Apress®

U0287787

Spring 5 攻略

Spring 5 Recipes

A Problem-Solution Approach

[美] 马腾·代伊纳姆（Marten Deinum）
[美] 丹尼尔·鲁比奥（Daniel Rubio） 著
[美] 乔希·朗（Josh Long）
张龙 译

人民邮电出版社
北京

图书在版编目（CIP）数据

Spring 5攻略 / (美) 马腾·代伊纳姆
(Marten Deinum) 著 ; (美) 丹尼尔·鲁比奥
(Daniel Rubio) , (美) 乔希·朗 (Josh Long) 著 ; 张
龙译. -- 北京 : 人民邮电出版社, 2021.6
ISBN 978-7-115-56138-1

Ⅰ. ①S… Ⅱ. ①马… ②丹… ③乔… ④张… Ⅲ. ①
JAVA语言—程序设计 Ⅳ. ①TP312.8

中国版本图书馆CIP数据核字(2021)第046453号

版权声明

◆ 著　　[美] 马腾·代伊纳姆（Marten Deinum）
　　　　[美] 丹尼尔·鲁比奥（Daniel Rubio）
　　　　[美] 乔希·朗（Josh Long）
译　　张 龙
责任编辑　傅道坤
责任印制　王 郁　焦志炜

◆ 人民邮电出版社出版发行　　北京市丰台区成寿寺路 11 号
邮编 100164　　电子邮件 315@ptpress.com.cn
网址 https://www.ptpress.com.cn
北京市艺辉印刷有限公司印刷

◆ 开本：787×1092 1/16
印张：35.25
字数：945 千字　　2021 年 6 月第 1 版
印数：1 – 2 200 册　　2021 年 6 月北京第 1 次印刷

著作权合同登记号　图字：01-2015-2822 号

定价：149.90 元

读者服务热线：(010) 81055410　印装质量热线：(010) 81055316
反盗版热线：(010) 81055315
广告经营许可证：京东市监广登字 20170147 号

内容提要

Spring 是一个开源的轻量级 Java 开发框架，主要用于解决企业应用开发的复杂性，简化应用程序的开发。

本书以“菜谱”的方式，介绍了 Spring 开发期间会遇到的各种需求、问题以及相应的解决方案。本书分为 17 章，主要内容有 Spring 开发工具的简单介绍；Spring 是什么、如何配置、如何使用；如何使用 Spring Web MVC 框架进行基于 Web 的开发；Spring 对 Restful Web Service 的支持；Spring MVC 的异步处理；使用 Spring Social 集成社交网络；使用 Spring Security 保护应用；使用 Spring Mobile 在应用中集成移动设备检测和使用功能；如何使用 Spring 访问数据；Spring 事务管理；使用 Spring Batch 框架对大型机领域的解决方案进行建模；Spring 与 NoSQL 和 Hadoop 的混合使用；Spring Java 企业服务与远程技术；Spring 消息机制；使用 Spring Integration 框架集成不同的服务与数据；Spring Framework 的单元测试；Grails 框架的简单介绍。

本书适合对 Java 开发和企业应用集成有一定了解，希望在实际开发中掌握一种全面、快速、可伸缩、可移植的工具平台的开发人员阅读。

译者序

毋庸置疑，Spring 现已成为 Java 企业级开发事实上的标准，围绕着 Spring 的生态圈也异常强大和繁荣。现如今的 Spring 已经不再只是一个框架，而是一整套解决方案。近些年，基于 Spring 发展起来的 Spring Boot、Spring Cloud 等微服务框架已经成为各大公司项目开发的首选。虽然这些微服务框架可以帮助我们快速创建项目并进行实际的业务开发，但其背后依然是使用 Spring 框架进行支撑。因此，从这个角度来看，Spring 实际上是这些框架的根基。若想在 Spring 之上进行项目开发，全方位地理解与掌握 Spring 就是必不可少的了。

本书可以看作是一本关于 Spring 的百科全书。众所周知，现如今的 Spring 家族异常庞大，拥有众多的子项目，每个子项目都是为了解决某个领域的特定问题而出现的。本书真正做到了适合不同层次的开发者阅读。无论你是一位 Spring 老手，还是刚刚接触 Spring 的初学者，都可以从本书中学到很多知识，而且能够立刻将所学应用到实际的项目开发当中。

本书首先从 Spring 开发环境讲起，详细介绍了 Spring 的核心。这部分内容是 Spring 的基础与精华所在。接下来对 Spring MVC、Spring REST 以及 Spring MVC 的异步处理进行了全方位的讲解。通过这部分内容的学习，读者可以对 Spring Web 开发相关的理论与实践拥有一个完整的理解。Spring Social、Spring Security 与 Spring Mobile 则是接下来介绍的重点内容。任何真实的应用都离不开对数据库的访问，因此本书对数据访问和 Spring 事务管理展开了深入的介绍与剖析，通过大量实例展现了 Spring 是如何简化数据访问与事务管理的。Spring Batch、Spring 与 NoSQL 及 Spring Java 企业服务与远程技术等章节则对 Java 企业级开发中常见的问题进行了论述并给出了相应的解决方案。最后，本书介绍了 Spring 与消息中间件、Spring Integration、Spring 测试，同时通过大量案例进行了针对性极强的剖析。值得一提的是，本书还单独拿出一章对快速开发框架 Grails 进行了讲解，旨在帮助读者迅速掌握这一基于 Groovy 的重要框架。

以电子版形式提供的两个附录则分别对云端部署和缓存这两个重要话题进行了讲解。通过实际案例介绍了如何通过 Pivotal 的 CloudFoundry 解决方案将 Java（Web）应用部署到云端，还介绍了 Spring 缓存抽象，包括如何配置以及如何透明地为应用添加缓存。

虽然目前基于约定优于配置理念的 Spring Boot、Spring Cloud 等快速开发框架在企业中得到了广泛的应用，但值得一提的是，它们的底层均是基于 Spring 框架。作为一名有追求的开发者，不应该仅仅停留在框架的使用层面，还需要关注框架底层的设计理念与核心执行逻辑。充分理解了框架的运作逻辑后，我们才能更有自信地编写代码；当出现问题时也能够快速定位问题并解决问题。本书可以帮助广大开发者从 Spring 框架自身展开探索，围绕 Spring 在各个领域所提供的解决方案，系统全面地学习 Spring 的诸多子项目以及解决问题的方式。

相信读者在学习完本书后，能够对 Spring 庞大的体系拥有一个清晰且透彻的认识，并可以将所学内容应用到自己在后续的工作与学习中，全面提升效率，直击问题本质，轻松解决问题。

在本书的翻译过程中，得到了人民邮电出版社傅道坤老师的大力支持与帮助。傅老师无论在专业领域还是个人魅力方面都对我产生了极大的影响。没有傅老师的指导与帮助，本译作是不可能完成的。在这里，再一次对傅老师的指导与教诲表示深深的谢意。

感谢我的家人，在本书翻译过程中，家人为我分担了生活中的各种琐事，能够让我将全部精力放在译稿的打磨上。这本译作的出版有你们一半的功劳。同时，我也会将本书作为送给你们的礼物。没有你们背后默默地鼓励与支持，我是无论如何也无法坚持下去的。

虽然在翻译过程中尽心尽力且小心翼翼，但奈何技术能力与文字功底有限，错误与疏漏之处在所难免。读者在阅读本书的过程中发现任何问题或是想与我交流，均可通过邮箱 zhanglong217@163.com 与我联系，非常高兴与大家一同学习探讨关于 Spring 的一切。

最后，衷心希望阅读本书的读者能够收获满满，深入且彻底地掌握 Spring 这一 Java 领域最为重要的框架，为今后的工作与学习打下坚实的基础。

张龙
2021 年 3 月

关于作者

Marten Deinum，Spring Framework 开源项目的提交者，也是 Conspect 公司的 Java/软件咨询师，为各种小型和大型公司开发并架构软件（主要基于 Java）。他是位热忱的开源用户，并且是 Spring Framework 的长期粉丝、用户与拥护者。他拥有多个角色，包括软件工程师、开发负责人和 Java 与 Spring 培训师。

Daniel Rubio，拥有 10 年以上的企业与 Web 软件开发经验，目前是 MashupSoft 公司的创始人与技术负责人。他已经为 Apress 编写了多本著作。Daniel 擅长 Java、Spring、Python、Django、JavaScript/CSS 和 HTML 等技术。

Josh Long，Spring 开发大使，目前就职于 Pivotal。Josh 是一位 Java 拥趸、5 本图书的作者（包括 O'Reilly 出版的 *Cloud Native Java*），也是 3 个销售极佳的培训视频的作者（包括与 Phil Webb 合作的 *Building Microservices with Spring Boot*），同时还是一位开源贡献者（Spring Boot、Spring Integration、Spring Cloud、Activiti 和 Vaadin）。

关于技术审稿人

Massimo Nardone，在安全、Web/移动开发、云计算和 IT 架构等领域拥有 23 年以上的经验。在 IT 领域中，他真正的热情在安全与 Android 上。

他目前是 Cargotec Oyj 的首席信息安全官（CISO），同时也是信息系统审计与控制协会（ISACA）芬兰理事会的成员。在漫长的职业生涯中，他担任过如下这些职位：项目经理、软件工程师、研究工程师、首席安全架构师、信息安全经理、PCI/SCADA 审计员，以及资深 IT 安全/云/SCADA 架构师。此外，他还是赫尔辛基理工大学（现在的阿尔托大学）网络实验室的客座讲师与练习主管。

Massimo 拥有意大利萨莱诺大学计算机科学专业的理学硕士学位，并且拥有 4 项国际专利（PKI、SIP、SAML 与代理领域）。

除了对本书进行审稿外，Massimo 还为不同出版公司审校了 40 多本 IT 图书，同时还是 *Pro Android Games*（Apress，2015）一书的合著者。

致谢

致谢可能是一本书中最难写的一部分内容。我无法说出想要感谢的所有人的名字，还可能会忘掉一些人。由于这个原因，我得先说声抱歉。虽然这是我编写的第三本书，但如果没有 Apress 卓越团队的帮助，本书是不可能出版的。非常感谢 Mark Powers 让我能够始终保持专注并促使我按时完成写作，感谢 Amrita 让我能够跟进最终的审校工作。

要感谢 Massimo Nardone，没有他的评论和建议，本书不可能是现在这样的。

感谢我的家人和朋友，能够接受我缺席的那些日子；感谢与我一同潜水的朋友，能够容忍我无法与他们一同参加潜水和旅游。

最后，感谢我的妻子 Djoke Deinum 和两个女儿 Geeske 与 Sietske。为了完成本书，我牺牲了无数个夜晚、周末与假期时光，谢谢你们对我无尽的支持、爱与奉献。没有你们的支持，我可能早就放弃了。

——Marten Deinum

前言

Spring Framework 在蓬勃发展。它总是与选择有关。Java EE 专注于少量技术，这在很大程度上损害了其他更好的解决方案。当 Spring Framework 横空出世时，没多少人会认为 Java EE 代表了当时最佳的架构。Spring 引起了人们的热议，因为它旨在简化 Java EE。从那时起，Spring 的每个版本都会引入一些新特性来简化并赋能解决方案。

从 2.0 版本开始，Spring Framework 开始瞄准了多个平台。框架一如既往地在既有平台之上提供了服务，不过在尽可能的情况下做到了与底层平台的解耦。Java EE 依然是个主要的参照点，不过并非唯一目标。此外，Spring Framework 可以在不同的云环境下运行。构建在 Spring 之上的框架不断涌现出来，可以支持应用集成、批处理、消息传递等。Spring Framework 5 是一个重大的升级，其基线升级到了 Java 8，添加了对基于注解的配置的更多支持，同时还引入了对 JUnit 5 的支持。一个新添加的特性是以 Spring WebFlux 的形式提供了对反应式编程的支持。

本书涵盖了更新后的框架，介绍了新特性以及不同的配置选项。

我们无法介绍 Spring 生态圈的每个项目，因此需要决定保留哪些内容、添加哪些内容以及更新哪些内容。这是个艰难的选择，不过我们认为已经包含了最重要的内容。

本书面向的读者

本书适合想要在 Java EE 平台范围外简化架构并解决问题的 Java 开发者阅读。如果你已经在项目中使用了 Spring，那么本书中更高阶的章节介绍了一些你可能还不太了解的新技术。如果你是框架初学者，那么本书将会带领你即刻起步。

本书假设你已经熟悉 Java 并且用过某个 IDE。虽然只使用 Java 开发客户端应用是可行也是有价值的，不过 Java 最大的社区还是在企业范畴内，这也是诸多技术产生价值之所在。因此，我们假设你已经熟悉了一些基本的企业编程概念，如 Servlet API。

本书的组织结构

第 1 章，“Spring 开发工具”，总体介绍了支持 Spring Framework 的工具以及如何使用它们。

第 2 章，“Spring 核心任务”，概览了 Spring Framework，包括 Spring Framework 是什么、如何配置以及如何使用 Spring Framework。

第 3 章，“Spring MVC”，介绍了如何通过 Spring Web MVC 框架进行基于 Web 的应用开发。

第 4 章，“Spring REST”，介绍了 Spring 对 RESTful Web Service 的支持。

第 5 章，“Spring MVC：异步处理”，介绍了如何通过 Spring MVC 进行异步处理。

第 6 章，“Spring Social”，介绍了 Spring Social，可以通过它轻松集成社交网络。

第 7 章，“Spring Security”，介绍了 Spring Security 项目，帮助你更好地保护应用。

第 8 章，“Spring Mobile”，介绍了 Spring Mobile，可以通过它在应用中集成并使用移动设备检测。

第 9 章，“数据访问”，介绍了如何通过 JDBC、Hibernate 和 JPA 等 API 来使用 Spring 访问数据存储。

第 10 章，“Spring 事务管理”，介绍了 Spring 健壮的事务管理设施背后的概念。

第 11 章，“Spring Batch”，介绍了 Spring Batch 框架，它提供了一种方式来对传统大型机领域的解决方案进行建模。

第 12 章，“Spring 与 NoSQL”，介绍了多个 Spring Data 项目，包括不同的 NoSQL 和 Hadoop 等大数据技术。

第 13 章，“Spring Java 企业服务与远程技术”，介绍了 JMX 支持、调度、邮件支持和各种 RPC 设施，包括 Spring Web Service 项目。

第 14 章，“Spring 消息机制”，介绍了如何通过 JMS 和 RabbitMQ 在 Spring 中使用面向消息的中间件，以及简化的 Spring 抽象。

第 15 章，“Spring Integration”，介绍了如何通过 Spring Integration 框架集成不同的服务与数据。

第 16 章，“Spring 测试”，介绍了 Spring Framework 的单元测试。

第 17 章，“Grails”，介绍了 Grails 框架，可以通过它提升生产力，方法是使用各种最佳组合并通过 Groovy 代码将它们黏合起来。

阅读本书的先决条件

由于 Java 编程语言独立于平台，因此可以随意选择任何支持该语言的操作系统。但是，本书中的一些示例使用管理特定于平台的路径。大家在输入这些示例之前，需要将这些路径转换为适合自己操作系统的格式。

为了充分使用本书，请安装 JDK 1.8 或更高的版本。为了降低开发难度，最好安装一个 Java IDE。本书中的代码示例以 Gradle 为基础。如果你运行的是 Eclipse 并安装了 Gradle 插件，则可以在 Eclipse 中直接打开代码，CLASSPATH 和依赖关系将由 Gradle 元数据进行填充。

资源与支持

本书由异步社区出品，社区（https://www.epubit.com/）为您提供相关资源和后续服务。

配套资源

本书提供如下资源：

- 本书附录文件；
- 本书源代码。

要获得配套资源，请在异步社区本书页面中单击 配套资源 ，跳转到下载界面，按提示进行操作即可。注意：为保证购书读者的权益，该操作会给出相关提示，要求输入提取码进行验证。

如果您是教师，希望获得教学配套资源，请在社区本书页面中直接联系本书的责任编辑。

提交勘误

作者和编辑尽最大努力来确保书中内容的准确性，但难免会存在疏漏。欢迎您将发现的问题反馈给我们，帮助我们提升图书的质量。

当您发现错误时，请登录异步社区，按书名搜索，进入本书页面，单击“提交勘误”，输入勘误信息，单击“提交”按钮即可。本书的作者和编辑会对您提交的勘误进行审核，确认并接受后，您将获赠异步社区的 100 积分。积分可用于在异步社区兑换优惠券、样书或奖品。

详细信息　写书评　提交勘误

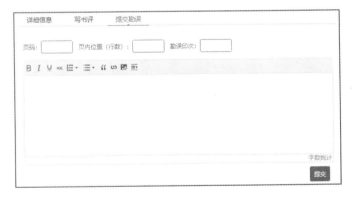

页码：　页内位置（行数）：　勘误印次：

字数统计

提交

扫码关注本书

扫描下方二维码，您将会在异步社区微信服务号中看到本书信息及相关的服务提示。

与我们联系

我们的联系邮箱是 contact@epubit.com.cn。

如果您对本书有任何疑问或建议，请您发邮件给我们，并请在邮件标题中注明本书书名，以便我们更高效地做出反馈。

如果您所在的学校、培训机构或企业，想批量购买本书或异步社区出版的其他图书，也可以发邮件给我们。

如果您在网上发现有针对异步社区出品图书的各种形式的盗版行为，包括对图书全部或部分内容的非授权传播，请您将怀疑有侵权行为的链接发邮件给我们。您的这一举动是对作者权益的保护，也是我们持续为您提供有价值的内容的动力之源。

关于异步社区和异步图书

“异步社区”是人民邮电出版社旗下 IT 专业图书社区，致力于出版精品 IT 技术图书和相关学习产品，为作译者提供优质出版服务。异步社区创办于 2015 年 8 月，提供大量精品 IT 技术图书和电子书，以及高品质技术文章和视频课程。更多详情请访问异步社区官网 https://www.epubit.com。

“异步图书”是由异步社区编辑团队策划出版的精品 IT 专业图书的品牌，依托于人民邮电出版社近 30 年的计算机图书出版积累和专业编辑团队，相关图书在封面上印有异步图书的 LOGO。异步图书的出版领域包括软件开发、大数据、AI、测试、前端、网络技术等。

异步社区

微信服务号

目录

第 1 章

Spring 开发工具

本章将会介绍如何搭建并使用最流行的开发工具来创建 Spring 应用。就像很多其他的软件框架一样，Spring 拥有大量的开发工具可供选择——从简单的命令行工具到名为集成开发环境（IDE）的复杂图形化工具。

无论你已经使用过一些 Java 开发工具还是一个新手，接下来的小节将会带领你搭建各种不同的工具箱来完成后续章节的练习，同时开发出 Spring 应用。

如下是 3 个工具箱以及对应的小节，跟着去做就可以开始创建 Spring 应用了。

- Spring Tool Suite：1-1 节。
- IntelliJ IDE：1-2 节（以及介绍 Maven 命令行界面的 1-3 节与 1-4 节，介绍 Gradle 命令行界面的攻略 1-5 节与 1-6 节）。
- 文本编辑器：1-3 节与 1-4 节介绍 Maven 命令行界面；1-5 节与 1-6 节介绍 Gradle 命令行界面。

请记住，使用 Spring 并不需要安装全部 3 个工具箱。可以都尝试一下，然后选择最喜欢的那个。

1-1 使用 Spring Tool Suite 构建 Spring 应用

问题提出

使用 Spring Tool Suite（STS）构建 Spring 应用。

解决方案

在工作站上安装 STS。打开 STS 并单击 Open Dashboard 链接。要想创建新的 Spring 应用，请单击 Create 表格中 Dashboard 里面的 Spring project 链接。要想打开使用了 Maven 的 Spring 应用，请从顶层的 File 菜单中选择 Import 选项，单击 Maven 图标，并选择 Existing Maven projects。接下来，从工作站中选择基于 Maven 的 Spring 应用。

要想在 STS 中安装 Gradle，请单击 Dashboard 窗口底部的 Extensions 选项卡。勾选 Gradle Support 复选框。继续进行 Gradle 扩展的安装，当安装完毕后请重启 STS。要想打开使用了 Gradle 的 Spring 应用，请从顶层的 File 菜单中选择 Import 选项，单击 Gradle 图标，并选择 Gradle project。接下来，在工作站中选择基于 Gradle 的 Spring 应用。单击 Build Model 按钮，最后单击 Finish 就可以开始项目的开发了。

解释说明

STS 是由 SpringSource 所开发的 IDE，而 SpringSource 是 Pivotal 的一个部门，Pivotal 则是 Spring 框架的创建者。STS 专门用于 Spring 应用的开发，这使得它成为了最为完善的工具之一。STS 是个基于 Eclipse 的工具，因此它与开源的 Eclipse IDE 拥有一致的外观。

STS 可以用在所有主流的操作系统版本之上：Windows（32 位或 64 位）、macOS（Cocoa，64 位）以及 Linux（GTK，32 位或 64 位）。此外，STS 自身也是有不同版本的，你可以下载最新的稳定版或是里程碑/开发版。请下载适合于你所使用的操作系统的版本。

下载好 STS 后，请确保系统中已经安装好了 Java SDK，这是安装 STS 的前提。请继续安装 STS。按照安装指南的指示，大约需要 5～10 分钟即可安装完毕。最后，一个名为 STS_<VERSION>的目录会在执行安装的用户的主目录下创建出来，用户也可以自行指定安装目录的位置。如果查看该目录，你会看到用于启动 STS 的 STS 可执行文件。

启动 STS。当启动时，STS 会要求你确定一个工作空间位置。所谓工作空间，就是 STS 放置所有项目信息的地方。你可以保持默认目录（位于主 STS 安装目录下）或是根据需要定义不同的目录。当启动完毕后，你会看到如图 1-1 所示的界面。

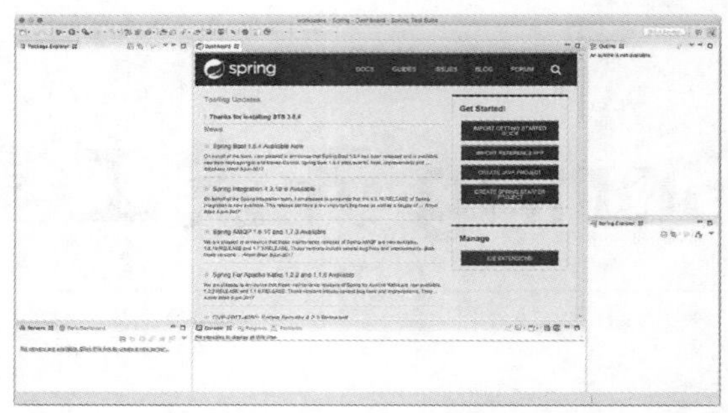

图 1-1　STS 启动界面

在 STS Dashboard 中，位于 Get Started!列的中间有一个名为 Create Spring Starter Project 的链接。你可以单击这个链接来新建一个 Spring 应用。如果喜欢，可以继续并创建一个空应用。你需要提供一个名字和一系列参数信息，这些参数信息也可以使用默认值。

与从头开始创建 Spring 应用相比，一个更为常见的场景是基于既有的 Spring 应用继续开发。在这种情况下，应用的所有者通常会通过构建脚本来分发应用源代码以简化正在进行的开发工作。

对于大多数 Java 应用来说，所选的构建脚本是针对于 Maven 构建工具的 pom.xml 文件，或是近来针对于 Gradle 构建工具的 build.gradle 文件。除了使用 Maven 构建文件的单个应用外，本书的源代码及其应用都是通过 Gradle 构建文件来提供的。

在一个 Java 应用中存在着大量繁琐的任务（比如说复制 JAR 或配置文件、设置 Java 的类路径以执行编译、下载 JAR 依赖等）。Java 构建工具可以执行 Java 应用中的这些任务。

Java 构建工具将会继续拥有自己的一席之地，这是因为使用构建文件分发的应用可以确保应用的创建者所指定的那些繁琐的任务会在使用该应用的任何用户身上得到精确的复制。如果应用使用 Ant 的 build.xml 文件、Maven 的 pom.xml 文件、Ivy 的 ivy.xml 文件或是 Gradle 的 build.gradle 文件进行分发，那么每种构建文件都会确保构建在不同用户以及不同系统上的一致性。

一些较新的 Java 构建工具更加强大，增强了早先那些构建工具的工作方式，每种构建文件都会使用自己的语法来定义动作、依赖，以及构建应用所需的任何任务。不过，你不应该被其遮蔽住双眼，因为 Java 构建工具只不过是完成任务的一种方式而已。它是由应用的创建者所选择的，旨在让构建过程变成流水化作业。如果你发现应用分发所使用的构建文件是最古老的 Ant 或是最新的 Gradle，都请不要惊慌失措；从最终用户的视角来看，你所需要做的只不过是按照创建者的意图下载并安装构建工具来创建应用而已。

由于很多 Spring 应用都还在使用 Maven，同时一些新的 Spring 应用在使用 Gradle，我们将会介绍如何将这两种类型的项目导入到 STS 中。

导入并构建 Maven 项目

下载本书的源代码并将其解压缩到本地目录中，单击 STS 顶部的 File 菜单并选择 Import 选项。这时会弹出一个窗口。在弹出的窗口中单击 Maven 图标并选择 Existing Maven Projects 选项，如图 1-2 所示。

单击 Next 按钮。在随后的界面中，单击 Browse 按钮并选择本书源码中 ch01 下名为 springintro_mvn 的目录，如图 1-3 所示。

注意到图 1-3 中的 Import Maven Projects 窗口更新了，包含了 pom.xml com.apress. springrecipes…一行，这反映了待导入的 Maven 项目。勾选上项目的复选框并单击 Finish 按钮导入项目。STS 中的所有项目都可以通过左侧的包浏览器窗口进行访问。对于该示例来说，项目名为 springintro_mvn。

如果单击包浏览器中的项目图标，你会看到项目结构（比如说 Java 类、依赖、配置文件等）。如果双击包浏览器中的任何项目文件，该文件就会在中央窗口中单独的选项卡中打开——在 Dashboard 旁边。文件打开后，你就可以查看、编辑或是删除其内容了。

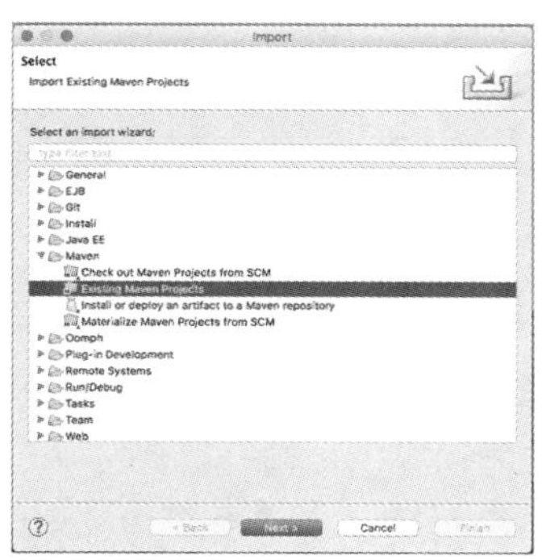

图 1-2 导入既有的 Maven 项目

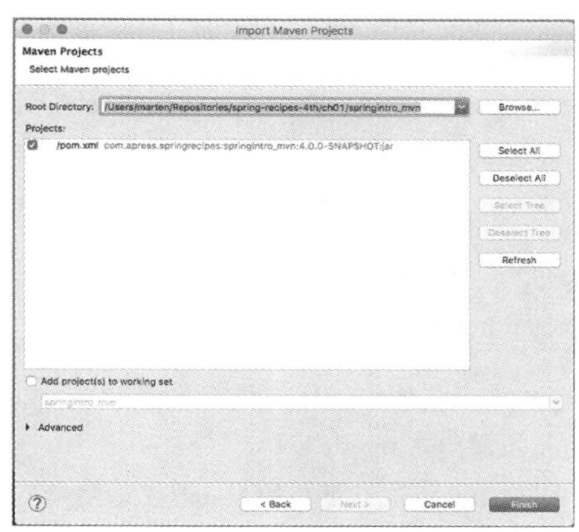

图 1-3 选择一个 Maven 项目

选中包浏览器中的项目图标并右键单击它。一个上下文菜单会出现并带有各种项目命令。选择 Run as 并接着选择 Maven build 选项。这时会出现一个弹出窗口，你可以编辑并配置项目构建。只需单击右下角的 Run 按钮即可。在 STS 中下部位置处，你会看到控制台窗口出现。在该示例中，控制台窗口显示了 Maven 所生成的一系列构建消息，如果构建过程失败还会显示出可能的错误消息。

你已经完成了应用的构建，现在来运行一下。再次从包浏览器中选中项目，按下 F5 键刷新项目目录。展开项目树。你会在靠近底部的位置处看到一个名为 target 的新目录，该目录包含了构建后的应用。单击图标展开 target 目录。接下来，选中文件 springintro_mvn-4.0.0- SNAPSHOT.jar，如图 1-4 所示。

选中文件后，单击右键打开带有各种项目命令的上下文菜单。选择 Run as，接下来再选择 Run configurations 选项。这时会出现一个弹出窗口，你可以对此次的运行信息进行编辑和配置。请确保选择了左侧的 Java application 选项。在 Main class 输入框中，输入 com.apress.springrecipes. hello.Main。这是该项目的主类，如图 1-5 所示。

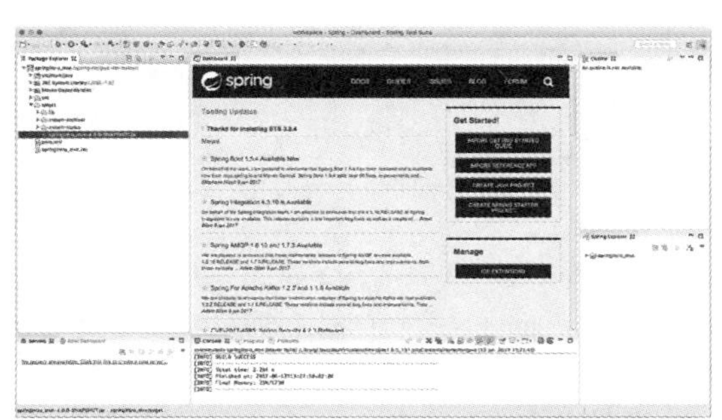

图 1-4 在 STS 中选中可执行文件

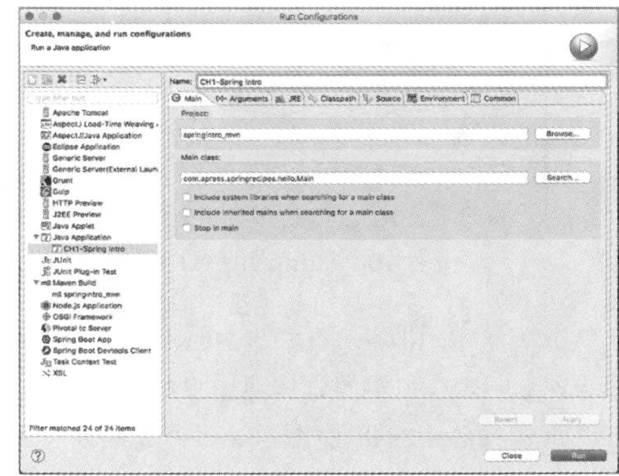

图 1-5 在 STS 中定义主执行类

单击右下角的 Run 按钮。在 STS 的中下部位置，你会看到控制台窗口。在该示例中，控制台窗口会展示出应用的日志消息，同时还有一条应用所定义的问候消息。

虽然已经借助于 STS 构建并运行了一个 Spring 应用，但你依旧什么都没做。刚刚在 STS 中所完成的过程几乎都是通过 Maven 构建工具来做的。接下来，我们来介绍如何通过更新的构建工具 Gradle 来导入 Spring 应用。

导入并构建 Gradle 项目

虽然 Gradle 是个相对较新的工具，但有迹象表明 Gradle 未来将会取代 Maven。比如说，很多大型 Java 项目（如 Spring 框架本身）现在就在使用 Gradle 而非 Maven，这是由 Gradle 的强大特性所决定的。考虑到这种趋势，我们有必要了解如何在 STS 中使用 Gradle。

> ■ **提示：** 如果你有 Maven 项目（即 pom.xml 文件），你可以使用 bootstrap 插件或是 maven2gradle 工具创建 Gradle 项目（即 build.gradle 文件）。bootstrap 插件包含在 Gradle 中，maven2gradle 工具可以从 GitHub 下载 https://github.com/jbaruch/maven2gradle.git。

要想在 STS 中安装 Gradle 支持，需要安装 BuildShip 扩展。请通过 Help 菜单打开 Eclipse Marketplace 并搜索 Gradle，如图 1-6 所示。

单击 BuildShip 集成右下角的 Install 按钮继续安装。

单击弹出窗口中的 Next 按钮。阅读完“许可并接受条款”后，单击弹出窗口中的 Finish 按钮。Gradle 扩展安装就会开始。安装过程完成后会提示你重启 STS 以让修改生效。请重启 STS 来完成 Gradle 的安装。

本书的源代码包含了大量使用 Gradle 构建的 Spring 应用，因此我们需要介绍如何将这些 Spring 应用导入到 STS 中。下载好本书的源代码并将其解压缩到本地目录后，在 STS 中单击顶层的 File 菜单并选择 Import 选项。这时会弹出一个窗口。在该弹出窗口中，单击 Gradle 图标并选择 Existing Gradle Project 选项，如图 1-7 所示。

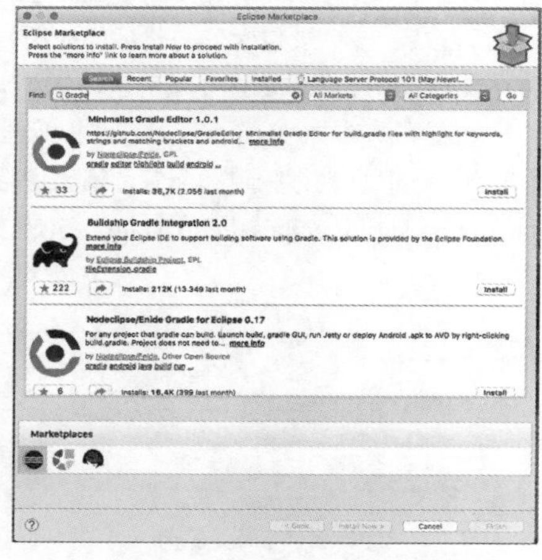

图 1-6　Buildship STS 安装

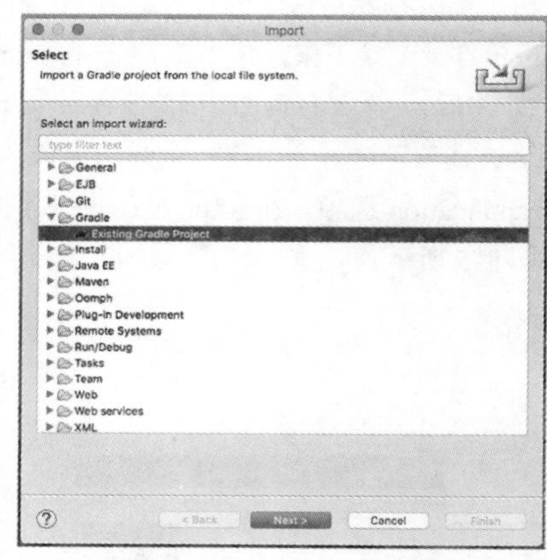

图 1-7　导入 Gradle 项目

单击 Next 按钮。在随后的界面中，单击 Browse 按钮并选择本书的 Ch01/springintro 目录。单击 Finish 按钮导入项目。在包浏览器中查看 STS 的左侧，你会看到名为 springintro 的项目被加载进来。单击项目图标则可以查看项目结构（比如说 Java 类、依赖与配置文件等）。

IDE 右侧有个 Gradle Tasks 选项卡。找到 springintro 项目，打开 Build 菜单并选择 Build。右键单击并选择 Run Gradle Tasks。你已经构建了应用，现在来运行它。

再次选中项目图标并单击 F5 键刷新项目目录。展开项目树。你会在中间位置看到一个名为 libs 的新目录，它包含了构建好的应用。单击图标展开 libs 目录。接下来，选中文件 springintro.jar。

选中文件后，从顶层菜单 Run 中选择 Run configurations 选项。这时会弹出一个窗口，你可以在这里编辑并配置运行信息。请确保选择了左侧的 Java application 选项。在 Main class 输入框中，输入 com.apress.springrecipes.hello.Main。这是项目的主类。单击右下角的 Run 按钮。在 STS 的中下部位置处，你会看到控制台窗口。在该示例中，控制台窗口会展示出应用的日志消息以及一条应用所定义的问候消息。

1-2 使用 IntelliJ IDE 构建 Spring 应用

问题提出

使用 IntelliJ IDE 构建 Spring 应用。

解决方案

要想在 IntelliJ Quick Start 窗口中开始新的 Spring 应用，请单击 Create New Project 链接。在接下来的窗口中，请为项目指定一个名字，选择一个运行时 JDK，并选择 Java Module 选项。在随后的窗口中，单击各种 Spring 复选框，这样 IntelliJ 就会为项目下载必要的 Spring 依赖。

要想打开使用了 Maven 的 Spring 应用，首先需要安装 Maven 以便在命令行界面中可以使用（参见 1-4 节）。从 IntelliJ 顶层的 File 菜单中选择 Import Project 选项。接下来，在工作站中选择基于 Maven 的 Spring 应用。在接下来的界面中，选择 Import project from external model 选项并选择 Maven 类型。

要想打开使用了 Gradle 的 Spring 应用，首先需要安装 Gradle 以便在命令行界面中可以使用（参见 1-5 节）。从 IntelliJ 顶层的 File 菜单中选择 Import Project 选项。接下来，在工作站中选择基于 Gradle 的 Spring 应用。在接下来的界面中，选择 Import Project from external model 选项并选择 Gradle 类型。

解释说明

IntelliJ 是市场上最为流行的商业 IDE 之一。不像其他那些由基金会所推出的 IDE，比如说 Eclipse，或是用于支持某公司旗舰软件的 IDE，比如说针对于 Spring 框架的 STS，IntelliJ 是由一家名为 JetBrains 的公司所推出的，该公司唯一的业务就是对开发工具进行商业化。正是这种专注方使得 IntelliJ 在企业环境下的专业开发者之间变得异常流行。

针对于本节，我们假设你已经安装好了 IntelliJ 旗舰版，只是想要上手并运行 Spring 应用。

警告： IntelliJ 提供了一个免费的社区版和一个拥有 30 天免费试用期的旗舰版。虽然免费的社区版为应用开发提供了不错的价值，不过它并不包含对于 Spring 应用的支持。后续的说明都假设你正在使用 IntelliJ 旗舰版。

创建 Spring 应用

要想开始一个 Spring 应用，请在 IntelliJ Quick Start 窗口中单击 Create New Project 链接。在 New Project 窗口中，选择 Spring 选项并勾选各种 Spring 复选框，如图 1-8 所示。

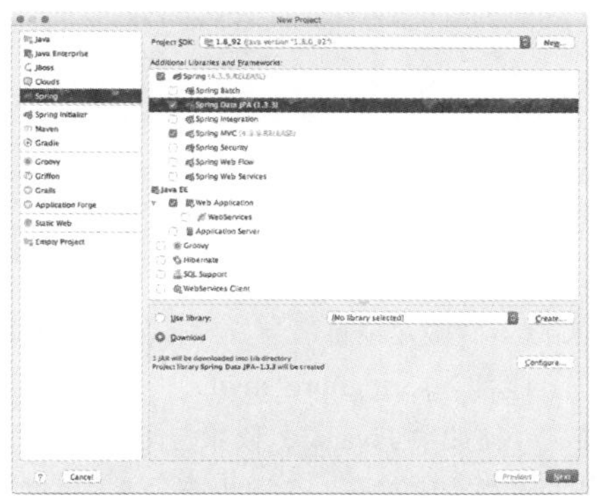

图 1-8 使用 IntelliJ 创建 Spring 项目

单击 Next 按钮。在接下来的窗口中，为项目起一个名字并单击 Finish。

导入并构建 Maven 项目

与从头开始创建 Spring 应用相比，一个更为常见的场景是基于既有的 Spring 应用继续开发。在这种情况下，应用的所有者通常会通过构建脚本来分发应用源代码以简化正在进行的开发工作。

对于大多数 Java 应用来说，所选的构建脚本是针对于 Maven 构建工具的 pom.xml 文件，或是近来针对 Gradle 构建工具的 build.gradle 文件。除了使用 Maven 构建文件的单个应用外，本书的源代码及其应用都是通过 Gradle 构建文件来提供的。

下载本书的源代码并将其解压缩到本地目录后，单击 IntelliJ 顶层的 File 菜单并选择 Import Project 选项。这时会弹出一个窗口，如图 1-9 所示。

在该窗口中，进入到子目录中，直到看到 ch01 目录为止，然后选择 springintro_mvn。单击 Open 按钮。在接下来的界面中，选择 Import project from external model 选项并选择 Maven 类型，如图 1-10 所示。

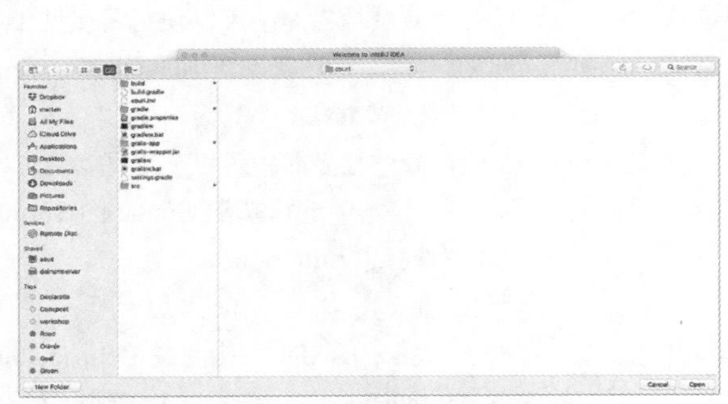

图 1-9　在 IntelliJ 中选择文件或目录进行导入

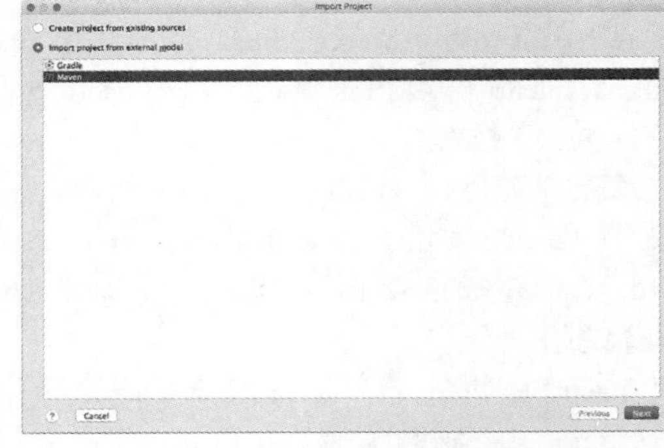

图 1-10　在 IntelliJ 中选择项目类型

在接下来的窗口中（见图 1-11），你可以细粒度地修改 Maven 项目的设置，比如说自动将修改导入到 pom.xml 中、下载依赖的源代码等。修改完毕后，请单击 Next。

请确保勾选项目复选框，单击 Next 按钮导入项目，如图 1-12 所示。

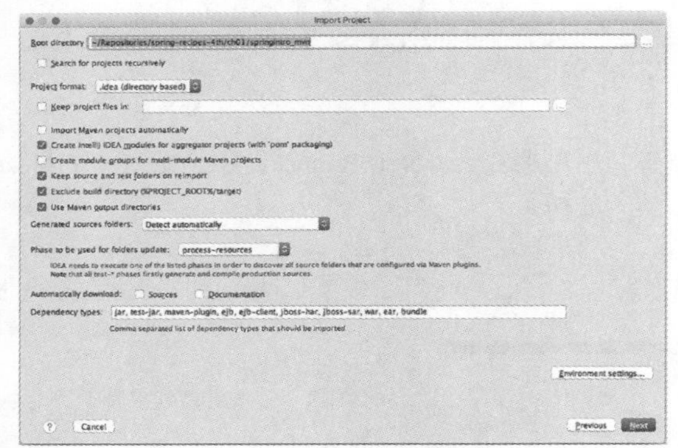

图 1-11　在 IntelliJ 中细粒度地修改 Maven 项目设置

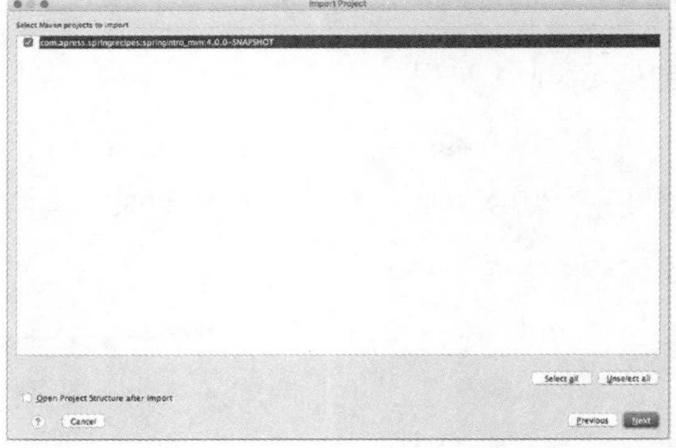

图 1-12　在 IntelliJ 中选择 Maven 项目

接下来，选择项目所用的 SDK 版本。确定项目名与位置并单击 Finish 按钮。IntelliJ 中的所有项目都位于项目窗口的左侧。在该示例中，项目名为 springintro_mvn。

单击项目图标可以看到项目结构（比如说 Java 类、依赖与配置文件等）。如果双击项目窗口中的任何项目文件，文件就会在中间窗口单独的选项卡中打开。你可以查看文件内容，也可以编辑或是删除其内容。

接下来，需要配置 Maven 来与 IntelliJ 协同使用。请按照 1-3 节的指示安装 Maven 以在命令行中使用。完成后，你就可以在 IntelliJ 中使用 Maven 了。

单击 IntelliJ 顶层的 File 菜单并选择 Settings 选项。这时会弹出一个窗口用于配置 IntelliJ 的设置。单击 Maven 选项，在 Maven home directory 中根据系统来设置 Maven 安装目录信息，如图 1-13 所示。单击 Apply 按钮，然后再单击 OK 按钮。

接下来，单击 IntelliJ 右侧的垂直选项卡 Maven Projects 显示出 Maven 项目窗格，如图 1-14 所示。

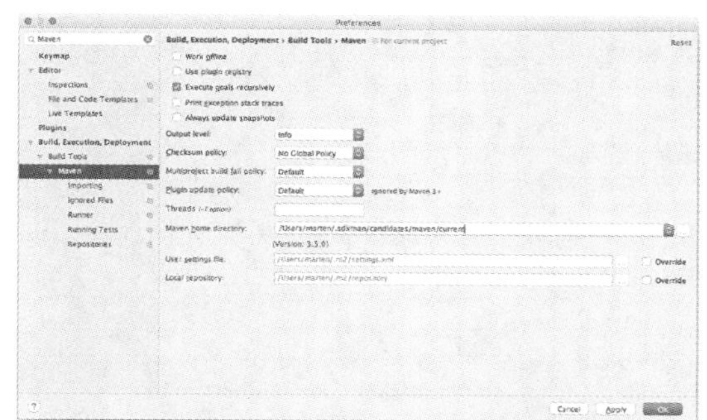

图 1-13 在 IntelliJ 中设置 Maven

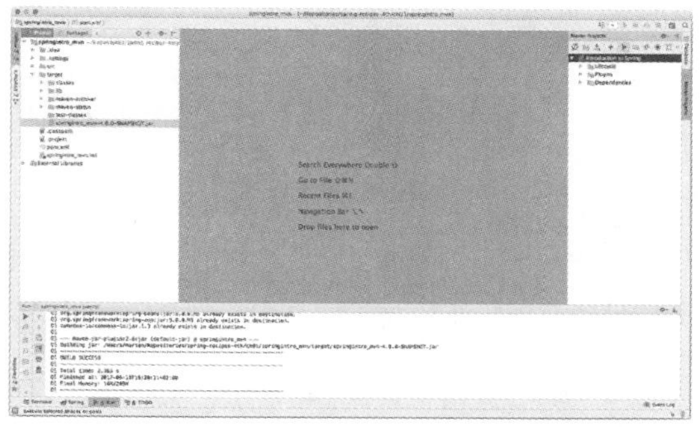

图 1-14 IntelliJ 中的 Maven 项目窗格

在 Maven 项目窗格中选中项目的 Introduction to Spring 这一行，单击右键打开上下文菜单，菜单中有关于项目的各种命令。选择 Run Maven Build 选项。在 IntelliJ 的中下部位置处，你会看到运行窗口。在该示例中，运行窗口会显示出 Maven 所生成的一系列构建信息，当构建过程失败时还会显示出可能的错误信息。

警告：如果看到错误消息“No valid Maven installation found. Either set the home directory in the configuration dialog or set the M2_HOME environment variable on your system”，那就表示 IntelliJ 找不到 Maven。请验证 Maven 的安装与配置过程。

你已经完成了应用的构建，现在来运行一下。如果看不到 target 目录，请按下 Ctrl+Alt+Y 组合键同步一下项目。单击 target 目录的图标将其展开。接下来，右键单击文件 springintro_mvn-4.0.0-SNAPSHOT.jar，并选择 Run 选项，如图 1-15 所示。

在 IntelliJ 中下部位置处的 Run 窗口中，你会看到应用的日志消息，还有一条应用所定义的问候消息。

图 1-15 在 IntelliJ 中运行应用

导入并构建 Gradle 项目

现在，我们在 IntelliJ 中构建一个 Gradle 应用。首先需要安装 Gradle。按照 1-4 节的说明从命令行中安装 Gradle。完成后，你就可以在 IntelliJ 中使用 Gradle 了。

单击 IntelliJ 顶层的 File 菜单并选择 Import Project 选项。这时会弹出一个窗口，如图 1-9 所示。从弹出窗口的目录树中进入到子目录，直到看到本书源代码的 ch01/springintro 目录，选择文件 build.gradle。

单击 Open 按钮。在接下来的界面中，选择 Import project from external model 选项并选择 Gradle。随后的界面中，在 Gradle home 输入框中根据系统中 Gradle 的安装信息输入 Gradle 主目录，如图 1-16 所示。

单击 Finish 按钮确认导入过程，然后单击 Finish 按钮完成导入过程。接下来，在项目窗口中右键单击 build.gradle 并选择 Run Build。

你已经完成了应用的构建。现在来运行一下。在项目窗口中，展开 build 目录并进入到 libs 目录。找到 springintro-all.jar，如图 1-17 所示。

注意：我们已经配置好了 build.gradle 文件，使之可以生成一个 shadow JAR，这意味着它包含了运行时所需的所有类和依赖。

现在，右键单击 springintro-all.jar 文件并选择 Run 选项。在 IntelliJ 下部位置的运行窗口中，你会看到应用日志消息，以及一条应用所定义的问候消息。

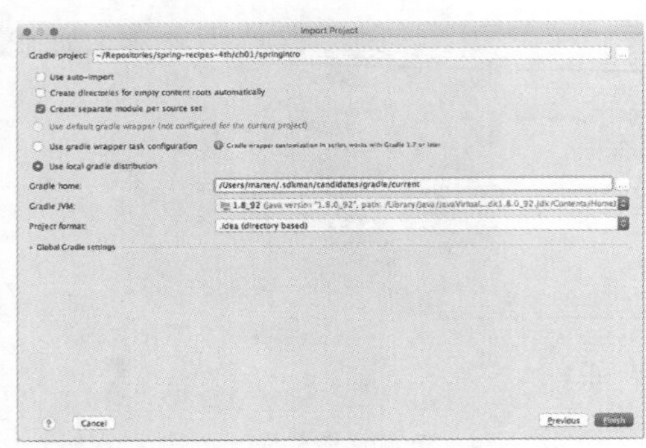

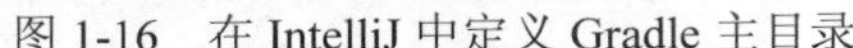

图 1-16　在 IntelliJ 中定义 Gradle 主目录

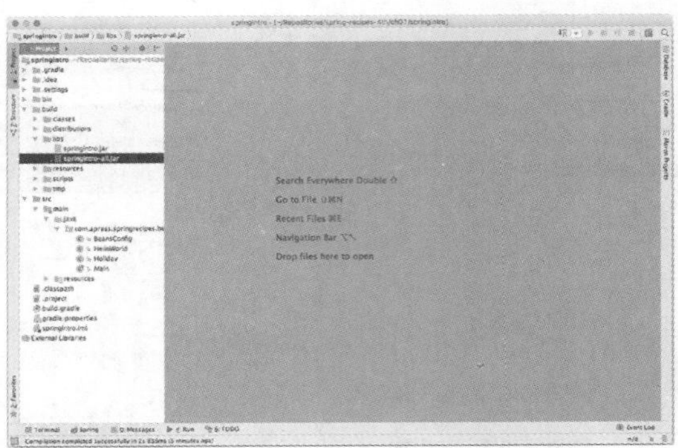

图 1-17　在 IntelliJ 中选择运行的应用

1-3　使用 Maven 命令行界面构建 Spring 应用

问题提出

通过命令行使用 Maven 构建 Spring 应用。

解决方案

从 Maven 官网下载 Maven。确保将 JAVA_HOME 环境变量设置为 Java SDK 主目录。修改 PATH 环境变量，使之包含 Maven 的 bin 目录。

解释说明

Maven 是以独立的命令行界面工具形式提供的。这样，我们就可以在各种不同的开发环境中使用它了。比如说，如果喜欢使用 Emacs 或是 Vi 等文本编辑器来编写应用代码，那么能够使用诸如 Maven 等构建工具来自动化与 Java 应用构建过程相关的工作（比如说复制文件、一步编译等）就是非常必要的了。

可以从 Maven 官网免费下载 Maven。它既提供了源代码版本，也提供了二进制版本。由于 Java 工具是跨平台的，因此我们推荐下载二进制版本以减少额外的编译步骤。在本书编写之际，Maven 最新的稳定版是 3.5.0。

下载好 Maven 后，请确保在系统上安装好了 Java SDK，因为 Maven 在运行期需要用到它。继续安装 Maven，将其解压缩，并定义 JAVA_HOME 与 PATH 环境变量。

执行如下命令来解压缩 Maven：

```
www@ubuntu:~$ tar -xzvf apache-maven-3.5.0-bin.tar.gz
```

执行如下命令来添加 JAVA_HOME 变量：

```
www@ubuntu:~$ export JAVA_HOME=/usr/lib/jvm/java-8-openjdk/
```

执行如下命令将 Maven 可执行文件添加到 PATH 变量中：

```
www@ubuntu:~$ export PATH=$PATH:/home/www/apache-maven-3.5.0/bin/
```

■ **提示：**如果按照上面的方式声明变量 JAVA_HOME 与 PATH，那么每次打开新的 shell 会话来使用 Maven 时都需要再这么做一遍。在 UNIX/Linux 系统中，你可以打开用户主目录下的.bashrc 文件并添加同样的 export 行以避免每次会话时都声明一次环境变量。在 Windows 系统中，你可以永久性地设置环境变量，方式是选中“我的电脑”图标、右键单击它，然后选择“属性”选项。在弹出窗口中选择“高级”选项卡并单击“环境变量”按钮。

Maven 可执行文件是通过 mvn 命令提供的。如果按照上面的方式正确设置了环境变量，那么在系统的任何目录下敲入 mvn 都会调用 Maven。对于 Maven 更为详尽的介绍已经超出了本节的范围。不过，接下来我们将会介绍如何通过 Maven 构建来自于本书源代码的一个 Spring 应用。

下载好本书的源代码并将其解压缩到本地目录后，请进入到名为 ch01/springintro_mvn 的目录。在 springintro_mvn 目录下输入 mvn 来调用 Maven 并构建应用。输出如图 1-18 所示。

图 1-18　Maven 构建输出

你已经完成了应用的构建，现在来运行一下。进入到 ch01/springintro_mvn 下由 Maven 所创建的目录 target。你会看到文件 springintro_mvn-4.0.0-SNAPSHOT.jar，这是构建好的应用。执行命令 java -jar springintro_mvn-1-0.SNAPSHOT.jar 来运行应用。你会看到应用日志消息以及一条由应用所定义的问候消息。

1-4　使用 Maven wrapper 构建 Spring 应用

问题提出

在命令行中使用 Maven wrapper 构建 Spring 应用。

解决方案

在命令行中运行 mvnw 脚本。

解释说明

虽然 Maven（参见 1-3 节）是以独立的命令行工具形式提供的，不过大量（开源）项目都在通过 Maven wrapper 的形式让使用者使用 Maven。这种方式的好处在于应用完全是自给自足的。作为开发者的你无须提前安装好 Maven，因为 Maven wrapper 会下载特定版本的 Maven 来构建项目。

当项目使用了 Maven wrapper 后，你只需在命令行上输入./mvnw package 就可以自动下载 Maven 并执行构建。唯一的前提就是要安装好 Java SDK，这是因为 Maven 在运行期会用到它，同时 Maven wrapper 也需要它才能运行。

下载好本书的源代码并将其解压缩到本地目录后，进入到名为 ch01/springintro_mvnw 的目录。输入./mvnw 来调用 Maven wrapper 并自动构建应用。输出如图 1-19 所示。

图 1-19　Maven wrapper 构建输出

注意到输出的第一部分是下载用于该项目的实际 Maven 版本。

1-5　使用 Gradle 命令行界面构建 Spring 应用

问题提出

在命令行中使用 Gradle 构建 Spring 应用。

解决方案

从 Gradle 官网下载 Gradle。请确保将 JAVA_HOME 环境变量设为 Java 的 SDK 主目录。修改 PATH 环境变量，使之包含 Gradle 的 bin 目录。

解释说明

Gradle 是以独立的命令行工具形式提供的。这样，我们就可以在各种不同的开发环境中使用它了。比如说，如果喜欢使用 Emacs 或是 Vi 等文本编辑器来编写应用代码，那么能够使用诸如 Gradle 等构建工具来自动化与 Java 应用构建过程相关的工作（比如说复制文件、一步编译等）就是非常必要的了。

可以从 Gradle 官网免费下载 Gradle。Gradle 官网既提供了源代码版本，也提供了二进制版本。由于 Java 工具是跨平台的，因此我们推荐下载二进制版本以减少额外的编译步骤。在本书编写之际，Gradle 最新的稳定版是 3.5。

下载好 Gradle 后，请确保在系统上安装好了 Java SDK，因为 Gradle 在运行期需要用到它。继续安装 Gradle，将其解压缩，并定义 JAVA_HOME 与 PATH 环境变量。

执行如下命令来解压缩 Gradle：

```
www@ubuntu:~$ unzip gradle-3.5-bin.zip
```

执行如下命令来添加 JAVA_HOME 变量：

```
www@ubuntu:~$ export JAVA_HOME=/usr/lib/jvm/java-8-openjdk/
```

执行如下命令将 Gradle 可执行文件添加到 PATH 变量中：

```
www@ubuntu:~$ export PATH=$PATH:/home/www/gradle-3.5/bin/
```

■ **提示：**如果按照上面的方式声明变量 JAVA_HOME 与 PATH，那么每次打开新的 shell 会话来使用 Gradle 时都需要再这么做一遍。在 UNIX/Linux 系统中，你可以打开用户主目录下的.bashrc 文件并添加同样的 export 行以避免每次会话时都声明一次环境变量。在 Windows 系统中，你可以永久性地设置环境变量，方式是选中“我的电脑”图标、右键单击它，然后选择“属性”选项。在弹出窗口中选择“高级”选项卡并单击“环境变量”按钮。

Gradle 可执行文件是通过 gradle 命令提供的。如果按照上面的方式正确设置了环境变量，那么在系统的任何目录下敲入 gradle 都会调用 Gradle。对于 Gradle 更为详尽的介绍已经超出了本节的范围。不过，由于本书的源代码包含了大量使用了 Gradle 的 Spring 应用，因此我们将会介绍如何使用 Gradle 构建其中一个 Spring 应用。

下载好本书的源代码并将其解压缩到本地目录后，请进入到名为 ch01/springintro 的目录。在 springintro 目录下输入 gradle 来调用 Gradle 并构建应用。输出如图 1-20 所示。

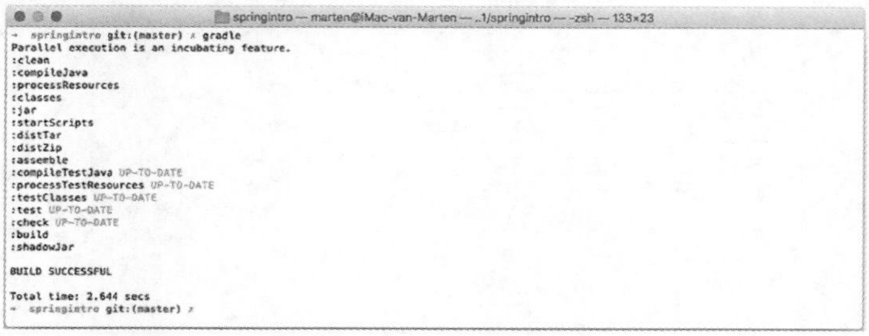

图 1-20　Gradle 构建输出

你已经完成了应用的构建，现在来运行一下。进入到 ch01/springintro 下由 Gradle 所创建的目录 libs。你会看到文件 springintro-all.jar，这是构建好的应用。执行命令 java -jar springintro-all.jar 来运行应用。你会看到应用日志消息以及一条由应用所定义的问候消息。

1-6 使用 Gradle wrapper 构建 Spring 应用

问题提出

在命令行中使用 Gradle wrapper 构建 Spring 应用。

解决方案

在命令行中运行 gradlew 脚本。

解释说明

虽然 Gradle（参见 1-5 节）是以独立的命令行工具形式提供的，不过大量（开源）项目都在通过 Gradle wrapper 的形式让使用者使用 Gradle。这种方式的好处在于应用完全是自给自足的。作为开发者的你无须提前安装好 Gradle，因为 Gradle wrapper 会下载特定版本的 Gradle 来构建项目。

当项目使用了 Gradle wrapper 后，你只需在命令行上输入./gradlew build 就可以自动下载 Gradle 并执行构建。唯一的前提就是要安装好 Java SDK，这是因为 Gradle 在运行期会用到它，同时 Gradle wrapper 也需要它才能运行。

下载好本书的源代码并将其解压缩到本地目录后，进入到名为 ch01/Recipe_1_6 的目录。输入./gradlew 来调用 Gradle wrapper 并自动在 Recipe_1_6 下构建应用。输出如图 1-21 所示。

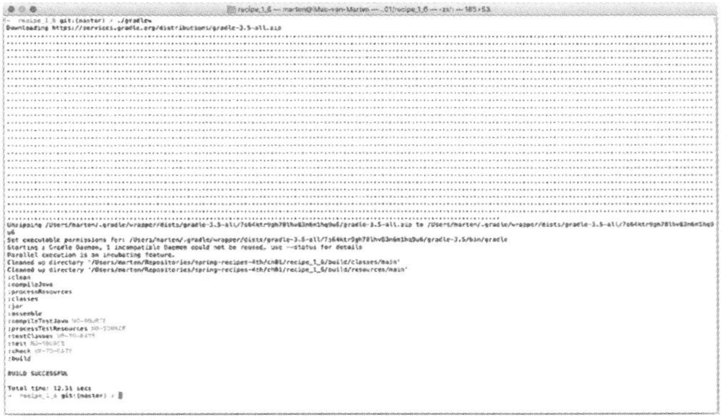

图 1-21 Gradle 构建输出

> **提示**：本书的源代码既可以使用普通的 Gradle 构建，也可以使用 Gradle wrapper 构建。后者是推荐的方式，因为代码在构建时使用的是与示例编写时相同版本的 Gradle。

小结

本章主要介绍了如何搭建最为流行的开发工具来创建 Spring 应用。我们讲解了如何使用 4 种工具箱来构建并运行 Spring 应用。其中两种工具箱用到了 IDE：由 Spring 框架创建者所推出的 Spring Tool Suite，以及由 JetBrains 所推出的 IntelliJ IDE。另外两种工具箱用到了命令行工具：Maven 构建工具以及更新的 Gradle 构建工具。相较于 Maven 来说，Gradle 正变得越来越流行。

第 2 章

Spring 核心任务

本章将会介绍与 Spring 相关的核心任务。Spring 框架的内核是 Spring 控制反转（IoC）容器。IoC 容器用于管理和配置 Plain Old Java Object（POJO）。由于 Spring 框架的主要吸引力在于使用 POJO 构建 Java 应用，因此很多 Spring 的核心任务都涉及在 IoC 容器中管理和配置 POJO。

无论你打算使用 Spring 框架进行 Web 应用、企业集成还是其他类型的项目开发，使用 POJO 与 IoC 容器都是需要掌握的第一个环节。本章的大部分内容都与全书涉及的任务相关，同时也可以作为开发 Spring 应用的基础。

■ **注意：** 在本书以及 Spring 文档中，术语 bean 都是可以与 POJO 实例进行互换的。它们指的都是通过 Java 类所创建的对象实例。此外，术语“组件”也是可以与 POJO 类进行互换的。它们指的都是实际的 Java 类，可以通过它创建对象实例。

■ **提示：** 下载的源代码是使用 Gradle 来组织的（通过 Gradle wrapper），可以构建相应的应用。Gradle 负责加载所有必要的 Java 类和依赖，并创建可执行的 JAR 文件。第 1 章已经介绍了如何搭建 Gradle 工具。此外，如果某示例介绍了多种方式，那么源代码是通过罗马字符来对各种示例进行分类的（比如说 Recipe_2_1_i、Recipe_2_1_ii、Recipe_2_1_iii 等）。

要想构建每个应用，请进入到 Recipe 目录（如 Ch2/Recipe_2_1_i/）并执行./gradlew 构建命令来编译源代码。源代码编译完毕后会创建出一个 build/libs 子目录，里面有应用的可执行文件。接下来就可以从命令行来运行应用的 JAR 文件了（如 java -jar Recipe_2_1_i-4.0.0.jar）。

2-1 使用 Java config 来配置 POJO

问题提出

使用注解和 Spring 的 IoC 容器来管理 POJO。

解决方案

设计一个 POJO 类。接下来，使用@Configuration 和@Bean 注解创建一个 Java 配置类来配置 POJO 实例值，或是使用@Component、@Repository、@Service 或@Controller 注解来创建 Java 组件，供后续创建 POJO 实例值所用。接下来，实例化 Spring IoC 容器来扫描带有注解的 Java 类。POJO 实例或是 bean 实例会变成可访问的，并成为应用的一部分。

解释说明

假设要开发一个生成序列号的应用，你需要很多序列号来满足不同的需求。每个序列都有自己的前缀、后缀与初始值。你需要为应用创建并维护多个生成器实例。使用 Java config 来创建一个 POJO 类，并通过它来创建 bean。

根据需求，你创建了一个 SequenceGenerator 类，它有 3 个属性：prefix、suffix 与 initial。还有一个私有

字段 counter 用于存储每个生成器的数值。每次调用生成器实例的 getSequence()方法时，你都会得到最后一个序列号，同时加上 prefix 与 suffix。

```
package com.apress.springrecipes.sequence;

import java.util.concurrent.atomic.AtomicInteger;

public class SequenceGenerator {

    private String prefix;
    private String suffix;
    private int initial;
    private final AtomicInteger counter = new AtomicInteger();

    public SequenceGenerator() {
    }

    public void setPrefix(String prefix) {
        this.prefix = prefix;
    }

    public void setSuffix(String suffix) {
        this.suffix = suffix;
    }

    public void setInitial(int initial) {
        this.initial = initial;
    }

    public String getSequence() {
        StringBuilder builder = new StringBuilder();
        builder.append(prefix)
               .append(initial)
               .append(counter.getAndIncrement())
               .append(suffix);
        return builder.toString();
    }
}
```

使用@Configuration 与@Bean 创建 Java config 并通过它创建 POJO

要想在 Spring IoC 容器中定义 POJO 类实例，你可以创建一个 Java 配置类并使用实例值。使用 POJO 或是 bean 定义所创建的 Java 配置类如下代码所示：

```
package com.apress.springrecipes.sequence.config;

import org.springframework.context.annotation.Bean;
import org.springframework.context.annotation.Configuration;
import com.apress.springrecipes.sequence.SequenceGenerator;

@Configuration
public class SequenceGeneratorConfiguration {

    @Bean
    public SequenceGenerator sequenceGenerator() {
        SequenceGenerator seqgen = new SequenceGenerator();
        seqgen.setPrefix("30");
        seqgen.setSuffix("A");
        seqgen.setInitial("100000");
        return seqgen;
    }
}
```

注意到 SequenceGeneratorConfiguration 类被@Configuration 注解所修饰；这告诉 Spring 它是个配置类。

当 Spring 遇到带有@Configuration 注解的类时，它会寻找类中的 bean 实例定义，这些定义是由@Bean 注解所修饰的 Java 方法。这些 Java 方法创建并返回一个 bean 实例。

任何使用@Bean 注解所修饰的方法声明都会根据方法名来生成一个 bean 名字。此外，你还可以在@Bean 注解中通过 name 属性显式指定 bean 名字。比如说，@Bean(name="mys1")会将 bean 名字指定为 mys1。

■ **注意：** 如果显式指定了 bean 名字，那么在 bean 创建时就会忽略掉方法名。

实例化 Spring IoC 容器来扫描注解

你需要实例化 Spring IoC 容器来扫描包含了注解的 Java 类。在这个过程中，Spring 会检测@Configuration 与@Bean 注解，这样后面就可以从 IoC 容器中获取到 bean 实例了。

Spring 提供了两种类型的 IoC 容器实现。基础的实现叫作 bean factory（bean 工厂），高级的实现叫作 application context（应用上下文），它兼容于 bean factory。注意到这两种类型的 IoC 容器的配置文件是一致的。

相较于 bean 工厂来说，应用上下文提供了更为高级的特性，同时又兼容基本的特性。因此，强烈建议你对于每个应用来说都使用应用上下文，除非应用资源受限（比如说在 applet 或是移动设备中运行 Spring）。bean 工厂与应用上下文的接口分别是 BeanFactory 与 ApplicationContext。ApplicationContext 是 BeanFactory 的子接口，这样可以保持兼容性。

由于 ApplicationContext 是个接口，因此你需要实例化它的一个实现才行。Spring 提供了几个应用上下文实现；推荐使用 AnnotationConfigApplicationContext，它是最新也是最灵活的一种实现。可以通过这个类加载 Java 配置文件。

```
ApplicationContext context = new AnnotationConfigApplicationContext
    (SequenceGeneratorConfiguration.class);
```

应用上下文实例化后，对象引用（该示例中就是 context）就会提供一个访问 POJO 实例或 bean 的入口。

从 IoC 容器中获取 POJO 实例或 bean

要想从 bean 工厂或应用上下文中获取声明的 bean，只需调用 getBean()方法并传递进唯一的 bean 名字即可。getBean()方法的返回类型是 java.lang.Object，因此在使用前需要将其转换为实际的类型。

```
SequenceGenerator generator =
    (SequenceGenerator) context.getBean("sequenceGenerator");
```

getBean()方法还支持另外一个版本，你可以提供 bean 的类名以避免类型转换。

```
SequenceGenerator generator = context.getBean("sequenceGenerator",SequenceGenerator.class);
```

如果只有一个 bean，还可以省略掉 bean 名字。

```
SequenceGenerator generator = context.getBean(SequenceGenerator.class);
```

到达这一步后，你就可以使用 POJO 或是 bean 了，就像在 Spring 之外使用构造方法创建的任何对象一样。

运行序列生成器应用的 Main 类代码如下所示：

```
package com.apress.springrecipes.sequence;

import com.apress.springrecipes.sequence.config.SequenceGeneratorConfiguration;
import org.springframework.context.ApplicationContext;
import org.springframework.context.annotation.AnnotationConfigApplicationContext;

public class Main {

    public static void main(String[] args) {
        ApplicationContext context =
            new AnnotationConfigApplicationContext(SequenceGeneratorConfiguration.class);

        SequenceGenerator generator = context.getBean(SequenceGenerator.class);

        System.out.println(generator.getSequence());
        System.out.println(generator.getSequence());
    }
}
```

如果一切都在 Java 类路径中（SequenceGenerator POJO 类与 Spring JAR 依赖），那么你会看到如下输出，还有一些日志信息：

```
30100000A
30100001A
```

使用@Component 注解创建 POJO 类并通过它来创建 DAO bean

到目前为止，Spring bean 实例化已经通过在 Java 配置类中以硬编码值的方式完成了。这是推荐的用于简化 Spring 示例的方式。

不过，对于大多数应用来说，POJO 实例化过程要么是通过数据库，要么是通过用户输入完成的。现在是时候往前一步，使用更加真实的场景了。对于本节来说，我们将会使用一个领域（Domain）类和一个数据访问对象（DAO）类来创建 POJO。你无须创建数据库（还是在 DAO 类中硬编码值），不过熟悉这类应用的结构是非常重要的，因为它是大多数真实应用与后续内容的基础。

假设你要开发一个序列生成器应用，就像上节所做的那样。你需要稍微修改一下类的结构，使之适合于领域类与 DAO 模式。首先，创建一个名为 Sequence 的领域类，其中包含了 id、prefix 与 suffix 属性。

```java
package com.apress.springrecipes.sequence;

public class Sequence {

    private final String id;
    private final String prefix;
    private final String suffix;

    public Sequence(String id, String prefix, String suffix) {
        this.id = id;
        this.prefix = prefix;
        this.suffix = suffix;
    }

    public String getId() {
        return id;
    }

    public String getPrefix() {
        return prefix;
    }

    public String getSuffix() {
        return suffix;
    }

}
```

接下来，为 DAO 创建一个接口，它负责访问数据库中的数据。getSequence()方法会从数据库中根据 ID 加载一个 POJO 或是 Sequence 对象，getNextValue()方法则会获取到特定数据库序列的下一个值。

```java
package com.apress.springrecipes.sequence;

public interface SequenceDao {

    public Sequence getSequence(String sequenceId);
    public int getNextValue(String sequenceId);
}
```

在生产应用中，你需要实现这个 DAO 接口以使用数据访问技术。不过为了简化，这里通过在 Map 中对值进行硬编码（存储序列实例与值）来实现 DAO。

```java
package com.apress.springrecipes.sequence;

import org.springframework.stereotype.Component;
```

```
import java.util.HashMap;
import java.util.Map;
import java.util.concurrent.atomic.AtomicInteger;

@Component("sequenceDao")
public class SequenceDaoImpl implements SequenceDao {

    private final Map<String, Sequence> sequences = new HashMap<>();
    private final Map<String, AtomicInteger> values = new HashMap<>();

    public SequenceDaoImpl() {
        sequences.put("IT", new Sequence("IT", "30", "A"));
        values.put("IT", new AtomicInteger(10000));
    }

    public Sequence getSequence(String sequenceId) {
        return sequences.get(sequenceId);
    }

    public int getNextValue(String sequenceId) {
        AtomicInteger value = values.get(sequenceId);
        return value.getAndIncrement();
    }
}
```

请注意 SequenceDaoImpl 类是如何被@Component("sequenceDao")注解修饰的。这会对类进行标记，这样 Spring 就可以通过它创建 POJO 了。@Component 注解中的值定义了 bean 实例 ID，在该示例中就是 sequenceDao。如果@Component 注解中没有提供 bean 值的名字，那么默认情况下，bean 的名字就是首字母小写的非全限定类名。比如说，对于 SequenceDaoImpl 类来说，默认的 bean 名字就是 sequenceDaoImpl。

调用 getSequence 方法会返回给定 sequenceID 的值。调用 getNextValue 方法会根据给定的 sequenceID 值创建一个新值并将这个新值返回。POJO 是在应用层中进行分类的。在 Spring 中有三层：持久化、服务与展示。@Component 是个通用的注解，用于修饰 POJO 以便让 Spring 能够检测到，而@Repository、@Service 与@Controller 则是@Component 的特化，它们针对具体的 POJO 层次，分别对应于持久化、服务与展示层。

如果不确定 POJO 的目的，你可以使用@Component 注解来修饰它。不过，在可能的情况下最好使用特化的注解，因为它们会根据 POJO 的目的提供额外的功能（比如说@Repository 会将异常包装为 DataAccessExceptions，这使得调试变得更加容易）。

使用过滤器实例化 Spring IoC 容器以扫描注解

默认情况下，Spring 会检测所有被@Configuration、@Bean、@Component、@Repository、@Service 与@Controller 注解所修饰的类。你可以定制扫描过程以包含一个或多个 include/exclude 过滤器。当一个 Java 包中有很多类时，这么做是很有帮助的。对于某些 Spring 应用上下文来说，排除或是包含拥有某些注解的 POJO 是必要的。

> ■ **警告：**扫描每个包会不必要地降低启动过程。

Spring 支持 4 类过滤器表达式。annotation 与 assignable 类型用于指定注解类型，类/接口则用于过滤。regex 与 aspectj 类型可以指定正则表达式以及 AspectJ 切点表达式以对类进行匹配。还可以通过 use-default-filters 属性禁用默认过滤器。

比如说，如下组件扫描包含了 com.apress.springrecipes.sequence 内所有名字中含有单词 Dao 或是 Service 的类，并且排除了带有@Controller 注解的类：

```
@ComponentScan(
    includeFilters = {
        @ComponentScan.Filter(
            type = FilterType.REGEX,
```

```
            pattern = {"com.apress.springrecipes.sequence.*Dao",
                       "com.apress.springrecipes.sequence.*Service"})
    },
    excludeFilters = {
        @ComponentScan.Filter(
            type = FilterType.ANNOTATION,
            classes = {org.springframework.stereotype.Controller.class}) }
    )
```

在应用包含过滤器检测名字中含有单词 Dao 或是 Service 的类时，甚至那些没有注解修饰的类也会被自动检测到。

从 IoC 容器中获取 POJO 实例或是 bean

接下来可以通过如下 Main 类测试之前的组件：

```
package com.apress.springrecipes.sequence;

import org.springframework.context.ApplicationContext;
import org.springframework.context.annotation.AnnotationConfigApplicationContext;

public class Main {

    public static void main(String[] args) {

        ApplicationContext context =
            new AnnotationConfigApplicationContext("com.apress.springrecipes.sequence");

        SequenceDao sequenceDao = context.getBean(SequenceDao.class);

        System.out.println(sequenceDao.getNextValue("IT"));
        System.out.println(sequenceDao.getNextValue("IT"));
    }
}
```

2-2 通过调用构造方法创建 POJO

问题提出

通过调用构造方法在 Spring IoC 容器中创建 POJO 实例或是 bean，这是最常见也是最直接的 bean 创建方式。它相当于在 Java 中使用 new 运算符创建对象。

解决方案

使用一个或多个构造方法定义 POJO 类。接下来，创建一个 Java 配置类，使用构造方法为 Spring IoC 容器配置 POJO 实例值。然后，实例化 Spring IoC 容器并扫描带有注解的 Java 类。POJO 实例或是 bean 实例会变成可访问的，并成为应用的一部分。

解释说明

假设你要开发一个商城应用以在线销售产品。首先，创建 Product POJO 类，它有几个属性，比如说产品名字和价格。由于商城中存在多种类型的产品，可将 Product 类设为抽象的，并从它继承下来不同的产品子类。

```
package com.apress.springrecipes.shop;

public abstract class Product {

    private String name;
    private double price;

    public Product() {}

    public Product(String name, double price) {
        this.name = name;
        this.price = price;
    }
```

```
    // Getters and Setters
    ...

    public String toString() {
        return name + " " + price;
    }
}
```

使用构造方法创建 POJO 类

接下来，创建两个产品子类：Battery 与 Disc。每个子类都有自己的属性。

```
package com.apress.springrecipes.shop;

public class Battery extends Product {

    private boolean rechargeable;

    public Battery() {
        super();
    }

    public Battery(String name, double price) {
        super(name, price);
    }

    // Getters and Setters
    ...
}

package com.apress.springrecipes.shop;

public class Disc extends Product {

    private int capacity;

    public Disc() {
        super();
    }

    public Disc(String name, double price) {
        super(name, price);
    }

    // Getters and Setters
    ...
}
```

为 POJO 创建 Java config

要想在 Spring IoC 容器中定义 POJO 类实例，你需要创建一个 Java 配置类并使用实例化值。拥有 POJO 或是 bean 定义并通过调用构造方法来定义的 Java 配置类如下代码所示：

```
package com.apress.springrecipes.shop.config;

import com.apress.springrecipes.shop.Battery;
import com.apress.springrecipes.shop.Disc;
import com.apress.springrecipes.shop.Product;
import org.springframework.context.annotation.Bean;
import org.springframework.context.annotation.Configuration;

@Configuration
public class ShopConfiguration {
```

```
    @Bean
    public Product aaa() {
        Battery p1 = new Battery("AAA", 2.5);
        p1.setRechargeable(true);
        return p1;
    }

    @Bean
    public Product cdrw() {
        Disc p2 = new Disc("CD-RW", 1.5);
        p2.setCapacity(700);
        return p2;
    }
}
```

接下来，编写如下的 Main 类，通过从 Spring IoC 容器中获取产品来进行测试：

```
package com.apress.springrecipes.shop;

import com.apress.springrecipes.shop.config.ShopConfiguration;
import org.springframework.context.ApplicationContext;
import org.springframework.context.annotation.AnnotationConfigApplicationContext;

public class Main {

    public static void main(String[] args) throws Exception {

        ApplicationContext context =
            new AnnotationConfigApplicationContext(ShopConfiguration.class);

        Product aaa = context.getBean("aaa", Product.class);
        Product cdrw = context.getBean("cdrw", Product.class);
        System.out.println(aaa);
        System.out.println(cdrw);
    }
}
```

2-3　使用 POJO 引用与自动装配和其他 POJO 进行交互

问题提出

构成应用的 POJO 实例或 bean 常常需要彼此协作来完成应用的功能。你想要通过注解来使用 POJO 引用与自动装配。

解决方案

对于定义在 Java 配置类中的 POJO 实例来说，你可以使用标准的 Java 代码在 bean 之间创建引用。要想自动装配 POJO 引用，你可以使用@Autowired 注解来标记字段、setter 方法、构造方法，甚至是任意方法。

解释说明

首先，我们会介绍使用构造方法、字段与属性来实现自动装配的不同方法。然后，你会看到如何解决自动装配中遇到的问题。

在 Java config 类中引用 POJO

当 POJO 实例定义在 Java 配置类中时（如 2-1 节与 2-2 节所示），POJO 引用的使用是很直接的，因为一切皆是 Java 代码。在如下示例中，一个 bean 属性引用了另外一个 bean：

```
package com.apress.springrecipes.sequence.config;

import com.apress.springrecipes.sequence.DatePrefixGenerator;
import com.apress.springrecipes.sequence.SequenceGenerator;
import org.springframework.context.annotation.Bean;
import org.springframework.context.annotation.Configuration;
```

```
@Configuration
public class SequenceConfiguration {

    @Bean
    public DatePrefixGenerator datePrefixGenerator() {
        DatePrefixGenerator dpg = new DatePrefixGenerator();
        dpg.setPattern("yyyyMMdd");
        return dpg;
    }

    @Bean
    public SequenceGenerator sequenceGenerator() {
        SequenceGenerator sequence = new SequenceGenerator();
        sequence.setInitial(100000);
        sequence.setSuffix("A");
        sequence.setPrefixGenerator(datePrefixGenerator());
        return sequence;
    }
}
```

SequenceGenerator 类的 prefixGenerator 属性是 DatePrefixGenerator bean 的实例。

第一个 bean 声明创建了一个 DatePrefixGenerator POJO。根据约定，可以通过 bean 名字 datePrefixGenerator（即方法名）来访问到它。不过，由于 bean 的实例化逻辑也是个标准的 Java 方法，因此还可以通过标准的 Java 调用来访问它。在设置 prefixGenerator 属性时（在第二个 bean 中，通过 setter），对方法 datePrefixGenerator()使用了标准的 Java 调用来引用该 bean。

使用@Autowired 注解自动装配 POJO 字段

接下来，我们对 2-1 节第二部分所引入的 DAO SequenceDaoImpl 类的 SequenceDao 字段进行自动装配。我们需要向应用添加一个服务类以说明 DAO 类的自动装配。

生成服务对象的服务类是另一个实际应用的最佳实践，它充当了访问 DAO 的外观，而非直接访问 DAO。在内部，服务对象会与 DAO 交互来处理序列生成请求。

```
package com.apress.springrecipes.sequence;
import org.springframework.beans.factory.annotation.Autowired;
import org.springframework.stereotype.Component;

@Component
public class SequenceService {

    @Autowired
    private SequenceDao sequenceDao;

    public void setSequenceDao(SequenceDao sequenceDao) {
        this.sequenceDao = sequenceDao;
    }

    public String generate(String sequenceId) {
        Sequence sequence = sequenceDao.getSequence(sequenceId);
        int value = sequenceDao.getNextValue(sequenceId);
        return sequence.getPrefix() + value + sequence.getSuffix();
    }
}
```

SequenceService 类使用了@Component 注解进行了修饰。这样，Spring 就可以检测到 POJO 了。由于@Component 注解没有名字，因此默认的 bean 名字是 sequenceService，这是根据类名生成的。

SequenceService 类的 sequenceDao 属性使用了@Autowired 注解进行了修饰。这样，Spring 就可以使用 sequenceDao bean 自动装配该属性了（即 SequenceDaoImpl 类）。

@Autowired 注解还可以用在数组类型的属性上，这样 Spring 就可以自动装配所有匹配的 bean 了。比如说，你可以对 PrefixGenerator[]属性使用@Autowired。这样，Spring 就会一次性自动装配其类型兼容于 PrefixGenerator 的所有 bean 了。

```
package com.apress.springrecipes.sequence;

import org.springframework.beans.factory.annotation.Autowired;

public class SequenceGenerator {

    @Autowired
    private PrefixGenerator[] prefixGenerators;
    ...
}
```

如果 IoC 容器中定义了多个类型兼容于 PrefixGenerator 的 bean，那么它们就会自动添加到 prefixGenerators 数组中。

与之类似，你可以将@Autowired 注解应用到类型安全的集合上。Spring 会读取该集合的类型信息，并自动装配类型兼容的所有 bean。

```
package com.apress.springrecipes.sequence;

import org.springframework.beans.factory.annotation.Autowired;

public class SequenceGenerator {

    @Autowired
    private List<PrefixGenerator> prefixGenerators;
    ...
}
```

如果 Spring 发现@Autowired 注解应用到了类型安全的 java.util.Map 上，并且 Map 的键是字符串，那么它会将所有类型兼容的 bean 添加到这个 map 中，并且将 bean 的名字作为键。

```
package com.apress.springrecipes.sequence;

import org.springframework.beans.factory.annotation.Autowired;

public class SequenceGenerator {

    @Autowired
    private Map<String, PrefixGenerator> prefixGenerators;
    ...
}
```

使用@Autowired 注解自动装配 POJO 方法与构造方法并且将自动装配配置为可选

@Autowired 注解还可以直接应用到 POJO 的 setter 方法上。比如说，你可以使用@Autowired 来注解 prefixGenerator 属性的 setter 方法。这样，Spring 就会尝试装配类型兼容于 prefixGenerator 的 bean。

```
package com.apress.springrecipes.sequence;

import org.springframework.beans.factory.annotation.Autowired;

public class SequenceGenerator {
    ...
    @Autowired
    public void setPrefixGenerator(PrefixGenerator prefixGenerator) {
        this.prefixGenerator = prefixGenerator;
    }
}
```

默认情况下，带有注解@Autowired 的所有属性都是必选的。当 Spring 无法找到匹配的 bean 进行自动装配时，它会抛出异常。如果想让某个属性变为可选的，请将@Autowired 的 required 属性设为 false。接下来，当 Spring 找不到匹配的 bean 时，它就不会再设置该属性了。

```
package com.apress.springrecipes.sequence;
```

```
import org.springframework.beans.factory.annotation.Autowired;

public class SequenceGenerator {
    ...
    @Autowired(required=false)
    public void setPrefixGenerator(PrefixGenerator prefixGenerator) {
        this.prefixGenerator = prefixGenerator;
    }
}
```

你也可以将@Autowired 注解应用到任意名称以及拥有任意数量参数的方法上；在这种情况下，Spring 会尝试对每个方法参数使用兼容类型来装配 bean。

```
package com.apress.springrecipes.sequence;

import org.springframework.beans.factory.annotation.Autowired;

public class SequenceGenerator {
    ...
    @Autowired
    public void myOwnCustomInjectionName(PrefixGenerator prefixGenerator) {
        this.prefixGenerator = prefixGenerator;
    }
}
```

最后，你还可以将@Autowired 注解应用到想要进行自动装配的构造方法上。构造方法可以拥有任意数量的参数，Spring 会尝试对构造方法的每个参数使用兼容类型来装配 bean。

```
@Service
public class SequenceService {

    private final SequenceDao sequenceDao;

    @Autowired
    public SequenceService(SequenceDao sequenceDao) {
        this.sequenceDao=sequenceDao;
    }

    public String generate(String sequenceId) {
        Sequence sequence = sequenceDao.getSequence(sequenceId);
        int value = sequenceDao.getNextValue(sequenceId);
        return sequence.getPrefix() + value + sequence.getSuffix();
    }
}
```

■ **提示：**从 Spring Framework 4.3 开始，如果只有一个构造方法，那么 Spring 会自动使用这个构造方法进行自动装配。在这种情况下，可以省略掉@Autowired 注解。

使用注解解决自动装配的模糊性问题

默认情况下，当 IoC 容器中具有兼容类型的 bean 数量多于一个且属性不是组类型（即数组、list 或 map 等）时，按类型自动装配将无法正常使用，这一点之前已经解释过了。不过，如果相同类型的 bean 数量多于一个，那么有两种方式可以解决按类型自动装配的问题：@Primary 注解与@Qualifier 注解。

使用@Primary 注解解决自动装配的模糊性问题

Spring 可以让我们通过为候选者使用@Primary 注解来根据类型指定一个候选 bean。当多个候选者都满足单值依赖的自动装配时，@Primary 注解可以让被注解的 bean 拥有更高的优先级。

```
package com.apress.springrecipes.sequence;
...
import org.springframework.stereotype.Component;
import org.springframework.context.annotation.Primary;
```

```
@Component
@Primary
public class DatePrefixGenerator implements PrefixGenerator {

    public String getPrefix() {
        DateFormat formatter = new SimpleDateFormat("yyyyMMdd");
        return formatter.format(new Date());
    }
}
```

注意到上面的 POJO 实现了 PrefixGenerator 接口并使用了@Primary 注解进行修饰。如果尝试使用 PrefixGenerator 类型进行 bean 的自动装配，即便 Spring 有多个具有相同类型 PrefixGenerator 的 bean 实例，它也会自动装配 DatePrefixGenerator，因为它被标记为了@Primary 注解。

使用@Qualifier 注解解决自动装配的模糊性问题

Spring 还可以通过在@Qualifier 注解中提供名字来根据类型指定一个候选 bean。

```
package com.apress.springrecipes.sequence;

import org.springframework.beans.factory.annotation.Autowired;
import org.springframework.beans.factory.annotation.Qualifier;

public class SequenceGenerator {

    @Autowired
    @Qualifier("datePrefixGenerator")
    private PrefixGenerator prefixGenerator;
    ...
}
```

加上了@Qualifier 注解后，Spring 会尝试在 IoC 容器中寻找具有该名字的 bean，并将其装配到属性中。

@Qualifier 注解还可以用在方法参数上进行自动装配。

```
package com.apress.springrecipes.sequence;

import org.springframework.beans.factory.annotation.Autowired;
import org.springframework.beans.factory.annotation.Qualifier;

public class SequenceGenerator {
    ...
    @Autowired
    public void myOwnCustomInjectionName(
        @Qualifier("datePrefixGenerator") PrefixGenerator prefixGenerator) {
        this.prefixGenerator = prefixGenerator;
    }
}
```

如果想要根据名字自动装配 bean 属性，你可以为 setter 方法、构造方法或是字段使用下一节将会介绍的 JSR-250 @Resource 注解。

从多个位置处解析 POJO 引用

随着应用规模的不断增长，在单个 Java 配置类中管理每一个 POJO 将会变得愈发困难。一种常见的做法是根据功能将 POJO 划分到多个 Java 配置类中。在创建多个 Java 配置类时，获取引用并自动装配定义在不同类中的 POJO 就不如单个 Java 配置类那么直接了。

一种方式是使用每个 Java 配置类的位置来初始化应用上下文。在这种方式下，每个 Java 配置类中的 POJO 都会被加载到上下文和引用中，在不同 POJO 中进行自动装配也就成为了可能。

```
AnnotationConfigApplicationContext context = new AnnotationConfigApplicationContext
    (PrefixConfiguration.class, SequenceGeneratorConfiguration.class);
```

另一种方式是使用@Import 注解，这样 Spring 就会让一个配置文件中的 POJO 对另外一个文件可见。

```
package com.apress.springrecipes.sequence.config;

import org.springframework.context.annotation.Bean;
import org.springframework.context.annotation.Import;
import org.springframework.beans.factory.annotation.Value;
import org.springframework.context.annotation.Configuration;
import com.apress.springrecipes.sequence.SequenceGenerator;
import com.apress.springrecipes.sequence.PrefixGenerator;

@Configuration
@Import(PrefixConfiguration.class)
public class SequenceConfiguration {
    @Value("#{datePrefixGenerator}")
    private PrefixGenerator prefixGenerator;

    @Bean
    public SequenceGenerator sequenceGenerator() {
        SequenceGenerator sequence= new SequenceGenerator();
        sequence.setInitial(100000);
        sequence.setSuffix("A");
        sequence.setPrefixGenerator(prefixGenerator);
        return sequence;
    }
}
```

sequenceGenerator bean 要求你设置一个 prefixGenerator bean。不过请注意，在该 Java 配置类中并未定义 prefixGenerator bean。prefixGenerator bean 定义在另一个名为 PrefixConfiguration 的 Java 配置类中。借助于@Import(PrefixConfiguration.class)注解，Spring 会将该 Java 配置类中的所有 POJO 都带入到当前配置类的作用域中。在 PrefixConfiguration 中的 POJO 都位于作用域中的情况下，你使用@Value 注解与 SpEL 将名为 datePrefixGenerator 的 bean 注入到了 prefixGenerator 字段中。当该 bean 被注入后，它就可以为 sequenceGenerator bean 设置 prefixGenerator bean 了。

2-4　使用@Resource 与@Inject 注解自动装配 POJO

问题提出

你想要使用 Java 标准的@Resource 与@Inject 注解通过自动装配来引用 POJO，而不是使用 Spring 专有的@Autowired 注解。

解决方案

JSR-250（Common Annotations for the Java Platform，Java 平台的通用注解）定义了@Resource 注解来根据名字自动装配 POJO 引用。JSR-330（Standard Annotations for Injection，Injection 的标准注解）定义了@Inject 注解来根据类型自动装配 POJO 引用。

解释说明

上一节介绍的@Autowired 注解来自于 Spring 框架，它位于 org.springframework.beans.factory.annotation 包下。这意味着它只能用在 Spring 框架上下文中。

在 Spring 支持@Autowired 注解后不久，Java 语言标准化了各种注解来满足@Autowired 注解的相同目的。这些注解有@Resource，它位于 javax.annotation 包下；还有@Inject，它位于 javax.inject 包下。

使用@Resource 注解自动装配 POJO

默认情况下，@Resource 注解的工作方式类似于 Spring 的@Autowired 注解，会根据类型进行自动装配。比如说，如下 POJO 属性使用@Resource 注解进行修饰，这样 Spring 就会尝试寻找匹配 PrefixGenerator 类型的 POJO。

```
package com.apress.springrecipes.sequence;

import javax.annotation.Resource;
```

```
public class SequenceGenerator {

    @Resource
    private PrefixGenerator prefixGenerator;
    ...
}
```

不过，与@Autowired 注解（需要使用@Qualifier 注解才能根据名字来自动装配 POJO）所不同的是，如果有多个相同类型的 POJO 存在，@Resource 会消除模糊性。本质上，@Resource 注解的功能相当于将@Autowired 注解和@Qualifier 注解合二为一。

使用@Inject 注解自动装配 POJO

@Inject 注解会根据类型进行自动装配，就像@Resource 与@Autowired 注解一样。比如说，如下 POJO 属性使用@Inject 注解进行修饰，这样 Spring 就会尝试寻找匹配 PrefixGenerator 类型的 POJO。

```
package com.apress.springrecipes.sequence;

import javax.inject.Inject;

public class SequenceGenerator {

    @Inject
    private PrefixGenerator prefixGenerator;
    ...
}
```

但就像@Resource 与@Autowired 注解一样，如果有多个相同类型的 POJO 存在，那就需要采取不同的方式来根据名字匹配 POJO 或是避免模糊性。使用@Inject 注解根据名字来进行自动装配的第一步是要创建一个自定义注解，用于标识 POJO 注入类与 POJO 注入点。

```
package com.apress.springrecipes.sequence;

import java.lang.annotation.Documented;
import java.lang.annotation.ElementType;
import java.lang.annotation.Retention;
import java.lang.annotation.RetentionPolicy;
import java.lang.annotation.Target;

import javax.inject.Qualifier;

@Qualifier
@Target({ElementType.TYPE, ElementType.FIELD, ElementType.PARAMETER})
@Documented
@Retention(RetentionPolicy.RUNTIME)
public @interface DatePrefixAnnotation {
}
```

注意到该自定义注解使用到了@Qualifier 注解。这个注解不同于 Spring 的@Qualifier 注解，因为它与@Inject 注解都位于相同的 Java 包中（即 javax.inject）。

定义好自定义注解后，接下来就要使用它来修饰生成 bean 实例的 POJO 注入类了，在该示例中就是 DatePrefixGenerator 类。

```
package com.apress.springrecipes.sequence;
...
@DatePrefixAnnotation
public class DatePrefixGenerator implements PrefixGenerator {
...
}
```

最后，POJO 属性或是注入点都会被相同的自定义注解所修饰，从而限定好 POJO 并消除任何模糊性。

```
package com.apress.springrecipes.sequence;

import javax.inject.Inject;
```

```
public class SequenceGenerator {

    @Inject @DataPrefixAnnotation
    private PrefixGenerator prefixGenerator;
    ...
}
```

如 2-3 节与 2-4 节所示，@Autowired、@Resource 与@Inject 这 3 个注解可以实现同样的效果。@Autowired 注解是基于 Spring 的解决方案，而@Resource 与@Inject 注解则是 Java 标准解决方案（即 JSR）。如果想要进行基于名字的自动装配，那么@Resource 注解提供了最为简单的语法。对于根据类型进行自动装配来说，这 3 个注解的使用都很直观，因为它们都只需要唯一一个注解即可。

2-5　使用@Scope 注解设置 POJO 的作用域

问题提出

在使用如@Component 这样的注解声明 POJO 实例时，实际上是为 bean 的创建而非 bean 实例定义了一个模板。当 bean 被 getBean()方法请求或是被其他 bean 引用时，Spring 会根据 bean 的作用域来决定应该返回哪个 bean 实例。有时，你需要为 bean 设置一个恰当的作用域而非使用默认作用域。

解决方案

Bean 的作用域是通过@Scope 注解来设置的。默认情况下，Spring 只会为 IoC 容器中所声明的 bean 创建唯一一个实例，该实例会在整个 IoC 容器的作用域中被共享。后续所有的 getBean()调用与 bean 引用都会返回这个唯一的 bean 实例。该作用域叫做 singleton（单例），它是所有 bean 的默认作用域。表 2-1 列出了 Spring 中所有有效的 bean 作用域。

表 2-1　Spring 中有效的 bean 作用域

作用域	说明
singleton	每个 Spring IoC 容器中创建单个 bean 实例
prototype	每次请求时都创建一个新的 bean 实例
request	每个 HTTP 请求都创建单个 bean 实例；只用在 Web 应用上下文环境中
session	每个 HTTP 会话都创建单个 bean 实例；只用在 Web 应用上下文环境中
globalSession	每个全局 HTTP 会话都创建单个 bean 实例；只用在 portal（门户）应用上下文环境中

解释说明

为了演示 bean 作用域的概念，我们考虑一个商城应用中的购物车示例。首先，创建一个 ShoppingCart 类，代码如下所示：

```
package com.apress.springrecipes.shop;
...
@Component
public class ShoppingCart {

    private List<Product> items = new ArrayList<>();

    public void addItem(Product item) {
        items.add(item);
    }

    public List<Product> getItems() {
        return items;
    }
}
```

接下来，在 Java 配置文件中声明几个产品 bean，稍后将其添加到购物车中。

```
package com.apress.springrecipes.shop.config;

import com.apress.springrecipes.shop.Battery;
import com.apress.springrecipes.shop.Disc;
import com.apress.springrecipes.shop.Product;
import org.springframework.context.annotation.Bean;
import org.springframework.context.annotation.ComponentScan;
import org.springframework.context.annotation.Configuration;

@Configuration
@ComponentScan("com.apress.springrecipes.shop")
public class ShopConfiguration {

    @Bean
    public Product aaa() {
        Battery p1 = new Battery();
        p1.setName("AAA");
        p1.setPrice(2.5);
        p1.setRechargeable(true);
        return p1;
    }

    @Bean
    public Product cdrw() {
        Disc p2 = new Disc("CD-RW", 1.5);
        p2.setCapacity(700);
        return p2;
    }

    @Bean
    public Product dvdrw() {
        Disc p2 = new Disc("DVD-RW", 3.0);
        p2.setCapacity(700);
        return p2;
    }
}
```

接下来，定义一个 Main 类，向购物车中添加几件产品进行测试。假设同时有两个客户在浏览你的商城。第一个客户通过 getBean()方法获得了一个购物车，并向其中添加了两件产品。接下来，第二个客户也通过 getBean()方法获得了一个购物车，并向其中添加了另外一件产品。

```
package com.apress.springrecipes.shop;

import com.apress.springrecipes.shop.config.ShopConfiguration;
import org.springframework.context.ApplicationContext;
import org.springframework.context.annotation.AnnotationConfigApplicationContext;

public class Main {

    public static void main(String[] args) throws Exception {
        ApplicationContext context =
            new AnnotationConfigApplicationContext(ShopConfiguration.class);

        Product aaa = context.getBean("aaa", Product.class);
        Product cdrw = context.getBean("cdrw", Product.class);
        Product dvdrw = context.getBean("dvdrw", Product.class);

        ShoppingCart cart1 = context.getBean("shoppingCart", ShoppingCart.class);
        cart1.addItem(aaa);
        cart1.addItem(cdrw);
        System.out.println("Shopping cart 1 contains " + cart1.getItems());


        ShoppingCart cart2 = context.getBean("shoppingCart", ShoppingCart.class);
```

```
        cart2.addItem(dvdrw);
        System.out.println("Shopping cart 2 contains " + cart2.getItems());
    }
}
```

上述 bean 声明的结果就是，你会看到这两个客户拿到的是同一个购物车实例。

```
Shopping cart 1 contains [AAA 2.5, CD-RW 1.5]
Shopping cart 2 contains [AAA 2.5, CD-RW 1.5, DVD-RW 3.0]
```

这是因为 Spring 默认的 bean 作用域是 singleton，这意味着针对于每个 IoC 容器来说，Spring 只会创建唯一一个购物车实例。

在商城应用中，你期望在调用 getBean()方法时，每个客户都能获取到不同的购物车实例。为了实现这种行为，需要将 shoppingCart 的作用域设为 prototype。这样，每次调用 getBean()方法时，Spring 都会创建一个新的 bean 实例。

```
package com.apress.springrecipes.shop;
...
import org.springframework.stereotype.Component;
import org.springframework.context.annotation.Scope;

@Component
@Scope("prototype")
public class ShoppingCart { … }
```

再次运行 Main 类，你会看到两个客户获取到了不同的购物车实例。

```
Shopping cart 1 contains [AAA 2.5, CD-RW 1.5]
Shopping cart 2 contains [DVD-RW 3.0]
```

2-6　使用来自于外部资源（文本文件、XML 文件、属性文件或图像文件）的数据

问题提出

有时，应用需要从不同位置（如文件系统、类路径或是 URL）读取外部资源（如文本文件、XML 文件、属性文件或是图像文件）。通常，你需要处理不同的 API 来加载不同位置的资源。

解决方案

Spring 提供了@PropertySource 注解来加载.properties 文件中的内容（即键值对）并设置 bean 属性。

Spring 还有一种资源加载器机制，它提供了一个统一的 Resource 接口，通过资源路径就可以获取到任意类型的外部资源。你可以通过@Value 注解为路径指定不同的前缀，以从不同的位置加载资源。要想从文件系统加载资源，可以使用 file 前缀。要想从类路径加载资源，可以使用 classpath 前缀。还可以在资源路径中指定 URL。

解释说明

要想读取属性文件（即键值对）的内容来设置 bean 属性，可以使用 Spring 的@PropertySource 注解以及 PropertySourcesPlaceholderConfigurer。要想读取任何文件的内容，可以使用 Spring 的 Resource 机制并使用@Value 注解进行修饰。

使用属性文件数据来设置 POJO 实例化值

假设在属性文件中有一系列值，你想要使用它们来设置 bean 属性。通常情况下，它们可能是数据库的配置属性或是键值对形式的其他应用值。比如说，获取存储在 discounts.properties 文件中的如下键值对：

```
specialcustomer.discount=0.1
summer.discount=0.15
endofyear.discount=0.2
```

■ **注意：**要想读取属性文件以进行国际化（i18n），请参考下一节内容。要想访问 discounts.properties 文件的内容来创建其他 bean，可以使用@PropertySource 注解并在 Java 配置类中将键值对转换为 bean。

```
package com.apress.springrecipes.shop.config;

import com.apress.springrecipes.shop.Battery;
import com.apress.springrecipes.shop.Disc;
import com.apress.springrecipes.shop.Product;
import org.springframework.beans.factory.annotation.Value;
import org.springframework.context.annotation.Bean;
import org.springframework.context.annotation.ComponentScan;
import org.springframework.context.annotation.Configuration;
import org.springframework.context.annotation.PropertySource;
import org.springframework.context.support.PropertySourcesPlaceholderConfigurer;

@Configuration
@PropertySource("classpath:discounts.properties")
@ComponentScan("com.apress.springrecipes.shop")
public class ShopConfiguration {

    @Value("${endofyear.discount:0}")
    private double specialEndofyearDiscountField;

    @Bean
    public static PropertySourcesPlaceholderConfigurer
        propertySourcesPlaceholderConfigurer() {
        return new PropertySourcesPlaceholderConfigurer();
    }

    @Bean
    public Product dvdrw() {
        Disc p2 = new Disc("DVD-RW", 3.0, specialEndofyearDiscountField);
        p2.setCapacity(700);
        return p2;
    }
}
```

这里定义了一个@PropertySource 注解并使用 classpath:discounts.properties 值来修饰 Java 配置类。classpath: 前缀告诉 Spring 在 Java 类路径中寻找 discounts.properties 文件

当定义了@PropertySource 注解来加载属性文件后，你还需要定义一个 PropertySourcePlaceholder-Configurer bean 并使用@Bean 注解修饰。Spring 会自动装配@PropertySource discounts. properties 文件，这样其属性就可以作为 bean 属性了。

接下来，你需要定义 Java 变量从 discount discounts.properties 文件中获取值。要想使用这些值定义 Java 变量值，你需要使用@Value 注解和一个占位符表达式。

语法是@Value("${key:default_value}")。Spring 会在所有加载的应用属性中搜索该键值。如果在属性文件中找到匹配的 key=value，那么相应的值就会被赋给 bean 属性。如果在加载的应用属性中找不到匹配的 key=value，那么默认值（即${key:之后的值）就会被赋给 bean 属性。

当 Java 变量被赋予 discount 值后，你就可以使用它为 bean 的 discount 属性赋值来创建 bean 实例了。

如果想要使用属性文件数据做其他事情而非设置 bean 属性，那么你就应该使用 Spring 的 Resource 机制，接下来就会介绍这一点。

在 POJO 中使用任意外部资源文件中的数据

假设你想在应用启动时展示一个条幅。该条幅包含了如下字符并存储在一个名为 banner.txt 的文本文件中。该文件位于应用的类路径下。

```
*************************
*  Welcome to My Shop!  *
*************************
```

接下来编写一个 BannerLoader POJO 类加载条幅并将其输出到控制台上。

```
package com.apress.springrecipes.shop;
```

```
import org.springframework.core.io.Resource;
...
import javax.annotation.PostConstruct;
public class BannerLoader {

    private Resource banner;
    public void setBanner(Resource banner) {
        this.banner = banner;
    }

    @PostConstruct
    public void showBanner() throws IOException {
        InputStream in = banner.getInputStream();

        BufferedReader reader = new BufferedReader(new InputStreamReader(in));
        while (true) {
            String line = reader.readLine();
            if (line == null)
                break;
            System.out.println(line);
        }
        reader.close();
    }
}
```

注意到 POJO banner 字段是个 Spring Resource 类型。一言以蔽之，在创建 bean 实例时，该字段值会通过 setter 注入进行装配。showBanner()方法会调用 getInputStream()方法从 Resource 字段中获取到输入流。在获取到 InputStream 后，你就可以使用标准的 Java 文件操作类了。在该示例中，通过 BufferedReader 逐行读取文件内容并输出到控制台。

此外，showBanner()方法使用了@PostConstruct 注解进行了修饰。由于要在启动时展示条幅，因此通过该注解告诉 Spring 在创建完毕后自动调用该方法。这可以确保 showBanner()方法成为应用运行的首批方法之一，因此也可以确保条幅一开始就会出现。

接下来，POJO BannerLoader 需要作为一个实例进行初始化。此外，BannerLoader 的 banner 字段也需要进行注入。下面就来创建一个 Java 配置类完成这些事情。

```
@Configuration
@PropertySource("classpath:discounts.properties")
@ComponentScan("com.apress.springrecipes.shop")
public class ShopConfiguration {

    @Value("classpath:banner.txt")
    private Resource banner;

    @Bean
    public static PropertySourcesPlaceholderConfigurer propertySourcesPlaceholderConfigurer() {
        return new PropertySourcesPlaceholderConfigurer();
    }

    @Bean
    public BannerLoader bannerLoader() {
        BannerLoader bl = new BannerLoader();
        bl.setBanner(banner);
        return bl;
    }
}
```

看一下 banner 属性是如何通过@Value("classpath:banner.txt")注解进行修饰的。这会告诉 Spring 在类路径中搜索 banner.txt 文件并将其注入进来。在注入到 bean 之前，Spring 会使用预先注册好的属性编辑器 ResourceEditor 将文件定义转换为 Resource 对象。

当 banner 属性被注入后，它会通过 setter 注入赋给 BannerLoader bean 实例。由于条幅文件位于 Java 类路径下，因此资源路径会以 classpath 前缀开始。之前的资源路径以文件系统相对路径的方式指定了资源（也可以指定一个绝对路径）。

```
file:c:/shop/banner.txt
```

当资源位于 Java 类路径下时，需要使用 classpath 前缀。如果没有路径信息，则会从类路径的根目录进行加载。

```
classpath:banner.txt
```

如果资源位于特定的包下面，则可以从类路径根目录处指定绝对路径。

```
classpath:com/apress/springrecipes/shop/banner.txt
```

除了支持从文件系统和类路径下加载外，还可以通过指定 URL 来加载资源。

```
http://springrecipes.apress.com/shop/banner.txt
```

由于 bean 类在 showBanner()方法上使用了@PostConstruct 注解，因此当 IoC 容器创建后条幅就会被发送至输出。由于这一点，现在就没必要使用应用上下文或是显式调用 bean 来输出条幅。不过，有时我们需要访问外部资源来与应用上下文进行交互。现在，假设你想在应用结束时展示一个说明。该说明包含了之前介绍的 discounts.properties 文件中的 discounts。要想访问到属性文件的内容，你还可以使用 Spring 的 Resource 机制。

接下来就来使用 Spring 的 Resource 机制，不过这次是在应用结束时直接在应用的 Main 类中输出说明。

```
import org.springframework.core.io.ClassPathResource;
import org.springframework.core.io.support.PropertiesLoaderUtils;
...
...
public class Main {

    public static void main(String[] args) throws Exception {
        ...
        Resource resource = new ClassPathResource("discounts.properties");
        Properties props = PropertiesLoaderUtils.loadProperties(resource);
        System.out.println("And don't forget our discounts!");
        System.out.println(props);
    }
}
```

Spring 的 ClassPathResource 类用于访问 discounts.properties 文件，它会将文件内容转换为 Resource 对象。接下来，Resource 对象会经由 Spring 的 PropertiesLoaderUtils 类处理为一个 Properties 对象。最后，Properties 对象的内容会被发送至控制台作为应用的最终输出。

由于说明文件（即 discounts.properties）位于 Java 类路径下，因此资源是通过 Spring 的 ClassPathResource 类访问的。如果外部资源位于文件系统路径下，那么资源将会由 Spring 的 FileSystemResource 加载。

```
Resource resource = new FileSystemResource("c:/shop/banner.txt")
```

如果外部资源是个 URL，那么资源将会由 Spring 的 UrlResource 加载。

```
Resource resource = new UrlResource("http://www.apress.com/")
```

2-7 针对不同地域的属性文件解析 i18n 文本信息

问题提出

希望应用通过注解方式支持国际化。

解决方案

MessageSource 是个接口，定义了用于解析资源包中信息的几个方法。ResourceBundleMessageSource 是最常用的一个 MessageSource 实现，可以针对不同地域解析资源包中的信息。在实现了 ResourceBundleMessageSource POJO 后，你就可以在 Java 配置文件中通过@Bean 注解在应用中使用 i18n 数据了。

解释说明

作为一个示例，针对美式英语创建一个名为 messages_en_US.properties 的资源包。该资源包是从类路径的根目录下加载的，这样可以确保它位于 Java 类路径下。在该文件中编写如下键值对：

```
alert.checkout=A shopping cart has been checked out.
alert.inventory.checkout=A shopping cart with {0} has been checked out at {1}.
```

要想从资源包中解析信息，我们需要通过 ReloadableResourceBundleMessageSource bean 实例创建一个 Java 配置文件。

```
package com.apress.springrecipes.shop.config;

import org.springframework.context.annotation.Bean;
import org.springframework.context.annotation.Configuration;
import org.springframework.context.support.ReloadableResourceBundleMessageSource;

@Configuration
public class ShopConfiguration {

    @Bean
    public ReloadableResourceBundleMessageSource messageSource() {
        ReloadableResourceBundleMessageSource messageSource =
            new ReloadableResourceBundleMessageSource();
        messageSource.setBasenames("classpath:messages");
        messageSource.setCacheSeconds(1);
        return messageSource;
    }
}
```

bean 实例必须命名为 messageSource，这样应用上下文才能检测到它。

在 bean 定义中，你通过 setBasenames 方法声明了一个 String 列表以便为 ResourceBundleMessageSource 定位到包资源。在该示例中，只是指定了默认的约定来查找 Java 类路径下名字以 messages 开头的文件。此外，setCacheSeconds 方法会将缓存时间设为 1 秒，从而避免读取到过期的消息。注意到在重新加载资源文件前，它会先检查属性文件的上次修改时间戳，这样如果文件没有变化，那么 setCacheSeconds 间隔时间就可以设置得更长一些，因为刷新并未进行实际的重新加载。

对于这个 MessageSource 定义来说，如果查找美国区域（首选语言是英语）的文本信息，那就会首先考虑 messages_en_US.properties 资源包。如果该资源包不存在或是找不到信息，那么就会考虑使用与语言匹配的 messages_en.properties 文件。如果依然找不到该文件，那么就会选择适用于所有地域的默认 messages.properties 文件。要想了解关于资源包加载的更多信息，可以参考 java.util.ResourceBundle 类的 Javadoc。

接下来，可以配置应用上下文以使用 getMessage()方法解析信息。第一个参数是对应于信息的键，第三个参数则是目标地域。

```
package com.apress.springrecipes.shop;

import com.apress.springrecipes.shop.config.ShopConfiguration;
import org.springframework.context.ApplicationContext;
import org.springframework.context.annotation.AnnotationConfigApplicationContext;

import java.util.Date;
import java.util.Locale;

public class Main {

    public static void main(String[] args) throws Exception {

        ApplicationContext context =
            new AnnotationConfigApplicationContext(ShopConfiguration.class);

        String alert = context.getMessage("alert.checkout", null, Locale.US);
        String alert_inventory = context.getMessage("alert.inventory.checkout", new Object[]
        {"[DVD-RW 3.0]", new Date()}, Locale.US);
        System.out.println("The I18N message for alert.checkout is: " + alert);
        System.out.println("The I18N message for alert.inventory.checkout is: " +
        alert_inventory);
    }
}
```

getMessage()方法的第二个参数是一个信息参数数组。在第一个 String 语句中，值是 null；在第二个 String

语句中，使用了一个对象数组来填充信息参数。

在 Main 类中，你可以解析文本信息，因为你可以直接访问到应用上下文。不过对于要解析文本信息的 bean 来说，你需要将一个 MessageSource 实现注入到想要解析文本信息的 bean 中。我们为商城应用实现一个 Cashier 类，它演示了如何解析信息。

```
package com.apress.springrecipes.shop;
...
@Component
public class Cashier {

    @Autowired
    private MessageSource messageSource;

    public void setMessageSource(MessageSource messageSource) {
        this.messageSource = messageSource;
    }

    public void checkout(ShoppingCart cart) throws IOException {
        String alert = messageSource.getMessage("alert.inventory.checkout",
                                                new Object[] { cart.getItems(),new Date() },
                                                Locale.US);
        System.out.println(alert);
    }
}
```

注意到 POJO messageSource 字段是个 Spring MessageSource 类型。该字段值使用@Autowired 注解进行了修饰，因此当 bean 实例创建时，它会通过注入来进行装配。接下来，checkout 方法就可以访问 messageSource 字段，它可以让 bean 访问到 getMessage 方法，从而根据 i18n 标准获取到文本信息。

2-8　使用注解自定义 POJO 初始化与销毁动作

问题提出

一些 POJO 需要在使用前执行某些初始化任务。这些任务有打开文件、打开网络/数据库连接、分配内存等。此外，同样的这些 POJO 还需要在生命周期结束时执行对应的销毁任务。因此，有时需要在 Spring IoC 容器中自定义 bean 的初始化与销毁动作。

解决方案

Spring 可以通过在 Java 配置类中设置@Bean 定义的 initMethod 与 destroyMethod 属性来识别出初始化与销毁回调方法。如果 POJO 方法使用了@PostConstruct 与@PreDestroy 注解修饰，Spring 同样可以识别出来。此外，Spring 还可以延迟 bean 的创建，直到需要这个 bean 的时候再创建（叫做延迟初始化），这可以通过@Lazy 注解来达成。Spring 还可以借助于@DependsOn 注解让某些 bean 的初始化发生在其他 bean 之前。

解释说明

使用@Bean 注解来定义在 POJO 初始化和销毁前运行的方法。我们来看一下商城应用，考虑一个涉及结账功能的示例。修改 Cashier 类以将购物车中的产品和结账时间记录到文本文件中。

```
package com.apress.springrecipes.shop;

import java.io.*;
import java.util.Date;

public class Cashier {

    private String fileName;
    private String path;
    private BufferedWriter writer;

    public void setFileName(String fileName) {
        this.fileName = fileName;
    }
```

```
    public void setPath(String path) {
        this.path = path;
    }

    public void openFile() throws IOException {

        File targetDir = new File(path);
        if (!targetDir.exists()) {
            targetDir.mkdir();
        }

        File checkoutFile = new File(path, fileName + ".txt");
        if (!checkoutFile.exists()) {
            checkoutFile.createNewFile();
        }

        writer = new BufferedWriter(new OutputStreamWriter(
                new FileOutputStream(checkoutFile, true)));
    }

    public void checkout(ShoppingCart cart) throws IOException {
        writer.write(new Date() + "\t" + cart.getItems() + "\r\n");
        writer.flush();
    }

    public void closeFile() throws IOException {
        writer.close();
    }
}
```

在 Cashier 类中，openFile()方法会先验证用于写入数据的目标目录与文件是否存在。接下来，打开指定系统路径下的文本文件并将其赋给 writer 字段。在每次调用 checkout()方法时都会将日期与购物车中的产品追加到这个文本文件中。最后，closeFile()方法会关闭文件以释放系统资源。

接下来，我们看看如何在 Java 配置类中创建这个 bean 定义，使之在 bean 创建前执行 openFile()方法，在 bean 销毁前执行 closeFile()方法。

```
@Configuration
public class ShopConfiguration {

    @Bean(initMethod = "openFile", destroyMethod = "closeFile")
    public Cashier cashier() {

        String path = System.getProperty("java.io.tmpdir") + "/cashier";
        Cashier c1 = new Cashier();
        c1.setFileName("checkout");
        c1.setPath(path);
        return c1;
    }
}
```

注意到 POJO 的初始化与销毁任务是通过定义在@Bean 注解中的 initMethod 与 destroyMethod 属性实现的。通过在 bean 声明中定义这两个属性，当 Cashier 类创建时，它会首先触发 openFile()方法，验证用于写入数据的目标目录与文件是否存在，并打开文件来追加记录。当 bean 销毁时，它会触发 closeFile()方法，确保关闭了文件引用，从而释放系统资源。

使用@PostConstruct 与@PreDestroy 定义 POJO 初始化与销毁前执行的方法

如果在 Java 配置类外部定义了 POJO 实例（比如说使用@Component 注解），那么可以直接在 POJO 类中使用@PostConstruct 与@PreDestroy 注解。

```
@Component
```

```
public class Cashier {

    @Value("checkout")
    private String fileName;
    @Value("c:/Windows/Temp/cashier")
    private String path;
    private BufferedWriter writer;

    public void setFileName(String fileName) {
        this.fileName = fileName;
    }

    public void setPath(String path) {
        this.path = path;
    }

    @PostConstruct
    public void openFile() throws IOException {
        File targetDir = new File(path);
        if (!targetDir.exists()) {
            targetDir.mkdir();
        }
        File checkoutFile = new File(path, fileName + ".txt");
        if(!checkoutFile.exists()) {
            checkoutFile.createNewFile();
        }
        writer = new BufferedWriter(new OutputStreamWriter(
                 new FileOutputStream(checkoutFile, true)));
    }

    public void checkout(ShoppingCart cart) throws IOException {
        writer.write(new Date() + "\t" +cart.getItems() + "\r\n");
        writer.flush();
    }

    @PreDestroy
    public void closeFile() throws IOException {
        writer.close();
    }
}
```

@Component 注解告诉 Spring 要管理该 POJO，就像前文中所用的那样。其中有两个 POJO 字段值使用到了@Value 注解，@Value 注解在前文中也介绍过了。openFile()方法使用@PostConstruct 注解进行修饰，这告诉 Spring 在 bean 构建完毕后就去执行这个方法。closeFile()方法使用@PreDestroy 注解进行修饰，这告诉 Spring 在 bean 销毁前执行该方法。

使用@Lazy 定义 POJO 的延迟初始化

默认情况下，Spring 会对所有 POJO 执行尽早的初始化。这意味着 POJO 会在启动时就进行初始化。不过在某些场景下，将 POJO 的初始化过程延迟到需要时再进行会更加方便，这就叫作延迟初始化。

延迟初始化有助于在启动时限制资源的消耗并能节省整体的系统资源。对于执行重量级操作（如网络连接、文件操作等）的 POJO 来说，延迟初始化尤其有帮助。要想将 bean 标记为延迟初始化，只需使用@Lazy 注解来修饰 bean 即可。

```
package com.apress.springrecipes.shop;
...
import org.springframework.stereotype.Component;
import org.springframework.context.annotation.Scope;
import org.springframework.context.annotation.Lazy;

@Component
```

```
@Scope("prototype")
@Lazy
public class ShoppingCart {

    private List<Product> items = new ArrayList<>();

    public void addItem(Product item) {
        items.add(item);
    }

    public List<Product> getItems() {
        return items;
    }
}
```

在上述声明中，POJO 使用了@Lazy 注解进行了修饰，如果这个 POJO 没有被应用所用到，也没有被其他 POJO 所引用，那么它就永远不会实例化。

使用@DependsOn 实现一个 POJO 在其他 POJO 前完成初始化

随着应用中 POJO 数量的不断增长，POJO 的初始化数量也会持续攀升。如果 POJO 之间相互引用并散落在不同的 Java 配置类中，那就会导致竞态条件。如果 bean C 需要用到 bean B 与 bean F 中的逻辑，会导致什么情况发生呢？如果 bean C 被先检测到，同时 Spring 尚未初始化 bean B 与 bean F，那就会导致难以发现的错误。为了确保某些 POJO 会在其他 POJOs 前完成初始化，并且在初始化过程失败时能够获得更具描述力的错误信息，Spring 提供了@DependsOn 注解。@DependsOn 注解可以确保给定的 bean 会在另一个 bean 之前完成初始化。

```
package com.apress.springrecipes.sequence.config;

import org.springframework.context.annotation.Bean;
import org.springframework.context.annotation.DependsOn;
import org.springframework.context.annotation.Configuration;
import com.apress.springrecipes.sequence.DatePrefixGenerator;
import com.apress.springrecipes.sequence.NumberPrefixGenerator;
import com.apress.springrecipes.sequence.SequenceGenerator;

@Configuration
public class SequenceConfiguration {
    @Bean
    @DependsOn("datePrefixGenerator")
    public SequenceGenerator sequenceGenerator() {
        SequenceGenerator sequence= new SequenceGenerator();
        sequence.setInitial(100000);
        sequence.setSuffix("A");
        return sequence;
    }
}
```

在上述代码片段中，声明@DependsOn("datePrefixGenerator")可以确保 datePrefixGenerator bean 会在 sequenceGenerator bean 之前创建出来。@DependsOn 属性还支持通过{}围起来的 CSV 列表来定义多个依赖 bean（即@DependsOn({"datePrefixGenerator,numberPrefixGenerator, randomPrefixGenerator"}这种形式)。

2-9 创建后置处理器来验证和修改 POJO

问题提出

你想在 bean 构建过程中将一些任务应用到所有 bean 实例或是特定类型的实例上，以根据特定标准来验证或是修改 bean 属性。

解决方案

可以通过 bean 后置处理器在初始化回调方法（即@Bean 注解中使用了 initMethod 属性的方法，或通过

@PostConstruct 注解进行修饰的方法）执行前后对 bean 进行处理。bean 后置处理器的主要特性在于它会处理 IoC 容器中所有的 bean 实例，而不仅仅只是单个 bean 实例。通常，bean 后置处理器用于检查 bean 属性的有效性，根据特定标准修改 bean 属性，或是对所有 bean 实例应用某些任务。

Spring 还支持@Required 注解，它就是由内建的 Spring 后置处理器 RequiredAnnotationBeanPostProcessor 所支持的。RequiredAnnotationBeanPostProcessor bean 后置处理器会检查所有带有@Required 注解的 bean 属性是否被设置了。

解释说明

假设你想要审计每个 bean 的创建，借此来调试应用，验证每个 bean 的属性，或是进行其他一些处理。bean 后置处理器就是实现该特性的一种理想选择，因为你无须修改既有的 POJO 代码。

创建 POJO 来处理每个 bean 实例

要想编写 bean 后置处理器，类需要实现 BeanPostProcessor。当 Spring 检测到实现了这个类的 bean 时，它就会为 Spring 管理的所有 bean 实例应用 postProcessBeforeInitialization()与 postProcessAfterInitialization()方法。你可以在这些方法中实现任何逻辑，比如说查看、修改或是验证 bean 的状态。

```
package com.apress.springrecipes.shop;

import org.springframework.beans.BeansException;
import org.springframework.beans.factory.config.BeanPostProcessor;

import org.springframework.stereotype.Component;

@Component
public class AuditCheckBeanPostProcessor implements BeanPostProcessor {

    public Object postProcessBeforeInitialization(Object bean, String beanName)
        throws BeansException {
            System.out.println("In AuditCheckBeanPostProcessor.
            postProcessBeforeInitialization, processing bean type: " + bean.getClass());
        return bean;
    }

    public Object postProcessAfterInitialization(Object bean, String beanName)
        throws BeansException {
        return bean;
    }
}
```

注意，即便没有在方法中做任何事情，postProcessBeforeInitialization()与 postProcessAfterInitialization()方法也必须要返回原始的 bean 实例。

要想将 bean 后置处理器注册到应用上下文中，你只需为类加上@Component 注解即可。应用上下文可以检测到哪个 bean 实现了 BeanPostProcessor 接口，并将其注册进来，从而可以处理容器中所有其他的 bean 实例。

创建 POJO 来处理选定的 bean 实例

在 bean 构建过程中，Spring IoC 容器会将所有 bean 实例逐个传递给 bean 后置处理器。这意味着如果只想将一个 bean 后置处理器应用到某些类型的 bean 上，你就需要通过检查其实例类型来过滤 bean。通过这种方式，你可以有选择地对 bean 应用处理逻辑。

假设你只想对 Product bean 实例应用 bean 后置处理器。如下示例展示了完成该功能的另一个 bean 后置处理器：

```
package com.apress.springrecipes.shop;

import org.springframework.beans.BeansException;
import org.springframework.beans.factory.config.BeanPostProcessor;
```

```
import org.springframework.stereotype.Component;

@Component
public class ProductCheckBeanPostProcessor implements BeanPostProcessor {

    public Object postProcessBeforeInitialization(Object bean, String beanName)
        throws BeansException {
        if (bean instanceof Product) {
            String productName = ((Product) bean).getName();
                System.out.println("In ProductCheckBeanPostProcessor.
                postProcessBeforeInitialization, processing Product: " + productName);
        }
        return bean;
    }

    public Object postProcessAfterInitialization(Object bean, String beanName)
        throws BeansException {
        if (bean instanceof Product) {
            String productName = ((Product) bean).getName();
                System.out.println("In ProductCheckBeanPostProcessor.
                postProcessAfterInitialization, processing Product: " + productName);
        }
        return bean;
    }
}
```

postProcessBeforeInitialization()与 postProcessAfterInitialization()方法都必须要返回待处理的 bean 实例。不过，这也意味着你可以在 bean 后置处理器中使用新的实例来代替原始的 bean 实例。

使用@Required 注解验证 POJO 属性

在某些情况下，我们需要检查特定的属性是否被设置了。相较于创建自定义后置处理器来验证 bean 的特定属性来说，我们可以使用@Required 注解来修饰属性。@Required 注解可以访问到 RequiredAnnotation-BeanPostProcessor 类，它是一个 Spring bean 后置处理器，可以检查某些 bean 属性是否被设置了。注意到该处理器只能检查属性是否被设置了，但无法检查其值是否为 null 或其他值。

假设序列生成器需要 prefixGenerator 与 suffix 属性。你可以使用@Required 注解来修饰其 setter 方法。

```
package com.apress.springrecipes.sequence;

import org.springframework.beans.factory.annotation.Required;

public class SequenceGenerator {

    private PrefixGenerator prefixGenerator;
    private String suffix;
    ...
    @Required
    public void setPrefixGenerator(PrefixGenerator prefixGenerator) {
        this.prefixGenerator = prefixGenerator;
    }

    @Required
    public void setSuffix(String suffix) {
        this.suffix = suffix;
    }
    ...
}
```

要想让 Spring 检查这些属性是否被设置了，你只需开启扫描，这样 Spring 就可以检查并应用@Required 注解了。如果没有设置被@Required 注解修饰的属性，那就会抛出 BeanInitializationException 错误。

2-10　使用工厂（静态工厂、实例方法与 Spring 的 FactoryBean）创建 POJO

问题提出

你想通过调用静态工厂方法或实例工厂方法在 Spring IoC 容器中创建 POJO 实例。这种做法的目的在于将对象创建过程封装到静态方法或另一个对象实例的方法中。请求对象的客户端只需调用这个方法即可，无须了解创建的细节信息。你想通过 Spring 的工厂 bean 在 Spring IoC 容器中创建 POJO 实例。工厂 bean 是这样一种 bean：它作为一个工厂，在 IoC 容器中创建其他 bean。从概念上来说，工厂 bean 类似于工厂方法，不过它是 Spring 特有的 bean，可以在 bean 创建过程中被 Spring IoC 容器识别出来。

解决方案

要想在 Java 配置类的@Bean 定义下通过调用静态工厂创建 POJO，你可以使用标准的 Java 语法来调用静态工厂方法。要想在 Java 配置类的@Bean 定义下通过调用实例工厂方法创建 POJO，你可以创建一个 POJO 来实例化工厂值，再创建另外一个 POJO 作为门面来访问这个工厂。

为了方便，Spring 提供了一个名为 AbstractFactoryBean 的抽象模板类来扩展 FactoryBean 接口。

解释说明

我们将会介绍在 Spring 中定义与使用工厂方法的不同方式。首先会介绍如何使用静态工厂方法，接下来介绍基于实例的工厂方法，最后会介绍 Spring FactoryBean。

通过调用静态工厂方法创建 POJO

比如说，你可以编写如下 createProduct 静态工厂方法，根据预先定义的产品 ID 来创建出产品。根据产品 ID，该方法会决定实例化哪个具体的产品类。如果没有与这个 ID 相匹配的产品，那就会抛出 IllegalArgumentException 异常。

```
package com.apress.springrecipes.shop;

public class ProductCreator {

    public static Product createProduct(String productId) {
        if ("aaa".equals(productId)) {
            return new Battery("AAA", 2.5);
        } else if ("cdrw".equals(productId)) {
            return new Disc("CD-RW", 1.5);
        } else if ("dvdrw".equals(productId)) {
            return new Disc("DVD-RW", 3.0);
        }
        throw new IllegalArgumentException("Unknown product");
    }
}
```

要想在 Java 配置类的@Bean 定义下通过静态工厂方法创建 POJO，你可以使用常规的 Java 语法来调用工厂方法。

```
package com.apress.springrecipes.shop.config;

import com.apress.springrecipes.shop.Product;
import com.apress.springrecipes.shop.ProductCreator;
import org.springframework.context.annotation.Bean;
import org.springframework.context.annotation.Configuration;

@Configuration
public class ShopConfiguration {
    @Bean
    public Product aaa() {
        return ProductCreator.createProduct("aaa");
    }
```

```
    @Bean
    public Product cdrw() {
        return ProductCreator.createProduct("cdrw");
    }

    @Bean
    public Product dvdrw() {
        return ProductCreator.createProduct("dvdrw");
    }
}
```

通过调用实例工厂方法创建 POJO

比如说，你可以编写如下 ProductCreator 类，使用可配置的 map 来存储预先定义好的产品。createProduct()实例工厂方法通过在 map 中查找所提供的 productId 值来找到产品。如果没有与这个 ID 相匹配的产品，那就会抛出 IllegalArgumentException 异常。

```
package com.apress.springrecipes.shop;
...
public class ProductCreator {

    private Map<String, Product> products;

    public void setProducts(Map<String, Product> products) {
        this.products = products;
    }

    public Product createProduct(String productId) {
        Product product = products.get(productId);
        if (product != null) {
            return product;
        }
        throw new IllegalArgumentException("Unknown product");
    }
}
```

要想从这个 ProductCreator 中创建出产品，首先需要声明一个@Bean 来实例化工厂值。接下来，再声明一个 bean 作为门面来访问工厂。最后，调用工厂并执行 createProduct()方法来实例化其他 bean。

```
package com.apress.springrecipes.shop.config;

import com.apress.springrecipes.shop.Battery;
import com.apress.springrecipes.shop.Disc;
import com.apress.springrecipes.shop.Product;
import com.apress.springrecipes.shop.ProductCreator;
import org.springframework.context.annotation.Bean;
import org.springframework.context.annotation.Configuration;

import java.util.HashMap;
import java.util.Map;

@Configuration
public class ShopConfiguration {

    @Bean
    public ProductCreator productCreatorFactory() {

        ProductCreator factory = new ProductCreator();
        Map<String, Product> products = new HashMap<>();
        products.put("aaa", new Battery("AAA", 2.5));
        products.put("cdrw", new Disc("CD-RW", 1.5));
        products.put("dvdrw", new Disc("DVD-RW", 3.0));
        factory.setProducts(products);
```

```
        return factory;
    }
    @Bean
    public Product aaa() {
        return productCreatorFactory().createProduct("aaa");
    }

    @Bean
    public Product cdrw() {
        return productCreatorFactory().createProduct("cdrw");
    }

    @Bean
    public Product dvdrw() {
        return productCreatorFactory().createProduct("dvdrw");
    }

}
```

使用 Spring 的 Factory Bean 创建 POJO

虽然很少需要编写自定义的工厂 bean，不过通过示例来理解其内部机制是很有用的。比如说，你可以编写一个工厂 bean 来创建产品，并在价格上应用折扣。它会接收 product 与 discount 属性，将折扣应用到产品上并将其作为新的 bean 返回。

```
package com.apress.springrecipes.shop;

import org.springframework.beans.factory.config.AbstractFactoryBean;

public class DiscountFactoryBean extends AbstractFactoryBean<Product> {

    private Product product;
    private double discount;

    public void setProduct(Product product) {
        this.product = product;
    }

    public void setDiscount(double discount) {
        this.discount = discount;
    }

    public Class<?> getObjectType() {
        return product.getClass();
    }

    protected Product createInstance() throws Exception {
        product.setPrice(product.getPrice() * (1 - discount));
        return product;
    }
}
```

通过继承 AbstractFactoryBean 类，工厂 bean 只需重写 createInstance()方法来创建目标 bean 实例即可。此外，你还需要在 getObjectType()方法中返回目标 bean 的类型，从而让自动装配特性能够正常工作。

接下来，使用常规的@Bean 注解来声明产品实例以应用 DiscountFactoryBean。

```
package com.apress.springrecipes.shop.config;

import com.apress.springrecipes.shop.Battery;
import com.apress.springrecipes.shop.Disc;
import com.apress.springrecipes.shop.DiscountFactoryBean;
import org.springframework.context.annotation.Bean;
import org.springframework.context.annotation.ComponentScan;
import org.springframework.context.annotation.Configuration;
```

```
@Configuration
@ComponentScan("com.apress.springrecipes.shop")
public class ShopConfiguration {
    @Bean
    public Battery aaa() {
        Battery aaa = new Battery("AAA", 2.5);
        return aaa;
    }

    @Bean
    public Disc cdrw() {
        Disc aaa = new Disc("CD-RW", 1.5);
        return aaa;
    }

    @Bean
    public Disc dvdrw() {
        Disc aaa = new Disc("DVD-RW", 3.0);
        return aaa;
    }

    @Bean
    public DiscountFactoryBean discountFactoryBeanAAA() {
        DiscountFactoryBean factory = new DiscountFactoryBean();
        factory.setProduct(aaa());
        factory.setDiscount(0.2);
        return factory;
    }

    @Bean
    public DiscountFactoryBean discountFactoryBeanCDRW() {
        DiscountFactoryBean factory = new DiscountFactoryBean();
        factory.setProduct(cdrw());
        factory.setDiscount(0.1);
        return factory;
    }

    @Bean
    public DiscountFactoryBean discountFactoryBeanDVDRW() {
        DiscountFactoryBean factory = new DiscountFactoryBean();
        factory.setProduct(dvdrw());
        factory.setDiscount(0.1);
        return factory;
    }
}
```

2-11 使用 Spring 环境与 profile 加载不同的 POJO

问题提出

你想要使用同一套 POJO 实例或 bean，但要针对不同的应用场景（比如说生产、开发与测试）使用不同的实例值。

解决方案

创建多个 Java 配置类并将 POJO 实例或 bean 分组到这些类中。根据分组的目的使用@Profile 注解为 Java 配置类指定一个 profile 名。获取应用上下文的环境信息并设置 profile 来加载特定的 POJO 分组。

解释说明

根据不同的应用场景，POJO 实例值也会不同。比如说，一个常见的场景就是应用从开发环境进入到测试环境再到生产环境。在每一种环境下，某些 bean 的属性会发生变化以适应环境的变更（比如说数据库用户名/密码、文件路径等）。

你可以创建多个 Java 配置类，每个都对应不同的 POJO（比如说 ShopConfigurationGlobal、ShopConfigurationStr、ShopConfigurationSumWin），在应用上下文中，根据场景只加载给定的配置类文件。

使用@Profile 注解创建 Java 配置类

使用@Profile 注解为前文介绍的商城应用创建多个 Java 配置类。

```
package com.apress.springrecipes.shop.config;

import com.apress.springrecipes.shop.Cashier;
import org.springframework.context.annotation.Bean;
import org.springframework.context.annotation.ComponentScan;
import org.springframework.context.annotation.Configuration;
import org.springframework.context.annotation.Profile;

@Configuration
@Profile("global")
@ComponentScan("com.apress.springrecipes.shop")
public class ShopConfigurationGlobal {

    @Bean(initMethod = "openFile", destroyMethod = "closeFile")
    public Cashier cashier() {
        final String path = System.getProperty("java.io.tmpdir") + "cashier";
        Cashier c1 = new Cashier();
        c1.setFileName("checkout");
        c1.setPath(path);
        return c1;
    }
}
package com.apress.springrecipes.shop.config;

import com.apress.springrecipes.shop.Battery;
import com.apress.springrecipes.shop.Disc;
import com.apress.springrecipes.shop.Product;
import org.springframework.context.annotation.Bean;
import org.springframework.context.annotation.Configuration;
import org.springframework.context.annotation.Profile;

@Configuration
@Profile({"summer", "winter"})
public class ShopConfigurationSumWin {
    @Bean
    public Product aaa() {
        Battery p1 = new Battery();
        p1.setName("AAA");
        p1.setPrice(2.0);
        p1.setRechargeable(true);
        return p1;
    }

    @Bean
    public Product cdrw() {
        Disc p2 = new Disc("CD-RW", 1.0);
        p2.setCapacity(700);
        return p2;
    }

    @Bean
    public Product dvdrw() {
        Disc p2 = new Disc("DVD-RW", 2.5);
        p2.setCapacity(700);
        return p2;
    }
}
```

@Profile 注解修饰了整个 Java 配置类，这样所有的@Bean 实例就都属于同一个 profile。要想为@Profile 指定一个名字，只需将名字放在""中即可。注意，还可以通过逗号分隔值（CSV）语法并放在{}中来指定多个@Profile 名（如{"summer","winter"}）。

将 profile 加载到环境中

要想将某个 profile 中的 bean 加载到应用中，需要激活一个 profile。可以一次加载多个 profile，也可以以编程的方式加载 profile，还可以借助于 Java 运行时标记，甚至作为 WAR 文件的初始化参数。

要想以编程的方式加载 profile（即通过应用上下文），你需要通过 setActiveProfiles()方法从能够加载到 profile 的地方获取到上下文环境。

```
AnnotationConfigApplicationContext context = new AnnotationConfigApplicationContext();
context.getEnvironment().setActiveProfiles("global", "winter");
context.scan("com.apress.springrecipes.shop");
context.refresh();
```

还可以通过 Java 运行时标记来指定加载哪个 Spring profile。在这种方式下，你可以传递如下运行时标记来加载属于 global 和 winter profile 的所有 bean：

```
-Dspring.profiles.active=global,winter
```

设置默认 profile

为了避免因没有 profile 被加载到应用中而可能导致的错误，你可以定义默认 profile。只有当 Spring 无法检测到任何激活的 profile 时才会使用默认 profile，默认 profile 可以通过编程的方式、Java 运行时标记，或是 Web 应用初始化参数的形式来定义。

要想设置默认 profile，你依然可以使用用于创建激活的 profile 的这 3 种方式。对于编程的方式来说，请将方法 setActiveProfiles()替换为 setDefaultProfiles()；对于 Java 运行时标记或是 Web 应用初始化参数来说，请将参数 spring.profiles.active 替换为 spring.profiles.default。

2-12　让 POJO 能够感知到 Spring 的 IoC 容器资源

问题提出

虽然设计良好的组件不应该直接依赖于 Spring 的 IoC 容器，但有时 bean 还是需要能够感知到容器的资源。

解决方案

可以通过实现某些“aware”接口来让 bean 能够感知到 Spring IoC 容器的资源。表 2-2 列出了大多数常见的接口。Spring 会通过定义在这些接口中的 setter 方法将相应的资源注入到实现了这些接口的 bean 中。

表 2-2　Spring 中常见的 aware 接口

aware 接口	目标资源类型
BeanNameAware	配置在 IoC 容器中的 bean 实例名
BeanFactoryAware	当前的 bean 工厂，可以通过它调用容器的服务
ApplicationContextAware	当前的应用上下文，可以通过它调用容器的服务
MessageSourceAware	消息源，可以通过它解析文本消息
ApplicationEventPublisherAware	应用事件发布器，可以通过它发布应用事件
ResourceLoaderAware	资源加载器，可以通过它加载外部资源
EnvironmentAware	与 ApplicationContext 接口相关联的 org.springframework.core.env.Environment 实例

■ **注意：** ApplicationContext 接口实际上继承了 MessageSource、ApplicationEventPublisher 与 ResourceLoader 接口，这样你只需要感知到应用上下文就可以访问到所有这些服务。不过，最佳实践是选择一个能够满足需求

的最小作用域的 aware 接口。

aware 接口中的 setter 方法会被 Spring 所调用，调用时机是在所有 bean 属性都被设置完毕之后但初始化回调方法执行之前，下面来详细说明一下。

1. 通过构造方法或是工厂方法创建 bean 实例。
2. 为 bean 属性设置值和 bean 引用。
3. 调用定义在 aware 接口中的 setter 方法。
4. 将 bean 实例传递给每个 bean 后置处理器的 postProcessBeforeInitialization()方法。调用初始化回调方法。
5. 将 bean 实例传递给每个 bean 后置处理器的 postProcessAfterInitialization()方法。这时的 bean 就可以使用了。
6. 当容器关闭时，调用销毁回调方法。

请记住，一旦类实现了 aware 接口，它们就会绑定到 Spring，在 Spring IoC 容器之外将无法正常使用。因此，请考虑清楚是否有必要实现这种私有接口。

注意： 在新版本的 Spring 中，实现 aware 接口通常是没必要的。比如说，还可以通过@Autowired 来访问 ApplicationContext。不过，如果你在编写框架或是库，那么最好实现这些接口。

解释说明

比如说，可以通过实现BeanNameAware接口让商城应用的Cashier类的POJO实例感知到它们对应的bean名字。通过实现该接口，Spring 会自动将 bean 名字注入到 POJO 实例中。除了实现该接口外，你还需要添加必要的 setter 方法来处理 bean 名字。

```
package com.apress.springrecipes.shop;
...
import org.springframework.beans.factory.BeanNameAware;

public class Cashier implements BeanNameAware {
    ...
    private String fileName;

    public void setBeanName(String beanName) {
        this.fileName = beanName;
    }
}
```

当注入了 bean 名字后，你就可以使用其值来完成需要 bean 名字的相关 POJO 任务。比如说，你可以使用其值来设置文件名以记录收银台的结账数据。通过这种方式，可以不必再配置 fileName 属性，也不需要 setFileName()方法。

```
@Bean(initMethod = "openFile", destroyMethod = "closeFile")
public Cashier cashier() {
    final String path = System.getProperty("java.io.tmpdir") + "cashier";
    Cashier cashier = new Cashier();
    cashier.setPath(path);
    return c1;
}
```

2-13　使用注解实现面向切面编程

问题提出

使用 Spring 与注解实现面向切面编程。

解决方案

通过@Aspect 注解修饰 Java 类来定义切面。借助于另外一个注解的帮助，类中的每个方法都会成为一个通知。可以使用 5 种通知注解：@Before、@After、@AfterReturning、@AfterThrowing 与@Around。

要想在 Spring IoC 容器中启用注解支持，你需要对配置类添加@EnableAspectJAutoProxy 注解。为了应用 AOP，Spring 会创建代理，默认情况下，它会创建 JDK 动态代理，这是基于接口的。当应用设计中没有使用接口时，代理的创建就会依赖于 CGLIB。要想启用 CGLIB，你需要在@EnableAspectJAutoProxy 注解中设置属性 proxyTargetClass=true。

解释说明

为了通过注解来支持面向切面编程，Spring 使用了与 AspectJ 相同的注解，使用了 AspectJ 提供的库来进行切点解析与匹配。不过，AOP 运行时依旧是纯粹的 Spring AOP，没有对 AspectJ 编译器或 weaver 产生任何依赖。

为了说明如何通过注解来实现面向切面编程，这里使用了如下的计算器接口来定义一组示例 POJO：

```
package com.apress.springrecipes.calculator;

public interface ArithmeticCalculator {

    public double add(double a, double b);
    public double sub(double a, double b);
    public double mul(double a, double b);
    public double div(double a, double b);
}
package com.apress.springrecipes.calculator;

public interface UnitCalculator {

    public double kilogramToPound(double kilogram);
    public double kilometerToMile(double kilometer);
}
```

接下来，为每个接口创建 POJO 类，类中通过 println 语句获悉每个方法的调用时机：

```
package com.apress.springrecipes.calculator;

import org.springframework.stereotype.Component;

@Component("arithmeticCalculator")
public class ArithmeticCalculatorImpl implements ArithmeticCalculator {

    public double add(double a, double b) {
        double result = a + b;
        System.out.println(a + " + " + b + " = " + result);
        return result;
    }

    public double sub(double a, double b) {
        double result = a - b;
        System.out.println(a + " - " + b + " = " + result);
        return result;
    }
    public double mul(double a, double b) {
        double result = a * b;
        System.out.println(a + " * " + b + " = " + result);
        return result;
    }

    public double div(double a, double b) {
        if (b == 0) {
            throw new IllegalArgumentException("Division by zero");
        }
        double result = a / b;
        System.out.println(a + " / " + b + " = " + result);
        return result;
    }
}
```

```
package com.apress.springrecipes.calculator;

import org.springframework.stereotype.Component;

@Component("unitCalculator")
public class UnitCalculatorImpl implements UnitCalculator {

    public double kilogramToPound(double kilogram) {
        double pound = kilogram * 2.2;
        System.out.println(kilogram + " kilogram = " + pound + " pound");
        return pound;
    }

    public double kilometerToMile(double kilometer) {
        double mile = kilometer * 0.62;
        System.out.println(kilometer + " kilometer = " + mile + " mile");
        return mile;
    }
}
```

注意，每个 POJO 实现都使用@Component 注解进行修饰来创建 bean 实例。

声明切面、通知与切点

切面是个 Java 类，它对一组关注点（比如说日志或是事务处理）进行了模块化，这些关注点跨越了多个类型与对象。模块化这些关注点的 Java 类使用了@Aspect 注解进行修饰。在 AOP 术语中，切面还会由通知进行补充，通知本身包含了切点。通知是个简单的 Java 方法，并拥有一个通知注解。AspectJ 支持 5 种通知注解：@Before、@After、@AfterReturning、@AfterThrowing 与@Around。切点是个表达式，它会寻找应用于切面通知的类型与对象。

使用@Before 通知实现的切面

要想创建前置通知以处理特定的程序执行点前的横切关注点，需要使用@Before 注解并将切点表达式作为注解值。

```
package com.apress.springrecipes.calculator;

import org.apache.commons.logging.Log;
import org.apache.commons.logging.LogFactory;
import org.aspectj.lang.annotation.Aspect;
import org.aspectj.lang.annotation.Before;
import org.springframework.stereotype.Component;

@Aspect
@Component
public class CalculatorLoggingAspect {

    private Log log = LogFactory.getLog(this.getClass());

    @Before("execution(* ArithmeticCalculator.add(..))")
    public void logBefore() {
        log.info("The method add() begins");
    }
}
```

该切点表达式会匹配 ArithmeticCalculator 接口的 add()方法的执行。表达式中前面的通配符会匹配任何修饰符（public、protected 与 private）与任何返回类型。参数列表中的两个点会匹配任意数量的参数。

要想让这个切面能够正常工作（比如说输出其消息），你需要创建日志。特别是创建一个 logback.xml 文件，其配置属性如下所示：

```
<?xml version="1.0" encoding="UTF-8"?>
```

```
<configuration>

    <appender name="STDOUT" class="ch.qos.logback.core.ConsoleAppender">
        <layout class="ch.qos.logback.classic.PatternLayout">
            <Pattern>%d [%15.15t] %-5p %30.30c - %m%n</Pattern>
        </layout>
    </appender>

    <root level="INFO">
        <appender-ref ref="STDOUT" />
    </root>

</configuration>
```

■ **注意**：@Aspect 注解不足以实现类路径中的自动检测。因此，你还需要添加一个@Component 注解以便让 POJO 能够检测到。

接下来，创建一个 Spring 配置来扫描所有 POJO，包括 POJO 计算器实现与切面，并在配置类上使用@EnableAspectJAutoProxy 注解。

```
@Configuration
@EnableAspectJAutoProxy
@ComponentScan
public class CalculatorConfiguration {
}
```

最后，使用如下的 Main 类测试切面：

```
package com.apress.springrecipes.calculator;

import org.springframework.context.ApplicationContext;
import org.springframework.context.annotation.AnnotationConfigApplicationContext;

public class Main {

    public static void main(String[] args) {

        ApplicationContext context =
            new AnnotationConfigApplicationContext(CalculatorConfiguration.class);

        ArithmeticCalculator arithmeticCalculator =
            context.getBean("arithmeticCalculator", ArithmeticCalculator.class);
        arithmeticCalculator.add(1, 2);
        arithmeticCalculator.sub(4, 3);
        arithmeticCalculator.mul(2, 3);
        arithmeticCalculator.div(4, 2);

        UnitCalculator unitCalculator = context.getBean("unitCalculator", UnitCalculator.class);
        unitCalculator.kilogramToPound(10);
        unitCalculator.kilometerToMile(5);
    }
}
```

切点所匹配的执行点叫做连接点。从这个角度看，切点是个匹配一组连接点的表达式，而通知则是在特定连接点处所采取的动作。

要想让通知能够访问到当前连接点的细节信息，你可以在通知方法中声明一个 JoinPoint 类型的参数。接下来就可以访问到连接点的细节信息，如方法名和参数值。现在，可以通过将类名与方法名修改为通配符使得切点能够匹配所有方法。

```
package com.apress.springrecipes.calculator;
...
import java.util.Arrays;

import org.aspectj.lang.JoinPoint;
```

```
import org.aspectj.lang.annotation.Aspect;
import org.aspectj.lang.annotation.Before;

@Aspect
@Component
public class CalculatorLoggingAspect {
    ...
    @Before("execution(* *.*(..))")
    public void logBefore(JoinPoint joinPoint) {
        log.info("The method " + joinPoint.getSignature().getName()
            + "() begins with " + Arrays.toString(joinPoint.getArgs()));
    }
}
```

使用@After 通知实现的切面

后置通知会在连接点完毕后执行，它由一个使用了@After 注解的方法来表示，无论该方法返回一个结果还是抛出异常。如下后置通知会在计算器方法执行完毕后记录日志：

```
package com.apress.springrecipes.calculator;
...
import org.aspectj.lang.JoinPoint;
import org.aspectj.lang.annotation.After;
import org.aspectj.lang.annotation.Aspect;

@Aspect
public class CalculatorLoggingAspect {
    ...
    @After("execution(* *.*(..))")
    public void logAfter(JoinPoint joinPoint) {
        log.info("The method " + joinPoint.getSignature().getName()
            + "() ends");
    }
}
```

使用@AfterReturning 通知实现的切面

无论连接点正常返回还是抛出异常，后置通知都会执行。如果希望只在连接点返回时才打印日志，那就应该使用返回后通知来代替后置通知。

```
package com.apress.springrecipes.calculator;
...
import org.aspectj.lang.JoinPoint;
import org.aspectj.lang.annotation.AfterReturning;
import org.aspectj.lang.annotation.Aspect;

@Aspect
public class CalculatorLoggingAspect {
    ...
    @AfterReturning("execution(* *.*(..))")
    public void logAfterReturning(JoinPoint joinPoint) {
        log.info("The method {}() ends with {}", joinPoint.getSignature().getName(), result);
    }
}
```

在返回后通知中，你可以通过向@AfterReturning 注解添加一个 returning 属性来获取连接点的返回值。该属性值应该是通知方法的参数名，这样返回值就会传递进来。接下来，你需要使用这个名字为通知方法签名添加一个参数。在运行时，Spring AOP 会通过该参数将返回值传递进来。还需要注意的是，原来的切点表达式需要赋给 pointcut 属性。

```
package com.apress.springrecipes.calculator;
...
import org.aspectj.lang.JoinPoint;
import org.aspectj.lang.annotation.AfterReturning;
```

```
import org.aspectj.lang.annotation.Aspect;

@Aspect
public class CalculatorLoggingAspect {
    ...
    @AfterReturning(
        pointcut = "execution(* *.*(..))",
        returning = "result")
    public void logAfterReturning(JoinPoint joinPoint, Object result) {
        log.info("The method " + joinPoint.getSignature().getName()
            + "() ends with " + result);
    }
}
```

使用@AfterThrowing 通知实现的切面

异常后通知只会在连接点抛出异常时才会执行。

```
package com.apress.springrecipes.calculator;
...
import org.aspectj.lang.JoinPoint;
import org.aspectj.lang.annotation.AfterThrowing;
import org.aspectj.lang.annotation.Aspect;

@Aspect
public class CalculatorLoggingAspect {
    ...
    @AfterThrowing("execution(* *.*(..))")
    public void logAfterThrowing(JoinPoint joinPoint) {
        log.error("An exception has been thrown in {}()", joinPoint.getSignature().getName());
    }
}
```

类似地，可以通过向@AfterThrowing 注解添加一个 throwing 属性来获取到连接点抛出的异常。类型 Throwable 是 Java 语言中所有错误与异常的父类型。因此，如下通知会捕获到连接点所抛出的所有错误与异常。

```
package com.apress.springrecipes.calculator;
...
import org.aspectj.lang.JoinPoint;
import org.aspectj.lang.annotation.AfterThrowing;
import org.aspectj.lang.annotation.Aspect;

@Aspect
public class CalculatorLoggingAspect {
    ...
    @AfterThrowing(
        pointcut = "execution(* *.*(..))",
        throwing = "e")
    public void logAfterThrowing(JoinPoint joinPoint, Throwable e) {
        log.error("An exception {} has been thrown in {}()", e, joinPoint.getSignature().
        getName());
    }
}
```

不过，如果只对某一特定类型的异常感兴趣，那就可以将其声明为异常的参数类型。接下来，只有当抛出了兼容类型的异常（即该类型和其子类型）时，通知才会执行。

```
package com.apress.springrecipes.calculator;
...
import java.util.Arrays;

import org.aspectj.lang.JoinPoint;
import org.aspectj.lang.annotation.AfterThrowing;
import org.aspectj.lang.annotation.Aspect;
```

```
@Aspect
public class CalculatorLoggingAspect {
    ...
    @AfterThrowing(
        pointcut = "execution(* *.*(..))",
        throwing = "e")
    public void logAfterThrowing(JoinPoint joinPoint, IllegalArgumentException e) {
        log.error("Illegal argument {} in {}()", Arrays.toString(joinPoint.getArgs()),
        joinPoint.getSignature().getName());
    }
}
```

使用@Around 通知实现的切面

最后一种通知类型是环绕（around）通知。它是所有通知类型中最为强大的一种。它能完全控制连接点，这样就可以将之前通知的所有动作合并到单个通知中。你甚至可以控制何时以及是否继续执行原来的连接点。

如下的环绕通知组合了之前所创建的前置通知、返回后通知与异常后通知。注意，针对于环绕通知来说，连接点的参数类型必须得是 ProceedingJoinPoint。它是 JoinPoint 的子接口，可以通过它控制何时继续执行原来的连接点。

```
package com.apress.springrecipes.calculator;

import org.aspectj.lang.ProceedingJoinPoint;
import org.aspectj.lang.annotation.Around;
import org.aspectj.lang.annotation.Aspect;
import org.slf4j.Logger;
import org.slf4j.LoggerFactory;
import org.springframework.stereotype.Component;

import java.util.Arrays;

@Aspect
@Component
public class CalculatorLoggingAspect {

    private Logger log = LoggerFactory.getLogger(this.getClass());

    @Around("execution(* *.*(..))")
    public Object logAround(ProceedingJoinPoint joinPoint) throws Throwable {

        log.info("The method {}() begins with {}", joinPoint.getSignature().getName(),
        Arrays.toString(joinPoint.getArgs()));
        try {
            Object result = joinPoint.proceed();
            log.info("The method {}() ends with ", joinPoint.getSignature().getName(),
            result);
            return result;
        } catch (IllegalArgumentException e) {
            log.error("Illegal argument {} in {}()", Arrays.toString(joinPoint.getArgs()),
            joinPoint.getSignature().getName());
            throw e;
        }
    }
}
```

环绕通知类型非常强大且灵活，因为你甚至可以修改原始的参数值并改变最终的返回值。使用该类型通知时请务必小心，因为很有可能就会忘记继续调用初始连接点。

提示： 选择通知类型的一般原则是使用可以满足需求的功能最少的那一个。

2-14 访问连接点信息

问题提出

在 AOP 中，通知会被应用到叫做连接点的不同的程序执行点处。要想让通知采取正确的动作，通常需要获取到关于连接点的详细信息。

解决方案

通知可以通过在方法签名中声明一个类型为 org.aspectj.lang.JoinPoint 的参数来获取到当前的连接点信息。

解释说明

比如说，你可以通过如下通知访问到连接点信息。信息包括连接点类型（在 Spring AOP 中只有方法执行）、方法签名（声明类型与方法名）与参数值，还有目标对象与代理对象。

```
package com.apress.springrecipes.calculator;

import org.aspectj.lang.JoinPoint;
import org.aspectj.lang.annotation.Aspect;
import org.aspectj.lang.annotation.Before;
import org.slf4j.Logger;
import org.slf4j.LoggerFactory;
import org.springframework.stereotype.Component;

import java.util.Arrays;

@Aspect
@Component
public class CalculatorLoggingAspect {

    private Logger log = LoggerFactory.getLogger(this.getClass());

    @Before("execution(* *.*(..))")
    public void logJoinPoint(JoinPoint joinPoint) {

        log.info("Join point kind : {}", joinPoint.getKind());
        log.info("Signature declaring type : {}", joinPoint.getSignature().
        getDeclaringTypeName());
        log.info("Signature name : {}", joinPoint.getSignature().getName());
        log.info("Arguments : {}", Arrays.toString(joinPoint.getArgs()));
        log.info("Target class : {}", joinPoint.getTarget().getClass().getName());
        log.info("This class : {}", joinPoint.getThis().getClass().getName());
    }
}
```

由代理所包装的原始 bean 叫作目标对象，而代理对象就是 this 对象。可以通过连接点的 getTarget()与 getThis()方法访问到它们。从下面的输出中可以看到，这两个对象的类是不同的：

```
Join point kind : method-execution
Signature declaring type : com.apress.springrecipes.calculator.ArithmeticCalculator
Signature name : add
Arguments : [1.0, 2.0]
Target class : com.apress.springrecipes.calculator.ArithmeticCalculatorImpl
This class : com.sun.proxy.$Proxy6
```

2-15 通过@Order 注解指定切面的顺序

问题提出

当有多个切面应用到同一个连接点时，如果不显式指定的话，那么切面的顺序是不确定的。

解决方案

可以通过实现 Ordered 接口或是使用@Order 注解来指定切面的顺序。

解释说明

假设你编写了另一个切面来验证计算器参数。在这个切面中只有一个前置通知。

```
package com.apress.springrecipes.calculator;

import org.aspectj.lang.JoinPoint;
import org.aspectj.lang.annotation.Aspect;
import org.aspectj.lang.annotation.Before;

import org.springframework.stereotype.Component;

@Aspect
@Component
public class CalculatorValidationAspect {

    @Before("execution(* *.*(double, double))")
    public void validateBefore(JoinPoint joinPoint) {
        for (Object arg : joinPoint.getArgs()) {
            validate((Double) arg);
        }
    }

    private void validate(double a) {
        if (a < 0) {
            throw new IllegalArgumentException("Positive numbers only");
        }
    }
}
```

如果应用这个切面和之前介绍的切面，那么你是无法保证哪个场面会先执行。为了确保一个切面在另一个切面之前执行，你需要指定顺序。为了指定顺序，你需要让这两个切面实现 Ordered 接口或是使用@Order 注解。

如果决定实现 Ordered 接口，那么 getOrder 方法所返回的值中较小的值将会拥有更高的优先级。如果希望验证切面先执行，那么它应该返回比日志切面更小的值。

```
package com.apress.springrecipes.calculator;
...
import org.springframework.core.Ordered;

@Aspect
@Component
public class CalculatorValidationAspect implements Ordered {
    ...
    public int getOrder() {
        return 0;
    }
}
package com.apress.springrecipes.calculator;
...
import org.springframework.core.Ordered;

@Aspect
@Component
public class CalculatorLoggingAspect implements Ordered {
    ...
    public int getOrder() {
        return 1;
    }
}
```

指定顺序的另一种方式是使用@Order 注解。序号应该写在注解值中。

```
package com.apress.springrecipes.calculator;
...
import org.springframework.core.annotation.Order;
```

```
@Aspect
@Component
@Order(0)
public class CalculatorValidationAspect { … }
package com.apress.springrecipes.calculator;
...
import org.springframework.core.annotation.Order;

@Aspect
@Component
@Order(1)
public class CalculatorLoggingAspect { … }
```

2-16 重用切面的切点定义

问题提出

在编写切面时，你可以直接将一个切点表达式嵌入到通知注解中。你希望在多个通知中使用相同的切点表达式而不必嵌入多次。

解决方案

可以使用@Pointcut 注解单独定义切点以便在多个通知中进行重用。

解释说明

在切面中，可以通过@Pointcut 注解将切点声明为一个简单的方法。切点的方法体通常为空，因为将切点定义与应用逻辑混合到一起是不合理的。切点方法的访问修饰符还会控制该切点的可见性。其他通知可以通过方法名来引用该切点。

```
package com.apress.springrecipes.calculator;
...
import org.aspectj.lang.annotation.Pointcut;

@Aspect
@Component
public class CalculatorLoggingAspect {
    ...
    @Pointcut("execution(* *.*(..))")
    private void loggingOperation() {}

    @Before("loggingOperation()")
    public void logBefore(JoinPoint joinPoint) {
        ...
    }

    @AfterReturning(
        pointcut = "loggingOperation()",
        returning = "result")
    public void logAfterReturning(JoinPoint joinPoint, Object result) {
        ...
    }

    @AfterThrowing(
        pointcut = "loggingOperation()",
        throwing = "e")
    public void logAfterThrowing(JoinPoint joinPoint, IllegalArgumentException e) {
        ...
    }

    @Around("loggingOperation()")
    public Object logAround(ProceedingJoinPoint joinPoint) throws Throwable {
        ...
    }
}
```

通常，如果切点在多个切面中共享，那么最好将其单独定义在一个公共类中。在这种情况下，必须将其声明为 public。

```
package com.apress.springrecipes.calculator;

import org.aspectj.lang.annotation.Aspect;
import org.aspectj.lang.annotation.Pointcut;

@Aspect
public class CalculatorPointcuts {

    @Pointcut("execution(* *.*(..))")
    public void loggingOperation() {}
}
```

当引用这个切点时，还需要使用到类名。如果类与切面不在同一个包下，那么还需要包含包名。

```
package com.apress.springrecipes.calculator;
...
@Aspect
public class CalculatorLoggingAspect {
    ...
    @Before("CalculatorPointcuts.loggingOperation()")
    public void logBefore(JoinPoint joinPoint) {
        ...
    }

    @AfterReturning(
        pointcut = "CalculatorPointcuts.loggingOperation()",
        returning = "result")
    public void logAfterReturning(JoinPoint joinPoint, Object result) {
        ...
    }

    @AfterThrowing(
        pointcut = "CalculatorPointcuts.loggingOperation()",
        throwing = "e")
    public void logAfterThrowing(JoinPoint joinPoint, IllegalArgumentException e) {
        ...
    }

    @Around("CalculatorPointcuts.loggingOperation()")
    public Object logAround(ProceedingJoinPoint joinPoint) throws Throwable {
        ...
    }
}
```

2-17　编写 AspectJ 切点表达式

问题提出

横切关注点会出现在名为连接点的不同的程序执行点处。由于连接点各不相同，你需要一种强大的表达式语言来帮助进行匹配。

解决方案

AspectJ 切点语言是一种强大的表达式语言，可用于匹配各种各样的连接点。不过，Spring AOP 只支持为 IoC 容器中声明的 bean 方法执行连接点。出于这个原因，本节只会介绍 Spring AOP 支持的那些切点表达式。要想了解对 AspectJ 切点语言的完整描述，请参考 AspectJ 语言指南，它位于 AspectJ 的网站。Spring AOP 使用 AspectJ 切点语言进行切点的定义并通过 AspectJ 所提供的库在运行时解释切点表达式。在为 Spring AOP 编写 AspectJ 切点表达式时，你要记住 Spring AOP 只支持为 IoC 容器中的 bean 方法执行连接点。使用其他的切点表达式会抛出 IllegalArgumentException 异常。

解释说明

我们来看一下 Spring 支持的用于编写切点表达式的模式。首先，我们会介绍如何基于消息签名、类型模式来编写切点以及如何使用（与访问）方法参数。

使用方法签名模式

最典型的切点表达式会根据方法签名来匹配多个方法。比如说，如下切点表达式会匹配 ArithmeticCalculator 接口中声明的所有方法。最开始的通配符会匹配拥有任意修饰符（public、protected 与 private）和任意返回类型的方法。参数列表中的两个点会匹配任意数量的参数。

```
execution(* com.apress.springrecipes.calculator.ArithmeticCalculator.*(..))
```

如果目标类或接口与切面位于同一个包下，那么就可以省略掉包名。

```
execution(* ArithmeticCalculator.*(..))
```

如下切点表达式会匹配声明在 ArithmeticCalculator 接口中的所有 public 方法：

```
execution(public * ArithmeticCalculator.*(..))
```

还可以限制方法的返回类型。比如说，如下切点会匹配返回 double 数值的方法：

```
execution(public double ArithmeticCalculator.*(..))
```

还可以限制方法的参数列表。比如说，如下切点会匹配第一个参数为原生 double 类型的方法。接下来的两个点会匹配后续任意数量的参数。

```
execution(public double ArithmeticCalculator.*(double, ..))
```

还可以指定方法签名中所有的参数类型以让切点进行匹配。

```
execution(public double ArithmeticCalculator.*(double, double))
```

虽然 AspectJ 切点语言在匹配各种连接点上是非常强大的，不过有时对于想要匹配的方法而言，你可能找不到任何共同点（比如说修饰符、返回类型、方法名模式或是参数等）。在这种情况下，可以考虑为其提供自定义注解。比如说，可以定义如下标识注解。该注解可用在方法与类型级别上。

```
package com.apress.springrecipes.calculator;

import java.lang.annotation.Documented;
import java.lang.annotation.ElementType;
import java.lang.annotation.Retention;
import java.lang.annotation.RetentionPolicy;
import java.lang.annotation.Target;

@Target( { ElementType.METHOD, ElementType.TYPE })
@Retention(RetentionPolicy.RUNTIME)
@Documented
public @interface LoggingRequired {
}
```

接下来，可以将该注解用在需要日志的所有方法上，也可以用在类上以将行为应用到其所有方法上。注意，需要将注解添加到实现类而非接口上，因为注解无法被继承。

```
package com.apress.springrecipes.calculator;

@LoggingRequired
public class ArithmeticCalculatorImpl implements ArithmeticCalculator {

    public double add(double a, double b) {
        ...
    }

    public double sub(double a, double b) {
        ...
    }

    ...
}
```

接下来就可以编写一个切点表达式，在@Pointcut 注解上使用 annotation 关键字匹配使用了@LoggingRequired

注解的类或是方法。

```
package com.apress.springrecipes.calculator;

import org.aspectj.lang.annotation.Aspect;
import org.aspectj.lang.annotation.Pointcut;

@Aspect
public class CalculatorPointcuts {

    @Pointcut("annotation(com.apress.springrecipes.calculator.LoggingRequired)")
    public void loggingOperation() {}

}
```

使用类型签名模式

另一种切点表达式会匹配某些类型下的所有连接点。在应用到 Spring AOP 上时，这些切点的作用域会收窄以匹配类型下的所有方法执行。比如说，如下切点会匹配 com.apress.springrecipes.calculator 包下所有的方法执行连接点：

```
within(com.apress.springrecipes.calculator.*)
```

要想匹配某个包及其子包下的连接点，需要在通配符前再添加一个点。

```
within(com.apress.springrecipes.calculator..*)
```

如下切点表达式会匹配特定类中的方法执行连接点：

```
within(com.apress.springrecipes.calculator.ArithmeticCalculatorImpl)
```

如果目标类与切面位于同一个包下，那么包名可以省略。

```
within(ArithmeticCalculatorImpl)
```

可以匹配实现了 ArithmeticCalculator 接口的所有类中的方法执行连接点，方法是添加一个加号。

```
within(ArithmeticCalculator+)
```

如前所述，自定义注解@LoggingRequired 可用在类或是方法级别上。

```
package com.apress.springrecipes.calculator;

@LoggingRequired
public class ArithmeticCalculatorImpl implements ArithmeticCalculator {
    ...
}
```

接下来，可以在@Pointcut 注解中通过 within 关键字匹配使用了@LoggingRequired 注解的类或是方法中的连接点。

```
@Pointcut("within(com.apress.springrecipes.calculator.LoggingRequired)")
public void loggingOperation() {}
```

组合切点表达式

在 AspectJ 中，可以通过运算符&&（与）、||（或）和!（非）来组合切点表达式。比如说，如下切点会匹配实现了 ArithmeticCalculator 或 UnitCalculator 接口的类中的连接点：

```
within(ArithmeticCalculator+) || within(UnitCalculator+)
```

这些运算符的操作数可以是任何切点表达式以及对其他切点的引用。

```
package com.apress.springrecipes.calculator;

import org.aspectj.lang.annotation.Aspect;
import org.aspectj.lang.annotation.Pointcut;

@Aspect
public class CalculatorPointcuts {

    @Pointcut("within(ArithmeticCalculator+)")
    public void arithmeticOperation() {}
```

```
    @Pointcut("within(UnitCalculator+)")
    public void unitOperation() {}

    @Pointcut("arithmeticOperation() || unitOperation()")
    public void loggingOperation() {}
}
```

声明切点参数

访问连接点信息的一种方式是使用反射（即通过通知方法中 org.aspectj.lang.JoinPoint 类型的参数）。此外，还可以通过某些特殊的切点表达式以声明的方式来访问连接点信息。比如说，表达式 target()与 args()会捕获到当前连接点的目标对象与参数值，并将其作为切点参数公开出来。这些参数会以相同名字的实参形式传递给通知方法。

```
package com.apress.springrecipes.calculator;
...
import org.aspectj.lang.annotation.Aspect;
import org.aspectj.lang.annotation.Before;

@Aspect
public class CalculatorLoggingAspect {
    ...
    @Before("execution(* *.*(..)) && target(target) && args(a,b)")
    public void logParameter(Object target, double a, double b) {
        log.info("Target class : {}", target.getClass().getName());
        log.info("Arguments : {}, {}", a,b);
    }
}
```

在声明公开了参数的独立切点时，还需要将其添加到切点方法的实参列表中。

```
package com.apress.springrecipes.calculator;

import org.aspectj.lang.annotation.Aspect;
import org.aspectj.lang.annotation.Pointcut;

@Aspect
public class CalculatorPointcuts {
    ...
    @Pointcut("execution(* *.*(..)) && target(target) && args(a,b)")
    public void parameterPointcut(Object target, double a, double b) {}
}
```

引用了该参数化切点的任何通知都可以通过同名的方法实参来访问切点参数。

```
package com.apress.springrecipes.calculator;
...
import org.aspectj.lang.annotation.Aspect;
import org.aspectj.lang.annotation.Before;

@Aspect
public class CalculatorLoggingAspect {
    ...
    @Before("CalculatorPointcuts.parameterPointcut(target, a, b)")
    public void logParameter(Object target, double a, double b) {
        log.info("Target class : {}", target.getClass().getName());
        log.info("Arguments : {}, {}"a,b);
    }
}
```

2-18　使用 AOP 为 POJO 添加引介

问题提出

有时，你可能会有一组共享某个公共行为的类。在 OOP 中，它们需要继承相同的父类或是实现相同的接口。这实际上是个横切关注点问题，可以通过 AOP 进行模块化。此外，Java 的单继承机制只允许一个类至多继承一个父类。这样，我们无法同时从多个实现类中继承行为。

解决方案

引介是 AOP 中一种特殊类型的通知。它可以让对象动态实现一个接口，方式是为该接口提供一个实现类。看起来好像是对象在运行时继承了一个实现类。此外，你可以同时为你的对象引入拥有多个实现类的多个接口。这就实现了与多继承相同的效果。

解释说明

假设有 MaxCalculator 与 MinCalculator 两个接口，它们分别定义了 max()与 min()操作。

```
package com.apress.springrecipes.calculator;

public interface MaxCalculator {

    public double max(double a, double b);
}
```

```
package com.apress.springrecipes.calculator;

public interface MinCalculator {

    public double min(double a, double b);
}
```

接下来，为每个接口提供一个实现类，使用 println 语句以便知道方法何时执行了。

```
package com.apress.springrecipes.calculator;

public class MaxCalculatorImpl implements MaxCalculator {

    public double max(double a, double b) {
        double result = (a >= b) ? a : b;
        System.out.println("max(" + a + ", " + b + ") = " + result);
        return result;
    }
}
```

```
package com.apress.springrecipes.calculator;

public class MinCalculatorImpl implements MinCalculator {

    public double min(double a, double b) {
        double result = (a <= b) ? a : b;
        System.out.println("min(" + a + ", " + b + ") = " + result);
        return result;
    }
}
```

现在，假设你想让 ArithmeticCalculatorImpl 执行 max()与 min()计算。由于 Java 语言只支持单继承，因此 ArithmeticCalculatorImpl 类无法同时继承 MaxCalculatorImpl 与 MinCalculatorImpl 类。唯一可行的方式就是继承其中一个类（比如 MaxCalculatorImpl）并实现另一个接口（比如 MinCalculator），然后要么将实现代码复制过来，要么将处理委托给实际的实现类。无论哪种情况，你都需要重复方法声明。

借助于引介，你可以让 ArithmeticCalculatorImpl 通过实现类 MaxCalculatorImp 与 MinCalculatorImpl 动态实现 MaxCalculator 与 MinCalculator 接口。它与同时继承 MaxCalculatorImpl 和 MinCalculatorImpl 这种多继承机制效果相同。引介背后的思想是无须修改 ArithmeticCalculatorImpl 类就可以引入新的方法。这意味着即便在没有源代码的情况下也可以向既有类中引入方法。

> **提示：**你可能想知道引介是如何在 Spring AOP 中做到这一点的。答案就是动态代理。回忆一下，你可以指定一组接口让一个动态代理去实现。引介则是向动态代理添加一个接口（比如 MaxCalculator）。当声明在该接口中的方法在代理对象上调用时，代理会将调用委托给后端的实现类（比如 MaxCalculatorImpl）。

就像通知一样，引介也必须要在切面中进行声明。你可以为此创建新的切面或是复用已有的切面。在该切面中，可以对任意字段应用@DeclareParents 注解来声明引介。

```
package com.apress.springrecipes.calculator;

import org.aspectj.lang.annotation.Aspect;
import org.aspectj.lang.annotation.DeclareParents;

import org.springframework.stereotype.Component;

@Aspect
@Component
public class CalculatorIntroduction {

    @DeclareParents(
        value = "com.apress.springrecipes.calculator.ArithmeticCalculatorImpl",
        defaultImpl = MaxCalculatorImpl.class)
    public MaxCalculator maxCalculator;

    @DeclareParents(
        value = "com.apress.springrecipes.calculator.ArithmeticCalculatorImpl",
        defaultImpl = MinCalculatorImpl.class)
    public MinCalculator minCalculator;
}
```

@DeclareParents 注解类型的 value 属性指定了引介的目标类。引入的接口是由被注解字段的类型所决定的。最后，新接口所用的实现类是通过 defaultImpl 属性指定的。

通过这两个引介，你可以动态向 ArithmeticCalculatorImpl 类引入两个接口。实际上，你可以在@DeclareParents 注解的 value 属性中指定 AspectJ 的类型匹配表达式，以将一个接口引入到多个类中。

在将 MaxCalculator 与 MinCalculator 接口引入到算术计算器后，你可以将其类型转换为对应的接口来执行 max()与 min()计算。

```
package com.apress.springrecipes.calculator;

public class Main {

    public static void main(String[] args) {
        ...
        ArithmeticCalculator arithmeticCalculator =
            (ArithmeticCalculator) context.getBean("arithmeticCalculator");
        ...
        MaxCalculator maxCalculator = (MaxCalculator) arithmeticCalculator;
        maxCalculator.max(1, 2);

        MinCalculator minCalculator = (MinCalculator) arithmeticCalculator;
        minCalculator.min(1, 2);
    }
}
```

2-19　使用 AOP 为 POJO 引入状态

问题提出

有时，你想向一组既有的对象添加新的状态以追踪其使用情况，比如说调用数量、最后的修改日期等。如果所有对象都有相同的父类，那就不是什么难事了。不过，如果它们不在同一个类层次体系中，那么向不同类添加这种状态就比较困难了。

解决方案

可以为对象引入一个新的接口，该接口有一个实现类，这个实现类持有状态字段。接下来，编写另一个通知根据具体情况来修改状态。

解释说明

假设你想要追踪每个计算器对象的调用次数。由于在原始的计算器类中并没有用于存储计数器值的字段，因此需要通过 Spring AOP 引入一个。首先，为计数器操作创建一个接口。

```
package com.apress.springrecipes.calculator;

public interface Counter {

    public void increase();
    public int getCount();
}
```

接下来，为该接口编写一个简单的实现类。该类有一个 count 字段用于存储计数器值。

```
package com.apress.springrecipes.calculator;

public class CounterImpl implements Counter {

    private int count;

    public void increase() {
        count++;
    }

    public int getCount() {
        return count;
    }
}
```

要想将 Counter 接口（CounterImpl 作为其实现）引入到所有计算器对象中，可以编写如下引介，使用类型匹配表达式来匹配所有计算器实现：

```
package com.apress.springrecipes.calculator;
...
import org.aspectj.lang.annotation.Aspect;
import org.aspectj.lang.annotation.DeclareParents;

@Aspect
@Component
public class CalculatorIntroduction {
    ...
    @DeclareParents(
        value = "com.apress.springrecipes.calculator.*CalculatorImpl",
        defaultImpl = CounterImpl.class)
    public Counter counter;
}
```

该引介向每个计算器对象引入了 CounterImpl。不过，这还不足以追踪调用次数。每次调用计算器方法时都需要增加计数器值。可以编写一个后置通知来实现这个目的。注意，你需要获取到 this 对象而非目标对象，因为只有代理对象才实现了 Counter 接口。

```
package com.apress.springrecipes.calculator;
...
import org.aspectj.lang.annotation.After;
import org.aspectj.lang.annotation.Aspect;

@Aspect
@Component
public class CalculatorIntroduction {
    ...
    @After("execution(* com.apress.springrecipes.calculator.*Calculator.*(..))"
        + " && this(counter)")
    public void increaseCount(Counter counter) {
        counter.increase();
    }
}
```

在 Main 类中，可以通过将计算器对象类型转换为 Counter 类型来输出它们的计数器值。

```
package com.apress.springrecipes.calculator;
```

```
public class Main {

    public static void main(String[] args) {
        ...
        ArithmeticCalculator arithmeticCalculator =
            (ArithmeticCalculator) context.getBean("arithmeticCalculator");
        ...

        UnitCalculator unitCalculator =
            (UnitCalculator) context.getBean("unitCalculator");
        ...

        Counter arithmeticCounter = (Counter) arithmeticCalculator;
        System.out.println(arithmeticCounter.getCount());

        Counter unitCounter = (Counter) unitCalculator;
        System.out.println(unitCounter.getCount());
    }
}
```

2-20 在 Spring 中使用加载期编织的 AspectJ 切面

问题提出

Spring AOP 框架只支持有限类型的 AspectJ 切点，并且可以让切点应用到 IoC 容器中所声明的 bean 上。如果想要使用其他切点类型或是在 Spring IoC 容器外将切面应用到对象上，你就需要在 Spring 应用中使用 AspectJ 框架了。

解决方案

编织指的是将切面应用到目标对象的过程。在 Spring AOP 中，编织是通过动态代理实现的，发生在运行期。与之相反，AspectJ 框架支持编译期与加载期编织。

AspectJ 编译期编织是通过一个名为 ajc 的特殊的 AspectJ 编译器实现的。它可以将切面织入到 Java 源文件中并输出编织后的二进制类文件。它还可以将切面织入到编译后的类文件或 JAR 文件中。这个过程叫作编译期后织入。你可以在 Spring IoC 容器中声明类之前对其执行编译期与编译期后织入。Spring 完全不会介入到织入过程。要想了解关于编译期与编译期后织入的更多信息，请参考 AspectJ 文档。

AspectJ 加载期织入（又叫作 LTW）发生在目标类被类加载器载入到 JVM 的时候。对于待织入的类来说，需要使用特殊的类加载器来增强目标类的字节码。AspectJ 与 Spring 都提供了加载期编织器来为类加载器增加加载期织入功能。你只需要简单的配置即可启用这些加载期编织器。

解释说明

为了理解在 Spring 应用中 AspectJ 加载期的编织过程，我们考虑一个用于复数的计算器。首先，创建一个 Complex 类表示复数。为该类定义 toString()方法，将复数转换为字符串表示形式（a + bi）。

```
package com.apress.springrecipes.calculator;

public class Complex {

    private int real;
    private int imaginary;

    public Complex(int real, int imaginary) {
        this.real = real;
        this.imaginary = imaginary;
    }

    // Getters and Setters
    ...

    public String toString() {
        return "(" + real + " + " + imaginary + "i)";
```

```
    }
}
```

接下来，为复数的操作定义一个接口。出于简化，接口只提供 add()与 sub()方法。

```
package com.apress.springrecipes.calculator;

public interface ComplexCalculator {

    public Complex add(Complex a, Complex b);
    public Complex sub(Complex a, Complex b);
}
```

该接口的实现代码如下所示。每次都会返回一个新的 Complex 对象作为结果。

```
package com.apress.springrecipes.calculator;

import org.springframework.stereotype.Component;

@Component("complexCalculator")
public class ComplexCalculatorImpl implements ComplexCalculator {

    public Complex add(Complex a, Complex b) {
        Complex result = new Complex(a.getReal() + b.getReal(),
            a.getImaginary() + b.getImaginary());
        System.out.println(a + " + " + b + " = " + result);
        return result;
    }

    public Complex sub(Complex a, Complex b) {
        Complex result = new Complex(a.getReal() - b.getReal(),
            a.getImaginary() - b.getImaginary());
        System.out.println(a + " - " + b + " = " + result);
        return result;
    }
}
```

现在，可以通过如下 Main 类代码来测试这个复数计算器：

```
package com.apress.springrecipes.calculator;

import org.springframework.context.ApplicationContext;
import org.springframework.context.annotation.AnnotationConfigApplicationContext;

public class Main {

    public static void main(String[] args) {
        ApplicationContext context =
            new AnnotationConfigApplicationContext(CalculatorConfiguration.class);

        ComplexCalculator complexCalculator =
            context.getBean("complexCalculator", ComplexCalculator.class);

        complexCalculator.add(new Complex(1, 2), new Complex(2, 3));
        complexCalculator.sub(new Complex(5, 8), new Complex(2, 3));
    }
}
```

到目前为止，这个复数计算器没什么问题。不过，你可能希望将复数对象缓存起来以改进计算器的性能。由于缓存是个众所周知的横切关注点，因此可以通过切面将其模块化。

```
package com.apress.springrecipes.calculator;

import java.util.Collections;
import java.util.HashMap;
import java.util.Map;

import org.aspectj.lang.ProceedingJoinPoint;
import org.aspectj.lang.annotation.Around;
```

```
import org.aspectj.lang.annotation.Aspect;

@Aspect
public class ComplexCachingAspect {

    private final Map<String, Complex> cache = new ConcurrentHashMap<>();

    @Around("call(public Complex.new(int, int)) && args(a,b)")
    public Object cacheAround(ProceedingJoinPoint joinPoint, int a, int b)
        throws Throwable {
        String key = a + "," + b;
        Complex complex = cache.get(key);
        if (complex == null) {
            System.out.println("Cache MISS for (" + key + ")");
            complex = (Complex) joinPoint.proceed();
            cache.put(key, complex);
        }
        else {
            System.out.println("Cache HIT for (" + key + ")");
        }
        return complex;
    }
}
```

在该切面中，你将复数对象缓存到 map 中，并将其实部与虚部值作为键。接下来，最适合查询缓存的时间点就是调用构造方法创建复数对象之际。你使用 AspectJ 切点表达式 call 来捕获调用 Complex(int,int)构造方法的连接点。

接下来，需要一个环绕通知来修改返回值。如果在缓存中找到具有相同值的复数对象，那么就直接将其返回给调用者。否则，继续之前的构造方法调用来创建新的复数对象。在将其返回给调用者之前，你将其缓存到 map 中供后续使用。

Spring AOP 并不支持 call 切点，如果让 Spring 扫描切点注解，你会收到错误“unsupported pointcut primitive call”。

由于 Spring AOP 不支持该类型的切点，你需要使用 AspectJ 框架来应用这个切面。AspectJ 框架的配置是通过类路径根目录下 META-INF 目录中名为 aop.xml 的文件实现的。

```
<!DOCTYPE aspectj PUBLIC "-//AspectJ//DTD//EN"
    "http://www.eclipse.org/aspectj/dtd/aspectj.dtd">

<aspectj>
    <weaver>
        <include within="com.apress.springrecipes.calculator.*" />
    </weaver>

    <aspects>
        <aspect
            name="com.apress.springrecipes.calculator.ComplexCachingAspect" />
    </aspects>
</aspectj>
```

在上述 AspectJ 配置文件中，你需要指定切面以及希望切面织入的类。这里指定了要将 ComplexCachingAspect 织入到 com.apress.springrecipes.calculator 包下的所有类中。

最后，要想实现这个加载期织入，你可以通过两种方式来运行应用，下面就来介绍一下。

使用 AspectJ 编织器实现加载期织入

AspectJ 提供了加载期编织代理来启用加载期织入。你只需要为运行应用的命令添加一个 VM 参数。接下来，当类被加载到 JVM 中时就会进行编织。

```
java -javaagent:lib/aspectjweaver-1.9.0.jar -jar Recipe_2_19_ii-4.0.0.jar
```

如果使用上述参数运行应用，你会看到如下输出与缓存状态。AspectJ 代理会为对 Complex(int,int)构造方法的所有调用应用切面。

```
Cache MISS for (1,2)
Cache MISS for (2,3)
Cache MISS for (3,5)
(1 + 2i) + (2 + 3i) = (3 + 5i)
Cache MISS for (5,8)
Cache HIT for (2,3)
Cache HIT for (3,5)
(5 + 8i) - (2 + 3i) = (3 + 5i)
```

使用 Spring 加载期编织器实现加载期织入

Spring 提供了几个用于不同运行时环境的加载期编织器。要想为 Spring 应用打开合适的加载期编织器，只需要为配置类添加@EnableLoadTimeWeaving 注解即可。

Spring 会为运行期环境寻找最适合的加载期编织器。一些 Java EE 应用服务器提供了支持 Spring 加载期编织机制的类加载器，这样就没必要在启动命令中指定 Java 代理了。

不过，对于简单的 Java 应用来说，你还是需要 Spring 提供的编织代理来启用加载期织入。你需要在启动命令的 VM 参数中指定 Spring 代理。

```
java -javaagent:lib/spring-instrument-5.0.0.jar -jar Recipe_2_19_iii-4.0.0.jar
```

不过，如果运行应用，你会看到如下输出与缓存状态：

```
Cache MISS for (3,5)
(1 + 2i) + (2 + 3i) = (3 + 5i)
Cache HIT for (3,5)
(5 + 8i) - (2 + 3i) = (3 + 5i)
```

这是因为 Spring 代理只会对声明在 Spring IoC 容器中的 bean 调用 Complex(int,int)构造方法时应用通知。由于复数操作数是在 Main 类中创建的，因此 Spring 代理并不会对构造方法调用应用通知。

2-21　在 Spring 中配置 AspectJ 切面

问题提出

AspectJ 框架中所用的切面是由 AspectJ 框架自身实例化的。因此，你需要从 AspectJ 框架获取到切面实例来对其进行配置。

解决方案

每个 AspectJ 切面都提供了一个名为 Aspects 的工厂类，它有一个叫做 aspectOf()的静态工厂方法，可以通过该方法访问到当前的切面实例。在 Spring IoC 容器中，可以声明由该工厂方法 Aspects.aspectOf（ComplexCachingAspect.class）创建的 bean。

解释说明

比如说，可以通过 setter 方法来预先配置 ComplexCachingAspect 中的 cache map。

```java
package com.apress.springrecipes.calculator;

import org.aspectj.lang.ProceedingJoinPoint;
import org.aspectj.lang.annotation.Around;
import org.aspectj.lang.annotation.Aspect;

import java.util.Map;
import java.util.concurrent.ConcurrentHashMap;

@Aspect
public class ComplexCachingAspect {

    private Map<String, Complex> cache = new ConcurrentHashMap<>();

    public void setCache(Map<String, Complex> cache) {
        this.cache.clear();
        this.cache.putAll(cache);
```

```
    }

    @Around("call(public Complex.new(int, int)) && args(a,b)")
    public Object cacheAround(ProceedingJoinPoint joinPoint, int a, int b) throws Throwable {
        String key = a + "," + b;
        Complex complex = cache.get(key);
        if (complex == null) {
            System.out.println("Cache MISS for (" + key + ")");
            complex = (Complex) joinPoint.proceed();
            cache.put(key, complex);
        } else {
            System.out.println("Cache HIT for (" + key + ")");
        }
        return complex;
    }

}
```

要想配置切面，请创建一个使用了@Bean 注解的方法，该方法会调用上面提及的工厂方法 Aspects.aspectOf；它会返回一个切面实例，之后可以配置该实例。

```
package com.apress.springrecipes.calculator;

import org.aspectj.lang.Aspects;
import org.springframework.context.annotation.Bean;
import org.springframework.context.annotation.ComponentScan;
import org.springframework.context.annotation.Configuration;

import java.util.HashMap;
import java.util.Map;

@Configuration
@ComponentScan
public class CalculatorConfiguration {

    @Bean
    public ComplexCachingAspect complexCachingAspect() {

        Map<String, Complex> cache = new HashMap<>();
        cache.put("2,3", new Complex(2,3));
        cache.put("3,5", new Complex(3,5));

        ComplexCachingAspect complexCachingAspect =
            Aspects.aspectOf(ComplexCachingAspect.class);
        complexCachingAspect.setCache(cache);
        return complexCachingAspect;
    }
}
```

请使用 AspectJ 的编织器运行应用。

```
java -javaagent:lib/aspectjweaver-1.9.0.jar -jar Recipe_2_20-4.0.0.jar
```

2-22　使用 AOP 将 POJO 注入到领域对象中

问题提出

声明在 Spring IoC 容器中的 bean 可以通过 Spring 的依赖注入能力实现彼此装配。不过，在 Spring IoC 容器外部创建的对象却无法通过配置将自身装配到 Spring bean 中。你需要通过程序代码手工实现装配。

解决方案

在 Spring IoC 容器外部创建的对象通常都是领域对象。它们常常通过 new 运算符或是根据数据库查询结果来创建。要想将 Spring bean 注入到在 Spring 外部创建的领域对象中，你需要借助于 AOP 的帮助。实际上，Spring bean 的注入也是一种横切关注点。由于领域对象并不是由 Spring 创建的，因此无法使用 Spring AOP 进行注入。Spring 提供了一个专门用于此目的的 AspectJ 切面。你可以在 AspectJ 框架中启用该切面。

解释说明

假设你有一个全局格式化器来格式化复数。该格式化器接收一个格式化模式，并使用标准的@Component与@Value 注解来实例化 POJO。

```
package com.apress.springrecipes.calculator;

@Component
public class ComplexFormatter {

    @Value("(a + bi)")
    private String pattern;

    public void setPattern(String pattern) {
        this.pattern = pattern;
    }

    public String format(Complex complex) {
        return pattern.replaceAll("a", Integer.toString(complex.getReal()))
                .replaceAll("b", Integer.toString(complex.getImaginary()));
    }
}
```

在 Complex 类中，你在 toString()方法中使用这个格式化器将复数转换为字符串。该类为 ComplexFormatter 公开了一个 setter 方法。

```
package com.apress.springrecipes.calculator;

public class Complex {

    private int real;
    private int imaginary;
    ...
    private ComplexFormatter formatter;

    public void setFormatter(ComplexFormatter formatter) {
        this.formatter = formatter;
    }

    public String toString() {
        return formatter.format(this);
    }
}
```

不过，由于 Complex 对象不是由 Spring IoC 容器实例化的，因此无法通过常规方式实现其依赖注入。Spring 在其切面库中提供了一个 AnnotationBeanConfigurerAspect，它可以配置任何对象的依赖，即便对象不是由 Spring IoC 容器创建的也可以。

首先，你需要为对象类型加上@Configurable 注解来声明该对象类型是可配置的。

```
package com.apress.springrecipes.calculator;

import org.springframework.beans.factory.annotation.Configurable;
import org.springframework.beans.factory.annotation.Configurable;
import org.springframework.context.annotation.Scope;

@Configurable
@Component
@Scope("prototype")
public class Complex {
    ...
    @Autowired
    public void setFormatter(ComplexFormatter formatter) {
        this.formatter = formatter;
    }

}
```

除了@Configurable 注解外，你还可使用标准的@Component、@Scope 与@Autowired 注解来修饰 POJO，这样 bean 就可以拥有标准的 Spring 行为了。不过，@Configurable 注解是最为重要的配置信息，针对这一点，Spring 提供了一个便捷注解@EnableSpringConfigured 来启用上面提及的切面。

```
@Configuration
@EnableSpringConfigured
@ComponentScan
public class CalculatorConfiguration {}
```

当使用了@Configurable 注解的类被实例化时，切面会寻找类型与该类相同且作用域为 prototype 的 bean 定义。接下来，它会根据 bean 定义来配置新的实例。如果 bean 定义中声明了属性，那么新实例中相同的属性也会被切面进行设置。

最后，为了运行应用，你需要使用 AspectJ 代理在加载时将切面织入到类中。

```
java -javaagent:lib/aspectjweaver-1.9.0.jar -jar Recipe_2_21-4.0.0.jar
```

2-23 使用 Spring 与 TaskExecutor 实现并发

问题提出

你想要通过 Spring 构建一个线程化的并发程序，但又不知道如何去做，因为并没有标准的方式。

解决方案

使用 Spring 的 TaskExecutor 抽象。该抽象提供了针对多种环境的各种实现，包括基础的 Java SE Executor 实现、CommonJ WorkManager 实现，以及自定义实现。

在 Spring 中，所有实现都是统一的，也都可以转换为 Java SE 的 Executor 接口。

解释说明

线程是个有难度的话题，在 Java SE 环境中使用标准的线程机制来实现是一件非常乏味的事情。并发则是服务器端组件的另一个重要方面，不过在企业级 Java 领域中，它却几乎没有什么标准化做法。实际上，Java 企业版规范的某些部分会禁止显式创建和操纵线程。

在 Java SE 领域中，随着时间的流逝，引入了很多选择来处理线程与并发。首先，从 Java Development Kit（JDK）1.0 开始就提供了标准的 java.lang.Thread 支持。Java 1.3 引入了 java.util.TimerTask 来支持周期性完成的某些工作。Java 5 则首次引入了 java.util.concurrent 包，以及用于构建线程池的改造后的层次体系，这些都是围绕着 java.util.concurrent.Executor 来进行的。

Executor 的应用编程接口（API）是很简单的。

```
package java.util.concurrent;
public interface Executor {
    void execute(Runnable command);
}
```

ExecutorService 是个子接口，提供了更多的功能来管理线程并为线程所增加的事件提供了支持，如 shutdown()。从 Java SE 5.0 开始，随 JDK 一同发布了 ExecutorService 的几种实现。很多都是通过 java.util.concurrent 包下的静态工厂方法来实现的。接下来就使用 Java SE 类来提供几个示例。

ExecutorService 类提供了一个 submit()方法，它返回一个 Future<T>对象。Future<T>的实例可用于追踪通常是异步执行的线程的执行过程。可以通过 Future.isDone()或 Future.isCancelled()方法来确定任务是完成了还是取消了。在使用 ExecutorService 时，如果向 submit()传递一个 Runnable 实例（其 run 方法没有返回类型），那么调用返回的 Future 的 get()方法就会返回 null，或是返回提交任务时所指定的值。

```
Runnable task = new Runnable(){
    public void run(){
        try{
            Thread.sleep( 1000 * 60 ) ;
            System.out.println("Done sleeping for a minute, returning! " );
        } catch (Exception ex) { /* ... */ }
    }
};
```

```
ExecutorService executorService  = Executors.newCachedThreadPool() ;

if(executorService.submit(task, Boolean.TRUE).get().equals( Boolean.TRUE ))
    System.out.println( "Job has finished!");
```

拥有了这些背景信息，我们就可以探索各种实现的特性了。比如说，下面这个类使用 Runnable 来标记经过的时间：

```
package com.apress.springrecipes.spring3.executors;

import java.util.Date;

public class DemonstrationRunnable implements Runnable {
    public void run() {
        try {
            Thread.sleep(1000);
            } catch (InterruptedException e) {
                e.printStackTrace();
            }
            System.out.println(Thread.currentThread().getName());
            System.out.printf("Hello at %s \n", new Date());
    }
}
```

在探究 Java SE Executor 与 Spring 的 TaskExecutor 支持时，将使用同样的实例。

```
package com.apress.springrecipes.spring3.executors;
import java.util.Date;
import java.util.concurrent.ExecutorService;
import java.util.concurrent.Executors;
import java.util.concurrent.ScheduledExecutorService;
import java.util.concurrent.TimeUnit;

public class ExecutorsDemo {

    public static void main(String[] args) throws Throwable {
        Runnable task = new DemonstrationRunnable();

        ExecutorService cachedThreadPoolExecutorService =
            Executors.newCachedThreadPool();
        if (cachedThreadPoolExecutorService.submit(task).get() == null)
            System.out.printf("The cachedThreadPoolExecutorService "
                + "has succeeded at %s \n", new Date());

        ExecutorService fixedThreadPool = Executors.newFixedThreadPool(100);
        if (fixedThreadPool.submit(task).get() == null)
            System.out.printf("The fixedThreadPool has " +
                "succeeded at %s \n",
                new Date());

        ExecutorService singleThreadExecutorService =
            Executors.newSingleThreadExecutor();
        if (singleThreadExecutorService.submit(task).get() == null)
            System.out.printf("The singleThreadExecutorService "
                + "has succeeded at %s \n", new Date());

        ExecutorService es = Executors.newCachedThreadPool();
        if (es.submit(task, Boolean.TRUE).get().equals(Boolean.TRUE))
            System.out.println("Job has finished!");

        ScheduledExecutorService scheduledThreadExecutorService =
            Executors.newScheduledThreadPool(10);
        if (scheduledThreadExecutorService.schedule(
            task, 30, TimeUnit.SECONDS).get() == null)
            System.out.printf("The scheduledThreadExecutorService "
                + "has succeeded at %s \n", new Date());
```

```
        scheduledThreadExecutorService.scheduleAtFixedRate(task, 0, 5,
            TimeUnit.SECONDS);

    }
}
```

如果使用 ExecutorService 子接口中接收 Callable<T>的 submit()方法，那么 submit()就会返回 Callable 中的主方法 call()的返回结果。如下是 Callable 接口的定义：

```
package java.util.concurrent;

public interface Callable<V> {
    V call() throws Exception;
}
```

Java EE 领域提供了用于解决这些问题的不同方法，因为 Java EE 从设计上就限制了对线程的处理。

Quartz（一个任务调度框架）是填补这种线程特性沟壑的第一个解决方案，它提供了调度与并发的解决方案。JCA 1.5（即 J2EE Connector Architecture）是另一个规范，它为集成功能提供了一个原生类型的网关，并支持即时并发。借助于 JCA，组件会收到关于到来的消息的通知并可以并发响应。JCA 1.5 提供了一种原生、有限的企业服务总线——类似于集成特性，但又无须像 SpringSource 的 Spring Integration 框架那样提供诸多处理。

不过，应用服务器厂商并未对并发需求视而不见。很多新厂商都冲在了最前线。比如说，IBM 与 BEA 在 2003 年就合作创建了 Timer 与 WorkManager API，并最终成为了 JSR-237，接下来又与 JSR-236 进行了合并以聚焦于如何在托管环境下实现并发。Service Data Object（SDO）规范（即 JSR-235）提供了类似的解决方案。此外，CommonJ API 的开源实现也在最近几年不断涌现并实现了同样的解决方案。

问题在于，对于托管环境下的组件来说，并没有可移植、标准、简单的方式来控制线程并提供并发处理，这一点与 Java SE 的解决方案不同。

Spring 通过 org.springframeworks.core.task.TaskExecutor 接口提供了一种统一的解决方案。TaskExecutor 抽象继承了 Java 1.5 中的 java.util.concurrent.Executor。

事实上，TaskExecutor 接口主要在 Spring 框架内部使用。比如说，对于 Spring Quartz 集成（支持线程）和消息驱动的 POJO 容器支持来说，它们都广泛使用了 TaskExecutor。

```
package org.springframework.core.task;

import java.util.concurrent.Executor;

public interface TaskExecutor extends Executor {
    void execute(Runnable task);
}
```

在有些地方，不同的解决方案会使用核心 JDK 所提供的功能。在其他地方，它们则提供了独有的方案并与 CommonJ WorkManager 等框架进行了集成。这些集成通常会使用目标框架中的类，但你可以像其他 TaskExecutor 抽象一样操纵它们。

虽然可以将既有的 Java SE Executor 或是 ExecutorService 适配为 TaskExecutor，但在 Spring 中这并非那么重要，因为 TaskExecutor 的父类就是一个 Executor。通过这种方式，Spring 中的 TaskExecutor 填平了 Java EE 与 Java SE 中各种解决方案的沟壑。

接下来，我们来看一个 TaskExecutor 的简单示例，这里使用了之前定义的相同的 Runnable。客户端代码是个简单的 Spring POJO，你将各种 TaskExecutor 实例注入其中，这么做的唯一目的在于提交 Runnable。

```
package com.apress.springrecipes.executors;

import org.springframework.beans.factory.annotation.Autowired;
import org.springframework.context.ApplicationContext;
import org.springframework.context.annotation.AnnotationConfigApplicationContext;
import org.springframework.core.task.SimpleAsyncTaskExecutor;
import org.springframework.core.task.SyncTaskExecutor;
```

```
import org.springframework.core.task.support.TaskExecutorAdapter;
import org.springframework.scheduling.concurrent.ThreadPoolTaskExecutor;
import org.springframework.stereotype.Component;

import javax.annotation.PostConstruct;

@Component
public class SpringExecutorsDemo {

    @Autowired
    private SimpleAsyncTaskExecutor asyncTaskExecutor;
    @Autowired
    private SyncTaskExecutor syncTaskExecutor;
    @Autowired
    private TaskExecutorAdapter taskExecutorAdapter;
    @Autowired
    private ThreadPoolTaskExecutor threadPoolTaskExecutor;
    @Autowired
    private DemonstrationRunnable task;

    @PostConstruct
    public void submitJobs() {
        syncTaskExecutor.execute(task);
        taskExecutorAdapter.submit(task);
        asyncTaskExecutor.submit(task);

        for (int i = 0; i < 500; i++)
            threadPoolTaskExecutor.submit(task);
    }

    public static void main(String[] args) {

        new AnnotationConfigApplicationContext(ExecutorsConfiguration.class)
            .registerShutdownHook();
    }
}
```

应用上下文演示了各种 TaskExecutor 实现的创建。大多数实现都很简单，甚至可以手工创建。只有一个地方委托给了一个工厂 bean 来自动触发执行，代码如下所示：

```
package com.apress.springrecipes.executors;

import org.springframework.context.annotation.Bean;
import org.springframework.context.annotation.ComponentScan;
import org.springframework.context.annotation.Configuration;
import org.springframework.core.task.SimpleAsyncTaskExecutor;
import org.springframework.core.task.SyncTaskExecutor;
import org.springframework.core.task.support.TaskExecutorAdapter;
import org.springframework.scheduling.concurrent.ScheduledExecutorFactoryBean;
import org.springframework.scheduling.concurrent.ScheduledExecutorTask;
import org.springframework.scheduling.concurrent.ThreadPoolTaskExecutor;

import java.util.concurrent.Executors;

@Configuration
@ComponentScan
public class ExecutorsConfiguration {

    @Bean
    public TaskExecutorAdapter taskExecutorAdapter() {
        return new TaskExecutorAdapter(Executors.newCachedThreadPool());
    }

    @Bean
    public SimpleAsyncTaskExecutor simpleAsyncTaskExecutor() {
```

```
        return new SimpleAsyncTaskExecutor();
    }

    @Bean
    public SyncTaskExecutor syncTaskExecutor() {
        return new SyncTaskExecutor();
    }
    @Bean
    public ScheduledExecutorFactoryBean scheduledExecutorFactoryBean(ScheduledExecutorTask
    scheduledExecutorTask) {
        ScheduledExecutorFactoryBean scheduledExecutorFactoryBean =
        new ScheduledExecutorFactoryBean();
        scheduledExecutorFactoryBean.setScheduledExecutorTasks(scheduledExecutorTask);
        return scheduledExecutorFactoryBean;
    }

    @Bean
    public ScheduledExecutorTask scheduledExecutorTask(Runnable runnable) {
        ScheduledExecutorTask scheduledExecutorTask = new ScheduledExecutorTask();
        scheduledExecutorTask.setPeriod(1000);
        scheduledExecutorTask.setRunnable(runnable);
        return scheduledExecutorTask;
    }

    @Bean
    public ThreadPoolTaskExecutor threadPoolTaskExecutor() {
        ThreadPoolTaskExecutor taskExecutor = new ThreadPoolTaskExecutor();
        taskExecutor.setCorePoolSize(50);
        taskExecutor.setMaxPoolSize(100);
        taskExecutor.setAllowCoreThreadTimeOut(true);
        taskExecutor.setWaitForTasksToCompleteOnShutdown(true);
        return taskExecutor;
    }
}
```

上述代码展示了 TaskExecutor 接口的不同实现。第一个 bean（TaskExecutorAdapter 实例）是个针对 java.util.concurrent.Executors 实例的简单包装器，因此你可以按照 Spring TaskExecutor 接口的形式使用它。这里使用 Spring 来配置一个 Executor 实例并将其作为构造方法参数传递进去。

SimpleAsyncTaskExecutor 为每个提交过来的任务创建了一个新线程。它并未使用线程池，也没有重用线程。每个提交的任务都异步运行在一个线程中。

SyncTaskExecutor 是最简单的一个 TaskExecutor 实现。任务的提交是同步的，相当于启动一个线程，运行它，然后立刻使用 join()连接它。这相当于手工执行调用线程的 run()方法，并完全忽略掉线程本身。

ScheduledExecutorFactoryBean 会自动触发以 ScheduledExecutorTask bean 形式定义的任务。你可以指定一个 ScheduledExecutorTask 实例列表来同时触发多个任务。ScheduledExecutorTask 实例可以接收一个周期参数将任务的执行间隔开来。

最后一个示例是个 ThreadPoolTaskExecutor，它是个基于 java.util.concurrent.ThreadPoolExecutor 的完全的线程池实现。

如果想要通过诸如 IBM WebSphere 等应用服务器所支持的 CommonJ WorkManager/TimerManager 来构建应用，那就可以使用 org.springframework.scheduling.commonj.WorkManagerTaskExecutor。这个类会委托给 WebSphere 中的 CommonJ Work Manager 引用。通常情况下，你需要以“对恰当资源的 JNDI 引用”的形式来提供它。

JEE 7 添加了 javax.enterprise.concurrent 包，特别是 ManagedExecutorService。JEE 7 兼容的服务器必须要提供这个 ManagedExecutorService 实例。如果想要在 Spring TaskExecutor 的支持下使用该机制，那么可以配置一个 DefaultManagedTaskExecutor，它会尝试检测默认的 ManagedExecutorService（规范所要求的），也可以显式进行配置。

TaskExecutor 支持提供了一种强大的方式来通过统一接口访问应用服务器中的调度服务。如果想要寻找可部署在任何应用服务器（比如说 Tomcat 和 Jetty）上的更为健壮（虽然也会更加重量级）的支持，那么可以考虑 Spring 提供的 Quartz 支持。

2-24 在 POJO 间实现应用事件通信

问题提出

在 POJO 间的典型通信中，发送者需要定位到接收者才能调用其方法。在这种情况下，发送者 POJO 必须要能感知到接收者组件。这种通信方式是直接且简单的，不过发送者与接收者 POJO 却是紧耦合的。

当使用了 IoC 容器时，POJO 可以通过接口（而非实现）进行通信。这种通信模型有助于降低耦合。不过，只有当一个发送者组件与一个接收者组件进行通信时才行。当一个发送者需要与多个接收者进行通信时，发送者就需要逐一调用接收者。

解决方案

Spring 的应用上下文支持 bean 之间的基于事件的通信。在基于事件的通信模型中，发送者 POJO 只需发送事件即可，无须知道接收者是谁，因为可能会有多个接收者。此外，接收者不一定知道谁在发送事件。它可以在同一时刻监听来自于不同发送者的多个事件。通过这种方式，发送者与接收者组件可以做到松耦合。

传统上，要想监听事件，bean 需要实现 ApplicationListener 接口并通过类型参数（如 ApplicationListener <CheckoutEvent>）来指定想要得到通知的事件类型。这种监听器只能监听继承自 ApplicationEvent 的事件，因为它是 ApplicationListener 接口的类型签名。

要想发布事件，bean 需要访问 ApplicationEventPublisher，对于事件的发送来说则需要调用 publishEvent 方法。要想访问 ApplicationEventPublisher，类可以实现 ApplicationEventPublisherAware，或是在 ApplicationEventPublisher 类型的字段上使用@Autowired 注解。

解释说明

首先，编写一个自定义的 ApplicationEvent，然后将其发布出去；然后编写一个组件来接收这些事件并对其进行响应。

使用 ApplicationEvent 定义事件

启用基于事件通信的第一步是定义事件。假设你想让一个 cashier bean 在购物车结账后发布一个 CheckoutEvent。该事件包含了一个结账时间的属性。

```
package com.apress.springrecipes.shop;

import org.springframework.context.ApplicationEvent;

import java.util.Date;

public class CheckoutEvent extends ApplicationEvent {

    private final ShoppingCart cart;
    private final Date time;

    public CheckoutEvent(ShoppingCart cart, Date time) {
        super(cart);
        this.cart=cart;
        this.time = time;
    }

    public ShoppingCart getCart() {
        return cart;
    }

    public Date getTime() {
```

```
        return this.time;
    }
}
```

发布事件

要想发布事件，只需创建一个事件实例并调用应用事件发布器的 publishEvent()方法即可，可以通过实现 ApplicationEventPublisherAware 接口来访问应用事件发布器。

```
package com.apress.springrecipes.shop;
...
import org.springframework.context.ApplicationEventPublisher;
import org.springframework.context.ApplicationEventPublisherAware;

public class Cashier implements ApplicationEventPublisherAware {
    ...
    private ApplicationEventPublisher applicationEventPublisher;

    public void setApplicationEventPublisher(
        ApplicationEventPublisher applicationEventPublisher) {
        this.applicationEventPublisher = applicationEventPublisher;
    }

    public void checkout(ShoppingCart cart) throws IOException {
        ...
        CheckoutEvent event = new CheckoutEvent(this, new Date());
        applicationEventPublisher.publishEvent(event);
    }
}
```

或是在字段属性上进行自动装配。

```
package com.apress.springrecipes.shop;
...
import org.springframework.context.ApplicationEventPublisher;

public class Cashier {
    ...
    @Autowired
    private ApplicationEventPublisher applicationEventPublisher;

    public void checkout(ShoppingCart cart) throws IOException {
        ...
        CheckoutEvent event = new CheckoutEvent(cart, new Date());
        applicationEventPublisher.publishEvent(event);
    }
}
```

监听事件

定义在应用上下文中且实现了 ApplicationListener 接口的任何 bean 都会收到与类型参数匹配的所有类型事件的通知（通过这种方式，可以监听某一组事件，如 ApplicationContextEvent）。

```
package com.apress.springrecipes.shop;
...
import org.springframework.context.ApplicationListener;

@Component
public class CheckoutListener implements ApplicationListener<CheckoutEvent> {

    public void onApplicationEvent(CheckoutEvent event) {
        // Do anything you like with the checkout amount and time
        System.out.println("Checkout event [" + event.getTime() + "]");
    }
}
```

在新版本的 Spring 中，还可以通过@EventListener 注解替代实现 ApplicationListener 接口来创建事件监听器。

```
package com.apress.springrecipes.shop;

...

@Component
public class CheckoutListener {

    @EventListener
    public void onApplicationEvent(CheckoutEvent event) {
        // Do anything you like with the checkout amount and time
        System.out.println("Checkout event [" + event.getTime() + "]");
    }
}
```

接下来，需要在应用上下文中注册监听器来监听所有事件。注册非常简单，只需声明监听器的一个 bean 实例或是让组件扫描能够检测到它即可。应用上下文会识别出实现了 ApplicationListener 接口的 bean 以及其方法使用了@EventListener 注解的 bean，并且当其感兴趣的事件到来时会通知它们。

使用@EventListener 注解还有另外一个很棒的特性，即事件无须再继承 ApplicationEvent 了。通过这种方式，事件不必再依赖于 Spring 框架中的类，它们只是普通的 POJO 而已。

```
package com.apress.springrecipes.shop;

import java.util.Date;

public class CheckoutEvent {

    private final ShoppingCart cart;
    private final Date time;

    public CheckoutEvent(ShoppingCart cart, Date time) {
        this.cart=cart;
        this.time = time;
    }

    public ShoppingCart getCart() {
        return cart;
    }

    public Date getTime() {
        return this.time;
    }
}
```

注意： 最后，请记住应用上下文本身也会发布容器事件，如 ContextClosedEvent、ContextRefreshedEvent 以及 RequestHandledEvent。如果有 bean 想要接收到这些事件通知，它们可以实现 ApplicationListener 接口。

小结

本章介绍了 Spring 的核心任务。我们介绍了 Spring 是如何支持@Configuration 与@Bean 注解来通过 Java 配置类实例化 POJO 的，还介绍了如何通过@Component 注解管理 POJO。此外，我们介绍了@Repository、@Service 与@Controller 注解，相较于@Component 注解来说，它们提供了更为具体的行为。

我们又介绍了如何在 POJO 中引用其他 POJO，以及如何使用@Autowired 注解，它会根据类型或是名字自动关联 POJO。此外，我们探究了标准的@Resource 与@Inject 注解是如何通过自动装配引用 POJO 的，从而替代 Spring 独有的@Autowired 注解。

接下来介绍了如何通过@Scope 注解来设置 Spring POJO 的作用域，还介绍了如何通过@PropertySource 与@Value 注解在 POJO 配置与创建的上下文中读取外部资源并使用所读取的数据。此外，我们介绍了 Spring

是如何使用 i18n 资源包以在 POJOs 中支持不同语言的。

接下来介绍了如何通过@Bean 注解的 initmethod 与 destroyMethod 属性以及@PostConstruct 与@PreDestroy 注解来自定义 POJOs 的初始化与销毁。此外，我们还介绍了如何通过@Lazy 注解实现延迟初始化，如何通过@DependsOn 注解定义初始化依赖。

接下来，我们介绍了如何通过 Spring 的后置处理器来验证并修改 POJO 值，包括如何使用@Required 注解。然后，我们探究了如何使用 Spring 环境与 profile 来加载不同的 POJO 集合，包括如何使用@Profile 注解。

接下来，我们介绍了 Spring 下的面向切面编程并介绍了如何创建切面、切点与通知。这包括对@Aspect、@Before、@After、@AfterReturning、@AfterThrowing 与@Around 注解的使用。

然后，我们介绍了如何访问 AOP 连接点信息并将其应用到不同的程序执行点。接下来介绍了如何通过@Order 注解来指定切面顺序，以及如何重用切面切点（aspect pointcut）的定义。

我们还介绍了如何编写 AspectJ 切点表达式，如何应用 AOP 引介的概念使得 POJO 可以在同一时刻从多个实现类中继承行为。我们还介绍了如何通过 AOP 将状态存储到 POJO 中，以及如何应用加载期编织技术。

最后，我们介绍了如何在 Spring 中配置 AspectJ 切面，如何将 POJO 注入到领域对象中，如何使用 Spring 与 TaskExecutor 处理并发，以及如何在 Spring 中创建、发布和监听事件。

第 3 章

Spring MVC

MVC 是 Spring 框架中最重要的模块之一。它构建在强大的 Spring IoC 容器之上，并广泛使用容器的特性来简化其配置。

模型—视图—控制器（MVC）是 UI 设计中一种常见的设计模式。它通过分离应用中的模型、视图与控制器角色实现了业务逻辑与 UI 的解耦。模型负责封装应用数据以展现在视图上。视图只应该展现这种数据，不会包含任何业务逻辑。控制器负责接收来自用户的请求并调用后端服务进行业务处理。处理完毕后，后端服务可以返回一些数据供视图展示。控制器会收集这些数据并准备好模型以供视图进行展示。MVC 模式的核心理念在于分离业务逻辑与 UI，使之可以独立变化而不会影响到彼此。

在 Spring MVC 应用中，模型通常会包含领域对象，这些领域对象由服务层进行处理并由持久层进行持久化。视图通常是由 Java 标准标签库（JSTL）编写的 JSP 模板。不过，也可以将视图定义为 PDF 文件、Excel 文件、RESTful Web Service，甚至是 Flex 接口，其中 Flex 通常被称作富互联网应用（RIA）。

学习完本章后，你可以使用 Spring MVC 开发 Java Web 应用，还会理解 Spring MVC 的常用控制器与视图类型，包括 Spring 3.0 发布后用于创建控制器的注解，这些注解已经成为事实上的用法。你还会理解 Spring MVC 的基本原则，这将作为后续章节所介绍的更高级话题的基础。

3-1 使用 Spring MVC 开发一个简单的 Web 应用

问题提出

使用 Spring MVC 开发一个简单的 Web 应用以了解该框架的基本概念与配置。

解决方案

Spring MVC 的核心组件是一个前端控制器。在最简单的 Spring MVC 应用中，该控制器是唯一需要在 Java Web 部署描述符（例如 web.xml 文件或是 ServletContainerInitializer）中配置的 servlet。Spring MVC 控制器（通常叫做 Dispatcher Servlet）实现了 Sun Java EE 设计模式中名为前端控制器（Front Controller）的模式。它充当 Spring MVC 框架中的前端控制器，每个 Web 请求都会流经该控制器，这样它就可以管理整个请求处理过程了。

当 Web 请求被发送给 Spring MVC 应用时，控制器首先会接收到请求。接下来，它会组织 Spring Web 应用上下文中配置的不同组件或是位于控制器上的注解来处理请求。图 3-1 所示为 Spring MVC 中请求处理的主要流程。

要想在 Spring 中定义控制器类，需要将类标记为@Controller 或@RestController 注解。

当标注了@Controller 注解的类（比如一个控制器类）接收到一个请求，它会寻找一个恰当的处理器方法来处理请求。这要求控制器类通过一个或多个处理器映射将每个请求映射到一个处理器方法上。为了做到这一点，控制器类的方法需要使用@RequestMapping 注解，这会将其标记为处理器方法。

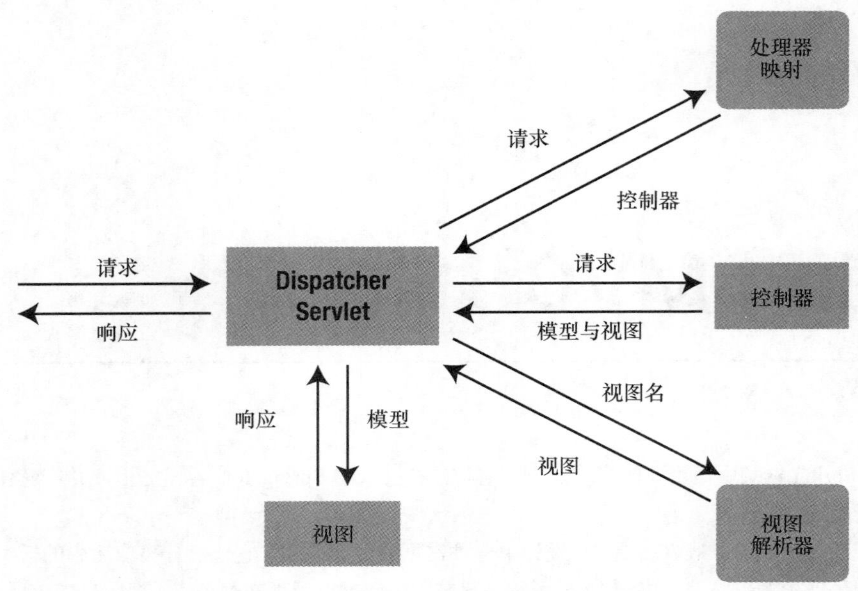

图 3-1 Spring MVC 中请求处理的主要流程

这些处理器方法的签名（就如任何标准的类一样）并没有什么限制。你可以为处理器方法指定任意名字，定义各种方法参数。同样，处理器方法可以返回各种值（如 String 或是 void），这取决于它所满足的应用逻辑。在阅读本书的过程中，你会在使用了@RequestMapping 注解的处理器方法中看到各种方法参数。如下列出了部分有效的参数类型，仅供参考。

- HttpServletRequest 或 HttpServleResponse。
- 使用@RequestParam 注解的任意类型的请求参数。
- 使用@ModelAttribute 注解的任意类型的模型属性。
- 使用@CookieValue 注解的传入请求中的 cookie 值。
- Map 或 ModelMap，用于处理器方法将属性添加到模型中。
- Errors 或 BindingResult，用于处理器方法访问命令对象的绑定与验证结果。
- SessionStatus，用于处理器方法通知会话处理的完成。

当控制器类选择好了恰当的处理器方法后，它会使用请求来调用处理器方法的逻辑。通常，控制器的逻辑会调用后端服务来处理请求。此外，处理器方法的逻辑很可能会从众多输入参数（如 HttpServletRequest、Map、Errors 或是 SessionStatus）中添加或移除一些信息，这些信息会成为正在进行的 Spring MVC 流程的一部分。

当处理器方法完成了请求处理后，它会将控制委托给视图，这表现为处理器方法的返回值。为了提供一种灵活的方式，处理器方法的返回值并不会表示视图的实现（比如说 user.jsp 或 report.pdf），而是表示为一种逻辑视图（如 user 或 report，注意，这里没有文件扩展名）。

处理器方法的返回值可以是一个字符串，表示一个逻辑视图名；也可以是 void，这时会根据处理器方法或是控制器的名字来确定一个默认的逻辑视图名。要想将信息从控制器传递给视图，无论处理器方法返回何种逻辑视图名（字符串或是 void）都可以，因为处理器方法的输入参数对于视图是可见的。比如说，如果处理器方法接收一个 Map 和 SessionStatus 对象作为输入参数（在处理器方法逻辑中修改其内容），那么由处理器方法所返回的视图是可以访问到相同对象的。

当控制器类接收到一个视图时，它会通过视图解析器将逻辑视图名解析为具体的视图实现（比如说 user.jsp 或 report.pdf）。视图解析器是个实现了 ViewResolver 接口并在 Web 应用上下文中配置的 bean，其职责是为逻辑视图名返回一个具体的视图实现（HTML、JSP 或是 PDF 等）。

当控制器类将视图名解析为视图实现后，根据视图实现的设计，它会渲染控制器的处理器方法所传递的对象（比如说HttpServletRequest、Map、Errors或是SessionStatus）。视图的职责是将处理器逻辑中所添加的对象呈现给用户。

解释说明

假设你要为一个运动中心开发一款场地预订系统。该应用的UI是基于Web的，这样用户就可以通过互联网在线预订了。你想要使用Spring MVC开发这款应用。首先，在domain子包中创建如下领域类：

```
package com.apress.springrecipes.court.domain;

public class Reservation {

    private String courtName;
    private Date date;
    private int hour;
    private Player player;
    private SportType sportType;

    // Constructors, Getters and Setters
    ...
}
package com.apress.springrecipes.court.domain;

public class Player {

    private String name;
    private String phone;

    // Constructors, Getters and Setters
    ...
}
package com.apress.springrecipes.court.domain;

public class SportType {

    private int id;
    private String name;

    // Constructors, Getters and Setters
    ...
}
```

接下来，在service子包中定义如下服务接口来向展示层提供预订服务：

```
package com.apress.springrecipes.court.service;

import com.apress.springrecipes.court.domain.Reservation;

import java.util.List;

public interface ReservationService {

    public List<Reservation> query(String courtName);
}
```

在生产应用中，你应该使用数据库持久化来实现这个接口。不过出于简化的目的，可以将预定记录保存到列表中并硬编码几条预订信息进行测试。

```
package com.apress.springrecipes.court.service;

import com.apress.springrecipes.court.domain.Player;
import com.apress.springrecipes.court.domain.Reservation;
import com.apress.springrecipes.court.domain.SportType;
import org.springframework.stereotype.Service;
```

```java
import java.time.LocalDate;
import java.util.ArrayList;
import java.util.List;
import java.util.Objects;
import java.util.stream.Collectors;

@Service
public class ReservationServiceImpl implements ReservationService {

    public static final SportType TENNIS = new SportType(1, "Tennis");
    public static final SportType SOCCER = new SportType(2, "Soccer");

    private final List<Reservation> reservations = new ArrayList<>();

    public ReservationServiceImpl() {

        reservations.add(new Reservation("Tennis #1", LocalDate.of(2008, 1, 14), 16,
            new Player("Roger", "N/A"), TENNIS));
        reservations.add(new Reservation("Tennis #2", LocalDate.of(2008, 1, 14), 20,
            new Player("James", "N/A"), TENNIS));
    }

    @Override
    public List<Reservation> query(String courtName) {

        return this.reservations.stream()
            .filter(reservation -> Objects.equals(reservation.getCourtName(), courtName))
            .collect(Collectors.toList());
    }
}
```

创建一个 Spring MVC 应用

接下来，你需要创建一个 Spring MVC 应用布局。一般来说，使用 Spring MVC 开发的 Web 应用的创建方式与标准的 Java Web 应用一样，只不过需要添加一些 Spring MVC 专属的配置文件与库。

Java EE 规范为构成 Web 归档（WAR 文件）的 Java Web 应用定义了有效的目录结构。比如说，你需要在 WEB-INF 根目录下提供一个 Web 部署描述符（即 web.xml），或是实现了 ServletContainerInitializer 的一个或多个类。Web 应用的类文件与 JAR 文件应该分别放在 WEB-INF/classes 与 WEB-INF/lib 目录下。

对于这个场地预订系统来说，创建如下目录结构。注意到高亮的文件是 Spring 专属配置文件。

注意：要想使用 Spring MVC 开发 Web 应用，你需要将常规的 Spring 依赖（请参考第 1 章了解更多信息）以及 Spring Web 与 Spring MVC 依赖添加到类路径下。如果使用 Maven，请将如下依赖添加到 Maven 项目中：

```xml
<dependency>
    <groupId>org.springframework</groupId>
    <artifactId>spring-webmvc</artifactId>
    <version>${spring.version}</version>
</dependency>
```

如果使用 Gradle，请添加如下依赖：

```groovy
dependencies {
    compile "org.springframework:spring-webmvc:$springVersion"
}
```

用户可以通过 URL 直接访问 WEB-INF 目录外的文件，因此 CSS 文件与图片文件必须要放在这里。在使用 Spring MVC 时，JSP 文件用作模板。它们会被框架读取并生成动态内容，因此 JSP 文件应该放在 WEB-INF 目录中以防止被直接访问。不过，一些应用服务器内部并不允许 Web 应用读取 WEB-INF 中的文件。在这种情况下，只能将它们放到 WEB-INF 目录外。

创建配置文件

Web 部署描述符（web.xml 或 ServletContainerInitializer）是 Java Web 应用必要的配置文件。在该文件中，你可以定义应用的 servlet 以及 Web 请求映射到 servlet 的方式。对于 Spring MVC 应用来说，你只需要定义一个 DispatcherServlet 实例来作为 Spring MVC 的前端控制器即可，不过如果需要，还可以定义其他 servlet。

在大型应用中，使用多个 DispatcherServlet 实例会很方便。这可以让 DispatcherServlet 针对特定的 URL，使得代码管理变得更加轻松，也可以让每个团队成员处理一个应用的逻辑时不会干扰到彼此。

```
package com.apress.springrecipes.court.web;

import com.apress.springrecipes.court.config.CourtConfiguration;
import org.springframework.web.context.support.AnnotationConfigWebApplicationContext;
import org.springframework.web.servlet.DispatcherServlet;

import javax.servlet.ServletContainerInitializer;
import javax.servlet.ServletContext;
import javax.servlet.ServletException;
import javax.servlet.ServletRegistration;
import java.util.Set;

public class CourtServletContainerInitializer implements ServletContainerInitializer {

    @Override
    public void onStartup(Set<Class<?>> c, ServletContext ctx) throws ServletException {

        AnnotationConfigWebApplicationContext applicationContext =
            new AnnotationConfigWebApplicationContext();
        applicationContext.register(CourtConfiguration.class);

        DispatcherServlet dispatcherServlet = new DispatcherServlet(applicationContext);

        ServletRegistration.Dynamic courtRegistration =
            ctx.addServlet("court", dispatcherServlet);
        courtRegistration.setLoadOnStartup(1);
        courtRegistration.addMapping("/");
    }
}
```

在这个 CourtServletContainerInitializer 中，你定义了一个 DispatcherServlet 类型的 servlet。这是 Spring MVC 中的核心 servlet 类，它接收 Web 请求并将其分发到恰当的处理器上。你将该 servlet 的名字设为 court 并使用斜杠（/）映射所有的 URL，这个斜杠表示根目录。注意，可以将 URL 模式设为更为细粒度的模式。在大型应用中，在多个 servlet 中来委托模式会更有意义，不过为了简化，应用中的所有 URL 都委托给了单个 court servlet。

要想检测到 CourtServletContainerInitializer，还需要向 META-INF/services 目录中添加一个名为 javax.servlet.ServletContainerInitializer 的文件。文件内容应该是 CourtServletContainerInitializer 的全名。该文件会被 servlet 容器加载并用于启动应用。

```
com.apress.springrecipes.court.web.CourtServletContainerInitializer
```

最后，添加 CourtConfiguration 类，它是个简单的@Configuration 类。

```
package com.apress.springrecipes.court.config;

import org.springframework.context.annotation.ComponentScan;
import org.springframework.context.annotation.Configuration;

@Configuration
@ComponentScan("com.apress.springrecipes.court")
public class CourtConfiguration {}
```

这定义了一个@ComponentScan 注解，它会扫描 com.apress.springrecipes.court 包（及其子包）并注册所有检测到的 bean（在该示例中就是 ReservationServiceImpl 以及将要创建的被@Controller 所注解的类）。

创建 Spring MVC 控制器

基于注解的控制器类可以是任意类，无须实现特定接口或是继承特定父类。你可以使用@Controller 注解来修饰它。一个控制器中会有一个或多个处理器方法来处理单个或多个动作。处理器方法的签名足够灵活，可以接收各种参数。

@RequestMapping 注解可以应用到类级别或是方法级别上。第一种映射策略是将特定的 URL 模式映射到控制器类上，然后将特定的 HTTP 方法映射到每个处理器方法上。

```
package com.apress.springrecipes.court.web;

import org.springframework.stereotype.Controller;
import org.springframework.ui.Model;
import org.springframework.web.bind.annotation.RequestMapping;
import org.springframework.web.bind.annotation.RequestMethod;

import java.util.Date;

@Controller
@RequestMapping("/welcome")
public class WelcomeController {

    @RequestMapping(method = RequestMethod.GET)
    public String welcome(Model model) {

        Date today = new Date();
        model.addAttribute("today", today);
        return "welcome";
    }
}
```

该控制器创建了一个 java.util.Date 对象来获取当前日期，然后作为属性将其添加到输入 Model 对象上，这样目标视图就可以展示了。

由于已经在 com.apress.springrecipes.court 包上激活了注解扫描，因此在部署时控制器类上的注解就会被检测到。

@Controller 注解会将类定义为 Spring MVC 控制器。@RequestMapping 注解更有意思一些，因为它包含了属性，并且可以声明在类或处理器方法层次上。该类所用的第一个值（"/welcome"）指定了一个 URL，控制器会对这个 URL 采取动作，这意味着/welcome URL 上所接收的任何请求都会被 WelcomeController 类所接收。

当请求被控制器类接收到后，它会将调用委托给控制器中声明的默认 HTTP GET 处理器方法。出现这种行为的原因在于对 URL 所发出的每个初始请求都是 HTTP GET 类型。这样，当控制器接收到/welcome URL 上的请求时，它会将其委托给默认的 HTTP GET 处理器方法进行处理。

注解@RequestMapping(method = RequestMethod.GET)用于将 welcome 方法修饰为控制器的默认 HTTP GET 处理器方法。值得注意的是，如果没有声明默认的 HTTP GET 处理器方法，那就会抛出 ServletException 异常。因此，对于 Spring MVC 控制器来说，重要的是指定一个最基础的 URL 路由和默认的 HTTP GET 处理器方法。

另外一种方式则是在方法层次上所用的@RequestMapping 注解中声明两个值（URL 路由与默认的 HTTP GET 处理器方法）。这种声明如下所示：

```
@Controller
public class WelcomeController {

    @RequestMapping(value = "/welcome", method=RequestMethod.GET)
    public String welcome(Model model) { … }

}
```

该声明等价于之前那个。value 属性指定了处理器方法所映射的 URL，method 属性则定义了作为控制器默认HTTP GET方法的处理器方法。最后，还有其他一些方便的注解（如@GetMapping和@PostMapping等）可以简化配置。如下映射与上面的声明等价：

```
@Controller
public class WelcomeController {

    @GetMapping("/welcome")
    public String welcome(Model model) { … }

}
```

@GetMapping注解让这个类更为简短，更容易阅读。

最后一个控制器演示了Spring MVC的基本原则。不过，典型的控制器会调用后端服务来进行业务处理。比如说，你可以创建一个控制器来查询特定场地的预订情况，代码如下所示：

```
package com.apress.springrecipes.court.web;

import com.apress.springrecipes.court.domain.Reservation;
import com.apress.springrecipes.court.service.ReservationService;
import org.springframework.beans.factory.annotation.Autowired;
import org.springframework.stereotype.Controller;
import org.springframework.ui.Model;
import org.springframework.web.bind.annotation.RequestMapping;
import org.springframework.web.bind.annotation.RequestMethod;
import org.springframework.web.bind.annotation.RequestParam;

import java.util.List;

@Controller
@RequestMapping("/reservationQuery")
public class ReservationQueryController {

    private final ReservationService reservationService;

    public ReservationQueryController(ReservationService reservationService) {
        this.reservationService = reservationService;
    }

    @GetMapping
    public void setupForm() {}

    @PostMapping
    public String sumbitForm(@RequestParam("courtName") String courtName, Model model) {

        List<Reservation> reservations = java.util.Collections.emptyList();
        if (courtName != null) {
            reservations = reservationService.query(courtName);
        }
            model.addAttribute("reservations", reservations);
        return "reservationQuery";
    }
}
```

如前所述，控制器接下来会寻找默认的HTTP GET处理器方法。由于为public void setupForm()方法指定了必要的@RequestMapping注解来实现这个目的，接下来就会调用它。

与之前默认的HTTP GET处理器方法不同，注意到该方法没有输入参数，没有逻辑，返回值为void。这意味着两点。由于没有输入参数也没有逻辑，视图只会在实现模板（如JSP）中展示出硬编码的数据，这是因为控制器中并未添加数据。由于返回值为void，因此会根据请求URL使用默认的视图名；由于请求URL是/reservationQuery，故名为reservationQuery的视图会返回。

剩下的处理器方法使用@PostMapping注解进行了修饰。乍看之下，两个处理器方法只有类级别的/reservationQuery URL语句，这令人感到困惑，不过其实很简单。一个方法会在向/reservationQuery URL发出 HTTP GET 请求时调用，另一个方法则会在向同样的 URL 发出 HTTP POST 请求时调用。

Web 应用中大多数请求都是 HTTP GET 类型的，而 HTTP POST 类型的请求通常都是在用户提交 HTML 表单时发出的。因此，揭示了对应用的更多视图（稍后将会介绍），当 HTML 表单初次加载时会调用一个方法（HTTP GET），当 HTML 表单提交时则会调用另一个方法（HTTP POST）。

仔细看一下 HTTP POST 默认的处理器方法，注意它的两个输入参数。首先看一下@RequestParam ("courtName") String courtName 声明，它用于抽取出名为 courtName 的请求参数。在该示例中，HTTP POST 请求的形式是/reservationQuery?courtName=<value>；该声明使得 value 成为方法中名为 courtName 的变量的值。再来看一下 Model 声明，它用于定义一个对象，该对象可以将数据传递给返回的视图中。

由处理器方法所执行的逻辑包含了使用控制器的 reservationService 并根据 courtName 变量来执行查询。该查询所得到的结果会被赋给 Model 对象，接下来返回的视图就可以将其展示出来了。

最后，注意到该方法返回了一个名为 reservationQuery 的视图。该方法也可以返回 void，就像默认的 HTTP GET 一样，而且由于请求 URL 的缘故，它会被赋予同样的 reservationQuery 默认视图。这两种方式是等价的。

在学习了 Spring MVC 控制器是如何构成的之后，现在探究一下视图，视图会接收控制器的处理器方法所传递的结果。

创建 JSP 视图

Spring MVC 针对不同的展示层技术支持多种类型的视图，包括 JSP、HTML、PDF、Excel 工作表（XLS）、XML、JSON、Atom 与 RSS feed、JasperReports 和其他第三方视图实现。

在 Spring MVC 应用中，视图大多都是使用 JSTL 编写的 JSP 模板。当 DispatcherServlet（定义在应用的 web.xml 文件中）接收到处理器所返回的视图名时，它会将逻辑视图名解析为视图实现进行渲染。比如说，你可以在 Web 应用上下文的 CourtConfiguration 中配置一个 InternalResourceViewResolver bean，将视图名解析为/WEB-INF/jsp/目录中的 JSP 文件。

```
@Bean
public InternalResourceViewResolver internalResourceViewResolver() {

    InternalResourceViewResolver viewResolver = new InternalResourceViewResolver();
    viewResolver.setPrefix("/WEB-INF/jsp/");
    viewResolver.setSuffix(".jsp");
    return viewResolver;
}
```

借助于上述配置，名为 reservationQuery 的逻辑视图会委托给位于/WEB-INF/jsp/reservationQuery.jsp 的视图实现。知道了这一点，你就可以为 Welcome 控制器创建如下 JSP 模板，将其命名为 welcome.jsp 并放在/WEB-INF/jsp/目录下：

```
<%@ taglib prefix="fmt" uri="http://java.sun.com/jsp/jstl/fmt" %>

<html>
<head>
    <title>Welcome</title>
</head>

<body>
<h2>Welcome to Court Reservation System</h2>
Today is <fmt:formatDate value="${today}" pattern="yyyy-MM-dd" />.
</body>
</html>
```

在该 JSP 模板中，使用 JSTL 中的 fmt 标签库将 today 模型属性格式化为 yyyy-MM-dd 模式。不要忘记在该 JSP 模板的顶部引入 fmt 标签库定义。

接下来，为预留查询控制器创建另一个 JSP 模板，将其命名为 reservationQuery.jsp 以匹配视图名。

```
<%@ taglib prefix="c" uri="http://java.sun.com/jsp/jstl/core" %>
<%@ taglib prefix="fmt" uri="http://java.sun.com/jsp/jstl/fmt" %>
```

```
<html>
<head>
<title>Reservation Query</title>
</head>

<body>
<form method="post">
Court Name
<input type="text" name="courtName" value="${courtName}" />
<input type="submit" value="Query" />
</form>

<table border="1">
    <tr>
        <th>Court Name</th>
        <th>Date</th>
        <th>Hour</th>
        <th>Player</th>
    </tr>
    <c:forEach items="${reservations}" var="reservation">
    <tr>
        <td>${reservation.courtName}</td>
        <td><fmt:formatDate value="${reservation.date}" pattern="yyyy-MM-dd" /></td>
        <td>${reservation.hour}</td>
        <td>${reservation.player.name}</td>
    </tr>
    </c:forEach>
</table>
</body>
</html>
```

在该 JSP 模板中，使用了一个表单供用户输入想要查询的场地名，然后使用<c:forEach>标签遍历 reservation 的模型属性来生成结果表格。

部署 Web 应用

在 Web 应用的开发过程中，强烈建议安装一个带有 Web 容器的本地 Java EE 应用服务器进行测试与调试。出于简化配置与部署的目的，我们选择 Apache Tomcat 8.5.x 作为 Web 容器。

该 Web 容器的部署目录位于 webapps 目录下。默认情况下，Tomcat 会监听 8080 端口，并部署到与应用 WAR 名字相同的上下文中。因此，如果将应用打包到名为 court.war 的 WAR 中，那就可以通过如下 URL 访问欢迎控制器与预订查询控制器：

```
http://localhost:8080/court/welcome
http://localhost:8080/court/reservationQuery
```

提示：该项目还可以创建一个 Docker 容器。运行../gradlew buildDocker 获取到一个带有 Tomcat 和应用的容器。接下来就可以启动 Docker 容器来测试应用了（docker run -p 8080:8080 spring-recipes-4th/court-web）。

使用 WebApplicationInitializer 启动应用

上一节创建了一个 CourtServletContainerInitializer 并在 META-INF/services 下放置一个文件来启动应用。

之前是实现了自己的启动类，现在我们使用便捷的 Spring 实现——SpringServletContainerInitializer。该类是 ServletContainerInitializer 接口的一个实现，它会扫描类路径寻找 WebApplicationInitializer 接口的实现。值得称道的是，Spring 提供了该接口的一些便捷实现，可以在应用中使用；其中一个就是 AbstractAnnotationConfigDispatcherServletInitializer。

```
package com.apress.springrecipes.court.web;

import com.apress.springrecipes.court.config.CourtConfiguration;
```

```
import org.springframework.web.servlet.support.
AbstractAnnotationConfigDispatcherServletInitializer;

public class CourtWebApplicationInitializer extends
AbstractAnnotationConfigDispatcherServletInitializer {

    @Override
    protected Class<?>[] getRootConfigClasses() {
        return null;
    }

    @Override
    protected Class<?>[] getServletConfigClasses() {
        return new Class[] {CourtConfiguration.class};
    }

    @Override
    protected String[] getServletMappings() {
        return new String[] { "/"};
    }
}
```

新创建的 CourtWebApplicationInitializer 已经创建了一个 DispatcherServlet，这样只需要在 getServlet-Mappings 方法中配置映射，以及在 getServletConfigClasses 方法中配置想要加载的类。除了 servlet 外，还有另一个待创建的组件，即 ContextLoaderListener。它是个 ServletContextListener，也会创建一个 Application-Context，并作为 DispatcherServlet 的父 ApplicationContext。如果有多个 servlet 需要访问相同的 bean（服务、数据源等），这么做就会很方便。

3-2 使用@RequestMapping 映射请求

问题提出

当 DispatcherServlet 接收到 Web 请求时，它会将请求分发给使用了@Controller 注解修饰的各种控制器类。分发过程取决于声明在控制器类中的各种@RequestMapping 注解及处理器方法。需要使用@RequestMapping 注解为映射请求定义一种策略。

解决方案

在 Spring MVC 应用中，我们是通过声明在控制器类中的一个或多个@RequestMapping 注解将 Web 请求映射到控制器上的。

处理器映射在匹配 URL 时是根据它们相对于上下文路径（即 Web 应用上下文的部署路径）与 servlet 路径（即映射到 DispatcherServlet 的路径）的路径来进行的。比如说，在 URL http://localhost:8080/court/welcome 中，待匹配的路径是/welcome，因为上下文路径是/court，并且没有 servlet 路径——因为 CourtWebApplicationInitializer 中将 servlet 路径声明为/。

解释说明

首先，你会看到应用到方法级别的请求映射，接下来探究一下类级别上的请求映射，并将其与方法级别的请求映射组合起来。最后，你会看到如何为请求映射方法指定 HTTP 方法。

根据方法映射请求

@RequestMapping 注解最简单的使用方式就是直接修饰处理器方法。要想做到这一点，需要使用@RequestMapping 注解（包含一个 URL 模式）来声明每个处理器方法。如果处理器的@RequestMapping 注解匹配了请求的 URL，那么 DispatcherServlet 就会将请求分发给该处理器，由它来处理这个请求。

```
package com.apress.springrecipes.court.web;

import com.apress.springrecipes.court.domain.Member;
import com.apress.springrecipes.court.service.MemberService;
```

```
import org.springframework.stereotype.Controller;
import org.springframework.ui.Model;
import org.springframework.web.bind.annotation.RequestMapping;
import org.springframework.web.bind.annotation.RequestMethod;
import org.springframework.web.bind.annotation.RequestParam;

@Controller
public class MemberController {

    private MemberService memberService;

    public MemberController(MemberService memberService) {
        this.memberService = memberService;
    }

    @RequestMapping("/member/add")
    public String addMember(Model model) {

        model.addAttribute("member", new Member());
        model.addAttribute("guests", memberService.list());
        return "memberList";
    }

    @RequestMapping(value = {"/member/remove", "/member/delete"}, method = RequestMethod.GET)
    public String removeMember(@RequestParam("memberName")String memberName) {
        memberService.remove(memberName);
        return "redirect:";
    }
}
```

上述代码演示了如何通过@RequestMapping 注解将每个处理器方法映射到特定的 URL 上。第二个处理器方法演示了如何指定多个 URL，这样/member/remove 与/member/delete 都会触发该处理器方法的执行。默认情况下，假设所有对 URL 的请求都是 HTTP GET 类型的。

根据类映射请求

@RequestMapping 注解还可以修饰控制器类。这样，处理器方法既可以不再使用@RequestMapping 注解（如 3-1 节中的 ReservationQueryController 控制器那样），也可以搭配上自己的@RequestMapping 注解以使用更细粒度的 URL。针对更为广泛的 URL 匹配，@RequestMapping 注解还支持使用通配符（即*）。

如下代码演示了@RequestMapping 注解中 URL 通配符的使用，以及处理器方法上@RequestMapping 注解更为细粒度的 URL 匹配：

```
package com.apress.springrecipes.court.web;

import com.apress.springrecipes.court.domain.Member;
import com.apress.springrecipes.court.service.MemberService;
import org.springframework.stereotype.Controller;

import org.springframework.ui.Model;
import org.springframework.web.bind.annotation.PathVariable;
import org.springframework.web.bind.annotation.RequestMapping;
import org.springframework.web.bind.annotation.RequestMethod;
import org.springframework.web.bind.annotation.RequestParam;

@Controller
@RequestMapping("/member/*")
public class MemberController {

    private final MemberService memberService;

    public MemberController(MemberService memberService) {
```

```
        this.memberService = memberService;
    }

    @RequestMapping("add")
    public String addMember(Model model) {

        model.addAttribute("member", new Member());
        model.addAttribute("guests", memberService.list());
        return "memberList";
    }

    @RequestMapping(value={"remove","delete"}, method=RequestMethod.GET)
    public String removeMember(@RequestParam("memberName") String memberName) {
        memberService.remove(memberName);
        return "redirect:";
    }

    @RequestMapping("display/{member}")
    public String displayMember(@PathVariable("member") String member, Model model) {
        model.addAttribute("member", memberService.find(member).orElse(null));
        return "member";
    }

    @RequestMapping
    public void memberList() {}

    public void memberLogic(String memberName) {}

}
```

注意到类级别的@RequestMapping 注解使用了一个 URL 通配符：/member/*。这会将/member/ URL 下的所有请求委托给该控制器的处理器方法。

前两个处理器方法使用了@RequestMapping 注解。在向/member/add URL 发出 HTTP GET 请求时会调用 addMember()方法，在向/member/remove 或是/member/delete URL 发出 HTTP GET 请求时则会调用 removeMember()方法。

第 3 个处理器方法使用了一个特殊符号{path_variable}来指定其@RequestMapping 的值。通过这种方式，URL 中的值就会作为输入传递给处理器方法。注意到处理器方法声明了@PathVariable ("user") String user。借助这种方式，如果请求形式为 member/display/jdoe，那么处理器方法就可以通过值 jdoe 访问到成员变量。这种方式可以省去处理器的请求对象的解析过程，同时在设计 RESTful Web Service 时颇为有用。

第 4 个处理器方法也使用了@RequestMapping 注解，不过它并没有指定 URL 值。由于类级别使用了/member/* URL 通配符，因此该处理器方法将会作为一个大容器。任何 URL 请求（如/member/abcdefg 或是/member/randomroute）都会触发该方法。注意到该方法的 void 返回值，它会让处理器方法根据其名字（即 memberList）默认转向一个视图。

最后一个方法 memberLogic 没有使用@RequestMapping 注解，这意味着该方法是类的一个辅助方法，与 Spring MVC 没有任何关系。

根据 HTTP 请求类型映射请求

默认情况下，@RequestMapping 注解会处理所有类型的传入请求。不过，在大多数情况下，你并不希望 GET 请求与 POST 请求都执行同样的方法。为了区分 HTTP 请求，有必要在@RequestMapping 注解中显式指定类型，代码如下所示：

```
@RequestMapping(value= "processUser", method = RequestMethod.POST)
public String submitForm(@ModelAttribute("member") Member member,
                         BindingResult result, Model model) {

}
```

到底在何种程度上指定处理器的 HTTP 类型取决于如何以及什么在与控制器进行交互。对于大多数情况来说，Web 浏览器主要通过 HTTP GET 与 POST 请求执行操作。不过，其他设备或是应用（如 RESTful Web Service）可能需要支持其他的 HTTP 请求类型。总的来说，一共有 9 种不同的 HTTP 请求类型，分别是 HEAD、GET、POST、PUT、DELETE、PATCH、TRACE、OPTIONS 与 CONNECT。不过，处理所有这些请求类型的支持已经超出了 MVC 控制器的范畴，因为 Web 服务器（以及请求方）需要支持这些 HTTP 请求类型。考虑到大多数 HTTP 请求都是 GET 或是 POST 类型，因此很少需要实现对其他 HTTP 请求类型的支持。

对于大多数常用的请求方法来说，Spring MVC 提供了专门的注解，如表 3-1 所示。

表 3-1　　请求方法与注解之间的映射

请求方法	注解
POST	@PostMapping
GET	@GetMapping
DELETE	@DeleteMapping
PUT	@PutMapping

这些便捷的注解都是@RequestMapping 注解的特化，可以使得编写请求处理方法更加紧凑一些。

```
@PostMapping("processUser")
public String submitForm(@ModelAttribute("member") Member member,
                         BindingResult result, Model model) {

}
```

你可能会注意到，在@RequestMapping 注解中指定的所有 URL 中，并没有.html 或是.jsp 等文件扩展名。这是符合 MVC 设计的好做法，不过并未得到广泛使用。

控制器不应该绑定到用于标识某项视图技术的任何类型的扩展上，比如说 HTML 或是 JSP。这正是控制器返回逻辑视图的原因，同时也是匹配的 URL 不应该声明任何扩展的原因。

有时，应用需要以不同格式来呈现相同的内容，比如说 XML、JSON、PDF 或是 XLS（Excel），我们应该让视图解析器检测请求中所提供的扩展，并决定应当使用哪种视图技术。

在这个部分，我们谈及了如何在 MVC 配置类中配置解析器将逻辑视图映射到 JSP 文件上，这里面都不会使用 URL 文件扩展名，如.jsp。

在接下来的正文中，我们将会介绍 Spring MVC 如何通过这种无扩展名的 URL 方式并使用不同的视图技术来提供内容。

3-3 使用处理器拦截器拦截请求

问题提出

Servlet API 所定义的 servlet 过滤器可以在每个 Web 请求被 servlet 处理前后进行预先处理和事后处理。你想要在 Spring 的 Web 应用上下文中配置一些具有与过滤器类似功能的组件以利用容器的特性。

此外，有时你想要对 Spring MVC 处理器所处理的 Web 请求进行预先处理和事后处理，并且操纵这些处理器所返回的模型属性，之后再将其传递给视图。

解决方案

Spring MVC 可以通过处理器拦截器来拦截 Web 请求并对其进行预先处理与事后处理。处理器拦截器是配置在 Spring 的 Web 应用上下文中的，这样它们就可以利用容器特性并引用容器中所声明的 bean 了。处理器拦截器可以针对特定的 URL 映射进行注册，这样它只会拦截映射到某些 URL 上的请求。

每个处理器拦截器都要实现 HandlerInterceptor 接口，它包含了 3 个回调方法供你实现：preHandle()、postHandle()与 afterCompletion()。前两个方法分别是在请求被处理器处理前后调用的。第二个方法还可以访问到所返回的 ModelAndView 对象，这样就可以操纵其中的模型属性了。最后一个方法是在所有请求处理都完毕后调用的（即视图被渲染之后）。

解释说明

假设你想要度量每个请求处理器对每个 Web 请求的处理时间并将这个时间展现给用户。可以为此创建一个自定义的处理器拦截器。

```
package com.apress.springrecipes.court.web;
...
import org.springframework.web.servlet.HandlerInterceptor;
import org.springframework.web.servlet.ModelAndView;

public class MeasurementInterceptor implements HandlerInterceptor {

    public boolean preHandle(HttpServletRequest request,
        HttpServletResponse response, Object handler) throws Exception {
        long startTime = System.currentTimeMillis();
        request.setAttribute("startTime", startTime);
        return true;
    }

    public void postHandle(HttpServletRequest request,
        HttpServletResponse response, Object handler,
        ModelAndView modelAndView) throws Exception {
        long startTime = (Long) request.getAttribute("startTime");
        request.removeAttribute("startTime");

        long endTime = System.currentTimeMillis();
        modelAndView.addObject("handlingTime", endTime - startTime);
    }

    public void afterCompletion(HttpServletRequest request,
        HttpServletResponse response, Object handler, Exception ex)
        throws Exception {
    }
}
```

在拦截器的 preHandle()方法中记录下了开始时间并将其保存到请求属性中。该方法应该返回 true，这样 DispatcherServlet 就可以继续处理请求了。否则，DispatcherServlet 会认为该方法已经处理完了请求，这样 DispatcherServlet 就会直接将响应返回给用户。接下来，在 postHandle()方法中从请求属性中加载了开始时间，并将其与当前时间进行比较。你可以计算出总持续时间，然后将其添加到模型中传递给视图。最后，由于 afterCompletion()方法没什么可做的，将方法体置空即可。

在实现接口时，你需要实现所有方法，即便有些方法没必要实现。更好的方式是继承拦截器适配器类。该类默认实现了所有拦截器方法，只需重写需要的方法即可。

```
package com.apress.springrecipes.court.web;
...
import org.springframework.web.servlet.ModelAndView;
import org.springframework.web.servlet.handler.HandlerInterceptorAdapter;

public class MeasurementInterceptor extends HandlerInterceptorAdapter {
    public boolean preHandle(HttpServletRequest request,
        HttpServletResponse response, Object handler) throws Exception {
        ...
    }

    public void postHandle(HttpServletRequest request,
        HttpServletResponse response, Object handler,
        ModelAndView modelAndView) throws Exception {
        ...
    }
}
```

要想注册拦截器，你需要修改 3-1 节中创建的 CourtConfiguration。你需要让其实现 WebMvcConfigurer

并重写 addInterceptors 方法。借助于该方法可以访问 InterceptorRegistry，可以通过它添加拦截器。修改后的类如下所示：

```
@Configuration
public class InterceptorConfiguration implements WebMvcConfigurer {

    @Override
    public void addInterceptors(InterceptorRegistry registry) {
        registry.addInterceptor(measurementInterceptor());
    }

    @Bean
    public MeasurementInterceptor measurementInterceptor() {
        return new MeasurementInterceptor();
    }

    ...
}
```

现在可以在 welcome.jsp 中展示这个时间以验证拦截器的功能了。由于 WelcomeController 并没有做太多事情，因此很可能看到的处理时间是 0 秒。如果是这种情况，你可以向类中添加一个睡眠语句，这样就可以看到更长的处理时间了。

```
<%@ taglib prefix="fmt" uri="http://java.sun.com/jsp/jstl/fmt" %>

<html>
<head>
<title>Welcome</title>
</head>

<body>
...
<hr />
Handling time : ${handlingTime} ms
</body>
</html>
```

默认情况下，HandlerInterceptors 会应用到所有@Controllers 上；然而，有时你想区分哪些控制器需要使用拦截器。命名空间与基于 Java 的配置都可以让拦截器映射到特定 URL 上。这只是个配置问题。如下代码展示了实现该功能的 Java 配置：

```
package com.apress.springrecipes.court.config;

import com.apress.springrecipes.court.web.ExtensionInterceptor;
import com.apress.springrecipes.court.web.MeasurementInterceptor;
import org.springframework.context.annotation.Bean;
import org.springframework.context.annotation.Configuration;
import org.springframework.web.servlet.config.annotation.InterceptorRegistry;
import org.springframework.web.servlet.config.annotation.WebMvcConfigurer;

@Configuration
public class InterceptorConfiguration implements WebMvcConfigurer {

    @Override
    public void addInterceptors(InterceptorRegistry registry) {

        registry.addInterceptor(measurementInterceptor());
        registry.addInterceptor(summaryReportInterceptor())
                .addPathPatterns("/reservationSummary*");
    }

    @Bean
    public MeasurementInterceptor measurementInterceptor() {
        return new MeasurementInterceptor();
    }
```

```
    @Bean
    public ExtensionInterceptor summaryReportInterceptor() {
        return new ExtensionInterceptor();
    }
}
```

首先，增加了拦截器 bean summaryReportInterceptor。该 bean 对应的类结构与 measurementInterceptor（实现了 HandlerInterceptor 接口）一样。不过，该拦截器所执行的逻辑只会限定到特定的控制器，即映射到 /reservationSummary URI 的控制器。在注册拦截器时，你可以指定映射到哪些 URL；默认情况下使用的是 Ant 风格的表达式。将该模式传递给 addPathPatterns 方法，还有一个 excludePathPatterns 方法可用于排除针对某些 URL 的拦截器。

3-4 解析用户地域

问题提出

要想让 Web 应用支持国际化，你需要识别出每个用户的首选地域，并根据该地域展示内容。

解决方案

在 Spring MVC 应用中，用户的地域是通过一个地域解析器来识别的，它实现了 LocaleResolver 接口。Spring MVC 自带几个 LocaleResolver 实现，可以根据不同标准解析地域信息。此外，可以通过实现该接口创建自定义的地域解析器。

可以通过在 Web 应用上下文中注册一个类型为 LocaleResolver 的 bean 来定义地域解析器。必须将地域解析器的 bean 名字设为 localeResolver，这样 DispatcherServlet 才会自动检测到。注意到每个 DispatcherServlet 只能注册一个地域解析器。

解释说明

我们会介绍 Spring MVC 中不同的 LocaleResolver，以及如何通过拦截器修改用户的地域信息。

根据 HTTP 请求头解析地域信息

Spring 默认的地域解析器是 AcceptHeaderLocaleResolver。它会通过检测 HTTP 请求头的 accept-language 来解析地域信息。这个头信息是用户浏览器根据底层操作系统的地域设置来设定的。注意，该地域解析器无法修改用户的地域信息，因为它不能改变用户操作系统的地域设置。

根据会话属性解析地域信息

另一种解析地域信息的方式是使用 SessionLocaleResolver。它会检测用户会话中一个预定义的属性来解析地域信息。如果该会话属性不存在，那么该地域解析器就会根据 accept-language HTTP 头来确定默认地域信息。

```
@Bean
public LocaleResolver localeResolver () {
    SessionLocaleResolver localeResolver = new SessionLocaleResolver();
    localeResolver.setDefaultLocale(new Locale("en"));
    return localeResolver;
}
```

可以在会话属性不存在时为该解析器设置 defaultLocale 属性。注意，该地域解析器可以改变用户的地域信息，方式是修改用于存储地域信息的会话属性。

根据 cookie 解析地域信息

还可以使用 CookieLocaleResolver 通过检测用户浏览器的 cookie 来解析地域信息。如果 cookie 不存在，那么该地域解析器就会根据 accept-language HTTP 头来确定默认地域信息。

```
@Bean
```

```
public LocaleResolver localeResolver() {
    return new CookieLocaleResolver();
}
```

该地域解析器所用的 cookie 可以通过设置 cookieName 与 cookieMaxAge 属性进行定制。cookieMaxAge 属性表示该 cookie 会持久化多少秒。值-1 表示当浏览器关闭后 cookie 就会失效。

```
@Bean
public LocaleResolver localeResolver() {
    CookieLocaleResolver cookieLocaleResolver = new CookieLocaleResolver();
    cookieLocaleResolver.setCookieName("language");
    cookieLocaleResolver.setCookieMaxAge(3600);
    cookieLocaleResolver.setDefaultLocale(new Locale("en"));
    return cookieLocaleResolver;
}
```

当用户浏览器中不存在 cookie 时，还可以为该解析器设置 defaultLocale 属性。该地域解析器可以改变用户的地域信息，方式是修改存储地域信息的 cookie。

修改用户的地域信息

除了显式调用 LocaleResolver.setLocale()修改用户的地域信息外，还可以将 LocaleChangeInterceptor 应用到处理器映射上来达成所愿。该拦截器会检测当前的 HTTP 请求中是否存在一个特殊的参数。可以通过拦截器的 paramName 属性来自定义参数名。如果该参数存在于当前请求中，那么拦截器就会根据参数值修改用户的地域信息。

```
package com.apress.springrecipes.court.web.config;

import org.springframework.web.servlet.i18n.CookieLocaleResolver;
import org.springframework.web.servlet.i18n.LocaleChangeInterceptor;
import org.springframework.web.servlet.view.InternalResourceViewResolver;

import java.util.Locale;

// Other imports omitted

@Configuration
public class I18NConfiguration implements WebMvcConfigurer {

    @Override
    public void addInterceptors(InterceptorRegistry registry) {
        registry.addInterceptor(measurementInterceptor());
        registry.addInterceptor(localeChangeInterceptor());
        registry.addInterceptor(summaryReportInterceptor())
                .addPathPatterns("/reservationSummary*");
    }

    @Bean
    public LocaleChangeInterceptor localeChangeInterceptor() {
          LocaleChangeInterceptor localeChangeInterceptor = new LocaleChangeInterceptor();
          localeChangeInterceptor.setParamName("language");
          return localeChangeInterceptor;
    }

    @Bean
    public CookieLocaleResolver localeResolver() {
        CookieLocaleResolver cookieLocaleResolver = new CookieLocaleResolver();
        cookieLocaleResolver.setCookieName("language");
        cookieLocaleResolver.setCookieMaxAge(3600);
        cookieLocaleResolver.setDefaultLocale(new Locale("en"));
        return cookieLocaleResolver;
    }
    ...
}
```

现在，用户的地域信息可以通过带有 language 参数的任何 URL 修改。比如说，如下两个 URL 会分别将用户的地域信息修改为美式英语和德语：

```
http://localhost:8080/court/welcome?language=en_US
http://localhost:8080/court/welcome?language=de
```

接下来，可以在 welcome.jsp 中展示 HTTP 响应对象的地域信息，进而验证地域拦截器的配置。

```
<%@ taglib prefix="fmt" uri="http://java.sun.com/jsp/jstl/fmt" %>

<html>
<head>
<title>Welcome</title>
</head>

<body>
...
<br />
Locale : ${pageContext.response.locale}
</body>
</html>
```

3-5　外部化地域相关的文本信息

问题提出

在开发国际化 Web 应用时，你需要根据用户的首选地域信息展示网页，并且不希望为不同地域创建同一个页面的不同版本。

解决方案

为了避免不同地域下为一个页面创建不同版本的情况，你应该将地域相关的文本信息外部化，从而让网页独立于地域信息。Spring 可以通过消息源来解析文本信息，这个消息源要实现 MessageSource 接口。接下来，JSP 文件可以使用定义在 Spring 标签库中的<spring:message>标签来解析给定 code 的消息。

解释说明

可以通过在 Web 应用上下文中注册一个 MessageSource 类型的 bean 来定义消息源。必须要将消息源的 bean 名字设为 messageSource，这样 DispatcherServlet 才能自动检测到。注意，每个 DispatcherServlet 只能注册一个消息源。ResourceBundleMessageSource 实现会针对不同地域的不同资源包来解析信息。比如说，你可以将其注册到 WebConfiguration 中来加载 base 名字为 messages 的资源包。

```
@Bean
public MessageSource messageSource() {
    ResourceBundleMessageSource messageSource = new ResourceBundleMessageSource();
    messageSource.setBasename("messages");
    return messageSource;
}
```

接下来创建 messages.properties 与 messages_de.properties 这两个资源包，分别存储默认地域与德语地域的信息。应该将这些资源包放到类路径的根目录下。

```
welcome.title=Welcome
welcome.message=Welcome to Court Reservation System

welcome.title=Willkommen
welcome.message=Willkommen zum Spielplatz-Reservierungssystem
```

现在，在 welcome.jsp 这样的 JSP 文件中，你可以使用<spring:message>标签解析指定 code 的信息。该标签会根据用户当前的地域自动解析信息。注意，该标签定义在 Spring 的标签库中，因此需要在 JSP 文件顶部声明它。

```
<%@ taglib prefix="spring" uri="http://www.springframework.org/tags" %>

<html>
<head>
```

```
<title><spring:message code="welcome.title" text="Welcome" /></title>
</head>

<body>
<h2><spring:message code="welcome.message"
    text="Welcome to Court Reservation System" /></h2>
...
</body>
</html>
```

在<spring:message>中，当无法解析给定 code 的信息时，你可以指定输出的默认文本。

3-6 根据名字解析视图

问题提出

当处理器处理完请求后，它会返回一个逻辑视图名，DispatcherServlet 会将控制委托给视图模板，这样信息才会渲染出来。你想要为 DispatcherServlet 定义一种策略来根据逻辑名解析视图。

解决方案

在 Spring MVC 应用中，视图是通过声明在 Web 应用上下文中的一个或多个视图解析器 bean 来解析的。这些 bean 需要实现 ViewResolver 接口，这样 DispatcherServlet 才能自动检测到它们。Spring MVC 自带了几个 ViewResolver 实现，用于根据不同策略来解析视图。

解释说明

我们将会介绍不同的视图解析策略，从使用前缀与后缀来生成实际名字的具名模板开始，再到根据名字来进行视图的解析，这些名字可能来自 XML 文件或 ResourceBundle。最后，我们将会介绍如何使用多个 ViewResolver。

根据模板名与位置解析视图

解析视图的基本策略是将它们直接映射到模板的名称和位置。视图解析器 InternalResourceViewResolver 会通过一个 prefix 和 suffix 声明的方式将每个视图名映射到一个应用的目录中。要想注册 InternalResourceViewResolver，你可以在 Web 应用上下文中声明一个该类型的 bean。

```
@Bean
public InternalResourceViewResolver viewResolver() {
    InternalResourceViewResolver viewResolver = new InternalResourceViewResolver();
    viewResolver.setPrefix("/WEB-INF/jsp/");
    viewResolver.setSuffix(".jsp");
    return viewResolver;
}
```

比如说，InternalResourceViewResolver 会按照如下方式解析视图名 welcome 与 reservationQuery：

```
welcome --> /WEB-INF/jsp/welcome.jsp
reservationQuery --> /WEB-INF/jsp/reservationQuery.jsp
```

可以通过 viewClass 属性指定解析的视图类型。默认情况下，如果 JSTL 库（即 jstl.jar）位于类路径下，那么 InternalResourceViewResolver 就会将视图名解析为类型为 JstlView 的视图对象。这样，如果视图是使用了 JSTL 标签的 JSP 模板，那么就可以省略掉 viewClass 属性。

InternalResourceViewResolver 很简单，但它只能解析可由 Servlet API 的 RequestDispatcher 转发的内部资源视图（如内部 JSP 文件或是 servlet）。对于 Spring MVC 支持的其他视图类型来说，你需要使用其他策略来解析。

从 XML 配置文件解析视图

解析视图的另一种策略是将其声明为 Spring bean，并通过其 bean 名字来解析。可以将视图 bean 声明在 Web 应用上下文相同的配置文件中，不过最好将其放到单独的配置文件中。默认情况下，XmlViewResolver 会从/WEB-INF/views.xml 中加载视图 bean，不过可以通过 location 属性改变这个位置。

```
Configuration
public class ViewResolverConfiguration implements WebMvcConfigurer, ResourceLoaderAware {

    private ResourceLoader resourceLoader;

    @Bean
    public ViewResolver viewResolver() {
        XmlViewResolver viewResolver = new XmlViewResolver();
        viewResolver.setLocation(resourceLoader.getResource("/WEB-INF/court-views.nl"));
        return viewResolver;
    }

    @Override
    public void setResourceLoader(ResourceLoader resourceLoader) {
        this.resourceLoader=resourceLoader;
    }
}
```

注意，在 ResourceLoaderAware 接口的实现中，你需要加载资源，因为 location 属性接收一个类型为 Resource 的参数。在 Spring XML 文件中，从 String 到 Resource 的转换已经被处理了；不过，在使用 Java 配置时，你需要做一些额外的工作。在 court-views.xml 配置文件中，可以通过设置类名与属性将每个视图声明为一个普通的 Spring bean。通过这种方式，可以声明任意类型的视图（比如说 RedirectView，甚至是自定义视图类型）。

```
<beans xmlns="http://www.springframework.org/schema/beans"
    xmlns:xsi="http://www.w3.org/2001/XMLSchema-instance"
    xsi:schemaLocation="http://www.springframework.org/schema/beans
        http://www.springframework.org/schema/beans/spring-beans.xsd">

    <bean id="welcome"
        class="org.springframework.web.servlet.view.JstlView">
        <property name="url" value="/WEB-INF/jsp/welcome.jsp" />
    </bean>

    <bean id="reservationQuery"
        class="org.springframework.web.servlet.view.JstlView">
        <property name="url" value="/WEB-INF/jsp/reservationQuery.jsp" />
    </bean>

    <bean id="welcomeRedirect"
        class="org.springframework.web.servlet.view.RedirectView">
        <property name="url" value="welcome" />
    </bean>
</beans>
```

从资源包解析视图

除了 XML 配置文件外，你可以在资源包中声明视图 bean。ResourceBundleViewResolver 会从类路径下的资源包中加载视图 bean。注意，ResourceBundleViewResolver 还可以利用资源包的能力针对不同地域从不同资源包中加载视图 bean。

```
@Bean
public ResourceBundleViewResolver viewResolver() {
    ResourceBundleViewResolver viewResolver = new ResourceBundleViewResolver();
    viewResolver.setBasename("court-views");
    return viewResolver;
}
```

由于将 court-views 作为 ResourceBundleViewResolver 的 base 名字，因此资源包就是 court-views.properties 了。在该资源包中，你可以以属性（properties）格式声明视图 bean。这种类型的声明等价于 XML bean 声明。

```
welcome.(class)=org.springframework.web.servlet.view.JstlView
welcome.url=/WEB-INF/jsp/welcome.jsp
reservationQuery.(class)=org.springframework.web.servlet.view.JstlView
```

```
reservationQuery.url=/WEB-INF/jsp/reservationQuery.jsp
welcomeRedirect.(class)=org.springframework.web.servlet.view.RedirectView
welcomeRedirect.url=welcome
```

使用多个解析器解析视图

如果 Web 应用中有大量视图，那么只使用一种视图解析策略常常是不够的。通常情况下，InternalResourceViewResolver 可以解析大多数内部 JSP 视图，不过通常还会有需要由 ResourceBundleViewResolver 解析的其他类型的视图。在这种情况下，你需要组合使用这两种策略进行视图解析。

```
@Bean
public ResourceBundleViewResolver viewResolver() {
    ResourceBundleViewResolver viewResolver = new ResourceBundleViewResolver();
    viewResolver.setOrder(0);
    viewResolver.setBasename("court-views");
    return viewResolver;
}

@Bean
public InternalResourceViewResolver internalResourceViewResolver() {
    InternalResourceViewResolver viewResolver = new InternalResourceViewResolver();
    viewResolver.setOrder(1);
    viewResolver.setPrefix("/WEB-INF/jsp/");
    viewResolver.setSuffix(".jsp");
    return viewResolver;
}
```

当同时使用多种策略时，需要指定解析优先级。你可以设置视图解析器 bean 的 order 属性来做到这一点。数值小代表优先级高。注意，应该为 InternalResourceViewResolver 设置最低的优先级，这是因为无论视图存在与否，它总是会进行解析。因此，如果其他解析器的优先级更低，那么它们将没有机会去解析视图。现在，资源包 court-views.properties 只应该包含不能被 InternalResourceViewResolver 解析的视图（比如说重定向视图）。

```
welcomeRedirect.(class)=org.springframework.web.servlet.view.RedirectView
welcomeRedirect.url=welcome
```

使用重定向前缀

如果在 Web 应用上下文中配置了 InternalResourceViewResolver，那么可以通过在视图名中使用 redirect: 前缀来解析重定向视图。视图名的其他部分就会作为重定向 URL。比如说，视图名 redirect:welcome 会触发对相对 URL welcome 的重定向。还可以在视图名中指定绝对 URL。

3-7 使用视图与内容协商

问题提出

控制器中使用的是无扩展名的 URL——welcome，而不是 welcome.html 或 welcome.pdf。你想要设计一种策略，使得所有请求都会返回正确的内容与类型。

解决方案

当请求到达 Web 应用时，它会包含一系列属性，可以让处理框架（这里就是 Spring MVC）确定返回给请求方的正确内容与类型。主要的两个属性是请求中所提供的 URL 扩展名和 HTTP Accept 头。比如说，如果请求 URL 是/reservationSummary.xml，那么控制器就可以检测到扩展名并将其委托给代表 XML 视图的逻辑视图。不过，请求 URL 还可能是/reservationSummary 这种形式。该请求应该委托给 XML 视图还是 HTML 视图呢？又或者是其他类型的视图呢？这一点无法通过 URL 判断出来。不过，相较于使用默认视图来处理这种请求，检测请求的 HTTP Accept 头来确定视图类型是更为妥当的做法。

在控制器中检测 HTTP Accept 头是比较麻烦的事情。因此，Spring MVC 通过 ContentNegotiatingViewResolver 来支持头的检测，可以根据 URL 文件扩展名或是 HTTP Accept 头的值来进行视图的委托。

解释说明

就 Spring MVC 内容协商来说，首先要知晓的是它被配置为一个解析器，就像 3-6 节所介绍的那些一样。Spring MVC 内容协商解析器基于 ContentNegotiatingViewResolver 类。不过，在介绍其原理之前，我们需要知道如何配置它并将其与其他解析器进行集成。

```
@Configuration
Public class ViewResolverConfiguration implements WebMvcConfigurer {

    @Autowired
    private ContentNegotiationManager contentNegotiationManager;

    @Override
    public void configureContentNegotiation(ContentNegotiationConfigurer configurer) {
        Map<String, MediaType> mediatypes = new HashMap<>();
        mediatypes.put("html", MediaType.TEXT_HTML);
        mediatypes.put("pdf", MediaType.valueOf("application/pdf"));
        mediatypes.put("xls", MediaType.valueOf("application/vnd.ms-excel"));
        mediatypes.put("xml", MediaType.APPLICATION_XML);
        mediatypes.put("json", MediaType.APPLICATION_JSON);
        configurer.mediaTypes(mediatypes);
    }

    @Bean
    public ContentNegotiatingViewResolver contentNegotiatingViewResolver() {
        ContentNegotiatingViewResolver viewResolver = new ContentNegotiatingViewResolver();
        viewResolver.setContentNegotiationManager(contentNegotiationManager);
        return viewResolver;
    }
}
```

首先需要配置内容协商。默认配置添加了一个 ContentNegotiationManager，可通过实现 configureContentNegotiation 方法对其进行配置。要想访问配置后的 ContentNegotiationManager，只需在配置类中对其进行自动装配即可。

将注意力转回到 ContentNegotiatingViewResolver 解析器上。该配置会将解析器设为最高优先级，这对于内容协商解析器来说是必要的。之所以要让该解析器拥有最高优先级，原因在于它并不会解析视图本身，而是将其委托给其他视图解析器（会自动检测）。考虑到“解析器并不会解析视图”这件事儿可能会令你感到困惑，我们用一个示例说明一下。

假设控制器接收到一个发往/reservationSummary.xml 的请求。当处理器方法处理完毕后，它会将控制转向一个名为 reservation 的逻辑视图。这时，Spring MVC 解析器就会粉墨登场，其中第一个就是 ContentNegotiatingViewResolver 解析器，因为它拥有最高优先级。

ContentNegotiatingViewResolver 解析器首先会根据如下标准确定请求的媒体类型：它会针对 ContentNegotiatingManager bean 配置中 mediaTypes map 所指定的默认媒体类型（如 text/html）检查请求的路径扩展名（如.html 或是.pdf）。如果请求路径有扩展名，但在默认的 mediaTypes 中却找不到匹配项，那就会使用 Java Activation Framework 中的 FileTypeMap 来确定扩展的媒体类型。如果请求路径中没有扩展名，那就会使用请求的 HTTP Accept 头。如果请求路径是/reservationSummary.xml，那么媒体类型就会由第一步确定为 application/xml。不过，如果请求路径为 reservationSummary，那么直到第三步才会确定好媒体类型。

HTTP Accept 头包含了诸如 Accept: text/html 或是 Accept: application/pdf 这样的值。这些值会帮助解析器确定请求所期望的媒体类型，即便请求 URL 中没有扩展名也没关系。

这个时候，ContentNegotiatingViewResolver 解析器拥有一个媒体类型和名为 reservation 的逻辑视图。根据该信息，会对其余解析器进行遍历（根据其顺序），从而根据所检测的媒体类型来确定哪个视图与逻辑视图名是最匹配的。

这个过程可以支持拥有相同名字的多个逻辑视图，每个都支持不同的媒体类型（如 HTML、PDF 或是 XLS），由 ContentNegotiatingViewResolver 来解析哪一个是最匹配的。在这种情况下，控制器的设计就得到了进一步简化，因为它无须再硬编码逻辑视图来创建某些媒体类型了（如 pdfReservation、xlsReservation 或是 htmlReservation），而是通过单个视图（如 reservation）就可以让 ContentNegotiatingViewResolver 解析器确定最佳的匹配。

这一过程的一系列结果如下所示。

- 媒体类型确定为 application/pdf。如果拥有最高优先级（更低的 order）的解析器包含了对名为 reservation 的逻辑视图的映射，但视图不支持 application/pdf，这就导致没有匹配的结果——查询过程会继续进入到剩余的解析器中。
- 媒体类型确定为 application/pdf。拥有最高优先级（更低的 order）的解析器包含了对名为 reservation 的逻辑视图的映射，并且支持 application/pdf，结果就是匹配。
- 媒体类型确定为 text/html。有 4 个解析器以及一个名为 reservation 的逻辑视图，不过与拥有最高优先级的两个解析器匹配的视图不支持 text/html。剩下的解析器中包含了一个对名为 reservation 的视图的映射，同时视图支持 text/html，那就会匹配该解析器。

这个视图搜索过程会自动发生在应用中所配置的所有解析器中。如果不想回退到 ContentNegotiatingViewResolver 解析器外的配置，那还可以配置（在 ContentNegotiatingViewResolver bean 中）默认视图与解析器。

3-11 节将要介绍的控制器就依赖于 ContentNegotiatingViewResolver 解析器来确定应用的视图。

3-8 将异常映射到视图

问题提出

当未知异常发生时，应用服务器通常会向用户展示出万恶的异常堆栈。用户完全不了解这个堆栈，而且还会抱怨应用对用户不友好。此外，这可能还是个潜在的安全风险，因为可能会向用户暴露出内部的方法调用层次。不过，可以配置 Web 应用的 web.xml，使得在 HTTP 错误或是类异常发生时展示友好的 JSP 页面。Spring MVC 支持一种更为健壮的方式来管理类异常时的视图。

解决方案

在 Spring MVC 应用中，可以在 Web 应用上下文中注册一个或多个异常解析器 bean 来解析未捕获的异常。这些 bean 需要实现 HandlerExceptionResolver 接口，这样 DispatcherServlet 才能自动检测到它们。Spring MVC 自带了一个简单的异常解析器将每一类异常映射到视图上。

解释说明

假设预订服务会因无法预订而抛出如下异常：

```java
package com.apress.springrecipes.court.service;
...
public class ReservationNotAvailableException extends RuntimeException {

    private String courtName;
    private Date date;
    private int hour;

    // Constructors and Getters
    ...
}
```

为了解析未捕获异常，可以通过实现 HandlerExceptionResolver 接口来编写自定义的异常解析器。通常情况下，需要将不同类别的异常映射到不同的错误页面。Spring MVC 自带了异常解析器 SimpleMappingExceptionResolver 供开发者在 Web 应用上下文中配置异常映射。比如说，可以在配置中注册如下异常解析器：

```java
@Override
public void configureHandlerExceptionResolvers(List<HandlerExceptionResolver>
    exceptionResolvers) {
    exceptionResolvers.add(handlerExceptionResolver());
```

```
}

@Bean
public HandlerExceptionResolver handlerExceptionResolver() {
    Properties exceptionMapping = new Properties();
    exceptionMapping.setProperty(
    ReservationNotAvailableException.class.getName(), "reservationNotAvailable");

    SimpleMappingExceptionResolver exceptionResolver = new SimpleMappingExceptionResolver();
    exceptionResolver.setExceptionMappings(exceptionMapping);
    exceptionResolver.setDefaultErrorView("error");
    return exceptionResolver;
}
```

在该异常解析器中，为 ReservationNotAvailableException 定义了逻辑视图名 reservationNotAvailable。可以通过 exceptionMappings 属性添加任意数量的异常类，以及更为通用的异常类 java.lang.Exception。在这种方式下，根据异常的类型，用户会看到与异常一致的视图。

属性 defaultErrorView 用于定义名为 error 的默认视图，当出现 exceptionMapping 元素中没有映射到的异常时就会使用该视图。

就对应的视图来说，如果 Web 应用上下文中配置了 InternalResourceViewResolver，那么当 reservation 不存在时就会展示如下 reservationNotAvailable.jsp 页面：

```
<%@ taglib prefix="fmt" uri="http://java.sun.com/jsp/jstl/fmt" %>
<html>
<head>
<title>Reservation Not Available</title>
</head>
<body>
Your reservation for ${exception.courtName} is not available on <fmt:formatDate
value="${exception.date}" pattern="yyyy-MM-dd" /> at ${exception.hour}:00.
</body>
```

在错误页面中，可以通过变量${exception}访问异常实例，这样就可以向用户展示关于该异常的详细信息了。

为任何未知异常定义默认的错误页面是一种最佳实践。可以通过属性 defaultErrorView 定义默认视图，或是将页面映射为 java.lang.Exception 以作为映射的最后一项，这样如果前面找不到其他匹配项就会展示这个页面。接下来可以创建视图的 JSP（error.jsp），如下所示：

```
<html>
<head>
<title>Error</title>
</head>
<body>
An error has occurred. Please contact our administrator for details.
</body>
</html>
```

使用@ExceptionHandler 映射异常

相较于配置 HandlerExceptionResolver，你可以对方法应用@ExceptionHandler 注解。它的工作方式类似于@RequestMapping 注解。

```
@Controller
@RequestMapping("/reservationForm")
@SessionAttributes("reservation")
public class ReservationFormController {

    @ExceptionHandler(ReservationNotAvailableException.class)
    public String handle(ReservationNotAvailableException ex) {
        return "reservationNotAvailable";
    }
```

```
    @ExceptionHandler
    public String handleDefault(Exception e) {
        return "error";
    }
    ...
}
```

有两个方法使用了@ExceptionHandler注解。第一个用于处理具体的异常ReservationNotAvailableException；第二个则是个通用（捕获所有）的异常处理方法。此外，无须在WebConfiguration中指定HandlerExceptionResolver。

使用了@ExceptionHandler注解的方法可拥有各种返回类型（就像@RequestMapping方法一样）；这里只是返回了需要渲染的视图名，不过还可以返回一个ModelAndView或是View等。

虽然使用了@ExceptionHandler注解的方法非常强大且灵活，不过在将其放到控制器中时却有一个缺憾。这些方法只适用于它们所在的控制器，如果另一个控制器出现了异常（比如说WelcomeController），那么这些方法将不会被调用。通用的异常处理方法需要放在单独的类中，这个类需要使用@ControllerAdvice注解。

```
@ControllerAdvice
public class ExceptionHandlingAdvice {

    @ExceptionHandler(ReservationNotAvailableException.class)
    public String handle(ReservationNotAvailableException ex) {
        return "reservationNotAvailable";
    }

    @ExceptionHandler
    public String handleDefault(Exception e) {
        return "error";
    }
}
```

这个类会应用到应用上下文中所有的控制器中，这也是它叫做@ControllerAdvice的原因所在。

3-9　使用控制器处理表单

问题提出

在Web应用中常常要处理表单。表单控制器会向用户展示表单，还需要处理表单提交。表单提交是一个复杂多变的任务。

解决方案

当用户与表单交互时，它需要控制器支持两种操作。首先，在第一次请求表单时，它会让控制器根据HTTP GET请求来展示表单，这会将表单视图渲染给用户。接下来，在提交表单时会发出一个HTTP POST请求来处理诸如表单中的数据验证与业务处理等事情。如果表单处理成功，那么它会向用户渲染成功视图。否则，它会使用错误信息再次渲染表单视图。

解释说明

假设用户可以通过填写表单来预订场地。为了能够更好地理解控制器所处理的数据，我们先来介绍一下控制器的视图（即表单）。

创建表单视图

我们来创建表单视图reservationForm.jsp。表单使用了Spring的表单标签库，因为它简化了表单的数据绑定，可以展示出错误消息，还能在错误发生时重新展示用户之前所输入的值。

```
<%@ taglib prefix="form" uri="http://www.springframework.org/tags/form"%>

<html>
<head>
<title>Reservation Form</title>
<style>
```

```
.error {
    color: #ff0000;
    font-weight: bold;
}
</style>
</head>

<body>
<form:form method="post" modelAttribute="reservation">
<form:errors path="*" cssClass="error" />
<table>
    <tr>
        <td>Court Name</td>
        <td><form:input path="courtName" /></td>
        <td><form:errors path="courtName" cssClass="error" /></td>
    </tr>
    <tr>
        <td>Date</td>
        <td><form:input path="date" /></td>
        <td><form:errors path="date" cssClass="error" /></td>
    </tr>
    <tr>
        <td>Hour</td>
        <td><form:input path="hour" /></td>
        <td><form:errors path="hour" cssClass="error" /></td>
    </tr>
    <tr>
        <td colspan="3"><input type="submit" /></td>
    </tr>
</table>
</form:form>
</body>
</html>
```

Spring <form:form>声明了两个属性。method="post"属性表示表单会执行一个 HTTP POST 提交请求。modelAttribute="reservation"属性表示表单数据会绑定到名为 reservation 的模型上。大家应该对第一个属性很熟悉，因为它会用在大多数 HTML 表单上。第二个属性在我们介绍处理表单的控制器时也会熟悉起来。

请记住，<form:form>在发送给用户前会被渲染为标准的 HTML，因此 modelAttribute ="reservation"并不是给浏览器使用的；该属性用于生成实际的 HTML 表单。

接下来，你会看到<form:errors>标签，它用于定义表单没有满足控制器所设置的规则时错误的显式位置。属性 path="*"表示会显示所有错误（这里用到了通配符），而属性 cssClass="error"则表示显示错误时所用的 CSS 格式化类。

再往后，你会看到表单的各种<form:input>标签，后面还会跟着另一组对应的<form:errors>标签。这些标签通过 path 属性来表示表单的字段，该示例中分别是 courtName、date 与 hour。

<form:input>标签通过 path 属性来绑定到对应 modelAttribute 的属性。它们会向用户显示字段的初始值——要么是绑定的属性值，要么是由于绑定错误而生成的拒绝值。它们需要用在<form:form>标签内，该标签所定义的表单会根据名字来绑定到 modelAttribute 上。

最后，你会看到标准的 HTML 标签<input type="submit" />，它会生成一个提交按钮并触发数据发向服务端，其后是</form:form>标签，它用于关闭表单。如果表单及其数据处理正确，那么需要创建一个成功视图来通知用户预订成功。接下来要介绍的 reservationSuccess.jsp 就是为了这个目的：

```
<html>
<head>
<title>Reservation Success</title>
</head>

<body>
```

```
Your reservation has been made successfully.
</body>
</html>
```

表单中提交了不合法的值也会导致错误的发生。比如说，如果日期格式不合法或是 hour 字段出现了字母，控制器就需要拒绝这种字段值。接下来，控制器会为每个错误生成一个选择好的错误码列表并返回给表单视图，其值会放在<form:errors>标签中。

比如说，如果 date 字段的输入值不合法，那么控制器就会生成如下错误码：

```
typeMismatch.command.date
typeMismatch.date
typeMismatch.java.time.LocalDate
typeMismatch
```

如果定义了 ResourceBundleMessageSource，那就可以针对恰当的地域信息在资源包中提供如下错误消息（比如说，messages.properties 是默认地域）：

```
typeMismatch.date=Invalid date format
typeMismatch.hour=Invalid hour format
```

在处理表单数据时，如果失败了，那么与之对应的错误码和值就会返回给用户。

在知道了表单视图的结构及其所处理的数据后，我们就来看一看处理表单中所提交数据（即 reservation）的逻辑。

创建表单的服务处理

并非是控制器来处理表单的数据 reservation，而是由控制器所使用的服务来处理的。首先在 ReservationService 接口中定义一个 make()方法。

```
package com.apress.springrecipes.court.service;
...
public interface ReservationService {
    ...
    void make(Reservation reservation)
        throws ReservationNotAvailableException;
}
```

接下来实现这个 make()方法，向存储预订信息的列表中添加一个 Reservation 条目。如果有重复的预订则抛出 ReservationNotAvailableException 异常。

```
package com.apress.springrecipes.court.service;
...
public class ReservationServiceImpl implements ReservationService {
    ...
    @Override
    public void make(Reservation reservation) throws ReservationNotAvailableException {
        long cnt = reservations.stream()
            .filter(made -> Objects.equals(made.getCourtName(), reservation.getCourtName()))
            .filter(made -> Objects.equals(made.getDate(), reservation.getDate()))
            .filter(made -> made.getHour() == reservation.getHour())
            .count();

        if (cnt > 0) {
            throw new ReservationNotAvailableException(reservation
                .getCourtName(), reservation.getDate(), reservation
                .getHour());
        } else {
            reservations.add(reservation);
        }
    }
}
```

在对与控制器交互的两个元素（表单的视图与预订服务类）有了进一步的理解后，我们创建一个控制器来处理场地预订表单。

创建表单控制器

用于处理表单的控制器使用了与前文所用的相同注解。我们直接看代码：

```
package com.apress.springrecipes.court.web;
...

@Controller
@RequestMapping("/reservationForm")

@SessionAttributes("reservation")
public class ReservationFormController {

    private final ReservationService reservationService;

    @Autowired
    public ReservationFormController(ReservationService reservationService) {
        this.reservationService = reservationService;
    }

    @RequestMapping(method = RequestMethod.GET)
    public String setupForm(Model model) {
        Reservation reservation = new Reservation();
        model.addAttribute("reservation", reservation);
        return "reservationForm";
    }

    @RequestMapping(method = RequestMethod.POST)
    public String submitForm(
        @ModelAttribute("reservation") Reservation reservation,
        BindingResult result, SessionStatus status) {
            reservationService.make(reservation);
            return "redirect:reservationSuccess";
    }
}
```

首先，控制器使用了标准的@Controller 注解和@RequestMapping 注解，可以通过如下 URL 访问控制器：

```
http://localhost:8080/court/reservationForm
```

当在浏览器输入这个 URL 时，它会向 Web 应用发出一个 HTTP GET 请求。这又会触发 setupForm 方法的执行，根据@RequestMapping 注解，该方法会处理该类型的请求。

setupForm 方法定义了一个 Model 对象作为输入参数，它用于将模型数据发送给视图（即表单）。在处理器方法中创建了一个空的 Reservation 对象并作为属性添加到了控制器的 Model 对象中。接下来，控制器会将执行流返回给 reservationForm 视图，在该示例中会被解析为 reservationForm.jsp（即表单）。

上面这个方法中最重要的部分是添加了一个空的 Reservation 对象。如果分析一下表单 reservationForm.jsp，会注意到<form:form>标签声明了属性 modelAttribute="reservation"。这意味着在渲染视图时，表单期望有一个名为 reservation 的对象，这是通过将其放在处理器方法的 Model 中实现的。事实上，我们可以进一步发现每个<form:input>标签的 path 值都对应于 Reservation 对象的一个字段名。由于表单是首次加载的，显然需要一个空的 Reservation 对象。

在分析其他控制器处理器方法之前还需要介绍一下@SessionAttributes("reservation")注解，它声明在控制器类的顶部。由于表单可能会包含错误，因此在随后的每次提交中，要是丢失了用户已经提供的有效数据就会变得极其不便。为了解决这一问题，@SessionAttributes 注解用于将一个 reservation 字段保存到用户的会话中，这样后面对 reservation 字段的引用实际上都是相同的，无论表单被提交了两次还是多次。这也是整个控制器中只创建一个 Reservation 对象并赋给 reservation 字段的原因所在。当创建好了空的 Reservation 对象后（在 HTTP GET 处理器方法中），所有的动作都会基于相同的对象，因为它被赋给了用户的会话。

现在将注意力转向到首次提交表单上。在填写完表单字段后，提交表单会触发一个 HTTP POST 请求，它又会调用 submitForm 方法，因为与该方法的@RequestMapping 值匹配。声明在 submitForm 方法中的输入字段有 3 个。@ModelAttribute("reservation") Reservation reservation 用于引用 reservation 对象。BindingResult 对象包含了用户新提交的数据。SessionStatus 对象可以将处理标记为“已完成”，这样 Reservation 对象就会从 HttpSession 中移除掉。

这时，处理器方法还没有进行验证或是执行对用户会话的访问，这正是 BindingResult 与 SessionStatus 对象的目的所在，接下来将会介绍它们。

处理器方法所执行的唯一操作就是 reservationService.make(reservation);。该操作会使用当前的 reservation 对象的状态调用预订服务。通常，在对其执行这类操作前需要先验证控制器对象。最后，注意到处理器方法返回了一个名为 redirect:reservationSuccess 的视图。在该示例中，视图的实际名字是 reservationSuccess，它会被解析为之前所创建的 reservationSuccess.jsp 页面。

视图名中的 redirect:前缀用于避免表单重复提交问题。

当在表单成功视图中刷新页面时，方才所提交的表单会再次提交。为了避免这个问题，可以使用 post/redirect/get 设计模式，推荐的做法是当成功进行表单处理后重定向到另外一个 URL 而不是直接返回 HTML 页面。这正是在视图名前加上 redirect:前缀的原因所在。

初始化 Model 属性对象并使用值来预先填充表单

表单的设计目的在于让用户进行场地预订。不过，如果分析一下 Reservation 领域类，就会注意到表单还是需要再增加两个字段才能创建出一个完整的 reservation 对象。其中一个是 player 字段，它对应于 Player 对象。根据 Player 类的定义，Player 对象有一个 name 字段和一个 phone 字段。

那么，是否可以将 player 字段纳入到表单视图和控制器中呢？我们先来分析一下表单视图。

```
<html>
<head>
<title>Reservation Form</title>
</head>
<body>
<form method="post" modelAttribute="reservation">
<table>
    ...
    <tr>
        <td>Player Name</td>
        <td><form:input path="player.name" /></td>
        <td><form:errors path="player.name" cssClass="error" /></td>
    </tr>
    <tr>
        <td>Player Phone</td>
        <td><form:input path="player.phone" /></td>
        <td><form:errors path="player.phone" cssClass="error" /></td>
    </tr>
    <tr>
        <td colspan="3"><input type="submit" /></td>
    </tr>
</table>
</form>
</body>
</html>
```

使用一种直接的方式，再添加两个<form:input>标签表示 Player 对象的字段。虽然这些表单声明很简单，不过还是需要修改控制器。回忆一下，当使用了<form:input>标签后，视图期望能够访问到控制器所传递的与<form:input>标签的 path 值匹配的模型对象。

虽然控制器的 HTTP GET 处理器方法向上面这个视图返回了一个名为 reservation 的空 reservation 对象，但 player 属性是 null，因此在渲染表单时会导致异常。为了解决这个问题，需要初始化一个空的 Player 对象并将其赋给返回给视图的 Reservation 对象。

```
@RequestMapping(method = RequestMethod.GET)
public String setupForm(
@RequestParam(required = false, value = "username") String username, Model model) {
    Reservation reservation = new Reservation();
    reservation.setPlayer(new Player(username, null));
    model.addAttribute("reservation", reservation);
    return "reservationForm";
}
```

在该示例中，当创建了空的 Reservation 对象后，setPlayer 方法用于将一个空的 Player 对象赋给它。此外，Person 对象的创建依赖于 username 值。这个值是从@RequestParam 输入值中获取到的，它也会被添加到处理器方法中。这样，Player 对象的创建就可以使用作为请求参数传递进来的具体的 username 值，所以 username 表单字段就会使用该值预先填充进去。

比如说，如果按照如下方式请求表单：

```
http://localhost:8080/court/reservationForm?username=Roger
```

处理器方法会抽取出 username 参数来创建 Player 对象，然后使用 Roger 值来预先填充表单的 username 字段。值得注意的是，username 参数的@RequestParam 注解使用了属性 required=false，这样即便该请求参数不存在，表单请求也会得到处理。

向表单提供引用数据

当表单控制器渲染表单视图时，它可能会向表单提供一些引用数据（比如说在 HTML 下拉列表中展示的条目）。假设在预订场地时可以让用户选择运动类型，这是 Reservation 类中不存在的字段。

```
<html>
<head>
<title>Reservation Form</title>
</head>
<body>
<form method="post" modelAttribute="reservation">
<table>
    ...
    <tr>
        <td>Sport Type</td>
        <td><form:select path="sportType" items="${sportTypes}"
            itemValue="id" itemLabel="name" /></td>
        <td><form:errors path="sportType" cssClass="error" /></td>
    </tr>
    <tr>
        <td colspan="3"><input type="submit" /></td>
    /<tr>
</table>
</form>
</body>
</html>
```

<form:select>标签可以根据控制器传递给视图的值生成一个下拉列表。这样，表单会将 sportType 字段表示为一组 HTML <select>元素而不是之前的<input>元素（这需要用户自己输入文本值）。

接下来，我们看看控制器如何将 sportType 字段指定为模型属性；这个过程与之前的字段有些许不同。

首先，在 ReservationService 接口中定义 getAllSportTypes()方法来获取所有可用的运动类型。

```
package com.apress.springrecipes.court.service;
...
public interface ReservationService {
    ...
    public List<SportType> getAllSportTypes();
}
```

接下来，实现该方法，返回一个硬编码的列表。

```
package com.apress.springrecipes.court.service;
...
public class ReservationServiceImpl implements ReservationService {
```

```
    ...
    public static final SportType TENNIS = new SportType(1, "Tennis");
    public static final SportType SOCCER = new SportType(2, "Soccer");

    public List<SportType> getAllSportTypes() {
        return Arrays.asList(TENNIS, SOCCER);
    }
}
```

在有了硬编码的返回 SportType 对象列表的实现之后，我们来看看控制器如何将该列表关联到返回的表单视图上。

```
package com.apress.springrecipes.court.service;
.....
    @ModelAttribute("sportTypes")
    public List<SportType> populateSportTypes() {
    return reservationService.getAllSportTypes();
}

@RequestMapping(method = RequestMethod.GET)
public String setupForm(
@RequestParam(required = false, value = "username") String username, Model model) {
    Reservation reservation = new Reservation();
    reservation.setPlayer(new Player(username, null));
    model.addAttribute("reservation", reservation);
    return "reservationForm";
}
```

注意，负责向表单视图返回空 Reservation 对象的 setupForm 处理器方法没有发生变化。

新添加的以及负责将 SportType 列表作为模型属性传递给表单视图的是被注解@ModelAttribute("sportTypes")修饰的方法。@ModelAttribute 注解用于定义全局模型属性，可用在处理器方法所返回的任何视图中。同样地，处理器方法声明了一个 Model 对象作为输入参数，并为其赋予了可在返回的视图中访问的属性。

由于被@ModelAttribute("sportTypes")注解所修饰的方法的返回类型是 List<SportType>，并且它会调用 reservationService.getAllSportTypes()，因此硬编码的 TENNIS 与 SOCCER SportType 对象就会被赋给名为 sportTypes 的模型属性。这个模型属性会用在表单视图中，用于装配下拉列表（即<form:select>标签）。

绑定自定义类型的属性

当表单提交时，控制器会将表单字段值绑定到模型对象的同名属性上，在该示例中就是 Reservation 对象。不过，对于自定义类型的属性来说，控制器就无法对其进行转换了，除非指定了相应的属性编辑器。

比如说，运动类型选择字段只会提交所选的运动类型 ID，因为这是 HTML <select>字段的操作方式。因此，需要使用属性编辑器将这个 ID 转换为 SportType 对象。首先，需要通过 ReservationService 中的 getSportType() 方法根据 ID 来获取一个 SportType 对象。

```
package com.apress.springrecipes.court.service;
...
public interface ReservationService {
    ...
    public SportType getSportType(int sportTypeId);
}
```

出于测试的目的，可以使用一个 switch/case 语句来实现该方法。

```
package com.apress.springrecipes.court.service;
...
public class ReservationServiceImpl implements ReservationService {
    ...
    public SportType getSportType(int sportTypeId) {
        switch (sportTypeId) {
        case 1:
            return TENNIS;
```

```
            case 2:
                return SOCCER;
            default:
                return null;
            }
        }
    }
```

接下来，创建 SportTypeConverter 类将运动类型 ID 转换为 SportType 对象。该转换器需要通过 ReservationService 来执行查询。

```
package com.apress.springrecipes.court.domain;

import com.apress.springrecipes.court.service.ReservationService;
import org.springframework.core.convert.converter.Converter;

public class SportTypeConverter implements Converter<String, SportType> {

    private ReservationService reservationService;

    public SportTypeConverter(ReservationService reservationService) {
        this.reservationService = reservationService;
    }

    @Override
    public SportType convert(String source) {
        int sportTypeId = Integer.parseInt(source);
        SportType sportType = reservationService.getSportType(sportTypeId);
        return sportType;
    }
}
```

在通过 SportTypeConverter 类将表单属性绑定到 SportType 这样的自定义类上之后，现在需要将其关联到控制器上。为了实现这个目的，可以使用 WebMvcConfigurer 的 addFormatters 方法。

通过在配置类中重写该方法，自定义类型就可以关联到控制器上。这包括 SportTypeConverter 和 Date 等其他自定义类型。虽然之前并未提及 date 字段，不过它与运动类型选择字段面临着相同的问题。用户会将日期字段作为文本值。要想让控制器将这些文本值赋给 Reservation 对象的 date 字段，需要将 date 字段关联到 Date 对象上。考虑到 Date 类是 Java 语言的一部分，因此没必要像 SportTypeConverter 那样创建一个特殊的类。出于这个目的，Spring 框架已经自带了一个自定义类。

由于需要将 SportTypeConverter 类与 Date 类绑定到底层的控制器上，如下代码展示了对于配置类的修改过程：

```
package com.apress.springrecipes.court.web.config;
...
import com.apress.springrecipes.court.domain.SportTypeConverter;
import com.apress.springrecipes.court.service.ReservationService;
import org.springframework.beans.factory.annotation.Autowired;
import org.springframework.format.FormatterRegistry;
import org.springframework.format.datetime.DateFormatter;

@Configuration
@EnableWebMvc
@ComponentScan("com.apress.springrecipes.court.web")
public class WebConfiguration implements WebMvcConfigurer {

    @Autowired
    private ReservationService reservationService;

    @Override
    public void addFormatters(FormatterRegistry registry) {
        registry.addConverter(new SportTypeConverter(reservationService));
    }
}
```

上述类中的唯一字段对应于 reservationService，它用于访问应用的 ReservationService bean。注意到这里使用了@Autowired 注解开启了 bean 的自动装配。接下来，你会看到使用 addFormatters 方法绑定了 Date 与 SportTypeConverter 类，然后通过两个调用注册转换器与格式化器。这些方法属于 FormatterRegistry 对象，该对象作为输入参数传递给了 addFormatters 方法。

第一个调用将 Date 类绑定到了 DateFormatter 类。DateFormatter 类由 Spring 框架提供，它提供了解析和打印 Date 对象的功能。

第二个调用注册了 SportTypeConverter 类。由于已经创建了 SportTypeConverter 类，你应该知道其唯一的输入参数是 ReservationService bean。通过这种方式，每个被注解的控制器（比如说使用了@Controller 注解的类）都可以在其处理器方法中访问相同的自定义转换器与格式化器。

验证表单数据

在提交表单时，标准的做法是在提交成功前验证用户所提供的数据。Spring MVC 支持使用实现了 Validator 接口的验证器对象进行验证。可以编写如下验证来检查必填的表单字段是否都填写了，以及预订时间在假期与工作日是否合法：

```java
package com.apress.springrecipes.court.domain;

import org.springframework.stereotype.Component;
import org.springframework.validation.Errors;
import org.springframework.validation.ValidationUtils;
import org.springframework.validation.Validator;

import java.time.DayOfWeek;
import java.time.LocalDate;

@Component
public class ReservationValidator implements Validator {

    public boolean supports(Class<?> clazz) {
        return Reservation.class.isAssignableFrom(clazz);
    }

    public void validate(Object target, Errors errors) {
        ValidationUtils.rejectIfEmptyOrWhitespace(errors, "courtName",
            "required.courtName", "Court name is required.");
        ValidationUtils.rejectIfEmpty(errors, "date",
            "required.date", "Date is required.");
        ValidationUtils.rejectIfEmpty(errors, "hour",
            "required.hour", "Hour is required.");
        ValidationUtils.rejectIfEmptyOrWhitespace(errors, "player.name",
            "required.playerName", "Player name is required.");
        ValidationUtils.rejectIfEmpty(errors, "sportType",
            "required.sportType", "Sport type is required.");

        Reservation reservation = (Reservation) target;
        LocalDate date = reservation.getDate();
        int hour = reservation.getHour();
        if (date != null) {
            if (date.getDayOfWeek() == DayOfWeek.SUNDAY) {
                if (hour < 8 || hour > 22) {
                    errors.reject("invalid.holidayHour", "Invalid holiday hour.");
                }
            } else {
                if (hour < 9 || hour > 21) {
                    errors.reject("invalid.weekdayHour", "Invalid weekday hour.");
                }
            }
        }
```

```
    }
}
```

在该验证器中，使用了来自于 ValidationUtils 类的 rejectIfEmptyOrWhitespace()与 rejectIfEmpty()方法验证表单字段。如果表单字段为空，那么这些方法会创建一个字段 error 并将其绑定到字段上。这些方法的第二个参数是属性名，第三个与第四个字段分别是错误码和默认的错误消息。

还可以检查预订时间在假期与工作日是否合法。如果不合法，那就应该使用 reject()方法创建一个对象 error 并绑定到 reservation 对象而不是字段上。

由于验证类使用了@Component 注解，因此 Spring 会尝试根据类名将类实例化为 bean，即 reservationValidator。

由于验证器会在验证过程中创建错误，因此应该为错误码定义消息以显示给用户。如果定义了 ResourceBundleMessageSource，就可以针对恰当的地域信息在资源包中定义如下错误消息（比如说，针对默认地域的 messages.properties）：

```
required.courtName=Court name is required
required.date=Date is required
required.hour=Hour is required
required.playerName=Player name is required
required.sportType=Sport type is required
invalid.holidayHour=Invalid holiday hour
invalid.weekdayHour=Invalid weekday hour
```

为了应用这个验证器，需要对控制器进行如下修改：

```
package com.apress.springrecipes.court.service;
.....
    private ReservationService reservationService;
    private ReservationValidator reservationValidator;
public ReservationFormController(ReservationService reservationService,
    ReservationValidator reservationValidator) {
    this.reservationService = reservationService;
    this.reservationValidator = reservationValidator;
}

@RequestMapping(method = RequestMethod.POST)
public String submitForm(
    @ModelAttribute("reservation") @Validated Reservation reservation,
    BindingResult result, SessionStatus status) {
    if (result.hasErrors()) {
        return "reservationForm";
    } else {
        reservationService.make(reservation);
        return "redirect:reservationSuccess";
    }
}

@InitBinder
public void initBinder(WebDataBinder binder) {
    binder.setValidator(reservationValidator);
}
```

首先向控制器添加了 ReservationValidator 字段，它可以让控制器访问验证器 bean 的实例。

接下来修改了 HTTP POST 处理器方法，当用户提交表单时会调用它。@ModelAttribute 注解旁边现在增加了一个@Validated 注解，它会触发对对象的验证。验证后，结果参数（BindingResult 对象）就会包含验证过程的结果。接下来会根据 result 的值进行 hasErrors()方法的条件判断。如果验证类检测到了错误，那么该值就为 true。

如果验证过程中检测到了错误，那么方法处理器就会返回视图 reservationForm，它对应于相同的表单，这样用户就可以重新提交信息了。如果验证过程没有检测到错误，那就会执行预订的调用（reservationService.make(reservation);），然后重定向到成功视图 reservationSuccess。

验证器的注册是在@InitBinder 所注解的方法中完成的，同时验证器是在 WebDataBinder 上设置的，这样绑定后就可以使用了。为了注册验证器，需要使用 setValidator 方法。还可以通过 addValidators 方法注册多个验证器，它接收一个可变参数来注册一个或多个 Validator 实例。

注意： WebDataBinder 还可针对类型转换注册额外的 ProperyEditor、Converter 与 Formatter 实例。可以通过这种方式取代全局的 ProperyEditor、Converter 与 Formatter 注册。

让控制器的会话数据过期

为了支持表单多次提交且不会丢失用户所提供的数据的场景，控制器使用了@SessionAttributes 注解。借助于它，在请求之间会保存一个对 Reservation 类型的 reservation 字段的引用。

不过，当表单成功提交且预订完毕后，我们就不需要在用户会话中继续保存 Reservation 对象了。实际上，当用户在短时间内再次访问表单时，如果没有清除掉会话数据，那么这个老的 Reservation 对象数据还有可能显示出来。

可以通过 SessionStatus 对象删除掉使用@SessionAttributes 注解所赋予的值，该对象可作为输入参数传递给处理器方法。如下代码演示了如何让控制器的会话数据过期：

```
package com.apress.springrecipes.court.web;

@Controller
@RequestMapping("/reservationForm")
@SessionAttributes("reservation")
public class ReservationFormController {

    @RequestMapping(method = RequestMethod.POST)
    public String submitForm(
        @ModelAttribute("reservation") Reservation reservation,
        BindingResult result, SessionStatus status) {

        if (result.hasErrors()) {
            return "reservationForm";
        } else {
            reservationService.make(reservation);
            status.setComplete();
            return "redirect:reservationSuccess";
        }
    }
}
```

处理器方法调用 reservationService.make(reservation);执行预订且在用户被重定向到成功页面前是让控制器会话数据过期的绝佳时刻。这是通过调用 SessionStatus 对象的 setComplete()方法实现的。

3-10 使用向导表单控制器处理多页面表单

问题提出

在 Web 应用中，有时需要处理跨越多个页面的复杂表单。这种表单通常叫做向导表单，因为用户需要逐页填写——就像使用软件向导一样。毫无疑问，可以创建一个或多个表单控制器来处理向导表单。

解决方案

由于一个向导表单有多个表单页面，因此需要为向导表单控制器定义多个页面视图。接下来，控制器会管理所有这些表单页面的表单状态。在向导表单中，还会有一个控制器处理器方法来处理表单提交，就像单个表单一样。不过，为了区分用户的动作，每个表单中都需要嵌入一个特殊的请求参数，这个参数通常会被指定为提交按钮的名字。

```
'_finish': Finish the wizard form.
'_cancel': Cancel the wizard form.
'_targetx': Step to the target page, where x is the zero-based page index.
```

借助于这些参数，控制器的处理器方法就可以根据表单与用户动作确定接下来该怎么做。

解释说明

假设你想要提供这样一个功能，即允许用户定期按照固定时间段来预订场地。首先在 domain 子包中定义 PeriodicReservation 类。

```
package com.apress.springrecipes.court.domain;
...
public class PeriodicReservation {

    private String courtName;
    private Date fromDate;
    private Date toDate;
    private int period;
    private int hour;
    private Player player;

    // Getters and Setters
    ...
}
```

接下来，向 ReservationService 接口中添加 makePeriodic()方法进行定期的预订。

```
package com.apress.springrecipes.court.service;
...
public interface ReservationService {
    ...
    public void makePeriodic(PeriodicReservation periodicReservation)
        throws ReservationNotAvailableException;
}
```

该方法的实现会从 PeriodicReservation 中生成一系列的 Reservation 对象，并将每个 reservation 传递给 make()方法。显然在这个简单的应用中是不支持事务管理的。

```
package com.apress.springrecipes.court.service;
...
public class ReservationServiceImpl implements ReservationService {
    ...
    @Override
    public void makePeriodic(PeriodicReservation periodicReservation)
        throws ReservationNotAvailableException {

        LocalDate fromDate = periodicReservation.getFromDate();
        while (fromDate.isBefore(periodicReservation.getToDate())) {
            Reservation reservation = new Reservation();
            reservation.setCourtName(periodicReservation.getCourtName());
            reservation.setDate(fromDate);
            reservation.setHour(periodicReservation.getHour());
            reservation.setPlayer(periodicReservation.getPlayer());
            make(reservation);

            fromDate = fromDate.plusDays(periodicReservation.getPeriod());

        }
    }
}
```

创建向导表单页面

假设要在 3 个不同的页面中向用户展示定期预订表单。每个页面都包含了部分表单字段。第一个页面是 reservationCourtForm.jsp，它只包含了定期预订的场地名字段。

```
<%@ taglib prefix="form" uri="http://www.springframework.org/tags/form"%>

<html>
<head>
<title>Reservation Court Form</title>
<style>
```

```
.error {
    color: #ff0000;
    font-weight: bold;
}
</style>
</head>

<body>
<form:form method="post" modelAttribute="reservation">
<table>
    <tr>
        <td>Court Name</td>
        <td><form:input path="courtName" /></td>
        <td><form:errors path="courtName" cssClass="error" /></td>
    </tr>
    <tr>
        <td colspan="3">
            <input type="hidden" value="0" name="_page" />
            <input type="submit" value="Next" name="_target1" />
            <input type="submit" value="Cancel" name="_cancel" />
        </td>
    </tr>
</table>
</form:form>
</body>
</html>
```

该页面中的表单与输入字段是通过 Spring 的<form:form>与<form:input>标签定义的。它们会绑定到模型属性 reservation 及其属性上。还有一个错误标签用于向用户展示字段错误消息。注意该页面有两个提交按钮。Next 按钮的名字是_target1。它会让向导表单控制器进入第二个页面，其页面索引是 1（从 0 开始）。Cancel 按钮的名字是_cancel。它会让控制器取消这个表单。此外，还有一个隐藏的表单字段，用于追踪用户所在的页面；在该示例中，它为 0。

第二个页面是 reservationTimeForm.jsp。它包含了定期预订的日期与时间字段。

```
<%@ taglib prefix="form" uri="http://www.springframework.org/tags/form"%>

<html>
<head>
<title>Reservation Time Form</title>
<style>
.error {
    color: #ff0000;
    font-weight: bold;
}
</style>
</head>

<body>
<form:form method="post" modelAttribute="reservation">
<table>
    <tr>
        <td>From Date</td>
        <td><form:input path="fromDate" /></td>
        <td><form:errors path="fromDate" cssClass="error" /></td>
    </tr>
    <tr>
        <td>To Date</td>
        <td><form:input path="toDate" /></td>
        <td><form:errors path="toDate" cssClass="error" /></td>
    </tr>
    <tr>
        <td>Period</td>
        <td><form:select path="period" items="${periods}" /></td>
```

```
            <td><form:errors path="period" cssClass="error" /></td>
        </tr>
        <tr>
            <td>Hour</td>
            <td><form:input path="hour" /></td>
            <td><form:errors path="hour" cssClass="error" /></td>
        </tr>
        <tr>
            <td colspan="3">
                <input type="hidden" value="1" name="_page"/>
                <input type="submit" value="Previous" name="_target0" />
                <input type="submit" value="Next" name="_target2" />
                <input type="submit" value="Cancel" name="_cancel" />
            </td>
        </tr>
    </table>
    </form:form>
    </body>
    </html>
```

该表单有 3 个提交按钮。Previous 与 Next 按钮的名字必须分别是_target0 与_target2。它们会让向导表单控制器进入第一个页面与第三个页面。Cancel 按钮会让控制器取消这个表单。此外，还有一个隐藏的表单字段来追踪用户当前所在的页面；在该示例中，它为 1。

第三个页面是 reservationPlayerForm.jsp。它包含了定期预订的选手信息。

```
<%@ taglib prefix="form" uri="http://www.springframework.org/tags/form"%>

<html>
<head>
<title>Reservation Player Form</title>
<style>
.error {
    color: #ff0000;
    font-weight: bold;
}
</style>
</head>

<body>
<form:form method="POST" commandName="reservation">
<table>
    <tr>
        <td>Player Name</td>
        <td><form:input path="player.name" /></td>
        <td><form:errors path="player.name" cssClass="error" /></td>
    </tr>
    <tr>
        <td>Player Phone</td>
        <td><form:input path="player.phone" /></td>
        <td><form:errors path="player.phone" cssClass="error" /></td>
    </tr>
    <tr>
        <td colspan="3">
            <input type="hidden" value="2" name="_page"/>
            <input type="submit" value="Previous" name="_target1" />
            <input type="submit" value="Finish" name="_finish" />
            <input type="submit" value="Cancel" name="_cancel" />
        <td>
    <tr>
</table>
</form:form>
</body>
</html>
```

该表单有 3 个提交按钮。Previous 按钮会让向导表单控制器回退到第二个页面。Finish 按钮的名字必须是_finish。它会让控制器完成这个表单。Cancel 按钮会让控制器取消这个表单。此外，还有一个隐藏的表单字段来追踪用户当前所在的页面；在该示例中，它为 2。

创建向导表单控制器

现在，我们创建向导表单控制器来处理这个定期预订表单。就像之前的 Spring MVC 控制器一样，该控制器有 4 个主要的处理器方法（一个针对 HTTP GET 请求，另外三个针对 HTTP POST 请求），也使用了与之前控制器相同的控制器元素（比如说注解、验证以及会话）。对于向导表单控制器来说，不同页面中的所有表单字段都会绑定到单个模型属性的 Reservation 对象上，它跨越多个请求存储在用户会话中。

```
package com.apress.springrecipes.court.web;

import com.apress.springrecipes.court.domain.PeriodicReservation;
import com.apress.springrecipes.court.domain.Player;
import com.apress.springrecipes.court.service.ReservationService;
import org.springframework.stereotype.Controller;
import org.springframework.ui.Model;
import org.springframework.validation.BindingResult;
import org.springframework.validation.annotation.Validated;
import org.springframework.web.bind.annotation.*;
import org.springframework.web.bind.support.SessionStatus;
import org.springframework.web.util.WebUtils;

import javax.annotation.PostConstruct;
import javax.servlet.http.HttpServletRequest;
import java.util.Enumeration;
import java.util.HashMap;
import java.util.Map;

@Controller
@RequestMapping("/periodicReservationForm")
@SessionAttributes("reservation")
public class PeriodicReservationController {

    private final Map<Integer, String> pageForms = new HashMap<>(3);
    private final ReservationService reservationService;

    public PeriodicReservationController(ReservationService reservationService) {
        this.reservationService = reservationService;
    }

    @PostConstruct
    public void initialize() {
        pageForms.put(0, "reservationCourtForm");
        pageForms.put(1, "reservationTimeForm");
        pageForms.put(2, "reservationPlayerForm");
    }

@GetMapping
public String setupForm(Model model) {
    PeriodicReservation reservation = new PeriodicReservation();
    reservation.setPlayer(new Player());
    model.addAttribute("reservation", reservation);
    return "reservationCourtForm";
}

@PostMapping(params = {"_cancel"})
public String cancelForm(@RequestParam("_page") int currentPage) {

    return pageForms.get(currentPage);
}
```

```
    @PostMapping(params = {"_finish"})
    public String completeForm(
        @ModelAttribute("reservation") PeriodicReservation reservation,
        BindingResult result, SessionStatus status,
        @RequestParam("_page") int currentPage) {

        if (!result.hasErrors()) {
            reservationService.makePeriodic(reservation);
            status.setComplete();
            return "redirect:reservationSuccess";
        } else {
            return pageForms.get(currentPage);
        }
    }

    @PostMapping
    public String submitForm(
        HttpServletRequest request,
        @ModelAttribute("reservation") PeriodicReservation reservation,
        BindingResult result, @RequestParam("_page") int currentPage) {

        int targetPage = getTargetPage(request, "_target", currentPage);
        if (targetPage < currentPage) {
            return pageForms.get(targetPage);
        }

        if (!result.hasErrors()) {
            return pageForms.get(targetPage);
        } else {
            return pageForms.get(currentPage);
        }
    }

    @ModelAttribute("periods")
    public Map<Integer, String> periods() {
            Map<Integer, String> periods = new HashMap<Integer, String>();
            periods.put(1, "Daily");
            periods.put(7, "Weekly");
            return periods;
        }

        private int getTargetPage(HttpServletRequest request, String paramPrefix, int currentPage) {

            Enumeration<String> paramNames = request.getParameterNames();
            while (paramNames.hasMoreElements()) {
                String paramName = paramNames.nextElement();
                if (paramName.startsWith(paramPrefix)) {
                    for (int i = 0; i < WebUtils.SUBMIT_IMAGE_SUFFIXES.length; i++) {
                        String suffix = WebUtils.SUBMIT_IMAGE_SUFFIXES[i];
                        if (paramName.endsWith(suffix)) {
                            paramName = paramName.substring(0, paramName.length() -
                            suffix.length());
                        }
                    }
                    return Integer.parseInt(paramName.substring(paramPrefix.length()));
                }
            }
            return currentPage;
        }
    }
```

该控制器使用了与之前的 ReservationFormController 控制器一些相同的元素，因此已经介绍过的内容这里就不再赘述了。总的来说，它使用@SessionAttributes 注解将 reservation 对象放到用户的会话中。它拥有相同的 HTTP GET 方法，在加载第一个表单视图时分配空的 Reservation 与 Player 对象。

接下来，控制器定义了一个 HashMap，通过它将页面序号与视图名关联到了一起。这个 HashMap 在控制器中会使用多次，因为控制器需要确定不同场景下（比如说验证或用户单击了 Cancal 或 Next）的用户视图。

你还会看到使用了@ModelAttribute("periods")注解的方法。在之前的控制器中对其做过介绍，该声明会将一个值的列表传递给控制器中所返回的任何视图中。查看之前的表单 reservationTimeForm.jsp，你会看到它会访问一个名为 periods 的模型属性。

接下来，如果到来的请求 URL 中包含了_cancel 参数，那么第一个@PostMapping 就会被调用。它还会通过@RequestParam("page")从请求的 page 属性中提取出 currentPage 值。当该方法被调用时，它会将控制返回给与 currentPage 值对应的视图。结果就是输入被重置为之前的值。

如果到来的请求 URL 中包含了_finish 参数，那么下一个@PostMapping(params= {"_finish"})就会被调用，这出现在用户单击 Finish 按钮时。由于这是流程的最后一步，因此需要验证 Reservation 对象，所以为属性应用了@Validated 注解。如果没有错误，那么处理器方法就会调用 reservationService.makePeriodic(reservation);进行预订并将用户重定向到 reservationSuccess 视图。

使用@PostMapping 注解的最后一个处理器方法会处理其他情况并为 HttpServletRequest 声明一个输入参数，这样处理器方法就可以访问该对象的内容了。之前的处理器方法使用了诸如@RequestParam 的参数来输入这些标准对象中的数据，这是一种简洁的机制。它表明可以在处理器方法中完全访问标准的 HttpServletRequest 与 HttpServletResponse 对象。大家应该对其余输入参数的名字与符号很熟悉了，它们来自于之前的控制器。如果该处理器方法被调用，那就意味着用户单击了表单的 Next 或是 Previous 按钮。这就意味着在 HttpServletRequest 对象中有一个名为_target 的参数。这是因为每个表单的 Next 与 Previous 按钮都会被赋予该参数。

借助于 getTargetPage 方法可以抽取出_target 参数的值，它对应于 target0、target1 或是 target2，并且会被解析为代表目标页面的 0、1 或是 2。

当有了目标页面号与当前页面号后，就可以确定用户单击的是 Next 还是 Previous 按钮。如果目标页面号小于当前页面号，那就表示用户单击的是 Previous 按钮。如果目标页面号大于当前页面号，那就表示用户单击的是 Next 按钮。

这时，你可能还不太清楚为何需要确定用户单击的是 Next 还是 Previous 按钮，尤其是总是返回与目标页面对应的视图。这个逻辑背后的原因是：如果用户单击的是 Next 按钮，则需要验证数据；如果用户单击的是 Previous 按钮，就没必要验证任何东西了。下一节会将验证融入到控制器中，到时就清楚了。

由于使用了@RequestMapping("/periodicReservationForm")注解来修饰 PeriodicReservationController 类，因此可以通过如下 URL 来访问该控制器：

```
http://localhost:8080/court/periodicReservation
```

验证向导表单数据

在简单的表单控制器中，当表单提交时，可以一次性验证整个模型属性对象。不过，由于向导表单控制器有多个表单页面，当表单提交时需要验证每个页面。出于这个原因，需要创建如下验证器，它将 validate()方法拆分为几个细粒度的验证方法，每个方法验证特定页面上的字段：

```
package com.apress.springrecipes.court.domain;

import org.springframework.validation.Errors;
import org.springframework.validation.ValidationUtils;
import org.springframework.validation.Validator;

public class PeriodicReservationValidator implements Validator {

    public boolean supports(Class clazz) {
        return PeriodicReservation.class.isAssignableFrom(clazz);
    }
```

```
    public void validate(Object target, Errors errors) {
        validateCourt(target, errors);
        validateTime(target, errors);
        validatePlayer(target, errors);
    }

    public void validateCourt(Object target, Errors errors) {
        ValidationUtils.rejectIfEmptyOrWhitespace(errors, "courtName",
            "required.courtName", "Court name is required.");
    }

    public void validateTime(Object target, Errors errors) {
        ValidationUtils.rejectIfEmpty(errors, "fromDate",
            "required.fromDate", "From date is required.");
        ValidationUtils.rejectIfEmpty(errors, "toDate", "required.toDate",
            "To date is required.");
        ValidationUtils.rejectIfEmpty(errors, "period",
            "required.period", "Period is required.");
        ValidationUtils.rejectIfEmpty(errors, "hour", "required.hour",
            "Hour is required.");
    }

    public void validatePlayer(Object target, Errors errors) {
        ValidationUtils.rejectIfEmptyOrWhitespace(errors, "player.name",
            "required.playerName", "Player name is required.");
    }
}
```

类似于之前的验证器示例，注意到该验证器也使用了@Component注解自动将验证器类注册为bean。当验证器bean注册好后，剩下的唯一事情就是将验证器融合到控制器中。

```
package com.apress.springrecipes.court.web;

import com.apress.springrecipes.court.domain.PeriodicReservation;
import com.apress.springrecipes.court.domain.PeriodicReservationValidator;
import com.apress.springrecipes.court.domain.Player;
import com.apress.springrecipes.court.service.ReservationService;
import org.springframework.stereotype.Controller;
import org.springframework.ui.Model;
import org.springframework.validation.BindingResult;
import org.springframework.validation.annotation.Validated;
import org.springframework.web.bind.WebDataBinder;
import org.springframework.web.bind.annotation.*;
import org.springframework.web.bind.support.SessionStatus;
import org.springframework.web.util.WebUtils;

import javax.annotation.PostConstruct;
import javax.servlet.http.HttpServletRequest;
import java.util.Enumeration;
import java.util.HashMap;
import java.util.Map;

@Controller
@RequestMapping("/periodicReservationForm")
@SessionAttributes("reservation")
public class PeriodicReservationController {

    private final Map<Integer, String> pageForms = new HashMap<>(3);
    private final ReservationService reservationService;
    private final PeriodicReservationValidator validator;

    public PeriodicReservationController(ReservationService reservationService,
                                         PeriodicReservationValidator
                                         periodicReservationValidator) {
        this.reservationService = reservationService;
```

```
        this.validator = periodicReservationValidator;
    }

    @InitBinder
    public void initBinder(WebDataBinder binder) {
        binder.setValidator(this.validator);
    }

    @PostMapping(params = {"_finish"})
    public String completeForm(
        @Validated @ModelAttribute("reservation") PeriodicReservation reservation,
        BindingResult result, SessionStatus status,
        @RequestParam("_page") int currentPage) {
        if (!result.hasErrors()) {
            reservationService.makePeriodic(reservation);
            status.setComplete();
            return "redirect:reservationSuccess";
        } else {
            return pageForms.get(currentPage);
        }
    }

    @PostMapping
    public String submitForm(
        HttpServletRequest request,
        @ModelAttribute("reservation") PeriodicReservation reservation,
        BindingResult result, @RequestParam("_page") int currentPage) {
        int targetPage = getTargetPage(request, "_target", currentPage);
        if (targetPage < currentPage) {
            return pageForms.get(targetPage);
        }
        validateCurrentPage(reservation, result, currentPage);
        if (!result.hasErrors()) {
            return pageForms.get(targetPage);
        } else {
            return pageForms.get(currentPage);
        }
    }
    private void validateCurrentPage(PeriodicReservation reservation,
        BindingResult result, int currentPage) {
        switch (currentPage) {
            case 0:
                validator.validateCourt(reservation, result);
                break;
            case 1:
                validator.validateTime(reservation, result);
                break;
            case 2:
                validator.validatePlayer(reservation, result);
                break;
        }
    }

    ...
}
```

首先向控制器添加的是 validator 字段，通过类的构造方法将 PeriodicReservationValidator 验证器 bean 的实例赋给它。接下来会看到控制器中有两个对 validator 的引用。

第一个是当用户提交完表单时。要想调用验证器，需要向方法的 Reservation 参数添加@Validated 注解。为了让其起作用，还需要添加一个被@InitBinder 注解的方法，它会将 PeriodicReservationValidator 注册到数据绑定器上。如果验证器没有返回错误，那么预订就会提交，用户的会话会被重置，用户也会被重定向到 reservationSuccess 视图。如果验证器返回了错误，那么用户会转向到当前的视图表单来修正错误（参见 3-9 节）。

当用户单击了表单的 Next 按钮时也会使用到验证器。由于用户想要进入下一个表单，因此需要对所提供的数据进行验证。考虑到有 3 个表单视图需要验证，这里使用了 case 语句来确定该调用哪个验证器方法。当验证器方法的执行返回时，如果检测到了错误，那么用户会被转向到 currentPage 视图来修复错误；如果没有检测到错误，那么用户会被转向到 targetPage 视图；注意到这些目标页面号会映射到控制器中的 Map。

3-11 使用注解进行 bean 验证（JSR-303）

问题提出

你想要基于 JSR-303 标准使用注解在 Web 应用中验证 Java bean。

解决方案

JSR-303（即 Bean Validation）是一个规范，旨在通过注解来标准化 Java bean 的验证。

在之前的示例中，你已经看到了 Spring 框架是如何通过专门的技术来验证 bean 的。这需要继承 Spring 框架中的类为特定类型的 Java bean 创建验证器类。

JSR-303 标准的目标是在 Java bean 类中直接使用注解。这样就可以直接在想要验证的代码中指定验证规则，而无须在单独的类中创建验证规则了——就像之前使用 Spring Validator 类所做的那样。

解释说明

首先需要使用必要的 JSR-303 注解来修饰 Java bean。如下代码演示了如何使用 JSR-303 注解来修饰场地预订应用中的 Reservation 领域类：

```
public class Reservation {

    @NotNull
    @Size(min = 4)
    private String courtName;

    @NotNull
    private Date date;

    @Min(9)
    @Max(21)
    private int hour;

    @Valid
    private Player player;

    @NotNull
    private SportType sportType;

    // Getter/Setter methods omitted for brevity
}
```

courtName 字段使用了两个注解：@NotNull 注解表示该字段不能为 null；@Size 注解表示该字段至少要包含 4 个字符。

date 与 sportType 字段使用了@NotNull 注解，因为它们是必填字段。

hour 字段使用了@Min 与@Max 注解，表示 hour 字段值的上下限。

player 字段使用了@Valid 注解来触发内嵌的 Player 对象的验证，Player 领域类中的两个字段也都使用了@NotNull 注解。

在知道如何通过 JSR-303 标准中的注解来修饰 Java bean 类后，我们就来看看控制器是如何使用这些验证器注解的。

```
package com.apress.springrecipes.court.service;
.....
    private final ReservationService reservationService;

    public ReservationFormController(ReservationService reservationService) {
        this.reservationService = reservationService;
```

```
    }

    @RequestMapping(method = RequestMethod.POST)
    public String submitForm(
        @ModelAttribute("reservation") @Valid Reservation reservation,
        BindingResult result, SessionStatus status) {
    if (result.hasErrors()) {
        return "reservationForm";
    } else {
        reservationService.make(reservation);
        return "redirect:reservationSuccess";
    }
}
```

控制器与 3-9 节中的非常类似，唯一的差别在于缺少了被@InitBinder 注解所修饰的方法。Spring MVC 会检测到类路径上的 javax.validation.Validator。我们向类路径添加 hibernate-validator，它是一个验证实现。

接下来，你会看到控制器的 HTTP POST 处理器方法，它用于处理用户数据的提交。由于处理器方法期望一个 Reservation 对象实例（使用 JSR-303 注解进行了修饰），因此可以验证其数据。

submitForm 方法的其余部分与 3-9 节相同。

注意： 要想在 Web 应用中使用 JSR-303 bean 验证，需要向类路径添加一个其实现的依赖。如果使用 Maven，请将如下依赖添加到 Maven 项目中：

```
<dependency>
    <groupId>org.hibernate</groupId>
    <artifactId>hibernate-validator</artifactId>
    <version>5.4.0.Final</version>
</dependency>
```

如果使用的是 Gradle，请添加如下依赖：

```
compile 'org.hibernate:hibernate-validator:5.4.0.Final'
```

3-12　创建 Excel 与 PDF 视图

问题提出

虽然 HTML 是最常用的展示 Web 内容的方式，不过有时用户可能需要将内容从 Web 应用中导出为 Excel 或是 PDF 格式。在 Java 中，有一些库可以用于生成 Excel 和 PDF 文件。不过，要想在 Web 应用中直接使用这些库，则需要在后台生成内容，并将其以二进制附件的形式返回给用户。你需要为此处理 HTTP 响应头与输出流。

解决方案

Spring 将 Excel 与 PDF 文件的生成集成到了 MVC 框架中。可以将 Excel 与 PDF 文件看作特殊的视图，这样就能以一致的方式在控制器中处理 Web 请求，并向模型添加数据以传递给 Excel 与 PDF 视图。通过这种方式，无须再处理 HTTP 响应头和输出流。Spring MVC 支持使用 Apache POI 库来生成 Excel 文件。对应的视图类是 AbstractExcelView、AbstractXlsxView 与 AbstractXlsxStreamingView。PDF 文件可以通过 iText 库生成，对应的视图类是 AbstractPdfView。

解释说明

假设用户想要生成某一天的预订总结报告。他希望这个报告使用 Excel、PDF 或是基本的 HTML 格式生成。对于这个报告生成功能来说，需要在 service 层声明一个方法，返回某一天所有的预订信息。

```
package com.apress.springrecipes.court.service;
...
public interface ReservationService {
    ...
    public List<Reservation> findByDate(LocalDate date);
}
```

接下来，为该方法提供一个简单的实现，遍历所有的预订信息。

```
package com.apress.springrecipes.court.service;
...
public class ReservationServiceImpl implements ReservationService {
    ...
    public List<Reservation> findByDate(LocalDate date) {
        return reservations.stream()
            .filter(r -> Objects.equals(r.getDate(), date))
            .collect(Collectors.toList());
    }
}
```

现在，编写一个简单的控制器从 URL 中获取到 date 参数。date 参数会被格式化为一个 date 对象并传递给 service 层来查询预订信息。控制器使用了 3-7 节所介绍的内容协商解析器。因此，控制器会返回一个逻辑视图，让解析器确定使用 Excel、PDF 还是默认的 HTML 页面来生成报告。

```
package com.apress.springrecipes.court.web;
...
@Controller
@RequestMapping("/reservationSummary*")
public class ReservationSummaryController {
    private ReservationService reservationService;

    @Autowired
    public ReservationSummaryController(ReservationService reservationService) {
        this.reservationService = reservationService;
    }

    @RequestMapping(method = RequestMethod.GET)
    public String generateSummary(
        @RequestParam(required = true, value = "date")
        String selectedDate,
        Model model) {
        List<Reservation> reservations = java.util.Collections.emptyList();
        try {
            Date summaryDate = new SimpleDateFormat("yyyy-MM-dd").parse(selectedDate);
            reservations = reservationService.findByDate(summaryDate);
        } catch (java.text.ParseException ex) {
            StringWriter sw = new StringWriter();
            PrintWriter pw = new PrintWriter(sw);
            ex.printStackTrace(pw);
            throw new ReservationWebException("Invalid date format for reservation
            summary",new Date(),sw.toString());
        }
        model.addAttribute("reservations",reservations);
        return "reservationSummary";
    }
}
```

该控制器只包含了一个默认的 HTTP GET 处理器方法。该方法执行的第一个动作是创建一个空的 Reservation 列表来存放从预订服务获取的结果。接下来，会看到一个 try/catch 块，它会根据 selectedDate @RequestParam 创建一个 Date 对象，并使用它调用预订服务。如果创建 Date 对象失败，那么就会抛出一个名为 ReservationWebException 的 Spring 异常。

如果 try/catch 块中没有错误，那么 Reservation 列表就会被放到控制器的 Model 对象中。接下来，方法会将控制转交给 reservationSummary 视图。

注意，虽然控制器支持 PDF、XLS 与 HTML 视图，但它只会返回一个视图。这是因为使用了 ContentNegotiatingViewResolver 解析器，它会根据视图名来确定使用哪个视图。请参见 3-7 节了解关于该解析器的更多信息。

创建 Excel 视图

可以通过继承 AbstractXlsView 或是 AbstractXlsxView 类（针对 Apache POI）来创建 Excel 视图。这里使

用了 AbstractXlsxView。在 buildExcelDocument()方法中，可以访问从控制器传递过来的模型以及预先创建的 Excel 工作簿。你的任务是使用模型中的数据装配这个工作簿。

注意：要想在 Web 应用中使用 Apache POI 生成 Excel 文件，需要在类路径中引入 Apache POI 依赖。如果使用的是 Apache Maven，请将如下依赖添加到 Maven 项目中：

```
<dependency>
    <groupId>org.apache.poi</groupId>
    <artifactId>poi</artifactId>
    <version>3.10-FINAL</version>
</dependency>
```

```
package com.apress.springrecipes.court.web.view;

import com.apress.springrecipes.court.domain.Reservation;
import org.apache.poi.ss.usermodel.Row;
import org.apache.poi.ss.usermodel.Sheet;
import org.apache.poi.ss.usermodel.Workbook;
import org.springframework.web.servlet.view.document.AbstractXlsxView;
import javax.servlet.http.HttpServletRequest;
import javax.servlet.http.HttpServletResponse;
import java.text.DateFormat;
import java.text.SimpleDateFormat;
import java.util.List;
import java.util.Map;

public class ExcelReservationSummary extends AbstractXlsxView {

    @Override
    protected void buildExcelDocument(Map<String, Object> model, Workbook workbook,
                                      HttpServletRequest request, HttpServletResponse
                                      response) throws Exception {
        @SuppressWarnings({"unchecked"})
        final List<Reservation> reservations =
            (List<Reservation>) model.get("reservations");
        final DateFormat dateFormat = new SimpleDateFormat("yyyy-MM-dd");
        final Sheet sheet = workbook.createSheet();

        addHeaderRow(sheet);

        reservations.forEach(reservation -> createRow(dateFormat, sheet, reservation));
    }

    private void addHeaderRow(Sheet sheet) {
        Row header = sheet.createRow(0);
        header.createCell((short) 0).setCellValue("Court Name");
        header.createCell((short) 1).setCellValue("Date");
        header.createCell((short) 2).setCellValue("Hour");
        header.createCell((short) 3).setCellValue("Player Name");
        header.createCell((short) 4).setCellValue("Player Phone");
    }

    private void createRow(DateFormat dateFormat, Sheet sheet, Reservation reservation) {
        Row row = sheet.createRow(sheet.getLastRowNum() + 1);
        row.createCell((short) 0).setCellValue(reservation.getCourtName());
        row.createCell((short) 1).setCellValue(dateFormat.format(reservation.getDate()));
        row.createCell((short) 2).setCellValue(reservation.getHour());
        row.createCell((short) 3).setCellValue(reservation.getPlayer().getName());
        row.createCell((short) 4).setCellValue(reservation.getPlayer().getPhone());
    }
}
```

在上面的 Excel 视图中，首先在工作簿中创建了一个工作表。该工作表的第一行展示了报告的头。接下来，遍历预订列表为每个预订创建一行。

由于控制器中配置了@RequestMapping("/reservationSummary*")，并且处理器方法需要一个日期作为请求参数，因此可以通过如下 URL 访问这个 Excel 视图：

```
http://localhost:8080/court/reservationSummary.xls?date=2009-01-14
```

创建 PDF 视图

PDF 视图是通过继承 AbstractPdfView 类创建的。在 buildPdfDocument()方法中，可以访问控制器传递过来的模型和预先创建的 PDF 文档。你的任务是使用模型中的数据装配该文档。

■ **注意**：要想在 Web 应用中使用 iText 生成 PDFl 文件，需要在类路径中引入 iText 库。如果使用的是 Apache Maven，请将如下依赖添加到 Maven 项目中：

```
<dependency>
    <groupId>com.lowagie</groupId>
    <artifactId>itext</artifactId>
    <version>4.2.1</version>
</dependency>
```

```
package com.apress.springrecipes.court.web.view;
...
import org.springframework.web.servlet.view.document.AbstractPdfView;

import com.lowagie.text.Document;
import com.lowagie.text.Table;
import com.lowagie.text.pdf.PdfWriter;

public class PdfReservationSummary extends AbstractPdfView {

    protected void buildPdfDocument(Map model, Document document,
        PdfWriter writer, HttpServletRequest request,
        HttpServletResponse response) throws Exception {
        List<Reservation> reservations = (List) model.get("reservations");
        DateFormat dateFormat = new SimpleDateFormat("yyyy-MM-dd");
        Table table = new Table(5);

        table.addCell("Court Name");
        table.addCell("Date");
        table.addCell("Hour");
        table.addCell("Player Name");
        table.addCell("Player Phone");

        for (Reservation reservation : reservations) {
            table.addCell(reservation.getCourtName());
            table.addCell(dateFormat.format(reservation.getDate()));
            table.addCell(Integer.toString(reservation.getHour()));
            table.addCell(reservation.getPlayer().getName());
            table.addCell(reservation.getPlayer().getPhone());
        }

        document.add(table);
    }
}
```

由于控制器中配置了@RequestMapping("/reservationSummary*")，并且处理器方法需要一个日期作为请求参数，因此可以通过如下 URL 访问这个 PDF 视图：

```
http://localhost:8080/court/reservationSummary.pdf?date=2009-01-14
```

为 Excel 与 PDF 视图创建解析器

3-6 节介绍了将逻辑视图名解析为具体视图实现的不同策略。其中一种策略就是从资源包中解析视图；这种策略更适合于将逻辑视图名映射为包含了 PDF 或 XLS 类的视图实现。

请确保在 Web 应用上下文中配置了 ResourceBundleViewResolver bean 作为视图解析器，可以在 Web 应用类路径的根目录下的 views.properties 文件中定义视图。

将如下映射添加到 views.properties 中，以将 XLS 视图类映射到逻辑视图名上：

```
reservationSummary.(class)=com.apress.springrecipes.court.web.view.ExcelReservationSummary
```

由于应用依赖于内容协商的过程，这意味着同样的视图名会映射到不同的视图技术上。此外，由于在同样的 views.properties 文件中无法定义相同的名字，因此需要创建一个名为 secondaryviews.properties 的文件将 PDF 视图类映射到逻辑视图名上，如下所示：

```
reservationSummary.(class)=com.apress.springrecipes.court.web.view.PdfReservationSummary
```

注意，secondaryviews.properties 文件需要配置在自己的 ResourceBundleViewResolver 解析器中。属性名 reservationSummary 对应于控制器所返回的视图名。ContentNegotiatingViewResolver 解析器负责根据用户请求来确定该使用哪个类。确定好之后，相应类的执行就会生成 PDF 或是 XLS 文件。

创建基于日期的 PDF 与 XLS 文件名

当用户使用如下 URL 发出针对 PDF 或是 XLS 文件的请求时：

```
http://localhost:8080/court/reservationSummary.pdf?date=2008-01-14
http://localhost:8080/court/reservationSummary.xls?date=2008-02-24
```

浏览器会询问用户一个问题："保存为 reservationSummary.pdf，还是保存为 reservationSummary.xls？"。这种约定是基于用户所请求资源的 URL。不过，考虑到用户还在 URL 中提供了一个日期，因此更好的自动提示形式是"保存为 ReservationSummary_2009_01_24.xls，还是保存为 ReservationSummary_2009_02_24.xls？"。这可以通过应用一个拦截器来重写返回的 URL 达成。如下代码展示了该拦截器：

```
package com.apress.springrecipes.court.web
...

public class ExtensionInterceptor extends HandlerInterceptorAdapter {
    public void postHandle(HttpServletRequest request,
        HttpServletResponse response, Object handler,
            ModelAndView modelAndView) throws Exception {
        // Report date is present in request
        String reportName = null;
        String reportDate = request.getQueryString().replace("date=","").replace("-","_");
        if(request.getServletPath().endsWith(".pdf")) {
            reportName= "ReservationSummary_" + reportDate + ".pdf";
        }
        if(request.getServletPath().endsWith(".xls")) {
            reportName= "ReservationSummary_" + reportDate + ".xls";
        }
        if (reportName != null) {
            response.setHeader("Content-Disposition","attachment; filename="+reportName);
        }
    }
}
```

如果 URL 包含了.pdf 或.xls 扩展名，那么该拦截器会提取出整个 URL。如果它检测到这样的扩展名，那就会以 ReservationSummary_<report_date>.<.pdf|.xls>的形式为返回的文件名创建出相应的值。为了确保用户能以这种形式接收到下载提示，需要将 HTTP 头 Content-Disposition 设为该文件名格式。

要想部署该拦截器且只应用到负责生成 PDF 与 XLS 文件的控制器所对应的 URL，建议回顾一下 3-3 节，它介绍了这种配置以及关于拦截器类的更多信息。

内容协商与在拦截器中设置 HTTP 头

虽然该应用使用了 ContentNegotiatingViewResolver 解析器来选择恰当的视图，不过修改返回 URL 的过程已经超出了视图解析器的范围。因此，有必要使用拦截器手工检测请求扩展名，同时设置必要的 HTTP 头来修改发出的 URL。

小结

本章介绍了如何通过 Spring MVC 框架开发 Java Web 应用。Spring MVC 的核心组件是 DispatcherServlet，它作为前端控制器将请求分发到恰当的处理器以便处理请求。在 Spring MVC 中，控制器是使用@Controller 注解修饰的标准 Java 类。在本章的各个小节中，我们介绍了如何使用 Spring MVC 控制器中的其他注解，这包括标识访问 URL 的@RequestMapping、自动注入 bean 引用的@Autowired，以及在用户会话中维护对象的@SessionAttributes 等。我们还介绍了如何将拦截器纳入到应用中；这可以修改控制器中的请求与响应对象。此外，还介绍了 Spring MVC 对表单处理的支持，包括使用 Spring 验证器和 JSR-303 bean 验证标准来验证数据。还介绍了 Spring MVC 如何结合 SpEL 来简化某些配置任务，以及 Spring MVC 如何针对不同的展示层技术来支持不同类型的视图。最后介绍了 Spring 是如何支持内容协商来根据请求扩展名确定视图的。

第 4 章

Spring REST

本章将介绍 Spring 是如何支持表述性状态转移（REST）的。自从 Roy Fielding 在 2000 年提出该术语以来，REST 对 Web 应用有着重要的影响。

基于 Web 的超文本传输协议（HTTP）的基础，REST 所提出的架构在 Web Service 实现中变得越发流行。Web Service 本身已经成为了 Web 上很多机器与机器之间通信的基础。正是很多公司所使用的技术（如 Java、Python、Ruby、.NET 等）不同，才使得填平这些不同环境之间沟壑的解决方案变得尤为必要。比如说，使用 Python 编写的应用要访问 Java 应用所提供的信息该怎么办？Java 应用要访问来自于.NET 所编写的应用的信息该怎么办？Web Service 填补了这个空白。

有很多技术可以实现 Web Service，不过 RESTful Web Service 已经成为 Web 应用中最常见的选择了。它们已经被一些大型互联网站点（如 Google 和 Yahoo!）用来提供对其信息的访问、支持浏览器的 Ajax 调用，并提供各种信息分发的基础，如新闻订阅（即 RSS）。

本章将会介绍 Spring 应用该如何使用 REST，这样就可以通过这种流行的方式访问并提供信息了。

4-1 使用 REST 服务发布 XML

问题提出

你想要通过 Spring 发布基于 XML 的 REST 服务。

解决方案

在 Spring 中设计 REST 服务存在两种可能。一种是将应用数据以 REST 服务的形式发布出去；另一种是访问来自于第三方 REST 服务的数据以在应用中使用。本节将会介绍如何以 REST 服务的形式将应用数据发布出去。4-2 节将会介绍如何访问第三方 REST 服务的数据。将应用数据以 REST 服务的形式发布出去又会使用到 Spring MVC 注解@RequestMapping 与@PathVariable。通过使用这些注解来修饰 Spring MVC 处理器方法，Spring 应用就能以 REST 服务的形式将应用数据发布出去。

此外，Spring 支持一系列机制来生成 REST 服务的负载。本节将会介绍最为简单的一种机制，它会用到 Spring 的 MarshallingView 类。随着讲解的不断推进，你将会了解到 Spring 支持的用于生成 REST 服务负载的高级机制。

解释说明

以 REST 服务的形式（用 Web Service 的术语来说是“创建端点”）发布 Web 应用的数据会紧紧绑定到 Spring MVC 上，第 3 章已经对此进行了介绍。由于 Spring MVC 依赖于注解@RequestMapping 来修饰处理器方法并定义访问点（即 URL），因此这是定义 REST 服务端点的首选方式。

使用 MarshallingView 生成 XML

如下代码展示了一个 Spring MVC 控制器，它有一个处理器方法定义了一个 REST 服务端点：

```
package com.apress.springrecipes.court.web;
```

```
import com.apress.springrecipes.court.domain.Members;
import com.apress.springrecipes.court.service.MemberService;
import org.springframework.beans.factory.annotation.Autowired;
import org.springframework.stereotype.Controller;
import org.springframework.ui.Model;
import org.springframework.web.bind.annotation.RequestMapping;

@Controller
public class RestMemberController {

    private final MemberService memberService;

    @Autowired
    public RestMemberController(MemberService memberService) {
        super();
        this.memberService=memberService;
    }

    @RequestMapping("/members")
    public String getRestMembers(Model model) {
        Members members = new Members();
        members.addMembers(memberService.findAll());
        model.addAttribute("members", members);
        return "membertemplate";
    }
}
```

通过使用@RequestMapping("/members")来修饰控制器的处理器方法，我们就可以通过 host_name/[app-name]/members 来访问 REST 服务端点。你会发现控制会转移给名为 membertemplate 的逻辑视图。如下代码展示的声明定义了一个名为 membertemplate 的逻辑视图：

```
@Configuration
@EnableWebMvc
@ComponentScan(basePackages = "com.apress.springrecipes.court")
public class CourtRestConfiguration {

    @Bean
    public View membertemplate() {
        return new MarshallingView(jaxb2Marshaller());
    }

    @Bean
    public Marshaller jaxb2Marshaller() {
        Jaxb2Marshaller marshaller = new Jaxb2Marshaller();
        marshaller.setClassesToBeBound(Members.class, Member.class);
        return marshaller;
    }

    @Bean
    public ViewResolver viewResolver() {
        return new BeanNameViewResolver();
    }
}
```

membertemplate 视图被定义为 MarshallingView 类型，这是个通用目的的类，可以通过编组器来渲染响应。编组指的是将对象的内存表示转换为某种数据格式的过程。因此，针对这个特定的场景，编组器负责将 Members 与 Member 对象转换为 XML 数据格式。MarshallingView 所用的编组器属于 Spring 提供的一系列 XML 编组器之一，即 Jaxb2Marshaller。Spring 提供的其他编组器还有 CastorMarshaller、JibxMarshaller、XmlBeansMarshaller 与 XStreamMarshaller。

编组器本身也需要配置。我们选择使用 Jaxb2Marshaller 编组器是因为其简洁性与 Java Architecture for XML Binding（JAXB）基础。不过，如果更想使用 Castor XML 框架，你会发现 CastorMarshaller 的使用很简单；如果使用 XStream，你会发现 XStreamMarshaller 的使用也很简单；对于其他编组器来说也是类似的。

需要通过名为 classesToBeBound 或 contextPath 的属性来配置 Jaxb2Marshaller 编组器。如果使用 classesToBeBound，那么赋给该属性的类就表示将要被转换为 XML 的类（即对象）结构。如下代码展示了赋给 Jaxb2Marshaller 编组器的 Members 与 Member 类：

```
package com.apress.springrecipes.court.domain;

import javax.xml.bind.annotation.XmlRootElement;

@XmlRootElement
public class Member {
    private String name;
    private String phone;
    private String email;

    public String getEmail() {
        return email;
    }

    public String getName() {
        return name;
    }

    public String getPhone() {
        return phone;
    }

    public void setEmail(String email) {
        this.email = email;
    }

    public void setName(String name) {
        this.name = name;
    }

    public void setPhone(String phone) {
        this.phone = phone;
    }
}
```

这是 Members 类：

```
package com.apress.springrecipes.court.domain;

import javax.xml.bind.annotation.XmlAccessType;
import javax.xml.bind.annotation.XmlAccessorType;
import javax.xml.bind.annotation.XmlElement;
import javax.xml.bind.annotation.XmlRootElement;
import java.util.ArrayList;
import java.util.Collection;
import java.util.List;

@XmlRootElement
@XmlAccessorType(XmlAccessType.FIELD)
public class Members {

    @XmlElement(name="member")
    private List<Member> members = new ArrayList<>();

    public List<Member> getMembers() {
        return members;
    }
```

```
    public void setMembers(List<Member> members) {
        this.members = members;
    }

    public void addMembers(Collection<Member> members) {
        this.members.addAll(members);
    }
}
```

注意到 Member 类是个使用了@XmlRootElement 注解修饰的 POJO。Jaxb2Marshaller 编组器可以通过该注解检测一个类（即对象）的字段并将其转换为 XML 数据（即 name=John 转换为<name>john</name>，email=john@doe.com 转换为<email>john@doe.com</email>）。

总结一下，当请求以 http://[host_name]//app-name]/members.xm l 的形式发送给一个 URL 时，对应的处理器会负责创建一个 Members 对象，接下来将其传递给名为 membertemplate 的逻辑视图。根据上面视图的定义，编组器会将一个 Members 对象转换为 XML 负载并返回给 REST 服务的请求方。如下代码展示了 REST 服务所返回的 XML 负载：

```
<?xml version="1.0" encoding="UTF-8" standalone="yes"?>
<members>
    <member>
        <email>marten@deinum.biz</email>
        <name>Marten Deinum</name>
        <phone>00-31-1234567890</phone>
    </member>
    <member>
        <email>john@doe.com</email>
        <name>John Doe</name>
        <phone>1-800-800-800</phone>
    </member>
    <member>
        <email>jane@doe.com</email>
        <name>Jane Doe</name>
        <phone>1-801-802-803</phone>
    </member>
</members>
```

该 XML 负载代表了一种生成 REST 服务响应的简单方式。随着本书内容的不断推进，你会学习到更加复杂的方式，比如说创建广泛使用的 REST 服务负载（如 RSS、Atom 和 JSON）的能力。

仔细观察上面介绍的 REST 服务端点或是 URL，你会发现它有一个.xml 扩展名。如果换成其他扩展名（甚至是省略掉扩展名），这个特定的 REST 服务就不会被触发。这种行为是直接绑定到 Spring MVC 及其处理视图解析的方式上的。它与 REST 服务本身没有什么关系。

默认情况下，由于与这个特定的 REST 服务处理器方法关联的视图返回了 XML，它会被.xml 扩展名所触发。这样，同样的处理器方法就可以支持多个视图了。比如说，诸如 http://[host_name]/[app-name]/members.pdf 这样的请求会以 PDF 文档的形式返回同样的信息，http://[host_name]/[app- name]/members.html 这样的请求会返回 HTML 内容，而 http://[host_name]/[app-name]/members.xml 这样的请求则会对 REST 请求返回 XML。

那么，如果请求没有 URL 扩展名呢？比如说 http://[host_name]/[app-name]/members。这也完全取决于 Spring MVC 的视图解析。针对这个目的，Spring MVC 支持一种叫做内容协商的过程，可以根据请求的扩展名或是 HTTP 头来确定视图。

由于 REST 服务请求中 HTTP 头通常是 Accept: application/xml 这种形式，因此 Spring MVC 可以配置为使用内容协商来确定向这种请求返回 XML（REST）负载，即便请求没有扩展名也可以。这样，也可以通过诸如 HTML、PDF 与 XLS 等格式来发送没有扩展名的请求，这些格式都基于 HTTP 头。3-7 节介绍了内容协商的相关内容。

使用@ResponseBody 生成 XML

使用 MarshallingView 生成 XML 是生成结果的一种方式；不过，如果希望同样的数据（一个 Member 对象列表）有多种表示（比如说 JSON），那么再添加一个视图可能会很麻烦。相反，可以使用 Spring MVC HttpMessageConverters 将对象转换为用户请求的表示形式。如下代码展示了对 RestMemberController 所做的修改：

```
@Controller
public class RestMemberController {
...
    @RequestMapping("/members")
    @ResponseBody
    public Members getRestMembers() {
        Members members = new Members();
        members.addMembers(memberService.findAll());
        return members;
    }
}
```

第一个变化是为控制器方法添加了@ResponseBody。该注解告诉 Spring MVC，方法的结果应该作为响应体来使用。由于需要 XML，因此编组是由 Spring 所提供的 Jaxb2RootElementHttpMessageConverter 类完成的。第二个变化是，由于使用了@ResponseBody 注解，现在就无须视图名了，只需返回 Members 对象即可。

提示：如果使用的是 Spring 4 或更高版本，除了使用@ResponseBody 来注解方法外，还可以使用@RestController 代替@Controller 来注解控制器，结果是相同的。如果一个控制器中有多个方法，那么这种方式会非常方便。

进行了上述这些修改后，可以清理一下配置了，因为不再需要 MarshallingView 与 Jaxb2Marshaller 了。

```
package com.apress.springrecipes.court.web.config;

import org.springframework.context.annotation.Bean;
import org.springframework.context.annotation.ComponentScan;
import org.springframework.context.annotation.Configuration;
import org.springframework.web.servlet.config.annotation.EnableWebMvc;

@Configuration
@EnableWebMvc
@ComponentScan(basePackages = "com.apress.springrecipes.court")
public class CourtRestConfiguration {}
```

部署好应用后，向 http://localhost:8080/court/members.xml 发出成员请求，这会得到与之前相同的结果。

```
<?xml version="1.0" encoding="UTF-8" standalone="yes"?>
<members>
    <member>
        <email>marten@deinum.biz</email>
        <name>Marten Deinum</name>
        <phone>00-31-1234567890</phone>
    </member>
    <member>
        <email>john@doe.com</email>
        <name>John Doe</name>
        <phone>1-800-800-800</phone>
    </member>
    <member>
        <email>jane@doe.com</email>
        <name>Jane Doe</name>
        <phone>1-801-802-803</phone>
    </member>
</members>
```

使用@PathVariable 限制结果

REST 服务请求常常会有参数，用于限制或是过滤服务的负载。比如说，http://[host_name]/ [app-name] /member/353/形式的请求可用于检索成员 353 的信息。http://[host_name]/[app-name]/reservations/07-07-2010/形式的请求可用于检索 2010 年 7 月 7 日的预订信息。

在 Spring 中要想使用参数来构建 REST 服务，可以使用@PathVariable 注解。根据 Spring MVC 的约定，@PathVariable 注解用作处理器方法的输入参数，这样就可以在处理器方法体中使用它了。如下代码片段展示了使用@PathVariable 注解的用作 REST 服务的处理器方法：

```
import org.springframework.web.bind.annotation.PathVariable;

@Controller
public class RestMemberController {
...
    @RequestMapping("/member/{memberid}")
    @ResponseBody
    public Member getMember(@PathVariable("memberid") long memberID) {
        return memberService.find(memberID);
    }
}
```

注意，@RequestMapping 值包含了{memberid}。由{ }所包围的值用于标识 URL 参数是变量。此外，处理器方法定义中使用了输入参数@PathVariable("memberid") long memberID。这最后一个声明将构成 URL 一部分的 memberid 值关联起来，并将其赋给名为 memberID 的变量，可以在处理器方法中使用它。因此，/member/353/与/member/777/形式的 REST 端点将会由最后一个处理器方法所处理，其 memberID 变量将会分别被赋予值 353 和 777。在处理器方法中，针对成员 353 与 777 进行了查询（通过 memberID 变量），并作为 REST 服务的负载返回。

对 http://localhost:8080/court/member/2 的请求会生成 ID 为 2 的成员的 XML 表示。

```
<?xml version="1.0" encoding="UTF-8" standalone="yes"?>
<member>
    <email>john@doe.com</email>
    <name>John Doe</name>
    <phone>1-800-800-800</phone>
</member>
```

除了支持{ }符号外，还可以使用通配符（*）来定义 REST 端点。当设计团队选择使用具有表现力的 URL（通常叫做整洁的 URL）或是使用搜索引擎优化（SEO）技术让 REST URL 对搜索引擎更加友好时会使用到它。如下代码片段展示了如何使用通配符来声明一个 REST 服务：

```
@RequestMapping("/member/*/{memberid}")
@ResponseBody
public Member getMember(@PathVariable("memberid") long memberID) { ... }
```

在该示例中，添加的通配符不会对 REST 服务所执行的逻辑产生任何影响。不过，它会匹配/member/John+Smith/353/与/member/Mary+Jones/353/形式的端点请求，这对最终用户的可读性和 SEO 来说具有很重要的影响。

值得一提的是，可以在 REST 端点的处理器方法的定义中使用数据绑定。如下代码片段展示了使用数据绑定的 REST 服务的声明：

```
@InitBinder
public void initBinder(WebDataBinder binder) {
    SimpleDateFormat dateFormat = new SimpleDateFormat("yyyy-MM-dd");
    binder.registerCustomEditor(Date.class, new CustomDateEditor(dateFormat, false));
}

@RequestMapping("/reservations/{date}")
public void getReservation(@PathVariable("date") Date resDate) { ... }
```

在该示例中，http://[host_name]/[app-name]/reservations/07-07-2010/形式的请求会被最后一个处理器方法所

匹配，值 07-07-2010 会被传递给处理器方法（作为变量 resDate），可以通过它过滤 REST Web Service 的负载。

使用 ResponseEntity 通知客户端

检索 Member 实例的端点要么返回一个有效的成员，要么什么都不返回。这两种情况都会使得请求向客户端返回 HTTP 响应码 200，表示 OK。不过，这可能并不是用户所期待的。在处理资源时，应该通知用户找不到资源这一事实。理想情况下，你希望返回 HTTP 响应码 404，表示“not found”。如下代码片段展示了修改后的 getMember 方法：

```
package com.apress.springrecipes.court.web;

import org.springframework.http.HttpStatus;
import org.springframework.http.ResponseEntity;
...

@Controller
public class RestMemberController {
...
    @RequestMapping("/member/{memberid}")
    @ResponseBody
    public ResponseEntity<Member> getMember(@PathVariable("memberid") long memberID) {
        Member member = memberService.find(memberID);
        if (member != null) {
            return new ResponseEntity<Member>(member, HttpStatus.OK);
        }
        return new ResponseEntity(HttpStatus.NOT_FOUND);
    }
}
```

方法的返回值修改为 ResponseEntity<Member>。Spring MVC 中的 ResponseEntity 类会将结果体对象与 HTTP 状态码包装起来。当找到了 Member 时，它会返回 HttpStatus.OK，对应于 HTTP 状态码 200。如果没有结果，那就会返回 HttpStatus.NOT_FOUND，对应于 HTTP 状态码 404，表示“not found”。

4-2 使用 REST 服务发布 JSON

问题提出

你想要通过 Spring 发布基于 JavaScript Object Notation（JSON）的 REST 服务。

解决方案

JSON 已经成为 REST 服务最受人喜欢的负载格式。然而，与大多数 REST 服务负载（依赖于 XML 标记）不同的是，JSON 中的内容是基于 JavaScript 语言的特殊符号。对于本节来说，除了依赖于 Spring 的 REST 支持外，你还会用到 MappingJackson2JsonView 类（作为 Spring 的一部分）来简化 JSON 内容的发布。

> **注意：**MappingJackson2JsonView 类依赖于 Jackson JSON 处理器库版本 2。如果使用 Maven 或是 Gradle，只需在项目构建文件中添加 Jackson 库依赖即可。

如果你的 Spring 应用集成了 Ajax 设计，那么你会发现自己已经在设计 REST 服务，且该服务发布 JSON 作为其负载。这主要是因为浏览器中的处理能力有限。虽然浏览器可以从发布 XML 负载的 REST 服务中处理并抽取出信息，但效率很低。通过以 JSON 格式来传递负载（基于 JavaScript 语言，浏览器提供了针对它的原生解释器），数据的处理与抽取变得更加高效。与 RSS 和 Atom 源（它们是标准）不同的是，JSON 并没有需要遵循的特定结构（除了语法），稍后将会对此进行介绍。这样，JSON 元素的负载结构需要与负责应用程序 Ajax 设计的团队成员共同协作方能确定下来。

解释说明

首先要确定想作为 JSON 负载发布的信息。该信息可能位于 RDBMS 或是文本文件中（通过 JDBC 或是 ORM 访问），或是 Spring bean 以及其他类型组件的一部分。介绍如何获取这些信息已经超出了本节的范围，所以假设你可以通过恰当的方式访问到它。倘若不熟悉 JSON，如下代码展示了该格式的一个片段：

```
{
    "glossary": {
        "title": "example glossary",
        "GlossDiv": {
            "title": "S",
            "GlossList": {
                "GlossEntry": {
                    "ID": "SGML",
                    "SortAs": "SGML",
                    "GlossTerm": "Standard Generalized Markup Language",
                    "Acronym": "SGML",
                    "Abbrev": "ISO 8879:1986",
                    "GlossDef": {
                        "para": "A meta-markup language, used to create markup
                        languages such as DocBook.",
                        "GlossSeeAlso": ["GML", "XML"]
                    },
                    "GlossSee": "markup"
                }
            }
        }
    }
}
```

可以看到，JSON 负载包含了文本与分隔符，如{、}、[、]、:与"。我们不会详细介绍每个分隔符，不过可以说的是，相较于 XML 类型格式来说，这种类型的语法可以让 JavaScript 引擎更轻松地访问和操纵数据。

使用 MappingJackson2JsonView 生成 JSON

由于 4-1 节与 4-3 节介绍了如何通过 REST 服务发布数据，因此我们直接来介绍 Spring MVC 控制器为了实现这一过程所需的实际的处理器方法。

```
@RequestMapping("/members")
public String getRestMembers(Model model) {

    Members members = new Members();
    members.addMembers(memberService.findAll());
    model.addAttribute("members", members);
    return "jsonmembertemplate";
}
```

你可能会注意到，它与 4-1 节所介绍的控制器方法颇为相似。唯一的差别在于为视图返回了不同的名字。这里所返回的视图名 jsonmembertemplate 是不同的，它会映射到 MappingJackson2JsonView 视图。需要在配置类中配置该视图。

```
@Configuration
@EnableWebMvc
@ComponentScan(basePackages = "com.apress.springrecipes.court")
public class CourtRestConfiguration {
    ...
    @Bean
    public View jsonmembertemplate() {
        MappingJackson2JsonView view = new MappingJackson2JsonView();
        view.setPrettyPrint(true);
        return view;
    }
}
```

MappingJackson2JsonView 视图使用 Jackson2 库实现了对象与 JSON 之间的相互转换。它使用了一个 Jackson2 ObjectMapper 实例进行转换。当请求发送给 http://localhost:8080/court/members.json 时，控制器方法会被调用并返回一个 JSON 表示。

```
{
    "members" : {
        "members" : [ {
```

```
            "name" : "Marten Deinum",
            "phone" : "00-31-1234567890",
            "email" : "marten@deinum.biz"
        }, {
            "name" : "John Doe",
            "phone" : "1-800-800-800",
            "email" : "john@doe.com"
        }, {
            "name" : "Jane Doe",
            "phone" : "1-801-802-803",
            "email" : "jane@doe.com"
        } ]
    }
}
```

实际上，每次调用/members 或/members.*时都会返回该 JSON（比如说，/members.xml 也会生成 JSON）。下面将 4-1 节中的方法与视图添加到该控制器中。

```
@Controller
public class RestMemberController {
...
    @RequestMapping(value="/members", produces=MediaType.APPLICATION_XML_VALUE)
    public String getRestMembersXml(Model model) {
        Members members = new Members();
        members.addMembers(memberService.findAll());
        model.addAttribute("members", members);
        return "xmlmembertemplate";
    }

    @RequestMapping(value="/members", produces= MediaType.APPLICATION_JSON_VALUE)
    public String getRestMembersJson(Model model) {
        Members members = new Members();
        members.addMembers(memberService.findAll());
        model.addAttribute("members", members);
        return "jsonmembertemplate";
    }
}
```

现在有了 getMembersXml 方法和 getMembersJson 方法；这两个方法基本一样，只不过返回不同的视图名。注意@RequestMapping 注解中的 produces 属性，它用于确定调用哪个方法：/members.xml 会生成 XML，而/members.json 则会生成 JSON。

虽然这种方式可以使用，不过对于企业应用来说，要为所支持的不同视图类型复制所有方法并不是一种灵活的解决方案。可以创建一个辅助方法来减少重复，但依然需要大量样板代码，因为@RequestMapping 注解存在差别。

使用@ResponseBody 生成 JSON

使用 MappingJackson2JsonView 生成 JSON 是一种生成结果的方式；不过，如上一节所述，这么做会很麻烦，特别是在想要支持多种视图类型的情况下更是如此。相反，你可以使用 Spring MVC 的 HttpMessageConverters 将对象转换为用户所请求的表示形式。如下代码展示了对 RestMemberController 所做的修改：

```
@Controller
public class RestMemberController {
...
    @RequestMapping("/members")
    @ResponseBody
    public Members getRestMembers() {
        Members members = new Members();
        members.addMembers(memberService.findAll());
        return members;
    }
}
```

第一个变化是为控制器方法添加了@ResponseBody 注解。该注解告诉 Spring MVC，方法的结果应该作为响应体来使用。由于需要 JSON，因此编组是由 Spring 所提供的 Jackson2JsonMessageConverter 类完成的。第二个变化是，由于使用了@ResponseBody 注解，现在就无须视图名了，只需返回 Members 对象即可。

■ **提示：** 如果使用的是 Spring 4 或更高版本，除了使用@ResponseBody 来注解方法外，还可以使用@RestController 代替@Controller 来注解控制器，结果是相同的。如果一个控制器中有多个方法，那么这种方式会非常方便。

进行了上述这些修改后，可以清理一下配置，因为不再需要 MappingJackson2JsonView 了。

```
package com.apress.springrecipes.court.web.config;

import org.springframework.context.annotation.Bean;
import org.springframework.context.annotation.ComponentScan;
import org.springframework.context.annotation.Configuration;
import org.springframework.web.servlet.config.annotation.EnableWebMvc;

@Configuration
@EnableWebMvc
@ComponentScan(basePackages = "com.apress.springrecipes.court")
public class CourtRestConfiguration {}
```

当部署了应用程序并从 http://localhost:8080/court/members.json 请求成员时，它将给出与之前相同的结果。

```
{
    "members" : {
        "members" : [ {
            "name" : "Marten Deinum",
            "phone" : "00-31-1234567890",
            "email" : "marten@deinum.biz"
        }, {
            "name" : "John Doe",
            "phone" : "1-800-800-800",
            "email" : "john@doe.com"
        }, {
            "name" : "Jane Doe",
            "phone" : "1-801-802-803",
            "email" : "jane@doe.com"
        } ]
    }
}
```

你可能会注意到 RestMemberController 和 CourtRestConfiguration 现在与 4-1 节中的一样。当调用 http://localhost:8080/court/members.xml 时，会得到 XML。

在没有添加任何额外配置的情况下这是如何做到的呢？Spring MVC 会检测类路径；当自动检测到 JAXB 2、Jackson 与 Rome（参见 4-4 节）时，它会针对可用的技术注册恰当的 HttpMessageConverter。

使用 GSON 生成 JSON

到现在为止，你一直在使用 Jackson 从对象生成 JSON；另一个流行的库是 GSON，Spring 对其提供了现成的支持。要想使用 GSON，要将其添加到类路径中（取代 Jackson），接下来就可以使用它生成 JSON 了。

如果使用 Maven，请添加如下依赖：

```
<dependency>
    <groupId>com.google.code.gson</groupId>
    <artifactId>gson</artifactId>
    <version>2.8.0</version>
</dependency>
```

如果使用 Gradle，请添加如下依赖：

```
compile 'com.google.code.gson:gson:2.8.0'
```

就像使用 Jackson 一样，这是使用 GSON 实现 JSON 序列化所需做的全部工作。如果开启应用并调用 http://localhost:8080/court/members.json，你依然会接收到 JSON，不过现在是通过 GSON 完成的。

4-3 使用 Spring 访问 REST 服务

问题提出

你想在 Spring 应用中访问第三方（如 Google、Yahoo!，或是其他业务方）的 REST 服务并使用其负载。

解决方案

在 Spring 应用中访问第三方的 REST 服务时，需要使用 Spring RestTemplate 类。RestTemplate 类的设计与其他很多 Spring *Template 类（如 JdbcTemplate 和 JmsTemplate 等）遵循了同样的原则，通过默认行为提供了一种简化的方式来执行冗长的任务。这意味着在 Spring 应用中对调用 REST 服务并使用其返回的负载的处理都实现了流水化作业。

解释说明

在介绍 RestTemplate 类的细节之前，我们有必要了解一下 REST 服务的生命周期，这样就能知晓 RestTemplate 类所执行的实际工作了。探索 REST 服务的生命周期最好通过浏览器来实现，在工作站中打开钟爱的浏览器来开始吧。首先，我们需要一个 REST 服务端点。这里重用 4-2 节中所创建的端点。该端点位于 http://localhost:8080/court/members.xml（或.json）。如果在浏览器中加载这个 REST 服务端点，浏览器就会执行一个 GET 请求，这是 REST 服务所支持的诸多流行的 HTTP 请求之一。在加载 REST 服务时，浏览器会展示出对应的负载，如下所示：

```
<?xml version="1.0" encoding="UTF-8" standalone="yes"?>
<members>
    <member>
        <email>marten@deinum.biz</email>
        <name>Marten Deinum</name>
        <phone>00-31-1234567890</phone>
    </member>
    <member>
        <email>john@doe.com</email>
        <name>John Doe</name>
        <phone>1-800-800-800</phone>
    </member>
    <member>
        <email>jane@doe.com</email>
        <name>Jane Doe</name>
        <phone>1-801-802-803</phone>
    </member>
</members>
```

上面的负载代表一个格式良好的 XML 片段，它符合大多数 REST 服务的响应。负载的实际含义完全取决于 REST 服务。在该示例中，XML 标签（<members>、<member>等）是自己定义的，每个 XML 标签中的字符数据则表示与 REST 服务请求相关的信息。

REST 服务消费者（比如说你自己）需要知道 REST 服务的负载结构（有时叫做词汇表），这样才能恰当地处理其信息。虽然 REST 服务依赖于自定义的词汇表，但一系列 REST 服务通常会依赖于标准化的词汇表（如 RSS），这可以统一处理 REST 服务负载。此外，值得注意的是，有些 REST 服务提供了 Web Application Description Language（WADL）契约来简化负载的发现与消费。

在通过浏览器了解了 REST 服务的生命周期后，现在来看看如何使用 Spring RestTemplate 类将 REST 服务负载纳入到 Spring 应用中。考虑到 RestTemplate 类设计的目的就是为了调用 REST 服务，因此其主要的方法都会紧密绑定到 REST 底层，即 HTTP 协议的方法：HEAD、GET、POST、PUT、DELETE 与 OPTIONS。表 4-1 所示为 RestTemplate 类所支持的主要方法。

表 4-1 基于 HTTP 协议请求方法的 RestTemplate 类的方法

方法	说明
headForHeaders(String, Object...)	执行 HTTP HEAD 操作
getForObject(String, Class, Object...)	执行 HTTP GET 操作并将结果作为给定的类型返回
getForObject(String, Class, Object...)	执行 HTTP GET 操作并返回一个 ResponseEntity
postForLocation(String, Object, Object...)	执行 HTTP POST 操作并返回 location 头的值
postForObject(String, Object, Class, Object...)	执行 HTTP POST 操作并将结果作为给定的类型返回
postForEntity(String, Object, Class, Object...)	执行 HTTP POST 操作并返回一个 ResponseEntity
put(String, Object, Object...)	执行 HTTP PUT 操作
delete(String, Object...)	执行 HTTP DELETE 操作
optionsForAllow(String, Object...)	执行 HTTP OPTIONS 操作
execute(String, HttpMethod, RequestCallback, ResponseExtractor, Object...)	可执行除了 CONNECT 外的任何 HTTP 操作

从表 4-1 中可以看到，RestTemplate 类中的方法都会使用一系列 HTTP 协议方法作为前缀，包括 HEAD、GET、POST、PUT、DELETE 和 OPTIONS。一方面，execute 方法作为一种通用目的的方法可以执行任何 HTTP 操作，包括不太常用的 HTTP 协议的 TRACE 方法，但 CONNECT 方法不行，因为 execute 方法所用的底层 HttpMethod 枚举不支持它。注意，到目前为止，REST 服务中最常用的 HTTP 方法是 GET，因为它表示一种获取信息的安全操作（即它并不会修改任何数据）。另一方面，诸如 PUT、POST 与 DELETE 等 HTTP 方法都会修改提供者的信息，这使得它们不太受 REST 服务提供者的支持。如果需要修改数据，很多提供者倾向于使用 SOAP 协议，这是 REST 服务的一种替代机制。

在了解了 RestTemplate 类的方法后，现在可以调用之前使用浏览器调用的相同的 REST 服务，只不过这次使用的是 Spring 框架的 Java 代码。如下代码所展示的类会访问 REST 服务并将其内容发送给 System.out：

```
package com.apress.springrecipes.court;

import org.springframework.web.client.RestTemplate;

public class Main {

    public static void main(String[] args) throws Exception {
        final String uri = "http://localhost:8080/court/members.json";
        RestTemplate restTemplate = new RestTemplate();
        String result = restTemplate.getForObject(uri, String.class);
        System.out.println(result);
    }
}
```

■ **警告：**一些 REST 服务提供者会根据请求方的不同限制对其数据源的访问。通常会根据请求中的数据（如 HTTP 头或是 IP 地址）来拒绝访问。因此，根据环境的不同，提供者可能会返回拒绝访问响应，即便数据源可能在另外一个环境中使用（比如说，你可能在浏览器中能够访问 REST 服务，但在 Spring 应用中访问同样的数据源就会得到拒绝访问响应）。这取决于 REST 提供者预设的使用条款。

上述代码的第一行声明了一条 import 语句以便在类中使用 RestTemplate 类。首先，需要创建一个 RestTemplate 类的实例。接下来，调用了 RestTemplate 类的 getForObject 方法，表 4-1 对此已经进行了介绍，它用于执行 HTTP GET 操作，就像通过浏览器所执行的获取 REST 服务的负载那样。这个方法有两个重要的方面：响应与参数。

调用 getForObject 方法的响应会被赋给一个 String 对象。这意味着你在浏览器中所看到的该 REST 服务（即 XML 结构）的输出现在被赋给了一个 String。即便从来没有在 Java 中处理过 XML，你也会认识到以 Java

String 的形式抽取和处理数据并不是那么容易的事情。换言之，有其他类要比 String 对象更适合处理 XML 数据（以及 REST 服务的负载）。现在请记住，本章的其他小节会介绍如何更好地抽取和操纵从 REST 服务获取到的数据。

传递给 getForObject 方法的参数包含了实际的 REST 服务端点。第一个参数对应于 URL（即端点）声明。注意，URL 与使用浏览器时所用的 URL 是一样的。

当执行后，输出与使用浏览器时是一样的，只不过现在会打印到控制台上。

从参数化 URL 中获取数据

上一节介绍了如何通过调用 URI 来获取数据，不过如果 URI 需要参数该怎么做呢？你并不想将参数硬编码到 URL 中。借助于 RestTemplate 类，可以通过占位符来使用 URL，在执行时，这些占位符会被替换为实际值。占位符是通过 { 和 } 定义的，就像请求映射一样（参见 4-1 节与 4-2 节）。

http://localhost:8080/court/member/{memberId}就是一种参数化 URI。要想调用该方法，你需要为占位符传递一个值。这可以使用一个 Map 并将其作为 RestTemplate 类的 getForObject 方法的第三个参数传递进去。

```
public class Main {

    public static void main(String[] args) throws Exception {
        final String uri = "http://localhost:8080/court/member/{memberId}";
        Map<String, String> params = new HashMap<>();
        params.put("memberId", "1");
        RestTemplate restTemplate = new RestTemplate();
        String result = restTemplate.getForObject(uri, String.class, params );
        System.out.println(result);
    }
}
```

上述代码片段使用了 HashMap 类（位于 Java 集合框架中）并使用相应的 REST 服务参数创建了一个实例，后面会将其传递给 RestTemplate 类的 getForObject 方法。向各种 RestTemplate 方法传递一系列 String 参数或是单个 Map 参数所得到的结果是一样的。

将数据作为映射对象返回

相较于返回一个 String 在应用中使用，还可以重用 Members 与 Member 类来映射结果。相较于将 String.class 作为第二个参数，还可以传递 Members.class，这时响应就会映射到该类上。

```
package com.apress.springrecipes.court;

import com.apress.springrecipes.court.domain.Members;
import org.springframework.web.client.RestTemplate;

public class Main {

    public static void main(String[] args) throws Exception {
        final String uri = "http://localhost:8080/court/members.xml";
        RestTemplate restTemplate = new RestTemplate();
        Members result = restTemplate.getForObject(uri, Members.class);
        System.out.println(result);
    }
}
```

当控制器使用了被@ResponseBody 标记的方法时，RestTemplate 类会使用同样的 HttpMessageConverter 基础设施。JAXB 2（以及 Jackson）会被自动检测到，因此映射到与 JAXB 对应的对象就是非常简单的事情了。

4-4 发布 RSS 与 Atom 源

问题提出

你想在 Spring 应用中发布 RSS 或是 Atom 源。

解决方案

RSS 与 Atom 源已经成为一种流行的发布信息的方式。访问这些类型的源是通过 REST 服务的方式来进行的，这意味着构建 REST 服务是发布 RSS 与 Atom 源的前置条件。除了依赖于 Spring 的 REST 支持外，使用专门用于处理 RSS 与 Atom 源的第三方库也会十分方便。这使得 REST 服务发布这种类型的 XML 负载变得更加轻松。针对这一点，你需要使用 Project Rome，这是一个开源库。

> ■ **提示**：虽然 RSS 与 Atom 源常常被看作是新闻源，不过它们已经越过了仅仅提供新闻的初始使用场景。时至今日，RSS 与 Atom 源用于以跨平台的方式（比如说使用 XML）发布与博客、天气、旅行等相关的信息。因此，如果需要发布能以跨平台的方式所访问的各种信息，RSS 或 Atom 源的方式会是一种完美选择，因为它们的使用非常广泛（很多应用都支持它们，很多开发者也知道它们的结构）。

解释说明

首先需要确定作为 RSS 或 Atom 新闻源发布的信息。该信息可能位于 RDBMS 或是文本文件中（通过 JDBC 或是 ORM 访问），或是 Spring bean 以及其他类型组件的一部分。介绍如何获取这些信息已经超出了本节的范围，所以假设大家可以通过恰当的方式访问到它。得到了想要发布的信息后，需要将其结构化为 RSS 或 Atom 源，这正是 Project Rome 的用武之地。

倘若你不熟悉 Atom 源的结构，如下代码展示了该格式的一个片段：

```xml
<?xml version="1.0" encoding="utf-8"?>
<feed xmlns="http://www.w3.org/2005/Atom">
    <title>Example Feed</title>
    <link href="http://example.org/"/>
    <updated>2010-08-31T18:30:02Z</updated>
    <author>
        <name>John Doe</name>
    </author>
    <id>urn:uuid:60a76c80-d399-11d9-b93C-0003939e0af6</id>
    <entry>
        <title>Atom-Powered Robots Run Amok</title>
        <link href="http://example.org/2010/08/31/atom03"/>
        <id>urn:uuid:1225c695-cfb8-4ebb-aaaa-80da344efa6a</id>
        <updated>2010-08-31T18:30:02Z</updated>
        <summary>Some text.</summary>
    </entry>
</feed>
```

如下代码展示了 RSS 源结构的一个片段：

```xml
<?xml version="1.0" encoding="utf-8"?>
<rss version="2.0">
    <channel>
        <title>RSS Example</title>
        <description>This is an example of an RSS feed</description>
        <link>http://www.example.org/link.htm</link>
        <lastBuildDate>Mon, 28 Aug 2006 11:12:55 -0400 </lastBuildDate>
        <pubDate>Tue, 31 Aug 2010 09:00:00 -0400</pubDate>
        <item>
            <title>Item Example</title>
            <description>This is an example of an Item</description>
            <link>http://www.example.org/link.htm</link>
            <guid isPermaLink="false"> 1102345</guid>
            <pubDate>Tue, 31 Aug 2010 09:00:00 -0400</pubDate>
        </item>
    </channel>
</rss>
```

从上面两个代码片段可以看到，RSS 与 Atom 源只不过是 XML 负载而已，它们依赖于一系列元素来发布信息。虽然深入介绍 RSS 或 Atom 源结构需要整本书的篇幅，不过这两种格式拥有不少共性，简单来说有下面这些。

- 它们都有元数据部分，用来描述源的内容（比如说 Atom 格式的<author>与<title>元素，以及 RSS 格式的<description>与<pubDate>元素）。
- 它们都有重复元素用来描述信息（比如说 Atom 源格式的<entry>元素，以及 RSS 源格式的<item>元素）。此外，每个重复元素也拥有自己的元素用来进一步描述信息。
- 它们都有多个版本。RSS 版本有 0.90、0.91 Netscape、0.91 Userland、0.92、0.93、0.94、1.0 与 2.0。Atom 版本有 0.3 和 1.0。可以通过 Project Rome 根据 Java 代码中的信息（比如说 Strings、Maps 等结构）创建源的元数据部分、重复元素，以及之前提到的各种版本。

在了解了 RSS 与 Atom 源的结构以及本节中 Project Rome 所扮演的角色后，现在就来看看负责向最终用户展示源的 Spring MVC 控制器：

```
//FINAL
package com.apress.springrecipes.court.web;

import com.apress.springrecipes.court.feeds.TournamentContent;
import org.springframework.stereotype.Controller;
import org.springframework.ui.Model;
import org.springframework.web.bind.annotation.RequestMapping;
import java.util.ArrayList;
import java.util.Date;
import java.util.List;

@Controller
public class FeedController {
    @RequestMapping("/atomfeed")
    public String getAtomFeed(Model model) {
        List<TournamentContent> tournamentList = new ArrayList<>();
        tournamentList.add(TournamentContent.of("ATP", new Date(), "Australian Open",
        "www.australianopen.com"));
        tournamentList.add(TournamentContent.of("ATP", new Date(), "Roland Garros",
        "www.rolandgarros.com"));
        tournamentList.add(TournamentContent.of("ATP", new Date(), "Wimbledon",
        "www.wimbledon.org"));
        tournamentList.add(TournamentContent.of("ATP", new Date(), "US Open",
        "www.usopen.org"));
        model.addAttribute("feedContent", tournamentList);

        return "atomfeedtemplate";
    }

    @RequestMapping("/rssfeed")
    public String getRSSFeed(Model model) {
        List<TournamentContent> tournamentList;
        tournamentList = new ArrayList<TournamentContent>();
        tournamentList.add(TournamentContent.of("FIFA", new Date(), "World Cup",
        "www.fifa.com/worldcup/"));
        tournamentList.add(TournamentContent.of("FIFA", new Date(), "U-20 World Cup",
        "www.fifa.com/u20worldcup/"));
        tournamentList.add(TournamentContent.of("FIFA", new Date(), "U-17 World Cup",
        "www.fifa.com/u17worldcup/"));
        tournamentList.add(TournamentContent.of("FIFA", new Date(), "Confederations Cup",
        "www.fifa.com/confederationscup/"));
        model.addAttribute("feedContent", tournamentList);

        return "rssfeedtemplate";
    }
}
```

这个 Spring MVC 控制器有两个处理器方法：一个是 getAtomFeed()，映射到 URL http://[host_name]/[app-name]/atomfeed；另一个是 getRSSFeed()，映射到 URL http://[host_name]/ [app-name]/rssfeed。

每个处理器方法都定义了一个 TournamentContent 对象的列表，TournamentContent 对象背后的类是个

POJO。接下来，该列表会被赋给处理器方法的 Model 对象，这样返回的视图就可以访问它了。处理器方法所返回的逻辑视图分别是 atomfeedtemplate 与 rssfeedtemplate。这些逻辑视图按照如下方式定义在 Spring 配置类中：

```
package com.apress.springrecipes.court.web.config;

import com.apress.springrecipes.court.feeds.AtomFeedView;
import com.apress.springrecipes.court.feeds.RSSFeedView;
import org.springframework.context.annotation.Bean;
import org.springframework.context.annotation.ComponentScan;
import org.springframework.context.annotation.Configuration;
import org.springframework.web.servlet.HandlerMapping;
import org.springframework.web.servlet.config.annotation.EnableWebMvc;
import org.springframework.web.servlet.handler.BeanNameUrlHandlerMapping;

@Configuration
@EnableWebMvc
@ComponentScan(basePackages = "com.apress.springrecipes.court")
public class CourtRestConfiguration {

    @Bean
    public AtomFeedView atomfeedtemplate() {
        return new AtomFeedView();
    }

    @Bean
    public RSSFeedView rssfeedtemplate() {
        return new RSSFeedView();
    }
...
}
```

可以看到，每个逻辑视图都会映射到一个类上。每个类负责实现必要的逻辑来构建 Atom 或 RSS 视图。回忆一下第 3 章，当时使用了同样的方式（使用类）来实现 PDF 与 Excel 视图。

对于 Atom 与 RSS 视图来说，Spring 自带了两个构建于 Project Rome 之上的类。这两个类分别是 AbstractAtomFeedView 与 AbstractRssFeedView。它们提供了构建 Atom 与 RSS 源的基础，无须处理每种格式的细节信息。

如下代码展示了实现 AbstractAtomFeedView 类的 AtomFeedView 类，用于支持 atomfeedtemplate 逻辑视图：

```
package com.apress.springrecipes.court.feeds;

import com.rometools.rome.feed.atom.Content;
import com.rometools.rome.feed.atom.Entry;
import com.rometools.rome.feed.atom.Feed;
import org.springframework.web.servlet.view.feed.AbstractAtomFeedView;

import javax.servlet.http.HttpServletRequest;
import javax.servlet.http.HttpServletResponse;
import java.util.List;
import java.util.Map;
import java.util.stream.Collectors;

public class AtomFeedView extends AbstractAtomFeedView {

    @Override
    protected void buildFeedMetadata(Map model, Feed feed, HttpServletRequest request) {
        feed.setId("tag:tennis.org");
        feed.setTitle("Grand Slam Tournaments");
        List<TournamentContent> tournamentList = (List<TournamentContent>)
        model.get("feedContent");

        feed.setUpdated(tournamentList.stream().map(TournamentContent::getPublicationDate).
```

```
            sorted().findFirst().orElse(null));

    }

    @Override
    protected List buildFeedEntries(Map model, HttpServletRequest request,
    HttpServletResponse response)
        throws Exception {
        List<TournamentContent> tournamentList = (List<TournamentContent>) model.
        get("feedContent");
        return tournamentList.stream().map(this::toEntry).collect(Collectors.toList());
    }

    private Entry toEntry(TournamentContent tournament) {
        Entry entry = new Entry();
        String date = String.format("%1$tY-%1$tm-%1$td", tournament.getPublicationDate());
        entry.setId(String.format("tag:tennis.org,%s:%d", date, tournament.getId()));
        entry.setTitle(String.format("%s - Posted by %s", tournament.getName(), tournament.
        getAuthor()));
        entry.setUpdated(tournament.getPublicationDate());

        Content summary = new Content();
        summary.setValue(String.format("%s - %s", tournament.getName(), tournament.
        getLink()));
        entry.setSummary(summary);
        return entry;
    }
}
```

首先注意到的是，除了实现 Spring 框架所提供的 AbstractAtomFeedView 类之外，这个类还从 com.sun.syndication.feed.atom 包中导入了几个 Project Rome 类。这样，接下来只需为从 AbstractAtomFeedView 类所继承的两个方法（buildFeedMetadata 与 buildFeedEntries）提供源的实现细节。

buildFeedMetadata 有 3 个输入参数：一个 Map 对象，表示用于构建源的数据（即在处理器方法中所指定的数据，该示例中就是一个 TournamentContent 对象列表）；一个基于 Project Rome 类的 Feed 对象，用于操纵源本身；一个 HttpServletRequest 对象，用于操纵 HTTP 请求。

在 buildFeedMetadata 方法中，你会看到调用了几次 Feed 对象的 setter 方法（即 setId、setTitle 与 setUpdated）。其中两个调用使用了硬编码的字符串，另外一个调用是根据遍历源的数据（即 Map 对象）之后所确定的值来进行的。所有这些调用都代表了 Atom 源的元数据信息的赋值。

> **注意：** 如果想为 Atom 源的元数据赋予更多的值以及指定特定的 Atom 版本（默认版本是 Atom 1.0），请查阅 Project Rome 的 API。

buildFeedEntries 方法也有 3 个输入参数：一个 Map 对象，表示用于构建源的数据（即在处理器方法中所指定的数据，该示例中就是一个 TournamentContent 对象列表）；一个 HttpServletRequest 对象，用于操纵 HTTP 请求；一个 HttpServletResponse 对象，用于操纵 HTTP 响应。值得注意的是，buildFeedEntries 方法返回一个对象列表，该示例中就是一个基于 Project Rome 类的 Entry 对象列表，它包含了一个 Atom 源的重复元素。

在 buildFeedEntries 方法中，通过 Map 对象获取到处理器方法中所分配的 feedContent 对象。接下来会创建一个空的 Entry 对象列表。然后对 feedContent 对象进行一次循环，它包含了 TournamentContent 对象列表的列表，针对每个元素会创建一个 Entry 对象，并将其赋给顶层的 Entry 对象列表。当循环完成后，方法会返回一个填充好的 Entry 对象列表。

> **注意：** 如果想为 Atom 源的重复元素赋予更多的值，请查阅 Project Rome 的 API。

部署好上面的类后，除了之前介绍的 Spring MVC 控制器外，访问 URL http://[host_name]/[app-name]/atomfeed.atom（或 http://[host_name]/atomfeed.xml）会得到如下响应：

```
<?xml version="1.0" encoding="UTF-8"?>
<feed xmlns="http://www.w3.org/2005/Atom">
    <title>Grand Slam Tournaments</title>
    <id>tag:tennis.org</id>
    <updated>2017-04-19T01:32:52Z</updated>
    <entry>
        <title>Australian Open - Posted by ATP</title>
        <id>tag:tennis.org,2017-04-19:5</id>
        <updated>2017-04-19T01:32:52Z</updated>
        <summary>Australian Open - www.australianopen.com</summary>
    </entry>
    <entry>
        <title>Roland Garros - Posted by ATP</title>
        <id>tag:tennis.org,2017-04-19:6</id>
        <updated>2017-04-19T01:32:52Z</updated>
        <summary>Roland Garros - www.rolandgarros.com</summary>
    </entry>
    <entry>
        <title>Wimbledon - Posted by ATP</title>
        <id>tag:tennis.org,2017-04-19:7</id>
        <updated>2017-04-19T01:32:52Z</updated>
        <summary>Wimbledon - www.wimbledon.org</summary>
    </entry>
    <entry>
        <title>US Open - Posted by ATP</title>
        <id>tag:tennis.org,2017-04-19:8</id>
        <updated>2017-04-19T01:32:52Z</updated>
        <summary>US Open - www.usopen.org</summary>
    </entry>
</feed>
```

现在将注意力从构建 RSS 源的 Spring MVC 控制器转向到剩下的处理器方法 getRSSFeed 上，你会看到过程与之前介绍的构建 Atom 源很类似。处理器方法也创建了一个 TournamentContent 对象列表，接下来赋给处理器方法的 Model 对象，这样就可以在返回的视图中访问它了。不过，现在返回的逻辑视图对应于 rssfeedtemplate。如前所述，该逻辑视图会映射到名为 RssFeedView 的类上。

如下代码展示了 RssFeedView 类，它实现了 AbstractRssFeedView 类：

```
package com.apress.springrecipes.court.feeds;

import com.rometools.rome.feed.rss.Channel;
import com.rometools.rome.feed.rss.Item;
import org.springframework.web.servlet.view.feed.AbstractRssFeedView;

import javax.servlet.http.HttpServletRequest;
import javax.servlet.http.HttpServletResponse;
import java.util.List;
import java.util.Map;
import java.util.stream.Collectors;

public class RSSFeedView extends AbstractRssFeedView {

    @Override
    protected void buildFeedMetadata(Map model, Channel feed, HttpServletRequest request) {
        feed.setTitle("World Soccer Tournaments");
        feed.setDescription("FIFA World Soccer Tournament Calendar");
        feed.setLink("tennis.org");

        List<TournamentContent> tournamentList = (List<TournamentContent>) model.
        get("feedContent");
        feed.setLastBuildDate(tournamentList.stream().map( TournamentContent::getPublication
        Date).sorted().findFirst().orElse(null) );
    }
```

```
    @Override
    protected List<Item> buildFeedItems(Map model, HttpServletRequest request,
    HttpServletResponse response)
        throws Exception {
        List<TournamentContent> tournamentList = (List<TournamentContent>) model.
        get("feedContent");

        return tournamentList.stream().map(this::toItem).collect(Collectors.toList());
    }

    private Item toItem(TournamentContent tournament) {
        Item item = new Item();
        item.setAuthor(tournament.getAuthor());
        item.setTitle(String.format("%s - Posted by %s", tournament.getName(),
        tournament.getAuthor()));
        item.setPubDate(tournament.getPublicationDate());
        item.setLink(tournament.getLink());
        return item;
    }
}
```

首先注意到的是，除了实现 Spring 框架所提供的 AbstractRssFeedView 类之外，这个类还从 com.sun.syndication.feed.rss 包中导入了几个 Project Rome 类。这样，接下来只需为从 AbstractRssFeedView 类所继承的 buildFeedMetadata 与 buildFeedItems 这两个方法提供源的实现细节。buildFeedMetadata 方法本质上类似于在构建 Atom 源时所用的同名方法。注意到 buildFeedMetadata 方法会基于 Project Rome 类操纵一个 Channel 对象而不是 Feed 对象，Channel 对象用于构建 RSS 源，而 Feed 对象则用于构建 Atom 源。对 Channel 对象所调用的 setter 方法（如 setTitle、setDescription 与 setLink）表示对 RSS 源的元数据信息进行赋值。buildFeedItems 方法不同于 Atom 实现中的 buildFeedEntries，之所以这样命名是因为 Atom 源的重复元素叫做 entries，而 RSS 源的重复元素叫做 items。除了命名约定外，其逻辑是类似的。

在 buildFeedItems 方法中会访问 Map 对象以获取在处理器方法中所分配的 feedContent 对象。接下来会创建一个空的 Item 对象列表。然后会对 feedContent 对象执行循环，它包含了一个 TournamentContent 对象的列表，针对每个元素都会创建一个 Item 对象，并将其赋给 Item 对象的顶层列表。循环完成后，方法会返回一个填充好的 Item 对象列表。

注意：如果想为 RSS 源的元数据和重复元素部分赋予更多的值以及指定特定的 RSS 版本（默认版本是 RSS 2.0），请查阅 Project Rome 的 API。

部署好上面的类后，除了之前介绍的 Spring MVC 控制器外，访问 URL http://[host_name]/ rssfeed.rss（或 http://[host_name]/rssfeed.xml）会得到如下响应：

```
<?xml version="1.0" encoding="UTF-8"?>
<rss version="2.0">
    <channel>
        <title>World Soccer Tournaments</title>
        <link>tennis.org</link>
        <description>FIFA World Soccer Tournament Calendar</description>
        <lastBuildDate>Wed, 19 Apr 2017 01:32:31 GMT</lastBuildDate>
        <item>
            <title>World Cup - Posted by FIFA</title>
            <link>www.fifa.com/worldcup/</link>
            <pubDate>Wed, 19 Apr 2017 01:32:31 GMT</pubDate>
            314861_4_EnFIFA</author>
        </item>
        <item>
            <title>U-20 World Cup - Posted by FIFA</title>
            <link>www.fifa.com/u20worldcup/</link>
            <pubDate>Wed, 19 Apr 2017 01:32:31 GMT</pubDate>
            314861_4_EnFIFA</author>
```

```
        </item>
        <item>
            <title>U-17 World Cup - Posted by FIFA</title>
            <link>www.fifa.com/u17worldcup/</link>
            <pubDate>Wed, 19 Apr 2017 01:32:31 GMT</pubDate>
            314861_4_EnFIFA</author>
        </item>
        <item>
            <title>Confederations Cup - Posted by FIFA</title>
            <link>www.fifa.com/confederationscup/</link>
            <pubDate>Wed, 19 Apr 2017 01:32:31 GMT</pubDate>
            314861_4_EnFIFA</author>
        </item>
    </channel>
</rss>
```

小结

本章介绍了如何通过 Spring 开发和访问 REST 服务。REST 服务是紧紧绑定到 Spring MVC 上的，通过控制器来分发向 REST 服务所发出的请求，同时还可以访问第三方 REST 服务以在应用中使用其信息。

本章介绍了 REST 服务如何使用 Spring MVC 控制器中的注解，这包括标识服务端点的@RequestMapping，以及指定访问参数以过滤请求负载的@PathVariable。此外，本章还介绍了 Spring 的 XML 编组器，如 Jaxb2Marshaller，它可以将应用对象转换为 XML 并以 REST 服务负载的形式输出。还介绍了 Spring 的 RestTemplate 类以及它是如何支持各种 HTTP 方法的，包括 HEAD、GET、POST、PUT 与 DELETE——可以通过这些方法从 Spring 应用上下文中直接访问第三方 REST 服务并执行操作。

最后，本章介绍了如何通过 Project Rome API 在 Spring 应用中发布 Atom 与 RSS 源。

第 5 章

Spring MVC：异步处理

当 Servlet API 发布时，主流的容器实现采取的做法是每个请求一个线程。这意味着线程会被阻塞，直到请求处理完成且响应发送给客户端。

不过，早期并不像现在这样有那么多设备连接到互联网上。由于设备数量不断增加，要处理的 HTTP 请求数也急剧攀升，因为这种攀升，对于大量 Web 应用来说，让线程阻塞就变得不那么灵活了。随着 Servlet 3 规范的出现，可以异步处理 HTTP 请求并释放最初处理 HTTP 请求的线程了。新的线程会在后台运行，当结果生成后，将其写入客户端。如果处理正确，这些都可以在 Servlet 3.1 兼容的 servlet 容器中以非阻塞的方式进行。当然，所有用到的资源也需要是非阻塞的。

在过去的几年间，反应式编程的使用越来越多，随着 Spring 5 的发布，我们可以编写反应式 Web 应用了。反应式 Spring 项目使用了 Project Reactor（与 Spring 一样，它也是由 Pivotal 维护的）作为 Reactive Streams API 的实现。完整介绍反应式编程已经超出了本书的讨论范围，不过一言以蔽之，这是一种非阻塞函数式编程方式。

传统上，在处理 Web 应用时会有一个请求；HTML 会在服务器端渲染，然后发送给客户端。近几年，渲染 HTML 的工作转移到了客户端，通信不再通过 HTML 进行，而是通过向客户端返回 JSON、XML 或是其他表示形式来实现。这依旧是一种请求响应循环，不过它是从客户端通过 XMLHttpRequest 对象以异步调用的方式驱动的。客户端与服务器端之间还有其他的通信方式；可以通过服务器端发送事件来实现服务器端到客户端的单向通信，对于全双工通信来说，可以使用 WebSocket 协议。

5-1 使用控制器与 TaskExecutor 异步处理请求

问题提出

为了减少 servlet 容器的负载，你想要异步处理请求。

解决方案

当请求到来时，它会被同步处理，这会阻塞 HTTP 请求处理线程。响应保持打开状态，等待着被写入。当调用需要花一些时间完成时，这么做是有意义的。相较于阻塞线程来说，你可以在后台进行处理，当完成时向用户返回值。

解释说明

如 3-1 节所述，Spring MVC 支持从方法返回各种类型。除了返回类型外，表 5-1 中列出的类型会以异步方式进行处理。

表 5-1　　异步返回类型

类型	说明
DeferredResult	从另外一个线程生成的异步结果
ListenableFuture<?>	从另外一个线程生成的异步结果；等价于 DeferredResult

续表

类型	说明
CompletableStage<?> / CompletableFuture<?>	从另外一个线程生成的异步结果；等价于 DeferredResult
Callable<?>	计算完成后对所生成的结果进行异步计算
ResponseBodyEmitter	可用于异步将多个对象写入到响应中
SseEmitter	可用于异步写入一个服务器端发送的事件
StreamingResponseBody	可用于异步写入到 OutputStream 中

通用的异步返回类型可以保存控制器的任何返回类型，包括添加到模型中的对象、视图的名字，甚至是 ModelAndView 对象。

配置异步处理

要想使用 Spring MVC 的异步处理特性，首先需要将其开启。异步请求处理的支持已经被添加到 Servlet 3.0 规范中。要想开启它，需要让过滤器与 servlet 异步执行。为了做到这一点，可以在注册过滤器或 servlet 时调用 setAsyncSupported()方法。

在编写 WebApplicationInitializer 时，需要这样做：

```
public class CourtWebApplicationInitializer implements WebApplicationInitializer {

    public void onStartup(ServletContext ctx) {

        DispatcherServlet servlet = new DispatcherServlet();
        ServletRegistration.Dynamic registration = ctx.addServlet("dispatcher", servlet);
        registration.setAsyncSupported(true);
    }

}
```

■ **注意**：*在进行异步处理时，应用中的所有 servlet 过滤器与 servlet 都需要将该属性设为 true，否则异步处理将无法使用！*

幸好 Spring 帮我们做到了，在将 AbstractAnnotationConfigDispatcherServletInitializer 作为父类时，针对注册的 DispatcherServlet 与过滤器，该属性默认就会启用。要想改变这一点，请重写 isAsyncSupported()并实现逻辑以确定到底是开启还是关闭该属性。

根据需要，可能还需要在 MVC 配置中配置一个 AsyncTaskExecutor 并装配它。

```
package com.apress.springrecipes.court.config;

import org.springframework.context.annotation.Bean;
import org.springframework.context.annotation.Configuration;
import org.springframework.core.task.AsyncTaskExecutor;
import org.springframework.scheduling.concurrent.ThreadPoolTaskExecutor;
import org.springframework.web.servlet.config.annotation.AsyncSupportConfigurer;
import org.springframework.web.servlet.config.annotation.WebMvcConfigurationSupport;

@Configuration
public class AsyncConfiguration extends WebMvcConfigurationSupport {

    @Override
    protected void configureAsyncSupport(AsyncSupportConfigurer configurer) {
        configurer.setDefaultTimeout(5000);
        configurer.setTaskExecutor(mvcTaskExecutor());
    }

    @Bean
    public ThreadPoolTaskExecutor mvcTaskExecutor() {
```

```
        ThreadPoolTaskExecutor taskExecutor = new ThreadPoolTaskExecutor();
        taskExecutor.setThreadGroupName("mvc-executor");
        return taskExecutor;
    }
}
```

要想配置异步处理，需要重写 WebMvcConfigurationSupport 的 configureAsyncSupport 方法；重写后的方法可以访问 AsyncSupportConfigurer，并且可以设置 defaultTimeout 与 AsyncTaskExecutor 的值。超时时间设为了 5 秒，执行器则使用了 ThreadPoolTaskExecutor（参见 2-23 节）。

编写异步控制器

编写控制器并异步处理请求是很简单的，只需修改控制器的处理器方法的返回类型即可。假设 ReservationService.query 调用需要花费较长时间，但我们不想让服务器因此而阻塞。

使用 Callable

如下代码展示了 Callable 的使用：

```
package com.apress.springrecipes.court.web;

import com.apress.springrecipes.court.Delayer;
import com.apress.springrecipes.court.domain.Reservation;
import com.apress.springrecipes.court.service.ReservationService;
import org.springframework.stereotype.Controller;
import org.springframework.ui.Model;
import org.springframework.web.bind.annotation.GetMapping;
import org.springframework.web.bind.annotation.PostMapping;
import org.springframework.web.bind.annotation.RequestMapping;
import org.springframework.web.bind.annotation.RequestParam;

import java.util.List;
import java.util.concurrent.Callable;

@Controller
@RequestMapping("/reservationQuery")
public class ReservationQueryController {

    private final ReservationService reservationService;

    public ReservationQueryController(ReservationService reservationService) {
        this.reservationService = reservationService;
    }

    @GetMapping
    public void setupForm() {}

    @PostMapping
    public Callable<String> sumbitForm(@RequestParam("courtName") String courtName, Model
    model) {
        return () -> {
            List<Reservation> reservations = java.util.Collections.emptyList();
            if (courtName != null) {
                Delayer.randomDelay(); // Simulate a slow service
                reservations = reservationService.query(courtName);
            }
            model.addAttribute("reservations", reservations);
            return "reservationQuery";
        };
    }
}
```

看一下 submitForm 方法，现在它返回了一个 Callable<String>而非直接返回 String。在新构建的

Callable<String>中有一个随机等待时间用于模拟调用 query 方法前的延迟。

现在在预订时，你会在日志中看到如下内容：

```
2017-06-20 10:37:04,836 [nio-8080-exec-2] DEBUG o.s.w.c.request.async.WebAsyncManager       :
Concurrent handling starting for POST [/court/reservationQuery]
2017-06-20 10:37:04,838 [nio-8080-exec-2] DEBUG o.s.web.servlet.DispatcherServlet           :
Leaving response open for concurrent processing
2017-06-20 10:37:09,954 [mvc-executor-1 ] DEBUG o.s.w.c.request.async.WebAsyncManager :
Concurrent result value [reservationQuery] - dispatching request to resume processing
2017-06-20 10:37:09,959 [nio-8080-exec-3] DEBUG o.s.web.servlet.DispatcherServlet           :
DispatcherServlet with name 'dispatcher' resumed processing POST request for [/court/
reservationQuery]
```

注意，请求处理是由某个线程来处理的（这里是 nio-8080-exec-2），它会被释放，接下来另一个线程会处理并返回结果（这里是 mvc-executor-1）。最后，请求会再次分发给 DispatcherServlet 以在另外一个线程上处理结果。

使用 DeferredResult

相较于 Callable<String>，你还可以使用 DeferredResult<String>。在使用 DeferredResult 时，需要构建出该类的一个实例，提交一个任务进行异步处理，在该任务中通过 setResult 方法填充 DeferredResult 的结果。当出现异常时，可以将异常传递给 DeferredResult 的 setErrorResult 方法。

```
package com.apress.springrecipes.court.web;

import com.apress.springrecipes.court.Delayer;
import com.apress.springrecipes.court.domain.Reservation;
import com.apress.springrecipes.court.service.ReservationService;

import org.springframework.core.task.AsyncTaskExecutor;
import org.springframework.core.task.TaskExecutor;
import org.springframework.stereotype.Controller;
import org.springframework.ui.Model;
import org.springframework.web.bind.annotation.GetMapping;
import org.springframework.web.bind.annotation.PostMapping;
import org.springframework.web.bind.annotation.RequestMapping;
import org.springframework.web.bind.annotation.RequestParam;
import org.springframework.web.context.request.async.DeferredResult;

import java.util.List;

@Controller
@RequestMapping("/reservationQuery")
public class ReservationQueryController {

    private final ReservationService reservationService;
    private final TaskExecutor taskExecutor;

    public ReservationQueryController(ReservationService reservationService,
                                      AsyncTaskExecutor taskExecutor) {
        this.reservationService = reservationService;
        this.taskExecutor = taskExecutor;
    }

    @GetMapping
    public void setupForm() {}

    @PostMapping
    public DeferredResult<String> sumbitForm(@RequestParam("courtName") String courtName,
    Model model) {
        final DeferredResult<String> result = new DeferredResult<>();

        taskExecutor.execute(() -> {
```

```
            List<Reservation> reservations = java.util.Collections.emptyList();
            if (courtName != null) {
                Delayer.randomDelay(); // Simulate a slow service
                reservations = reservationService.query(courtName);
            }
            model.addAttribute("reservations", reservations);
            result.setResult("reservationQuery");
        });
        return result;
    }
}
```

方法现在返回了一个 DeferredResult<String>，它依然是待渲染的视图名。实际的结果是通过一个 Runnable 来设置的，它会被传递给注入进来的 TaskExecutor 的 execute 方法。返回 DeferredResult 与返回 Callable 之间的主要差别在于前者需要创建自己的线程（或是委托给 TaskExecutor），后者则不需要。

使用 CompletableFuture

修改方法的签名，使之返回一个 CompletableFuture<String>，使用 TaskExecutor 异步执行代码。

```
package com.apress.springrecipes.court.web;

import com.apress.springrecipes.court.Delayer;
import com.apress.springrecipes.court.domain.Reservation;
import com.apress.springrecipes.court.service.ReservationService;

import org.springframework.core.task.TaskExecutor;
import org.springframework.stereotype.Controller;
import org.springframework.ui.Model;
import org.springframework.web.bind.annotation.GetMapping;
import org.springframework.web.bind.annotation.PostMapping;
import org.springframework.web.bind.annotation.RequestMapping;
import org.springframework.web.bind.annotation.RequestParam;

import java.util.List;
import java.util.concurrent.CompletableFuture;

@Controller
@RequestMapping("/reservationQuery")
public class ReservationQueryController {

    private final ReservationService reservationService;
    private final TaskExecutor taskExecutor;

    public ReservationQueryController(ReservationService reservationService,
    TaskExecutor taskExecutor) {
        this.reservationService = reservationService;
        this.taskExecutor = taskExecutor;
    }

    @GetMapping
    public void setupForm() {}

    @PostMapping
    public CompletableFuture<String> sumbitForm(@RequestParam("courtName")
    String courtName, Model model) {

        return CompletableFuture.supplyAsync(() -> {
            List<Reservation> reservations = java.util.Collections.emptyList();
            if (courtName != null) {
                Delayer.randomDelay(); // Simulate a slow service
                reservations = reservationService.query(courtName);
            }
            model.addAttribute("reservations", reservations);
```

```
            return "reservationQuery";
        }, taskExecutor);
    }
}
```

在调用 supplyAsync（如果需要 void，那么可以使用 runAsync）时，你提交一个任务并得到一个 CompletableFuture。这里使用了 supplyAsync 方法，它接收一个 Supplier 和一个 Executor，这样就可以重用 TaskExecutor 进行异步处理了。如果使用只接收一个 Supplier 的 supplyAsync 方法，那就会使用 JVM 中默认的 fork/join 池。

当返回 CompletableFuture 时，可以充分利用它的特性，比如说组合并链式调用多个 CompletableFuture 实例。

使用 ListenableFuture

Spring 提供了 ListenableFuture 接口，它是一个 Future 实现，当 Future 完成时会执行回调。要想创建 ListenableFuture，你需要向 AsyncListenableTaskExecutor 提交一个任务，它会返回一个 ListenableFuture。之前配置的 ThreadPoolTaskExecutor 就是 AsyncListenableTaskExecutor 接口的一个实现。

```
// FINAL
package com.apress.springrecipes.court.web;

import com.apress.springrecipes.court.Delayer;
import com.apress.springrecipes.court.domain.Reservation;
import com.apress.springrecipes.court.service.ReservationService;
import org.springframework.core.task.AsyncListenableTaskExecutor;
import org.springframework.stereotype.Controller;
import org.springframework.ui.Model;
import org.springframework.util.concurrent.ListenableFuture;
import org.springframework.web.bind.annotation.GetMapping;
import org.springframework.web.bind.annotation.PostMapping;
import org.springframework.web.bind.annotation.RequestMapping;
import org.springframework.web.bind.annotation.RequestParam;

import java.util.List;

@Controller
@RequestMapping("/reservationQuery")
public class ReservationQueryController {

    private final ReservationService reservationService;
    private final AsyncListenableTaskExecutor taskExecutor;

    public ReservationQueryController(ReservationService reservationService,
    AsyncListenableTaskExecutor taskExecutor) {
        this.reservationService = reservationService;
        this.taskExecutor = taskExecutor;
    }

    @GetMapping
    public void setupForm() {}

    @PostMapping
    public ListenableFuture<String> sumbitForm(@RequestParam("courtName")
    String courtName, Model model) {

        return taskExecutor.submitListenable(() -> {
            List<Reservation> reservations = java.util.Collections.emptyList();
            if (courtName != null) {
                Delayer.randomDelay(); // Simulate a slow service
                reservations = reservationService.query(courtName);
            }
```

```
                model.addAttribute("reservations", reservations);
                return "reservationQuery";
            });
        }
    }
```

这里使用 submitListenable 方法向 taskExecutor 提交了一个任务，这会返回一个 ListenableFuture，可用作方法的结果。

你可能想知道对于所创建的 ListenableFuture 来说，成功回调与失败回调在哪里。Spring MVC 会将 ListenableFuture 适配为一个 DeferredResult，当成功完成时会调用 DeferredResult.setResult，当出现错误时会调用 DeferredResult.setErrorResult。通过 Spring 自带的 HandlerMethodReturnValueHandler 实现帮我们完成了所有的处理；在该示例中是由 DeferredResultMethodReturnValueHandler 处理的。

5-2 使用响应写入器

问题提出

你有一个服务或多个调用，希望以块的形式将响应发送给客户端。

解决方案

使用 ResponseBodyEmitter（或 SseEmitter）以块的形式发送响应。

解释说明

Spring 支持通过 HttpMessageConverter 基础设施以普通对象的形式来写入对象，结果是发送给客户端的块（或流）列表。除了对象，还可以以事件的形式发送，这就是所谓的服务端发送事件。

在一个响应中发送多个结果

Spring MVC 提供了一个名为 ResponseBodyEmitter 的类，相较于单个结果（如视图名或 ModelAndView）来说，如果希望向客户端返回多个对象，那么它就能大显身手。在发送对象时，会通过 HttpMessageConverter 将对象转换为一个结果（参见 4-2 节）。要想使用 ResponseBodyEmitter，你需要从请求处理方法中将其返回。

修改 ReservationQueryController 的 find 方法，使其返回一个 ResponseBodyEmitter，并将结果逐一发送给客户端。

```
// FINAL
package com.apress.springrecipes.court.web;

import com.apress.springrecipes.court.Delayer;
import com.apress.springrecipes.court.domain.Reservation;
import com.apress.springrecipes.court.service.ReservationService;
import org.springframework.core.task.TaskExecutor;
import org.springframework.http.MediaType;
import org.springframework.stereotype.Controller;
import org.springframework.ui.Model;
import org.springframework.web.bind.annotation.GetMapping;
import org.springframework.web.bind.annotation.PostMapping;
import org.springframework.web.bind.annotation.RequestMapping;
import org.springframework.web.bind.annotation.RequestParam;
import org.springframework.web.servlet.mvc.method.annotation.ResponseBodyEmitter;

import java.io.IOException;
import java.util.Collection;
import java.util.List;
import java.util.concurrent.Callable;

@Controller
@RequestMapping("/reservationQuery")
public class ReservationQueryController {

    private final ReservationService reservationService;
```

```
    private final TaskExecutor taskExecutor;

    ...

    @GetMapping(params = "courtName")
    public ResponseBodyEmitter find(@RequestParam("courtName") String courtName) {
        final ResponseBodyEmitter emitter = new ResponseBodyEmitter();
        taskExecutor.execute(() -> {
            Collection<Reservation> reservations = reservationService.query(courtName);
            try {
                for (Reservation reservation : reservations) {
                    emitter.send(reservation);
                }
                emitter.complete();
            } catch (IOException e) {
                emitter.completeWithError(e);
            }
        });
        return emitter;
    }
}
```

上述代码中，首先创建了一个 ResponseBodyEmitter，并在方法最后返回。接下来执行一个任务，它会使用 ReservationService.query 方法查询预订信息。该调用的所有结果会通过 ResponseBodyEmitter 的 send 方法逐一返回。当所有对象都发送后需要调用 complete()方法，这样负责发送响应的线程就会完成请求并开始处理接下来的响应。当出现异常且需要通知用户时，就需要调用 completeWithError。异常会经过 Spring MVC 常规的处理（参见 3-8 节），之后响应完成。

当使用 httpie 或 curl 等工具时，调用 URL http://localhost:8080/court/reservationQuerycourtName=='Tennis #1'会生成如图 5-1 所示的结果。结果是块状的，同时状态码为 200（OK）。

如果想要修改状态码或是添加自定义的头，可以将 ResponseBodyEmitter 包装到 ResponseEntity 中，它可以自定义返回码、头等信息（参见 4-1 节）。

```
@GetMapping(params = "courtName")
public ResponseEntity<ResponseBodyEmitter> find(@RequestParam("courtName") String  courtName)
{
    final ResponseBodyEmitter emitter = new ResponseBodyEmitter();
    ....
    return ResponseEntity.status(HttpStatus.I_AM_A_TEAPOT)
        .header("Custom-Header", "Custom-Value")
        .body(emitter);
}
```

现在的状态码变成了 418，还包含了一个自定义头，如图 5-2 所示。

图 5-1　块状结果

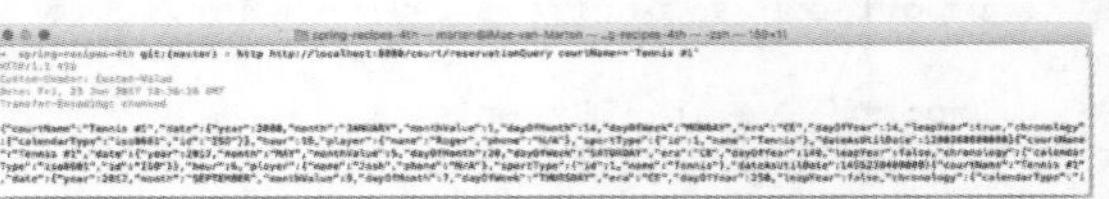

图 5-2　修改后的块状结果

以事件的形式发送多个结果

ResponseBodyEmitter 的一个替代者是 SseEmitter，它可以通过服务器端发送事件的方式将事件从服务器端发送给客户端。服务器端事件指的是从服务器端发向客户端的消息，它们有一个 text/event-stream 内容类型头。它们是非常轻量级的，可以定义 4 个字段，如表 5-2 所示。

表 5-2　　服务端发送事件可使用的字段

字段	说明
id	事件 ID
event	事件类型
data	事件数据
retry	事件流的重连时间

要想从请求处理方法中发送事件，你需要创建一个 SseEmitter 实例并将其从请求处理方法中返回。接下来，使用 send 方法将各个元素发送给客户端。

```
@GetMapping(params = "courtName")
public SseEmitter find(@RequestParam("courtName") String courtName) {
    final SseEmitter emitter = new SseEmitter();
    taskExecutor.execute(() -> {
        Collection<Reservation> reservations = reservationService.query(courtName);
        try {
            for (Reservation reservation : reservations) {
                Delayer.delay(125);
                emitter.send(reservation);
            }
            emitter.complete();
        } catch (IOException e) {
            emitter.completeWithError(e);
        }
    });
    return emitter;
}
```

注意：每次将条目发送给客户端时会有一个延时，这样就可以看到进来的不同事件了。实际代码中不要这么做。

当使用 curl 等工具调用 URL http://localhost:8080/court/reservationQuery courtName=='Tennis #1'时，你会看到事件一个接一个地到来，如图 5-3 所示。

图 5-3　服务端发送事件的结果

注意，Content-Type 头有一个 text/event-stream 值来标识你得到了一个事件流。你可以保持流的打开状态，持续接收事件通知。此外，每个被写入的对象都转换为了 JSON；这是通过 HttpMessageConverter 实现的，就像使用普通的 ResponseBodyEmitter 一样。每个对象都写在了 data 标签中，作为事件数据。

如果想要向事件添加更多信息（换句话说，添加表 5-2 中的其他字段），可以使用 SseEventBuilder。要想获得 SseEventBuilder 的实例，可以调用 SseEmitter 的 event()工厂方法。我们使用它来填入 id 字段以及 Reservation 的 hash code。

```
@GetMapping(params = "courtName")
public SseEmitter find(@RequestParam("courtName") String courtName) {
```

```
        final SseEmitter emitter = new SseEmitter();
        taskExecutor.execute(() -> {
            Collection<Reservation> reservations = reservationService.query(courtName);
            try {
                for (Reservation reservation : reservations) {
                    Delayer.delay(120);
                    emitter.send(emitter.event().id(String.valueOf(reservation.hashCode())).
                    data(reservation));
                }
                emitter.complete();
            } catch (IOException e) {
                emitter.completeWithError(e);
            }
        });
        return emitter;
    }
```

现在，当使用 curl 等工具调用 URL http://localhost:8080/court/reservationQuery courtName== 'Tennis #1'，你会看到事件逐一到来，它们包含了 id 与 data 字段。

5-3　使用异步拦截器

问题提出

Servlet API 所定义的 servlet 过滤器可以在每个 Web 请求被 servlet 处理前后对其进行预先处理与事后处理。你想在 Spring 的 Web 应用上下文中配置类似于过滤器的功能来充分利用容器的特性。

此外，有时你想对 Spring MVC 处理器所处理的 Web 请求进行预先处理与事后处理，并操纵这些处理器所返回的模型属性，然后再将其传递给视图。

解决方案

Spring MVC 可以通过处理器拦截器来拦截 Web 请求以进行预先处理和事后处理。处理器拦截器是配置在 Spring 的 Web 应用上下文中的，这样它们可以使用容器特性并引用容器中声明的 bean。处理器拦截器可以针对特定的 URL 映射进行注册，这样它就只会拦截与某些 URL 匹配的请求了。

如 3-3 节所介绍的，Spring 提供了 HandlerInterceptor 接口，它包含了 3 个需要你实现的回调方法：preHandle()、postHandle()与 afterCompletion()。前两个方法分别是在请求被处理器处理之前和之后调用的。第 2 个方法还可以访问到返回的 ModelAndView 对象，这样就可以操纵里面的模型属性了。最后一个方法会在所有请求处理都完成后调用（即视图渲染之后）。

对于异步处理来说，Spring 提供了 AsyncHandlerInterceptor，它包含了另外一个需要你实现的回调方法：afterConcurrentHandlingStarted。当异步处理开始时会调用这个方法，而不是 postHandle 或是 afterCompletion。当异步处理完成时，正常的流程会再次调用。

解释说明

3-3 节中创建了一个 MeasurementInterceptor，它会根据每个请求处理器来度量每个 Web 请求的处理时间，并将其添加到 ModelAndView 中。下面对其进行修改，使其记录请求与响应的处理时间，包括用于处理请求的线程。

```
package com.apress.springrecipes.court.web;

import org.springframework.web.servlet.AsyncHandlerInterceptor;
import org.springframework.web.servlet.ModelAndView;

import javax.servlet.http.HttpServletRequest;
import javax.servlet.http.HttpServletResponse;
public class MeasurementInterceptor implements AsyncHandlerInterceptor {

    public static final String START_TIME = "startTime";

    public boolean preHandle(HttpServletRequest request, HttpServletResponse response,
```

```
        Object handler) throws Exception {

        if (request.getAttribute(START_TIME) == null) {
            request.setAttribute(START_TIME, System.currentTimeMillis());
        }
        return true;
    }

    public void postHandle(HttpServletRequest request, HttpServletResponse response,
        Object handler, ModelAndView modelAndView) throws Exception {

        long startTime = (Long) request.getAttribute(START_TIME);
        request.removeAttribute(START_TIME);
        long endTime = System.currentTimeMillis();
        System.out.println("Response-Processing-Time: " + (endTime - startTime) + "ms.");
        System.out.println("Response-Processing-Thread: " + Thread.currentThread().
        getName());
    }

    @Override
    public void afterConcurrentHandlingStarted(HttpServletRequest request,
        HttpServletResponse response, Object handler) throws Exception {

        long startTime = (Long) request.getAttribute(START_TIME);
        request.setAttribute(START_TIME, System.currentTimeMillis());
        long endTime = System.currentTimeMillis();

        System.out.println("Request-Processing-Time: " + (endTime - startTime) + "ms.");
        System.out.println("Request-Processing-Thread: " + Thread.currentThread().
        getName());
    }
}
```

在该拦截器的 preHandle()方法中，记录下了开始时间并将其保存到请求属性中。该方法应该返回 true，以便 DispatcherServlet 能够继续请求处理。否则，DispatcherServlet 会认为该方法已经处理完了请求，这样 DispatcherServlet 就会直接将响应返回给用户。接下来，在 afterConcurrentHandlingStarted 中，获取到注册的时间并计算开始异步处理所花费的时间。之后，重置了开始时间并将请求处理时间和线程打印到控制台上。

接下来，在 postHandle()方法中从请求属性中获取到开始时间并将其与当前时间进行比较。然后计算出总的间隔时间，并将该时间和当前的线程名一同打印到控制台上。

要想注册拦截器，你需要修改 AsyncConfiguration，它是在 5-1 节中创建的。你需要让其实现 WebMvcConfigurer 并重写 addInterceptors 方法。该方法可以访问到 InterceptorRegistry，你可以通过它添加拦截器。修改后的类如下所示：

```
package com.apress.springrecipes.court.config;

import com.apress.springrecipes.court.web.MeasurementInterceptor;
import org.springframework.context.annotation.Bean;
import org.springframework.context.annotation.Configuration;
import org.springframework.scheduling.concurrent.ThreadPoolTaskExecutor;
import org.springframework.web.servlet.config.annotation.AsyncSupportConfigurer;
import org.springframework.web.servlet.config.annotation.InterceptorRegistry;
import org.springframework.web.servlet.config.annotation.WebMvcConfigurer;

import java.util.concurrent.TimeUnit;

@Configuration
public class AsyncConfiguration implements WebMvcConfigurer {

    ...

    @Override
```

```
    public void addInterceptors(InterceptorRegistry registry) {
        registry.addInterceptor(new MeasurementInterceptor());
    }
}
```

现在运行应用，当处理完请求后，日志如图 5-4 所示。

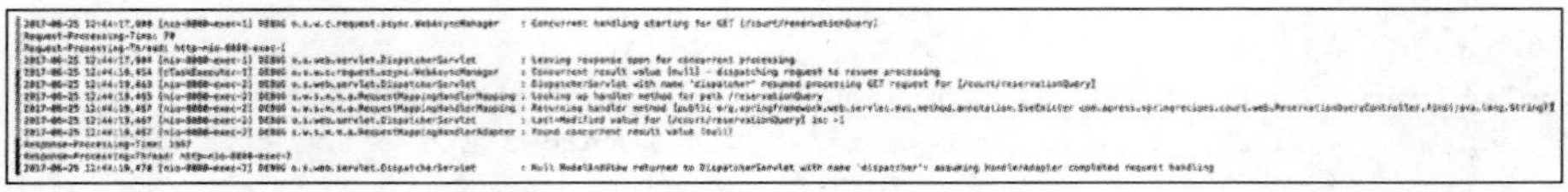

图 5-4 请求/响应处理时间

> **注意**：微软的浏览器（Internet Explorer 和 Edge）不支持服务器端发送事件。要想在这些浏览器中能够正常使用这些事件，你需要使用一个 polyfill 来添加支持。

5-4 使用 WebSocket

问题提出

你想在 Web 上实现客户端到服务器之间的双向通信。

解决方案

使用 WebSocket 实现客户端到服务器的通信，反之亦然。与 HTTP 不同，WebSocket 技术提供了一种全双工通信。

解释说明

对 WebSocket 技术的全面介绍超出了本节的范围；不过值得一提的是，HTTP 与 WebSocket 技术之间的关系实际上并不多。对于 WebSocket 技术来说，HTTP 唯一的用处在于初始的握手会用到它。这会将连接从普通的 HTTP 升级到 TCP socket 连接。

配置 WebSocket 支持

要想启用 WebSocket 技术，只需向配置类添加@EnableWebSocket 注解即可。

```
@Configuration
@EnableWebSocket
public class WebSocketConfiguration {}
```

对于 WebSocket 引擎的进一步配置，可以通过添加一个 ServletServerContainerFactoryBean 对象来配置缓冲区大小、超时时间等属性。

```
@Bean
public ServletServerContainerFactoryBean configureWebSocketContainer() {
    ServletServerContainerFactoryBean factory = new ServletServerContainerFactoryBean();
    factory.setMaxBinaryMessageBufferSize(16384);
    factory.setMaxTextMessageBufferSize(16384);
    factory.setMaxSessionIdleTimeout(TimeUnit.MINUTES.convert(30, TimeUnit.MILLISECONDS));
    factory.setAsyncSendTimeout(TimeUnit.SECONDS.convert(5, TimeUnit.MILLISECONDS));
    return factory;
}
```

这会将文本与二进制缓冲区大小设为 16KB，将 asyncSendTimeout 设为 5 秒，将会话超时时间设为 30 分钟。

创建 WebSocketHandler

要想创建 WebSocket 消息与生命周期事件（握手、连接建立等），你需要创建一个 WebSocketHandler 并将其注册到端点 URL 上。

WebSocketHandler 定义了 5 个方法（见表 5-3），如果直接实现该接口，那么就需要实现全部这 5 个方法。不过，Spring 提供了一个很棒的类层次体系，你可以使用它。在编写自定义的处理器时，继承

TextWebSocketHandler 或 BinaryWebSocketHandler 就足够了，顾名思义，它们分别可以处理文本消息与二进制消息。

表 5-3 WebSocketHandler 方法

方法	说明
afterConnectionEstablished	当 WebSocket 连接打开并准备使用时调用
handleMessage	当 WebSocket 消息到达处理器时调用
handleTransportError	当出现错误时调用
afterConnectionClosed	当 WebSocket 连接关闭时调用
supportsPartialMessages	当处理器支持部分消息时会调用。如果设为 true，那么 WebSocket 消息就会在多次调用中到达

创建 EchoHandler，让它继承 TextWebSocketHandler，然后实现 afterConnectionEstablished 与 handleMessage 方法。

```
package com.apress.springrecipes.websocket;

import org.springframework.web.socket.CloseStatus;
import org.springframework.web.socket.TextMessage;
import org.springframework.web.socket.WebSocketSession;
import org.springframework.web.socket.handler.TextWebSocketHandler;

public class EchoHandler extends TextWebSocketHandler {

    @Override
    public void afterConnectionEstablished(WebSocketSession session) throws Exception {
        session.sendMessage(new TextMessage("CONNECTION ESTABLISHED"));
    }

    @Override
    protected void handleTextMessage(WebSocketSession session, TextMessage message)
        throws Exception {
        String msg = message.getPayload();
        session.sendMessage(new TextMessage("RECEIVED: " + msg));
    }
}
```

当连接建立后，一条 TextMessage 会发送回客户端，告诉客户端连接已经建立。当接收到 TextMessage 时，会抽取出负载（实际的消息）并添加 RECEIVED:前缀，然后发送回客户端。

接下来，需要将这个处理器注册到 URI 上。为了做到这一点，创建一个@Configuration 类并实现 WebSocketConfigurer，在 registerWebSocketHandlers 方法中完成注册。我们将这个接口添加到 WebSocketConfiguration 类上，如下所示：

```
package com.apress.springrecipes.websocket;

import org.springframework.context.annotation.Bean;
import org.springframework.context.annotation.Configuration;
import org.springframework.web.socket.config.annotation.EnableWebSocket;
import org.springframework.web.socket.config.annotation.WebSocketConfigurer;
import org.springframework.web.socket.config.annotation.WebSocketHandlerRegistry;
import java.util.concurrent.TimeUnit;

@Configuration
@EnableWebSocket
public class WebSocketConfiguration implements WebSocketConfigurer {

    @Bean
    public EchoHandler echoHandler() {
        return new EchoHandler();
```

```
    }

    @Override
    public void registerWebSocketHandlers(WebSocketHandlerRegistry registry) {
        registry.addHandler(echoHandler(), "/echo");
    }
}
```

首先，将 EchoHandler 注册为一个 bean，这样就可以将其附加到一个 URI 上了。在 registerWebSocketHandlers 中，可以使用 WebSocketHandlerRegistry 注册处理器。借助于 addHandler 方法，你可以将处理器注册到 URI 上，该示例就是/echo。通过该配置，你可以使用 ws://localhost:8080/echo-ws/echo URL 从客户端打开一个 WebSocket 连接。

在服务器端就绪后，你需要一个客户端连接到 WebSocket 端点。这需要一些 JavaScript 和 HTML。编写如下 app.js：

```
var ws = null;
var url = "ws://localhost:8080/echo-ws/echo";

function setConnected(connected) {
    document.getElementById('connect').disabled = connected;
    document.getElementById('disconnect').disabled = !connected;
    document.getElementById('echo').disabled = !connected;
}

function connect() {
    ws = new WebSocket(url);

    ws.onopen = function () {
        setConnected(true);
    };

    ws.onmessage = function (event) {
        log(event.data);
    };

    ws.onclose = function (event) {
        setConnected(false);
        log('Info: Closing Connection.');
    };
}
function disconnect() {
    if (ws != null) {
        ws.close();
        ws = null;
    }
    setConnected(false);
}
function echo() {
    if (ws != null) {
        var message = document.getElementById('message').value;
        log('Sent: ' + message);
        ws.send(message);
    } else {
        alert('connection not established, please connect.');
    }
}

function log(message) {
    var console = document.getElementById('logging');
    var p = document.createElement('p');
    p.appendChild(document.createTextNode(message));
    console.appendChild(p);
    while (console.childNodes.length > 12) {
```

```
        console.removeChild(console.firstChild);
    }
    console.scrollTop = console.scrollHeight;
}
```

这里有几个函数。当单击 Connect 按钮时会调用第一个函数 connect。这会打开到 ws://localhost:8080/echo-ws/echo 的一个 WebSocket 连接，该 URL 对应于之前创建和注册的处理器。连接到服务器会创建一个 WebSocket JavaScript 对象，这样就可以在客户端监听消息与事件了。这里定义了 onopen、onmessage 与 onclose 回调。最重要的是 onmessage 回调，因为当来自于服务器端的消息到来时会调用它；该方法仅仅调用了 log 函数，它会将接收到的消息添加到屏幕上的 logging 元素中。

接下来是 disconnect，它会关闭 WebSocket 连接并清理 JavaScript 对象。最后还有一个 echo 函数，当单击 Echo Message 按钮时会调用它。给定的消息会发送到服务器端（最后还会返回）。

添加 index.html 文件以使用 app.js，代码如下所示：

```
<!DOCTYPE html>
<html>
<head>
    <link type="text/css" rel="stylesheet" href="https://cdnjs.cloudflare.com/ajax/libs/
    semantic-ui/2.2.10/semantic.min.css" />
    <script type="text/javascript" src="app.js"></script>
</head>
<body>
<div>
    <div id="connect-container" class="ui centered grid">
        <div class="row">
            <button id="connect" onclick="connect();" class="ui green button ">
            Connect</button>
            <button id="disconnect" disabled="disabled" onclick="disconnect();" class="ui
            red button">Disconnect</button>
        </div>
        <div class="row">
            <textarea id="message" style="width: 350px" class="ui input"
            placeholder="Message to Echo"></textarea>
        </div>
        <div class="row">
            <button id="echo" onclick="echo();" disabled="disabled" class="ui button">
            Echo message</button>
        </div>
    </div>
    <div id="console-container">
        <h3>Logging</h3>
        <div id="logging"></div>
    </div>
</div>
</body>
</html>
```

部署好应用后就可以连接到 echo WebSocket 服务并发送一些消息，这些消息又会返回，如图 5-5 所示。

使用 STOMP 与 MessageMapping

在使用 WebSocket 技术创建应用时，或多或少都会涉及消息。虽然可以使用 WebSocket 协议本身，不过还可以使用该协议的子协议。STOMP 就是 Spring WebSocket 支持的一种子协议。

STOMP 是一种简单的文本协议，面向 Ruby 和 Python 等脚本语言，用于连接到消息代理。STOMP 可用在任何可靠的双向网络协议上，如 TCP 和 WebSocket。协议本身是基于文本的，不过消息的负载却并不是严格地绑定到文本，也可以包含二进制数据。

当使用 Spring WebSocket 支持来配置和使用 STOMP 时，WebSocket 应用作为所有已连接客户端的代理。这可以是内存代理，也可以是实际的、功能完善的支持 STOMP 协议的企业解决方案（如 RabbitMQ 或 ActiveMQ）。对于后者来说，Spring WebSocket 应用会作为实际代理的中继。为了通过 WebSocket 协议添加消息，Spring 使用了 Spring Messaging（参见第 14 章了解关于消息的更多信息）。

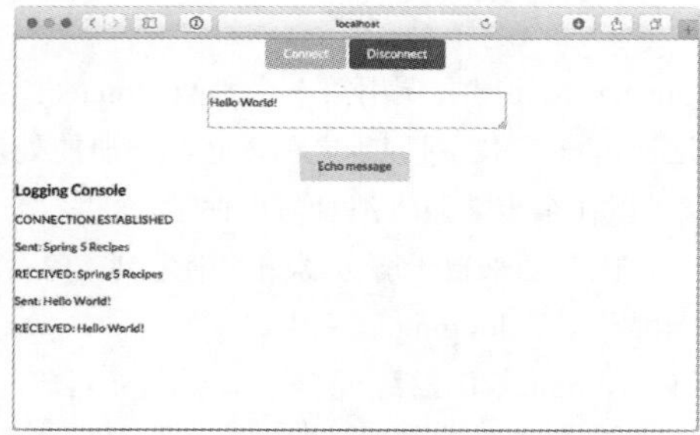

图 5-5　WebSocket 客户端输出

要想接收消息，你需要将@Controller 中的方法标记为@MessageMapping 注解，并告诉它从哪里接收消息。我们现在来修改 EchoHandler 以使用注解：

```
package com.apress.springrecipes.websocket;

import org.springframework.messaging.handler.annotation.MessageMapping;
import org.springframework.messaging.handler.annotation.SendTo;
import org.springframework.stereotype.Controller;

@Controller
public class EchoHandler {

    @MessageMapping("/echo")
    @SendTo("/topic/echo")
    public String echo(String msg) {
        return "RECEIVED: " + msg;
    }
}
```

当在/app/echo 上接收到消息时，消息会被传递给@MessageMapping 所注解的方法。注意方法上的@SendTo("/topic/echo")注解，这会让 Spring 将这个字符串结果放到所述的主题中。

现在需要配置消息代理并添加端点来接收消息。为了做到这一点，将@EnableWebSocketMessageBroker 注解添加到 WebSocketConfiguration 上，并让其继承 AbstractWebSocketMessageBrokerConfigurer（实现了 WebSocketMessageBrokerConfigurer，用于对 WebSocket 消息做进一步的配置）。

```
package com.apress.springrecipes.websocket;

import org.springframework.context.annotation.ComponentScan;
import org.springframework.context.annotation.Configuration;
import org.springframework.messaging.simp.config.MessageBrokerRegistry;
import org.springframework.web.socket.config.annotation.
AbstractWebSocketMessageBrokerConfigurer;
import org.springframework.web.socket.config.annotation.EnableWebSocketMessageBroker;
import org.springframework.web.socket.config.annotation.StompEndpointRegistry;
@Configuration
@EnableWebSocketMessageBroker
@ComponentScan
public class WebSocketConfiguration extends AbstractWebSocketMessageBrokerConfigurer {

    @Override
    public void configureMessageBroker(MessageBrokerRegistry registry) {
        registry.enableSimpleBroker("/topic");
        registry.setApplicationDestinationPrefixes("/app");
    }

    @Override
    public void registerStompEndpoints(StompEndpointRegistry registry) {
        registry.addEndpoint("/echo-endpoint");
```

```
    }
}
```

被@EnableWebSocketMessageBroker 注解后就可以在 WebSocket 上使用 STOMP 了。代理是在 configureMessageBroker 方法中配置的；这里使用的是简单的消息代理。要想连接到企业代理，请使用 registry.enableStompBrokerRelay 连接到实际的代理。为了区分代理与应用所处理的消息，这里使用了不同的前缀。以/topic 开头的消息会被传递给代理，以/app 开头的消息会被传递给消息处理器（即被@MessageMapping 所注解的方法）。

最后一部分是监听到传入 STOMP 消息的 WebSocket 端点的注册；在该示例中，端点映射到了/echo-endpoint。注册是在 registerStompEndpoints 方法中完成的，该方法可被重写。

最后，需要修改客户端以使用 STOMP 而非普通的 WebSocket。HTML 基本上是一样的；你还需要一个库才能在浏览器中使用 STOMP。这里使用的是 webstomp-client（https://github.com/ JSteunou/webstomp-client），也可以使用其他库。

```
<head>
    <link type="text/css" rel="stylesheet" href="https://cdnjs.cloudflare.com/ajax/
    libs/semantic-ui/2.2.10/semantic.min.css" />

    <script type="text/javascript" src="webstomp.js"></script>
    <script type="text/javascript" src="app.js"></script>

</head>
```

最大的变化是 app.js 文件。

```
var ws = null;
var url = "ws://localhost:8080/echo-ws/echo-endpoint";

function setConnected(connected) {
    document.getElementById('connect').disabled = connected;
    document.getElementById('disconnect').disabled = !connected;
    document.getElementById('echo').disabled = !connected;
}

function connect() {
    ws = webstomp.client(url);
    ws.connect({}, function(frame) {
        setConnected(true);
        log(frame);
        ws.subscribe('/topic/echo', function(message){
            log(message.body);
        })
    });
}

function disconnect() {
    if (ws != null) {
        ws.disconnect();
        ws = null;
    }
    setConnected(false);
}
function echo() {
    if (ws != null) {
        var message = document.getElementById('message').value;
        log('Sent: ' + message);
        ws.send("/app/echo", message);
    } else {
        alert('connection not established, please connect.');
    }
}
```

```
function log(message) {
    var console = document.getElementById('logging')
    var p = document.createElement('p');
    p.appendChild(document.createTextNode(message));
    console.appendChild(p);
    while (console.childNodes.length > 12) {
        console.removeChild(console.firstChild);
    }
    console.scrollTop = console.scrollHeight;
}
```

connect 函数现在使用 webstomp.client 创建一个面向代理的 STOMP 客户端连接。当连接后，客户端会订阅/topic/echo 并接收该主题上的消息。echo 函数被修改为使用客户端的 send 方法向/app/echo 目标发送消息，消息又会传递给被@MessageMapping 所注解的方法。

当启动应用并打开客户端时，依旧可以发送并接收消息，不过现在使用的是 STOMP 子协议。甚至可以连接多个浏览器，每个浏览器都会看到/topic/echo 目标上的消息，因为它本身是作为主题存在的。

在编写被@MessageMapping 所注解的方法时，你可以使用多种方法参数和注解（见表 5-4）来接收关于消息的种种信息。默认情况下，单个参数会映射到消息的负载，MessageConverter 用于将消息负载转换为所需的类型（参见 14-2 节以了解消息转换）。

表 5-4　　支持的方法参数与注解

类型	说明
Message	实际的底层消息，包括消息头与消息体
@Payload	消息的负载（默认情况下）；可以使用@Validated 注解来验证参数
@Header	从 Message 中获取指定的消息头
@Headers	可以用在 Map 参数上来获取所有的 Message 消息头
MessageHeaders	所有的 Message 消息头
Principal	当前用户（如果设置了的话）

5-5　使用 Spring WebFlux 开发反应式应用

问题提出

你想要通过 Spring WebFlux 开发一个简单的反应式 Web 应用，从而学习关于该框架的基本概念与配置。

解决方案

Spring WebFlux 最底层的组件是 HttpHandler。这是个接口，只有一个 handle 方法。

```
public interface HttpHandler {

   Mono<Void> handle(ServerHttpRequest request, ServerHttpResponse response);

}
```

handle 方法返回 Mono<Void>，这是以反应式方式来表示返回 void。它接收 org.springframework.http.server.reactive 包中的 ServerHttpRequest 对象与 ServerHttpResonse 对象。它们都是接口，根据所用的容器来创建其实例。针对这一点，提供了针对容器的几个适配器和桥接器。当运行在 Servlet 3.1 容器（支持非阻塞 I/O）中时，会使用 ServletHttpHandlerAdapter（或是其子类）来适配普通的 servlet 与反应式编程。当运行在如 Netty 等原生反应式引擎中时，会使用 ReactorHttpHandlerAdapter。

当 Web 请求发送给 Spring WebFlux 应用时，HandlerAdapter 首先会接收该请求。然后，它会组织好 Spring 应用上下文中所配置的不同组件来处理请求。

要想在 Spring WebFlux 中定义控制器类，需要使用@Controller 或@RestController 注解来标识类（就像 Spring MVC 一样；参见第 3 章与第 4 章）。

当被@Controller 注解的类（即控制器类）接收到请求时，它会寻找恰当的处理器方法来处理请求。这需要控制器类通过一个或多个处理器映射将每个请求映射到处理器方法上。为了做到这一点，控制器类的方法需要使用@RequestMapping 注解进行修饰，这使得它们成为了处理器方法。

这些处理器方法的签名（如任何标准的类一样）是无限多的。可以为处理器方法指定任何名字，定义各种方法参数。同样，处理器方法可以返回各种值（如 String 或 void)，这取决于它所满足的应用逻辑。下文列出了部分合法的参数类型，仅供参考：

- ServerHttpRequest 或 ServerHttpResponse；
- 任意类型的 URL 请求参数，使用注解@RequestParam；
- 任意类型的模型属性，使用注解@ModelAttribute；
- 请求中包含的 cookie 值，使用注解@CookieValue；
- 任意类型的请求消息头值，使用注解@RequestHeader；
- 任意类型的请求属性，使用注解@RequestAttribute；
- Map 或 ModelMap，可以让处理器方法向模型中添加属性；
- Errors 或 BindingResult，可以让处理器方法访问命令对象的绑定与验证结果；
- WebSession，用于会话。

当处理器类选择好了恰当的处理器方法后，它会使用请求来调用处理器方法的逻辑。通常，控制器的逻辑会调用后端服务来处理请求。此外，处理器方法的逻辑一般会涉及从各种输入参数（如 ServerHttpRequest、Map 或是 Errors）中添加或是移除信息，这构成了整个流程的一部分。

当处理器方法完成了请求处理后，它会将控制委托给视图，视图则表示为处理器方法的返回值。为了提供一种灵活的方式，控制器方法的返回值并不表示视图的实现（如 user.html 或 report.pdf)，而是一种逻辑视图（如 user 或 report)，注意这里没有文件扩展名。

处理器方法的返回值可以是 String，表示逻辑视图名；也可以是 void，这会根据处理器方法或控制器的名字来确定出默认的逻辑视图名。

要想将信息从控制器传递给视图，处理器方法返回什么逻辑视图名（String 或 void）是不重要的，因为处理器方法的输入参数是可以被视图使用的。

比如说，如果处理器方法接收 Map 与 Model 对象作为输入参数（在处理器方法的逻辑中修改其内容），那么处理器方法所返回的视图就可以访问到它们。

当控制器类接收到一个视图时，它会通过视图解析器将逻辑视图名解析为特定的视图实现（比如说 user.html 或 report.fmt)。视图解析器是配置在 Web 应用上下文中的 bean，它实现了 ViewResolver 接口，其职责是为逻辑视图名返回具体的视图实现。

当控制器类将视图名解析为视图实现后，根据视图实现的设计，它会渲染控制器的处理器方法所传递过来的对象（如 ServerHttpRequest、Map、Errors 或 WebSession)。视图的职责是将处理器方法的逻辑中所添加的对象展示给用户。

解释说明

我们来编写一个第 3 章介绍的场地预订系统的反应式版本。首先需要编写如下领域类，它就是普通的类（并不涉及反应式)：

```
package com.apress.springrecipes.reactive.court;

public class Reservation {

    private String courtName;

    @DateTimeFormat(iso = DateTimeFormat.ISO.DATE)
    private LocalDate date;
    private int hour;
    private Player player;
```

```
    private SportType sportType;

    // Constructors, Getters and Setters
    ...
}

package com.apress.springrecipes.court.domain;

public class Player {

    private String name;
    private String phone;

    // Constructors, Getters and Setters
    ...
}

package com.apress.springrecipes.court.domain;

public class SportType {

    private int id;
    private String name;

    // Constructors, Getters and Setters
    ...
}
```

接下来，定义如下服务接口向展示层提供预订服务：

```
package com.apress.springrecipes.reactive.court;

import reactor.core.publisher.Flux;

public interface ReservationService {

    Flux<Reservation> query(String courtName);
}
```

注意，query 方法返回的是一个 Flux<Reservation>类型，这表示零个或多个预订。

在生产应用中，你应该通过数据存储持久化来实现该接口，而且该接口最好拥有反应式支持。不过为了简化，可以将预订记录存储到一个列表中，并硬编码几条预订记录以便测试之用。

```
package com.apress.springrecipes.reactive.court;

import reactor.core.publisher.Flux;

import java.time.LocalDate;
import java.util.ArrayList;
import java.util.List;
import java.util.Objects;

public class InMemoryReservationService implements ReservationService {
    public static final SportType TENNIS = new SportType(1, "Tennis");
    public static final SportType SOCCER = new SportType(2, "Soccer");

    private final List<Reservation> reservations = new ArrayList<>();

    public InMemoryReservationService() {

        reservations.add(new Reservation("Tennis #1", LocalDate.of(2008, 1, 14), 16,
            new Player("Roger", "N/A"), TENNIS));
        reservations.add(new Reservation("Tennis #2", LocalDate.of(2008, 1, 14), 20,
            new Player("James", "N/A"), TENNIS));
    }
```

```
    @Override
    public Flux<Reservation> query(String courtName) {
        return Flux.fromIterable(reservations)
            .filter(r -> Objects.equals(r.getCourtName(), courtName));
    }
}
```

query 方法根据嵌入的 Reservation 列表返回一个 Flux，该 Flux 会过滤掉不匹配的预订记录。

创建 Spring WebFlux 应用

要想以反应式的方式来处理请求，你需要启用 WebFlux。这是通过向@Configuration 类添加@EnableWebFlux 注解实现的（类似于针对常规请求处理的@EnableWebMvc 注解）。

```
package com.apress.springrecipes.websocket;

import org.springframework.context.annotation.ComponentScan;
import org.springframework.context.annotation.Configuration;
import org.springframework.web.reactive.config.EnableWebFlux;
import org.springframework.web.reactive.config.WebFluxConfigurer;
@Configuration
@EnableWebFlux
@ComponentScan
public class WebFluxConfiguration implements WebFluxConfigurer { … }
```

@EnableWebFlux 注解会开启反应式处理。要想配置更多的 WebFlux，可以实现 WebFluxConfigurer 并添加更多的转换器。

启动应用

就像常规的 Spring MVC 应用一样，你需要启动应用。如何启动应用在一定程度上取决于所选择的运行时。对于所有支持的容器来说（见表 5-5），它们提供了不同的处理器适配器，这样运行时就可以使用针对 Spring WebFlux 的 HttpHandler 抽象了。

表 5-5 支持的运行时与 HandlerAdapter

运行时	适配器
Servlet 3.1 容器	ServletHttpHandlerAdapter
Tomcat	ServletHttpHandlerAdapter 或 TomcatHttpHandlerAdapter
Jetty	ServletHttpHandlerAdapter 或 JettyHttpHandlerAdapter
Reactor Netty	ReactorHttpHandlerAdapter
RxNetty	RxNettyHttpHandlerAdapter
Undertow	UndertowHttpHandlerAdapter

在适配运行时前，你需要通过 AnnotationConfigApplicationContext 启动应用并配置 HttpHandler。可以通过 WebHttpHandlerBuilder.applicationContext 工厂方法来创建它。这会创建一个 HttpHandler 并使用传递进来的 ApplicationContext 对其进行配置。

```
AnnotationConfigApplicationContext context =
    new AnnotationConfigApplicationContext(WebFluxConfiguration.class);
HttpHandler handler = WebHttpHandlerBuilder.applicationContext(context).build();
```

接下来需要将 HttpHandler 适配到运行时。

对于 Reactor Netty 来说，代码如下所示：

```
ReactorHttpHandlerAdapter adapter = new ReactorHttpHandlerAdapter(handler);
HttpServer.create(host, port).newHandler(adapter).block();
```

首先创建一个 ReactorHttpHandlerAdapter 组件，该组件知道如何将 Reactor Netty 处理适配到内部的 HttpHandler。接下来，将该适配器作为一个处理器注册到新创建的 Reactor Netty 服务器上。

将应用部署到 servlet 容器中时，你可以创建一个实现了 WebApplicationInitializer 的类并进行手工启动。

```
public class WebFluxInitializer implements WebApplicationInitializer {

    public void onStartup(ServletContext servletContext) throws ServletException {}
        AnnotationConfigApplicationContext context =
            new AnnotationConfigApplicationContext(WebFluxConfiguration.class);
        HttpHandler handler = WebHttpHandlerBuilder.applicationContext(context).build();
        ServletHttpHandlerAdapter handlerAdapter = new ServletHttpHandlerAdapter(httpHandler)
        ServletRegistration.Dynamic registration =
            servletContext.addServlet("dispatcher-handler", handlerAdapter);
        registration.setLoadOnStartup(1);
        registration.addMapping("/");
        registration.setAsyncSupported(true);
    }
}
```

首先创建一个 AnnotationConfigApplicationContext，因为你想要使用注解进行配置并将 WebFluxConfiguration 类传递给它。接下来需要一个 HttpHandler 来处理和分发请求。需要将这个 HttpHandler 作为 servlet 注册到所用的 servlet 容器中。为了做到这一点，将其包装到 ServletHttpHandlerAdapter 中。要想进行反应式编程，需要将 asyncSupported 设为 true。

为了简化配置，Spring WebFlux 提供了几个方便的实现供你扩展。在该示例中，可以使用 AbstractAnnotationConfigDispatcherHandlerInitializer 作为父类。现在的配置如下所示：

```
package com.apress.springrecipes.websocket;

import org.springframework.web.reactive.support.AbstractAnnotationConfigDispatcherHandler
Initializer;

public class WebFluxInitializer extends AbstractAnnotationConfigDispatcherHandlerInitializer {

    @Override
    protected Class<?>[] getConfigClasses() {
        return new Class<?>[] {WebFluxConfiguration.class};
    }
}
```

唯一需要的是 getConfigClasses 方法；所有的配置现在都是由 Spring WebFlux 所提供的基础实现来处理的。现在就可以在常规的 servlet 容器中运行应用了。

创建 Spring WebFlux 控制器

基于注解的控制器类可以是任何类，无须实现特定接口或是继承特定父类。你可以使用注解@Controller 或@RestController。一个控制器中可以有一个或多个处理器方法来处理单个或多个动作。处理器方法的签名非常灵活，可以接收很多参数（参见 3-2 节了解关于请求映射的更多信息）。

@RequestMapping 注解可以使用在类级别或方法级别上。第一个映射策略是将特定的 URL 模式映射到控制类上，然后再将特定的 HTTP 方法映射到每个处理器方法上。

```
package com.apress.springrecipes.reactive.court.web;

import org.springframework.stereotype.Controller;
import org.springframework.ui.Model;
import org.springframework.web.bind.annotation.GetMapping;
import org.springframework.web.bind.annotation.RequestMapping;
import reactor.core.publisher.Mono;

import java.time.LocalDate;

@Controller
@RequestMapping("/welcome")
public class WelcomeController {
```

```
    @GetMapping
    public String welcome(Model model) {
        model.addAttribute("today", Mono.just(LocalDate.now()));
        return "welcome";
    }

}
```

该控制器创建了一个 java.util.Date 对象来接收当前日期，然后将其添加到 Model 输入中作为一个属性，这样目标视图就可以展示它了。虽然这个控制器看起来就像是个普通的控制器一样，但主要差别在于属性添加到模型的方式上。相较于直接添加到模型中，当前的日期最终会出现在模型中，原因是使用了 Mono.just(...)。

由于已经在 com.apress.springrecipes.reactive.court 包上激活了注解扫描，因此部署后控制器类上的注解就会被检测到。

@Controller 注解会将类定义为一个控制器。@RequestMapping 注解更加有趣，因为它包含了可声明在类级别或方法级别上的属性。该类中的第一个值（“/welcome”）用于指定一个 URL，控制器会对该 URL 进行响应，这意味着在/welcome URL 上所接收到的请求会被 WelcomeController 类所处理。

当请求被控制器处理后，它会将调用委托给控制器中声明的默认 HTTP GET 处理器方法。该行为背后的原因在于针对 URL 所发出的每个初始请求都是 HTTP GET 类型的。当控制器接收到/welcome URL 的请求后，它会委托给默认的 HTTP GET 处理器方法进行处理。

注解@GetMapping 用于将 welcome 方法修饰为控制器默认的 HTTP GET 处理器方法。值得一提的是，如果没有声明默认的 HTTP GET 处理器方法，则会抛出 ServletException 异常。这正是 Spring MVC 控制器为何要有至少一个 URL 路由和默认 HTTP GET 处理器方法的原因所在。

另外一种做法可以在方法级别的@GetMapping 注解中声明两个值（URL 路由与默认的 HTTP GET 处理器方法）。该声明如下所示：

```
@Controller
public class WelcomeController {

    @GetMapping("/welcome")
    public String welcome(Model model) { … }

}
```

上面的控制器展示了 Spring MVC 的基本原则。不过，典型的控制器会调用后端服务进行业务处理。比如说，你可以创建一个控制器查询特定场地的预订情况，代码如下所示：

```
package com.apress.springrecipes.reactive.court.web;

import com.apress.springrecipes.reactive.court.Reservation;
import com.apress.springrecipes.reactive.court.ReservationService;
import org.springframework.stereotype.Controller;
import org.springframework.ui.Model;
import org.springframework.web.bind.annotation.GetMapping;
import org.springframework.web.bind.annotation.PostMapping;
import org.springframework.web.bind.annotation.RequestMapping;
import org.springframework.web.server.ServerWebExchange;
import reactor.core.publisher.Flux;

@Controller
@RequestMapping("/reservationQuery")
public class ReservationQueryController {

    private final ReservationService reservationService;

    public ReservationQueryController(ReservationService reservationService) {
        this.reservationService = reservationService;
    }

    @GetMapping
```

```
    public void setupForm() {
    }

    @PostMapping
    public String sumbitForm(ServerWebExchange serverWebExchange, Model model) {

        Flux<Reservation> reservations =
            serverWebExchange.getFormData()
                .map(form -> form.get("courtName"))
                .flatMapMany(Flux::fromIterable)
                .concatMap(courtName -> reservationService.query(courtName));
        model.addAttribute("reservations", reservations);
        return "reservationQuery";
    }
}
```

如前所述，控制器接下来会寻找默认的 HTTP GET 处理器方法。由于已经为 public void setupForm()方法指定了必要的@GetMapping 注解，因此会调用它。

与之前默认的 HTTP GET 处理器方法不同，该方法没有输入参数，没有逻辑，返回值为 void。这意味着两点。由于没有输入参数，没有逻辑，因此视图只会展示实现模板（比如说 JSP）中硬编码的数据，因为控制器没有添加数据进去。由于返回值为 void，因此会根据请求 URL 使用默认的视图名；这样，由于请求 URL 是/reservationQuery，因此会使用名为 reservationQuery 的视图。

剩下的处理器方法使用了注解@PostMapping。乍一看，两个处理器方法只拥有一个类级别的/reservationQuery URL 语句，这会让人感到困惑，不过这其实很简单。当 HTTP GET 请求发向/reservationQuery URL 时会调用一个方法；当 HTTP POST 请求发向同样的 URL 时会调用另一个方法。

在 Web 应用中大多数请求都是 HTTP GET 类型的，而 HTTP POST 类型的请求则主要用在用户提交 HTML 表单的时候。这样，当应用的视图越来越多时（稍后将会介绍），一个方法是在 HTML 表单首次加载时调用（即 HTTP GET），而另一个方法则是在 HTML 表单提交时调用（即 HTTP POST）。

仔细查看 HTTP POST 默认的处理器方法，注意到它有两个输入参数。第一个是 ServerWebExchange 声明，它用于提取出名为 courtName 的请求参数。在该示例中，HTTP POST 请求的形式是/reservationQuery?courtName=<value>。该声明使得上面的值可以在方法中通过名为 courtName 的变量访问。第二个是 Model 声明，它用于定义一个可以将数据传递给返回视图的对象。在常规的 Spring MVC 控制器中，可以通过@RequestParam("courtName") String courtName（见 3-1 节）获取到参数，不过对于 Spring WebFlux 来说，它无法处理作为表单数据传递过来的参数；它只能处理作为 URL 一部分的参数。因此，需要通过 ServerWebExchange 来获取表单数据，获取参数，并调用服务。

处理器方法所执行的逻辑包括使用控制器的 reservationService 并通过 courtName 变量执行查询。该查询所获取到的结果会被赋给 Model 对象，接下来可供返回的视图进行展示。

最后，注意方法返回了一个名为 reservationQuery 的视图。该方法也可以返回 void，就像默认的 HTTP GET 一样，并根据请求 URL 赋给同样的 reservationQuery 默认视图。这两种方式是等价的。

在了解了 Spring MVC 控制器是如何构成的之后，现在就来看看控制器的处理器方法将其结果所委托的视图。

创建 Thymeleaf 视图

Spring WebFlux 支持不同展示层技术的多种视图类型，包括 HTML、XML、JSON、Atom 与 RSS 源、JasperReports 以及其他第三方视图实现。接下来将会使用 Thymeleaf 编写几个简单的 HTML 模板。你需要向 WebFluxConfiguration 类增加一些配置来设置 Thymeleaf 并注册 ViewResolver，它会将控制器返回的视图名返回至实际待加载的资源。

如下是 Thymeleaf 配置：

```
@Bean
```

```
public SpringResourceTemplateResolver thymeleafTemplateResolver() {

    final SpringResourceTemplateResolver resolver = new SpringResourceTemplateResolver();
    resolver.setPrefix("classpath:/templates/");
    resolver.setSuffix(".html");
    resolver.setTemplateMode(TemplateMode.HTML);
    return resolver;
}

@Bean
public ISpringWebFluxTemplateEngine thymeleafTemplateEngine(){

    final SpringWebFluxTemplateEngine templateEngine = new SpringWebFluxTemplateEngine();
    templateEngine.setTemplateResolver(thymeleafTemplateResolver());
    return templateEngine;
}
```

Thymeleaf 通过一个模板引擎将模板转换为实际的 HTML。接下来需要配置 ViewResolver，它知道如何使用 Thymeleaf。ThymeleafReactiveViewResolver 是 ViewResolver 的反应式实现。最后，你需要让 WebFlux 配置感知到新的视图解析器。这是通过重写 configureViewResolvers 方法并将其添加到 ViewResolverRegistry 实现的。

```
@Bean
public ThymeleafReactiveViewResolver thymeleafReactiveViewResolver() {

    final ThymeleafReactiveViewResolver viewResolver = new ThymeleafReactiveViewResolver();
    viewResolver.setTemplateEngine(thymeleafTemplateEngine());
    viewResolver.setResponseMaxChunkSizeBytes(16384);
    return viewResolver;
}

@Override
public void configureViewResolvers(ViewResolverRegistry registry) {
    registry.viewResolver(thymeleafReactiveViewResolver());
}
```

模板是通过模板解析器来解析的；这里的 SpringResourceTemplateResolver 通过 Spring 的资源加载机制来加载模板。模板放在了 src/main/resources/templates 目录中。比如说，对于 welcome 视图来说，实际的 src/main/resources/templates/welcome.html 文件会被模板引擎加载并解析。

下面编写 welcome.html 模板。

```
<!DOCTYPE html>
<html lang="en" xmlns:th="http://www.thymeleaf.org">
<head>
    <meta charset="UTF-8">
    <title>Welcome</title>
</head>
<body>
<h2>Welcome to Court Reservation System</h2>
Today is <strong th:text="${#temporals.format(today, 'dd-MM-yyyy')}">21-06-2017</strong>
</body>
</html>
```

在该模板中，你使用了 EL 中的 temporals 对象将 today 模型属性格式化为 dd-MM-yyyy 模式。

接下来，为预订查询控制器再创建一个 JSP 模板，将其命名为 reservationQuery.html 以与视图名相匹配。

```
<!DOCTYPE html>
<html lang="en" xmlns:th="http://www.thymeleaf.org">
<head>
    <meta charset="UTF-8">
    <title>Reservation Query</title>
</head>
<body>
<form method="post">
    Court Name
```

```
        <input type="text" name="courtName" value="${courtName}"/>
        <input type="submit" value="Query"/>
    </form>

    <table border="1">
        <thead>
        <tr>
            <th>Court Name</th>
            <th>Date</th>
            <th>Hour</th>
            <th>Player</th>
        </tr>
        </thead>
        <tbody>
        <tr th:each="reservation : ${reservations}">
            <td th:text="${reservation.courtName}">Court</td>
            <td th:text="${#temporals.format(reservation.date, 'dd-MM-yyyy')}">21-06-2017</td>
            <td th:text="${reservation.hour}">22</td>
            <td th:text="${reservation.player.name}">Player</td>
        </tr>
        </tbody>
    </table>
    </body>
    </html>
```

在该模板中，你创建了一个表单，用户可以输入想要查询的场地名，然后通过 th:each 标签循环遍历 reservations 模型属性来生成结果表格。

运行 Web 应用

根据运行时的不同，你可以通过执行 main 方法来运行应用，也可以构建一个 WAR 档案并将其部署到 servlet 容器中。这里采用后面的方式，使用 Apache Tomcat 8.5.x 作为 Web 容器。

■ **提示：**该项目还可以创建一个带有应用的 Docker 容器。运行../gradlew buildDocker 获取一个带有 Tomcat 和应用的容器。接下来启动 Docker 容器来测试应用（docker run -p 8080:8080 spring-recipes-4th/court-rx/welcome）。

5-6　使用反应式控制器处理表单

问题提出

在 Web 应用中，你经常需要处理表单。表单控制器需要向用户展示表单，还需要处理表单提交。表单处理是一件复杂且多变的任务。

解决方案

当用户与表单交互时，需要控制器支持两种操作。首先，当首次请求表单时会通过一个 HTTP GET 请求让控制器展示出表单，这会将表单视图渲染给用户。接下来，当表单提交时会发出一个 HTTP POST 请求来处理诸如验证以及对表单中的数据进行业务处理等事项。如果表单成功处理，那么它会向用户渲染成功视图。否则，它会使用错误消息再次渲染表单视图。

解释说明

假设用户可以通过填写表单来进行场地预订。为了能够更好地理解控制器所处理的数据，我们先来介绍控制器的视图（即表单）。

创建表单视图

创建名为 reservationForm.html 的表单视图。表单使用了 Thymeleaf 表单标签库，因为它简化了表单的数据绑定、错误消息显示，还会在出错时重新显示用户之前输入的值。

```
<!DOCTYPE html>
<html lang="en" xmlns:th="http://www.thymeleaf.org">
<head>
    <title>Reservation Form</title>
    <style>
        .error {
            color: #ff0000;
            font-weight: bold;
        }
    </style>
</head>

<body>
<form method="post" th:object="${reservation}">

</form>
<table>
    <tr>
        <td>Court Name</td>
        <td><input type="text" th:field="*{courtName}" required/></td>
        <td><span class="error" th:if="${#fields.hasErrors('courtName')}"
        th:errors="*{courtName}"></span></td>
    </tr>
    <tr>
        <td>Date</td>
        <td><input type="date" th:field="*{date}" required/></td>
        <td><span class="error" th:if="${#fields.hasErrors('date')}" th:errors="*{date}">
        </span></td>
    </tr>
    <tr>
        <td>Hour</td>
        <td><input type="number" min="8" max="22" th:field="*{hour}" /></td>
        <td><span class="error" th:if="${#fields.hasErrors('hour')}" th:errors="*{hour}">
        </span></td>
    </tr>
    <tr>
        <td colspan="3"><input type="submit" /></td>
    </tr>
</table>

</form>
</body>
</html>
```

由于 form 标签上存在 th:object=${reservation}标签，因此该表单使用 Thymeleaf 将所有的表单字段绑定到了名为 reservation 的模型属性上。每个字段都会绑定到 Reservation 对象中的实际字段（并显示其值），这正是 th:field 标签的作用所在。如果字段存在错误，它们就会通过 th:errors 标签显示出来。

最后，你会看到标准的 HTML 标签<input type="submit" />，它会生成一个提交按钮并触发数据向服务器端的发送。

如果表单及其数据处理正确，那就需要创建一个成功视图来通知用户预订成功。接下来要介绍的 reservationSuccess.html 文件就是为了这个目的：

```
<html>
<head>
<title>Reservation Success</title>
</head>

<body>
Your reservation has been made successfully.
</body>
</html>
```

如果表单提交了不合法的值，那就会出现错误。比如说，如果日期格式不合法或是 hour 字段使用了字母，那么控制器就会拒绝这些字段值。接下来，控制器会为每一个错误生成一个选择好的错误码列表并返回给表单视图；这些值会放在 th:errors 标签中。

比如说，如果 date 字段出现了不合法的输入，那么数据绑定就会生成如下错误码：

```
typeMismatch.command.date
typeMismatch.date
typeMismatch.java.time.LocalDate
typeMismatch
```

如果定义了 ResourceBundleMessageSource 对象，那就可以针对恰当的地域（如针对默认地域的 messages.properties）在资源包中加入如下错误消息；参见 3-5 节来了解如何外部化地域信息：

```
typeMismatch.date=Invalid date format
typeMismatch.hour=Invalid hour format
```

在处理表单数据时，如果出现了错误，那么对应的错误码及其值就会返回给用户。

在知道了表单视图的结构以及它所处理的数据之后，现在就来看看处理表单中所提交数据（即预订信息）的逻辑。

创建表单的服务处理

表单的数据预订信息并不是由控制器来处理的，而是由控制器所使用的服务来处理的。首先在 ReservationService 接口中定义一个 make()方法。

```
package com.apress.springrecipes.court.service;
...
public interface ReservationService {
    ...
    Mono<Reservation> make(Mono<Reservation> reservation)
        throws ReservationNotAvailableException;
}
```

接下来实现该 make()方法，向存储预订信息的列表中添加一个 Reservation 条目。如果出现重复预订则抛出 ReservationNotAvailableException 异常。

```
package com.apress.springrecipes.reactive.court;
...
public class InMemoryReservationService implements ReservationService {
    ...
    @Override
    public Mono<Reservation> make(Reservation reservation) {

        long cnt = reservations.stream()
            .filter(made -> Objects.equals(made.getCourtName(), reservation.getCourtName()))
            .filter(made -> Objects.equals(made.getDate(), reservation.getDate()))
            .filter(made -> made.getHour() == reservation.getHour())
            .count();

        if (cnt > 0) {
            return Mono.error(new ReservationNotAvailableException(reservation
                .getCourtName(), reservation.getDate(), reservation
                .getHour()));
        } else {
            reservations.add(reservation);
            return Mono.just(reservation);
        }
    }
}
```

在对与控制器交互的两个元素（表单视图与预订服务类）有了更好的了解之后，现在就来创建一个处理场地预订表单的控制器。

创建表单控制器

用于处理表单的控制器使用了与上一节中相同的注解。我们直接看代码。

```
package com.apress.springrecipes.reactive.court.web;
...

@Controller
@RequestMapping("/reservationForm")
public class ReservationFormController {

    private final ReservationService reservationService;

    @Autowired
    public ReservationFormController(ReservationService reservationService) {
        this.reservationService = reservationService;
    }

    @RequestMapping(method = RequestMethod.GET)
    public String setupForm(Model model) {
        Reservation reservation = new Reservation();
        model.addAttribute("reservation", reservation);
        return "reservationForm";
    }

    @RequestMapping(method = RequestMethod.POST)
    public String submitForm(
        @ModelAttribute("reservation") Reservation reservation,
        BindingResult result) {
            reservationService.make(reservation);
            return "redirect:reservationSuccess";
    }
}
```

首先，控制器使用了标准的@Controller 注解以及@RequestMapping 注解，可以通过如下 URL 访问控制器：

```
http://localhost:8080/court-rx/reservationForm
```

当在浏览器中输入这个 URL 时，它会向 Web 应用发送一个 HTTP GET 请求。这又会触发 setupForm 方法的执行，根据其@GetMapping 注解，该方法会处理这类请求。

setupForm 方法定义了一个 Model 对象作为输入参数，旨在向视图（即表单）发送模型数据。在该处理器方法中创建了一个空的 Reservation 对象，并作为属性添加到控制器的 Model 对象中。接下来，控制器将执行流程返回到 reservationForm 视图，在该示例中会被解析为 reservationForm.jsp（即表单）。

上面这个方法最重要的地方是添加了一个空的 Reservation 对象。分析一下表单 reservationForm.html，你会发现 form 标签声明了 th:object="${reservation}"属性。这意味着在渲染视图时，表单需要一个名为 reservation 的对象，这是通过将其放到处理器方法的 Model 中做到的。实际上，进一步观察会发现，每个 input 标签的 th:field=*{expression}值都对应于 Reservation 对象的字段名。由于表单是首次加载，显然需要一个空的 Reservation 对象。

现在将注意力放到首次提交表单上来。在填写完表单字段后，提交表单会触发一个 HTTP POST 请求，这又会调用 submitForm 方法——由于方法的@PostMapping 值的缘故。

submitForm 方法中声明的输入字段@ModelAttribute("reservation") Reservation reservation 用于引用 reservation 对象，BindingResult 对象则包含了用户新提交的数据。

这时的处理器方法还没有包含验证，它是 BindingResult 对象的目的。

处理器方法所执行的唯一操作是 reservationService.make(reservation);。该操作会使用预订对象的当前状态来调用预订服务。

一般来说，在执行这类操作前需要先验证控制器对象。

最后，注意到处理器方法返回了一个名为 redirect:reservationSuccess 的视图。在该示例中，视图实际的名字是 reservationSuccess，它会被解析为之前所创建的 reservationSuccess.html 页面。

视图名中的 redirect:前缀用于避免表单重复提交的问题。

如果在表单成功视图中刷新页面，那么方才所提交的表单就会再次提交。为了避免这个问题，可以使用

post/redirect/get 设计模式，该模式推荐在表单提交处理成功后重定向到另一个 URL 而不是直接返回 HTML 页面。这正是在视图名前使用 redirect:前缀的目的所在。

初始化 Model 属性对象并使用值来预先填充表单

表单的设计目的在于让用户进行场地预订。不过，如果分析一下 Reservation 领域类，你会注意到表单还是需要再增加两个字段才能创建出一个完整的 reservation 对象。其中一个是 player 字段，它对应于 Player 对象。根据 Player 类的定义，Player 对象有一个 name 字段和一个 phone 字段。

那么，是否可以将 player 字段纳入到表单视图和控制器中呢？我们先来分析一下表单视图。

```
<!DOCTYPE html>
<html lang="en" xmlns:th="http://www.thymeleaf.org">
<body>
<form method="post" th:object="${reservation}">

    <table>
      ...
        <tr>
            <td>Player Name</td>
            <td><input type="text" th:field="*{player.name}" required/></td>
            <td><span class="error" th:if="${#fields.hasErrors('player.name')}"
            th:errors="*{player.name}"></span></td>
        </tr>
        <tr>
            <td>Player Phone</td>
            <td><input type="text" th:field="*{player.phone}" required/></td>
            <td><span class="error" th:if="${#fields.hasErrors('player.phone')}"
            th:errors="*{player.phone}"></span>
            </td>
        </tr>
        <tr>
            <td colspan="3"><input type="submit"/></td>
        </tr>
    </table>

</form>
</body>
</html>
```

我们使用一种直接的方式，再添加两个<input>标签来表示 Player 对象的字段。虽然这些表单声明都很简单，不过还是需要修改一下控制器。还记得，通过使用<input>标签，视图就可以访问控制器所传递的模型对象，它匹配了<input>标签的 path 值。

控制器的 HTTP GET 处理器方法会向上面的视图返回一个空的 Reservation 对象，所以 player 属性为 null，这会导致渲染表单时出现异常。为了解决这个问题，需要初始化一个空的 Player 对象并将其赋给返回给视图的 Reservation 对象。

```
@RequestMapping(method = RequestMethod.GET)
public String setupForm(
@RequestParam(required = false, value = "username") String username, Model model) {
    Reservation reservation = new Reservation();
    reservation.setPlayer(new Player(username, null));
    model.addAttribute("reservation", reservation);
    return "reservationForm";
}
```

在该示例中，当创建完空的 Reservation 对象后，setPlayer 方法会将一个空的 Player 对象赋给它。此外，Person 对象的创建要用到 username 值。这个值是通过@RequestParam 所注解的输入值来获取的，它也被添加到了处理器方法中。这样，我们就可以通过作为请求参数而传递进来的特定 username 值来创建 Player 对象，username 表单字段会使用该值进行重新装配。

比如说，如果通过如下方式发起对表单的请求：

```
http://localhost:8080/court/reservationForm?username=Roger
```

这样，处理器方法就可以提取出username参数来创建Player对象，并使用Roger值来重新装配表单的username字段。值得注意的是，username参数的@RequestParam注解使用了属性required=false；这样，即便该请求参数不存在，表单请求也会得到处理。

向表单提供引用数据

当表单控制器渲染表单视图时，它可能会向表单提供一些引用数据（比如说在HTML下拉列表中展示的条目）。假设在预订场地时可以让用户选择运动类型，这是Reservation类中不存在的字段。

```
<!DOCTYPE html>
<html lang="en" xmlns:th="http://www.thymeleaf.org">
<body>
<form method="post" th:object="${reservation}">

    <table>
        ...
        <tr>
            <td>Sport Type</td>
            <td>
                <select th:field="*{sportType}">
                    <option th:each="sportType : ${sportTypes}" th:value="${sportType.id}"
                    th:text="${sportType.name}"/>
                </select>
            </td>
            <td><span class="error" th:if="${#fields.hasErrors('sportType')}"
            th:errors="*{sportType}"></span></td>
        </tr>
        <tr>
            <td colspan="3"><input type="submit"/></td>
        </tr>
    </table>

</form>
</body>
</html>
```

<form:select>标签可以根据控制器传递给视图的值生成一个下拉列表。这样，表单会将sportType字段表示为一组HTML <select>元素而不是之前的<input>元素（这需要用户自己输入文本值）。

接下来，我们看看控制器如何将sportType字段指定为模型属性；这个过程与之前的字段有些许不同。

首先，在ReservationService接口中定义getAllSportTypes()方法来获取所有可用的运动类型。

```
package com.apress.springrecipes.reactive.court;
...
public interface ReservationService {
    ...
    Flux<SportType> getAllSportTypes();
}
```

接下来，实现该方法，返回一个硬编码的列表。

```
package com.apress.springrecipes.reactive.court;
...
public class InMemoryReservationService implements ReservationService {
    ...
    public static final SportType TENNIS = new SportType(1, "Tennis");
    public static final SportType SOCCER = new SportType(2, "Soccer");

    public Flux<SportType> getAllSportTypes() {
        return Flux.fromIterable(Arrays.asList(TENNIS, SOCCER));
    }
}
```

在有了硬编码的返回SportType对象列表的实现后，我们来看看控制器如何将该列表关联到返回的表单视图上。

```
package com.apress.springrecipes.court.service;
.....
    @ModelAttribute("sportTypes")
    public Flux<SportType> populateSportTypes() {
        return reservationService.getAllSportTypes();
    }

    @RequestMapping(method = RequestMethod.GET)
    public String setupForm(
    @RequestParam(required = false, value = "username") String username, Model model) {
        Reservation reservation = new Reservation();
        reservation.setPlayer(new Player(username, null));
        model.addAttribute("reservation", reservation);
        return "reservationForm";
    }
```

注意，负责向表单视图返回空 Reservation 对象的 setupForm 处理器方法没有发生变化。

新添加的方法，即负责将 SportType 列表作为模型属性传递给表单视图的方法是被注解@ModelAttribute("sportTypes")修饰的方法。@ModelAttribute 注解用于定义全局模型属性，可用在处理器方法所返回的任何视图中。同样，处理器方法声明了一个 Model 对象作为输入参数，并为其赋予了可在返回的视图中访问的属性。

由于被@ModelAttribute("sportTypes")注解所修饰的方法的返回类型是 Flux<SportType>，并且它会调用 reservationService.getAllSportTypes()，因此硬编码的 TENNIS 与 SOCCER SportType 对象就会被赋给名为 sportTypes 的模型属性。这个模型属性会用在表单视图中，用于装配下拉列表（即<select>标签）。

绑定自定义类型的属性

当表单提交时，控制器会将表单字段值绑定到模型对象的同名属性上，在该示例中就是 Reservation 对象。不过，对于自定义类型的属性来说，控制器就无法对其进行转换了，除非指定了相应的属性编辑器。

比如说，运动类型选择字段只会提交所选的运动类型 ID，因为这是 HTML <select>字段的操作方式。因此，你需要使用属性编辑器将这个 ID 转换为 SportType 对象。首先，需要通过 ReservationService 中的 getSportType()方法根据 ID 来获取一个 SportType 对象。

```
package com.apress.springrecipes.court.service;
...
public interface ReservationService {
    ...
    public SportType getSportType(int sportTypeId);
}
```

出于测试的目的，可以使用一个 switch/case 语句来实现该方法。

```
package com.apress.springrecipes.court.service;
...
public class ReservationServiceImpl implements ReservationService {
    ...
    public SportType getSportType(int sportTypeId) {
        switch (sportTypeId) {
        case 1:
            return TENNIS;
        case 2:
            return SOCCER;
        default:
            return null;
        }
    }
}
```

接下来，创建 SportTypeConverter 类将运动类型 ID 转换为 SportType 对象。该转换器需要通过 ReservationService 来执行查询。

```
package com.apress.springrecipes.reactive.court.domain;
```

```
import com.apress.springrecipes.court.service.ReservationService;
import org.springframework.core.convert.converter.Converter;

public class SportTypeConverter implements Converter<String, SportType> {

    private final ReservationService reservationService;

    public SportTypeConverter(ReservationService reservationService) {
        this.reservationService = reservationService;
    }

    @Override
    public SportType convert(String source) {
        int sportTypeId = Integer.parseInt(source);
        SportType sportType = reservationService.getSportType(sportTypeId);
        return sportType;
    }
}
```

既然已经可以通过 SportTypeConverter 类将表单属性绑定到 SportType 这样的自定义类上，现在需要将其关联到控制器上。为了实现这个目的，可以使用 WebFluxConfigurer 的 addFormatters 方法。

通过在配置类中重写该方法，自定义类型就可以关联到控制器上。这包括 SportTypeConverter 和 Date 等其他自定义类型。虽然之前并未提及日期字段，不过它与运动类型选择字段面临着相同的问题。用户会将日期字段作为文本值。要想让控制器将这些文本值赋给 Reservation 对象的日期字段，则需要将日期字段关联到 Date 对象上。考虑到 Date 类是 Java 语言的一部分，因此没必要像 SportTypeConverter 那样创建一个特殊的类。出于这个目的，Spring 框架已经自带了一个自定义类。

在知道了需要将 SportTypeConverter 类与 Date 类绑定到底层的控制器上后，如下代码展示了配置类的修改过程：

```
package com.apress.springrecipes.reactive.court;

import org.springframework.beans.factory.annotation.Autowired;
import org.springframework.context.annotation.Bean;
import org.springframework.context.annotation.ComponentScan;
import org.springframework.context.annotation.Configuration;
import org.springframework.format.FormatterRegistry;
import org.springframework.web.reactive.config.EnableWebFlux;
import org.springframework.web.reactive.config.ViewResolverRegistry;
import org.springframework.web.reactive.config.WebFluxConfigurer;
import org.thymeleaf.extras.java8time.dialect.Java8TimeDialect;
import org.thymeleaf.spring5.ISpringWebFluxTemplateEngine;
import org.thymeleaf.spring5.SpringWebFluxTemplateEngine;
import org.thymeleaf.spring5.templateresolver.SpringResourceTemplateResolver;
import org.thymeleaf.spring5.view.reactive.ThymeleafReactiveViewResolver;
import org.thymeleaf.templatemode.TemplateMode;

@Configuration
@EnableWebFlux
@ComponentScan
public class WebFluxConfiguration implements WebFluxConfigurer {

    @Autowired
    private ReservationService reservationService;

    ...

    @Override
    public void addFormatters(FormatterRegistry registry) {
        registry.addConverter(new SportTypeConverter(reservationService));
    }
}
```

上述类中的唯一字段对应于 reservationService，它用于访问应用的 ReservationService bean。注意，这里使用了@Autowired 注解开启了 bean 的自动装配。接下来，可以重写用于绑定 Date 与 SportTypeConverter 类的 addFormatters 方法。然后通过两个调用注册了转换器与格式化器。这些方法属于 FormatterRegistry 对象，该对象作为输入参数传递给了 addFormatters 方法。

第一个调用将 Date 类绑定到了 DateFormatter 类。DateFormatter 类由 Spring 框架提供，它提供了解析和打印 Date 对象的功能。

第二个调用注册了 SportTypeConverter 类。由于已经创建了 SportTypeConverter 类，你应该知道其唯一的输入参数是 ReservationService bean。通过这种方式，每个被注解的控制器（比如说使用了@Controller 注解的类）都可以在其处理器方法中访问到相同的自定义转换器与格式化器。

验证表单数据

在提交表单时，标准的做法是在提交成功前验证用户所提供的数据。就像 Spring MVC 一样，Spring WebFlux 支持使用实现了 Validator 接口的验证器对象进行验证。你可以编写如下验证来检查必填的表单字段是否都填写了，以及预订时间在假期与工作日是否合法：

```
package com.apress.springrecipes.reactive.court;

import org.springframework.stereotype.Component;
import org.springframework.validation.Errors;
import org.springframework.validation.ValidationUtils;
import org.springframework.validation.Validator;
import java.time.DayOfWeek;
import java.time.LocalDate;

@Component
public class ReservationValidator implements Validator {

    public boolean supports(Class<?> clazz) {
        return Reservation.class.isAssignableFrom(clazz);
    }

    public void validate(Object target, Errors errors) {
        ValidationUtils.rejectIfEmptyOrWhitespace(errors, "courtName",
            "required.courtName", "Court name is required.");
        ValidationUtils.rejectIfEmpty(errors, "date",
            "required.date", "Date is required.");
        ValidationUtils.rejectIfEmpty(errors, "hour",
            "required.hour", "Hour is required.");
        ValidationUtils.rejectIfEmptyOrWhitespace(errors, "player.name",
            "required.playerName", "Player name is required.");
        ValidationUtils.rejectIfEmpty(errors, "sportType",
            "required.sportType", "Sport type is required.");

        Reservation reservation = (Reservation) target;
        LocalDate date = reservation.getDate();
        int hour = reservation.getHour();
        if (date != null) {
            if (date.getDayOfWeek() == DayOfWeek.SUNDAY) {
                if (hour < 8 || hour > 22) {
                    errors.reject("invalid.holidayHour", "Invalid holiday hour.");
                }
            } else {
                if (hour < 9 || hour > 21) {
                    errors.reject("invalid.weekdayHour", "Invalid weekday hour.");
                }
            }
        }
    }
}
```

在该验证器中，使用了来自于 ValidationUtils 类的 rejectIfEmptyOrWhitespace()与 rejectIfEmpty()方法来验证表单字段。如果表单字段为空，那么这些方法会创建一个字段 error 并将其绑定到字段上。这些方法的第二个参数是属性名，第三个与第四个字段分别是错误码和默认的错误消息。

还可以检查预订时间在假期与工作日是否合法。如果不合法，就应该使用 reject()方法创建一个对象 error 并绑定到 reservation 对象而不是字段上。

由于验证类使用了@Component 注解，因此 Spring 会尝试根据类名将类实例化为 bean，即 reservationValidator。

由于验证器会在验证过程中创建错误，因此应该为错误码定义消息以显示给用户。如果定义了 ResourceBundleMessageSource，那就可以针对恰当的地域信息在资源包中定义如下错误消息（比如说，针对默认地域的 messages.properties）；还可以参考 3-5 节：

```
required.courtName=Court name is required
required.date=Date is required
required.hour=Hour is required
required.playerName=Player name is required
required.sportType=Sport type is required
invalid.holidayHour=Invalid holiday hour
invalid.weekdayHour=Invalid weekday hour
```

为了应用这个验证器，你需要对控制器进行如下修改：

```
package com.apress.springrecipes.court.service;
.....
    private final ReservationService reservationService;
    private final ReservationValidator reservationValidator;

    public ReservationFormController(ReservationService reservationService,
        ReservationValidator reservationValidator) {
        this.reservationService = reservationService;
        this.reservationValidator = reservationValidator;
    }

    @RequestMapping(method = RequestMethod.POST)
    public String submitForm(
        @ModelAttribute("reservation") @Validated Reservation reservation,
        BindingResult result, SessionStatus status) {
        if (result.hasErrors()) {
            return "reservationForm";
        } else {
            reservationService.make(reservation);
            return "redirect:reservationSuccess";
        }
    }

    @InitBinder
    public void initBinder(WebDataBinder binder) {
        binder.setValidator(reservationValidator);
    }
```

首先向控制器添加了 ReservationValidator 字段，它可以让控制器访问验证器 bean 的实例。

接下来修改了 HTTP POST 处理器方法，当用户提交表单时会调用它。@ModelAttribute 注解旁边现在增加了一个@Validated 注解，它会触发对对象的验证。验证后，结果参数（BindingResult 对象）就会包含验证过程的结果。接下来会根据 result 的值进行 hasErrors()方法的条件判断。如果验证类检测到了错误，那么该值就为 true。

如果验证过程中检测到了错误，那么方法处理器就会返回视图 reservationForm，它对应于相同的视图，这样用户就可以重新提交信息了。如果验证过程没有检测到错误，那就会执行预订的调用（reservationService.make(reservation);），然后重定向到成功视图 reservationSuccess。

验证器的注册是在@InitBinder 所注解的方法中完成的，同时验证器是在 WebDataBinder 上设置的，这样绑定后就可以使用了。为了注册验证器，你需要使用 setValidator 方法。还可以通过 addValidators 方法注册多个验证器，它接收一个可变参数来注册一个或多个 Validator 实例。

■ **注意：**WebDataBinder 还可针对类型转换注册额外的 ProperyEditor、Converter 与 Formatter 实例。可以通过这种方式取代全局的 ProperyEditor、Converter 与 Formatter 注册。

■ **提示：**相较于编写自定义的 Spring Validator 实例，还可以使用 JSR-303 验证并为字段添加注解来对其进行验证。

5-7 使用反应式 REST 服务发布和消费 JSON

问题提出

你想要以反应式的方式发布 XML 或 JSON 服务。

解决方案

使用 4-1 节与 4-2 节中所介绍的相同的声明，可以编写反应式端点。

解释说明

要想发布 JSON，可以使用@ResponseBody 或@RestController 注解。通过返回一个反应式类型（Mono 或 Flux），可以实现块状响应。如何处理结果取决于所请求的表述形式。在消费 JSON 时，可以通过@ResponseBody 对 Mono 或 Flux 类型的反应式方法参数进行注解以实现反应式消费。

发布 JSON

通过对请求处理方法应用注解@ResponseBody，输出会以 JSON 或 XML（取决于请求返回类型与类路径下的库）的格式返回。除了对方法应用注解@ResponseBody，还可以在类级别上使用@RestController 注解，这会自动对所有请求处理方法生效。

我们来编写一个 REST 控制器，它会返回系统中的所有预订信息。在类上使用注解@RestController，并声明一个使用了@GetMapping 注解的方法，该方法返回一个 Flux<Reservation>对象。

```java
package com.apress.springrecipes.reactive.court.web;

import com.apress.springrecipes.reactive.court.Reservation;
import com.apress.springrecipes.reactive.court.ReservationService;
import org.springframework.web.bind.annotation.GetMapping;
import org.springframework.web.bind.annotation.RequestMapping;
import org.springframework.web.bind.annotation.RestController;
import reactor.core.publisher.Flux;

@RestController
@RequestMapping("/reservations")
public class ReservationRestController {

    private final ReservationService reservationService;

    public ReservationRestController(ReservationService reservationService) {
        this.reservationService = reservationService;
    }

    @GetMapping
    public Flux<Reservation> listAll() {
        return reservationService.findAll();
    }
}
```

当返回类似于此的反应式类型时，结果会以流式 JSON/XML 或服务器端发送事件（见 5-2 节）的形式返回给客户端。结果取决于客户端的 Accept-Header 请求头。使用 httpie 并执行 http http://localhost:8080/court-rx/reservations --stream 会返回 JSON。当添加了 Accept:text/event-stream 时，结果会以服务器端发送事件的形式发布。

消费 JSON

除了生产 JSON 外，还可以消费它。要想做到这一点，请添加一个方法参数并对其使用注解 @RequestBody。传入的 JSON 请求体会被映射到该对象上。对于反应式控制器来说，可以将单个与多个结果分别包装到 Mono 与 Flux 中。

首先创建一个简单的 POJO，它接收一个 courtName，这样就可以查询预订信息了。

```
package com.apress.springrecipes.reactive.court.web;

public class ReservationQuery {

    private String courtName;

    public String getCourtName() {
        return courtName;
    }

    public void setCourtName(String courtName) {
        this.courtName = courtName;
    }
}
```

这是个基本的 POJO，将会通过 JSON 进行属性填充。现在来看看控制器。向控制器添加一个方法，该方法接收 Mono<ReservationQuery>作为参数。

```
@PostMapping
public Flux<Reservation> find(@RequestBody Mono<ReservationQuery> query) {
    return query.flatMapMany(q -> reservationService.query(q.getCourtName()));
}
```

现在，当带有 JSON 体的请求到来时，它会被反序列化为 ReservationQuery 对象。为了做到这一点，Spring WebFlux 使用了（就像 Spring MVC 一样）一个转换器。转换会被委托给 HttpMessageReader 的一个实例，在该示例中就是 DecoderHttpMessageReader。该类会将反应式流解码为对象。这再一次会被委托给一个 Decoder 对象。由于想要使用 JSON（确保 Jackson 2 JSON 库位于类路径上），它会使用 Jackson2JsonDecoder。HttpMessageReader 与 Decoder 实现是常规的 Spring MVC 所用的 HttpMessageConverter 的反应式对等组件。

使用 httpie 并执行请求 http POST http://localhost:8080/court-rx/reservations courtName="Tennis #1" --stream，你会得到场地 Tennis #1 的所有结果。该命令会向服务器发送如下 JSON：

```
{ courtName: "Tennis #1"}
```

5-8　使用异步 Web 客户端

问题提出

你想在 Spring 应用中访问第三方（如 Google、Yahoo!，或是其他业务方）的 REST 服务并使用其负载。

解决方案

在 Spring 应用中访问第三方的 REST 服务时需要使用 Spring WebClient 类。WebClient 类的设计与其他很多 Spring *Template 类（如 JdbcTemplate 和 JmsTemplate 等）遵循了同样的原则，通过默认行为提供了一种简化的方式来执行冗长的任务。

这意味着在 Spring 应用中对调用 REST 服务并使用其返回的负载的处理都实现了流水化作业。

> **注意：**在 Spring 5 之前，可以使用 AsyncRestTemplate；不过，从 Spring 5 开始，AsyncRestTemplate 已经被 WebClient 替代了。

解释说明

在介绍 WebClient 类的细节之前，我们有必要了解一下 REST 服务的生命周期，这样就能知晓 WebClient 类所执行的实际工作了。探索 REST 服务的生命周期最好通过浏览器来实现，在工作站中打开钟爱的浏览器来开始吧。

首先，我们需要一个 REST 服务端点。这里重用 5-7 节中所创建的端点。该端点位于 http://localhost:8080/court-rx/reservations。如果在浏览器中加载这个 REST 服务端点，浏览器就会执行一个 GET 请求，这是 REST 服务所支持的诸多流行的 HTTP 请求之一。在加载 REST 服务时，浏览器会展示出对应的负载，如图 5-6 所示。

REST 服务消费者（比如说你自己）需要知道 REST 服务的负载结构（有时叫做词汇表），这样才能恰当地处理其信息。虽然 REST 服务依赖于自定义的词汇表，但一系列 REST 服务通常会依赖于标准化的词汇表（如 RSS），这可以统一处理 REST 服务负载。此外，值得注意的是，有些 REST 服务提供了 Web 应用描述语言（Web Application Description Language，WADL）契约来简化负载的发现与消费。

在通过浏览器了解了 REST 服务的生命周期后，现在可以看看如何使用 Spring WebClient 类将 REST 服务负载纳入到 Spring 应用中。考虑到 WebClient 类设计的目的就是为了调用 REST 服务，因此其主要的方法都会紧密绑定到 REST 底层，即 HTTP 协议的方法：HEAD、GET、POST、PUT、DELETE 与 OPTIONS。表 5-6 所示为 WebClient 类所支持的主要方法。

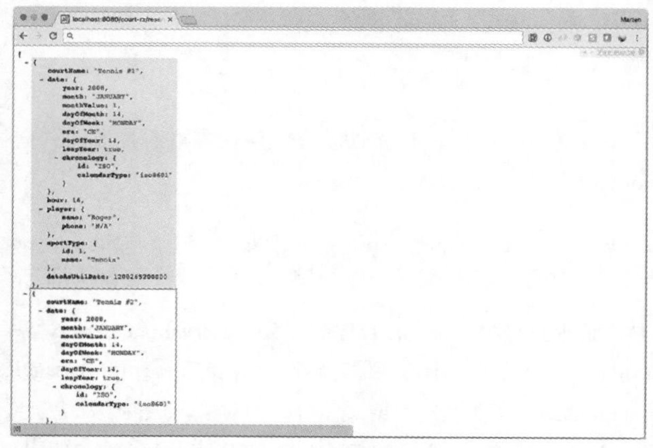

图 5-6 生成的 JSON

表 5-6 基于 HTTP 协议请求方法的 WebClient 类的方法

方法	说明
create	创建一个 WebClient，还可以指定一个默认 URL
head()	准备一个 HTTP HEAD 操作
get()	准备一个 HTTP GET 操作
post()	准备一个 HTTP POST 操作
put()	准备一个 HTTP PUT 操作
options()	准备一个 HTTP OPTIONS 操作
patch()	准备一个 HTTP PATCH 操作
delete()	准备一个 HTTP DELETE 操作

从表 5-6 中可以看到，WebClient 类的构建器方法都是根据 HTTP 协议方法建模的，包括 HEAD、GET、POST、PUT、DELETE 和 OPTIONS。

■ **注意**：到目前为止，REST 服务中最常用的 HTTP 方法是 GET，因为它表示一种获取信息的安全操作（即它并不会修改任何数据）。另外，诸如 PUT、POST 与 DELETE 等 HTTP 方法都会修改提供者的信息，这使得它们不太受 REST 服务提供者的支持。如果需要修改数据，很多提供者倾向于 SOAP 协议，这是 REST 服务的一种替代机制。

在了解了 WebClient 类基本的构建器方法后，现在可以调用之前在浏览器中调用的同一个 REST 服务，只不过这次使用的是 Spring 框架的 Java 代码。如下代码所展示的类会访问 REST 服务并将其内容发送给 System.out：

```
package com.apress.springrecipes.reactive.court;

import org.springframework.http.MediaType;
import org.springframework.web.reactive.function.client.WebClient;

import java.io.IOException;

public class Main {

    public static void main(String[] args) throws IOException {
        final String url = "http://localhost:8080/court-rx";
        WebClient.create(url)
            .get()
            .uri("/reservations")
            .accept(MediaType.APPLICATION_STREAM_JSON)
            .exchange()
            .flatMapMany(cr -> cr.bodyToFlux(String.class)).subscribe(System.out::println);

        System.in.read();
    }
}
```

警告：一些 REST 服务提供者会根据请求方的不同限制对其数据源的访问。通常会根据请求中的数据（如 HTTP 头或是 IP 地址）来拒绝访问。因此，根据环境的不同，提供者可能会返回拒绝访问响应，即便数据源可能在另外一个环境中使用（比如说，你可能在浏览器中能够访问 REST 服务，但在 Spring 应用中访问同样的数据源就会得到拒绝访问响应）。这取决于 REST 提供者预设的使用条款。

第一行声明了一条 import 语句以便在类中使用 WebClient 类，首先，需要通过 WebClient.create 创建一个 WebClient 类的实例。接下来，调用了 WebClient 类的 get()方法，表 5-6 对此已经进行了介绍，它用于准备一个 HTTP GET 操作，就像通过浏览器所执行的获取 REST 服务的负载那样。接下来，扩展了要调用的 URL，因为想要调用 http://localhost:8080/courtrx/reservations 且需要一个 JSON 流，这正是使用 accept(MediaType.APPLICATION_STREAM_JSON)的原因所在。

接下来，调用 exchange()会将配置从设置请求切换至定义响应处理。由于可能得到零个或多个元素，因此需要将 ClientResponse 体转换为 Flux。可以调用 ClientResponse 的 bodyToFlux 方法来做到这一点（如果想要执行自定义转换，可以使用普通的 body 方法，或者使用 bodyToMono 方法转换为单元素的结果）。由于要将每个元素写到 System.out，因此订阅它。

执行应用后，输出与使用浏览器时是一样的，只不过现在会打印到控制台上。

从参数化 URL 中获取数据

上一节介绍了如何通过调用 URI 来获取数据，不过如果 URI 需要参数该怎么做呢？你并不想将参数硬编码到 URL 中。借助于 WebClient 类，可以通过占位符来使用 URL，在执行时，这些占位符会被替换为实际值。占位符是通过 { 和 } 定义的，就像请求映射一样（参见 4-1 节与 4-2 节）。

http://localhost:8080/court-rx/reservations/{courtName}就是一种参数化 URI。要想调用该方法，你需要为占位符传递一个值。这可以通过向 WebClient 类的 uri 方法传递参数来实现。

```
public class Main {

    public static void main(String[] args) throws Exception {
        WebClient.create(url)
            .get()
            .uri("/reservations/{courtName}", "Tennis")
```

```
                .accept(MediaType.APPLICATION_STREAM_JSON)
                .exchange()
                .flatMapMany(cr -> cr.bodyToFlux(String.class))
                .subscribe(System.out::println);

            System.in.read();
        }
    }
```

将数据作为映射对象返回

相较于返回一个 String 在应用中使用，你还可以重用 Reservation、Player 与 SportType 类来映射结果。相较于传递 String.class 作为 bodyToFlux 方法的参数，可以传递 Reservation.class，这时响应就会映射到该类上。

```
package com.apress.springrecipes.reactive.court;

import org.springframework.http.MediaType;
import org.springframework.web.reactive.function.client.WebClient;

import java.io.IOException;

public class Main {

    public static void main(String[] args) throws IOException {
        final String url = "http://localhost:8080/court-rx";

        WebClient.create(url)
            .get()
            .uri("/reservations")
            .accept(MediaType.APPLICATION_STREAM_JSON)
            .exchange()
            .flatMapMany(cr -> cr.bodyToFlux(Reservation.class))
            .subscribe(System.out::println);

        System.in.read();
    }
}
```

当控制器使用了被@ResponseBody 标记的方法时，WebClient 类会使用同样的 HttpMessageReader 基础设施。由于 JAXB 2（以及 Jackson）会被自动检测到，因此映射到对象就是非常简单的事情了。

5-9　编写反应式处理器函数

问题提出

你想要编写可以对请求进行响应的函数。

解决方案

你可以编写一个接收 ServerRequest 并返回一个 Mono<ServerResponse>的方法，并将其映射为路由函数。

解释说明

相较于使用@RequestMapping 将请求映射到方法上，还可以借助于 HandlerFunction 接口来编写函数。

```
package org.springframework.web.reactive.function.server;

import reactor.core.publisher.Mono;

@FunctionalInterface
public interface HandlerFunction<T extends ServerResponse> {

    Mono<T> handle(ServerRequest request);

}
```

如上代码所示，HandlerFunction 方法接收一个 ServerRequest 参数并返回一个 Mono<ServerResponse>。

ServerRequest 与 ServerResponse 都能以完全的反应式的方式访问底层的请求与响应；这是通过将其各部分以 Mono 或是 Flux 流的方式向上公开而实现的。

编写好函数后，可以通过 RouterFunctions 将其映射到传入的请求上。可以对 URL、请求头、方法或是自定义的 RequestPredicate 类进行映射。可以通过 RequestPredicates 类来访问默认的请求谓词。

编写处理器函数

重写 ReservationRestController 来实现简单的请求处理函数。

为此，删除所有的请求映射注解并向该类添加一个简单的@Component 注解。接下来，重写其中的方法，使之符合 HandlerFunction 接口的签名要求。

```
package com.apress.springrecipes.reactive.court.web;

import com.apress.springrecipes.reactive.court.Reservation;
import com.apress.springrecipes.reactive.court.ReservationService;
import org.springframework.stereotype.Component;
import org.springframework.web.bind.annotation.*;
import org.springframework.web.reactive.function.server.ServerRequest;
import org.springframework.web.reactive.function.server.ServerResponse;
import reactor.core.publisher.Flux;
import reactor.core.publisher.Mono;

@Component
public class ReservationRestController {

    private final ReservationService reservationService;

    public ReservationRestController(ReservationService reservationService) {
        this.reservationService = reservationService;
    }

    public Mono<ServerResponse> listAll(ServerRequest request) {
        return ServerResponse.ok().body(reservationService.findAll(), Reservation.class);
    }

    public Mono<ServerResponse> find(ServerRequest request) {
        return ServerResponse
            .ok()
            .body(
                request.bodyToMono(ReservationQuery.class)
                    .flatMapMany(q -> reservationService.query(q.getCourtName())),
                    Reservation.class);
    }
}
```

该类需要依赖于 ReservationService。注意对 listAll 与 find 方法的修改。它们现在都返回 Mono<ServerResposne>并接收一个 ServerRequest 作为参数。由于想要返回 HTTP 状态 OK（200），可以使用 ServerResponse.ok()来构建该响应。你需要添加一个响应体 Flux<Reservation>，并且指定元素类型 Reservation.class。之所以需要后者是因为在组合函数时无法读取反应式与泛型类型信息。

在 find 方法中，有些内容是类似的，不过首先通过 bodyToMono 将传入的请求体映射到 ReservationQuery 上。该结果最终用于调用 ReservationService 的 query 方法。

将请求路由到处理器函数

由于现在使用了简单的函数而不是基于注解的请求处理方法，因此路由的使用也是不同的。可以通过 RouterFunctions 进行映射。

```
@Bean
public RouterFunction<ServerResponse> reservationsRouter(ReservationRestController handler)
{
```

```
        return RouterFunctions
            .route(GET("/*/reservations"), handler::listAll)
            .andRoute(POST("/*/reservations"), handler::find);
    }
```

当 HTTP GET 请求了/court-rx/reservations 地址后，listAll 方法会调用，当 HTTP POST 请求到来时，则会调用 find 方法。

使用 RequestPredicates.GET 与 RequestPredicates.method(HttpMethod.GET.)and(RequestPredicates.path("/*/reservations"))的效果是一样的。可以根据需要组合多个 RequestPredicate 语句。表 5-7 列出的方法都是 RequestPredicates 类中的。

表 5-7　默认的 RequestPredicates

方法	说明
method	针对 HTTP 方法的 RequestPredicate
path	针对 URL 或部分 URL 的 RequestPredicate
accept	针对 Accept 头的 RequestPredicate，用于匹配请求的媒体类型
queryParam	用于检查请求参数存在与否的 RequestPredicate
headers	用于检查请求头存在与否的 RequestPredicate

RequestPredicates 还为 GET、POST、PUT、DELETE、HEAD、PATCH 与 OPTIONS 提供了简洁的方法。这样就不必组合两个表达式了。

小结

本章介绍了实现异步处理的各种方式。传统方式是使用 Servlet 3.1 的异步支持并让控制器返回一个 DeferredResult 或 Future。

关于通信，本章介绍了服务器端发送事件与 WebSocket 通信。这可以实现客户端与服务器之间的异步通信。

接下来，我们介绍了如何编写反应式控制器，这与第 3 章和第 4 章所介绍的内容并不是完全不同的。同时，这也展现了 Spring 强大的抽象能力；可以将几乎相同的编程模型用于完全不同的技术中。编写完反应式控制器后，我们又介绍了如何编写反应式处理器函数，它几乎能实现反应式控制器所做的一切，只不过是以一种更加函数式的方式来实现的。

我们还介绍了如何通过 WebClient 类来异步消费 REST API。

第 6 章

Spring Social

社交网络无处不在，大多数互联网用户都有一个或多个社交网络账号。人们发推文分享自己的见闻或是对一件事情的看法；在 Facebook 和 Instagram 上分享照片，使用 Tumblr 撰写博客。每天都有很多社交网络涌现出来。作为网站的站长，集成这些社交网络的好处是很多的，可以让用户轻松发布链接，展现每个人的观点。

Spring Social 旨在提供一个统一的 API 来连接这些不同的网络并提供一种扩展模型。Spring Social 本身集成了 Facebook、Twitter 和 LinkedIn；不过，还有很多社区项目提供了对不同社交网络的支持（比如说 Tumblr、Weibo 和 Instagram 等）。Spring Social 可以分为 3 部分。第一部分是 Connect Framework，处理底层社交网络的认证与连接流程。第二部分是 ConnectController，它是个控制器，执行服务提供者、消费者（应用）与应用的用户之间的 OAuth 交换。最后是 SocialAuthenticationFilter，它集成了 Spring Social 与 Spring Security（见第 7 章），使得用户能够使用自己的社交网络账号登录。

6-1 搭建 Spring Social

问题提出

你想在自己的应用中使用 Spring Social。

解决方案

将 Spring Social 添加到依赖中并在应用中启用 Spring Social。

解释说明

Spring Social 包含了几个核心模块，并为每个服务提供者提供了扩展模块（如 Twitter、Facebook 和 GitHub 等）。要想使用 Spring Social，你需要将其添加到应用依赖中。表 6-1 所示为可用的模块。

表 6-1 Spring Social 模块概览

模块	说明
spring-social-core	Spring Social 核心模块；包含了主要与共享的基础设施类
spring-social-config	Spring Social 配置模块；可以简化 Spring Social 的配置
spring-social-web	Spring Social 的 Web 集成，包含了过滤器与控制器，可以简化使用
spring-social-security	集成了 Spring Security（见第 7 章）

依赖位于 org.springframework.social 组中。本章会介绍每个模块（core、config、web 与 security）。添加好依赖后就可以搭建 Spring Social 了。

```
package com.apress.springrecipes.social.config;

import com.apress.springrecipes.social.StaticUserIdSource;
import org.springframework.context.annotation.*;
```

```
import org.springframework.core.env.Environment;
import org.springframework.social.config.annotation.EnableSocial;
import org.springframework.social.config.annotation.SocialConfigurerAdapter;

@Configuration
@EnableSocial
@PropertySource("classpath:/application.properties")
public class SocialConfig extends SocialConfigurerAdapter {

    @Override
    public StaticUserIdSource getUserIdSource() {
        return new StaticUserIdSource();
    }
}
```

要想启用 Spring Social，只需向被@Configuration 所注解的类添加@EnableSocial 注解即可。该注解会触发 Spring Social 配置的加载。它会检测 SocialConfigurer bean 的实例，该 bean 用于进一步配置 Spring Social。它们用于添加一个或多个服务提供者的配置。

SocialConfig 继承了 SocialConfigurerAdapter，它是 SocialConfigurer 的一个实现。可以看到，它有一个名为 getUserIdSource 的重写方法，该方法返回一个 StaticUserIdSource 对象。Spring Social 需要一个 UserIdSource 实例来确定当前用户。该用户用于查询与服务提供者的连接。这些连接会根据用户存储到每个 ConnectionRepository 中。UsersConnectionRepository 则决定该使用哪个 ConnectionRepository，它会用到当前用户。默认配置的 UsersConnectionRepository 是 InMemoryUsersConnectionRepository。

最后，从类路径中加载一个属性文件。该属性文件包含了应用要用到的针对服务提供者的 API key。除了将它们放到一个属性文件中，还可以将其硬编码到程序中。

现在使用 StaticUserIdSource 来确定当前用户。

```
package com.apress.springrecipes.social;

import org.springframework.social.UserIdSource;

public class StaticUserIdSource implements UserIdSource {

    private static final String DEFAULT_USERID = "anonymous";
    private String userId = DEFAULT_USERID;

    @Override
    public String getUserId() {
        return this.userId;
    }

    public void setUserId(String userId) {
        this.userId = userId;
    }
}
```

StaticUserIdSource 实现了 UserIdSource 并返回一个预设的 userId。虽然现在可以这么做，但在实际应用中，你需要根据每个用户来存储连接信息。

6-2 连接到 Twitter

问题提出

你想让应用能够访问 Twitter。

解决方案

将应用注册到 Twitter 上，配置 Spring Social 以使用应用凭证来访问 Twitter。

解释说明

在应用使用 Twitter 前，你需要将应用注册到 Twitter 上。注册后，就拥有了识别应用的凭证了（API key

与 API secret）。

将应用注册到 Twitter 上

要想将应用注册到 Twitter 上，请访问 https://dev.twitter.com 并在页面右上角找到你的头像；在下拉菜单中选择 My apps，如图 6-1 所示。

图 6-1　在 Twitter 上选择 My apps

选择好 My apps 后会出现 Application Management 页面。在该页面上有个按钮，单击它就可以创建新的应用了，如图 6-2 所示。

在该页面上，单击按钮打开新的界面（见图 6-3）来注册应用。

在该界面上，你需要输入应用的名字与说明信息以及应用所用的网站的 URL。当使用 Spring Social 时，还需要填写回调 URL 字段，因为需要用到回调；实际的值并不重要（除非使用非常老的版本的 OAuth）。

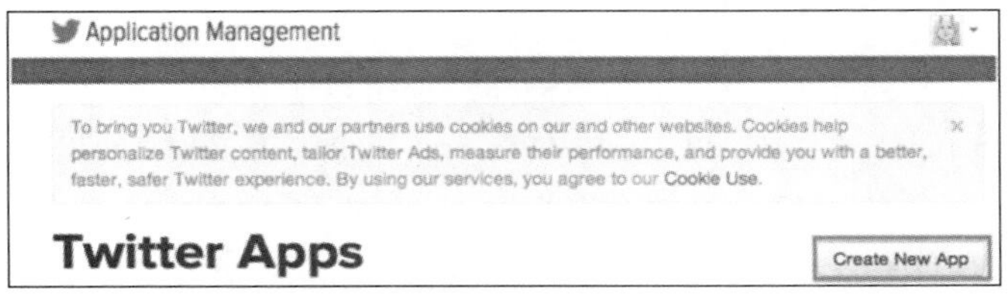

图 6-2　Application Management 页面

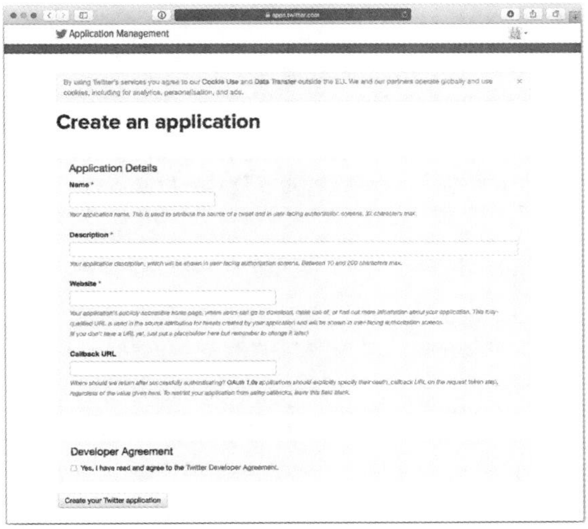

图 6-3　注册新的应用

接受条款后，单击最后的创建按钮，你会进入到应用设置页面。这意味着应用已经创建成功了。

要想将 Spring Social 连接到 Twitter 上，你需要知道 API key 与 API secret。可以在应用设置页面的 API Keys 选项卡下找到它们（见图 6-4 与图 6-5）。

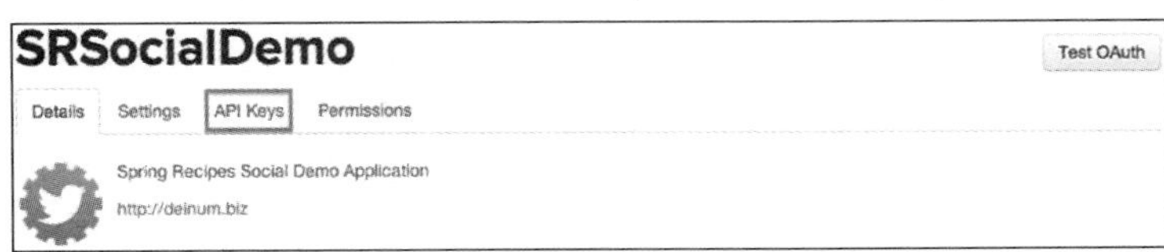

图 6-4　应用设置页面

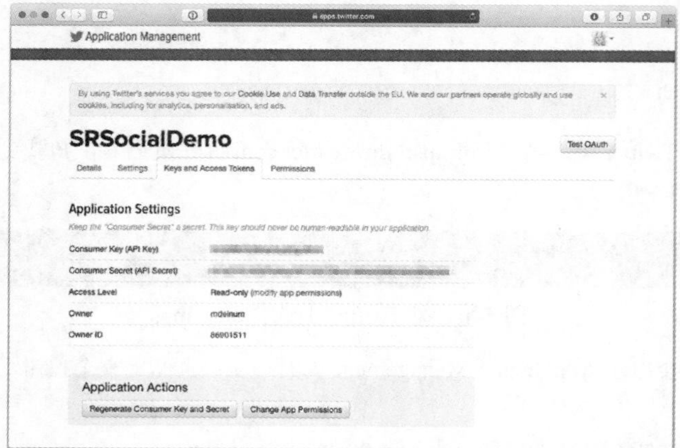

图 6-5　连接 Spring Social 所需的 API key 与 API secret

配置 Spring Social 以连接到 Twitter

在有了 API key 与 API secret 后，就可以配置 Spring Social 以连接到 Twitter 了。首先创建一个属性文件（比如说 application.properties）来存放 API key 与 API secret，这样在需要时就可以轻松获取到它们了。

```
twitter.appId=<your-twitter-API-key-here>
twitter.appSecret=<your-twitter-API-secret-here>
```

要想连接到 Twitter，你需要添加一个 TwitterConnectionFactory，在连接到 Twitter 时它会使用应用 ID 与 secret。

```
package com.apress.springrecipes.social.config;

import org.springframework.core.env.Environment;
import org.springframework.social.config.annotation.ConnectionFactoryConfigurer;
import org.springframework.social.connect.Connection;
import org.springframework.social.connect.ConnectionRepository;
import org.springframework.social.twitter.api.Twitter;
import org.springframework.social.twitter.connect.TwitterConnectionFactory;

@Configuration
@EnableSocial
@PropertySource("classpath:/application.properties")
public class SocialConfig extends SocialConfigurerAdapter {
...
    @Configuration
    public static class TwitterConfigurer extends SocialConfigurerAdapter {

        @Override
        public void addConnectionFactories(
            ConnectionFactoryConfigurer connectionFactoryConfigurer,
            Environment env) {

            connectionFactoryConfigurer.addConnectionFactory(
                new TwitterConnectionFactory(
                    env.getRequiredProperty("twitter.appId"),
                    env.getRequiredProperty("twitter.appSecret")));
        }

        @Bean
        @Scope(value = "request", proxyMode = ScopedProxyMode.INTERFACES)
        public Twitter twitterTemplate(ConnectionRepository connectionRepository) {

            Connection<Twitter> connection = connectionRepository.
            findPrimaryConnection(Twitter.class);
```

```
                return connection != null ? connection.getApi() : null;
            }
        }
    }
```

SocialConfigurer 接口拥有回调方法 addConnectionFactories，可以通过它添加 ConnectionFactory 实例来使用 Spring Social。对于 Twitter 来说，有 TwitterConnectionFactory，它接收两个参数。第一个是 API key，第二个是 API secret。这两个构造方法参数都来自于属性文件。当然，还可以将值硬编码到配置中。这样，与 Twitter 的连接就创建好了。虽然可以使用原生的底层连接，但并不推荐这么做。相反，请使用 TwitterTemplate，这可以更加轻松地使用 Twitter API。之前的配置向应用上下文添加了一个 TwitterTemplate。

注意@Scope 注解。这是一个请求作用域的 bean，对于每个请求来说，到 Twitter 的实际连接可能都是不同的，因为每个请求代表的是不同的用户，这就是需要使用请求作用域的原因所在。向方法中注入的 ConnectionRepository 是根据当前用户的 ID 来确定的，它是通过之前配置的 UserIdSource 来获取的。

注意： 虽然示例使用了单独的配置类将 Twitter 配置为一个服务提供者，但还可以将其添加到 SocialConfig 类中。不过，最好将全局的 Spring Social 配置与具体的服务提供者安装信息分隔开来。

6-3 连接到 Facebook

问题提出

你想让应用能够访问 Facebook。

解决方案

将应用注册到 Facebook 上，配置 Spring Social 以使用应用凭证来访问 Facebook。

解释说明

在应用使用 Facebook 前，首先需要将应用注册到 Facebook 上。注册后，就拥有了识别应用的凭证了（API key 与 API secret）。要想将应用注册到 Facebook 上，你需要拥有一个 Facebook 账号且需要注册成为开发者（这里假设你已经注册成为了 Facebook 开发者。如果没有，请访问 http://developers.facebook.com，单击 Register Now 按钮，然后填写好各个字段信息）。

将应用注册到 Facebook 上

访问 http://developers.facebook.com，单击页面顶部的 My Apps 菜单，选择 Add a New App，如图 6-6 所示。

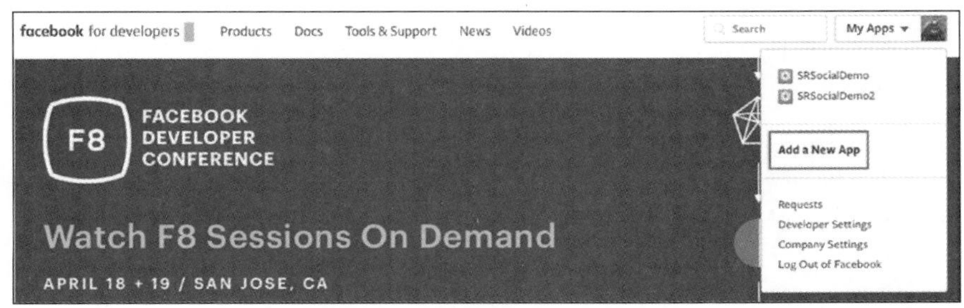

图 6-6　注册新应用的第一步

这会打开一个界面，如图 6-7 所示，请填写关于应用的一些细节信息。

应用名可以填写任何内容，只要不包含单词 face 或 book 即可。还需要填写一个电子邮件地址，这样 Facebook 就知道如何联系你了。表单填写完毕后，单击 Create App ID 按钮，这会进入到应用页面，如图 6-8 所示。在该页面中，单击 Settings 选项卡。

Create a New App ID
Get started integrating Facebook into your app or website
Display Name
The name you want to associate with this App ID
Contact Email
Used for important communication about your app
By proceeding, you agree to the Facebook Platform Policies
Cancel　Create App ID

图 6-7　创建新 App ID 窗口

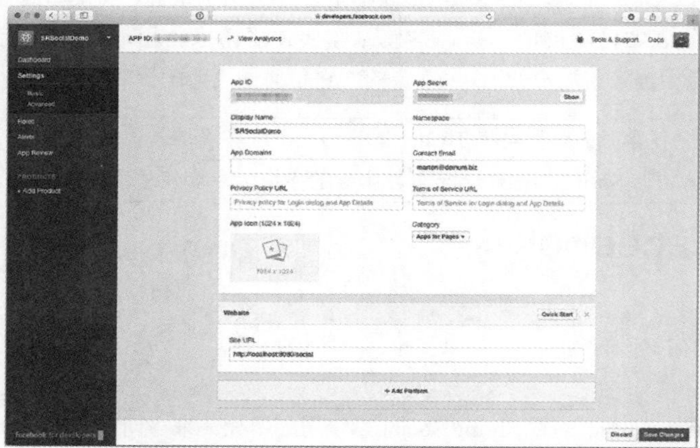

图 6-8　Facebook 设置页面

在设置页面中，单击 Add Platform 按钮并选择 Website。输入应用所属网站的 URL。在该练习中，URL 就是 http://localhost:8080/social。如果不写 URL，那就无法授权，也无法创建连接。

配置 Spring Social 以连接到 Facebook

Facebook 设置页面也有应用连接到 Facebook 所需的应用 ID 与 secret。将它们存储到 application.properties 文件中。

```
facebook.appId=<your app-id here>
facebook.appSecret=<your app-secret here>
```

假设 Spring Social 已经配置完毕（参见 6-1 节），只需要添加一个 FacebookConnectionFactory 和 FacebookTemplate 就可以轻松访问了。

```
package com.apress.springrecipes.social.config;

import org.springframework.social.facebook.api.Facebook;
import org.springframework.social.facebook.connect.FacebookConnectionFactory;

@Configuration
@EnableSocial
@PropertySource("classpath:/application.properties")
public class SocialConfig extends SocialConfigurerAdapter {

...

    @Configuration
    public static class FacebookConfiguration extends SocialConfigurerAdapter {

        @Override
        public void addConnectionFactories(
            ConnectionFactoryConfigurer connectionFactoryConfigurer,
            Environment env) {
```

```
            connectionFactoryConfigurer.addConnectionFactory(
                new FacebookConnectionFactory(
                    env.getRequiredProperty("facebook.appId"),
                    env.getRequiredProperty("facebook.appSecret")));
        }

        @Bean
        @Scope(value = "request", proxyMode = ScopedProxyMode.INTERFACES)
        public Facebook facebookTemplate(ConnectionRepository connectionRepository) {
            Connection<Facebook> connection = connectionRepository.
            findPrimaryConnection(Facebook.class);
            return connection != null ? connection.getApi() : null;
        }
    }
}
```

FacebookConnectionFactory 需要应用 ID 和 secret。这两个属性都被存储到了 application.properties 文件中，并且可以通过 Environment 对象获取。

之前的 bean 配置向应用上下文添加了一个名为 facebookTemplate 的 bean。注意@Scope 注解。这是一个请求作用域的 bean，对于每个请求来说，到 Facebook 的实际连接可能都是不同的，因为每个请求代表的是不同的用户，这就是需要使用请求作用域的原因所在。向方法中注入的 ConnectionRepository 是根据当前用户的 ID 来确定的，它是通过之前配置的 UserIdSource 来获取的（参见 6-1 节）。

注意： 虽然示例使用了单独的配置类将 Facebook 配置为了一个服务提供者，但还可以将其添加到 SocialConfig 类中。不过，最好将全局的 Spring Social 配置与具体的服务提供者安装信息分隔开来。

6-4 展示服务提供者的连接状态

问题提出

你想要展示所用的服务提供者的连接状态。

解决方案

配置 ConnectController，通过它向用户展示状态。

解释说明

Spring Social 自带 ConnectController，它负责创建与断开与服务提供者之间的连接，不过还可以展示所用服务提供者下当前用户的状态（连接还是断开）。ConnectController 使用几个 REST URL 来展示、添加或是移除给定用户的连接（见表 6-2）。

表 6-2 ConnectController URL 映射

URL	方法	说明
/connect	GET	展示可用服务提供者的连接状态。会将 connect/status 作为视图名返回给用户
	POST	开始建立与给定提供者的连接
	DELETE	移除给定提供者下当前用户的连接

要想使用该控制器，首先需要配置 Spring MVC（参见第 4 章）。对于该示例来说，请添加如下配置：

```
package com.apress.springrecipes.social.config;

import org.springframework.context.annotation.Bean;
import org.springframework.context.annotation.ComponentScan;
import org.springframework.context.annotation.Configuration;
import org.springframework.web.servlet.ViewResolver;
import org.springframework.web.servlet.config.annotation.EnableWebMvc;
import org.springframework.web.servlet.config.annotation.ViewControllerRegistry;
```

```
import org.springframework.web.servlet.config.annotation.WebMvcConfigurer;
import org.springframework.web.servlet.view.InternalResourceViewResolver;

@Configuration
@EnableWebMvc
@ComponentScan({"com.apress.springrecipes.social.web"})
public class WebConfig implements WebMvcConfigurer {

    @Bean
    public ViewResolver internalResourceViewResolver() {
        InternalResourceViewResolver viewResolver = new InternalResourceViewResolver();
        viewResolver.setPrefix("/WEB-INF/views/");
        viewResolver.setSuffix(".jsp");
        return viewResolver;
    }

    @Override
    public void addViewControllers(ViewControllerRegistry registry) {
        registry.addViewController("/").setViewName("index");
        registry.addViewController("/signin").setViewName("signin");
    }
}
```

需要通过@EnableWebMvc 注解启用 Spring MVC 并添加一个 ViewResolver，这样就会选择使用 JSP 页面。最后，在应用启动后想要展示 index.jsp 页面。向 WebConfig 类添加 ConnectController。该控制器需要 ConnectionFactoryLocator 和 ConnectionRepository 作为构造方法参数。要想访问它们，只需将其作为方法参数即可。

```
@Bean
public ConnectController connectController(
ConnectionFactoryLocator connectionFactoryLocator,
ConnectionRepository connectionRepository) {

    return new ConnectController(connectionFactoryLocator, connectionRepository);
}
```

ConnectController 会监听表 6-2 中的 URL。现在向/WEB-INF/views 目录下添加两个视图。第一个是主页，第二个是状态概览页。首先创建 index.jsp 文件。

```
<%@ taglib prefix="spring" uri="http://www.springframework.org/tags" %>
<html>
<head>
    <title>Hello Spring Social</title>
</head>
<body>

<h3>Connections</h3>
    Click <a href="<spring:url value='/connect'/>">here</a> to see your Social Network
    Connections.
</body>
</html>
```

接下来在/WEB-INF/views/connect 目录下创建 status.jsp 文件。

```
<%@ taglib prefix="spring" uri="http://www.springframework.org/tags" %>
<%@ taglib prefix="c" uri="http://java.sun.com/jsp/jstl/core" %>
<html>
<head>
    <title>Spring Social - Connections</title>
</head>
<body>
<h3>Spring Social - Connections</h3>
<c:forEach items="${providerIds}" var="provider">
    <h4>${provider}</h4>
    <c:if test="${not empty connectionMap[provider]}">
        You are connected to ${provider} as ${connectionMap[provider][0].displayName}
```

```
        </c:if>

        <c:if test="${empty connectionMap[provider]}">
            <div>
                You are not yet connected to ${provider}. Click <a href="<spring:url
                value="/connect/${provider}"/>">here</a> to connect to ${provider}.
            </div>
        </c:if>
    </c:forEach>
    </body>
    </html>
```

状态页会遍历所有可用的提供者，确定某个服务提供者（Twitter、Facebook 等）下的当前用户是否有存在的连接。ConnectController 会在 providerIds 属性下指定可用的提供者列表，connectionMap 会持有当前用户的连接。现在启动应用，你需要创建一个 WebApplicationInitializer，它会注册一个 ContextLoaderListener 和 DispatcherServlet 来处理请求。

```
package com.apress.springrecipes.social;

import com.apress.springrecipes.social.config.SocialConfig;
import com.apress.springrecipes.social.config.WebConfig;
import org.springframework.web.filter.DelegatingFilterProxy;

import org.springframework.web.servlet.support.
AbstractAnnotationConfigDispatcherServletInitializer;

import javax.servlet.Filter;

public class SocialWebApplicationInitializer extends
AbstractAnnotationConfigDispatcherServletInitializer {

    @Override
    protected Class<?>[] getRootConfigClasses() {

        return new Class<?>[]{SocialConfig.class};
    }

    @Override
    protected Class<?>[] getServletConfigClasses() {

        return new Class<?>[] {WebConfig.class, };
    }

    @Override
    protected String[] getServletMappings() {

        return new String[] {"/"};
    }
}
```

这会启动应用。SocialConfig 类会被 ContextLoaderListener 加载，WebConfig 类会被 DispatcherServlet 加载。要想处理请求，需要有一个 servlet 映射。对于该示例来说，映射地址为/。

现在一切配置就绪，应用就可以部署并通过 URL http://localhost:8080/social 访问了。这会展示主页，单击链接会展示连接状态页面，最初会显示当前用户尚未连接。

连接到服务提供者

当单击链接来连接到服务提供者时，用户将会发送/connect/{provider} URL。如果连接不存在，就会渲染 connect/{provider}Connect 页面，否则就展示 connect/{provider}Connected 页面。要想通过 ConnectController 连接到 Twitter，你需要添加 twitterConnect.jsp 与 twitterConnected.jsp 页面。对于 Facebook 来说，你需要添加 facebookConnect.jsp 与 facebookConnected.jsp 页面。这种模式也适用于 Spring Social 的其他所有服务提供者连

接器（比如说 GitHub、FourSquare 与 LinkedIn 等）。首先向/WEB-INF/views/connect 目录添加 twitterConnect.jsp。

```
<%@ taglib prefix="spring" uri="http://www.springframework.org/tags" %>
<html>
<head>
    <title>Spring Social - Connect to Twitter</title>
</head>
<body>
<h3>Connect to Twitter</h3>
<form action="<spring:url value='/connect/twitter'/>" method="POST">
    <div class="formInfo">
        <p>You aren't connected to Twitter yet. Click the button to connect this application
        with your Twitter account.</p>
    </div>
    <p><button type="submit">Connect to Twitter</button></p>
</form>
</body>
</html>
```

注意，表单标签向同样的 URL 发出了 POST 请求。当单击 Submit 按钮时会重定向到 Twitter，它会征求你的许可来允许应用访问你的 Twitter 个人信息（将其替换为 Facebook 就可以连接到 Facebook）。

接下来向/WEB-INF/views/connect 目录添加 twitterConnected.jsp。当已经连接到 Twitter，或是对应用授权后并从 Twitter 返回时，都会展示该页面。

```
<%@ taglib prefix="spring" uri="http://www.springframework.org/tags" %>
<html>
<head>
    <title>Spring Social - Connected to Twitter</title>
</head>

<body>
<h3>Connected to Twitter</h3>
<p>
    You are now connected to your Twitter account.
    Click <a href="<spring:url value='/connect'/>">here</a> to see your Connection Status.
</p>
</body>
</html>
```

添加好这些页面后，重启应用并转向状态页面。现在，单击 Connect to Twitter 链接，你会进入到 twitterConnect.jsp 页面。单击 Connect to Twitter 按钮后，你会看到 Twitter 授权应用页面，如图 6-9 所示。

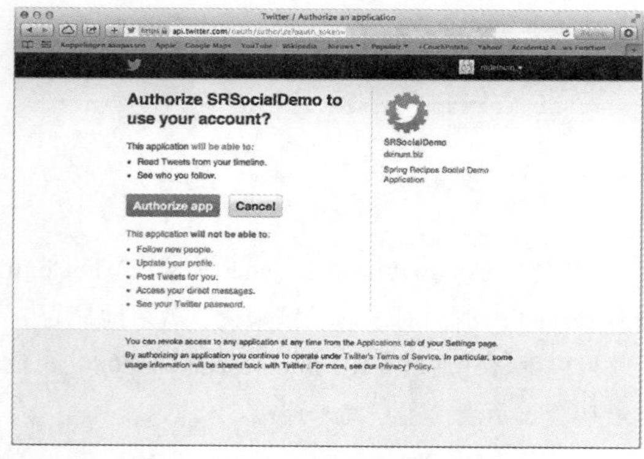

图 6-9　Twitter 授权页面

对应用授权后会返回到 twitterConnect.jsp 页面，告诉你已经成功连接到了 Twitter。当返回到状态页面时，你会看到已经使用你的昵称连接到了 Twitter。

对于 Facebook 或是其他服务提供者来说，请遵循同样的步骤，添加{provider}Connect 与{provider}Connected 页面，Spring Social 就可以连接到该提供者了，请确保添加了正确的服务提供者连接器与配置。

6-5 使用 Twitter API

问题提出

你想要使用 Twitter API。

解决方案

使用 Twitter 对象访问 Twitter API。

解释说明

每个服务提供者都提供了自己的 API 来使用 Twitter。有个实现了 Twitter 接口的对象，它代表 Java 中的 Twitter API；对于 Facebook 来说，有个实现了 Facebook 接口的对象。在 6-2 节中，你已经创建了到 Twitter 的连接以及 TwitterTemplate。TwitterTemplate 公开了 Twitter API 的各个部分，如表 6-3 所示。

表 6-3　Twitter API 公开的操作

操作	说明
blockOperations()	屏蔽与取消屏蔽用户
directMessageOperations()	读取并发送私信
friendOperations()	获取用户的好友与粉丝列表以及关注/取消关注用户
geoOperations()	使用位置
listOperations()	维护、订阅与取消订阅用户列表
searchOperations()	搜索推文并查看搜索趋势
streamingOperations()	通过 Twitter 的 Streaming API 在推文创建后获取推文
timelineOperations()	阅读时间线与发布推文
userOperations()	获取用户个人信息
restOperations()	如果部分 API 没有公开，那么通过其他 API 获取到的底层 RestTemplate

有时候，除了只读操作外，应用还需要其他类型的操作。如果想要发送推文或是访问私信，那就需要读写权限了。

要想发布状态更新，你需要使用 timelineOperations()方法，然后再使用 updateStatus()方法。根据需要，updateStatus 方法接收一个简单的 String（表示状态），或是叫做 TweetData 的值对象，它持有状态和其他信息，如位置、是否是对另一个推文的回复，以及其他资源（如图片等）。

一个简单的控制器的代码如下所示：

```
package com.apress.springrecipes.social.web;

import org.springframework.social.twitter.api.Twitter;
import org.springframework.stereotype.Controller;
import org.springframework.web.bind.annotation.RequestMapping;
import org.springframework.web.bind.annotation.RequestMethod;
import org.springframework.web.bind.annotation.RequestParam;

@Controller
@RequestMapping("/twitter")
public class TwitterController {

    private final Twitter twitter;
    public TwitterController(Twitter twitter) {
        this.twitter = twitter;
    }

    @RequestMapping(method = RequestMethod.GET)
```

```
    public String index() {
        return "twitter";
    }

    @RequestMapping(method = RequestMethod.POST)
    public String tweet(@RequestParam("status") String status) {
        twitter.timelineOperations().updateStatus(status);
        return "redirect:/twitter";
    }
}
```

控制器需要通过 TwitterTemplate 访问 Twitter API。TwitterTemplate 实现了 Twitter 接口。还记得在 6-2 节中，API 是请求作用域的。你会得到一个作用域代理，这也是需要使用 Twitter 接口的原因所在。tweet 方法接收一个参数并将其传递给 Twitter。

6-6　使用持久化的 UsersConnectionRepository

问题提出

你想要持久化用户的连接数据以便服务器重启后还可以继续使用。

解决方案

使用 JdbcUsersConnectionRepository 而非默认的 InMemoryUsersConnectionRepository。

解释说明

默认情况下，Spring Social 会自动配置一个 InMemoryUsersConnectionRepository 来存储用户的连接信息。不过，在集群环境下或是服务器重启时这种方式就行不通了。为了解决这个问题，可以使用数据库来存储连接信息。这是通过启用 JdbcUsersConnectionRepository 来实现的。

JdbcUsersConnectionRepository 需要用到数据库，而且数据库中需要有一张名为 UserConnection 的表，这张表会有几列。幸好，Spring Social 提供了一个 DDL 脚本 JdbcUsersConnectionRepository.sql，可以通过它来创建表。

首先，添加一个指向你所选择的数据库的数据源。在该示例中使用的是 PostgreSQL，不过其他数据库也是可以的。

提示： 在 bin 目录中有一个 postgres.sh 文件，它会开启一个 Docker 化的 PostgreSQL 实例供你使用。

```
@Bean
public DataSource dataSource() {

    DriverManagerDataSource dataSource = new DriverManagerDataSource();
    dataSource.setUrl(env.getRequiredProperty("datasource.url"));
    dataSource.setUsername(env.getRequiredProperty("datasource.username"));
    dataSource.setPassword(env.getRequiredProperty("datasource.password"));
    dataSource.setDriverClassName(env.getProperty("datasource.driverClassName"));
    return dataSource;
}
```

注意 dataSource.*属性，它用于配置 URL、JDBC 驱动以及用户名/密码。将这些属性添加到 application.properties 文件中。

```
dataSource.password=app
dataSource.username=app
dataSource.driverClassName=org.apache.derby.jdbc.ClientDriver
dataSource.url=jdbc:derby://localhost:1527/social;create=true
```

如果想要自动创建所选择的数据库中的表，你需要添加一个 DataSourceInitializer 并让其执行 JdbcUsersConnectionRepository.sql 文件。

```
@Bean
public DataSourceInitializer databasePopulator() {

    ResourceDatabasePopulator populator = new ResourceDatabasePopulator();
    populator.addScript(
        new ClassPathResource(
```

```
        "org/springframework/social/connect/jdbc/JdbcUsersConnectionRepository.sql"));
    populator.setContinueOnError(true);
    DataSourceInitializer initializer = new DataSourceInitializer();
    initializer.setDatabasePopulator(populator);
    initializer.setDataSource(dataSource());
    return initializer;
}
```

这个 DataSourceInitializer 会在应用启动时执行，并且会执行交给它的所有脚本。默认情况下，当出现错误时它会停止应用的启动。要想禁止这一点，请将 continueOnError 属性设为 true。在数据源创建并配置完毕后，最后一步就是向 SocialConfig 类添加 JdbcUsersConnectionRepository 了。

```
package com.apress.springrecipes.social.config;

import org.springframework.social.connect.jdbc.JdbcUsersConnectionRepository;
...

@Configuration
@EnableSocial
@PropertySource("classpath:/application.properties")
public class SocialConfig extends SocialConfigurerAdapter {

    @Override
    public UsersConnectionRepository getUsersConnectionRepository(ConnectionFactoryLocator
    connectionFactoryLocator) {
        return new JdbcUsersConnectionRepository(dataSource(), connectionFactoryLocator,
        Encryptors.noOpText());
    }

...
}
```

JdbcUsersConnectionRepository 接收 3 个构造方法参数。第一个是数据源，第二个是传递进来的 ConnectionFactoryLocator，最后一个是 TextEncryptor。TextEncryptor 类来自于 Spring Security crypto 模块，用于加密访问令牌、secret 并刷新令牌（如果有的话）。之所以需要加密是因为当数据存储为普通文本时很容易被窃取。令牌可用于访问你的个人信息。

不过，出于测试的目的，使用 noOpText 加密器会很方便，顾名思义，它并不会对数据进行加密。对于实际的产品来说，你需要使用 TextEncryptor，它会使用密码与盐对值进行加密。

配置好 JdbcUsersConnectionRepository 并启动数据库后，可以重启应用。乍一看，没什么不一样的；不过，在授权访问 Twitter 时，它并不受应用重启的影响。你还可以查看数据库，并且会发现信息存储到了 USERCONNECTION 表中。

6-7 集成 Spring Social 与 Spring Security

问题提出

你希望网站的用户可以连接到他们的社交网络账号。

解决方案

使用 spring-social-security 项目来集成这两个框架。

解释说明

我们来搭建 Spring Security。对 Spring Security 的详细介绍超出了本节的范围。要想了解关于它的更多信息，请查看第 7 章。本节的搭建如下所示：

```
@Configuratio
@EnableWebMvcSecurity
public class SecurityConfig extends WebSecurityConfigurerAdapter {

    @Override
    protected void configure(HttpSecurity http) throws Exception {
```

```
        http.authorizeRequests()
            .anyRequest().authenticated()
            .and()
                .formLogin()
                .loginPage("/signin")
                .failureUrl("/signin?param.error=bad_credentials")
                .loginProcessingUrl("/signin/authenticate").permitAll()
                .defaultSuccessUrl("/connect")
            .and()
                .logout().logoutUrl("/signout").permitAll();
    }

    @Bean
    public UserDetailsManager userDetailsManager(DataSource dataSource) {
        JdbcUserDetailsManager userDetailsManager = new JdbcUserDetailsManager();
        userDetailsManager.setDataSource(dataSource);
        userDetailsManager.setEnableAuthorities(true);
        return userDetailsManager;
    }

    @Override
    protected void configure(AuthenticationManagerBuilder auth) throws Exception {
        auth.userDetailsService(userDetailsManager(null));
    }

}
```

@EnableWebMvcSecurity 注解会为 Spring MVC 应用开启安全。它注册了 Spring Security 所需的 bean。要想做进一步的配置，比如说设置安全规则等，请添加一个或多个 WebSecurityConfigurers。为了简化这一点，你可以继承 WebSecurityConfigurerAdapter。

configure(HttpSecurity http)方法负责设置安全。这个配置使得每次调用时都会对用户进行认证。如果用户没有认证（即登录到应用），那么用户就会看到一个登录表单。还会注意到 loginPage、loginProcessingUrl 与 logoutUrl 被修改了。这么做是为了匹配 Spring Social 的默认 URL。

> **注意：** 如果想要保持 Spring Security 的默认配置，请显式配置 SocialAuthenticationFilter 并设置 signupUrl 与 defaultFailureUrl 属性。

借助于 configure(AuthenticationManagerBuilder auth)，你添加了一个 AuthenticationManager，它用于确定用户是否存在以及是否输入了正确的凭证。使用的 UserDetailsService 是个 JdbcUserDetailsManager，除了 UserDetailsService 的功能外，它还可以从仓库中添加和移除用户。当在应用中添加了社交登录页面时就会用到它。

JdbcUserDetailsManager 通过一个 DataSource 来读写数据，并将 enableAuthorities 属性设为 true，这样用户从应用中所获得的任何角色都会被添加到数据库中。要想启动数据库，请将 create_users.sql 脚本添加到上一节配置的数据库中。

```
@Bean
public DataSourceInitializer databasePopulator() {
    ResourceDatabasePopulator populator = new ResourceDatabasePopulator();
    populator.addScript(
    new ClassPathResource("org/springframework/social/connect/jdbc/JdbcUsersConnection
    Repository.sql"));
    populator.addScript(new ClassPathResource("sql/create_users.sql"));
    populator.setContinueOnError(true);

    DataSourceInitializer initializer = new DataSourceInitializer();
    initializer.setDatabasePopulator(populator);
    initializer.setDataSource(dataSource());
    return initializer;
}
```

接下来，要想渲染自定义登录或是注册页面，需要将其作为视图控制器添加到WebConfig类中。这样，对/signin的请求就会渲染signin.jsp页面。

```
package com.apress.springrecipes.social.config;

import org.springframework.context.annotation.Bean;
import org.springframework.context.annotation.ComponentScan;
import org.springframework.context.annotation.Configuration;
import org.springframework.web.servlet.config.annotation.ViewControllerRegistry;
import org.springframework.web.servlet.config.annotation.WebMvcConfigurer;
...

@Configuration
@EnableWebMvc
@ComponentScan({"com.apress.springrecipes.social.web"})
public class WebConfig implements WebMvcConfigurer {

    @Override
    public void addViewControllers(ViewControllerRegistry registry) {
        registry.addViewController("/").setViewName("index");
        registry.addViewController("/signin").setViewName("signin");
    }
    ...
}
```

signin.jsp是个简单的JSP页面，它会渲染一个用户名和一个密码输入框以及一个提交按钮。

```
<%@ taglib prefix="c" uri="http://java.sun.com/jsp/jstl/core" %>
<!DOCTYPE html>
<html>
<body>
    <c:url var="formLogin" value="/signin/authenticate" />
    <c:if test="${param.error eq 'bad_credentials'}">
        <div class="error">
            The login information was incorrect please try again.
        </div>
    </c:if>
    <form method="post" action="${formLogin}">
        <input type="hidden" name="_csrf" value="${_csrf.token}" />
        <table>
            <tr>
                <td><label for="username">Username</label></td>
                <td><input type="text" name="username"/></td>
            </tr>
            <tr>
                <td><label for="password">Password</label></td>
                <td><input type="password" name="password"/></td>
            </tr>
            <tr><td colspan="2"><button>Login</button></td> </tr>
        </table>
    </form>
</body>
</html>
```

注意隐藏的输入域，它包含了一个跨站点请求伪造（Cross-Site Request Forgery，CSRF）令牌。它用于防止恶意网站或JavaScript代码请求你的URL。在使用Spring Security时，默认情况下它会被启用。可以通过SecurityConfig类中的http.csfr().disable()将其禁用。

还剩下两个配置项。首先，这个配置需要能被加载；其次，需要注册一个过滤器将安全应用到请求上。基于此，修改SocialWebApplicationInitializer类。

```
package com.apress.springrecipes.social;

import com.apress.springrecipes.social.config.SecurityConfig;
import com.apress.springrecipes.social.config.SocialConfig;
import com.apress.springrecipes.social.config.WebConfig;
```

```
import org.springframework.web.filter.DelegatingFilterProxy;
import org.springframework.web.servlet.support.
AbstractAnnotationConfigDispatcherServletInitializer;

import javax.servlet.Filter;

public class SocialWebApplicationInitializer extends
AbstractAnnotationConfigDispatcherServletInitializer {

    @Override
    protected Class<?>[] getRootConfigClasses() {
        return new Class<?>[]{SecurityConfig.class, SocialConfig.class};
    }

    @Override
    protected Filter[] getServletFilters() {
        DelegatingFilterProxy springSecurityFilterChain = new DelegatingFilterProxy();
        springSecurityFilterChain.setTargetBeanName("springSecurityFilterChain");
        return new Filter[]{springSecurityFilterChain};
    }
...
}
```

首先注意到，SecurityConfig 类被添加到了 getRootConfigClasses 方法中。它负责加载配置类。接下来，添加了 getServletFilters 方法。该方法用于将过滤器注册到将要被 DispatcherServlet 所处理的请求上。默认情况下，Spring Security 会在应用上下文中注册一个名为 springSecurityFilterChain 的过滤器。为了执行它，你需要添加一个 DelegatingFilterProxy。DelegatingFilterProxy 会为指定的 targetBeanName 查找类型为 Filter 的 bean。

使用 Spring Security 获取用户名

在前文中，我们使用了一个 UserIdSource 实现，它返回一个静态的用户名。如果有已经在使用 Spring Security 的应用，那就可以使用 AuthenticationNameUserIdSource，它通过 SecurityContext（来自 Spring Security）获取已认证的当前用户的用户名。该用户名反过来又会用于存储和查找不同服务提供者的用户连接。

```
@Configuration
@EnableSocial
@PropertySource("classpath:/application.properties")
public class SocialConfig extends SocialConfigurerAdapter {

    @Override
    public UserIdSource getUserIdSource() {
        return new AuthenticationNameUserIdSource();
    }
    ...
}
```

■ **提示**：在使用 SpringsocialConfigurer 时，你可以省略掉这个 AuthenticationNameUserIdSource，因为默认就会创建并使用它。

注意 AuthenticationNameUserIdSource 的创建。这是从 Spring Security 中获取用户名所需做的全部工作。它会从 SecurityContext 中查找 Authentication 对象并返回该对象的 name 属性。当应用重启后，你会看到一个登录表单。现在请使用用户名 user1 和密码 user1 来登录。

使用 Spring Social 进行登录

让当前用户连接到社交网络是很不错。不过，如果用户能够使用自己的社交网络账号登录到应用上就更好了。Spring Social 提供了与 Spring Security 的紧密集成来达成这一点。为了实现这一点，还需要做一些额外的工作才行。

首先，Spring Social 需要集成 Spring Security。为此，需要使用 SpringSocialConfigurer 并将其应用到 Spring

Security 配置中。

```
@Configuration
@EnableWebMvcSecurity
public class SecurityConfig extends WebSecurityConfigurerAdapter {

    @Override
    protected void configure(HttpSecurity http) throws Exception {
        ...
        http.apply(new SpringSocialConfigurer());
    }
    ...
}
package com.apress.springrecipes.social.security;

import org.springframework.dao.DataAccessException;
import org.springframework.security.core.userdetails.UserDetails;
import org.springframework.security.core.userdetails.UserDetailsService;
import org.springframework.security.core.userdetails.UsernameNotFoundException;
import org.springframework.social.security.SocialUser;
import org.springframework.social.security.SocialUserDetails;
import org.springframework.social.security.SocialUserDetailsService;
import org.springframework.util.Assert;

public class SimpleSocialUserDetailsService implements SocialUserDetailsService {

    private final UserDetailsService userDetailsService;

    public SimpleSocialUserDetailsService(UserDetailsService userDetailsService) {
        Assert.notNull(userDetailsService, "UserDetailsService cannot be null.");
        this.userDetailsService = userDetailsService;    }

    @Override
    public SocialUserDetails loadUserByUserId(String userId) throws
    UsernameNotFoundException, DataAccessException {

        UserDetails user = userDetailsService.loadUserByUsername(userId);
        return new SocialUser(user.getUsername(), user.getPassword(), user.getAuthorities());
    }
}
```

接下来，为配置好的服务提供者添加登录页面链接。

```
<%@ taglib prefix="c" uri="http://java.sun.com/jsp/jstl/core" %>
<!DOCTYPE html>
<html>
<body>

...

<!-- TWITTER SIGNIN -->

<c:url var="twitterSigin" value="/auth/twitter"/>

<p><a href="${twitterSigin}">Sign in with Twitter</a></p>

<!-- FACEBOOK SIGNIN -->
<c:url var="facebookSigin" value="/auth/facebook"/>
<p><a href="${facebookSigin}">Sign in with Facebook</a></p>
</body>
</html>
```

SimpleSocialUserDetailsService 会将实际的查找委托给 UserDetailsService，后者是通过构造方法传递进来的。当检索到用户时，它会使用检索到的信息构建一个 SocialUser 实例。最后，该 bean 需要添加到配置中。

```
@Configuration
@EnableWebMvcSecurity
```

```
public class SecurityConfig extends WebSecurityConfigurerAdapter {

    @Bean
    public SocialUserDetailsService socialUserDetailsService(UserDetailsService
    userDetailsService) {
        return new SimpleSocialUserDetailsService(userDetailsService);
    }

    ...
}
```

这样，用户就可以使用自己的社交网络账号登录了；不过，应用需要知道账号是属于哪个用户的。如果用户在指定的社交网络中不存在，那就需要创建用户。基本上，应用需要为用户提供一种方式，使之能够注册到应用上。默认情况下，SocialAuthenticationFilter 会将用户重定向到/signup URL。你可以创建一个控制器映射到该 URL 上，并且渲染一个表单，让用户可以创建账号。

```
package com.apress.springrecipes.social.web;

import org.springframework.security.authentication.UsernamePasswordAuthenticationToken;
import org.springframework.security.core.GrantedAuthority;
import org.springframework.security.core.authority.SimpleGrantedAuthority;
import org.springframework.security.core.context.SecurityContextHolder;
import org.springframework.security.provisioning.UserDetailsManager;
import org.springframework.social.connect.Connection;
import org.springframework.social.connect.web.ProviderSignInUtils;
import org.springframework.social.security.SocialUser;
import org.springframework.stereotype.Controller;
import org.springframework.validation.BindingResult;
import org.springframework.validation.annotation.Validated;
import org.springframework.web.bind.annotation.GetMapping;
import org.springframework.web.bind.annotation.PostMapping;
import org.springframework.web.bind.annotation.RequestMapping;
import org.springframework.web.context.request.WebRequest;

import java.util.Collections;
import java.util.List;

@Controller
@RequestMapping("/signup")
public class SignupController {

    private static final List<GrantedAuthority> DEFAULT_ROLES = Collections.
    singletonList(new SimpleGrantedAuthority("USER"));
    private final ProviderSignInUtils providerSignInUtils;
    private final UserDetailsManager userDetailsManager;

    public SignupController(ProviderSignInUtils providerSignInUtils,
                            UserDetailsManager userDetailsManager) {
        this.providerSignInUtils = providerSignInUtils;
        this.userDetailsManager = userDetailsManager;
    }

    @GetMapping
    public SignupForm signupForm(WebRequest request) {
        Connection<?> connection = providerSignInUtils.getConnectionFromSession(request);
        if (connection != null) {
            return SignupForm.fromProviderUser(connection.fetchUserProfile());
        } else {
            return new SignupForm();
        }
    }

    @PostMapping
    public String signup(@Validated SignupForm form, BindingResult formBinding,
```

```
                        WebRequest request) {
        if (!formBinding.hasErrors()) {
            SocialUser user = createUser(form);
            SecurityContextHolder.getContext().setAuthentication(new UsernamePassword
            AuthenticationToken(user.getUsername(), null, user.getAuthorities()));
            providerSignInUtils.doPostSignUp(user.getUsername(), request);
            return "redirect:/";
        }
        return null;
    }

    private SocialUser createUser(SignupForm form) {
        SocialUser user = new SocialUser(form.getUsername(), form.getPassword(),
        DEFAULT_ROLES);
        userDetailsManager.createUser(user);
        return user;
    }
}
```

首先会调用 signupForm 方法，因为一开始是向/signup URL 发出的 GET 请求。signupForm 方法会发起连接请求。这是通过委托给 Spring Social 所提供的 ProviderSignInUtils 来实现的。如果连接成功，那么就会用获取到的 UserProfile 来预先填充 SignupForm。

```
package com.apress.springrecipes.social.web;

import org.springframework.social.connect.UserProfile;

public class SignupForm {

    private String username;
    private String password;

    public String getUsername() {
        return username;
    }

    public void setUsername(String username) {
        this.username = username;
    }

    public String getPassword() {
        return password;
    }

    public void setPassword(String password) {
        this.password = password;
    }

    public static SignupForm fromProviderUser(UserProfile providerUser) {

        SignupForm form = new SignupForm();
        form.setUsername(providerUser.getUsername());
        return form;
    }
}
```

下面是用于填写两个字段的 HTML 表单：

```
<%@ taglib prefix="form" uri="http://www.springframework.org/tags/form" %>
<%@ page contentType="text/html;charset=UTF-8" language="java" %>
<html>
<head>
    <title>Sign Up</title>
</head>

<body>
```

```
<h3>Sign Up</h3>

<form:form modelAttribute="signupForm" method="POST">
    <table>
        <tr><td><form:label path="username" /></td><td><form:input path="username"/></td></tr>
        <tr><td><form:label path="password" /></td><td><form:password path="password"/>
        </td></tr>
        <tr><td colspan="2"><button>Sign Up</button></td></tr>
    </table>
</form:form>
</body>
</html>
```

注意：这里并没有针对 CSFR 标签的隐藏输入域。Spring Security 紧密集成了 Spring MVC，当使用 Spring 框架的表单标签时，该字段会自动添加进来。

当用户填写完表单后，会调用 signup 方法。这会根据给定的用户名和密码创建一个用户。用户创建好之后，会为传递进来的用户名添加一个 Connection。在连接建立后，用户就可以登录到应用，并且在随后可以使用社交网络连接来登录到应用。

控制器使用 ProviderSignInUtils 来重用 Spring Social 的逻辑。可以在 SocialConfig 类中创建一个实例。

```
@Bean
public ProviderSignInUtils providerSignInUtils(ConnectionFactoryLocator
connectionFactoryLocator, UsersConnectionRepository usersConnectionRepository) {
    return new ProviderSignInUtils(connectionFactoryLocator, usersConnectionRepository);
}
```

配置的最后一部分是允许所有用户访问/signup URL。在 SecurityConfig 类中添加如下代码：

```
@Override
protected void configure(HttpSecurity http) throws Exception {

    http
        .authorizeRequests()
            .antMatchers("/signup").permitAll()
            .anyRequest().authenticated().and()
    ...
}
```

小结

本章介绍了 Spring Social。首先将应用注册到服务提供者，并使用生成的 API key 和 secret 将应用连接到该服务提供者。接下来介绍了如何将用户账号连接到应用，这样就可以访问用户信息了，还可以使用服务提供者的 API。对于 Twitter 来说，可以查询时间线或是查看某人的好友。

为了使得对服务提供者的连接更具价值，需要将其存储到基于 JDBC 的存储中。

最后，我们介绍了 Spring Social 如何集成 Spring Security，以及如何通过它让服务提供者登录到你的应用上。

第 7 章

Spring Security

本章将会介绍如何使用 Spring Security 框架保护你的应用，它是 Spring 框架的一个子项目。Spring Security 一开始名为 Acegi Security，后来在加入到 Spring 项目后更名为 Spring Security。Spring Security 可用于保护任何 Java 应用，不过主要用在 Web 应用中。对于 Web 应用特别是可通过互联网访问的 Web 应用来说，如果安全保护措施不到位，很容易遭到黑客攻击。

如果大家之前从未在应用中使用过安全，那么首先需要理解一些术语和概念。认证指的是对主体身份和其所声明的身份进行验证的过程。主体可以是一个用户、一台设备或是一个系统，不过一般来说指的是一个用户。主体需要提供身份证据来进行认证。这个证据叫做凭证，当目标主体是一个用户时，凭证通常就是密码。

授权指的是对已认证用户授予权利的过程，这样用户就可以访问目标应用特定的资源了。授权过程必须要在认证过程之后进行。通常情况下，授权会根据角色来进行。

访问控制指的是控制对应用资源的访问。它需要决定是否允许用户访问某个资源。这个决定叫做访问控制决定，它是通过比较资源的访问属性和用户的授权或是其他属性来进行的。

学习完本章内容后，你将会理解基本的安全概念，并知道如何在 URL 访问级别、方法调用级别、视图渲染级别以及领域对象级别保护你的 Web 应用。

注意：在开始本章的学习前，请看一下 recipe_7_1_i 的应用。这是初始的未使用安全的应用，本章后面将会使用到它。它是个基本的待办事项应用，你可以通过列表展示、创建待办事项以及将待办事项标记为已完成。在部署应用时，你会看到它的内容，如图 7-1 所示。

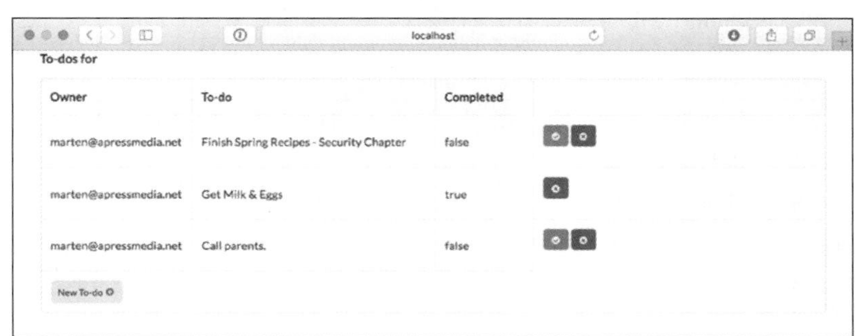

图 7-1　初始的待办事项应用

7-1　保护 URL 访问

问题提出

很多 Web 应用都有一些非常重要且私密的特定 URL。你需要保护这些 URL，防止遭受未授权的访问。

解决方案

Spring Security 可以通过简单配置以声明的方式保护 Web 应用的 URL 访问安全。它通过将 servlet 过滤器应用到 HTTP 请求来处理安全问题。为了注册过滤器并检测配置，Spring Security 提供了一个便捷的父类供你继承：AbstractSecurityWebApplicationInitializer。

Spring Security 通过 WebSecurityConfigurerAdapter 配置适配器的各种 configure 方法来配置 Web 应用的安全。如果 Web 应用的安全需求是直接且典型的，那就可以保持原有的配置并使用默认开启的安全设置，具体如下。

- 基于表单的登录服务：提供了一个默认页面，包括一个登录表单，供用户登录到应用中。
- HTTP Basic 认证：可以处理 HTTP 请求头中的 HTTP Basic 认证凭证，还可用于远程协议与 Web Service 的认证请求。
- 登出服务：提供了一个映射到 URL 的处理器，供用户登出应用。
- 匿名登录：为匿名用户指定一个主体和授权，这样就可以像正常用户一样处理匿名用户了。
- Servlet API 集成：可以通过标准的 Servlet API 访问 Web 应用中的安全信息，比如说 HttpServletRequest.isUserInRole()与 HttpServletRequest.getUserPrincipal()。
- CSRF：通过创建一个令牌并将其放到 HttpSession 中来实现跨站请求伪造防护。
- 安全头：类似于禁用安全包的缓存，这提供了 XSS 防护、传输安全与 X-Frame 安全。

注册好了这些安全服务后，可以指定需要特定授权访问的 URL 模式。Spring Security 会根据配置执行安全检查。用户在访问安全的 URL 前需要登录到应用中，除非这些 URL 对匿名访问是开放的。Spring Security 提供了一套认证提供器。认证提供器会对用户进行认证并返回对该用户的授权信息。

解释说明

首先，需要注册 Spring Security 所用的过滤器。最简单的方式是继承上面提到的 AbstractSecurityWebApplicationInitializer。

```
package com.apress.springrecipes.board.security;

import org.springframework.security.web.context.AbstractSecurityWebApplicationInitializer;

public class TodoSecurityInitializer extends AbstractSecurityWebApplicationInitializer {

    public TodoSecurityInitializer() {
        super(TodoSecurityConfig.class);
    }
}
```

AbstractSecurityWebApplicationInitializer 有一个接收一个或多个配置类的构造方法。这些配置类用于启用安全。

■ **注意**：如果已经有一个类继承了 AbstractAnnotationConfigDispatcherServletInitializer，那么请向它添加安全配置，否则会在启动时抛出异常。

虽然可以在与 Web 和服务层同样的配置类中配置 Spring Security，不过最好在单独的类（比如说 TodoSecurityConfig）中进行安全配置。在 WebApplicationInitializer 中（即 TodoWebInitializer），你需要将该配置类添加到类的列表中完成配置。

首先需要的是安全配置。为了做到这一点，需要创建一个类 TodoSecurityConfig，代码如下所示：

```
package com.apress.springrecipes.board.security;

import org.springframework.context.annotation.Configuration;
import org.springframework.security.config.annotation.web.configuration.EnableWebSecurity;
import org.springframework.security.config.annotation.web.configuration.
WebSecurityConfigurerAdapter;

@Configuration
```

```
@EnableWebSecurity
public class TodoSecurityConfig extends WebSecurityConfigurerAdapter {}
```

构建并部署应用，然后访问 http://localhost:8080/todos/todos，你会看到默认的 Spring Security 登录页面，如图 7-2 所示。

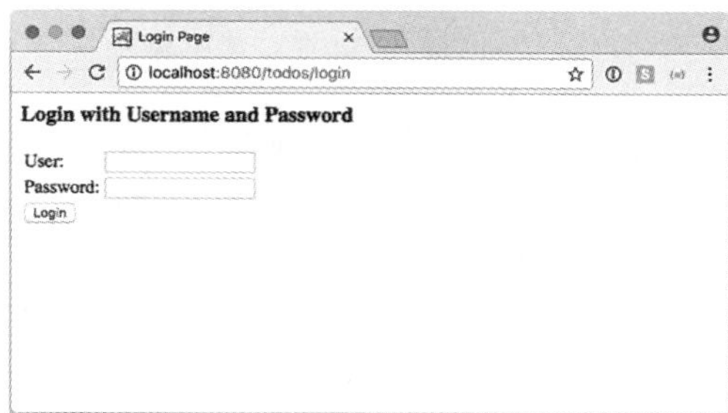

图 7-2　默认的 Spring Security 登录页面

保护 URL 访问

查看 org.springframework.security.config.annotation.web.configuration.WebSecurityConfigurerAdapter 类的 configure 方法，你会发现它里面调用了 anyRequest().authenticated()。这会告诉 Spring Security，对于每个到来的请求，都需要进行认证。你还会看到在默认情况下，HTTP Basic 认证与基于表单的登录是开启的。基于表单的登录还包含了一个默认的登录页面创建器，如果没有显式指定登录页面，那么就会使用它。

```
protected void configure(HttpSecurity http) throws Exception {
    http
        .authorizeRequests()
            .anyRequest().authenticated()
            .and()
        .formLogin().and()
        .httpBasic();
}
```

我们来编写几个安全规则。相较于仅仅需要登录来说，你可以为 URL 编写一些强大的访问规则。

```
package com.apress.springrecipes.board.security;

import org.springframework.context.annotation.Configuration;
import org.springframework.http.HttpMethod;
import org.springframework.security.config.annotation.authentication.builders.
AuthenticationManagerBuilder;
import org.springframework.security.config.annotation.web.builders.HttpSecurity;
import org.springframework.security.config.annotation.web.configuration.EnableWebSecurity;
import org.springframework.security.config.annotation.web.configuration.
WebSecurityConfigurerAdapter;

@Configuration
@EnableWebSecurity
public class TodoSecurityConfig extends WebSecurityConfigurerAdapter {

    @Override
    protected void configure(AuthenticationManagerBuilder auth) throws Exception {
        auth.inMemoryAuthentication()
            .withUser("marten@ya2do.io").password("user").authorities("USER")
            .and()
            .withUser("admin@ya2do.io").password("admin").authorities("USER", "ADMIN");
    }

    @Override
    protected void configure(HttpSecurity http) throws Exception {
```

```
        http.authorizeRequests()
            .antMatchers("/todos*").hasAuthority("USER")
            .antMatchers(HttpMethod.DELETE, "/todos*").hasAuthority("ADMIN")
            .and()
                .formLogin()
            .and()
                .csrf().disable();
    }
}
```

可以通过重写configure(HttpSecurity http)方法（还有其他的configure方法）来配置授权规则等内容。

从authorizeRequests()伊始，你开始对URL进行保护。接下来可以使用匹配器；在上述代码中，使用了antMatchers来定义匹配规则以及用户需要拥有哪些授权。记住，需要在URL模式的末尾加上一个通配符。如果不这么做就会导致URL模式无法匹配拥有请求参数的URL。这样，黑客就可以通过附加任意请求参数轻松跳过安全检查。我们对/todos的所有访问进行了保护，只有拥有USER授权的用户才能访问。要想通过DELETE请求访问/todos，用户需要拥有ADMIN角色。

注意：由于重写了默认访问规则和登录配置，因此需要再次启用formLogin。还有一个调用禁用了CSRF防护，因为CSRF防护会导致表单无法正常使用；稍后将会介绍如何启用它。

可以在重写的configure(AuthenticationManagerBuilder auth)方法中配置认证服务。Spring Security支持几种用户认证方式，包括使用数据库或是LDAP仓库进行认证。此外，它还支持直接定义用户的详细信息以实现简单的安全需求。你可以为每个用户指定一个用户名、一个密码和一组授权。

现在，可以重新部署应用来测试其安全配置。必须使用正确的用户名和密码登录到应用中才能看到待办事项。最后，要想删除待办事项，需要以管理员的身份登录。

使用CSRF防护

通常情况下，默认开启CSRF是个好的做法，因为这会降低CSRF攻击的风险。Spring Security默认是开启CSRF的，csfr().disable()这一行可以从配置中删掉了。当开启了CSRF防护后，Spring Security会向用于防护的过滤器列表添加一个CsrfFilter。该过滤器会使用CsrfTokenRepository的一个实现来生成并存储令牌；默认情况下，该实现是HttpSessionCsrfTokenRepository类，顾名思义，它会将生成的令牌存储到HttpSession接口中。还有一个CookieCsrfTokenRepository类，它会将令牌信息存储到cookie中。如果想要改变所用的CsrfTokenRepository实现，可以通过csrfTokenRepository()配置方法来达成。此外，还可以使用它来显式配置一个HttpSessionCsrfTokenRepository。

```
@Override
protected void configure(HttpSecurity http) throws Exception {

    HttpSessionCsrfTokenRepository repo = new HttpSessionCsrfTokenRepository();
    repo.setSessionAttributeName("csfr_token");
    repo.setParameterName("csfr_token")

    http.csrf().csrfTokenRepository(repo);
}
```

当启用了CSRF时，登录后尝试完成或是删除一个待办事项时会失败，原因是缺少了CSRF令牌。要想解决这个问题，你需要随着修改内容的请求向服务器一同传递CSRF令牌才行。可以通过在表单中使用一个隐藏输入来做到这一点。HttpSessionCsrfTokenRepository在会话的属性_csrf（默认属性，除非显式做了配置）中公开了该令牌。对于表单来说，你可以使用parameterName和token属性来创建恰当的input标签。

将如下代码添加到用于完成和删除待办事项的两个表单中：

```
<input type="hidden" name="${_csrf.parameterName}" value="${_csrf.token}"/>
```

现在，当提交表单时，令牌将会作为请求的一部分，你又可以完成或是删除待办事项了。

todo-create.jsp页面中还有一个表单；不过，由于它使用了Spring MVC表单标签，因此无须对其进行修改。

当使用了 Spring MVC 表单标签时，CSRF 会被自动添加到表单中。为了能够实现这一点，Spring Security 注册了一个 CsrfRequestDataValueProcessor 类，它负责将令牌添加到表单中。

7-2　登录到 Web 应用

问题提出

安全的应用会要求用户登录后才能访问某些安全的功能。这对于运行在开放互联网上的 Web 应用来说尤为重要，因为黑客可以轻松触达到它们。大多数 Web 应用都需要为用户提供一种输入凭证进行登录的方式。

解决方案

Spring Security 支持多种方式可以让用户登录到 Web 应用。它支持基于表单的登录，方式是提供一个包含了登录表单的默认页面。你还可以提供自定义的页面作为登录页面。此外，Spring Security 支持 HTTP Basic 认证，这是通过处理 HTTP 请求头中的 Basic 认证凭证来实现的。HTTP Basic 认证还可用于远程协议与 Web Service 的认证请求。

部分应用可能允许匿名访问（比如说欢迎页面）。Spring Security 提供一个匿名登录服务，可以为匿名用户分配一个主体并授权，这样在定义安全策略时就可以像正常用户一样处理匿名用户了。

Spring Security 还支持“记住我”登录，它可以跨越多个浏览器会话记住用户的身份，这样用户在首次登录后就无须再次登录了。

解释说明

为了更好地理解各种登录机制，我们先禁用默认的安全配置。

> **警告**：一般来说，我们会使用默认配置，仅仅禁用掉不需要的配置，比如说 httpBasic().disable()，而不是禁用所有的安全默认配置。

```
@Configuration
@EnableWebSecurity
public class TodoSecurityConfig extends WebSecurityConfigurerAdapter {

    public TodoSecurityConfig() {
        super(true);
    }
}
```

注意，如果启用了 HTTP 自动配置的话，那么接下来将会介绍的登录服务就会自动注册进来。不过，如果禁用了默认配置或是想要自定义这些服务，那就需要显式配置对应的特性了。

在启用认证特性之前，你需要开启基本的 Spring Security 功能，至少需要配置异常处理与安全上下文集成。

```
protected void configure(HttpSecurity http) {

    http.securityContext()
        .and()
        .exceptionHandling();
}
```

如果没有这些基本的配置，那么登录后 Spring Security 就不会将用户存储起来，也不会对安全相关的异常进行恰当的异常转换（异常只会向上抛出，这可能会将一些内部信息暴露给外面）。你还需要启用 Servlet API 集成，这样就可以使用 HttpServletRequest 的方法在视图中进行检查了。

```
protected void configure(HttpSecurity http) {
    http.servletApi();
}
```

使用 HTTP Basic 认证

可以通过 httpBasic()方法来配置 HTTP Basic 认证支持。当需要 HTTP Basic 认证时，浏览器通常会弹出一个登录对话框或是特定的登录页面让用户登录。

```
@Configuration
@EnableWebSecurity
public class TodoSecurityConfig extends WebSecurityConfigurerAdapter {

    @Override
    protected void configure(HttpSecurity http) throws Exception {
        http
            ...
            .httpBasic();
    }
}
```

注意：如果同时开启了 HTTP Basic 认证与基于表单的登录，那么将会使用后者。因此，如果希望 Web 应用的用户使用 HTTP Basic 认证的方式登录，那就不应该启用基于表单的登录。

使用基于表单的登录

基于表单的登录服务会渲染一个网页，该网页包含了一个登录表单供用户输入登录信息，同时该服务还会处理登录表单的提交。这是通过 formLogin 方法配置的。

```
@Configuration
@EnableWebSecurity
public class TodoSecurityConfig extends WebSecurityConfigurerAdapter {

    @Override
    protected void configure(HttpSecurity http) throws Exception {
        http
            ...
            .formLogin();
    }
}
```

默认情况下，Spring Security 会自动创建一个登录页面并将其映射到/login URL。因此，你可以向应用（比如说在 todos.jsp 中）添加一个链接，指向用于登录的该 URL。

```
<a href="<c:url value="/login" />">Login</a>
```

如果不想使用默认登录页面，那么可以提供自己的自定义登录页面。比如说，可以在 Web 应用的根目录下创建如下的 login.jsp 文件。注意，不要将该文件放到 WEB-INF 中，因为用户无法直接访问该目录下的文件。

```
<%@ taglib prefix="c" uri="http://java.sun.com/jsp/jstl/core" %>

<html>
<head>
    <title>Login</title>
    <link type="text/css" rel="stylesheet"
        href="https://cdnjs.cloudflare.com/ajax/libs/semantic-ui/2.2.10/semantic.min.css">
    <style type="text/css">
        body {
            background-color: #DADADA;
        }
        body > .grid {
            height: 100%;
        }
        .column {
            max-width: 450px;
        }
    </style>
</head>
<body>
<div class="ui middle aligned center aligned grid">
    <div class="column">
        <h2 class="ui header">Log-in to your account</h2>
        <form method="POST" action="<c:url value="/login" />" class="ui large form">
```

```
            <input type="hidden" name="${_csrf.parameterName}" value="${_csrf.token}"/>
            <div class="ui stacked segment">
                <div class="field">
                    <div class="ui left icon input">
                        <i class="user icon"></i>
                        <input type="text" name="username" placeholder="E-mail address">
                    </div>
                </div>
                <div class="field">
                    <div class="ui left icon input">
                        <i class="lock icon"></i>
                        <input type="password" name="password" placeholder="Password">
                    </div>
                </div>
                <button class="ui fluid large submit green button">Login</button>
            </div>
        </form>
    </div>
</div>
</body>
</html>
```

要想在登录时让 Spring Security 能够显示出自定义的登录页面，需要在 loginPage 配置方法中指定其 URL。

```
@Configuration
@EnableWebSecurity
public class TodoSecurityConfig extends WebSecurityConfigurerAdapter {

    @Override
    protected void configure(HttpSecurity http) throws Exception {
        http
            ...
            .formLogin().loginPage("/login.jsp");
    }
}
```

当用户请求一个安全的 URL 时，如果 Spring Security 显示了登录页面，那么当登录成功后用户会被重定向到目标 URL。不过，如果用户直接通过 URL 请求了登录页面，那么默认情况下当登录成功后用户会被重定向到上下文的根（比如说 http://localhost:8080/todos/）。如果没有在 Web 部署描述符中定义欢迎页面，那么当登录成功后你可能希望将用户重定向到默认的目标 URL。

```
@Configuration
@EnableWebSecurity
public class TodoSecurityConfig extends WebSecurityConfigurerAdapter {

    @Override
    protected void configure(HttpSecurity http) throws Exception {
        http
            ...
            .formLogin().loginPage("/login.jsp").defaultSuccessUrl("/todos");
    }
}
```

如果使用的是 Spring Security 创建的默认登录页面，那么当登录失败时，Spring Security 会使用错误消息再次渲染登录页面。不过，如果指定了自定义登录页面，那就需要配置 authentication-failure-url 值来指定当登录失败时将用户重定向到哪个 URL。比如说，可以使用错误请求参数将用户重定向到自定义的登录页面，如下所示：

```
@Configuration
@EnableWebSecurity
public class TodoSecurityConfig extends WebSecurityConfigurerAdapter {

    @Override
    protected void configure(HttpSecurity http) throws Exception {
        http
```

```
            ...
            .formLogin()
                .loginPage("/login.jsp")
                .defaultSuccessUrl("/messageList")
                .failureUrl("login.jsp?error=true");
    }
}
```

接下来，登录页面应该检查是否有错误请求参数。如果出现了错误，那就需要展示出错误消息，这是通过访问会话作用域属性 SPRING_SECURITY_LAST_EXCEPTION 来实现的，它存储了当前用户的上一次异常信息。

```
<form>
    ...
    <c:if test="${not empty param.error}">
        <div class="ui error message" style="display: block;">
            Authentication Failed<br/>
            Reason : ${sessionScope["SPRING_SECURITY_LAST_EXCEPTION"].message}
            </font>
        </div>
    </c:if>
</form>
```

使用登出服务

登出服务提供了一个用于处理登出请求的处理器。可以通过 logout()配置方法来配置。

```
@Configuration
@EnableWebSecurity
public class TodoSecurityConfig extends WebSecurityConfigurerAdapter {

    @Override
    protected void configure(HttpSecurity http) throws Exception {
        http
            ...
            .and()
                .logout();
    }
}
```

默认情况下，它会匹配到/logout URL 并且只会响应 POST 请求。你可以向页面中添加一个 HTML 表单实现登出。

```
<form action="<c:url value="/logout"/>" method="post"><button>Logout</button><form>
```

■ **注意：**当使用了 CSRF 防护时，不要忘记将 CSRF 令牌添加到表单中（参见 7-1 节），否则登出会失败。

默认情况下，当成功登出时，用户会被重定向到上下文路径的根页面，不过有时你想将用户重定向到别的 URL，这可以通过 logoutSuccessUrl 配置方法实现。

```
@Configuration
@EnableWebSecurity
public class TodoSecurityConfig extends WebSecurityConfigurerAdapter {

    @Override
    protected void configure(HttpSecurity http) throws Exception {
        http
            ...
            .and()
                .logout().logoutSuccessUrl("/logout-success.jsp");
    }
}
```

登出后，你会注意到当使用浏览器的后退按钮时，依然能够看到之前的页面，即便成功登出也是如此。这是因为浏览器缓存了页面。通过启用安全头（使用 headers()配置方法），浏览器就不会再缓存页面了。

```
@Configuration
```

```
@EnableWebSecurity
public class TodoSecurityConfig extends WebSecurityConfigurerAdapter {

    @Override
    protected void configure(HttpSecurity http) throws Exception {
        http
            ...
            .and()
                .headers();
    }
}
```

除了 no-cache 头之外，这还会禁用内容嗅探并启用 X-Frame 防护（参见 7-1 节了解更多信息）。启用它之后再使用浏览器的后退按钮时，你会被重定向到登录页面。

实现匿名登录

可以通过在 Java 配置中使用 anonymous()方法来配置匿名登录服务，你可以定制匿名用户的用户名和授权信息，其默认值分别是 anonymousUser 和 ROLE_ANONYMOUS。

```
@Configuration
@EnableWebSecurity
public class TodoSecurityConfig extends WebSecurityConfigurerAdapter {

    @Override
    protected void configure(HttpSecurity http) throws Exception {
        http
            ...
            .and()
                .anonymous().principal("guest").authorities("ROLE_GUEST");
    }
}
```

实现“记住我”支持

可以通过在 Java 配置中使用 rememberMe()方法配置“记住我”支持。默认情况下，它会将用户名、密码、“记住我”的过期时间与私钥编码为一个令牌并将其作为 cookie 存储到用户的浏览器中。用户下一次访问同一个 Web 应用时，该令牌会被检测到，这样用户就可以自动登录了。

```
@Configuration
@EnableWebSecurity
public class TodoSecurityConfig extends WebSecurityConfigurerAdapter {

    @Override
    protected void configure(HttpSecurity http) throws Exception {
        http
            ...
            .and()
                .rememberMe();
    }
}
```

不过，静态的“记住我”令牌会导致安全问题，因为它们有可能会被黑客捕获到。Spring Security 支持滚动令牌以满足更为高级的安全需求，不过这需要一个数据库来持久化令牌。要想了解关于“记住我”令牌滚动部署的详细信息，请查阅 Spring Security 参考文档。

7-3 对用户进行认证

问题提出

当用户尝试登录应用以访问安全资源时，你需要对用户的主体进行认证并对该用户进行授权。

解决方案

在 Spring Security 中，认证是通过将一个或多个 AuthenticationProviders 链接到一起来完成的。如果所有提供器都成功对用户进行了认证，那么该用户就可以登录到应用。如果有任何提供器发现用户被禁用、锁定、凭证不正确，或是没有提供器能够认证用户，那么该用户就无法登录到应用。

Spring Security 支持多种认证用户的方式，并且包含了内建的提供器实现。你可以通过内建的 XML 元素轻松配置这些提供器。大多数常见的认证提供器都会使用存储了用户信息的用户仓库来认证用户（比如说应用的内存、关系型数据库或是 LDAP 仓库）。

在将用户信息存储到仓库中时，不应该以明文形式存储用户密码，因为这会导致黑客攻击。相反，应该在仓库中存储加密后的密码信息。对密码进行加密的典型做法是使用单向的哈希函数对密码进行编码。当用户输入了密码进行登录时，你要对该密码应用同样的哈希函数并将结果与存储在仓库中的值进行比较。Spring Security 支持多种密码编码的算法（如 MD5 和 SHA），并为这些算法提供了内建的密码编码器。

如果在用户每次登录时都从用户仓库中获取用户的信息，那么应用就会有性能上的影响。这是因为用户仓库通常存储在远程，它需要执行某些查询来响应请求。出于这个原因，Spring Security 支持将用户信息存储到本地内存和本地存储中，从而减少执行远程查询的开销。

解释说明

我们将会探索不同的认证机制，首先来看看内存实现，接下来再介绍基于数据库的实现，最后再来讲一下 LDAP。最后一小节将会介绍如何为不同的认证机制启用缓存。

使用内存定义来认证用户

如果应用的用户不多并且很少会修改其信息，那就可以考虑在 Spring Security 的配置文件中定义用户信息，这样它们就会被加载到应用内存中。

```
@Configuration
@EnableWebSecurity
public class TodoSecurityConfig extends WebSecurityConfigurerAdapter {

...
    @Override
    protected void configure(AuthenticationManagerBuilder auth) throws Exception {
        auth.inMemoryAuthentication()
            .withUser("admin@ya2do.io").password("secret").authorities("ADMIN","USER").and()
            .withUser("marten@@ya2do.io").password("user").authorities("USER").and()
            .withUser("jdoe@does.net").password("unknown").disabled(true).
            authorities("USER");
    }
}
```

可以通过 inMemoryAuthentication()方法定义用户信息。借助于 withUser 方法，可以定义用户。对于每个用户来说，可以指定用户名、密码、禁用状态以及一组授权信息。被禁用的用户将无法登录到应用。

使用数据库来认证用户

更为典型的情况是，用户信息应当存储到数据库中以便更容易进行维护。Spring Security 对从数据库中查询用户信息提供了内建的支持。默认情况下，它会通过如下 SQL 语句查询用户信息，也包括授权信息：

```
SELECT username, password, enabled
FROM   users
WHERE  username = ?

SELECT username, authority
FROM   authorities
WHERE  username = ?
```

要想让 Spring Security 能够通过这些 SQL 语句查询用户信息，你需要在数据库中创建对应的表。比如说，可以通过如下 SQL 语句在 todo 模式中进行创建：

```
CREATE TABLE USERS (
    USERNAME     VARCHAR(50)  NOT NULL,
    PASSWORD     VARCHAR(50)  NOT NULL,
    ENABLED      SMALLINT     NOT NULL,
    PRIMARY KEY (USERNAME)
);

CREATE TABLE AUTHORITIES (
    USERNAME     VARCHAR(50)    NOT NULL,
    AUTHORITY    VARCHAR(50)    NOT NULL,
    FOREIGN KEY (USERNAME) REFERENCES USERS
);
```

接下来，可以向这些表输入一些用户信息以进行测试。表 7-1 和表 7-2 所示为这两张表的数据。

表 7-1　　USERS 表的测试用户数据

用户名	密码	启用
admin@ya2do.io	secret	1
marten@ya2do.io	user	1
jdoe@does.net	unknown	0

表 7-2　　AUTHORITIES 表的测试用户数据

用户名	权限
admin@ya2do.io	ADMIN
admin@ya2do.io	USER
marten@ya2do.io	USER
jdoe@does.net	USER

要想让 Spring Security 能够访问这些表，你需要声明一个数据源以便可以创建到该数据库的连接。对于 Java 配置来说，请使用 jdbcAuthentication()配置方法并将一个 DataSource 传递给它。

```
@Configuration
@EnableWebSecurity
public class TodoSecurityConfig extends WebSecurityConfigurerAdapter {

    @Bean
    public DataSource dataSource() {
        return new EmbeddedDatabaseBuilder()
            .setType(EmbeddedDatabaseType.H2)
            .setName("board")
            .addScript("classpath:/schema.sql")
            .addScript("classpath:/data.sql")
            .build();
    }

    @Override
    protected void configure(AuthenticationManagerBuilder auth) throws Exception {
        auth.jdbcAuthentication().dataSource(dataSource());
    }
}
```

注意： *用于 DataSource 的@Bean 方法从 TodoWebConfig 移动到了 TodoSecurityConfig 中。*

不过，在某些情况下，你可能已经将自己的用户仓库定义在了遗留的数据库中。比如说，假设通过如下 SQL 语句创建了表，并且 MEMBER 表中的所有用户都拥有启用状态：

```
CREATE TABLE MEMBER (
    ID           BIGINT       NOT NULL,
    USERNAME     VARCHAR(50)  NOT NULL,
    PASSWORD     VARCHAR(32)  NOT NULL,
```

```
    PRIMARY KEY (ID)
);

CREATE TABLE MEMBER_ROLE (
    MEMBER_ID    BIGINT         NOT NULL,
    ROLE         VARCHAR(10)    NOT NULL,
    FOREIGN KEY (MEMBER_ID)     REFERENCES MEMBER
);
```

假设遗留用户数据存储在了这些表中，如表 7-3 和表 7-4 所示。

表 7-3　　MEMBER 表中的遗留用户数据

ID	用户名	密码
1	admin@ya2do.io	secret
2	marten@ya2do.io	user

表 7-4　　MEMBER_ROLE 表中的遗留用户数据

MEMBER_ID	角色
1	ROLE_ADMIN
1	ROLE_USER
2	ROLE_USER

幸好，Spring Security 支持使用自定义的 SQL 语句查询遗留数据库来获取用户信息。你可以通过 usersByUsernameQuery()和 authoritiesByUsernameQuery()配置方法指定查询用户信息与授权的语句。

```
@Configuration
@EnableWebSecurity
public class TodoSecurityConfig extends WebSecurityConfigurerAdapter {

...

    @Override
    protected void configure(AuthenticationManagerBuilder auth) throws Exception {
        auth.jdbcAuthentication()
             .dataSource(dataSource)
             .usersByUsernameQuery(
                 "SELECT username, password, 'true' as enabled FROM member WHERE username = ?")
             .authoritiesByUsernameQuery(
                 "SELECT member.username, member_role.role as authorities " +
                 "FROM member, member_role " +
                 "WHERE member.username = ? AND member.id = member_role.member_id");
    }
}
```

对密码进行加密

到现在为止，我们一直在使用明文密码来存储用户信息。但这种方式很容易遭受黑客攻击，因此在存储之前需要对密码进行加密。Spring Security 支持多种算法来实现密码的加密。比如说，可以选择使用 BCrypt（一种单向哈希算法）来对密码进行加密。

■ **注意：**你可能需要一个帮手来计算密码的 BCrypt 哈希值。这可以在线搞定，还可以创建一个带有 main 方法的类，使用 Spring Security 的 BCryptPasswordEncoder 来完成。

现在，你可以将加密后的密码存储到用户仓库中了。比如说，如果使用内存用户定义，那就可以在 password 属性中指定加密后的密码了。接下来，可以通过 AuthenticationManagerBuilder 的 passwordEncoder()方法来指定密码编码器。

```
@Configuration
@EnableWebSecurity
```

```
public class TodoSecurityConfig extends WebSecurityConfigurerAdapter {

...

    @Bean
    public BCryptPasswordEncoder passwordEncoder() {
        return new BCryptPasswordEncoder();
    }

    @Override
    protected void configure(AuthenticationManagerBuilder auth) throws Exception {
        auth .jdbcAuthentication()
                .passwordEncoder(passwordEncoder())
                .dataSource(dataSource());
    }
}
```

当然，你需要将加密后的密码而不是明文密码存储到数据库表中，如表 7-5 所示。要想在 password 字段中存储 BCrypt 哈希值，字段长度至少得是 60 字符（这是 BCrypt 哈希值的长度）。

表 7-5　　使用了加密密码的 USERS 表中的测试用户数据

用户名	密码	启用
admin@ya2do.io	$2a$10$E3mPTZb50e7sSW15fDx8Ne7hDZpfDjrmMPTTUp8wVjLTu.G5oPYCO	1
marten@ya2do.io	$2a$10$5VWqjwoMYnFRTTmbWCRZT.iY3WW8ny27kQuUL9yPK1/WJcPcBLFWO	1
jdoe@does.net	$2a$10$cFKh0.XCUOA9L.in5smIiO2QIOT8.6ufQSwIIC.AVz26WctxhSWC6	0

使用 LDAP 仓库来认证用户

Spring Security 还支持通过访问 LDAP 仓库来认证用户。首先，你需要准备一些用户数据来填充 LDAP 仓库。我们以 LDAP 数据交换格式（LDAP Data Interchange Format，LDIF）来准备用户数据，这是一种标准的文本数据格式，用于导入和导出 LDAP 目录数据。比如说，创建包含如下内容的 users.ldif 文件：

```
dn: dc=springrecipes,dc=com
objectClass: top
objectClass: domain
dc: springrecipes

dn: ou=groups,dc=springrecipes,dc=com
objectclass: top
objectclass: organizationalUnit
ou: groups

dn: ou=people,dc=springrecipes,dc=com
objectclass: top
objectclass: organizationalUnit
ou: people

dn: uid=admin,ou=people,dc=springrecipes,dc=com
objectclass: top
objectclass: uidObject
objectclass: person
uid: admin
cn: admin
sn: admin
userPassword: secret

dn: uid=user1,ou=people,dc=springrecipes,dc=com
```

```
objectclass: top
objectclass: uidObject
objectclass: person
uid: user1
cn: user1
sn: user1
userPassword: 1111

dn: cn=admin,ou=groups,dc=springrecipes,dc=com
objectclass: top
objectclass: groupOfNames
cn: admin
member: uid=admin,ou=people,dc=springrecipes,dc=com

dn: cn=user,ou=groups,dc=springrecipes,dc=com
objectclass: top
objectclass: groupOfNames
cn: user
member: uid=admin,ou=people,dc=springrecipes,dc=com
member: uid=user1,ou=people,dc=springrecipes,dc=com
```

如果不太理解这个 LDIF 文件也不必担心。你可能不需要常常使用该文件格式来定义 LDAP 数据，因为大多数 LDAP 服务器都支持基于 GUI 的配置。这个 users.ldif 文件包含如下内容：

- 默认的 LDAP 域，dc=springrecipes,dc=com；
- 用于存储分组和用户的 groups 和 people 组织单元；
- 密码分别是 secret 与 1111 的 admin 与 user1 用户；
- admin 分组（包含了 admin 用户）与 user 分组（包含了 admin 与 user1 用户）

出于测试的目的，你可以在本地机器上安装一个 LDAP 服务器来托管这个用户仓库。为了能够更加轻松地安装与配置，推荐安装 OpenDS，这是一个支持 LDAP 的基于 Java 的开源目录服务引擎。

■ **提示：**在 bin 目录中有一个 ldap.sh 脚本，它会启动 Docker 版本的 OpenDS，并且会导入之前提到的 users.ldif。注意到该 LDAP 服务器的 root 用户与密码分别是 cn=Directory Manager 与 ldap。稍后将会使用该用户连接到服务器。

LDAP 服务器启动后，你可以配置 Spring Security 使用其仓库来认证用户。

需要通过 ldapAuthentication()配置方法来配置 LDAP 仓库。可以通过几个回调方法指定搜索过滤器和搜索范围来搜索用户与分组，其值必须要与仓库的目录结构保持一致。借助于前述的属性值，Spring Security 会使用特定的用户 ID 从 people 组织单元搜索用户，以及从 groups 组织单元搜索用户的分组。Spring Security 会自动向每个分组插入 ROLE_前缀作为授权。

```
@Configuration
@EnableWebSecurity
public class TodoSecurityConfig extends WebSecurityConfigurerAdapter {
...
    @Override
    protected void configure(AuthenticationManagerBuilder auth) throws Exception {
        auth
            .ldapAuthentication()
                .contextSource()
                .url("ldap://localhost:1389/dc=springrecipes,dc=com")
                .managerDn("cn=Directory Manager").managerPassword("ldap")
            .and()
                .userSearchFilter("uid={0}").userSearchBase("ou=people")
                .groupSearchFilter("member={0}").groupSearchBase("ou=groups")

            .passwordEncoder(new LdapShaPasswordEncoder())
            .passwordCompare().passwordAttribute("userPassword");
    }
}
```

由于 OpenDS 默认使用了盐渍安全哈希算法（Salted Secure Hash Algorithm，SSHA）对用户密码进行加密，因此需要指定 LdapShaPasswordEncoder 作为密码编码器。注意到这个值与 sha 是不同的，因为它是特定于 LDAP 的密码编码。你还需要指定 passwordAttribute 值，因为密码编码器需要知道 LDAP 中的哪个字段是密码。

最后，你需要引用 LDAP 服务器定义，它定义了如何创建去往 LDAP 服务器的连接。你可以指定 root 用户的用户名与密码来连接到运行在 localhost 上的 LDAP 服务器，这是通过使用 contextSource 方法来配置的。

缓存用户信息

<jdbc-user-service>与<ldap-user-service>都支持缓存用户信息，不过首先需要选择一种提供了缓存服务的缓存实现。由于 Spring 与 Spring Security 都内建了对 Ehcache 的支持，因此可以将其作为缓存实现，并为其创建一个配置文件（比如说类路径根目录下的 ehcache.xml），其内容如下所示：

```
<ehcache>
    <diskStore path="java.io.tmpdir"/>

    <defaultCache
        maxElementsInMemory="1000"
        eternal="false"
        timeToIdleSeconds="120"
        timeToLiveSeconds="120"
        overflowToDisk="true"
        />

    <cache name="userCache"
        maxElementsInMemory="100"
        eternal="false"
        timeToIdleSeconds="600"
        timeToLiveSeconds="3600"
        overflowToDisk="true"
        />
</ehcache>
```

这个 Ehcache 配置文件定义了两种类型的缓存配置。一个是默认的缓存配置，另一个则缓存了用户信息。如果使用了用户缓存配置，那么缓存实例就会在内存中至多缓存 100 个用户信息。如果超出了这个限制，那么被缓存的用户就会保存到磁盘上。如果被缓存的用户闲置了 10 分钟或是创建后存活了 1 小时，那么它就会过期。

Spring Security 自带两个 UserCache 实现：EhCacheBasedUserCache，它需要引用一个 Ehcache 实例；SpringCacheBasedUserCache，它使用了 Spring 的缓存抽象。

在基于 Java 的配置中，在本书撰写之际，只有 jdbcAuthentication()方法可以轻松实现对用户缓存的配置。对于 Spring 的基于缓存的解决方案来说（依旧是委托给 Ehcache），你需要配置一个 CacheManager 实例并让其感知到 Ehcache。

```
@Configuration
public class MessageBoardConfiguration {
...
    @Bean
    public EhCacheCacheManager cacheManager() {
        EhCacheCacheManager cacheManager = new EhCacheCacheManager();
        cacheManager.setCacheManager(ehCacheManager().getObject());
        return cacheManager;
    }

    @Bean
    public EhCacheManagerFactoryBean ehCacheManager() {
        return new EhCacheManagerFactoryBean();
    }
}
```

最好将其添加到服务的配置中，因为缓存还可用于其他方式（参见关于 Spring 缓存的内容）。在创建好了 CacheManager 实例后，现在需要配置一个 SpringCacheBasedUserCache 类。

```
@Configuration
@EnableWebMvcSecurity
public class TodoSecurityConfig extends WebSecurityConfigurerAdapter {

    @Autowired
    private CacheManager cacheManager;

    @Bean
    public SpringCacheBasedUserCache userCache() throws Exception {
        Cache cache = cacheManager.getCache("userCache");
        return new SpringCacheBasedUserCache(cache);
    }

    @Override
    protected void configure(AuthenticationManagerBuilder auth) throws Exception {
        auth.jdbcAuthentication()
            .userCache(userCache())
    ...
    }
}
```

注意，CacheManager 是被自动装配到配置类中的。之所以需要访问它是因为你需要获取一个 Cache 实例并将其传递给 SpringCacheBasedUseCache 的构造方法。你要使用名为 userCache 的缓存（在 ehcache.xml 文件中配置的）。最后，将配置好的 UserCache 传递给 jdbcAuthentications.userCache()方法。

7-4 做出访问控制决策

问题提出

在认证过程中，应用会对成功认证的用户进行授权。当该用户尝试访问应用中的资源时，应用需要确定凭借这些授权或是其他特性可否访问该资源。

解决方案

决定用户是否允许访问应用中资源的行为叫作访问控制决策。它是基于用户的认证状态与资源的特性以及访问属性做出的。在 Spring Security 中，访问控制决策是由访问决策管理器来做出的，它需要实现 AccessDecisionManager 接口。你可以通过实现该接口创建自己的访问决策管理器，不过 Spring Security 自带 3 个便捷的访问决策管理器，它们是基于投票方式实现的，如表 7-6 所示。

表 7-6 Spring Security 自带的访问决策管理器

访问决策管理器	指定何时授权访问
AffirmativeBased	至少一个投票者投票授权
ConsensusBased	一半以上的投票者投票授权
UnanimousBased	所有投票者都投票拒绝或是授权访问

所有这些访问决策管理器都需要配置一组投票者来对访问控制决策进行投票。每个投票者都需要实现 AccessDecisionVoter 接口。一个投票者可以授权、弃权或是拒绝对一个资源的访问。投票结果是通过定义在 AccessDecisionVoter 接口中的 ACCESS_GRANTED、ACCESS_DENIED 与 ACCESS_ABSTAIN 常量字段表示的。

默认情况下，如果没有显式指定访问决策管理器，那么 Spring Security 会通过如下配置的两个投票者自动配置一个 AffirmativeBased 访问决策管理器。

- RoleVoter 会基于用户的角色对访问控制决策进行投票。它只会处理以 ROLE_前缀开头的访问属性，不过该前缀是可以定制的。如果用户拥有访问资源所需的角色，那么它就会授权访问；如果用户缺少访问资源所需的任何角色，那么它就会拒绝访问。如果资源没有以 ROLE_开头的访问属性，那么它就会弃权。

- AuthenticatedVoter 会基于用户的认证级别对访问控制决策进行投票。它只会处理 IS_AUTHENTICATED_、FULLY、IS_AUTHENTICATED_REMEMBERED 与 IS_AUTHENTICATED_ANONYMOUSLY 访问属性。如果用户的认证级别高于所需的属性，那么它就会授权访问。认证级别从最高到最低分别是完全授权、记住授权与匿名授权。

解释说明

默认情况下，如果没有指定，那么 Spring Security 会自动配置一个访问决策管理器。这个默认的访问决策管理器等价于如下配置所定义的访问决策管理器：

```
@Bean
public AffirmativeBased accessDecisionManager() {
    List<AccessDecisionVoter> decisionVoters = Arrays.asList(new RoleVoter(), new
    AuthenticatedVoter());
    return new AffirmativeBased(decisionVoters);
}

@Override
protected void configure(HttpSecurity http) throws Exception {

    http.authorizeRequests()
        .accessDecisionManager(accessDecisionManager())
    ...
}
```

这个默认的访问决策管理器和它的决策投票者应该能够满足大多数典型的授权需求。不过，如果不满足，则可以自己创建。在大多数情况下，你只需创建一个自定义投票者即可。比如说，你可以创建一个投票者，使其基于用户的 IP 地址进行决策。

```
package com.apress.springrecipes.board.security;

import org.springframework.security.access.AccessDecisionVoter;
import org.springframework.security.access.ConfigAttribute;
import org.springframework.security.core.Authentication;
import org.springframework.security.web.authentication.WebAuthenticationDetails;

import java.util.Collection;
import java.util.Objects;

public class IpAddressVoter implements AccessDecisionVoter<Object> {

    private static final String IP_PREFIX = "IP_";
    private static final String IP_LOCAL_HOST = "IP_LOCAL_HOST";

    public boolean supports(ConfigAttribute attribute) {
        return (attribute.getAttribute() != null) && attribute.getAttribute().startsWith
        (IP_PREFIX);
    }

    @Override
    public boolean supports(Class<?> clazz) {
        return true;
    }

    public int vote(Authentication authentication, Object object,
    Collection<ConfigAttribute> configList) {
        if (!(authentication.getDetails() instanceof WebAuthenticationDetails)) {
            return ACCESS_DENIED;
        }

        WebAuthenticationDetails details = (WebAuthenticationDetails) authentication.
```

```
        getDetails();
        String address = details.getRemoteAddress();

        int result = ACCESS_ABSTAIN;

        for (ConfigAttribute config : configList) {
            result = ACCESS_DENIED;

            if (Objects.equals(IP_LOCAL_HOST, config.getAttribute())) {
                if (address.equals("127.0.0.1") || address.equals("0:0:0:0:0:0:0:1")) {
                    return ACCESS_GRANTED;
                }
            }
        }

        return result;
    }
}
```

注意，该投票者只会处理以 IP_前缀开头的访问属性。目前，它只支持 IP_LOCAL_HOST 访问属性。如果用户是个 Web 客户端，IP 地址是 127.0.0.1 或 0:0:0:0:0:0:0:1（无网络的 Linux 工作站返回的最后一个值），那么该投票者就会投票以授权访问。否则，它会投票以拒绝访问。如果资源没有以 IP_开头的访问属性，那么它就会弃权。

接下来，你需要自定义一个访问决策管理器来包含该投票者。

```
@Bean
public AffirmativeBased accessDecisionManager() {
    List<AccessDecisionVoter> decisionVoters = Arrays.asList(new RoleVoter(), new
    AuthenticatedVoter(), new IpAddressVoter());
    return new AffirmativeBased(decisionVoters);
}
```

现在，假设允许运行 Web 容器的机器上的用户（即服务器管理员）无须登录就可以删除待办事项。你需要在配置中引用该访问决策管理器，并将访问属性 IP_LOCAL_HOST 添加到删除 URL 映射中。

```
http.authorizeRequests()
    .accessDecisionManager()
    .antMatchers(HttpMethod.DELETE, "/todos*").access("ADMIN,IP_LOCAL_HOST");
```

当直接调用 URL 时，待办事项会被删除。要是通过 Web 界面访问，则还是需要登录。

使用表达式来做出访问控制决策

虽然 AccessDecisionVoters 考虑到了一定程度的灵活性，但有时还是希望更复杂的访问控制规则能够更加灵活一些。借助于 Spring Security，我们可以通过 Spring 表达式语言（SpEL）创建强大的访问控制规则。Spring Security 支持不少“开箱即用”的表达式（见表 7-7）。通过使用 and、or 和 not 等表达式，你可以创建非常强大且灵活的表达式。Spring Security 会通过 WebExpressionVoter 自动配置访问决策管理器。该访问决策管理器等价于使用如下 bean 配置所定义的访问决策管理器：

```
@Bean
public AffirmativeBased accessDecisionManager() {
    List<AccessDecisionVoter> decisionVoters = Arrays.asList(new WebExpressionVoter());
    return new AffirmativeBased(decisionVoters);
}
```

表 7-7 Spring Security 内建的表达式

表达式	说明
hasRole(role)或 hasAuthority(authority)	如果当前用户拥有给定角色则返回 true
hasAnyRole(role1,role2) / hasAnyAuthority(auth1,auth2)	如果当前用户至少拥有给定角色之一就返回 true

续表

表达式	说明
hasIpAddress(ip-address)	如果当前用户拥有给定 IP 地址就返回 true
principal	当前用户
Authentication	访问 Spring Security 认证对象
permitAll	总是为 true
denyAll	总是为 false
isAnonymous()	如果当前是匿名用户就返回 true
isRememberMe()	如果当前用户通过"记住我"功能登录就返回 true
isAuthenticated()	如果不是匿名用户就返回 true
isFullyAuthenticated()	如果用户不是匿名用户也没有使用"记住我"就返回 true

警告： 虽然角色与授权几乎是一样的，但在处理方式上还是存在一个细微但重要的差别。在使用 hasRole 时，如果传递进来的角色值以 ROLE_（默认的角色前缀）开头，那么它就会被检查。如果不是，那么在检查授权前它会被添加进来。这样，hasRole('ADMIN') 实际上会检查当前用户是否拥有 ROLE_ADMIN 授权。在使用 hasAuthority 时，它会检查这个值本身。

如果用户拥有 ADMIN 角色或是在本地机器上登录，那么之前的表达式就可以删除待办事项。在上一节中，你需要创建自定义的 AccessDecisionVoter。现在只需要编写一个表达式即可。在定义匹配器时，可以通过访问方法而非 has*方法来编写表达式。

```
@Configuration
@EnableWebSecurity
public class TodoSecurityConfig extends WebSecurityConfigurerAdapter {

    @Override
    protected void configure(HttpSecurity http) throws Exception {
        http
            .authorizeRequests()
                .antMatchers("/messageList*").hasAnyRole("USER", "GUEST")
                .antMatchers("/messagePost*").hasRole("USER")
                .antMatchers("/messageDelete*")
            .access("hasRole('ROLE_ADMIN') or hasIpAddress('127.0.0.1') or
            hasIpAddress('0:0:0:0:0:0:0:1')")
            ...
    }
...
}
```

虽然 Spring Security 已经内建了一些函数可以在创建表达式时使用，不过也可以根据需要扩展功能。要想做到这一点，你需要创建一个实现了 SecurityExpressionOperations 接口的类并将其注册到 Spring Security。虽然可以创建一个类来实现该接口的所有方法，不过在添加表达式时通常可以继承默认实现，这样会简单很多。

```
package com.apress.springrecipes.board.security;

import org.springframework.security.core.Authentication;
import org.springframework.security.web.FilterInvocation;
import org.springframework.security.web.access.expression.WebSecurityExpressionRoot;

public class ExtendedWebSecurityExpressionRoot extends WebSecurityExpressionRoot {

    public ExtendedWebSecurityExpressionRoot(Authentication a, FilterInvocation fi) {
        super(a, fi);
    }
```

```
    public boolean localAccess() {
        return hasIpAddress("127.0.0.1") || hasIpAddress("0:0:0:0:0:0:0:1");

    }
}
```

这里继承了 WebSecurityExpressionRoot，它提供了默认实现，同时你又添加了方法 localAccess()。该方法会检查是否从本地机器登录。为了让该类能被 Spring Security 使用，你需要创建 SecurityExpressionHandler 接口。

```
package com.apress.springrecipes.board.security;

import org.springframework.security.access.expression.SecurityExpressionOperations;
import org.springframework.security.authentication.AuthenticationTrustResolver;
import org.springframework.security.authentication.AuthenticationTrustResolverImpl;
import org.springframework.security.core.Authentication;
import org.springframework.security.web.FilterInvocation;
import org.springframework.security.web.access.expression.
DefaultWebSecurityExpressionHandler;
import org.springframework.security.web.access.expression.WebSecurityExpressionRoot;
public class ExtendedWebSecurityExpressionHandler extends
DefaultWebSecurityExpressionHandler {

    private AuthenticationTrustResolver trustResolver = new
    AuthenticationTrustResolverImpl();

    @Override
    protected SecurityExpressionOperations
        createSecurityExpressionRoot(Authentication authentication, FilterInvocation fi)
    {
        ExtendedWebSecurityExpressionRoot root =
            new ExtendedWebSecurityExpressionRoot(authentication, fi);
        root.setPermissionEvaluator(getPermissionEvaluator());
        root.setTrustResolver(trustResolver);
        root.setRoleHierarchy(getRoleHierarchy());
        return root;
    }

    @Override
    public void setTrustResolver(AuthenticationTrustResolver trustResolver) {
        this.trustResolver=trustResolver;
        super.setTrustResolver(trustResolver);
    }
}
```

这里继承了 DefaultWebSecurityExpressionHandler，它提供了默认实现。你重写了 createSecurityExpressionRoot 方法，在该方法中创建了 ExtendedWebSecurityExpressionRoot 类的一个实例。由于需要添加一些协作者，因此调用了父类的 get 方法。因为父类没有 getTrustResolver 方法，所以需要自己创建它的一个新实例并实现 setter 方法。

```
@Configuration
@EnableWebSecurity
public class TodoSecurityConfig extends WebSecurityConfigurerAdapter {

    @Override
    protected void configure(HttpSecurity http) throws Exception {
        http
            .authorizeRequests()
                .expressionHandler(new ExtendedWebSecurityExpressionHandler())
                .antMatchers("/todos*").hasAuthority("USER")
                .antMatchers(DELETE, "/todos*").access("hasRole('ROLE_ADMIN) or
                localAccess()")
    }
}
```

你通过 expressionHandler 方法设置了自定义表达式处理器。现在可以使用 localAccess()表达式重写自己的表达式了。

在表达式中使用 Spring bean 来做出访问控制决策

虽然可以通过这些方法扩展 Spring Security，不过并不推荐这么做。与之相反，建议的做法是编写一个类并在表达式中使用它。借助于表达式中的@语法，可以调用应用上下文中的任何 bean。这样，你可以编写类似于@accessChecker.hasLocalAccess(authentication)这样的表达式并提供一个名为 accessChecker 的 bean，它拥有一个 hasLocalAccess 方法，该方法接收一个 Authentication 对象。

```
package com.apress.springrecipes.board.security;

import org.springframework.security.core.Authentication;
import org.springframework.security.web.authentication.WebAuthenticationDetails;

public class AccessChecker {

    public boolean hasLocalAccess(Authentication authentication) {
        boolean access = false;
        if (authentication.getDetails() instanceof WebAuthenticationDetails) {
            WebAuthenticationDetails details = (WebAuthenticationDetails) authentication.
            getDetails();
            String address = details.getRemoteAddress();
            access = address.equals("127.0.0.1") || address.equals("0:0:0:0:0:0:0:1");
        }
        return access;
    }
}
```

AccessChecker 依然与之前的 IpAddressVoter 和自定义表达式处理器那样做同样的检查，不过它并未继承 Spring Security 的类。

```
@Bean
public AccessChecker accessChecker() {
    return new AccessChecker();
}

@Override
protected void configure(HttpSecurity http) throws Exception {

    http.authorizeRequests()
        .antMatchers("/todos*").hasAuthority("USER")
        .antMatchers(HttpMethod.DELETE, "/todos*").access("hasAuthority('ADMIN') or
        @accessChecker.hasLocalAccess(authentication)")
    ...
}
```

7-5 保护方法调用

问题提出

作为保护 Web 层 URL 访问的另外一种方式或是补充，你需要在服务层保护方法调用。比如说，如果一个控制器需要调用服务层的多个方法，你需要对这些方法进行细粒度的安全控制。

解决方案

Spring Security 可以以一种声明式的方式来保护方法调用。对声明在 bean 接口或实现类中的方法应用注解@Secured、@PreAuthorize/@PostAuthorize 或@PreFilter/@PostFilter，然后通过@EnableGlobalMethodSecurity 注解为其启用安全。

解释说明

首先，我们将会介绍如何通过注解来保护方法调用以及如何编写安全表达式。然后，我们会介绍如何通过注解和表达式来过滤方法的输入参数与输出。

使用注解来保护方法

保护方法是通过为其添加注解@Secured 实现的。比如说，你可以为 MessageBoardServiceImpl 中的方法应用注解@Secured 并指定其访问属性作为注解的值，值的类型是 String[]，用于接收一个或多个授权，拥有授权就可以访问该方法。

```
package com.apress.springrecipes.board.service;
...
import org.springframework.security.access.annotation.Secured;

public class MessageBoardServiceImpl implements MessageBoardService {
    ...
    @Secured({"ROLE_USER", "ROLE_GUEST"})
    public List<Message> listMessages() {
        ...
    }

    @Secured("ROLE_USER")
    public synchronized void postMessage(Message message) {
        ...
    }

    @Secured({"ROLE_ADMIN", "IP_LOCAL_HOST"})
    public synchronized void deleteMessage(Message message) {
        ...
    }

    @Secured({"ROLE_USER", "ROLE_GUEST"})
    public Message findMessageById(Long messageId) {
        return messages.get(messageId);
    }
}
```

最后，需要启用方法安全。要想做到这一点，需要向配置类添加@EnableGlobalMethodSecurity 注解。由于使用了@Secured，因此需要将 securedEnabled 属性设为 true。

```
@Configuration
@EnableGlobalMethodSecurity(securedEnabled = true)
public class TodoWebConfiguration { ... }
```

■ **注意：**需要将@EnableGlobalMethodSecurity 注解添加到应用上下文配置上，而该配置应该包含了你想要保护的 bean，这是非常重要的。

使用注解与表达式来保护方法

如果需要更为复杂的安全规则，那么可以像 URL 防护一样，使用基于 SpEL 的安全表达式来保护应用。要想做到这一点，可以使用@PreAuthorize 与@PostAuthorize 注解。借助于它们，你可以像基于 URL 的安全一样编写基于安全的表达式。要想开启对这些注解的处理，你需要将@EnableGlobalMethodSecurity 注解的 prePostEnabled 属性设为 true。

```
@Configuration
@EnableGlobalMethodSecurity(prePostEnabled = true)
public class TodoWebConfiguration { ... }
```

现在，可以使用@PreAuthorize 与@PostAuthorize 注解来保护应用的安全了。

```
package com.apress.springrecipes.board;

import org.springframework.security.access.prepost.PreAuthorize;
import org.springframework.stereotype.Service;
```

```java
import javax.transaction.Transactional;
import java.util.List;

@Service
@Transactional
class TodoServiceImpl implements TodoService {

    private final TodoRepository todoRepository;

    TodoServiceImpl(TodoRepository todoRepository) {
        this.todoRepository = todoRepository;
    }

    @Override
    @PreAuthorize("hasAuthority('USER')")
    public List<Todo> listTodos() {
        return todoRepository.findAll();
    }

    @Override
    @PreAuthorize("hasAuthority('USER')")
    public void save(Todo todo) {
        this.todoRepository.save(todo);
    }

    @Override
    @PreAuthorize("hasAuthority('USER')")
    public void complete(long id) {
        Todo todo = findById(id);
        todo.setCompleted(true);
        todoRepository.save(todo);
    }

    @Override
    @PreAuthorize("hasAnyAuthority('USER', 'ADMIN')")
    public void remove(long id) {
        todoRepository.remove(id);
    }

    @Override
    @PreAuthorize("hasAuthority('USER')")
    @PostAuthorize("returnObject.owner == authentication.name")
    public Todo findById(long id) {
        return todoRepository.findOne(id);
    }
}
```

@PreAuthorize 注解会在实际的方法调用前触发，@PostAuthorize 注解会在方法调用后触发。你还可以编写安全表达式，通过 returnObject 表达式来使用方法调用的结果。看一下 findById 方法上的表达式：如果其他人尝试访问 Todo 对象，那就会抛出安全异常。

使用注解与表达式进行过滤

除了@PreAuthorize 与@PostAuthorize 注解外，还有@PreFilter 与@PostFilter 注解。这两组注解之间的主要差别在于，如果安全规则不满足，那么@*Authorize 注解会抛出异常，而@*Filter 注解只会过滤你无权访问的元素的输入与输出变量。

现在，当调用 listTodos 时，数据库会返回全部信息。你想要对拥有 ADMIN 权限的用户所能获取的元素进行限制，其他人只能看到自己的待办事项列表。这可以通过@PostFilter 注解轻松实现。添加@PostFilter("hasAuthority('ADMIN') or filterObject.owner == authentication.name")就可以实现这个规则。

```
@PreAuthorize("hasAuthority('USER')")
@PostFilter("hasAnyAuthority('ADMIN') or filterObject.owner == authentication.name")
public List<Todo> listTodos() {
    return todoRepository.findAll();
}
```

重新部署应用并登录后，你只会看到自己的待办事项，当使用拥有 ADMIN 权限的用户时，你依然会看到所有的待办事项。参见 7-7 节来了解关于@*Filter 注解更为复杂的用法。

警告：虽然@PostFilter 与@PreFilter 注解是一种简单的过滤方法的输入/输出的方式，不过使用时要小心。在使用它们处理大量结果时，应用的性能会受到严重影响。

7-6　处理视图安全

问题提出

有时，你想在 Web 应用的视图中展示用户的认证信息，比如说主体名和授权。此外，你还想根据用户的授权有条件地渲染视图内容。

解决方案

虽然可以在 JSP 文件中编写 JSP 脚本，通过 Spring Security API 获取认证与授权信息，不过这并非高效的解决方案。Spring Security 提供了一个 JSP 标签库来处理 JSP 视图中的安全。它所提供的标签可以展示用户的认证信息，还可以根据用户的授权有条件地渲染视图内容。

解释说明

我们首先会介绍如何通过 Spring Security 标签展示当前已认证用户的信息。接下来将会介绍如何根据当前已认证用户的授权有条件地隐藏部分页面信息。

展示认证信息

假设你想在待办事项列表页面（即 todos.jsp）的头部展示用户的主体名与授权信息。首先，你需要导入 Spring Security 的标签库定义。

```
<%@ taglib prefix="c" uri="http://java.sun.com/jsp/jstl/core" %>
<%@ taglib prefix="sec" uri="http://www.springframework.org/security/tags" %>
```

<sec:authentication>标签会公开当前用户的 Authentication 对象供你渲染其属性。你可以在其属性中指定属性名或属性路径。比如说，可以通过 name 属性渲染用户的主体名。

```
<h4>Todos for <sec:authentication property="name" /></h4>
```

除了直接渲染认证属性外，该标签还支持将属性存储到 JSP 变量中，其名字是通过 var 属性来指定的。比如说，你可以将 authorities 属性（包含了用户的授权）存储到 JSP 变量 authorities 中，并通过<c:forEach>标签逐一渲染它们。还可以通过 scope 属性进一步指定变量的作用域。

```
<sec:authentication property="authorities" var="authorities" />
<ul>
    <c:forEach items="${authorities}" var="authority">
        <li>${authority.authority}</li>
    </c:forEach>
</ul>
```

有条件地渲染视图内容

如果想要根据用户的授权有条件地渲染视图内容，那就可以使用<sec:authorize>标签。比如说，你可以根据用户的授权决定是否渲染作者信息。

```
<td>
    <sec:authorize ifAllGranted="ROLE_ADMIN,ROLE_USER">${todo.owner}</sec:authorize>
</td>
```

如果希望只有当用户同时被授予了某些权限时才渲染被包围起来的内容，那就需要在 ifAllGranted 属性中指定它们。如果用户拥有任意授权就渲染被包围起来的内容，你就需要在 ifAnyGranted 属性中指定它们。

```
<td>
    <sec:authorize ifAnyGranted="ROLE_ADMIN,ROLE_USER">${todo.owner}</sec:authorize>
</td>
```

还可以在用户没有 ifNotGranted 属性中所指定的任何授权时渲染被包围的内容。

```
<td>
    <sec:authorize ifNotGranted="ROLE_ADMIN,ROLE_USER">${todo.owner}</sec:authorize>
</td>
```

7-7 处理领域对象的安全

问题提出

有时，你可能会有复杂的安全需求，需要在领域对象层次处理安全。这意味着每个领域对象对不同的主体有着不同的访问属性。

解决方案

Spring Security 提供了一个名为 ACL 的模块，可以让每个领域对象有自己的访问控制列表（ACL）。一个 ACL 包含了与对象关联的领域对象的对象身份，还持有多个访问控制项（ACE），每个访问控制项都包含了如下两个核心组成部分。

- 权限：ACE 的权限是由特定的位掩码表示的，每个位值表示特定类型的权限。BasePermission 类预先定义了 5 种基本的权限作为常量值供你使用：READ（位 0 或整数 1）、WRITE（位 1 或整数 2）、CREATE（位 2 或整数 4）、DELETE（位 3 或整数 8）以及 ADMINISTRATION（位 4 或整数 16）。还可以使用其他未使用的位定义自己的权限。
- 安全身份（SID）：每个 ACE 都包含了针对特定 SID 的权限。SID 可以是个主体（PrincipalSid），也可以是关联到权限的授权（GrantedAuthoritySid）。除了定义 ACL 对象模型外，Spring Security 还定义了用于读取和维护模型的 API，它为这些 API 提供了高性能的 JDBC 实现。为了简化 ACL 的使用，Spring Security 还提供了一些基础设施，如访问决策投票者与 JSP 标签，可以让你在使用 ACL 时能够与应用中的其他安全设施保持一致。

解释说明

首先，我们将会介绍如何创建 ACL 服务以及如何维护 ACL 权限。然后，我们会介绍如何通过存储的 ACL 权限来使用安全表达式保护对资源的访问。

创建 ACL 服务

Spring Security 对将 ACL 数据存储到关系型数据库并通过 JDBC 对其访问提供了内建的支持。首先，你需要在数据库中创建如下表来存储 ACL 数据：

```
CREATE TABLE ACL_SID(
    ID          BIGINT        NOT NULL GENERATED BY DEFAULT AS IDENTITY,
    SID         VARCHAR(100)  NOT NULL,
    PRINCIPAL   SMALLINT      NOT NULL,
    PRIMARY KEY (ID),
    UNIQUE (SID, PRINCIPAL)
);

CREATE TABLE ACL_CLASS(
    ID      BIGINT        NOT NULL GENERATED BY DEFAULT AS IDENTITY,
    CLASS   VARCHAR(100)  NOT NULL,
    PRIMARY KEY (ID),
    UNIQUE (CLASS)
);

CREATE TABLE ACL_OBJECT_IDENTITY(
    ID                    BIGINT    NOT NULL GENERATED BY DEFAULT AS IDENTITY,
    OBJECT_ID_CLASS       BIGINT    NOT NULL,
    OBJECT_ID_IDENTITY    BIGINT    NOT NULL,
```

```
    PARENT_OBJECT        BIGINT,
    OWNER_SID            BIGINT,
    ENTRIES_INHERITING   SMALLINT  NOT NULL,
    PRIMARY KEY (ID),
    UNIQUE (OBJECT_ID_CLASS, OBJECT_ID_IDENTITY),
    FOREIGN KEY (PARENT_OBJECT)   REFERENCES ACL_OBJECT_IDENTITY,
    FOREIGN KEY (OBJECT_ID_CLASS) REFERENCES ACL_CLASS,
    FOREIGN KEY (OWNER_SID)       REFERENCES ACL_SID
);

CREATE TABLE ACL_ENTRY(
    ID                  BIGINT     NOT NULL GENERATED BY DEFAULT AS IDENTITY,
    ACL_OBJECT_IDENTITY BIGINT     NOT NULL,
    ACE_ORDER           INT        NOT NULL,
    SID                 BIGINT     NOT NULL,
    MASK                INTEGER    NOT NULL,
    GRANTING            SMALLINT   NOT NULL,
    AUDIT_SUCCESS       SMALLINT   NOT NULL,
    AUDIT_FAILURE       SMALLINT   NOT NULL,
    PRIMARY KEY (ID),
    UNIQUE (ACL_OBJECT_IDENTITY, ACE_ORDER),
    FOREIGN KEY (ACL_OBJECT_IDENTITY) REFERENCES ACL_OBJECT_IDENTITY,
    FOREIGN KEY (SID)                 REFERENCES ACL_SID
);
```

Spring Security 定义了 API 并提供了高性能的 JDBC 实现供你访问存储在这些表中的 ACL 数据，因此你很少需要直接访问数据库中的 ACL 数据。由于每个领域对象都有自己的 ACL，因此应用中可能会有大量的 ACL。幸好，Spring Security 支持缓存 ACL 对象。你可以继续将 Ehcache 作为缓存实现，并在 ehcache.xml（位于类路径的根目录下）中为 ACL 缓存创建新的配置。

```
<ehcache>
    ...
    <cache name="aclCache"
        maxElementsInMemory="1000"
        eternal="false"
        timeToIdleSeconds="600"
        timeToLiveSeconds="3600"
        overflowToDisk="true"
        />
</ehcache>
```

接下来，你需要为应用创建 ACL 服务。不过，由于 Spring Security 尚不支持使用基于 Java 的配置来配置 ACL 模块，因此需要使用一组常规的 Spring bean 来配置该模块。出于这个原因，我们来创建一个名为 TodoAclConfig 的单独的 bean 配置类，它用于存储特定于 ACL 的配置，然后将其位置添加到部署描述符中。

```
package com.apress.springrecipes.board.security;

import org.springframework.security.web.context.AbstractSecurityWebApplicationInitializer;

public class TodoSecurityInitializer extends AbstractSecurityWebApplicationInitializer
{
    public TodoSecurityInitializer() {
        super(TodoSecurityConfig.class, TodoAclConfig.class);
    }
}
```

在 ACL 配置文件中，核心的 bean 是个 ACL 服务。在 Spring Security 中有两个接口定义了 ACL 服务的操作：AclService 与 MutableAclService。AclService 定义了读取 ACL 的操作。MutableAclService 是 AclService 的子接口，定义了创建、更新与删除 ACL 的操作。如果应用只需要读取 ACL，那就选择一个 AclService 实现即可，比如说 JdbcAclService。否则，应该选择一个 MutableAclService 实现，如 JdbcMutableAclService。

```
package com.apress.springrecipes.board.security;

import org.springframework.cache.CacheManager;
```

```java
import org.springframework.cache.ehcache.EhCacheManagerFactoryBean;
import org.springframework.context.annotation.Bean;
import org.springframework.context.annotation.Configuration;
import org.springframework.security.acls.AclEntryVoter;
import org.springframework.security.acls.domain.*;
import org.springframework.security.acls.jdbc.BasicLookupStrategy;
import org.springframework.security.acls.jdbc.JdbcMutableAclService;
import org.springframework.security.acls.jdbc.LookupStrategy;
import org.springframework.security.acls.model.AclCache;
import org.springframework.security.acls.model.AclService;
import org.springframework.security.acls.model.Permission;
import org.springframework.security.acls.model.PermissionGrantingStrategy;
import org.springframework.security.core.authority.SimpleGrantedAuthority;

import javax.sql.DataSource;

@Configuration
public class TodoAclConfig {

    private final DataSource dataSource;

    public TodoAclConfig(DataSource dataSource) {
        this.dataSource = dataSource;
    }

    @Bean
    public AclEntryVoter aclEntryVoter(AclService aclService) {
        return new AclEntryVoter(aclService, "ACL_MESSAGE_DELETE", new Permission[]
        {BasePermission.ADMINISTRATION, BasePermission.DELETE});
    }

    @Bean
    public EhCacheManagerFactoryBean ehCacheManagerFactoryBean() {
        return new EhCacheManagerFactoryBean();
    }

    @Bean
    public AuditLogger auditLogger() {
        return new ConsoleAuditLogger();
    }

    @Bean
    public PermissionGrantingStrategy permissionGrantingStrategy() {
        return new DefaultPermissionGrantingStrategy(auditLogger());
    }

    @Bean
    public AclAuthorizationStrategy aclAuthorizationStrategy() {
        return new AclAuthorizationStrategyImpl(new SimpleGrantedAuthority("ADMIN"));
    }

    @Bean
    public AclCache aclCache(CacheManager cacheManager) {
        return new SpringCacheBasedAclCache(cacheManager.getCache("aclCache"),
        permissionGrantingStrategy(), aclAuthorizationStrategy());
    }

    @Bean
    public LookupStrategy lookupStrategy(AclCache aclCache) {
        return new BasicLookupStrategy(this.dataSource, aclCache,
        aclAuthorizationStrategy(), permissionGrantingStrategy());
    }

    @Bean
```

```
    public AclService aclService(LookupStrategy lookupStrategy, AclCache aclCache) {
        return new JdbcMutableAclService(this.dataSource, lookupStrategy, aclCache);
    }
}
```

该 ACL 配置文件中的核心 bean 定义就是个 ACL 服务，它是 JdbcMutableAclService 的一个实例，可用于维护 ACL。这个类需要 3 个构造方法参数。第一个是个数据源，用于创建到存储了 ACL 数据的数据库的连接。应该提前定义好数据源，这样这里就只需引用它即可（假设已经在相同的数据库中建好了 ACL 表）。第 3 个构造方法参数是 ACL 所用的一个缓存实例，你可以使用 Ehcache 作为后端的缓存实现。

Spring Security 自带的唯一实现是 BasicLookupStrategy，它会使用标准且兼容的 SQL 语句执行基本的查询。如果想要使用高级的数据库特性来增强查询性能，你可以通过实现 LookupStrategy 接口创建自己的查询策略。BasicLookupStrategy 实例也需要一个数据源和一个缓存实例。除此之外，它还需要一个类型为 AclAuthorizationStrategy 的构造方法参数。该对象会确定某个主体是否拥有权限修改 ACL 的某些属性，这通常是通过对每一类属性指定一个必需的权限来实现的。对于之前的配置来说，只有拥有 ADMIN 权限的用户才能修改 ACL 的所有者、ACE 的审计信息或是其他 ACL 与 ACE 的详细信息。最后，它需要一个类型为 PermissionGrantingStrategy 的构造方法参数。该对象的作用是检查 ACL 是否对给定的 Sid 授予了它所拥有的 Permissions 值。

最后，JdbcMutableAclService 嵌入了标准的 SQL 语句以维护关系型数据库中的 ACL 数据。不过，这些 SQL 语句可能并不会与所有数据库产品都兼容。比如说，你需要为 Apache Derby 定制身份查询语句。

为领域对象维护 ACL

在后端服务与 DAO 中，你可以通过依赖注入使用之前定义的 ACL 服务为领域对象维护 ACL。对于留言板来说，当待办事项发布后你需要为其创建一个 ACL，当删除该待办事项时需要将其 ACL 一并删除。

```
package com.apress.springrecipes.board;

import org.springframework.security.access.prepost.PostFilter;
import org.springframework.security.access.prepost.PreAuthorize;
import org.springframework.security.acls.domain.*;
import org.springframework.security.acls.model.MutableAcl;
import org.springframework.security.acls.model.MutableAclService;
import org.springframework.security.acls.model.ObjectIdentity;
import org.springframework.stereotype.Service;

import javax.transaction.Transactional;
import java.util.List;

import static org.springframework.security.acls.domain.BasePermission.DELETE;
import static org.springframework.security.acls.domain.BasePermission.READ;
import static org.springframework.security.acls.domain.BasePermission.WRITE;

@Service
@Transactional
class TodoServiceImpl implements TodoService {

    private final TodoRepository todoRepository;
    private final MutableAclService mutableAclService;

    TodoServiceImpl(TodoRepository todoRepository, MutableAclService mutableAclService)
{
        this.todoRepository = todoRepository;
        this.mutableAclService = mutableAclService;
    }

    @Override
    @PreAuthorize("hasAuthority('USER')")
    public void save(Todo todo) {
```

```
        this.todoRepository.save(todo);
        ObjectIdentity oid = new ObjectIdentityImpl(Todo.class, todo.getId());
        MutableAcl acl = mutableAclService.createAcl(oid);
        acl.insertAce(0, READ, new PrincipalSid(todo.getOwner()), true);
        acl.insertAce(1, WRITE, new PrincipalSid(todo.getOwner()), true);
        acl.insertAce(2, DELETE, new PrincipalSid(todo.getOwner()), true);

        acl.insertAce(3, READ, new GrantedAuthoritySid("ADMIN"), true);
        acl.insertAce(4, WRITE, new GrantedAuthoritySid("ADMIN"), true);
        acl.insertAce(5, DELETE, new GrantedAuthoritySid("ADMIN"), true);

    }

    @Override
    @PreAuthorize("hasAnyAuthority('USER', 'ADMIN')")
    public void remove(long id) {
        todoRepository.remove(id);

        ObjectIdentity oid = new ObjectIdentityImpl(Todo.class, id);
        mutableAclService.deleteAcl(oid, false);
    }

    ...
}
```

当用户创建了待办事项时，你要同时为该消息创建一个新的 ACL，并使用 ID 作为 ACL 的对象身份。当用户删除了待办事项时，也要删除对应的 ACL。对于新的待办事项来说，需要将如下 ACE 插入到其 ACL 中：

- 待办事项的所用者可以执行 READ、WRITE 和 DELETE 待办事项；
- 拥有 ADMIN 权限的用户也可以执行 READ、WRITE 和 DELETE 待办事项。

JdbcMutableAclService 要求调用方法启用事务，这样其 SQL 语句就可以运行在事务中了。因此，你需要为涉及 ACL 维护的两个方法加上@Transactional 注解，然后在 TodoWebConfig 中定义一个事务管理器并为其加上@EnableTransactionManagement 注解。此外，不要忘记将 ACL 服务注入到 TodoService 中使其可以维护 ACL。

```
package com.apress.springrecipes.board.web;

import org.springframework.context.annotation.Configuration;
import org.springframework.jdbc.datasource.DataSourceTransactionManager;
import org.springframework.transaction.annotation.EnableTransactionManagement;
...

import javax.sql.DataSource;

@Configuration
@EnableTransactionManagement
...
public class TodoWebConfig implements WebMvcConfigurer {

    ...

    @Bean
    public DataSourceTransactionManager transactionManager(DataSource dataSource) {
        return new DataSourceTransactionManager(dataSource);
    }
}
```

使用表达式做出访问控制决策

由于每个领域对象都有一个 ACL，因此可以使用对象的 ACL 来为对象上的方法做出访问控制决策。比如说，当用户要删除一个待办事项时，你可以查看该消息的 ACL，判断该用户是否允许删除它。

配置 ACL 是个很艰巨的任务。幸好，你可以使用注解与表达式来简化这一点。可以通过@PreAuthorize

与@PreFilter 注解检查用户是否允许执行方法或是使用某些方法参数。@PostAuthorize 与@PostFilter 注解可用于检查用户是否允许访问结果或是根据 ACL 来过滤结果。要想开启对这些注解的处理，你需要将@EnableGlobalMethodSecurity 注解的 prePostEnabled 属性设为 true。

```
@EnableGlobalMethodSecurity(prePostEnabled=true)
```

此外，你需要配置基础设施组件以便可以进行决策。这需要创建一个 AclPermissionEvaluator，用于计算对象的权限。这是在 TodoWebConfig 中完成的，之所以要在这里面配置是因为它是启用全局方法安全的配置类。由于要通过表达式使用 ACL 保护方法，因此需要自定义的权限计算器。

```
package com.apress.springrecipes.board.web.config;

import org.springframework.beans.factory.annotation.Autowired;
import org.springframework.cache.Cache;
import org.springframework.cache.CacheManager;
import org.springframework.context.annotation.Bean;
import org.springframework.context.annotation.Configuration;
import org.springframework.security.acls.AclPermissionEvaluator;
import org.springframework.security.acls.domain.AclAuthorizationStrategyImpl;
import org.springframework.security.acls.domain.ConsoleAuditLogger;
import org.springframework.security.acls.domain.DefaultPermissionGrantingStrategy;
import org.springframework.security.acls.domain.SpringCacheBasedAclCache;
import org.springframework.security.acls.jdbc.BasicLookupStrategy;
import org.springframework.security.acls.jdbc.JdbcMutableAclService;
import org.springframework.security.core.GrantedAuthority;
import org.springframework.security.core.authority.SimpleGrantedAuthority;

import javax.sql.DataSource;

@Configuration
public class TodoWebConfig {

    ...

    @Bean
    public AclPermissionEvaluator permissionEvaluator() {
        return new AclPermissionEvaluator(jdbcMutableAclService());
    }

}
```

AclPermissionEvaluator 需要一个 AclService 来获取待检查对象的 ACL。使用基于 Java 的配置就足够了，因为 PermissionEvaluator 会被自动检测并装配到 DefaultMethodSecurityExpressionHandler。现在，我们就可以使用注解和表达式来控制访问了。

```
package com.apress.springrecipes.board;

...

@Service
@Transactional
class TodoServiceImpl implements TodoService {

    @Override
    @PreAuthorize("hasAuthority('USER')")
    @PostFilter("hasAnyAuthority('ADMIN') or hasPermission(filterObject, 'read')")
    public List<Todo> listTodos() { … }

    @Override
    @PreAuthorize("hasAuthority('USER')")
    public void save(Todo todo) { … }

    @Override
    @PreAuthorize("hasPermission(#id, 'com.apress.springrecipes.board.Todo', 'write')")
```

```
    public void complete(long id) { … }

    @Override
    @PreAuthorize("hasPermission(#id, 'com.apress.springrecipes.board.Todo', 'delete')")
    public void remove(long id) { … }

    @Override
    @PostFilter("hasPermission(filterObject, 'read')")
    public Todo findById(long id) { … }
}
```

你可能会注意到这些注解中的不同注解与表达式。@PreAuthorize 注解可用于检查用户是否具有执行方法的正确权限。表达式使用了#message，它会引用名为 message 的方法参数。hasPermission 是来自于 Spring Security 的内建表达式（见表 7-7）。

@PostFilter 注解可用于过滤集合并移除不允许用户读取的元素。在表达式中，关键字 filterObject 引用了集合中的元素。要想保留在集合中，登录用户需要拥有读取权限。

@PostAuthorize 注解可用于检查单个返回值是否可以使用（比如说，用户是否拥有正确的权限）。要想在表达式中使用返回值，请使用关键字 returnObject。

7-8 向 WebFlux 应用中添加安全

问题提出

你有一个使用 Spring WebFlux 构建的应用（见第 5 章），想要为其添加安全。

解决方案

通过向配置中添加@EnableWebFluxSecurity 注解并创建一个包含安全配置的 SecurityWebFilterChain 来启用安全。

解释说明

Spring WebFlux 应用在本质上与常规的 Spring MVC 应用之间存在着很大的区别。不过，Spring Security 已经尽力使配置变得非常轻松，它让 WebFlux 应用的配置与常规的 Web 配置尽可能保持一致。

保护 URL 访问

首先，创建一个 SecurityConfiguration 类并为它添加@EnableWebFluxSecurity 注解。

```
@Configuration
@EnableWebFluxSecurity
public class SecurityConfiguration { … }
```

@EnableWebFluxSecurity 注解注册一个 WebFluxConfigurer（见 5-5 节）来添加 AuthenticationPrincipal-ArgumentResolver，它可以将 Authentication 对象注入到 Spring WebFlux 处理器方法中。它还会注册 Spring Security 的 WebFluxSecurityConfiguration 类，这个类会检测 SecurityWebFilterChain（包含了安全配置）的实例，该类被包装为一个 WebFilter（相当于常规的 servlet 过滤器），它反过来又会被 WebFlux 所用，用于将行为添加到请求中（就像正常的 servlet 过滤器一样）。

你的配置现在只开启了安全，下面来添加一些安全规则。

```
@Bean
SecurityWebFilterChain springWebFilterChain(HttpSecurity http) throws Exception {
    return http
        .authorizeExchange()
            .pathMatchers("/welcome", "/welcome/**").permitAll()
            .pathMatchers("/reservation*").hasRole("USER")
            .anyExchange().authenticated()
        .and()
        .build();
}
```

org.springframework.security.config.annotation.web.reactive.HttpSecurity 看起来应该很熟悉（见 7-1 节），

它用于添加安全规则并做进一步的配置（如添加/移除头，以及配置登录方法）。借助于 authorizeExchange，我们可以编写规则。这里保护的是 URL；/welcome URL 允许任何人访问，/reservation URL 只允许角色为 USER 的用户访问。对于其他请求来说，你需要进行认证。最后，你需要调用 build()来实际构建 SecurityWebFilterChain。

除了 authorizeExchange 外，还可以使用 headers()配置方法向请求添加安全头（见 7-2 节），比如说跨站点脚本防护、缓存头等。

登录到 WebFlux 应用

现在，只有 httpBasic()认证机制被 Spring Security WebFlux 所支持，默认情况下它是启用的。你可以通过显式配置来重写部分默认配置，可以重写所用的认证管理器以及用于存储安全上下文的仓库。认证管理器会被自动检测到；你只需注册一个 ReactiveAuthenticationManager 类型或是 UserDetailsRepository 类型的 bean 即可。

还可以通过配置 SecurityContextRepository 来配置 SecurityContext 值所存储的位置。所用的默认实现是 WebSessionSecurityContextRepository，它会将上下文存储到 WebSession 中。另一个默认实现 ServerWebExchangeAttributeSecurityContextRepository 会将 SecurityContext 作为当前交换（即请求）的一个属性。

```
@Bean
SecurityWebFilterChain springWebFilterChain(HttpSecurity http) throws Exception {
    return http.httpBasic().
        .authenticationManager(new CustomReactiveAuthenticationManager())
        .securityContextRepository(new
        ServerWebExchangeAttributeSecurityContextRepository()).and().build();
}
```

这会使用 CustomReactiveAuthenticationManager 与无状态的 ServerWebExchangeAttribute SecurityContext-Repository 覆盖默认值。不过，对于该应用来说，要使用默认值。

对用户进行认证

在基于 Spring WebFlux 的应用中，对用户的认证是通过 ReactiveAuthenticationManager 完成的。这是一个接口，只有一个 authenticate 方法。你可以提供自己的实现，也可以使用两个既有实现之一。第一个是 UserDetailsRepositoryAuthenticationManager，它包装了一个 UserDetailsRepository 实例。

> **注意：**UserDetailsRepository 只有一个实现，即 MapUserDetailsRepository，它是个基于内存的实现。当然，你可以基于反应式数据存储（比如说 MongoDB 或 Couchbase）提供自己的实现。

另一个实现 ReactiveAuthenticationManagerAdapter 实际上是对常规的 AuthenticationManager（见 7-3 节）的一个包装器。它会包装一个常规的 AuthenticationManager 实例，由于这一点，你可以以反应式的方式使用阻塞实现。这并不会使其成为反应式的；它们依旧是阻塞的，不过这种方式下它们是可重用的。这样，你可以为你的反应式应用使用 JDBC、LDAP 等。

在 Spring WebFlux 应用中配置 Spring Security 时，你可以向 Java 配置类中添加一个 ReactiveAuthentication-Manager 或是 UserDetailsRepository。当检测到后者时，它会被自动包装到 UserDetailsRepository-AuthenticationManager 中。

```
@Bean
public MapUserDetailsRepository userDetailsRepository() {
    UserDetails marten = User.withUsername("marten").password("secret").roles("USER").build();
    UserDetails admin = User.withUsername("admin").password("admin").roles("USER","ADMIN").build();
    return new MapUserDetailsRepository(marten, admin);
}
```

部署应用后（或是运行 ReactorNettyBootstrap 类），你可以自由访问/welcome 页面，但在访问以/reservation 开头的 URL 时，你会看到浏览器弹出一个 Basic 认证窗口，如图 7-3 所示。

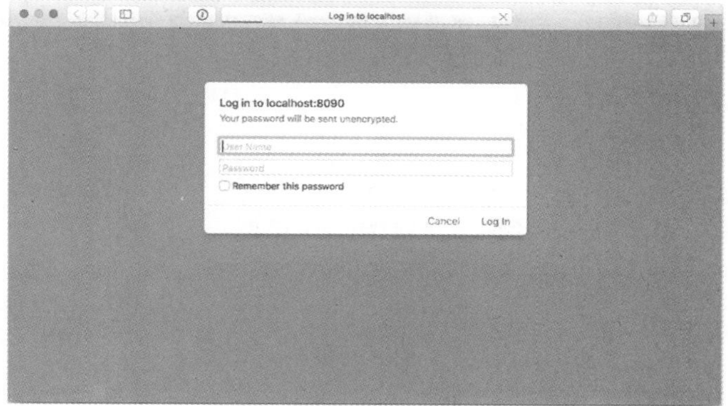

图 7-3　Basic 认证登录界面

做出访问控制决策

表 7-8 列出了 Spring Security WebFlux 内建的表达式。

表 7-8　Spring Security WebFlux 内建的表达式

表达式	说明
hasRole(role)或 hasAuthority(authority)	如果当前用户拥有给定角色则返回 true
permitAll()	总是为 true
denyAll()	总是为 false
authenticated()	如果用户经过认证则返回 true
access()	使用一个函数来确定访问是否已授权

警告：虽然角色与授权几乎是一样的，但在处理方式上还是存在一个细微但重要的差别。在使用 hasRole 时，如果传递进来的角色值以 ROLE_（默认的角色前缀）开头，那么它就会被检查。如果不是，那么在检查授权前它会被添加进来。这样，hasRole('ADMIN')实际上会检查当前用户是否拥有 ROLE_ADMIN 授权。在使用 hasAuthority 时，它会检查这个值本身。

```
@Bean
SecurityWebFilterChain springWebFilterChain(HttpSecurity http) throws Exception {
    return http
        .authorizeExchange()
            .pathMatchers("/users/{user}/**").access(this::userEditAllowed)
            .anyExchange().authenticated()
        .and()
        .build();
}

private Mono<AuthorizationDecision> userEditAllowed(Mono<Authentication> authentication,
AuthorizationContext context) {
    return authentication
        .map( a -> context.getVariables().get("user").equals(a.getName()) ||
        a.getAuthorities().contains(new SimpleGrantedAuthority("ROLE_ADMIN")))
        .map( granted -> new AuthorizationDecision(granted));
}
```

access()表达式可用于编写强大的表达式。上面的代码片段在 URL {user}中使用了一个路径参数，如果当前用户就是实际用户或是拥有 ROLE_ADMIN 授权，那么就允许访问。AuthorizationContext 包含了解析后的变量，你可以使用它与 URI 中的名字进行比较。Authentication 包含了 GrantedAuthorities 的集合，你可以使用它来检查 ROLE_ADMIN 授权。当然，你可以根据需要编写很多复杂的表达式，可以检查 IP 地址、请求头等。

小结

本章介绍了如何通过 Spring Security 保护应用。Spring Security 可用于保护任何 Java 应用，但它主要还是用于 Web 应用。认证、授权与访问控制是安全领域中的重要概念，你应该对它们有清晰的理解。

我们经常要通过防止对关键 URL 的未授权访问来保护它们。Spring Security 可以以一种声明式的方式帮你做到这一点。它会通过 servlet 过滤器来处理安全，这可以通过简单的基于 Java 的配置来实现。默认情况下，Spring Security 会自动配置基本的安全服务，并尽可能确保安全。

Spring Security 支持多种方式来让用户登录到 Web 应用，比如说基于表单的登录与 HTTP Basic 认证。它还提供了匿名登录服务，让你能像正常用户一样处理匿名用户。“记住我”支持可以让应用跨越多个浏览器会话记住用户的身份。

Spring Security 支持多种用户认证方式，并且为其提供了内建的提供器实现。比如说，它支持通过内存定义、关系型数据库和 LDAP 仓库来认证用户。我们应该总是将加密后的密码存储到用户仓库中，因为明文密码很容易遭受到黑客攻击。Spring Security 还支持将用户信息缓存到本地，从而降低执行远程查询的开销。

判定是否允许用户访问给定的资源是由访问决策管理器来实现的。Spring Security 自带 3 个基于投票方式的访问决策管理器。它们都需要一组配置好的投票者，并就访问控制决策进行投票。

Spring Security 可以以一种声明式的方式保护方法调用，这是通过两种方式来实现的，一种是将安全拦截器嵌入到 bean 定义中，另一种是使用 AspectJ 切点表达式或注解来匹配多个方法。Spring Security 还可以在 JSP 视图中展示用户的认证信息，并根据用户的授权有条件地渲染视图内容。

Spring Security 提供了一个 ACL 模块，可以让每个领域对象都有一个控制访问的 ACL。你可以通过 Spring Security 的高性能 API（使用 JDBC 实现）来读取和维护每个领域对象的 ACL。Spring Security 还提供了诸如访问决策投票者与 JSP 标签等设施，可以让你像使用其他安全设施一样使用 ACL。

Spring Security 还可以保护基于 Spring WebFlux 的应用。在 7-8 节中，我们介绍了如何向这类应用添加安全。

第 8 章

Spring Mobile

时至今日，移动设备的数量与日俱增。大多数移动设备都可以访问互联网和网站。不过，一些移动设备上的浏览器可能缺少网站上所用的某些 HTML 或 JavaScript 特性；你还可能想为移动用户展示不同的网站或是让他们自己选择查看网站的移动版本。在这些情况下，你可以自己编写所有的设备检测程序，不过 Spring Mobile 提供了检测所用设备的诸多方式。

8-1 不使用 Spring Mobile 来检测设备

问题提出

你想要检测连接到网站的设备类型。

解决方案

创建一个过滤器，检测传入请求的 User-Agent 值并设置一个请求属性，这样控制器就可以获取到它。

解释说明

如下是基于 User-Agent 的设备检测的 Filter 实现：

```
package com.apress.springrecipes.mobile.web.filter;

import org.springframework.util.StringUtils;
import org.springframework.web.filter.OncePerRequestFilter;

import javax.servlet.FilterChain;
import javax.servlet.ServletException;
import javax.servlet.http.HttpServletRequest;
import javax.servlet.http.HttpServletResponse;
import java.io.IOException;

public class DeviceResolverRequestFilter extends OncePerRequestFilter {

    public static final String CURRENT_DEVICE_ATTRIBUTE = "currentDevice";

    public static final String DEVICE_MOBILE = "MOBILE";
    public static final String DEVICE_TABLET = "TABLET";
    public static final String DEVICE_NORMAL = "NORMAL";

    @Override
    protected void doFilterInternal(HttpServletRequest request, HttpServletResponse
    response,
                                    FilterChain filterChain) throws ServletException,
                                    IOException {
        String userAgent = request.getHeader("User-Agent");
        String device = DEVICE_NORMAL;

        if (StringUtils.hasText(userAgent)) {
            userAgent = userAgent.toLowerCase();
```

```
            if (userAgent.contains("android")) {
                device = userAgent.contains("mobile") ? DEVICE_NORMAL : DEVICE_TABLET;
            } else if (userAgent.contains("ipad") || userAgent.contains("playbook") ||
            userAgent.contains("kindle")) {
                device = DEVICE_TABLET;
            } else if (userAgent.contains("mobil") || userAgent.contains("ipod") ||
            userAgent.contains("nintendo DS")) {
                device = DEVICE_MOBILE;
            }
        }
        request.setAttribute(CURRENT_DEVICE_ATTRIBUTE, device);
        filterChain.doFilter(request, response);
    }
}
```

该实现首先会从请求中获取 User-Agent 头。如果有值，那么过滤器就需要检查这个值。这个头中使用了几个 if/else 语句来对设备类型进行基本的检测。对于 Android 来说还有个特殊的情况，因为可能是平板或是移动设备。当过滤器确定了设备类型后，该类型会被存储为一个请求属性，这样其他组件就可以使用它了。接下来，有一个控制器和 JSP 页面来显示接下来的信息。控制器只是转向到 home.jsp 页面，它位于 WEB-INF/views 目录下。配置好的 InternalResourceViewResolver 负责将这个名字解析为实际的 JSP 页面（请参考第 4 章了解更多信息）。

```
package com.apress.springrecipes.mobile.web;

import org.springframework.stereotype.Controller;
import org.springframework.web.bind.annotation.RequestMapping;

import javax.servlet.http.HttpServletRequest;

@Controller
public class HomeController {

    @RequestMapping("/home")
    public String index(HttpServletRequest request) {
        return "home";
    }

}
```

如下是 home.jsp 页面：

```
<%@ taglib uri="http://java.sun.com/jsp/jstl/core" prefix="c" %>
<!doctype html>
<html>
<body>

<h1>Welcome</h1>
<p>
    Your User-Agent header: <c:out value="${header['User-Agent']}" />
</p>
<p>
    Your type of device: <c:out value="${requestScope.currentDevice}" />
</p>

</body>
</html>
```

JSP 页面展示了 User-Agent 头（如果有的话）和设备类型，这是由自己的 DeviceResolverRequestFilter 所确定的。

最后，如下是配置与启动逻辑：

```
package com.apress.springrecipes.mobile.web.config;

import org.springframework.context.annotation.Bean;
import org.springframework.context.annotation.ComponentScan;
```

```
import org.springframework.context.annotation.Configuration;
import org.springframework.web.servlet.ViewResolver;
import org.springframework.web.servlet.config.annotation.EnableWebMvc;
import org.springframework.web.servlet.view.InternalResourceView;
import org.springframework.web.servlet.view.InternalResourceViewResolver;

@Configuration
@ComponentScan("com.apress.springrecipes.mobile.web")
public class MobileConfiguration {

    @Bean
    public ViewResolver viewResolver() {

        InternalResourceViewResolver viewResolver = new InternalResourceViewResolver();
        viewResolver.setPrefix("/WEB-INF/views/");
        viewResolver.setSuffix(".jsp");
        return viewResolver;
    }
}
```

控制器是由@ComponentScan 注解所确定的。为了启动应用，这里使用了 MobileApplicationInitializer，它会启动 DispatcherServlet 以及 ContextLoaderListener（可选）。

```
package com.apress.springrecipes.mobile.web;

import com.apress.springrecipes.mobile.web.config.MobileConfiguration;
import com.apress.springrecipes.mobile.web.filter.DeviceResolverRequestFilter;
import org.springframework.web.servlet.support.
AbstractAnnotationConfigDispatcherServletInitializer;

import javax.servlet.Filter;

public class MobileApplicationInitializer extends
AbstractAnnotationConfigDispatcherServletInitializer {

    @Override
    protected Class<?>[] getRootConfigClasses() {
        return null;
    }

    @Override
    protected Class<?>[] getServletConfigClasses() {
        return new Class[] { MobileConfiguration.class };
    }

    @Override
    protected Filter[] getServletFilters() {
        return new Filter[] {new DeviceResolverRequestFilter()};
    }

    @Override
    protected String[] getServletMappings() {
        return new String[] {"/"};
    }
}
```

这里有两点值得注意。首先，之前提及的配置类会被传递给 DispatcherServlet，这是通过实现 getServletConfigClasses 方法做到的。其次，getServletFilters 方法的实现负责注册过滤器并将其映射到 DispatcherServlet。当应用部署完毕后，访问 http://localhost:8080/mobile/home 会展示出 User-Agent 值以及过滤器所判定的设备类型，如图 8-1 所示。

在 iMac 上使用 Chrome 访问会得到如图 8-1 所示的结果。当使用 iPhone 时，结果如图 8-2 所示。

注意：要想测试不同的浏览器，既可以在内网上使用平板或是移动设备，也可以使用 Chrome 或 Firefox 的浏览器插件，如 User-Agent Switcher。

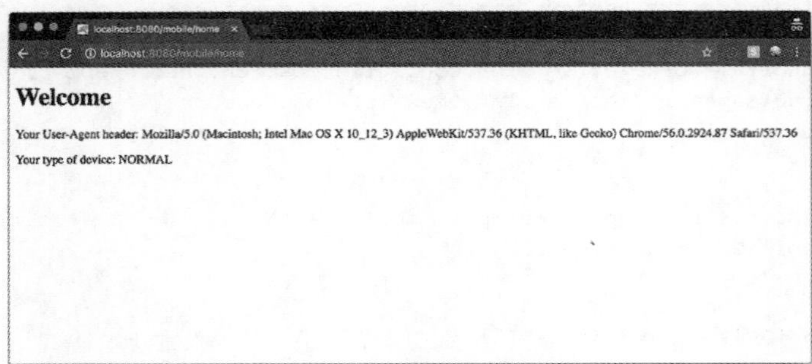

图 8-1 在 Chrome 中查看应用

虽然过滤器可以实现设备检测功能，但远非完美。比如说，有些移动设备并不匹配规则（Kindle Fire 与普通的 Kindle 设备的头就不同）。此外，维护规则与设备列表是一件非常困难的事情，测试诸多设备也绝非易事。使用 Spring Mobile 之类的库要比自己实现轻松得多。

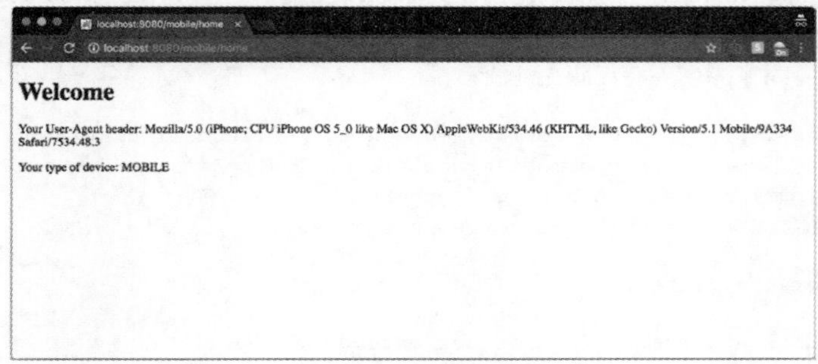

图 8-2 使用 iPhone 4 查看应用

8-2 使用 Spring Mobile 来检测设备

问题提出

你想要检测连接到网站的设备类型，并且想用 Spring Mobile 来实现。

解决方案

使用 Spring Mobile DeviceResolver 与辅助类，通过配置 DeviceResolverRequestFilter 或 DeviceResolverHandlerInterceptor 来确定设备类型。

解释说明

DeviceResolverRequestFilter 与 DeviceResolverHandlerInterceptor 都会将设备类型的检测工作委托给 DeviceResolver 来做。Spring Mobile 提供了该接口的一个实现，叫做 LiteDeviceResolver。DeviceResolver 会返回一个 Device 对象，它标识了类型。Device 对象会存储为一个请求属性，这样它就可以在后续环节中使用了。Spring Mobile 自带了 Device 接口的一个默认实现，叫做 LiteDevice。

使用 DeviceResolverRequestFilter

使用 DeviceResolverRequestFilter 时需要将其添加到 Web 应用中并将其映射到想让其处理的 servlet 或请求上。对于你的应用来说，这意味着将其添加到 getServletFilters 方法中。使用该过滤器的好处在于即便不是 Spring 应用也可以使用它。比如说，它可以用在 JSF 应用中。

```
package com.apress.springrecipes.mobile.web;
...
import org.springframework.mobile.device.DeviceResolverRequestFilter;

public class MobileApplicationInitializer extends
```

```
AbstractAnnotationConfigDispatcherServletInitializer {
...
    @Override
    protected Filter[] getServletFilters() {
        return new Filter[] {new DeviceResolverRequestFilter()};
    }
}
```

该配置会注册 DeviceResolverRequestFilter 并自动将其附加到 DispatcherServlet 所处理的请求上。要想测试，请向 http://localhost:8080/mobile/home 发出一个请求，这会展示出图 8-3 所示的结果：

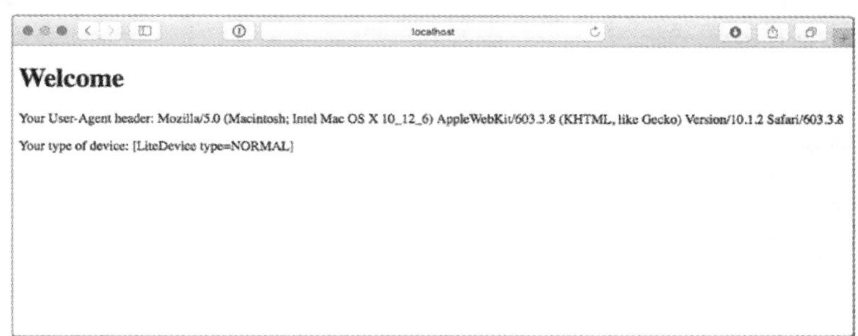

图 8-3　测试结果

该设备的输出是由 Spring Mobile 的 LiteDevice 类的 toString 方法所产生的文本。

使用 DeviceResolverHandlerInterceptor

当在基于 Spring MVC 的应用中使用 Spring Mobile 时，DeviceResolverHandlerInterceptor 会更加简单。这需要在配置类中对其进行配置，并使用 addInterceptors 辅助方法进行注册。

```
package com.apress.springrecipes.mobile.web.config;

import org.springframework.mobile.device.DeviceResolverHandlerInterceptor;
import org.springframework.web.servlet.config.annotation.InterceptorRegistry;
import org.springframework.web.servlet.config.annotation.WebMvcConfigurerAdapter;

@Configuration
@EnableWebMvc
@ComponentScan("com.apress.springrecipes.mobile.web")
public class MobileConfiguration extends WebMvcConfigurerAdapter {
...
    @Override
    public void addInterceptors(InterceptorRegistry registry) {
        registry.addInterceptor(new DeviceResolverHandlerInterceptor());
    }
}
```

MobileConfiguration 类继承了 WebMvcConfigurerAdapter，你可以重写 addInterceptors 方法。添加到注册表的所有拦截器都会被添加到应用上下文的 HandlerMapping beans 中。部署应用并向 http://localhost:8080/mobile/home 发出一个请求，结果应该与使用过滤器时一样。

8-3　使用站点首选项

问题提出

可以让用户选择其设备所访问的站点类型，并将其存储起来供后续使用。

解决方案

使用 Spring Mobile 所提供的 SitePreference 支持。

解释说明

SitePreferenceRequestFilter 与 SitePreferenceHandlerInterceptor 都会将当前 SitePreference 的获取委托给 SitePreferenceHandler 对象。默认实现使用 SitePreferenceRepository 类来存储首选项；默认情况下，这是存储在 cookie 中的。

```
package com.apress.springrecipes.mobile.web;

import org.springframework.mobile.device.site.SitePreferenceRequestFilter;
...

public class MobileApplicationInitializer extends
AbstractAnnotationConfigDispatcherServletInitializer {

    @Override
    protected Filter[] getServletFilters() {
        return new Filter[] {
            new DeviceResolverRequestFilter(),
            new SitePreferenceRequestFilter()};
    }

...
}
```

在注册了 SitePreferenceRequestFilter 之后，它就会检测传入的请求。如果请求有名为 site_preference 的参数，那么它就会使用传递进来的值（NORMAL、MOBILE 或 TABLET）来设置 SitePreference 值。确定后的值会存储在 cookie 中并供后续使用；如果检测到了新值，那么 cookie 中的值就会被重置。修改 home.jsp 页面，添加如下代码以展示当前的 SitePreference 值：

```
<p>
    Your site preferences <c:out value="${requestScope.currentSitePreference}" />
</p>
```

现在，访问 URL http://localhost:8080/mobile/home?site_preference=TABLET 则会将 SitePreference 值设为 TABLET。

使用 SitePreferenceHandlerInterceptor

当在基于 Spring MVC 的应用中使用 Spring Mobile 时，SitePreferenceHandlerInterceptor 会更加简单。这需要在配置类中对其进行配置，并使用 addInterceptors 辅助方法进行注册。

```
package com.apress.springrecipes.mobile.web.config;

import org.springframework.mobile.device.DeviceResolverHandlerInterceptor;
import org.springframework.mobile.device.site. SitePreferenceHandlerInterceptor;
import org.springframework.web.servlet.config.annotation.InterceptorRegistry;
import org.springframework.web.servlet.config.annotation.WebMvcConfigurerAdapter;

@Configuration
@EnableWebMvc
@ComponentScan("com.apress.springrecipes.mobile.web")
public class MobileConfiguration extends WebMvcConfigurerAdapter {
...
    @Override
    public void addInterceptors(InterceptorRegistry registry) {
        registry.addInterceptor(new DeviceResolverHandlerInterceptor());
        registry.addInterceptor(new SitePreferenceHandlerInterceptor());
    }
}
```

MobileConfiguration 类继承了 WebMvcConfigurerAdapter，你可以重写 addInterceptors 方法。添加到注册表的所有拦截器都会被添加到应用上下文的 HandlerMapping bean 中。部署应用并向 http://localhost:8080/mobile/home?site_preference=TABLET 发出一个请求，结果应该与上一节中使用过滤器时一样。

8-4　使用设备信息来渲染视图

问题提出

你想要根据设备或站点首选项渲染不同的视图。

解决方案

使用当前的 Device 与 SitePreferences 对象来确定渲染哪个视图。这可以手工实现，也可以使用 LiteDeviceDelegatingViewResolver 实现。

解释说明

在知道了设备类型后，就可以使用该信息了。首先，为支持的每种设备类型创建视图，并分别将其放到 WEB-INF/views 下的 mobile 或 tablet 目录下。如下是 mobile/home.jsp 的代码：

```
<%@ taglib uri="http://java.sun.com/jsp/jstl/core" prefix="c" %>
<!doctype html>
<html>
<body>

<h1>Welcome Mobile User</h1>
<p>
    Your User-Agent header: <c:out value="${header['User-Agent']}" />
</p>
<p>
    Your type of device: <c:out value="${requestScope.currentDevice}" />
</p>
<p>
    Your site preferences <c:out value="${requestScope.currentSitePreference}" />
</p>
</body>
</html>
```

如下是 tablet/home.jsp 的代码：

```
<%@ taglib uri="http://java.sun.com/jsp/jstl/core" prefix="c" %>
<!doctype html>
<html>
<body>

<h1>Welcome Tablet User</h1>
<p>
    Your User-Agent header: <c:out value="${header['User-Agent']}" />
</p>
<p>
    Your type of device: <c:out value="${requestScope.currentDevice}" />
</p>
<p>
    Your site preferences <c:out value="${requestScope.currentSitePreference}" />
</p>
</body>
</html>
```

在不同的视图就绪后，你需要根据检测出来的设备值来渲染它们。一种方式是手工从请求中获取到当前设备，并使用它来确定应当渲染哪个视图。

```
package com.apress.springrecipes.mobile.web;

import org.springframework.mobile.device.Device;
import org.springframework.mobile.device.DeviceUtils;
import org.springframework.stereotype.Controller;
import org.springframework.web.bind.annotation.RequestMapping;

import javax.servlet.http.HttpServletRequest;

@Controller
```

```
public class HomeController {

    @RequestMapping("/home")
    public String index(HttpServletRequest request) {
        Device device = DeviceUtils.getCurrentDevice(request);
        if (device.isMobile()) {
            return "mobile/home";
        } else if (device.isTablet()) {
            return "tablet/home";
        } else {
            return "home";
        }
    }
}
```

Spring Mobile 有一个 DeviceUtils 类，可用于获取当前设备。当前设备是从请求属性（currentDevice）中获取的，该属性是由过滤器或拦截器设置的。设备的值可用于确定该渲染哪个视图。

在需要设备值的每个方法中获取这个值是非常不方便的。如果能将该值作为方法参数传递给控制器方法就会变得轻松很多。要想做到这一点，你可以使用 DeviceHandlerMethodArgumentResolver，它会被注册进来并将方法参数解析为当前设备。要想获取到当前的 SitePreference 值，你可以添加一个 SitePreferenceHandlerMethodArgumentResolver 类。

```
package com.apress.springrecipes.mobile.web.config;

import org.springframework.mobile.device.DeviceHandlerMethodArgumentResolver;
...
import java.util.List;

@Configuration
@EnableWebMvc
@ComponentScan("com.apress.springrecipes.mobile.web")
public class MobileConfiguration extends WebMvcConfigurerAdapter {
...
    @Override
    public void addArgumentResolvers(List<HandlerMethodArgumentResolver> argumentResolvers)
{
        argumentResolvers.add(new DeviceHandlerMethodArgumentResolver());
        argumentResolvers.add(new SitePreferenceHandlerMethodArgumentResolver());
    }
}
```

在注册好了这些组件后，控制器方法就会简化很多，Device 值可作为方法参数传递进来。

```
package com.apress.springrecipes.mobile.web;

import org.springframework.mobile.device.Device;
import org.springframework.stereotype.Controller;
import org.springframework.web.bind.annotation.RequestMapping;

import javax.servlet.http.HttpServletRequest;

@Controller
public class HomeController {

    @RequestMapping("/home")
    public String index(Device device) {
        if (device.isMobile()) {
            return "mobile/home";
        } else if (device.isTablet()) {
            return "tablet/home";
        } else {
            return "home";
        }
    }
}
```

方法签名发生了变化，参数类型由 HttpServletRequest 改为了 Device。它负责查找并传递进来当前的设备。不过，虽然这要比手工获取方便很多，但在判断该渲染哪个视图时所采取的方式依然很烦琐。现在，我们并未使用首选项，不过可以将其添加到方法中，它可用于确定首选项。然而，这么做会使得检测算法变得很复杂。想象一下，这些代码会散布在很多控制器方法中，这很快就会成为一个维护噩梦。

Spring Mobile 自带一个 LiteDeviceDelegatingViewResolver 类，在将视图名传递给实际的视图解析器前，它可以向视图名添加额外的前缀与/或后缀。它还考虑到了可选的用户站点首选项。

```
package com.apress.springrecipes.mobile.web.config;

import org.springframework.mobile.device.view.LiteDeviceDelegatingViewResolver;
...

@Configuration
@EnableWebMvc
@ComponentScan("com.apress.springrecipes.mobile.web")
public class MobileConfiguration extends WebMvcConfigurerAdapter {
...
    @Bean
    public ViewResolver viewResolver() {
        InternalResourceViewResolver viewResolver = new InternalResourceViewResolver();
        viewResolver.setPrefix("/WEB-INF/views/");
        viewResolver.setSuffix(".jsp");
        viewResolver.setOrder(2);
        return viewResolver;
    }

    @Bean
    public ViewResolver mobileViewResolver() {
        LiteDeviceDelegatingViewResolver delegatingViewResolver =
            new LiteDeviceDelegatingViewResolver(viewResolver());
        delegatingViewResolver.setOrder(1);
        delegatingViewResolver.setMobilePrefix("mobile/");
        delegatingViewResolver.setTabletPrefix("tablet/");
        return delegatingViewResolver;

    }
}
```

LiteDeviceDelegatingViewResolver 接收一个委托的视图解析器作为构造方法参数；之前配置好的 InternalResourceViewResolver 会被传递进来作为委托。此外，注意视图解析器的顺序；你需要确保 LiteDeviceDelegatingViewResolver 在其他视图解析器之前执行。通过这种方式，它就可以确定是否存在针对特定设备的自定义视图了。接下来，注意在配置中，针对移动设备的视图位于 mobile 目录下，针对于平板设备的视图位于 tablet 目录下。要想将这些目录添加到视图名中，需要将这些设备类型的前缀设为各自的目录。现在，当控制器返回 home 作为视图名来选择移动设备时，它会被转换为 mobile/home。这个修改后的名字会被传递给 InternalResourceViewResolver，它会将其转换为/WEB-INF/views/mobile/home.jsp，即实际要渲染的页面。

```
package com.apress.springrecipes.mobile.web;

import org.springframework.stereotype.Controller;
import org.springframework.web.bind.annotation.RequestMapping;

@Controller
public class HomeController {

    @RequestMapping("/home")
    public String index() {
        return "home";
    }
}
```

现在的控制器变得相当整洁了。它只负责返回视图名。确定渲染哪个视图的工作留给了配置好的视图解析器。LiteDeviceDelegatingViewResolver 会负责处理任何找到的 SitePreferences 值。

8-5　实现站点切换

问题提出

移动站点的 URL 与正常的网站是不同的。

解决方案

使用 Spring Mobile 的站点切换支持设备重定向到网站的恰当页面上。

解释说明

Spring Mobile 自带一个 SiteSwitcherHandlerInterceptor 类，可以使用它基于检测到的 Device 值切换到移动版本的网站上。要想配置 SiteSwitcherHandlerInterceptor，有很多工厂方法可以使用，这些方法提供了现成的设置，如表 8-1 所示。

表 8-1　SiteSwitcherHandlerInterceptor 工厂方法一览

工厂方法	说明
mDot	重定向到以 m. 开头的域名；比如说 http://www.yourdomain.com 会重定向到 http://m.yourdomain.com
dotMobi	重定向到以.mobi 结尾的域名。向 http://www.yourdomain.com 发出的请求会被重定向到 http://www.yourdomain.mobi
urlPath	为不同设备创建不同的上下文根。这会重定向到为该设备所配置的 URL 路径。比如说，http://www.yourdomain.com 会被重定向到 http://www.yourdomain.com/mobile
standard	这是最为灵活的可配置的工厂方法，你可以为网站的移动版本、平板版本以及正常版本指定要重定向的域名

SiteSwitcherHandlerInterceptor 还提供了使用站点首选项的能力。当使用 SiteSwitcherHandlerInterceptor 时，不必再注册 SitePreferencesHandlerInterceptor，因为 SiteSwitcherHandlerInterceptor 提供了它的功能。配置非常简单，只需将其添加到你想要应用的拦截器列表中即可；唯一要注意的是需要将其放到 DeviceResolverHandlerInterceptor 之后，因为需要获取到设备信息才能得到重定向 URL。

```
package com.apress.springrecipes.mobile.web.config;

...
import org.springframework.mobile.device.switcher.SiteSwitcherHandlerInterceptor;

@Configuration
@EnableWebMvc
@ComponentScan("com.apress.springrecipes.mobile.web")
public class MobileConfiguration extends WebMvcConfigurerAdapter {

    @Override
    public void addInterceptors(InterceptorRegistry registry) {
        registry.addInterceptor(new DeviceResolverHandlerInterceptor());
        registry.addInterceptor(siteSwitcherHandlerInterceptor());
    }

    @Bean
    public SiteSwitcherHandlerInterceptor siteSwitcherHandlerInterceptor() {
        return SiteSwitcherHandlerInterceptor.mDot("yourdomain.com", true);
    }
...
}
```

注意在 SiteSwitcherHandlerInterceptor 的 bean 声明中，工厂方法 mDot 用于创建它的一个实例。该方法接收两个参数。第一个参数是要使用的根域名，第二个参数是个 boolean 值，表示是否将平板看作是移动设备，其默认值为 false。该配置会将由移动设备发起的对常规网站的请求重定向为 m.yourdomain.com。

```
@Bean
public SiteSwitcherHandlerInterceptor siteSwitcherHandlerInterceptor() {
    return SiteSwitcherHandlerInterceptor.dotMobi("yourdomain.com", true);
}
```

上述配置使用了 dotMobi 工厂方法，它接收两个参数。第一个参数是要使用的根域名，第二个参数是个 boolean 值，表示是否将平板看作是移动设备，其默认值为 false。该配置会将由移动设备发起的对常规网站的请求重定向为 yourdomain.mobi。

```
@Bean
public SiteSwitcherHandlerInterceptor siteSwitcherHandlerInterceptor() {
    return SiteSwitcherHandlerInterceptor.urlPath("/mobile", "/tablet", "/home");
}
```

上述配置使用了 urlPath 工厂方法，它接收 3 个参数。第一个参数是针对移动设备的上下文根，第二个参数是针对平板的上下文根，最后一个参数是应用的根路径。urlPath 工厂方法还有另外两个版本：一个只接收移动设备的路径；另外一个接收针对移动设备的路径和一个根路径。上述配置会将来自移动设备的请求重定向到 yourdomain.com/home/mobile，将来自平板的请求重定向到 yourdomain.com/home/tablet。

最后介绍一下 standard 工厂方法，它是最灵活的，配置也是最复杂的。

```
@Bean
public SiteSwitcherHandlerInterceptor siteSwitcherHandlerInterceptor() {
    return SiteSwitcherHandlerInterceptor
        .standard("yourdomain.com", "mobile.yourdomain.com",
            "tablet.yourdomain.com", "*.yourdomain.com");
}
```

上述配置使用了 standard 工厂方法。它为网站的正常版本、移动版本与平板版本指定了不同的域名。最后，它指定了用于存储站点首选项的 cookie 的域名。之所以需要它是因为指定了不同的子域名。

standard 工厂方法还有其他几个版本，可以指定前述配置的一个子集。

小结

本章介绍了 Spring Mobile 的使用方式，它可以检测请求页面的设备，并且可以让用户基于首选项选择某个页面。我们介绍了如何通过 DeviceResolverRequestFilter 或 DeviceResolverHandlerInterceptor 检测用户的设备，还介绍了如何通过 SitePreferences 让用户改变所检测到的设备。接下来，我们介绍了如何使用设备信息和首选项针对特定设备来渲染视图。最后，我们介绍了如何基于用户的设备或站点首选项将用户重定向到网站的不同页面。

第 9 章

数据访问

本章将会介绍如何通过 Spring 简化数据库访问任务（Spring 还可以简化 NoSQL 与大数据访问任务，第 12 章将会对此进行介绍）。数据访问对于大多数企业应用来说都是一个常见的需求，这通常需要访问存储在关系型数据库中的数据。作为 Java SE 的重要组成部分，Java Database Connectivity（JDBC）定义了一套标准 API，这些 API 能以厂商独立的形式访问关系型数据库。

JDBC 的目的在于提供一套 API，你可以通过它们对数据库执行 SQL 语句。不过，在使用 JDBC 时，需要自己管理数据库相关的资源，并显式处理数据库异常。为了简化 JDBC 的使用，Spring 为 JDBC 的交互提供了一个抽象框架。作为 Spring JDBC 框架的核心，JDBC 模板的设计目的旨在为不同类型的 JDBC 操作提供一套模板方法。每个模板方法负责控制整个过程，你也可以重写特定的处理任务。

如果原生 JDBC 无法满足你的需求，或是觉得应用应该使用一些更高层次的组件，那么 Spring 对于对象关系映射（Object-Relational Mapping，ORM）解决方案的支持应该会对你很有吸引力。本章还会介绍如何将 ORM 框架集成到 Spring 应用中。Spring 支持大多数流行的 ORM（或数据映射）框架，包括 Hibernate、JDO、iBATIS 和 Java Persistence API（JPA）。Spring 从 3.0 开始已经不再支持传统的 TopLink 了（当然，JPA 实现依然支持）。不过，JPA 支持对于很多 JPA 实现来说也是不同的，包括基于 Hibernate 与基于 TopLink 的版本。本章主要关注的是 Hibernate 与 JPA。不过，Spring 对 ORM 框架的支持是一致的，这样你可以轻松将本章所介绍的技术应用到其他 ORM 框架上。

ORM 是一种将对象持久化到关系型数据库中的现代化技术。ORM 框架会根据你所提供的映射元数据（基于 XML 或注解）持久化对象，比如说类与表之间的映射、属性与列之间的映射等。它会在运行期为对象持久化生成 SQL 语句，这样就无须编写特定于数据库的 SQL 语句了，除非你想使用特定于数据库的特性或是提供优化的 SQL 语句。这样，应用是独立于数据库的，可以在未来轻松迁移到其他数据库。相较于直接使用 JDBC 来说，ORM 框架可以极大降低应用的数据访问工作量。

Hibernate 是 Java 社区中一款流行的开源、高性能 ORM 框架。它支持大多数 JDBC 兼容的数据库，可以通过特定的语言访问特定数据库。除了基本的 ORM 特性外，Hibernate 还支持更为高级的特性，如缓存、级联、延迟加载等。它还定义了一个名为 Hibernate 查询语言（Hibernate Query Language，HQL）的查询语言，可以使用它编写简单但强大的对象查询。

JPA 在 Java SE 与 Java EE 平台中为对象持久化定义了一套标准的注解和 API。它在 JSR-220 中是作为 EJB 规范的一部分而定义的。JPA 只是一套标准的 API，它需要兼容 JPA 的引擎来提供持久化服务。你可以将 JPA 类比为 JDBC API，将 JPA 引擎类比为 JDBC 驱动。可以通过名为 Hibernate EntityManager 的扩展模块将 Hibernate 配置为兼容 JPA 的引擎。本章将主要介绍以 Hibernate 作为底层引擎时的 JPA。

直接使用 JDBC 的问题

假设你要开发一个车辆登记应用，其主要功能是对车辆记录进行基本的创建、读取、更新与删除（CRUD）

操作。这些记录存储在关系型数据库中并通过 JDBC 访问。首先，设计如下的 Vehicle 类，它在 Java 中代表一个车辆：

```
package com.apress.springrecipes.vehicle;

public class Vehicle {

    private String vehicleNo;
    private String color;
    private int wheel;
    private int seat;

    // Constructors, Getters and Setters
    ...
}
```

创建应用数据库

在开发车辆登记应用前，你需要为其创建数据库。我们选择 PostgreSQL 作为数据库引擎。PostgreSQL 是个开源的关系型数据库引擎（表 9-1 列出了连接属性）。

注意： 本章的示例代码在 bin 目录下提供了脚本，可以开启并连接到基于 Docker 的 PostgreSQL 实例。要想开启实例并创建数据库，请遵循如下步骤。

1. 执行 bin\postgres.sh，这会下载并启动 Postgres Docker 容器。
2. 执行 bin\psql.sh，这会连接到运行着的 Postgres 容器。
3. 执行 CREATE DATABASE vehicle 创建示例所用的数据库。
4. 接下来，需要创建 VEHICLE 表来存储车辆记录，SQL 语句如下所示。

```
CREATE TABLE VEHICLE (
    VEHICLE_NO    VARCHAR(10)  NOT NULL,
    COLOR         VARCHAR(10),
    WHEEL         INT,
    SEAT          INT,
    PRIMARY KEY (VEHICLE_NO)
);
```

表 9-1　连接应用数据库的 JDBC 属性

属性	值
驱动类	org.postgresql.Driver
URL	jdbc:postgresql://localhost:5432/vehicle
用户名	postgres
密码	password

理解数据访问对象设计模型

一种典型的设计错误是将不同类型的逻辑（比如说展示逻辑、业务逻辑与数据访问逻辑）放到一个大模块当中。这么做会降低模块的可重用性与可维护性，因为导致了紧耦合。数据访问对象（Data Access Object，DAO）模式的目的旨在通过分离数据访问逻辑和展示逻辑来避免这些问题。该模式推荐将数据访问逻辑封装到名为数据访问对象的独立模块中。

对于这个车辆登记应用来说，你可以将数据访问操作抽象为插入、更新、删除与查询车辆。这些操作应该声明在 DAO 接口中，可由不同的 DAO 实现技术来实现。

```
package com.apress.springrecipes.vehicle;

import java.util.List;
```

```
public interface VehicleDao {

    void insert(Vehicle vehicle);
    void insert(Iterable<Vehicle> vehicles);
    void update(Vehicle vehicle);
    void delete(Vehicle vehicle);
    Vehicle findByVehicleNo(String vehicleNo);
    List<Vehicle> findAll();
}
```

JDBC API 的大部分地方都会声明抛出 java.sql.SQLException 异常。不过，由于该接口旨在抽象出数据访问操作，因此它不应该依赖于实现技术。因此，让这个通用接口声明抛出特定于 JDBC 的 SQLException 异常是不明智的行为。在实现 DAO 接口时，一种常见的做法是将这种异常包装为运行期异常（自己的业务 Exception 子类或是一般性异常）。

使用 JDBC 实现 DAO

要想使用 JDBC 访问数据库，你需要为该 DAO 接口（如 JdbcVehicleDao）创建一个实现。由于 DAO 实现需要连接到数据库以执行 SQL 语句，因此需要通过指定驱动类名、数据库 URL、用户名与密码来建立数据库连接。不过，可以从预先配置好的 javax.sql.DataSource 对象获取数据库连接而无须知晓连接的细节信息。

```
package com.apress.springrecipes.vehicle;

import javax.sql.DataSource;
import java.sql.Connection;
import java.sql.PreparedStatement;
import java.sql.ResultSet;
import java.sql.SQLException;
import java.util.ArrayList;
import java.util.Collection;
import java.util.List;

public class PlainJdbcVehicleDao implements VehicleDao {

    private static final String INSERT_SQL  = "INSERT INTO VEHICLE (COLOR, WHEEL,
    SEAT, VEHICLE_NO) VALUES (?, ?, ?, ?)";
    private static final String UPDATE_SQL  = "UPDATE VEHICLE SET COLOR=?,WHEEL=?,SEAT=?
    WHERE VEHICLE_NO=?";
    private static final String SELECT_ALL_SQL = "SELECT * FROM VEHICLE";
    private static final String SELECT_ONE_SQL = "SELECT * FROM VEHICLE WHERE VEHICLE_NO = ?";
    private static final String DELETE_SQL  = "DELETE FROM VEHICLE WHERE VEHICLE_NO=?";

    private final DataSource dataSource;

    public PlainJdbcVehicleDao(DataSource dataSource) {
        this.dataSource = dataSource;
    }

    @Override
    public void insert(Vehicle vehicle) {
        try (Connection conn = dataSource.getConnection();
            PreparedStatement ps = conn.prepareStatement(INSERT_SQL)) {
            prepareStatement(ps, vehicle);
            ps.executeUpdate();
        } catch (SQLException e) {
            throw new RuntimeException(e);
        }
    }

    @Override
public void insert(Collection<Vehicle> vehicles) {
```

```
        vehicles.forEach(this::insert);
    }

    @Override
    public Vehicle findByVehicleNo(String vehicleNo) {
        try (Connection conn = dataSource.getConnection();
            PreparedStatement ps = conn.prepareStatement(SELECT_ONE_SQL)) {
            ps.setString(1, vehicleNo);

            Vehicle vehicle = null;
            try (ResultSet rs = ps.executeQuery()) {
                if (rs.next()) {
                    vehicle = toVehicle(rs);
                }
            }
            return vehicle;
        } catch (SQLException e) {
            throw new RuntimeException(e);
        }
    }

    @Override
    public List<Vehicle> findAll() {
        try (Connection conn = dataSource.getConnection();
            PreparedStatement ps = conn.prepareStatement(SELECT_ALL_SQL);
            ResultSet rs = ps.executeQuery()) {

            List<Vehicle> vehicles = new ArrayList<>();
            while (rs.next()) {
                vehicles.add(toVehicle(rs));
            }
            return vehicles;
        } catch (SQLException e) {
            throw new RuntimeException(e);
        }
    }

    private Vehicle toVehicle(ResultSet rs) throws SQLException {
        return new Vehicle(rs.getString("VEHICLE_NO"),
            rs.getString("COLOR"), rs.getInt("WHEEL"),
            rs.getInt("SEAT"));
    }

    private void prepareStatement(PreparedStatement ps, Vehicle vehicle) throws SQLException
     {
        ps.setString(1, vehicle.getColor());
        ps.setInt(2, vehicle.getWheel());
        ps.setInt(3, vehicle.getSeat());
        ps.setString(4, vehicle.getVehicleNo());
    }

    @Override
    public void update(Vehicle vehicle) { … }

    @Override
    public void delete(Vehicle vehicle) { … }
}
```

车辆插入操作是一种典型的 JDBC 更新场景。每次调用该方法时，都需要从数据源获取一个连接并在该连接上执行 SQL 语句。DAO 接口没有声明抛出任何检查异常，这样如果出现了 SQLException，你就需要将其包装为一个未检查的 RuntimeException（本章后面将会详细介绍如何处理 DAO 中的异常）。这里的代码使用了一种所谓的 try-with-resources 机制，它会自动关闭所用的资源（比如说 Connection、PreparedStatement 与 ResultSet）。如果没有使用 try-with-resources 块，那就要记得正确关闭所用的资源；否则会导致连接泄漏。

这里就不再介绍更新与删除操作了，因为从技术视角来看，它们与插入操作大同小异。对于查询操作来说，除了执行 SQL 语句外，你还需要从返回的结果集中抽取出数据来构建车辆对象。toVehicle 是个简单的辅助方法，可以重用映射逻辑；prepareStatement 也是个辅助方法，可用于设置 insert 与 update 方法的参数。

在 Spring 中配置数据源

javax.sql.DataSource 是由 JDBC 规范定义的标准接口，用于以工厂方式获取 Connection 实例。不同厂商与项目提供了很多数据源实现；HikariCP 与 Apache Commons DBCP 都是流行的开源选择，大多数应用服务器也都会提供自己的实现。切换不同的数据源实现是很容易的，因为它们都实现了公共的 DataSource 接口。作为一个 Java 应用框架，Spring 也提供了几个便捷但功能并没有那么强大的数据源实现。最简单的一个是 DriverManagerDataSource，它会在每次请求时开启一个新的连接。

```
package com.apress.springrecipes.vehicle.config;

import com.apress.springrecipes.vehicle.PlainJdbcVehicleDao;
import com.apress.springrecipes.vehicle.VehicleDao;
import org.apache.derby.jdbc.ClientDriver;
import org.springframework.context.annotation.Bean;
import org.springframework.context.annotation.Configuration;
import org.springframework.jdbc.datasource.DriverManagerDataSource;

import javax.sql.DataSource;

@Configuration
public class VehicleConfiguration {

    @Bean
    public VehicleDao vehicleDao() {
        return new PlainJdbcVehicleDao(dataSource());
    }

    @Bean
    public DataSource dataSource() {
        DriverManagerDataSource dataSource = new DriverManagerDataSource();
        dataSource.setDriverClassName(ClientDriver.class.getName());
        dataSource.setUrl("jdbc:derby://localhost:1527/vehicle;create=true");
        dataSource.setUsername("app");
        dataSource.setPassword("app");
        return dataSource;

    }
}
```

DriverManagerDataSource 并不是一个高效的数据源实现，因为每次请求时它都会为客户端打开一个新的连接。Spring 提供的另一个数据源实现是 SingleConnectionDataSource（DriverManagerDataSource 的子类）。顾名思义，它只会维护单个连接，该连接会一直被重用，永远不会关闭。显然，在多线程环境下它是不适合的。

Spring 自己的数据源实现主要用于测试的目的。很多产品级的数据源实现都支持连接池。比如说，HikariCP 提供了 HikariDataSource，它与 DriverManagerDataSource 接受同样的连接属性，并且还可以为连接池指定最小的连接数量以及最大的活动连接数。

```
@Bean
public DataSource dataSource() {
    HikariDataSource dataSource = new HikariDataSource();
    dataSource.setUsername("postgres");
    dataSource.setPassword("password");
    dataSource.setJdbcUrl("jdbc:postgresql://localhost:5432/vehicle");
    dataSource.setMinimumIdle(2);
    dataSource.setMaximumPoolSize(5);
    return dataSource;
}
```

注意：要想使用 HikariCP 所提供的数据源实现，你需要将其添加到类路径中。如果使用 Maven，请将如下依赖添加到项目中：

```
<dependency>
    <groupId>com.zaxxer</groupId>
    <artifactId>HikariCP</artifactId>
    <version>2.6.1</version>
</dependency>
```

如果使用 Gradle，请使用如下信息：

```
compile 'com.zaxxer:HikariCP:2.6.1'
```

很多 Java EE 应用服务器自带数据源实现，你可以通过服务器控制台或是配置文件对其进行配置。如果在应用服务器中配置了数据源并针对 JNDI 查找进行了公开，那就可以使用 JndiDataSourceLookup 查找它了。

```
@Bean
public DataSource dataSource() {
    return new JndiDataSourceLookup().getDataSource("jdbc/VehicleDS");
}
```

运行 DAO

如下的 Main 类会使用 DAO 将一个新的车辆插入到数据库中来对其进行测试。如果成功，你就可以立刻从数据库中查询到该车辆。

```
package com.apress.springrecipes.vehicle;

import org.springframework.context.ApplicationContext;
import org.springframework.context.support.ClassPathXmlApplicationContext;

public class Main {

    public static void main(String[] args) {
        ApplicationContext context =
            new AnnotationConfigApplicationContext(VehicleConfiguration.class);

        VehicleDao vehicleDao = context.getBean(VehicleDao.class);
        Vehicle vehicle = new Vehicle("TEM0001", "Red", 4, 4);
        vehicleDao.insert(vehicle);

        vehicle = vehicleDao.findByVehicleNo("TEM0001");
        System.out.println(vehicle);
    }
}
```

现在，可以直接使用 JDBC 实现 DAO 了。不过，从之前的 DAO 实现中可以看到，大多数 JDBC 代码都是相似的，每个数据库操作都需要重写一遍。这种冗余的代码会使得 DAO 方法变得过长且可读性很差。

更进一步

另一种方式是使用对象关系映射（ORM）工具，这样你只需要编写将领域模型中的实体映射到数据库表的逻辑就可以了。ORM 会解决将类数据持久化到数据库中的逻辑。这会彻底解放你：你只需要编写业务与领域模型，而不是数据库的 SQL 解析器。当然，另外一方面就是你不再能完全控制客户端与数据库之间的通信了——你得相信 ORM 层会做正确的事情。

9-1 使用 JDBC 模板来更新数据库

问题提出

使用 JDBC 是乏味的，充斥着烦琐的 API 调用，其中很多都不应该由你自己来管理。要想实现 JDBC 的更新操作，你需要执行如下任务，且大多数都是很烦琐的。

1. 从数据源获取一个数据库连接。
2. 从连接创建一个 PreparedStatement 对象。
3. 将参数绑定到 PreparedStatement 对象上。
4. 执行 PreparedStatement 对象。
5. 处理 SQLException。
6. 清理 statement 对象与连接。

JDBC 是个非常底层的 API，不过借助于 JDBC 模板，你要使用的 API 将会变得更具表现力（在杂事儿上花费的时间更少，从而将更多时间放在应用逻辑上），也更加易于使用且安全。

解决方案

org.springframework.jdbc.core.JdbcTemplate 类声明了大量重载的 update()模板方法来控制整个更新过程。不同版本的 update()方法可以重写默认过程的不同任务子集。Spring JDBC 框架预先定义了一些回调接口来封装不同的任务子集。你可以实现这些回调接口并将实例传递给对应的 update()方法来完成这个过程。

解释说明

我们会介绍通过 JdbcTemplate 的不同选项来更新数据库的不同方式。这里将会介绍 PreparedStatementCreators、PreparedStatementSetters，然后将会介绍 JdbcTemplate 本身的 update 方法。

使用 Statement Creator 更新数据库

要介绍的第一个回调接口是 PreparedStatementCreator。你将实现该接口以重写整个更新过程中的 statement 创建任务（任务 2）与参数绑定任务（任务 3）。要想将车辆插入到数据库中，需要实现 PreparedStatementCreator 接口，如下所示：

```
package com.apress.springrecipes.vehicle;

import java.sql.Connection;
import java.sql.PreparedStatement;
import java.sql.SQLException;

import org.springframework.jdbc.core.PreparedStatementCreator;

public class JdbcVehicleDao implements VehicleDao {

    private class InsertVehicleStatementCreator implements PreparedStatementCreator {

        private final Vehicle vehicle;

        InsertVehicleStatementCreator(Vehicle vehicle) {
            this.vehicle = vehicle;
        }

        public PreparedStatement createPreparedStatement(Connection con) throws SQLException
    {
            PreparedStatement ps = con.prepareStatement(INSERT_SQL);
            prepareStatement(ps, this.vehicle);
            return ps;
        }
    }
}
```

在实现 PreparedStatementCreator 接口时，需要获取到数据库连接并将其作为 createPreparedStatement()方法的参数。在该方法中需要基于该连接创建一个 PreparedStatement 对象并将参数绑定到它上面。最后，需要将 PreparedStatement 对象作为方法的返回值。注意该方法声明抛出了 SQLException 异常，这意味着你无须自己处理这类异常。由于这个类是 DAO 的一个内部类，因此可以在实现中调用 prepareStatement 辅助方法。

现在，你可以使用这个 statement creator 简化车辆的插入操作了。首先，需要创建 JdbcTemplate 类的一个实例并将数据源传递进去以从中获取到一个连接。接下来，只需调用 update()方法并将与模板对应的 statement creator 传递进来就可以完成更新过程了。

```
package com.apress.springrecipes.vehicle;
...
import org.springframework.jdbc.core.JdbcTemplate;

public class JdbcVehicleDao implements VehicleDao {
    ...
    public void insert(Vehicle vehicle) {
        JdbcTemplate jdbcTemplate = new JdbcTemplate(dataSource);
        jdbcTemplate.update(new InsertVehicleStatementCreator(vehicle));
    }
}
```

通常情况下，如果只在一个方法中使用，那么最好以内部类的形式来实现 PreparedStatementCreator 接口与其他回调接口。这是因为可以从内部类中直接访问到局部变量与方法参数，而不必通过构造方法参数来传递。在使用局部变量时，需要将其标记为 final。

```
package com.apress.springrecipes.vehicle;
...
import org.springframework.jdbc.core.JdbcTemplate;
import org.springframework.jdbc.core.PreparedStatementCreator;

public class JdbcVehicleDao implements VehicleDao {
    ...
    public void insert(Vehicle vehicle) {
        JdbcTemplate jdbcTemplate = new JdbcTemplate(dataSource);

        jdbcTemplate.update(new PreparedStatementCreator() {

            public PreparedStatement createPreparedStatement(Connection conn)
                throws SQLException {
                PreparedStatement ps = conn.prepareStatement(INSERT_SQL);
                prepareStatement(ps, vehicle);
                return ps;
            }
        });
    }
}
```

如果使用的是 Java 8，还可以通过 lambda 表达式来实现。

```
@Override
public void insert(final Vehicle vehicle) {
    JdbcTemplate jdbcTemplate = new JdbcTemplate(this.dataSource);
    jdbcTemplate.update(con -> {
        PreparedStatement ps = con.prepareStatement(INSERT_SQL);
        prepareStatement(ps, vehicle);
        return ps;
    });
}
```

现在可以删掉之前的 InsertVehicleStatementCreator 内部类了，因为它已不再使用。

使用 Statement Setter 更新数据库

顾名思义，第 2 个回调接口 PreparedStatementSetter 只会执行整个更新过程中的参数绑定任务（任务 3）。

另一个版本的 update()模板方法会接受一个 SQL 语句和一个 PreparedStatementSetter 对象作为参数。该方法会根据这个 SQL 语句创建一个 PreparedStatement 对象。对于该接口来说，只需将参数绑定到 PreparedStatement 对象上即可（可以再次委托给 prepareStatement 方法）。

```
package com.apress.springrecipes.vehicle;
...
```

```
import org.springframework.jdbc.core.JdbcTemplate;
import org.springframework.jdbc.core.PreparedStatementSetter;

public class JdbcVehicleDao implements VehicleDao {
    ...
    public void insert(final Vehicle vehicle) {
        JdbcTemplate jdbcTemplate = new JdbcTemplate(dataSource);

        jdbcTemplate.update(INSERT_SQL, new PreparedStatementSetter() {

            public void setValues(PreparedStatement ps)
                throws SQLException {
                prepareStatement(ps, vehicle);
            }
        });
    }
}
```

用 Java 8 中的 lambda 表达式实现会更加简洁，如下所示：

```
@Override
public void insert(Vehicle vehicle) {
    JdbcTemplate jdbcTemplate = new JdbcTemplate(this.dataSource);
    jdbcTemplate.update(INSERT_SQL, ps -> prepareStatement(ps, vehicle));
}
```

使用 SQL 语句和参数值更新数据库

最后，update()方法最简单的一个版本会接受一个 SQL 语句和一个对象数组作为语句的参数。它会根据这个 SQL 语句创建一个 PreparedStatement 对象并绑定参数。因此，你无须重写更新过程中的任何任务。

```
package com.apress.springrecipes.vehicle;
...
import org.springframework.jdbc.core.JdbcTemplate;

public class JdbcVehicleDao implements VehicleDao {
    ...
    public void insert(final Vehicle vehicle) {
        JdbcTemplate jdbcTemplate = new JdbcTemplate(dataSource);

        jdbcTemplate.update(INSERT_SQL, vehicle.getColor(),vehicle.getWheel(), vehicle.getSeat(),
        vehicle.getVehicleNo() );
    }
}
```

在介绍的 3 个不同版本的 update()方法中，最后一个是最简单的，因为你无须实现任何回调接口。此外，针对参数化查询，我们也不必再使用 setX（setInt、setString 等）风格的方法了。而第一个版本是最灵活的，因为可以在执行前对 PreparedStatement 对象进行任何预处理。在实际开发中，应该选择可以满足所有需求的最简单的版本。

JdbcTemplate 类还提供了其他重载的 update()方法。请参考 Javadoc 了解详情。

批量更新数据库

假设你想向数据库插入一批车辆。如果多次调用 update()方法，那么更新就会变得非常慢，因为 SQL 语句会重复编译。因此，更好的实现方式是使用批更新来插入一批车辆。

JdbcTemplate 类还提供了几个 batchUpdate()模板方法来实现批更新操作。下面将会使用到的 batchUpdate()方法接收一个 SQL 语句、一个条目集合、一个批大小和一个 ParameterizedPreparedStatementSetter。

```
package com.apress.springrecipes.vehicle;
...
import org.springframework.jdbc.core.BatchPreparedStatementSetter;
import org.springframework.jdbc.core.JdbcTemplate;
```

```
public class JdbcVehicleDao implements VehicleDao {
    ...
    @Override
    public void insert(Collection<Vehicle> vehicles) {
        JdbcTemplate jdbcTemplate = new JdbcTemplate(this.dataSource);
        jdbcTemplate.batchUpdate(INSERT_SQL, vehicles, vehicles.size(), new Parameterized
        PreparedStatementSetter<Vehicle>() {
            @Override
            public void setValues(PreparedStatement ps, Vehicle argument) throws
            SQLException {
                prepareStatement(ps, argument);
            }
        });
    }

}
```

下面是 Java 8 中的 lambda 的实现：

```
@Override
public void insert(Collection<Vehicle> vehicles) {
    JdbcTemplate jdbcTemplate = new JdbcTemplate(this.dataSource);
    jdbcTemplate.batchUpdate(INSERT_SQL, vehicles, vehicles.size(), this::prepareStatement);
}
```

你可以在 Main 类中使用以下代码测试批插入操作：

```
package com.apress.springrecipes.vehicle;
...
public class Main {

    public static void main(String[] args) {
        ...
        VehicleDao vehicleDao = (VehicleDao) context.getBean("vehicleDao");
        Vehicle vehicle1 = new Vehicle("TEM0022", "Blue", 4, 4);
        Vehicle vehicle2 = new Vehicle("TEM0023", "Black", 4, 6);
        Vehicle vehicle3 = new Vehicle("TEM0024", "Green", 4, 5);
        vehicleDao.insertBatch(Arrays.asList(vehicle1, vehicle2, vehicle3));
    }
}
```

9-2 使用 JDBC 模板查询数据库

问题提出

要想实现 JDBC 查询操作，你需要执行如下任务，任务 5 和任务 6 是比更新操作多出来的两个任务。

1. 从数据源获取一个数据库连接。
2. 从连接创建一个 PreparedStatement 对象。
3. 将参数绑定到 PreparedStatement 对象上。
4. 执行 PreparedStatement 对象。
5. 遍历返回的结果集。
6. 从结果集中提取出数据。
7. 处理 SQLException。
8. 清理 statement 对象与连接。

不过，与业务逻辑相关的步骤只有查询的定义和从结果集中提取出结果！其他最好都由 JDBC 模板来处理。

解决方案

JdbcTemplate 类声明了大量重载的 query()模板方法来控制整个查询过程。你可以通过实现 PreparedStatementCreator 与 PreparedStatementSetter 接口来重写 statement 创建（任务 2）与参数绑定（任务 3）任务，就像更新操作一样。此外，Spring JDBC 框架支持多种方式来重写数据提取任务（任务 6）。

解释说明

Spring 提供了 RowCallbackHandler 和 RowMapper 接口来在查询方法中处理结果。我们首先介绍这两个接口以及它们的不同使用场景，接下来将会介绍如何使用不同的查询方法来获取多个和单个结果。

使用 RowCallbackHandler 提取数据

RowCallbackHandler 是处理结果集中当前行的一个主要接口。query()方法会遍历结果集并针对每一行调用 RowCallbackHandler。这样，返回的结果集的每一行都会调用 processRow()方法一次。

```
package com.apress.springrecipes.vehicle;
...
import org.springframework.jdbc.core.JdbcTemplate;
import org.springframework.jdbc.core.RowCallbackHandler;

public class JdbcVehicleDao implements VehicleDao {
    ...
    @Override
    public Vehicle findByVehicleNo(String vehicleNo) {
        JdbcTemplate jdbcTemplate = new JdbcTemplate(dataSource);

        final Vehicle vehicle = new Vehicle();
        jdbcTemplate.query(SELECT_ONE_SQL,
            new RowCallbackHandler() {
                public void processRow(ResultSet rs) throws SQLException {
                    vehicle.setVehicleNo(rs.getString("VEHICLE_NO"));
                    vehicle.setColor(rs.getString("COLOR"));
                    vehicle.setWheel(rs.getInt("WHEEL"));
                    vehicle.setSeat(rs.getInt("SEAT"));
                }
            }, vehicleNo);
        return vehicle;
    }
}
```

使用 Java 8 中的 lambda 表达式的代码会更简洁一些。

```
@Override
public Vehicle findByVehicleNo(String vehicleNo) {
    JdbcTemplate jdbcTemplate = new JdbcTemplate(dataSource);

    final Vehicle vehicle = new Vehicle();
    jdbcTemplate.query(SELECT_ONE_SQL,
        rs -> {
            vehicle.setVehicleNo(rs.getString("VEHICLE_NO"));
            vehicle.setColor(rs.getString("COLOR"));
            vehicle.setWheel(rs.getInt("WHEEL"));
            vehicle.setSeat(rs.getInt("SEAT"));
        }, vehicleNo);
    return vehicle;
}
```

由于 SQL 查询至多返回一行，因此可以创建一个车辆对象作为局部变量并通过提取出结果集中的数据来设置其属性。对于有多行的结果集来说，应该以列表的形式来收集对象。

使用 RowMapper 提取数据

RowMapper<T>接口要比 RowCallbackHandler 更加通用，其目的在于将结果集中的一行映射到一个自定义对象上，这样它就可以应用到单行结果集与多行结果集上了。

从重用的角度来看，最好以普通类而非内部类的形式来实现 RowMapper<T>接口。在接口的 mapRow()方法中，你需要构建出代表一行的对象，并将其作为方法返回值返回。

```
package com.apress.springrecipes.vehicle;
```

```
import java.sql.ResultSet;
import java.sql.SQLException;

import org.springframework.jdbc.core.RowMapper;

public class JdbcVehicleDao implements VehicleDao {

    private class VehicleRowMapper implements RowMapper<Vehicle> {
        @Override
        public Vehicle mapRow(ResultSet rs, int rowNum) throws SQLException {
            return toVehicle(rs);
        }
    }
}
```

如前所述，RowMapper<T>既可用于单行结果集，也可用于多行结果集。在如 findByVehicleNo()方法这样查询唯一的对象时，你需要调用 JdbcTemplate 的 queryForObject()方法。

```
package com.apress.springrecipes.vehicle;
...
import org.springframework.jdbc.core.JdbcTemplate;

public class JdbcVehicleDao implements VehicleDao {
    ...
    public Vehicle findByVehicleNo(String vehicleNo) {

        JdbcTemplate jdbcTemplate = new JdbcTemplate(dataSource);
        return jdbcTemplate.queryForObject(SELECT_ONE_SQL, new VehicleRowMapper(),
        vehicleNo);
    }
}
```

Spring 自带一个便捷的 RowMapper<T>实现，名为 BeanPropertyRowMapper<T>，它可以自动将一行映射到指定类的一个新实例。注意，所指定的类必须是一个顶层类，同时要有一个默认或是无参构造方法。它首先会实例化这个类，然后通过匹配名字将每一列的值映射到属性上。它支持将属性名（如 vehicleNo）匹配到同名的列上，也支持匹配到带有下划线的列名上（如 VEHICLE_NO）。

```
package com.apress.springrecipes.vehicle;
...
import org.springframework.jdbc.core.BeanPropertyRowMapper;
import org.springframework.jdbc.core.JdbcTemplate;

public class JdbcVehicleDao implements VehicleDao {

    ...

    public Vehicle findByVehicleNo(String vehicleNo) {

        JdbcTemplate jdbcTemplate = new JdbcTemplate(dataSource);
        return jdbcTemplate.queryForObject(SELECT_ONE_SQL, BeanPropertyRowMapper.
        newInstance(Vehicle.class), vehicleNo);
    }
}
```

多行查询

现在，我们来看看如何查询拥有多行的结果集。比如说，假设 DAO 接口中需要一个 findAll()方法来获取到所有车辆。

```
package com.apress.springrecipes.vehicle;
...
public interface VehicleDao {
    ...
    public List<Vehicle> findAll();
}
```

如果不使用 RowMapper<T>，那么你依然可以调用 queryForList()方法并传递进一个 SQL 语句。返回的结果是个包含 map 的 list。每个 map 都存储了结果集中的一行并使用列名作为键。

```
package com.apress.springrecipes.vehicle;
...
import org.springframework.jdbc.core.JdbcTemplate;

public class JdbcVehicleDao implements VehicleDao {
    ...
    @Override
    public List<Vehicle> findAll() {
        JdbcTemplate jdbcTemplate = new JdbcTemplate(dataSource);

        List<Map<String, Object>> rows = jdbcTemplate.queryForList(SELECT_ALL_SQL);
        return rows.stream().map(row -> {
            Vehicle vehicle = new Vehicle();
            vehicle.setVehicleNo((String) row.get("VEHICLE_NO"));
            vehicle.setColor((String) row.get("COLOR"));
            vehicle.setWheel((Integer) row.get("WHEEL"));
            vehicle.setSeat((Integer) row.get("SEAT"));
            return vehicle;
        }).collect(Collectors.toList());
    }
}
```

可以通过 Main 类中的如下代码片段来测试 findAll()方法：

```
package com.apress.springrecipes.vehicle;
...
public class Main {

    public static void main(String[] args) {
        ...
        VehicleDao vehicleDao = (VehicleDao) context.getBean("vehicleDao");
        List<Vehicle> vehicles = vehicleDao.findAll();
        vehicles.forEach(System.out::println);
    }
}
```

如果使用 RowMapper<T>对象来映射结果集中的行，那么 query()方法将会得到一个被映射对象的列表。

```
package com.apress.springrecipes.vehicle;
...
import org.springframework.jdbc.core.BeanPropertyRowMapper;
import org.springframework.jdbc.core.JdbcTemplate;
public class JabcVehicleDao implements VehicleDao {
  ...
  public List<Vehicle> findAll() {
      JdbcTemplate jdbcTemplate = new JdbcTemplate(dataSource);
      return jdbcTemplate.query (SELECT_ALL_SQL,
                                 BeanPropertyRowMapper.newInstance(Vehicle.class));
  }
}
```

查询单个值

最后，我们来看看查询单行与单列结果集的情况。作为一个例子，我们将如下操作添加到 DAO 接口中：

```
package com.apress.springrecipes.vehicle;
...
public interface VehicleDao {
    ...
    public String getColor(String vehicleNo);
    public int countAll();
}
```

要想查询单个字符串值，你可以调用重载的 queryForObject()方法，它接收一个 java.lang.Class 类型的参数。该方法会将结果值映射为你所指定的类型。

```
package com.apress.springrecipes.vehicle;
...
import org.springframework.jdbc.core.JdbcTemplate;

public class JdbcVehicleDao implements VehicleDao {

    private static final String COUNT_ALL_SQL = "SELECT COUNT(*) FROM VEHICLE";
    private static final String SELECT_COLOR_SQL = "SELECT COLOR FROM VEHICLE WHERE
    VEHICLE_NO=?";

    ...
    public String getColor(String vehicleNo) {

        JdbcTemplate jdbcTemplate = new JdbcTemplate(dataSource);
        return jdbcTemplate.queryForObject(SELECT_COLOR_SQL, String.class, vehicleNo);
    }

    public int countAll() {

        JdbcTemplate jdbcTemplate = new JdbcTemplate(dataSource);
        return jdbcTemplate.queryForObject(COUNT_ALL_SQL, Integer.class);
    }
}
```

可以通过 Main 类中的如下代码片段测试这两个方法：

```
package com.apress.springrecipes.vehicle;
...
public class Main {

    public static void main(String[] args) {
        ...
        VehicleDao vehicleDao = context.getBean(VehicleDao.class);
        int count = vehicleDao.countAll();
        System.out.println("Vehicle Count: " + count);
        String color = vehicleDao.getColor("TEM0001");
        System.out.println("Color for [TEM0001]: " + color);
    }
}
```

9-3 简化 JDBC 模板的创建

问题提出

在每次需要时都创建 JdbcTemplate 的新实例是很低效的做法，因为需要重复执行创建语句，而创建新对象则会增加成本。

解决方案

JdbcTemplate 类是线程安全的，因此可以在 IoC 容器中声明它的单个实例并将其注入到所有 DAO 实例中。此外，Spring JDBC 框架提供了一个便捷的类 org.springframework.jdbc. core.support.JdbcDaoSupport 来简化 DAO 实现。该类声明了一个 jdbcTemplate 属性，可以从 IoC 容器中进行注入，也可以从数据源自动创建，比如说 JdbcTemplate jdbcTemplate = new JdbcTemplate(dataSource)。DAO 可以继承该类以继承这个属性。

解释说明

相较于需要时创建新的 JdbcTemplate，你还可以创建单个实例作为 bean，并将该实例注入到需要它的 DAO 中以实现重用。另一种方式则是继承 Spring 的 JdbcDaoSupport 类，它提供了对 JdbcTemplate 的访问器方法。

注入 JDBC 模板

到现在为止，你在每个 DAO 方法中都创建了 JdbcTemplate 的新实例。实际上，你可以在类级别上进行注入，并在所有 DAO 方法中使用注入的实例。出于简化，如下代码只展示了对 insert()方法的修改：

```
package com.apress.springrecipes.vehicle;
...
import org.springframework.jdbc.core.JdbcTemplate;
public class JdbcVehicleDao implements VehicleDao {

    private final JdbcTemplate jdbcTemplate;

    public JdbcVehicleDao (JdbcTemplate jdbcTemplate) {
        this.jdbcTemplate = jdbcTemplate;
    }

    public void insert(final Vehicle vehicle) {
        jdbcTemplate.update(INSERT_SQL, vehicle.getVehicleNo(), vehicle.getColor(),
        vehicle.getWheel(), vehicle.getSeat());
    }
    ...
}
```

JDBC 模板需要设置一个数据源。你可以通过 setter 方法或是构造方法参数来注入该属性。接下来就可以将这个 JDBC 模板注入到 DAO 中了。

```
@Configuration
public class VehicleConfiguration {

    @Bean
    public VehicleDao vehicleDao(JdbcTemplate jdbcTemplate) {
        return new JdbcVehicleDao(jdbcTemplate);
    }

    @Bean
    public JdbcTemplate jdbcTemplate(DataSource dataSource) {
        return new JdbcTemplate(dataSource);
    }
}
```

继承 JdbcDaoSupport 类

org.springframework.jdbc.core.support.JdbcDaoSupport 类有一个 setDataSource()方法和一个 setJdbcTemplate()方法。DAO 类可以继承这个类以便继承这两个方法。接下来，你可以直接注入一个 JDBC 模板或是注入一个数据源来创建一个 JDBC 模板。如下代码片段来自于 Spring 的 JdbcDaoSupport 类：

```
package org.springframework.jdbc.core.support;
...
public abstract class JdbcDaoSupport extends DaoSupport {

    private JdbcTemplate jdbcTemplate;

    public final void setDataSource(DataSource dataSource) {
        if( this.jdbcTemplate == null || dataSource != this.jdbcTemplate.getDataSource() ){
            this.jdbcTemplate = createJdbcTemplate(dataSource);
            initTemplateConfig();
        }
    }
    ...

    public final void setJdbcTemplate(JdbcTemplate jdbcTemplate) {
        this.jdbcTemplate = jdbcTemplate;
        initTemplateConfig();
    }

    public final JdbcTemplate getJdbcTemplate() {
        return this.jdbcTemplate;
    }
    ...
}
```

在 DAO 方法中，你只需调用 getJdbcTemplate()方法就可以得到 JDBC 模板。你还需要从 DAO 类中将 dataSource 与 jdbcTemplate 属性及其 setter 方法删除，因为它们已经从 JdbcDaoSupport 类继承下来了。出于简化，下面只展示了对 insert()方法的修改：

```
package com.apress.springrecipes.vehicle;
...
import org.springframework.jdbc.core.support.JdbcDaoSupport;

public class JdbcVehicleDao extends JdbcDaoSupport implements VehicleDao {

    public void insert(final Vehicle vehicle) {

        getJdbcTemplate().update(INSERT_SQL, vehicle.getVehicleNo(),
            vehicle.getColor(), vehicle.getWheel(), vehicle.getSeat());
    }
    ...
}
```

通过继承 JdbcDaoSupport，DAO 类会继承 setDataSource()方法。你可以将一个数据源注入到 DAO 实例中来创建一个 JDBC 模板。

```
@Configuration
public class VehicleConfiguration {
...
    @Bean
    public VehicleDao vehicleDao(DataSource dataSource) {
        JdbcVehicleDao vehicleDao = new JdbcVehicleDao();
        vehicleDao.setDataSource(dataSource);
        return vehicleDao;
    }
}
```

9-4 在 JDBC 模板中使用具名参数

问题提出

在传统的 JDBC 使用过程中，SQL 参数是通过占位符“?”表示的，并且会根据位置进行绑定。位置参数的问题在于当参数顺序改变后，还需要修改参数绑定。对于拥有很多参数的 SQL 语句来说，根据位置来匹配参数是非常麻烦的事情。

解决方案

在 Spring JDBC 框架中绑定 SQL 参数的另一种方式是使用具名参数。顾名思义，具名 SQL 参数是根据名字（以冒号开头）而非位置来指定的。具名参数更易于维护，可读性也更好。在运行期，框架类会使用占位符替换掉具名参数。具名参数是由 NamedParameterJdbcTemplate 所支持的。

解释说明

当在 SQL 语句中使用具名参数时，你可以以 map 的形式提供参数值，键则是参数名。

```
package com.apress.springrecipes.vehicle;
...
import org.springframework.jdbc.core.namedparam.NamedParameterJdbcDaoSupport;

public class JdbcVehicleDao extends NamedParameterJdbcDaoSupport implements
    VehicleDao {

    private static final String INSERT_SQL = "INSERT INTO VEHICLE (COLOR, WHEEL, SEAT,
    VEHICLE_NO) VALUES (:color, :wheel, :seat, :vehicleNo)";

    public void insert(Vehicle vehicle) {

        getNamedParameterJdbcTemplate().update(INSERT_SQL, toParameterMap(vehicle));
    }

    private Map<String, Object> toParameterMap(Vehicle vehicle) {
```

```
        Map<String, Object> parameters = new HashMap<>();
        parameters.put("vehicleNo", vehicle.getVehicleNo());
        parameters.put("color", vehicle.getColor());
        parameters.put("wheel", vehicle.getWheel());
        parameters.put("seat", vehicle.getSeat());
        return parameters;
    }

    ...
}
```

还可以提供一个 SQL 参数源，其职责是为具名 SQL 参数提供 SQL 参数值。SqlParameterSource 接口有 3 个实现。基本的实现是 MapSqlParameterSource，它将一个 map 包装为参数源。相较于之前的示例，该示例会有一些损耗，因为引入了一个额外对象——SqlParameterSource。

```
package com.apress.springrecipes.vehicle;
...
import org.springframework.jdbc.core.namedparam.MapSqlParameterSource;
import org.springframework.jdbc.core.namedparam.SqlParameterSource;
import org.springframework.jdbc.core.namedparam.NamedParameterJdbcDaoSupport;

public class JdbcVehicleDao extends NamedParameterJdbcDaoSupport implements
    VehicleDao {

    public void insert(Vehicle vehicle) {

        SqlParameterSource parameterSource =
            new MapSqlParameterSource(toParameterMap(vehicle));

        getNamedParameterJdbcTemplate().update(INSERT_SQL, parameterSource);
    }
    ...
}
```

当要在传递给 update 方法的参数与其值的来源之间构建额外一个间接层时，它的优势就体现出来了。比如说，如果想要从一个 JavaBean 中获取属性该怎么做呢？这正是 SqlParameterSource 大显身手的地方！SqlParameterSource 是个 BeanPropertySqlParameterSource，它将一个普通的 Java 对象包装为 SQL 参数源。对于每个具名参数来说，同名属性将会用作参数值。

```
package com.apress.springrecipes.vehicle;

import org.springframework.jdbc.core.namedparam.BeanPropertySqlParameterSource;
import org.springframework.jdbc.core.namedparam.SqlParameterSource;
import org.springframework.jdbc.core.namedparam.NamedParameterJdbcDaoSupport;

public class JdbcVehicleDao extends NamedParameterJdbcDaoSupport implements
    VehicleDao {

    public void insert(Vehicle vehicle) {

        SqlParameterSource parameterSource =
            new BeanPropertySqlParameterSource(vehicle);

        getNamedParameterJdbcTemplate ().update(INSERT_SQL, parameterSource);
    }
}
```

具名参数还可用在批更新上。你可以为参数值提供一个 map、数组或是 SqlParameterSource 数组。

```
package com.apress.springrecipes.vehicle;
...
import org.springframework.jdbc.core.namedparam.BeanPropertySqlParameterSource;
import org.springframework.jdbc.core.namedparam.SqlParameterSource;
import org.springframework.jdbc.core.namedparam.NamedParameterJdbcDaoSupport;

public class JdbcVehicleDao extends NamedParameterJdbcDaoSupport implements VehicleDao
```

```
{
    ...
    @Override
    public void insert(Collection<Vehicle> vehicles) {

        SqlParameterSource[] sources = vehicles.stream()
            .map(v -> new BeanPropertySqlParameterSource(v))
            .toArray(size -> new SqlParameterSource[size]);
        getNamedParameterJdbcTemplate().batchUpdate(INSERT_SQL, sources);
    }
}
```

9-5 在 Spring JDBC 框架中处理异常

问题提出

很多 JDBC API 都声明抛出 java.sql.SQLException，这是个必须要捕获的检查异常。每次执行数据库操作时处理这种异常都是一件非常麻烦的事情。你需要定义自己的策略来处理这种异常。不这么做就可能会导致不一致的异常处理。

解决方案

Spring 为数据访问模块（包括 JDBC 框架）提供了一种一致的数据访问异常处理机制。一般来说，Spring JDBC 框架抛出的所有异常都是 org.springframework.dao.DataAccessException 的子类，这是一种无须强制捕获的 RuntimeException。它是 Spring 数据访问模块中所有异常的根异常类。

图 9-1 只展示了 Spring 数据访问模块中 DataAccessException 的层次体系。总体来说，它为不同类别的数据访问异常定义了 30 多个异常类。

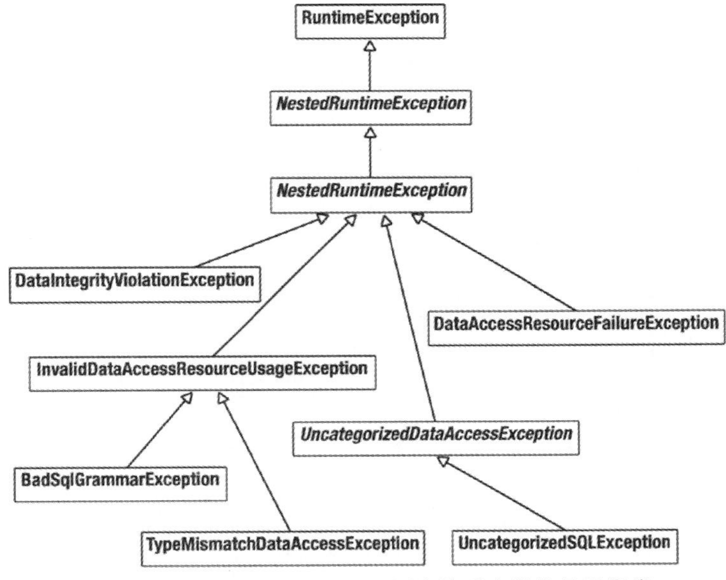

图 9-1 DataAccessException 层次体系中的常见异常类

解释说明

首先，我们将会介绍 Spring JDBC 中的异常处理方式，接下来会介绍如何通过创建自定义异常和映射来充分利用它。

理解 Spring JDBC 框架中的异常处理

到现在为止，在使用 JDBC 模板或 JDBC 操作对象时还没有显式处理过 JDBC 异常。为了帮助理解 Spring JDBC 框架的异常处理机制，我们看看如下 Main 类中的代码片段，它会插入一个车辆。如果使用重复车辆编码插入车辆，会发生什么呢？

```
package com.apress.springrecipes.vehicle;
...
public class Main {

    public static void main(String[] args) {
        ...
        VehicleDao vehicleDao = context.getBean(VehicleDao.class);
        Vehicle vehicle = new Vehicle("EX0001", "Green", 4, 4);
        vehicleDao.insert(vehicle);
    }
}
```

如果运行该方法两次或是该车辆之前已经被插入到了数据库中，那么它就会抛出DuplicateKeyException，这是DataAccessException的一个间接子类。在DAO方法中，你既不需要使用try/catch块将代码包围起来，也不需要在方法签名中声明抛出异常。这是因为DataAccessException（及其子类，包括DuplicateKeyException）是个无须强制捕获的未检查异常。DataAccessException的直接父类是NestedRuntimeException，这是一个核心的Spring异常类，它在RuntimeException中包装了另外一个异常。

在使用Spring JDBC框架中的类时，它们会捕获SQLException并使用DataAccessException的一个子类将其包装起来。由于该异常是个RuntimeException，因此无须捕获它。

不过，Spring JDBC框架如何知道该抛出DataAccessException层次体系中的哪个具体异常呢？这是通过查找被捕获异常的errorCode与SQLState属性做到的。由于DataAccessException会将底层SQLException包装为根本原因，因此可以通过如下catch块查看到errorCode与SQLState属性：

```
package com.apress.springrecipes.vehicle;
...
import java.sql.SQLException;

import org.springframework.dao.DataAccessException;

public class Main {

    public static void main(String[] args) {
        ...
        VehicleDao vehicleDao = context.getBean(VehicleDao.class);
        Vehicle vehicle = new Vehicle("EX0001", "Green", 4, 4);
        try {
            vehicleDao.insert(vehicle);
        } catch (DataAccessException e) {
            SQLException sqle = (SQLException) e.getCause();
            System.out.println("Error code: " + sqle.getErrorCode());
            System.out.println("SQL state: " + sqle.getSQLState());
        }
    }
}
```

当重复插入车辆时，注意到PostgreSQL会返回如下错误码和SQL状态：

```
Error code : 0
SQL state : 23505
```

查看PostgreSQL参考手册，你会发现表9-2所示的错误码说明。

表 9-2　　PostgreSQL 错误码说明

SQL 状态	消息文本
23505	unique_violation

Spring JDBC框架如何知道状态23505应该映射到DuplicateKeyException呢？错误码与SQL状态是特定于数据库的，这意味着不同的数据库产品可能会对同样的错误返回不同的错误码。此外，一些数据库产品会在errorCode属性中指定错误，而另外一些（如PostgreSQL）则会在SQLState属性中指定。

作为一个开放的 Java 应用框架，Spring 知道大多数流行的数据库产品的错误码。不过，由于错误码数量非常多，因此它只维护了最常遇到的错误的映射。映射定义在 sql-error-codes.xml 文件中，该文件位于 org.springframework.jdbc.support 包中。如下针对 PostgreSQL 的片段就来自于该文件：

```
<?xml version="1.0" encoding="UTF-8"?>
<!DOCTYPE beans PUBLIC "-//SPRING//DTD BEAN 3.0//EN"
    "http://www.springframework.org/dtd/spring-beans-3.0.dtd">

<beans>
    ...

    <bean id="PostgreSQL" class="org.springframework.jdbc.support.SQLErrorCodes">
        <property name="useSqlStateForTranslation">
            <value>true</value>
        </property>
        <property name="badSqlGrammarCodes">
            <value>03000,42000,42601,42602,42622,42804,42P01</value>
        </property>
        <property name="duplicateKeyCodes">
            <value>23505</value>
        </property>
        <property name="dataIntegrityViolationCodes">
            <value>23000,23502,23503,23514</value>
        </property>
        <property name="dataAccessResourceFailureCodes">
            <value>53000,53100,53200,53300</value>
        </property>
        <property name="cannotAcquireLockCodes">
            <value>55P03</value>
        </property>
        <property name="cannotSerializeTransactionCodes">
            <value>40001</value>
        </property>
        <property name="deadlockLoserCodes">
            <value>40P01</value>
        </property>
    </bean>
    ...
</beans>
```

useSqlStateForTranslation 属性表示会使用 SQLState 属性而不是 errorCode 属性来匹配错误码。最后，SQLErrorCodes 类定义了几个类别来映射数据库错误码。错误码 23505 位于 dataIntegrityViolationCodes 与 cdataIntegrityViolationCodes 类别中。

自定义数据访问异常处理

Spring JDBC 框架只会映射已知的错误码。有时，你想要自定义映射。比如说，你可能想向已有的类别中添加更多错误码，或是为特定的错误码自定义异常。

在表 9-2 中，错误码 23505 表示 PostgreSQL 中的重复键错误。它默认会被映射到 DataIntegrityViolationException。假设想要为该错误创建自定义的异常类型 MyDuplicateKeyException。它应该继承 DataIntegrityViolationException，因为它还是一种数据完整性违背错误。记住，对于 Spring JDBC 框架所抛出的异常来说，它必须要兼容于根异常类 DataAccessException。

```
package com.apress.springrecipes.vehicle;

import org.springframework.dao.DataIntegrityViolationException;

public class MyDuplicateKeyException extends DataIntegrityViolationException {

    public MyDuplicateKeyException(String msg) {
        super(msg);
```

```
    }

    public MyDuplicateKeyException(String msg, Throwable cause) {
        super(msg, cause);
    }
}
```

默认情况下，Spring 会从 org.springframework.jdbc.support 包中的 sql-error-codes.xml 文件查找异常。不过，你可以覆盖一些映射，方式是在类路径的根目录下提供一个同名文件。如果 Spring 能够找到你的自定义文件，那么它就会先从你的映射中寻找异常。但是，如果找不到合适的异常，Spring 就会寻找默认映射。

假设你想将自定义的 DuplicateKeyException 类型映射为错误码 23505。你需要通过 CustomSQLErrorCodesTranslation bean 添加绑定，然后将这个 bean 添加到 customTranslations 类别中。

```
<?xml version="1.0" encoding="UTF-8"?>
<!DOCTYPE beans PUBLIC "-//SPRING//DTD BEAN 2.0//EN"
    "http://www.springframework.org/dtd/spring-beans-2.0.dtd">

<beans>
    <bean id="PostgreSQL"
        class="org.springframework.jdbc.support.SQLErrorCodes">
        <property name="useSqlStateForTranslation">
            <value>true</value>
        </property>
        <property name="customTranslations">
            <list>
                <ref bean="myDuplicateKeyTranslation" />
            </list>
        </property>
    </bean>
    <bean id="myDuplicateKeyTranslation"
        class="org.springframework.jdbc.support.CustomSQLErrorCodesTranslation">
        <property name="errorCodes">
            <value>23505</value>
        </property>
        <property name="exceptionClass">
            <value>
                com.apress.springrecipes.vehicle.MyDuplicateKeyException
            </value>
        </property>
    </bean>
</beans>
```

现在，如果将包围着车辆插入操作的 try/catch 块删除，然后插入一个重复的车辆，那么 Spring JDBC 框架就会抛出一个 MyDuplicateKeyException 异常。

不过，如果对 SQLErrorCodes 类所用的基本的代码到异常的映射策略不太满意，那就可以进一步实现 SQLExceptionTranslator 接口并将其实例通过 setExceptionTranslator()方法注入到 JDBC 模板中。

9-6 直接使用 ORM 框架来避免问题

问题提出

你已经决定步入下一个层次——你有大量复杂的领域模型，为每个实体手工编写所有代码是非常乏味的事情，因此你准备探索其他方案，比如说 Hibernate。你会对它们的强大和简单所震惊到！

解决方案

让 Spring 来帮忙吧。它为 ORM 层的处理提供了相应的设施，而 ORM 层则超越了 JDBC 访问。

解释说明

假设你要为一家培训中心开发一个课程管理系统。该系统要创建的第一个类是 Course。这个类叫做实体类或是持久化类，因为它代表一个真实世界的实体，其实例会被持久化到数据库中。记住，要想让每个实体类都能被 ORM 框架持久化，实体类需要一个不带参数的默认构造方法。

```
package com.apress.springrecipes.course;
...
public class Course {

    private Long id;
    private String title;
    private Date beginDate;
    private Date endDate;
    private int fee;

    // Constructors, Getters and Setters
    ...
}
```

对于每个实体类来说，你必须要定义一个标识符属性来唯一标识一个实体。最佳实践是定义一个自动生成的标识符，因为它没有业务含义，因此在任何情况下都不会被修改。此外，这个标识符会被 ORM 框架用来确定实体的状态。如果标识符值为 null，那么这个实体就会被看作是新创建且尚未保存的。当这个实体被持久化时，会执行一个插入 SQL 语句；否则，会使用更新语句。为了让标识符可以为 null，应该为标识符选择原生包装类型，如 java.lang.Integer 和 java.lang.Long。

课程管理系统需要一个 DAO 接口来封装数据访问逻辑。我们在 CourseDao 接口中定义如下操作：

```
package com.apress.springrecipes.course;
...
public interface CourseDao {

    Course store(Course course);
    void delete(Long courseId);
    Course findById(Long courseId);
    List<Course> findAll();
}
```

通常情况下，在使用 ORM 持久化对象时，插入与更新会被合并为单个操作（即存储）。这是让 ORM 框架（不是你）来决定该插入还是更新对象。对于 ORM 框架来说，若想将对象持久化到数据库中，它必须得知道实体类的映射元数据。你需要以它支持的格式向其提供映射元数据。过去，Hibernate 使用 XML 来提供映射元数据。不过，由于每个 ORM 框架都有自己的定义映射元数据的格式，因此 JPA 定义了一套持久化注解，从而以标准格式来定义映射元数据，该格式可以在其他 ORM 框架中重用。

Hibernate 也支持使用 JPA 注解定义映射元数据，这样 Hibernate 与 JPA 一共有 3 种不同的策略来映射和持久化对象：

- 使用 Hibernate API 和 Hibernate XML 映射来持久化对象；
- 使用 Hibernate API 和 JPA 注解来持久化对象；
- 使用 JPA 和 JPA 注解来持久化对象。

Hibernate、JPA 和其他 ORM 框架的核心编程元素类似于 JDBC，表 9-3 对此进行了总结。

表 9-3　　不同数据访问策略的核心编程元素

概念	JDBC	Hibernate	JPA
资源	Connection	Session	EntityManager
资源工厂	DataSource	SessionFactory	EntityManagerFactory
异常	SQLException	HibernateException	PersistenceException

在 Hibernate 中，对象持久化的核心接口是 Session，可通过 SessionFactory 实例获取到它的实例。在 JPA 中，对应的接口是 EntityManager，可通过 EntityManagerFactory 实例获取到它的实例。Hibernate 抛出的异常类型是 HibernateException，JPA 抛出的异常类型是 PersistenceException 或其他的 Java SE 异常，如 IllegalArgumentException、IllegalStateException 等。注意，所有这些异常都是 RuntimeException 的子类，并不会强制你对其进行捕获和处理。

使用 Hibernate API 和 Hibernate XML 映射来持久化对象

要想使用 Hibernate XML 映射来映射实体类，可以为每个类提供一个映射文件，也可以为几个类提供一个大的文件。具体操作时，应该为每个类定义一个映射文件，文件名是类名加上作为文件扩展名的.hbm.xml，这样便于维护。中间的扩展名 hbm 代表“Hibernate metadata”。

Course 类的映射文件应当命名为 Course.hbm.xml，放在与实体类相同的包下。

```
<!DOCTYPE hibernate-mapping
    PUBLIC "-//Hibernate/Hibernate Mapping DTD 3.0//EN"
    "http://hibernate.sourceforge.net/hibernate-mapping-3.0.dtd">

<hibernate-mapping package="com.apress.springrecipes.course">
    <class name="Course" table="COURSE">
        <id name="id" type="long" column="ID">
            <generator class="identity" />
        </id>
        <property name="title" type="string">
            <column name="TITLE" length="100" not-null="true" />
        </property>
        <property name="beginDate" type="date" column="BEGIN_DATE" />
        <property name="endDate" type="date" column="END_DATE" />
        <property name="fee" type="int" column="FEE" />
    </class>
</hibernate-mapping>
```

在映射文件中，可以为实体类指定表名，为每个简单属性指定表中的列。还可以指定列的详细信息，比如说列的长度、not-null 约束和唯一性约束。此外，每个实体必须要有一个标识符，它可以自动生成，也可以手工赋值。在该示例中，标识符是通过表的身份字段生成的。

现在，我们在 hibernate 子包下使用普通的 Hibernate API 实现 DAO 接口。在调用 Hibernate API 进行对象持久化之前，你需要初始化一个 Hibernate session 工厂（比如说在构造方法中）。

```
package com.apress.springrecipes.course.hibernate;

import com.apress.springrecipes.course.Course;
import com.apress.springrecipes.course.CourseDao;
import org.hibernate.Session;
import org.hibernate.SessionFactory;
import org.hibernate.Transaction;
import org.hibernate.cfg.AvailableSettings;
import org.hibernate.cfg.Configuration;
import org.hibernate.dialect.PostgreSQL95Dialect;

import java.util.List;

public class HibernateCourseDao implements CourseDao {

    private final SessionFactory sessionFactory;

    public HibernateCourseDao() {

        Configuration configuration = new Configuration()
            .setProperty(AvailableSettings.URL, "jdbc:postgresql://localhost:5432/course")
            .setProperty(AvailableSettings.USER, "postgres")
            .setProperty(AvailableSettings.PASS, "password")
            .setProperty(AvailableSettings.DIALECT, PostgreSQL95Dialect.class.getName())
            .setProperty(AvailableSettings.SHOW_SQL, String.valueOf(true))
            .setProperty(AvailableSettings.HBM2DDL_AUTO, "update")
            .addClass(Course.class);
        sessionFactory = configuration.buildSessionFactory();
    }
```

```
    @Override
    public Course store(Course course) {
        Session session = sessionFactory.openSession();
        Transaction tx = session.getTransaction();
        try {
            tx.begin();
            session.saveOrUpdate(course);
            tx.commit();
            return course;
        } catch (RuntimeException e) {
            tx.rollback();
            throw e;
        } finally {
            session.close();
        }

    }

    @Override
    public void delete(Long courseId) {
        Session session = sessionFactory.openSession();
        Transaction tx = session.getTransaction();
        try {
            tx.begin();
            Course course = session.get(Course.class, courseId);
            session.delete(course);
            tx.commit();
        } catch (RuntimeException e) {
            tx.rollback();
            throw e;
        } finally {
            session.close();
        }
    }

    @Override
    public Course findById(Long courseId) {
        Session session = sessionFactory.openSession();
        try {
            return session.get(Course.class, courseId);
        } finally {
            session.close();
        }
    }

    @Override
    public List<Course> findAll() {
        Session session = sessionFactory.openSession();
        try {
            return session.createQuery("SELECT c FROM Course c", Course.class).list();
        } finally {
            session.close();
        }
    }
}
```

使用 Hibernate 的第一步是创建一个 Configuration 对象并配置属性，如数据库设置（JDBC 连接属性或是数据源的 JNDI 名字）、数据库方言、映射元数据的位置等。在使用 XML 映射文件定义映射元数据时，可以使用 addClass 方法告诉 Hibernate 它要管理的类是什么；根据约定，接下来它会加载 Course.hbm.xml 文件。然后，从这个 Configuration 对象构建一个 Hibernate 会话工厂。会话工厂的目的在于生成用于持久化对象的会话。

在持久化对象前，你需要在数据库模式中创建表来存储对象数据。当使用如 Hibernate 之类的 ORM 框架时，通常不需要自己设计表。如果将 hibernate.hbm2ddl.auto 属性设为 update，那么 Hibernate 就会帮助你更新数据库模式并在必要时创建表。

> ■ **提示：** 显然，不应该在生产环境下开启这个属性，不过它可以极大提升开发的速度。

在之前的 DAO 方法中，首先从会话工厂中打开一个会话。对于涉及数据库更新的任何操作（如 saveOrUpdate()和 delete()）来说，必须要在该会话上开启一个 Hibernate 事务。如果操作成功完成，那就提交事务。否则，如果出现任何 RuntimeException，那就需要回滚事务。对于只读操作（如 get()和 HQL 查询）来说，没必要开启事务。最后，必须要记得关闭会话来释放其所持有的资源。

可以创建如下 Main 类来测试并运行所有 DAO 方法。它还展示了一个实体的典型生命周期。

```
package com.apress.springrecipes.course;

import com.apress.springrecipes.course.hibernate.HibernateCourseDao;

import java.util.GregorianCalendar;

public class Main {
    public static void main(String[] args) {

        CourseDao courseDao = new HibernateCourseDao();

        Course course = new Course();
        course.setTitle("Core Spring");
        course.setBeginDate(new GregorianCalendar(2007, 8, 1).getTime());
        course.setEndDate(new GregorianCalendar(2007, 9, 1).getTime());
        course.setFee(1000);

        System.out.println("\nCourse before persisting");
        System.out.println(course);

        courseDao.store(course);

        System.out.println("\nCourse after persisting");
        System.out.println(course);

        Long courseId = course.getId();
        Course courseFromDb = courseDao.findById(courseId);

        System.out.println("\nCourse fresh from database");
        System.out.println(courseFromDb);

        courseDao.delete(courseId);

        System.exit(0);
    }
}
```

使用 Hibernate API 和 JPA 注解来持久化对象

JPA 注解是在 JSR-220 规范中标准化的，因此它们得到了所有兼容于 JPA 的 ORM 框架的支持，包括 Hibernate。此外，使用注解更加便于在同一个源文件中编辑映射元数据。

如下 Course 类展示了如何通过 JPA 注解来定义映射元数据：

```
package com.apress.springrecipes.course;
...
import javax.persistence.Column;
import javax.persistence.Entity;
import javax.persistence.GeneratedValue;
import javax.persistence.GenerationType;
import javax.persistence.Id;
```

```
import javax.persistence.Table;

@Entity
@Table(name = "COURSE")
public class Course {

    @Id
    @GeneratedValue(strategy = GenerationType.IDENTITY)
    @Column(name = "ID")
    private Long id;

    @Column(name = "TITLE", length = 100, nullable = false)
    private String title;

    @Column(name = "BEGIN_DATE")
    private Date beginDate;

    @Column(name = "END_DATE")
    private Date endDate;

    @Column(name = "FEE")
    private int fee;

    // Constructors, Getters and Setters
    ...
}
```

每个实体类必须使用@Entity 注解进行修饰。你可以在该注解中为实体类指定一个表名。对于每个属性来说，可以通过@Column 注解指定列名和列的详细信息。

每个实体类必须要有一个被@Id 注解所定义的标识符。你可以通过@GeneratedValue 注解为标识符的生成选择一种策略。这里，标识符是通过表的身份列来生成的。

DAO 与之前代码示例中所用的几乎一样，只不过在配置上有一点儿小变化。

```
public HibernateCourseDao() {

    Configuration configuration = new Configuration()
        .setProperty(AvailableSettings.URL, "jdbc:postgresql://localhost:5432/course")
        .setProperty(AvailableSettings.USER, "postgres")
        .setProperty(AvailableSettings.PASS, "password")
        .setProperty(AvailableSettings.DIALECT, PostgreSQL95Dialect.class.getName())
        .setProperty(AvailableSettings.SHOW_SQL, String.valueOf(true))
        .setProperty(AvailableSettings.HBM2DDL_AUTO, "update")
        .addAnnotatedClass(Course.class);
    sessionFactory = configuration.buildSessionFactory();
}
```

由于现在使用注解来指定元数据，因此需要使用 addAnnotatedClass 方法来代替 addClass。这会让 Hibernate 从类本身来读取映射元数据，而不是去寻找 hbm.xml 文件。可以使用相同的 Main 类来运行该示例。

将 Hibernate 作为 JPA 引擎来持久化对象

除了持久化注解外，JPA 还为对象持久化定义了一套编程接口。不过，JPA 并不是持久化实现；你需要选择一个兼容于 JPA 的引擎来提供持久化服务。Hibernate 可以通过 Hibernate EntityManager 模块兼容于 JPA。借助于它，Hibernate 可以作为底层 JPA 引擎来持久化对象。这一方面可以让你保留在 Hibernate 上的投资（速度更快，或是能更好地处理某些操作），另一方面可以编写出兼容于 JPA 且可以移植到其他 JPA 引擎的代码。这还是一种将代码基迁移至 JPA 上的很有价值的方式。新代码是严格按照 JPA API 编写的，老代码则会迁移至 JPA 接口。

在 Java EE 环境下，你可以在 Java EE 容器中配置 JPA 引擎。但在 Java SE 应用中则需要在本地创建引擎。JPA 的配置使用了中央化的 XML 文件 persistence.xml，它位于类路径根目录的 META-INF 目录下。在该文件中，你可以为底层引擎配置设置任何特定于厂商的属性。在使用 Spring 配置 EntityManagerFactory 时，这是不需要的，配置可以通过 Spring 完成。

现在，我们在类路径根目录的 META-INF 目录下创建 JPA 配置文件 persistence.xml。每个 JPA 配置文件都包含了一个或多个<persistence-unit>元素。一个持久化单元定义了一组持久化类及其持久化方式。每个持久化单元都需要一个名字来标识自己，这里为该持久化单元指定了名字 course。

```
<persistence xmlns="http://xmlns.jcp.org/xml/ns/persistence"
    xmlns:xsi=http://www.w3.org/2001/XMLSchema-instance
    xsi:schemaLocation="http://xmlns.jcp.org/xml/ns/persistence
    http://xmlns.jcp.org/xml/ns/persistence/persistence_2_1.xsd"
    version="2.1">

    <persistence-unit name="course" transaction-type="RESOURCE_LOCAL">
        <class>com.apress.springrecipes.course.Course</class>

        <properties>
            <property name="javax.persistence.jdbc.url"
                      value="jdbc:postgresql://localhost:5432/course" />
            <property name="javax.persistence.jdbc.user" value="postgres" />
            <property name="javax.persistence.jdbc.password" value="password" />

            <property name="hibernate.dialect"
                      value="org.hibernate.dialect.PostgreSQL95Dialect" />
            <property name="hibernate.show_sql" value="true" />
            <property name="hibernate.hbm2ddl.auto" value="update" />
        </properties>
    </persistence-unit>
</persistence>
```

在该 JPA 配置文件中，你将 Hibernate 配置为了底层 JPA 引擎。注意这里有几个 javax.persistence 开头的属性，它们用于配置数据库的位置以及使用的用户名/密码组合。接下来是一些 Hibernate 专有属性来配置方言和 hibernate.hbm2ddl.auto 属性。最后，有一个<class>元素来指定要对哪些类进行映射。

在 Java EE 环境中，Java EE 容器可以帮你管理实体管理器并直接将其注入到 EJB 组件中。不过，在 Java EE 容器外使用 JPA 时（比如说在 Java SE 应用中），需要自己创建并维护实体管理器。

现在，我们在 Java SE 应用中通过 JPA 来实现 CourseDao 接口。在调用 JPA 进行对象持久化之前，你需要初始化一个实体管理器工厂。它的目的在于生成实体管理器以持久化对象。

```
package com.apress.springrecipes.course.jpa;
...
import javax.persistence.EntityManager;
import javax.persistence.EntityManagerFactory;
import javax.persistence.EntityTransaction;
import javax.persistence.Persistence;
import javax.persistence.Query;

public class JpaCourseDao implements CourseDao {

    private EntityManagerFactory entityManagerFactory;

    public JpaCourseDao() {
        entityManagerFactory = Persistence.createEntityManagerFactory("course");
    }

    public void store(Course course) {
        EntityManager manager = entityManagerFactory.createEntityManager();
        EntityTransaction tx = manager.getTransaction();
        try {
            tx.begin();
            manager.merge(course);
            tx.commit();
        } catch (RuntimeException e) {
```

```
                tx.rollback();
                throw e;
            } finally {
                manager.close();
            }
        }

        public void delete(Long courseId) {
            EntityManager manager = entityManagerFactory.createEntityManager();
            EntityTransaction tx = manager.getTransaction();
            try {
                tx.begin();
                Course course = manager.find(Course.class, courseId);
                manager.remove(course);
                tx.commit();
            } catch (RuntimeException e) {
                tx.rollback();
                throw e;
            } finally {
                manager.close();
            }
        }

        public Course findById(Long courseId) {
            EntityManager manager = entityManagerFactory.createEntityManager();
            try {
                return manager.find(Course.class, courseId);
            } finally {
                manager.close();
            }
        }

        public List<Course> findAll() {
            EntityManager manager = entityManagerFactory.createEntityManager();
            try {
                Query query = manager.createQuery("select course from Course course");
                return query.getResultList();
            } finally {
                manager.close();
            }
        }
    }
```

实体管理器工厂是通过 javax.persistence.Persistence 类的静态方法 createEntityManagerFactory()创建的。你需要传进一个定义在 persistence.xml 中的实体管理器工厂的持久化单元名。

在之前的 DAO 方法中，首先从实体管理器工厂创建了一个实体管理器。对于涉及数据库更新的任何操作（如 merge()和 delete()），必须要在实体管理器上开启一个 JPA 事务。对于只读操作（如 find()和 JPA 查询），没必要开启事务。最后，必须要关闭实体管理器以释放资源。

可以使用类似的 Main 类测试该 DAO，不过这次需要实例化 JPA DAO 实现。

```
package com.apress.springrecipes.course;
...
public class Main {

    public static void main(String[] args) {
        CourseDao courseDao = new JpaCourseDao();
        ...
    }
}
```

在上述针对 Hibernate 与 JPA 的 DAO 实现中，每个 DAO 方法只有一两行代码不同。其他都是不得不重复编写的样板代码。此外，每个 ORM 框架的本地事务管理都有自己的 API。

9-7　在 Spring 中配置 ORM 资源工厂

问题提出

在单独使用 ORM 框架时，你需要使用其 API 配置资源工厂。对于 Hibernate 与 JPA 来说，你需要从原生 Hibernate API 与 JPA 构建会话工厂和实体管理器工厂。在没有 Spring 支持的情况下，除了手工管理这些对象外，我们别无选择。

解决方案

Spring 提供了几个工厂 bean 来创建 Hibernate 会话工厂和 JPA 实体管理器工厂，并将创建出来的工厂作为单例 bean 存放到 IoC 容器中。这些工厂可通过依赖注入在多个 bean 之间共享。此外，这还可以让会话工厂和实体管理器工厂集成其他的 Spring 数据访问设施，如数据源和事务管理器。

解释说明

对于 Hibernate 来说，Spring 提供了一个 LocalSessionFactoryBean 来创建普通的 Hibernate SessionFactory；对于 JPA 来说，Spring 提供了几种方式来构建 EntityManagerFactory。我们将会介绍如何从 JNDI 中获取 EntityManagerFactory，以及如何使用 LocalEntityManagerFactoryBean 和 LocalContainerEntityManagerFactoryBean，同时还会谈及每种方式之间的差别。

在 Spring 中配置 Hibernate 会话工厂

首先，我们修改一下 HibernateCourseDao，使之通过依赖注入接受一个会话工厂，而不是在构造方法中直接通过原生 Hibernate API 创建。

```
package com.apress.springrecipes.course.hibernate;
...
import org.hibernate.SessionFactory;

public class HibernateCourseDao implements CourseDao {

    private final SessionFactory sessionFactory;

    public HibernateCourseDao(SessionFactory sessionFactory) {
        this.sessionFactory = sessionFactory;
    }

    ...
}
```

接下来，创建一个配置类，将 Hibernate 作为 ORM 框架。此外，还在 Spring 的管理下声明了一个 HibernateCourseDao 实例。

```
package com.apress.springrecipes.course.config;

import com.apress.springrecipes.course.Course;
import com.apress.springrecipes.course.CourseDao;
import com.apress.springrecipes.course.hibernate.HibernateCourseDao;
import org.hibernate.SessionFactory;
import org.hibernate.cfg.AvailableSettings;
import org.hibernate.dialect.PostgreSQL95Dialect;
import org.springframework.context.annotation.Bean;
import org.springframework.context.annotation.Configuration;
import org.springframework.orm.hibernate5.LocalSessionFactoryBean;

import java.util.Properties;

@Configuration
public class CourseConfiguration {

    @Bean
    public CourseDao courseDao(SessionFactory sessionFactory) {
```

```
        return new HibernateCourseDao(sessionFactory);
    }

    @Bean
    public LocalSessionFactoryBean sessionFactory() {

        LocalSessionFactoryBean sessionFactoryBean = new LocalSessionFactoryBean();
        sessionFactoryBean.setHibernateProperties(hibernateProperties());
        sessionFactoryBean.setAnnotatedClasses(Course.class);
        return sessionFactoryBean;
    }

    private Properties hibernateProperties() {

        Properties properties = new Properties();
        properties.setProperty(AvailableSettings.URL,
                               "jdbc:postgresql://localhost:5432/course");
        properties.setProperty(AvailableSettings.USER, "postgres");
        properties.setProperty(AvailableSettings.PASS, "password");
        properties.setProperty(AvailableSettings.DIALECT,
                               PostgreSQL95Dialect.class.getName());
        properties.setProperty(AvailableSettings.SHOW_SQL, String.valueOf(true));
        properties.setProperty(AvailableSettings.HBM2DDL_AUTO, "update");
        return properties;
    }
}
```

之前在 Configuration 对象上设置的所有属性现在都被转换为一个 Properties 对象并添加到 LocalSessionFactoryBean 中。被注解的类会通过 setAnnotatedClasses 方法传递进来，这样最终 Hibernate 就知道它们了。创建好的 SessionFactory 会通过 HibernateCourseDao 的构造方法传递进来。

如果项目还在使用 Hibernate 映射文件，那么可以使用 mappingLocations 属性指定映射文件。凭借 LocalSessionFactoryBean，你可以利用 Spring 的资源加载支持从各种位置加载映射文件。可以在 mappingLocations 属性中指定映射文件的资源路径，其类型是 Resource[]。

```
@Bean
public LocalSessionFactoryBean sessionfactory() {
    LocalSessionFactoryBean sessionFactoryBean = new LocalSessionFactoryBean();
    sessionFactoryBean.setDataSource(dataSource());
    sessionFactoryBean.setMappingLocations(
        new ClassPathResource("com/apress/springrecipes/course/Course.hbm.xml"));
    sessionFactoryBean.setHibernateProperties(hibernateProperties());
    return sessionFactoryBean;
}
```

借助于 Spring 的资源加载支持，还可以在资源路径中使用通配符匹配多个映射文件，这样在每次添加新的实体类时就无须配置其位置了。要想做到这一点，你需要在配置类中配置一个 ResourcePatternResolver。可以通过使用 ResourcePatternUtils 和 ResourceLoaderAware 接口获得它。实现 ResourceLoaderAware 接口并使用 getResourcePatternResolver 方法来根据 ResourceLoader 获得一个 ResourcePatternResolver。

```
@Configuration
public class CourseConfiguration implements ResourceLoaderAware {

    private ResourcePatternResolver resourcePatternResolver;
...
    @Override
    public void setResourceLoader(ResourceLoader resourceLoader) {
        this.resourcePatternResolver =
            ResourcePatternUtils.getResourcePatternResolver(resourceLoader);
    }
}
```

现在，可以通过 ResourecePatternResolver 将资源模式解析为资源了。

```
@Bean
```

```
public LocalSessionFactoryBean sessionfactory() throws IOException {
    LocalSessionFactoryBean sessionFactoryBean = new LocalSessionFactoryBean();
    Resource[] mappingResources =
        resourcePatternResolver.getResources("classpath:com/apress/springrecipes/course/*.
        hbm.xml");
    sessionFactoryBean.setMappingLocations(mappingResources);
    ...
    return sessionFactoryBean;
}
```

现在，修改 Main 类，从 Spring IoC 容器中获取 HibernateCourseDao 实例。

```
package com.apress.springrecipes.course;
...
import org.springframework.context.ApplicationContext;
import org.springframework.context.support.ClassPathXmlApplicationContext;

public class Main {

    public static void main(String[] args) {
        ApplicationContext context =
            new AnnotationConfigApplicationContext(CourseConfiguration.class);
        CourseDao courseDao = context.getBean(CourseDao.class);
        ...
    }
}
```

上述工厂 bean 通过加载 Hibernate 配置文件创建了一个会话工厂，它包含了数据库设置（JDBC 连接属性或是数据源的 JNDI 名）。现在，假设 Spring IoC 容器中定义了一个数据源。如果想要为会话工厂使用该数据源，你可以将其注入到 LocalSessionFactoryBean 的 dataSource 属性中。该属性所指定的数据源会覆盖 Hibernate 配置的数据库设置。如果设置了，那么 Hibernate 设置就不应该再定义连接提供者了，以避免无意义的重复配置。

```
@Configuration
public class CourseConfiguration {
    ...
    @Bean
    public DataSource dataSource() {

        HikariDataSource dataSource = new HikariDataSource();
        dataSource.setUsername("postgres");
        dataSource.setPassword("password");
        dataSource.setJdbcUrl("jdbc:postgresql://localhost:5432/course");
        dataSource.setMinimumIdle(2);
        dataSource.setMaximumPoolSize(5);
        return dataSource;
    }

    @Bean
    public LocalSessionFactoryBean sessionFactory(DataSource dataSource) {

        LocalSessionFactoryBean sessionFactoryBean = new LocalSessionFactoryBean();
        sessionFactoryBean.setDataSource(dataSource);
        sessionFactoryBean.setHibernateProperties(hibernateProperties());
        sessionFactoryBean.setAnnotatedClasses(Course.class);
        return sessionFactoryBean;
    }

    private Properties hibernateProperties() {

        Properties properties = new Properties();
        properties.setProperty(AvailableSettings.DIALECT,
                               PostgreSQL95Dialect.class.getName());
        properties.setProperty(AvailableSettings.SHOW_SQL, String.valueOf(true));
        properties.setProperty(AvailableSettings.HBM2DDL_AUTO, "update");
```

```
        return properties;
    }}
```

甚至还可以将所有配置合并到 LocalSessionFactoryBean 中来忽略掉 Hibernate 配置文件。比如说，可以在 packagesToScan 属性中指定被 JPA 注解的类所在的包，还可以指定其他 Hibernate 属性，如 hibernateProperties 属性中的数据库方言等。

```
@Configuration
public class CourseConfiguration {
...
    @Bean
    public LocalSessionFactoryBean sessionfactory() {
        LocalSessionFactoryBean sessionFactoryBean = new LocalSessionFactoryBean();
        sessionFactoryBean.setDataSource(dataSource());
        sessionFactoryBean.setPackagesToScan("com.apress.springrecipes.course");
        sessionFactoryBean.setHibernateProperties(hibernateProperties());
        return sessionFactoryBean;
    }

    private Properties hibernateProperties() {
        Properties properties = new Properties();
        properties.put("hibernate.dialect",
                        org.hibernate.dialect.DerbyTenSevenDialect.class.getName());
        properties.put("hibernate.show_sql", true);
        properties.put("hibernate.hbm2dll.auto", "update");
        return properties;
    }
}
```

现在可以删除 Hibernate 配置文件（即 hibernate.cfg.xml）了，因为它的配置已经交由 Spring 了。

在 Spring 中配置 JPA 实体管理器工厂

首先，修改 JpaCourseDao，使之通过依赖注入接收一个实体管理器工厂，而不是直接在构造方法中创建。

```
package com.apress.springrecipes.course;
...
import javax.persistence.EntityManagerFactory;
import javax.persistence.Persistence;

public class JpaCourseDao implements CourseDao {

    private final EntityManagerFactory entityManagerFactory;

    public JpaCourseDao (EntityManagerFactory entityManagerFactory) {
        this.entityManagerFactory = entityManagerFactory;
    }
    ...
}
```

JPA 规范定义了如何在 Java SE 与 Java EE 环境中获取实体管理器工厂。在 Java SE 环境中，实体管理器工厂是通过调用 Persistence 类的 createEntityManagerFactory()静态方法手工创建的。

我们创建一个 bean 配置文件来使用 JPA。Spring 提供了一个工厂 bean，名为 LocalEntityManagerFactoryBean，以在 IoC 容器中创建实体管理器工厂。你需要指定定义在 JPA 配置文件中的持久化单元名，还可以在 Spring 管理下声明一个 JpaCourseDao 实例。

```
package com.apress.springrecipes.course.config;

import com.apress.springrecipes.course.CourseDao;
import com.apress.springrecipes.course.jpa.JpaCourseDao;
import org.springframework.context.annotation.Bean;
import org.springframework.context.annotation.Configuration;
import org.springframework.orm.jpa.LocalEntityManagerFactoryBean;

import javax.persistence.EntityManagerFactory;
```

```
@Configuration
public class CourseConfiguration {

    @Bean
    public CourseDao courseDao(EntityManagerFactory entityManagerFactory) {
        return new JpaCourseDao(entityManagerFactory);
    }

    @Bean
    public LocalEntityManagerFactoryBean entityManagerFactory() {

        LocalEntityManagerFactoryBean emf = new LocalEntityManagerFactoryBean();
        emf.setPersistenceUnitName("course");
        return emf;
    }
}
```

现在，可以使用 Main 类从 Spring IoC 容器中获取 JpaCourseDao 实例并对其进行测试了。

```
package com.apress.springrecipes.course;
...
import org.springframework.context.ApplicationContext;
import org.springframework.context.support.ClassPathXmlApplicationContext;

public class Main {

    public static void main(String[] args) {
        ApplicationContext context =
            new AnnotationConfigApplicationContext(CourseConfiguration.class);
        CourseDao courseDao = context.getBean(CourseDao.class);
        ...
    }
}
```

在 Java EE 环境中，可以通过 JNDI 从 Java EE 容器中查找实体管理器工厂。在 Spring 中，可以通过 JndiLocatorDelegate 对象（要比创建 JndiObjectFactoryBean 简单一些，它也可以实现同样的功能）执行 JNDI 查找。

```
@Bean
public EntityManagerFactory entityManagerFactory() throws NamingException {
    return JndiLocatorDelegate.createDefaultResourceRefLocator()
        .lookup("jpa/coursePU", EntityManagerFactory.class);
}
```

LocalEntityManagerFactoryBean 通过加载 JPA 配置文件（即 persistence.xml）创建一个实体管理器工厂。Spring 通过名为 LocalContainerEntityManagerFactoryBean 的另外一个工厂 bean 支持一种更为灵活的创建实体管理器工厂的方式。它可以覆盖 JPA 配置文件中的一些配置，比如说数据源和数据库方言。这样，你就可以利用 Spring 的数据访问设施来配置实体管理器工厂了。

```
@Configuration
public class CourseConfiguration {
    ...
    @Bean
    public LocalContainerEntityManagerFactoryBean entityManagerFactory(DataSourcedataSource)
    {

        LocalContainerEntityManagerFactoryBean emf =
            new LocalContainerEntityManagerFactoryBean();
        emf.setPersistenceUnitName("course");
        emf.setDataSource(dataSource);
        emf.setJpaVendorAdapter(jpaVendorAdapter());
        return emf;
    }

    private JpaVendorAdapter jpaVendorAdapter() {
```

```
        HibernateJpaVendorAdapter jpaVendorAdapter = new HibernateJpaVendorAdapter();
        jpaVendorAdapter.setShowSql(true);
        jpaVendorAdapter.setGenerateDdl(true);
        jpaVendorAdapter.setDatabasePlatform(PostgreSQL95Dialect.class.getName());
        return jpaVendorAdapter;
    }

    @Bean
    public DataSource dataSource() {

        HikariDataSource dataSource = new HikariDataSource();
        dataSource.setUsername("postgres");
        dataSource.setPassword("password");
        dataSource.setJdbcUrl("jdbc:postgresql://localhost:5432/course");
        dataSource.setMinimumIdle(2);
        dataSource.setMaximumPoolSize(5);
        return dataSource;
    }
}
```

在上述 bean 配置中，将一个数据源注入到了这个实体管理器工厂中。它会覆盖 JPA 配置文件中的数据库设置。你可以为 LocalContainerEntityManagerFactoryBean 设置一个 JPA 厂商适配器以指定特定于 JPA 引擎的属性。当使用 Hibernate 作为底层 JPA 引擎时，你应该选择 HibernateJpaVendorAdapter。适配器不支持的其他属性可以在 jpaProperties 属性中指定。

现在，JPA 配置文件（即 persistence.xml）可以简化为如下内容，因为其配置已经迁移到了 Spring 中：

```
<persistence xmlns=http://xmlns.jcp.org/xml/ns/persistence
    xmlns:xsi=http://www.w3.org/2001/XMLSchema-instance
    xsi:schemaLocation="http://xmlns.jcp.org/xml/ns/persistence
    http://xmlns.jcp.org/xml/ns/persistence/persistence_2_1.xsd"
    version="2.1">

    <persistence-unit name="course" transaction-type="RESOURCE_LOCAL">
        <class>com.apress.springrecipes.course.Course</class>
    </persistence-unit>

</persistence>
```

Spring 还可以在不使用 persistence.xml 文件的情况下配置 JPA EntityManagerFactory。如果想这么做，你可以在 Spring 配置文件中进行完全的配置。相较于 persistenceUnitName，你需要指定 packagesToScan 属性。接下来就可以完全移除 persistence.xml 文件了。

```
@Bean
public LocalContainerEntityManagerFactoryBean entityManagerFactory() {
    LocalContainerEntityManagerFactoryBean emf =
        new LocalContainerEntityManagerFactoryBean();
    emf.setDataSource(dataSource());
    emf.setPackagesToScan("com.apress.springrecipes.course");
    emf.setJpaVendorAdapter(jpaVendorAdapter());
    return emf;
}
```

9-8 使用 Hibernate 的上下文会话持久化对象

问题提出

你想要基于普通的 Hibernate API 编写 DAO 并依旧想使用 Spring 管理的事务。

解决方案

从 Hibernate 3 开始，会话工厂可以管理上下文会话，并且可以通过 org.hibernate.SessionFactory 的 getCurrentSession()方法获取到它们。在单个事务中，每次调用 getCurrentSession()方法时都会得到相同的会话。这可以确保每个事务只有一个 Hibernate 会话，因此它可以与 Spring 的事务管理支持完美结合。

解释说明

要想使用上下文会话方式，DAO 方法需要访问会话工厂，这可以通过 setter 方法或是构造方法参数进行注入。接下来，在每个 DAO 方法中，都可以从会话工厂中获取上下文会话并使用它进行对象持久化。

```
package com.apress.springrecipes.course.hibernate;

import com.apress.springrecipes.course.Course;
import com.apress.springrecipes.course.CourseDao;
import org.hibernate.Query;
import org.hibernate.Session;
import org.hibernate.SessionFactory;
import org.springframework.transaction.annotation.Transactional;

import java.util.List;

public class HibernateCourseDao implements CourseDao {

    private final SessionFactory sessionFactory;

    public HibernateCourseDao(SessionFactory sessionFactory) {
        this.sessionFactory=sessionFactory;
    }

    @Transactional
    public Course store(Course course) {
        Session session = sessionFactory.getCurrentSession();
        session.saveOrUpdate(course);
        return course;
    }

    @Transactional
    public void delete(Long courseId) {
        Session session = sessionFactory.getCurrentSession();
        Course course = session.get(Course.class, courseId);
        session.delete(course);
    }

    @Transactional(readOnly=true)
    public Course findById(Long courseId) {
        Session session = sessionFactory.getCurrentSession();
        return session.get(Course.class, courseId);
    }

    @Transactional(readOnly=true)
    public List<Course> findAll() {
        Session session = sessionFactory.getCurrentSession();
        return session.createQuery("from Course", Course.class).list();
    }
}
```

注意，所有 DAO 方法都必须是带有事务的。这是因为 Spring 是通过 Hibernate 的上下文会话支持集成 Hibernate 的。Spring 对 Hibernate 的 CurrentSessionContext 接口提供了自己的实现。它会尝试寻找事务并失败，然后提示没有 Hibernate 会话绑定到线程上。你可以通过对每个方法或是整个类使用@Transactional 注解来实现这一点。这确保了一个 DAO 方法中的持久化操作会在同一个事务也就是同一个会话中执行。此外，如果服务层组件的方法调用了多个 DAO 方法，同时它会将自己的事务传播给这些方法，那么所有这些 DAO 方法也都会在同一个会话中运行。

■ **警告：**在使用 Spring 配置 Hibernate 时，请确保不要设置 hibernate.current_session_context_class 属性，因为这会对 Spring 正确管理事务造成影响。只有在需要 JTA 事务时才应该设置该属性。

在 bean 配置文件中，你需要为该应用声明一个 HibernateTransactionManager 实例，并通过@Enable-

TransactionManagement 开启声明式事务管理。

```
@Configuration
@EnableTransactionManagement
public class CourseConfiguration {

    @Bean
    public CourseDao courseDao(SessionFactory sessionFactory) {
        return new HibernateCourseDao(sessionFactory);
    }
    @Bean
    public HibernateTransactionManager transactionManager(SessionFactory sessionFactory)
{
        return new HibernateTransactionManager(sessionFactory);
    }
}
```

不过，在调用 Hibernate 会话的原生方法时，抛出的异常是原生类型 HibernateException。如果想将 Hibernate 异常转换为 Spring 的 DataAccessException 以便进行一致的异常处理，你要对需要进行异常转换的 DAO 类使用@Repository 注解。

```
package com.apress.springrecipes.course.hibernate;
...
import org.springframework.stereotype.Repository;

@Repository
public class HibernateCourseDao implements CourseDao {
    ...
}
```

PersistenceExceptionTranslationPostProcessor 负责将原生 Hibernate 异常转换为 Spring DataAccess-Exception 体系中的数据访问异常。这个 bean 后置处理器只会对使用了@Repository 注解的 bean 进行异常转换。在使用基于 Java 的配置时，该 bean 会自动注册到 AnnotationConfigApplicationContext 中；因此不需要再显式声明它了。

在 Spring 中，@Repository 是个模板（stereotype）注解。通过使用该注解，组件类会通过组件扫描被自动检测到。你可以在该注解中指定一个组件名，让 Spring IoC 容器自动装配会话工厂。

```
package com.apress.springrecipes.course.hibernate;
...
import org.hibernate.SessionFactory;
import org.springframework.beans.factory.annotation.Autowired;
import org.springframework.stereotype.Repository;

@Repository("courseDao")
public class HibernateCourseDao implements CourseDao {

    private final SessionFactory sessionFactory;

    public HibernateCourseDao (SessionFactory sessionFactory) {
        this.sessionFactory = sessionFactory;
    }
    ...
}
```

接下来，只需添加@ComponentScan 注解并删除之前的 HibernateCourseDao bean 声明。

```
@Configuration
@EnableTransactionManagement
@ComponentScan("com.apress.springrecipes.course")
public class CourseConfiguration { ... }
```

9-9 使用 JPA 的上下文注入来持久化对象

问题提出

在 Java EE 环境中，Java EE 容器可以帮助你管理实体管理器并直接将它们注入到 EJB 组件中。EJB 组件只需使用注入的实体管理器执行持久化操作，而无须关注实体管理器的创建与事务管理问题。

解决方案

最初，@PersistenceContext 注解用于在 EJB 组件中进行实体管理器的注入。Spring 可以通过 bean 后置处理器来解释这个注解。它会通过该注解将实体管理器注入到属性中。Spring 可以确保单个事务中的所有持久化操作都会由同一个实体管理器来处理。

解释说明

要想使用上下文注入方式，你可以在 DAO 中声明一个实体管理器字段并将其注解为@PersistenceContext。Spring 会将实体管理器注入到该字段来持久化对象。

```
package com.apress.springrecipes.course.jpa;

import com.apress.springrecipes.course.Course;
import com.apress.springrecipes.course.CourseDao;
import org.springframework.transaction.annotation.Transactional;

import javax.persistence.EntityManager;
import javax.persistence.PersistenceContext;
import javax.persistence.TypedQuery;
import java.util.List;

public class JpaCourseDao implements CourseDao {

    @PersistenceContext
    private EntityManager entityManager;

    @Transactional
    public Course store(Course course) {
        return entityManager.merge(course);
    }

    @Transactional
    public void delete(Long courseId) {
        Course course = entityManager.find(Course.class, courseId);
        entityManager.remove(course);
    }

    @Transactional(readOnly = true)
    public Course findById(Long courseId) {
        return entityManager.find(Course.class, courseId);
    }

    @Transactional(readOnly = true)
    public List<Course> findAll() {
        TypedQuery<Course> query =
            entityManager.createQuery("select c from Course c", Course.class);
        return query.getResultList();
    }
}
```

你可以对每个 DAO 方法或是整个 DAO 类使用@Transactional 注解，以使所有方法都是事务性的。这确保了单个方法中的持久化操作会在同一个事务也就是同一个实体管理器中执行。

在 bean 配置文件中，你需要声明一个 JpaTransactionManager 实例并通过@EnableTransactionManagement 开启声明式事务管理。在使用基于 Java 的配置将实体管理器注入到被@PersistenceContext 所注解的属性时，PersistenceAnnotationBeanPostProcessor 实例会自动进行注册。

```
package com.apress.springrecipes.course.config;

import com.apress.springrecipes.course.CourseDao;
import com.apress.springrecipes.course.JpaCourseDao;
import org.apache.derby.jdbc.ClientDriver;
```

```
import org.hibernate.dialect.DerbyTenSevenDialect;
import org.springframework.context.annotation.Bean;
import org.springframework.context.annotation.Configuration;
import org.springframework.jdbc.datasource.SimpleDriverDataSource;
import org.springframework.orm.jpa.JpaTransactionManager;
import org.springframework.orm.jpa.JpaVendorAdapter;
import org.springframework.orm.jpa.LocalContainerEntityManagerFactoryBean;
import org.springframework.orm.jpa.vendor.HibernateJpaVendorAdapter;
import org.springframework.transaction.PlatformTransactionManager;
import org.springframework.transaction.annotation.EnableTransactionManagement;

import javax.sql.DataSource;

@Configuration
@EnableTransactionManagement
public class CourseConfiguration {

    @Bean
    public CourseDao courseDao() {
        return new JpaCourseDao();
    }

    @Bean
    public LocalContainerEntityManagerFactoryBean entityManagerFactory() {
        LocalContainerEntityManagerFactoryBean emf =
            new LocalContainerEntityManagerFactoryBean();
        emf.setDataSource(dataSource());
        emf.setJpaVendorAdapter(jpaVendorAdapter());
        return emf;
    }

    private JpaVendorAdapter jpaVendorAdapter() {
        HibernateJpaVendorAdapter jpaVendorAdapter = new HibernateJpaVendorAdapter();
        jpaVendorAdapter.setShowSql(true);
        jpaVendorAdapter.setGenerateDdl(true);
        jpaVendorAdapter.setDatabasePlatform(DerbyTenSevenDialect.class.getName());
        return jpaVendorAdapter;
    }

    @Bean
    public JpaTransactionManager transactionManager(EntityManagerFactory
    entityManagerFactory) {
        return new JpaTransactionManager(entityManagerFactory);
    }

    @Bean
    public DataSource dataSource() { ... }

}
```

PersistenceAnnotationBeanPostProcessor 还可以将实体管理器工厂注入到使用了@PersistenceUnit 注解的属性中。这样就可以自己创建实体管理器并管理事务了。这与通过 setter 方法注入实体管理器工厂没有区别。

```
package com.apress.springrecipes.course;
...
import javax.persistence.EntityManagerFactory;
import javax.persistence.PersistenceUnit;

public class JpaCourseDao implements CourseDao {
    @PersistenceContext
    private EntityManager entityManager;

    @PersistenceUnit
    private EntityManagerFactory entityManagerFactory;
    ...
}
```

当调用 JPA 实体管理器的原生方法时，抛出的异常是原生类型 PersistenceException 或其他 Java SE 异常，如 IllegalArgumentException 和 IllegalStateException。如果想将 JPA 异常转换为 Spring 的 DataAccessException，需要对 DAO 类使用@Repository 注解。

```
package com.apress.springrecipes.course;
...
import org.springframework.stereotype.Repository;

@Repository("courseDao")
public class JpaCourseDao implements CourseDao {
    ...
}
```

PersistenceExceptionTranslationPostProcessor 实例会将原生 JPA 异常转换为 Spring DataAccessException 异常体系中的异常。在使用基于 Java 的配置时，这个 bean 会自动注册到 AnnotationConfigApplicationContext 中；因此不需要显式声明它。

9-10 使用 Spring Data JPA 简化 JPA 操作

问题提出

编写数据访问代码是一件单调且重复性很高的任务，即便是使用 JPA 也是如此。你经常需要访问 EntityManager 或 EntityManagerFactory，并且要创建查询，更不必说当有大量 DAO 时为不同实体重复声明 findById 和 findAll 方法了。

解决方案

Spring Data JPA 可以让你像使用 Spring 本身一样专注在重要的地方，而不是完成任务的样板代码。它还为常用的数据访问方法提供了默认实现（如 findAll、delete 和 save 等）。

解释说明

要想使用 Spring Data JPA，你需要继承它的一个接口。这些接口会被检测到，并且仓库的一个默认实现会在运行期生成。在大多数情况下，继承 CrudRepository<T, ID>接口就足够了。

```
package com.apress.springrecipes.course;

import com.apress.springrecipes.course.Course;
import org.springframework.data.repository.CrudRepository;

public interface CourseRepository extends CrudRepository<Course, Long>{}
```

这足以为 Course 实体完成所有必要的 CRUD 动作。在继承 Spring Data 接口时，你需要指定类型 Course 和主键类型 Long。在运行期生成仓库时需要这些信息。

■ **注意：**还可以继承 JpaRepository，它添加了一些特定于 JPA 的方法（flush 和 saveAndFlush 等）并提供了带有分页/排序能力的查询方法。

接下来需要开启 Spring Data 仓库的检测。要想做到这一点，你可以使用 Spring Data JPA 提供的@EnableJpaRepositories 注解。

```
@Configuration
@EnableTransactionManagement
@EnableJpaRepositories("com.apress.springrecipes.course")
public class CourseConfiguration { ... }
```

这会启用 Spring Data JPA 并构建出一个可用的仓库。默认情况下，所有的仓库方法都会被标记为@Transactional，因此无须再添加其他注解了。

现在，可以使用 Main 类从 Spring IoC 容器中取出 CourseRepository 实例并对其进行测试了。

```
package com.apress.springrecipes.course.datajpa;
...
import org.springframework.context.ApplicationContext;
import org.springframework.context.support.ClassPathXmlApplicationContext;
```

```java
public class Main {

    public static void main(String[] args) {
        ApplicationContext context =
            new AnnotationConfigApplicationContext(CourseConfiguration.class);

        CourseRepository repository = context.getBean(CourseRepository.class);
        ...
    }
}
```

诸如异常转换、事务管理和 EntityManagerFactory 的轻松配置等事项依然会应用到基于 Spring Data JPA 的仓库上。它只是简化了你的工作并能让你专注在重要的地方。

小结

本章介绍了如何使用 Spring 针对 JDBC、Hibernate 和 JPA 的支持。我们介绍了如何配置 DataSource 对象来连接数据库，以及如何使用 Spring 的 JdbcTemplate 与 NamedParameterJdbcTemplate 对象减少单调的样板代码的处理。我们还介绍了如何使用辅助父类通过 JDBC 和 Hibernate 来构建 DAO 类，以及如何通过 Spring 对模板（stereotype）注解和组件扫描的支持来轻松构建新的 DAO 与服务。最后一节介绍了如何凭借 Spring Data JPA 的能力来极大简化数据访问代码。第 10 章将会介绍如何在 Spring 中通过事务（比如说针对 JMS 或数据库）来确保服务中状态的一致性。

第 10 章

Spring 事务管理

本章将会介绍事务的基本概念以及 Spring 在事务管理领域提供的能力。事务管理在企业应用中是一项必要的技术，它可以确保数据的完整性与一致性。作为企业应用框架的 Spring，在不同的事务管理 API 之上提供了一个抽象层。作为应用开发者，你可以使用 Spring 的事务管理设施而无须更多地了解底层的事务管理 API。

就像 EJB 中的 bean 管理事务（BMT）与容器管理事务（CMT）方式一样，Spring 支持编程式与声明式事务管理。Spring 的事务管理的目标在于通过向 POJO 添加事务能力来提供 EJB 事务的一种替代方案。

编程式事务管理是通过将事务管理代码嵌入到业务方法中以控制事务的提交与回滚来实现的。如果方法正常完成就需要提交事务，如果方法抛出了某种异常就需要回滚事务。借助于编程式事务管理，可以定义自己的规则来提交和回滚事务。

不过，在以编程的方式管理事务时，需要在每个事务性操作中引入事务管理代码。这样，每种操作都会有样板式的事务代码。此外，针对不同应用开启和关闭事务管理是很困难的事情。如果对 AOP 理解得很好的话，你可能已经发现事务管理就是一种横切关注点。

在大多数情况下，声明式事务管理要优于编程式事务管理。这是通过声明的方式分离事务管理代码与业务方法来实现的。作为一种横切关注点，事务管理可通过 AOP 的方式进行模块化。Spring 是通过 Spring AOP 框架来支持声明式事务管理的。这可以帮助你更加轻松地为应用开启事务并定义一致的事务策略。声明式事务管理的灵活性要比编程式事务管理差一些。

编程式事务管理可以通过代码来控制事务——显式进行开启、提交并将它们合并到一起。你可以指定一套事务属性以一种较好的粒度来定义事务。Spring 支持的事务属性包括传播行为、隔离级别、回滚规则、事务超时，以及是否是只读事务。这些属性可以让你进一步定制事务的行为。

学习完本章内容后，你就可以在应用中使用不同的事务管理策略了。此外，你还会熟悉不同的事务属性并且可以使用它们来定义事务。

在某些情况下，编程式事务管理是一种不错的做法，比如说你觉得添加 Spring 代理是得不偿失的，或是无法忍受它所带来的微小的性能损失。你可以自己访问原生的事务并手工控制事务。避免 Spring 代理的代价的一种更为便捷的做法是使用 TransactionTemplate 类，它提供了一个模板方法，事务边界会围绕着这个方法开启，然后提交。

10-1 使用事务管理来避免问题

事务管理是企业应用开发中的一项必要技术，它可以确保数据的完整性与一致性。如果没有事务管理，那么数据和资源可能会损坏并处于不一致的状态。事务管理对于在并发和分布式环境下从意外错误中进行恢复是至关重要的。

简而言之，所谓事务，就是将一系列动作看作是一个工作单元。这些动作要么全部完成，要么完全不生效。如果所有动作都没问题，那么事务就会永久提交。相反，如果任何一个动作出现了问题，那么事务就会回滚到初始状态，就好像什么都没有发生一样。

可以用 4 个关键的属性来描述事务的概念：原子性、一致性、隔离性与持久性（ACID）。

- 原子性：事务是个包含了一系列动作的原子性操作。事务的原子性可以确保动作要么全部完成，要么完全不生效。
- 一致性：一旦事务的所有动作都完成，事务就会提交。接下来，数据与资源将会处于符合业务规则的一致性状态。
- 隔离性：由于同一时刻可能会有多个事务在处理相同的数据集，因此事务之间会彼此隔离以防止数据损坏。
- 持久性：一旦事务完成，其结果就应该是持久的，可以经受任何系统失败（想象一下，如果在事务提交的过程中机器电源突然断掉了）。通常情况下，事务的结果会被写到持久化存储中。

为了理解事务管理的重要性，我们从一个示例开始，这个示例是关于从在线书店购买图书的。首先，你需要在数据库中为该应用创建一个新的模式。我们选择使用 PostgreSQL 作为数据库供这些示例使用。本章的源代码包含了一个 bin 目录，里面有两个脚本：一个（postgres.sh）会下载 Docker 容器并开启一个默认 Postgres 实例；另一个（psql.sh）会连接到运行着的 Postgres 实例。表 10-1 列出了 Java 应用中用到的连接属性。

注意：本章的示例代码在 bin 目录中提供了脚本来启动并连接到基于 Docker 的 PostgreSQL 实例。要想启动实例并创建数据库，请按照如下步骤进行。

1. 执行 bin\postgres.sh，它会下载并启动 Postgres Docker 容器。
2. 执行 bin\psql.sh，它会连接到运行着的 Postgres 容器。
3. 执行 CREATE DATABASE bookstore，创建示例所用的数据库。

表 10-1　连接到应用数据库所需的 JDBC 属性

属性	值
驱动类	org.postgresql.Driver
URL	jdbc:postgresql://localhost:5432/bookstore
用户名	postgres
密码	password

对于这个书店应用来说，你需要一个地方来存储数据。我们会创建一个简单的数据库来管理图书和账户。图 10-1 展示了表的实体关系（ER）图。

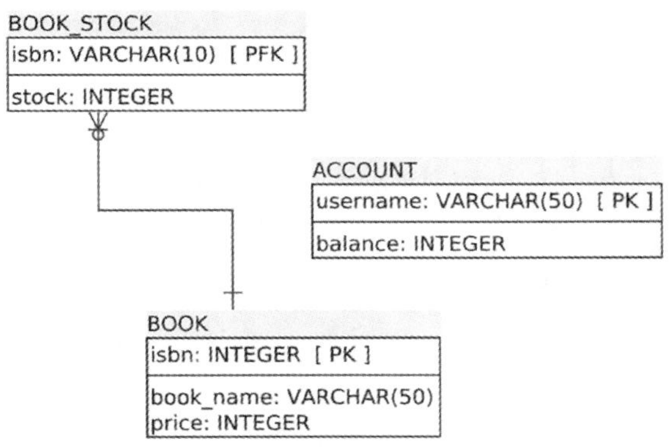

图 10-1　BOOK_STOCK 描述了有多少给定的 BOOK

现在，我们为上述模型创建 SQL。执行 bin\psql.sh 命令，连接到运行着的容器并打开 psql 工具。将如下 SQL 粘贴到 shell 中并验证是否成功：

```
CREATE TABLE BOOK (
    ISBN         VARCHAR(50)    NOT NULL,
    BOOK_NAME    VARCHAR(100)   NOT NULL,
    PRICE        INT,
    PRIMARY KEY (ISBN)
);

CREATE TABLE BOOK_STOCK (
    ISBN      VARCHAR(50)     NOT NULL,
    STOCK     INT             NOT NULL,
    PRIMARY KEY (ISBN),
    CONSTRAINT positive_stock CHECK (STOCK >= 0)
);

CREATE TABLE ACCOUNT (
    USERNAME    VARCHAR(50)    NOT NULL,
    BALANCE     INT            NOT NULL,
    PRIMARY KEY (USERNAME),
    CONSTRAINT positive_balance CHECK (BALANCE >= 0)
);
```

真实世界的这种应用可能会将 price 字段设为 decimal 类型，不过使用 int 会让程序更简单一些，所以就使用 int 吧。

BOOK 表存储了基本的图书信息，如名字和价格，并且将图书的 ISBN 作为主键。BOOK_STOCK 表会记录每本书的库存情况。stock 值使用了一个 CHECK 约束，它被限制为一个正数。虽然 CHECK 约束类型是在 SQL-99 中定义的，但并不是所有的数据库引擎都支持它。在本书编写之际，只有 MySQL 尚不支持，Sybase、Derby、HSQL、Oracle、DB2、SQL Server、Access、PostgreSQL 和 FireBird 都已经支持它了。如果你所用的数据库引擎不支持 CHECK 约束，那么请查阅文档了解类似的约束支持。最后，ACCOUNT 表存储了客户账户和余额。balance 也被限制为正数。

对书店的操作定义在如下 BookShop 接口中。现在只有一个操作：purchase()。

```
package com.apress.springrecipes.bookshop;

public interface BookShop {

    void purchase(String isbn, String username);
}
```

由于要使用 JDBC 实现该接口，因此需要创建如下 JdbcBookShop 类。为了更好地理解事务的本质，我们不借助 Spring 的 JDBC 支持来实现这个类。

```
package com.apress.springrecipes.bookshop;

import java.sql.Connection;
import java.sql.PreparedStatement;
import java.sql.ResultSet;
import java.sql.SQLException;

import javax.sql.DataSource;

public class JdbcBookShop implements BookShop {
    private DataSource dataSource;

    public void setDataSource(DataSource dataSource) {
        this.dataSource = dataSource;
    }

    public void purchase(String isbn, String username) {
        Connection conn = null;
        try {
            conn = dataSource.getConnection();
```

```
            PreparedStatement stmt1 = conn.prepareStatement(
                "SELECT PRICE FROM BOOK WHERE ISBN = ?");
            stmt1.setString(1, isbn);
            ResultSet rs = stmt1.executeQuery();
            rs.next();
            int price = rs.getInt("PRICE");
            stmt1.close();

            PreparedStatement stmt2 = conn.prepareStatement(
                "UPDATE BOOK_STOCK SET STOCK = STOCK - 1 "+
                "WHERE ISBN = ?");
            stmt2.setString(1, isbn);
            stmt2.executeUpdate();
            stmt2.close();

            PreparedStatement stmt3 = conn.prepareStatement(
                "UPDATE ACCOUNT SET BALANCE = BALANCE - ? "+
                "WHERE USERNAME = ?");
            stmt3.setInt(1, price);
            stmt3.setString(2, username);
            stmt3.executeUpdate();
            stmt3.close();
        } catch (SQLException e) {
            throw new RuntimeException(e);
        } finally {
            if (conn != null) {
                try {
                    conn.close();
                } catch (SQLException e) {}
            }
        }
    }
}
```

对于 purchase()操作来说，一共需要执行 3 条 SQL 语句。第一条是查询图书价格。第二条与第三条会更新图书库存与相应的账户余额。接下来，你可以在 Spring IoC 容器中声明一个 bookshop 实例来提供购买服务。为了简化，你可以使用 DriverManagerDataSource，它会在每次请求时打开一个新的数据库连接。

注意： 要想访问 PostgreSQL 数据库，你需要将 Postgres 客户端库添加到 CLASSPATH 中。

```
package com.apress.springrecipes.bookshop.config;

import com.apress.springrecipes.bookshop.BookShop;
import com.apress.springrecipes.bookshop.JdbcBookShop;
import org.springframework.context.annotation.Bean;
import org.springframework.context.annotation.Configuration;
import org.springframework.jdbc.datasource.DriverManagerDataSource;

import javax.sql.DataSource;

@Configuration
public class BookstoreConfiguration {

    @Bean
    public DataSource dataSource() {
        DriverManagerDataSource dataSource = new DriverManagerDataSource();
        dataSource.setDriverClassName(org.postgresql.Driver.class.getName());
        dataSource.setUrl("jdbc:postgresql://localhost:5432/bookstore");
        dataSource.setUsername("postgres");
        dataSource.setPassword("password");
        return dataSource;
    }
```

```
    @Bean
    public BookShop bookShop() {
        JdbcBookShop bookShop = new JdbcBookShop();
        bookShop.setDataSource(dataSource());
        return bookShop;
    }
}
```

为了说明不使用事务管理会出现的问题，假设将表 10-2～表 10-4 中的数据输入到书店数据库中。

表 10-2　　用于测试事务的 BOOK 表中的示例数据

ISBN	BOOK_NAME	PRICE
0001	The First Book	30

表 10-3　　用于测试事务的 BOOK_STOCK 表中的示例数据

ISBN	STOCK
0001	10

表 10-4　　用于测试事务的 ACCOUNT 表中的示例数据

USERNAME	BALANCE
user1	20

接下来，编写如下 Main 类，让用户 user1 购买 ISBN 为 0001 的图书。由于用户的账户中只有$20，因此金额是不够买书的。

```
package com.apress.springrecipes.bookshop;

import com.apress.springrecipes.bookshop.config.BookstoreConfiguration;
import org.springframework.context.ApplicationContext;
import org.springframework.context.annotation.AnnotationConfigApplicationContext;

public class Main {

    public static void main(String[] args) throws Throwable {

        ApplicationContext context =
            new AnnotationConfigApplicationContext(BookstoreConfiguration.class);

        BookShop bookShop = context.getBean(BookShop.class);
        bookShop.purchase("0001", "user1");

    }
}
```

当运行这个应用时，你会遇到一个 SQLException，这是因为违背了 ACCOUNT 表的 CHECK 约束。这是我们所期望的结果，因为使用的金额超出了账户余额。

不过，如果在 BOOK_STOCK 表中检查这本书的库存，你会发现库存意外地被这个失败的操作扣减了！原因在于在第三条语句抛出异常前已经执行第二条 SQL 语句扣减了库存。

可以看到，缺少事务管理会导致数据处于不一致的状态。为了避免这种不一致性，purchase()操作的 3 条 SQL 语句应该在一个事务中执行。一旦事务中的任何一个动作失败，整个事务应该回滚，以取消已执行动作所做的所有变更。

使用 JDBC 提交与回滚来管理事务

在使用 JDBC 更新数据库时，默认情况下，每条 SQL 语句都会在执行后立刻提交。这种行为叫做自动提交。不过，它不允许你管理操作的事务。JDBC 支持原生的事务管理策略，可以通过显式调用连接的 commit()与 rollback()方法来实现。不过，在这么做之前，需要关闭自动提交；默认情况下自动提交是开启的。

```
package com.apress.springrecipes.bookshop;
...
public class JdbcBookShop implements BookShop {
    ...
    public void purchase(String isbn, String username) {
        Connection conn = null;
        try {
            conn = dataSource.getConnection();
            conn.setAutoCommit(false);
            ...
            conn.commit();
        } catch (SQLException e) {
            if (conn != null) {
                try {
                    conn.rollback();
                } catch (SQLException e1) {}
            }
            throw new RuntimeException(e);
        } finally {
            if (conn != null) {
                try {
                    conn.close();
                } catch (SQLException e) {}
            }
        }
    }
}
```

可以通过调用 setAutoCommit()方法来修改数据库连接的自动提交行为。默认情况下，自动提交是开启的，它会在每条 SQL 语句执行后立刻提交。要想启用事务管理，你需要关闭这种默认行为，只在所有 SQL 语句都成功执行时才提交连接。如果任何一条语句出错了，你必须要回滚该连接所做的所有修改。

现在，如果再次运行该应用，当用户的余额不足以购买图书时，图书库存并不会扣减。

虽然可以通过显式提交与回滚 JDBC 连接来管理事务，但需要为不同的方法重复编写不少样板代码。此外，代码是特定于 JDBC 的，如果选择了其他数据访问技术还需要修改代码。Spring 的事务支持提供了一套不依赖于特定技术的设施，包括事务管理（如 org.springframework.transaction. PlatformTransactionManager）、事务模板（如 org.springframework.transaction.support.Transaction Template）和事务声明支持，从而简化了事务管理任务。

10-2 选择一种事务管理器实现

问题提出

通常，如果应用只涉及单个数据源，那就只需调用数据库连接的 commit()与 rollback()方法来管理事务。不过，如果事务跨越了多个数据源或是你想要使用 Java EE 应用服务器所提供的事务管理能力，那么可以选择 Java Transaction API（JTA）。此外，你可能还需要针对不同的对象关系映射框架（如 Hibernate 和 JPA）调用不同的私有事务 API。

这样，你需要针对不同技术使用不同的事务 API。从一套 API 切换至另外一套 API 是一件很困难的事情。

解决方案

Spring 从不同的事务管理 API 中抽象出了一套通用的事务设施。作为应用开发者，你只需使用 Spring 的事务设施即可，无须了解关于底层事务 API 的更多信息。借助于这些设施，事务管理代码将会独立于任何具体的事务技术。

Spring 的核心事务管理抽象基于接口 PlatformTransactionManager。它为事务管理封装了一套独立于技术的方法。请记住一点，在 Spring 中，无论选择哪一种事务管理策略（编程式或声明式），都需要一个事务管理器。PlatformTransactionManager 接口提供了如下 3 个方法来处理事务。

- TransactionStatus getTransaction(TransactionDefinition definition) throws TransactionException

- void commit(TransactionStatus status) throws TransactionException
- void rollback(TransactionStatus status) throws TransactionException

解释说明

PlatformTransactionManager 是面向所有 Spring 事务管理器的通用接口。Spring 内建了该接口的几个实现，这些实现使用了不同的事务管理 API。

- 如果应用中只使用了单个数据源并使用 JDBC 访问，那么 DataSourceTransactionManager 会满足你的需求。
- 如果在 Java EE 应用服务器中使用 JTA 进行事务管理，那就应该使用 JtaTransactionManager 从应用服务器查找事务。此外，JtaTransactionManager 还适合于分布式事务（跨越多个资源的事务）。注意，虽然我们经常使用 JTA 事务管理器来集成应用服务器的事务管理器，但也可以使用独立的 JTA 事务管理器，如 Atomikos。
- 如果使用对象关系映射框架访问数据库，那就应该选择该框架对应的事务管理器，如 HibernateTransactionManager 或 JpaTransactionManager。

图 10-2 所示为 Spring 中 PlatformTransactionManager 接口的常见实现。

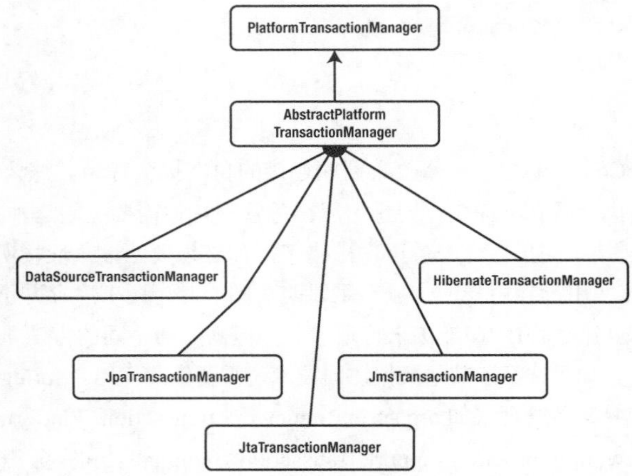

图 10-2　PlatformTransactionManager 接口的常见实现

事务管理器是作为一个普通的 bean 声明在 Spring IoC 容器中的。比如说，如下 bean 配置声明了一个 DataSourceTransactionManager 实例。需要设置它的 dataSource 属性，这样它就可以管理该数据源所创建的连接的事务了。

```
@Bean
public DataSourceTransactionManager transactionManager() {
    DataSourceTransactionManager transactionManager = new DataSourceTransactionManager()
    transactionManager.setDataSource(dataSource());
    return transactionManager;
}
```

10-3　使用事务管理器 API 以编程的方式管理事务

问题提出

你需要在业务方法中精确控制何时提交和回滚事务，但又不想直接使用底层事务 API。

解决方案

Spring 的事务管理器提供了一个独立于具体技术的 API，你可以通过调用 getTransaction()方法开启新的事务（或是获取到当前活动的事务），并通过调用 commit()和 rollback()方法来管理事务。由于 PlatformTransactionManager 是个事务管理的抽象单元，因此你对事务管理所调用的方法可以确保是独立于具体技术的。

解释说明

为了说明如何使用事务管理器 API，我们新建一个类 TransactionalJdbcBookShop，它会使用 Spring JDBC 模板。由于它需要用到事务管理器，因此我们向该类添加一个类型为 PlatformTransactionManager 的属性，并通过 setter 方法进行注入。

```
package com.apress.springrecipes.bookshop;

import org.springframework.dao.DataAccessException;
import org.springframework.jdbc.core.support.JdbcDaoSupport;
import org.springframework.transaction.PlatformTransactionManager;
import org.springframework.transaction.TransactionDefinition;
import org.springframework.transaction.TransactionStatus;
import org.springframework.transaction.support.DefaultTransactionDefinition;

public class TransactionalJdbcBookShop extends JdbcDaoSupport implements BookShop {

    private PlatformTransactionManager transactionManager;

    public void setTransactionManager(PlatformTransactionManager transactionManager) {
        this.transactionManager = transactionManager;
    }

    public void purchase(String isbn, String username) {
        TransactionDefinition def = new DefaultTransactionDefinition();
        TransactionStatus status = transactionManager.getTransaction(def);

        try {
            int price = getJdbcTemplate().queryForObject(
                "SELECT PRICE FROM BOOK WHERE ISBN = ?", Integer.class, isbn);

            getJdbcTemplate().update(
                "UPDATE BOOK_STOCK SET STOCK = STOCK - 1 WHERE ISBN = ?", isbn);

            getJdbcTemplate().update(
                "UPDATE ACCOUNT SET BALANCE = BALANCE - ? WHERE USERNAME = ?",
                price, username);
            transactionManager.commit(status);
        } catch (DataAccessException e) {
            transactionManager.rollback(status);
            throw e;
        }
    }

}
```

在开启新事务前，需要在类型为 TransactionDefinition 的事务定义对象中指定事务属性。对于该示例来说，只需创建一个 DefaultTransactionDefinition 的实例并使用默认的事务属性。

拥有了事务定义后就可以通过调用 getTransaction()方法让事务管理器使用该定义开启一个新事务了。接下来，它会返回一个 TransactionStatus 对象来记录事务状态。如果所有语句都成功执行，那就通过传递事务状态让事务管理器提交该事务。由于 Spring JDBC 模板抛出的所有异常都是 DataAccessException 的子类，因此当捕获到这种异常时就需要让事务管理器回滚事务。

该类中声明了通用类型 PlatformTransactionManager 的事务管理器属性。现在，你需要注入一个恰当的事务管理器实现。由于现在只使用了单个数据源并通过 JDBC 访问，因此应当选择 DataSourceTransactionManager。这里还装配了一个 dataSource 对象，因为 TransactionalJdbcBookShop 类是 Spring JdbcDaoSupport 类的一个子类，需要用到这个属性。

```
@Configuration
public class BookstoreConfiguration {
...
```

```
    @Bean
    public DataSourceTransactionManager transactionManager() {
        DataSourceTransactionManager transactionManager =
            new DataSourceTransactionManager();
        transactionManager.setDataSource(dataSource());
        return transactionManager;
    }

    @Bean
    public BookShop bookShop() {
        TransactionalJdbcBookShop bookShop = new TransactionalJdbcBookShop();
        bookShop.setDataSource(dataSource());
        bookShop.setTransactionManager(transactionManager());
        return bookShop;
    }
}
```

10-4 使用事务模板以编程的方式管理事务

问题提出

假设一个业务方法代码块（而非整个方法体）有如下的事务需求：

- 在代码块起始位置处开启新的事务；
- 在代码块成功完成后提交事务；
- 如果代码块抛出异常则回滚事务。

如果直接调用 Spring 的事务管理器 API，那么事务管理代码可以以一种独立于具体技术的方式编写。不过，你并不想在每个类似的代码块中都重复这种样板代码。

解决方案

与 JDBC 模板一样，Spring 还提供了一个 TransactionTemplate 来帮助你控制整个事务管理过程和事务异常处理。你只需将代码块封装到一个实现了 TransactionCallback<T>接口的回调类中并将其传递给 TransactionTemplate 的 execute 方法执行即可。通过这种方式，你无须再重复编写样板事务管理代码。Spring 提供的模板对象是轻量级的，通常可被丢弃或是重新创建而不会产生性能上的影响。比如说，我们可以随时通过 DataSource 引用来重新创建 JDBC 模板，也可以通过事务管理器引用来重新创建 TransactionTemplate，当然，也可以在 Spring 应用上下文中创建。

解释说明

TransactionTemplate 是通过事务管理器创建的，就像 JDBC 模板是通过数据源创建的一样。事务模板会执行封装了事务性代码块的事务回调对象。可以通过单独一个类或是内部类的形式实现回调接口。如果以内部类来实现，那就需要将方法参数标记为 final 才可以访问。

```
package com.apress.springrecipes.bookshop;

import org.springframework.jdbc.core.support.JdbcDaoSupport;
import org.springframework.transaction.PlatformTransactionManager;
import org.springframework.transaction.TransactionStatus;
import org.springframework.transaction.support.TransactionCallbackWithoutResult;
import org.springframework.transaction.support.TransactionTemplate;

public class TransactionalJdbcBookShop extends JdbcDaoSupport implements BookShop {

    private PlatformTransactionManager transactionManager;

    public void setTransactionManager(PlatformTransactionManager transactionManager) {
        this.transactionManager = transactionManager;
    }

    public void purchase(final String isbn, final String username) {
```

```
        TransactionTemplate transactionTemplate =
            new TransactionTemplate(transactionManager);

        transactionTemplate.execute(new TransactionCallbackWithoutResult() {

            protected void doInTransactionWithoutResult(
                TransactionStatus status) {

                int price = getJdbcTemplate().queryForObject(
                    "SELECT PRICE FROM BOOK WHERE ISBN = ?", Integer.class, isbn);

                getJdbcTemplate().update(
                    "UPDATE BOOK_STOCK SET STOCK = STOCK - 1 WHERE ISBN = ?", isbn );

                getJdbcTemplate().update(
                    "UPDATE ACCOUNT SET BALANCE = BALANCE - ? WHERE USERNAME = ?",
                    price, username);
            }
        });
    }
}
```

TransactionTemplate 可以接受一个事务回调对象，该对象要么实现 TransactionCallback<T>接口，要么实现框架所提供的 TransactionCallback<T>接口的子接口 TransactionCallbackWithoutResult。对于 purchase()方法中用于扣减图书库存和账户余额的代码块来说，不需要返回结果，因此使用 TransactionCallbackWithoutResult 会很恰当。对于拥有返回值的代码块来说，应该实现 TransactionCallback<T>接口。回调对象的返回值最终通过模板的 T execute()方法返回。这么做主要的好处在于剥离了开启、回滚与提交事务的职责。

在回调对象的执行过程中，如果抛出未检查异常（比如说 RuntimeException 和 DataAccessException）或是在 doInTransactionWithoutResult 方法中显式调用了 TransactionStatus 参数的 setRollbackOnly()，那么事务就会回滚，否则，当回调对象执行完毕后事务就会提交。

在 bean 配置文件中，bookshop bean 依然需要一个事务管理器来创建 TransactionTemplate。

```
@Configuration
public class BookstoreConfiguration {
...
    @Bean
    public DataSourceTransactionManager transactionManager() {
        DataSourceTransactionManager transactionManager =
            new DataSourceTransactionManager();
        transactionManager.setDataSource(dataSource());
        return transactionManager;
    }

    @Bean
    public BookShop bookShop() {
        TransactionalJdbcBookShop bookShop = new TransactionalJdbcBookShop();
        bookShop.setDataSource(dataSource());
        bookShop.setTransactionManager(transactionManager());
        return bookShop;
    }
}
```

还可以让 IoC 容器注入事务模板而不是直接创建。由于事务模板会处理所有事务，因此类中无须再引用事务管理器。

```
package com.apress.springrecipes.bookshop;
...
import org.springframework.transaction.support.TransactionTemplate;

public class TransactionalJdbcBookShop extends JdbcDaoSupport implements
    BookShop {
```

```
    private TransactionTemplate transactionTemplate;

    public void setTransactionTemplate(
        TransactionTemplate transactionTemplate) {
        this.transactionTemplate = transactionTemplate;
    }

    public void purchase(final String isbn, final String username) {
        transactionTemplate.execute(new TransactionCallbackWithoutResult() {
            protected void doInTransactionWithoutResult(TransactionStatus status) {
                ...
            }
        });
    }
}
```

接下来在 bean 配置文件中定义一个事务模板并将其（而不是事务管理器）注入到 bookshop bean 中。注意，事务模板实例可用在多个事务性 bean 上，因为它是线程安全的对象。最后，不要忘记为事务模板设置事务管理器属性。

```
package com.apress.springrecipes.bookshop.config;
...
import org.springframework.transaction.support.TransactionTemplate;

@Configuration
public class BookstoreConfiguration {
...
    @Bean
    public DataSourceTransactionManager transactionManager() { ... }

    @Bean
    public TransactionTemplate transactionTemplate() {
        TransactionTemplate transactionTemplate = new TransactionTemplate();
        transactionTemplate.setTransactionManager(transactionManager());
        return transactionTemplate;
    }

    @Bean
    public BookShop bookShop() {
        TransactionalJdbcBookShop bookShop = new TransactionalJdbcBookShop();
        bookShop.setDataSource(dataSource());
        bookShop.setTransactionTemplate(transactionTemplate());
        return bookShop;
    }
}
```

10-5 使用@Transactional 注解以声明的方式管理事务

问题提出

在 bean 配置文件中声明事务时，需要掌握 AOP 相关的概念，如切点、通知和通知器。缺乏这方面知识的开发者会觉得开启声明式事务管理是一件很难的事情。

解决方案

Spring 可以通过对事务性方法使用@Transactional 注解并将@EnableTransactionManagement 注解添加到配置类上来轻松声明事务。

解释说明

要想让方法成为事务性的，只需为其添加注解@Transactional 即可。注意，只应该注解 public 方法，这是由 Spring AOP 基于代理的限制所导致的。

```
package com.apress.springrecipes.bookshop;

import org.springframework.jdbc.core.support.JdbcDaoSupport;
```

```
import org.springframework.transaction.annotation.Transactional;

public class JdbcBookShop extends JdbcDaoSupport implements BookShop {

    @Transactional
    public void purchase(final String isbn, final String username) {

        int price = getJdbcTemplate().queryForObject(
            "SELECT PRICE FROM BOOK WHERE ISBN = ?", Integer.class, isbn);

        getJdbcTemplate().update(
            "UPDATE BOOK_STOCK SET STOCK = STOCK - 1 WHERE ISBN = ?", isbn);

        getJdbcTemplate().update(
            "UPDATE ACCOUNT SET BALANCE = BALANCE - ? WHERE USERNAME = ?", price, username);
    }
}
```

注意，由于继承了 JdbcDaoSupport，因此不需要为 DataSource 提供 setter，将其从 DAO 类中移除。

可以在方法级别或是类级别上应用@Transactional 注解。在将该注解应用到类上时，该类中的所有 public 方法都会是事务性的。虽然也可以将@Transactional 注解应用到接口或是接口中的方法声明上，但并不推荐这么做，因为在基于类的代理（比如说 CGLIB 代理）中这么做是不行的。

在 Java 配置类中，只需要添加@EnableTransactionManagement 注解即可，这就是你要做的全部工作。对于 IoC 容器所声明的 bean 来说，Spring 会对使用了@Transactional 注解的方法，以及使用@Transactional 注解的类中的方法应用通知。这样，Spring 就可以为这些方法管理事务。

```
@Configuration
@EnableTransactionManagement
public class BookstoreConfiguration { ... }
```

10-6 设置传播事务属性

问题提出

当一个事务性方法被另一个方法调用时，我们就需要指定事务是如何传播的。比如说，方法可以在既有事务中继续运行，或是开启一个新事务并在自己的事务中运行。

解决方案

可以通过传播事务属性来指定事务的传播行为。Spring 定义了 7 种传播行为，如表 10-5 所示。这些行为定义在 org.springframework.transaction.TransactionDefinition 接口中。注意，并非所有类型的事务管理器都支持全部这些传播行为。它们的行为取决于底层资源。比如说，数据库会支持不同的隔离级别，这约束了事务管理器所能支持的传播行为。

表 10-5　Spring 支持的传播行为

传播	说明
REQUIRED	如果过程中存在事务，那么当前方法就应该运行在该事务中；否则，它应该开启一个新事务并运行在自己的事务中
REQUIRES_NEW	当前方法必须要开启一个新事务并运行在自己的事务中。如果过程中存在事务，那么应该将其挂起
SUPPORTS	如果过程中存在事务，那么当前方法可以运行在该事务中；否则，没必要运行在事务中
NOT_SUPPORTED	当前方法不应该运行在事务中。如果过程中存在事务，那么应该将其挂起
MANDATORY	当前方法必须运行在事务中。如果过程中没有事务，那就会抛出异常
NEVER	当前方法不应该运行在事务中。如果过程中存在事务，那就会抛出异常

续表

传播	说明
NESTED	如果过程中存在事务，那么当前方法应该运行在该事务的嵌套事务中（由 JDBC 3.0 安全点特性支持）；否则，应该开启一个新事务并运行在自己的事务中。该特性只有 Spring 中才存在（之前的传播行为在 Java EE 事务传播中都有）。该行为对于批处理等情况很有用，假如你有一个长时间运行的处理（比如说处理 100 万条记录），你想要对批处理的提交进行截断。这样，每 1 万条记录提交一次。如果出错了，你可以回滚嵌套事务，这将只会丢失 1 万条记录（而不是全部的 100 万条记录）

解释说明

事务传播发生在一个事务性方法被另一个方法调用的时候。比如说，假设客户在书店收银员那里对购买的所有书结账。为了支持这个操作，定义如下 Cashier 接口：

```
package com.apress.springrecipes.bookshop;
...
public interface Cashier {

    public void checkout(List<String> isbns, String username);
}
```

可以将购买委托给 bookshop bean，多次调用其 purchase()方法来实现该接口。注意，由于使用了@Transactional 注解，因此 checkout()方法是事务性的。

```
package com.apress.springrecipes.bookshop;
...
import org.springframework.transaction.annotation.Transactional;

public class BookShopCashier implements Cashier {

    private BookShop bookShop;

    public void setBookShop(BookShop bookShop) {
        this.bookShop = bookShop;
    }

    @Transactional
    public void checkout(List<String> isbns, String username) {
        for (String isbn : isbns) {
            bookShop.purchase(isbn, username);
        }
    }
}
```

接下来，在 bean 配置文件中定义一个 cashier bean，并引用 bookshop bean 来购买图书。

```
@Configuration
@EnableTransactionManagement()
public class BookstoreConfiguration {
...

    @Bean
    public Cashier cashier() {
        BookShopCashier cashier = new BookShopCashier();
        cashier.setBookShop(bookShop());
        return cashier;
    }
}
```

为了说明事务的传播行为，在 bookshop 数据库中输入表 10-6～表 10-8 中的数据。

表 10-6　　用于测试传播行为的 BOOK 表中的示例数据

ISBN	BOOK_NAME	PRICE
0001	The First Book	30
0002	The Second Book	50

表 10-7　　用于测试传播行为的 BOOK_STOCK 表中的示例数据

ISBN	STOCK
0001	10
0002	10

表 10-8　　用于测试传播行为的 ACCOUNT 表中的示例数据

USERNAME	BALANCE
user1	40

使用 REQUIRED 传播行为

当用户 user1 从收银员那里结账两本书时，余额够买第一本，但不够第二本。

```
package com.apress.springrecipes.bookshop.spring;
...
public class Main {

    public static void main(String[] args) {
        ...
        Cashier cashier = context.getBean(Cashier.class);
        List<String> isbnList = Arrays.asList(new String[] { "0001", "0002"});
        cashier.checkout(isbnList, "user1");
    }
}
```

当 bookshop 的 purchase()方法被另一个事务性方法调用时（比如说 checkout()），默认情况下它会运行在既有事务中。这个默认的传播行为叫作 REQUIRED。这意味着只会有一个事务，其边界是 checkout()方法的起始位置和结束位置。该事务只会在 checkout()方法结束位置处被提交。这样，用户一本书也买不了。

图 10-3 所示为 REQUIRED 传播行为。

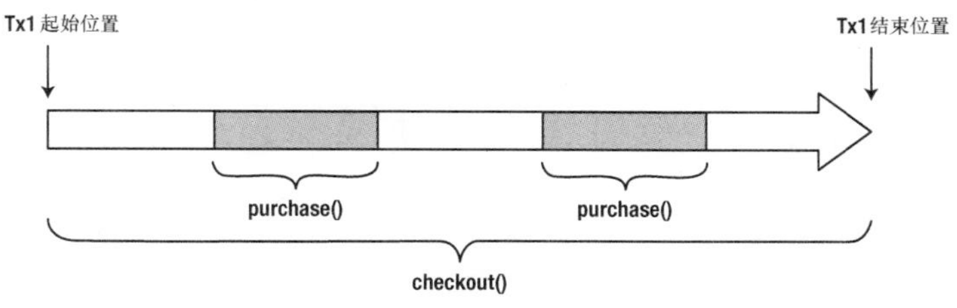

图 10-3　REQUIRED 事务传播行为

不过，如果 purchase()方法被另一个非事务性方法调用，同时过程中并不存在既有事务，那么它就会开启一个新事务并运行在自己的事务中。传播事务属性可以定义在@Transactional 注解中。比如说，你可以像下面这样设置该属性为 REQUIRED 行为。实际上，没必要这么做，因为 REQUIRED 就是默认行为。

```
package com.apress.springrecipes.bookshop.spring;
...
import org.springframework.transaction.annotation.Propagation;
import org.springframework.transaction.annotation.Transactional;

public class JdbcBookShop extends JdbcDaoSupport implements BookShop {
    @Transactional(propagation = Propagation.REQUIRED)
    public void purchase(String isbn, String username) {
        ...
```

```
        }
    }
    package com.apress.springrecipes.bookshop.spring;
    ...
    import org.springframework.transaction.annotation.Propagation;
    import org.springframework.transaction.annotation.Transactional;

    public class BookShopCashier implements Cashier {
        ...
        @Transactional(propagation = Propagation.REQUIRED)
        public void checkout(List<String> isbns, String username) {
            ...
        }
    }
```

使用 REQUIRES_NEW 传播行为

另一个常见的传播行为是 REQUIRES_NEW。这表示方法必须要开启一个新事务并运行在这个新事务中。如果过程中存在事务，那么它首先会被挂起（比如说，在 BookShopCashier 的 checkout 方法中被调用，其传播行为是 REQUIRED）。

```
package com.apress.springrecipes.bookshop.spring;
...
import org.springframework.transaction.annotation.Propagation;
import org.springframework.transaction.annotation.Transactional;

public class JdbcBookShop extends JdbcDaoSupport implements BookShop {
    @Transactional(propagation = Propagation.REQUIRES_NEW)
    public void purchase(String isbn, String username) {
        ...
    }
}
```

在这种情况下，一共会开启 3 个事务。第一个事务是由 checkout()方法开启的，不过当第一个 purchase()方法被调用时，第一个事务会被挂起，新的事务会开启。在第一个 purchase()方法的结束位置处，新的事务会完成并提交。当第二个 purchase()方法被调用时，另一个新事务会开启；然而，这个事务会失败并回滚。这样，第一本书会购买成功，而第二本则会失败。图 10-4 所示为 REQUIRES_NEW 事务传播行为。

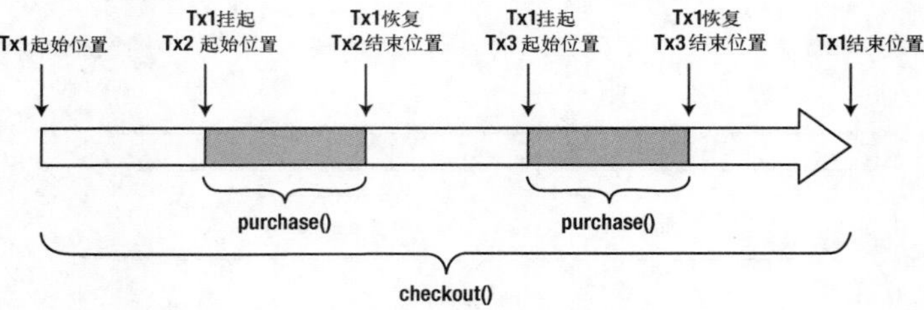

图 10-4 REQUIRES_NEW 事务传播行为

10-7 设置隔离事务属性

问题提出

当同一个应用或不同应用的多个事务同时操纵同一个数据集时，很多意料之外的问题就会出现。你需要指定事务之间的隔离方式。

解决方案

并发事务所导致的问题可以分为如下 4 类。

- 脏读：对于 T1 和 T2 两个事务来说，T1 读取到了 T2 已经更新但尚未提交的字段。接下来，如果 T2 回滚，那么 T1 所读取的字段将会变成临时且无效的。

- 不可重复读：对于 T1 和 T2 两个事务来说，T1 读取了一个字段，然后 T2 更新了该字段。接下来，如果 T1 再次读取这个字段，值将是不同的。
- 幻读：对于 T1 和 T2 两个事务来说，T1 从表中读取了一些行，然后 T2 向表中插入了一些新行。接下来，如果 T1 再次读取这张表，那么表中会有一些新行。
- 丢失更新：对于 T1 和 T2 两个事务来说，它们都选择了一行进行更新，并根据该行的状态对其进行更新。这样，一个事务就会覆盖掉另一个，第二个待提交的事务本应该在执行选择前等待第一个事务提交完成。

从理论上来说，事务应该做到彼此之间完全隔离（即序列化）从而避免上述问题。不过，这种隔离级别会对性能产生极大影响，因为这需要事务串行执行。实际上，事务可以运行在低一些的隔离级别中以改进性能。

事务的隔离级别可以通过隔离事务属性来指定。Spring 支持 5 种隔离级别，如表 10-9 所示。这些隔离级别定义在 org.springframework.transaction.TransactionDefinition 接口中。

表 10-9　Spring 支持的隔离级别

隔离级别	说明
DEFAULT	使用底层数据库默认的隔离级别。对于大多数数据库来说，默认隔离级别是 READ_COMMITTED
READ_UNCOMMITTED	一个事务可以读到其他事务尚未提交的更改；可能会出现脏读、不可重复读和幻读问题
READ_COMMITTED	一个事务只能读到其他事务已经提交的更改。这可以避免脏读，但依然会出现不可重复读和幻读问题
REPEATABLE_READ	确保一个事务能够多次从一个字段中读到相同的值。在该事务执行期间禁止其他事务对该字段进行更新。这可以避免脏读和不可重复读问题，但依然会出现幻读问题
SERIALIZABLE	确保一个事务能够多次从一张表中读到相同的行。在该事务执行期间禁止其他事务对该表进行插入、更新和删除操作。这可以避免所有并发问题，但性能也是最差的

注意：事务隔离是由底层数据库引擎支持的，并不是由应用或框架支持的。不过，并非所有数据库引擎都支持所有这些隔离级别。可以通过调用 java.sql.Connection 接口的 setTransactionIsolation()方法修改 JDBC 连接的隔离级别。

解释说明

为了说明并发事务导致的问题，我们向 bookshop 新增两个操作，用于增加和检查图书库存。

```
package com.apress.springrecipes.bookshop;

public interface BookShop {
    ...
    public void increaseStock(String isbn, int stock);
    public int checkStock(String isbn);
}
```

接下来，实现这两个操作，代码如下所示。注意，这两个操作应该声明为事务性的。

```
package com.apress.springrecipes.bookshop;

import org.springframework.jdbc.core.support.JdbcDaoSupport;
import org.springframework.transaction.annotation.Isolation;
import org.springframework.transaction.annotation.Transactional;

public class JdbcBookShop extends JdbcDaoSupport implements BookShop {

    @Transactional
    public void purchase(String isbn, String username) {
```

```
        int price = getJdbcTemplate().queryForObject(
            "SELECT PRICE FROM BOOK WHERE ISBN = ?", Integer.class, isbn);

        getJdbcTemplate().update(
            "UPDATE BOOK_STOCK SET STOCK = STOCK - 1 WHERE ISBN = ?", isbn );

        getJdbcTemplate().update(
            "UPDATE ACCOUNT SET BALANCE = BALANCE - ? WHERE USERNAME = ?",
            price, username);
    }

    @Transactional
    public void increaseStock(String isbn, int stock) {
        String threadName = Thread.currentThread().getName();
        System.out.println(threadName + " - Prepare to increase book stock");

        getJdbcTemplate().update("UPDATE BOOK_STOCK SET STOCK = STOCK + ?
        WHERE ISBN = ?", stock, isbn);

        System.out.println(threadName + " - Book stock increased by " + stock);
        sleep(threadName);

        System.out.println(threadName + " - Book stock rolled back");
        throw new RuntimeException("Increased by mistake");
    }

    @Transactional(isolation = Isolation.READ_UNCOMMITTED)
    public int checkStock(String isbn) {
        String threadName = Thread.currentThread().getName();
        System.out.println(threadName + " - Prepare to check book stock");

        int stock = getJdbcTemplate().queryForObject("SELECT STOCK FROM BOOK_STOCK
        WHERE ISBN = ?", Integer.class, isbn);

        System.out.println(threadName + " - Book stock is " + stock);
        sleep(threadName);

        return stock;
    }

    private void sleep(String threadName) {
        System.out.println(threadName + " - Sleeping");

        try {
            Thread.sleep(10000);
        } catch (InterruptedException e) {
        }

        System.out.println(threadName + " - Wake up");
    }
}
```

为了模拟并发，操作需要在多个线程中执行。可以通过 println 语句追踪操作的当前状态。对于每个操作来说，围绕着 SQL 语句的执行向控制台打印了两条消息。消息中包含了线程名，这样就可知道当前正在执行操作的是哪个线程。

在每个操作执行完 SQL 语句后，线程会睡眠 10 秒钟。众所周知，当操作完成后，事务会立刻提交或回滚。插入一个 sleep 语句可以延迟提交或回滚。对于 increase()操作来说，最终会抛出一个 RuntimeException 让事务回滚。我们来看看运行这些示例的简单客户端。

在开始运行隔离级别的示例前，请将表 10-10 和表 10-11 中的数据输入到 bookshop 数据库中（注意，该示例不需要 ACCOUNT 表）。

表 10-10 用于测试隔离级别的 BOOK 表中的示例数据

ISBN	BOOK_NAME	PRICE
0001	The First Book	30

表 10-11 用于测试隔离级别的 BOOK_STOCK 表中的示例数据

ISBN	STOCK
0001	10

使用 READ_UNCOMMITTED 和 READ_COMMITTED 隔离级别

READ_UNCOMMITTED 是最低的隔离级别，可以让一个事务读到其他事务尚未提交的变更。可以在 checkStock()方法的@Transactional 注解中设置该隔离级别。

```
package com.apress.springrecipes.bookshop.spring;
...
import org.springframework.transaction.annotation.Isolation;
import org.springframework.transaction.annotation.Transactional;

public class JdbcBookShop extends JdbcDaoSupport implements BookShop {
    ...
    @Transactional(isolation = Isolation.READ_UNCOMMITTED)
    public int checkStock(String isbn) {
        ...
    }
}
```

可以创建几个线程来对该事务的隔离级别进行试验。在如下的 Main 类中会创建两个线程。线程 1 会增加图书库存，线程 2 会检查图书库存。线程 1 比线程 2 先启动 5 秒。

```
package com.apress.springrecipes.bookshop.spring;
...
public class Main {

    public static void main(String[] args) {
        ...
        final BookShop bookShop = context.getBean(BookShop.class);

        Thread thread1 = new Thread(() -> {
            try {
                bookShop.increaseStock("0001", 5);
            } catch (RuntimeException e) {}
        }, "Thread 1");

        Thread thread2 = new Thread(() -> {
            bookShop.checkStock("0001");
        }, "Thread 2");

        thread1.start();
        try {
            Thread.sleep(5000);
        } catch (InterruptedException e) {}
            thread2.start();
    }
}
```

运行该应用会得到如下结果：

```
Thread 1-Prepare to increase book stock
Thread 1-Book stock increased by 5
Thread 1-Sleeping
Thread 2-Prepare to check book stock
Thread 2-Book stock is 15
Thread 2-Sleeping
Thread 1-Wake up
```

```
Thread 1-Book stock rolled back
Thread 2-Wake up
```

首先，线程 1 增加了图书库存，然后睡眠。这时，线程 1 的事务尚未回滚。当线程 1 睡眠时，线程 2 启动并读取图书库存。在 READ_UNCOMMITTED 隔离级别下，线程 2 可以读到尚未提交的事务所更新的库存值。

不过，当线程 1 醒来时，它的事务会因为 RuntimeException 而回滚，因此线程 2 所读取的值就是临时且无效的了。该问题叫做脏读，因为事务可能会读到“脏”值。

为了避免脏读问题，应该将 checkStock()的隔离级别提高至 READ_COMMITTED。

```
package com.apress.springrecipes.bookshop.spring;
...
import org.springframework.transaction.annotation.Isolation;
import org.springframework.transaction.annotation.Transactional;

public class JdbcBookShop extends JdbcDaoSupport implements BookShop {
    ...
    @Transactional(isolation = Isolation.READ_COMMITTED)
    public int checkStock(String isbn) {
        ...
    }
}
```

再次运行该应用，直到线程 1 回滚事务后线程 2 才能读到图书库存。以这种方式，脏读问题通过防止一个事务读到另一个未提交事务所更新的字段得以避免。

```
Thread 1-Prepare to increase book stock
Thread 1-Book stock increased by 5
Thread 1-Sleeping
Thread 2-Prepare to check book stock
Thread 1-Wake up
Thread 1-Book stock rolled back
Thread 2-Book stock is 10
Thread 2-Sleeping
Thread 2-Wake up
```

对于支持 READ_COMMITTED 隔离级别的底层数据库来说，它需要获取到已更新但尚未提交的行的更新锁。接下来，其他事务在读取该行时需要等待，直到更新锁释放为止。更新锁的释放出现在加锁的事务提交或回滚的时候。

使用 REPEATABLE_READ 隔离级别

现在，我们重新组织线程来演示另一个并发问题。交换两个线程的任务，让线程 1 先检查图书库存，然后再让线程 2 增加图书库存。

```
package com.apress.springrecipes.bookshop.spring;
...
public class Main {

    public static void main(String[] args) {
        ...
        final BookShop bookShop = (BookShop) context.getBean("bookShop");

        Thread thread1 = new Thread(() -> {
            public void run() {
                bookShop.checkStock("0001");
            }
        }, "Thread 1");

        Thread thread2 = new Thread(() -> {
            try {
                bookShop.increaseStock("0001", 5);
            } catch (RuntimeException e) {}
        }, "Thread 2");
```

```
        thread1.start();
        try {
            Thread.sleep(5000);
        } catch (InterruptedException e) {}
            thread2.start();
    }
}
```

运行应用会得到如下结果：

```
Thread 1-Prepare to check book stock
Thread 1-Book stock is 10
Thread 1-Sleeping
Thread 2-Prepare to increase book stock
Thread 2-Book stock increased by 5
Thread 2-Sleeping
Thread 1-Wake up
Thread 2-Wake up
Thread 2-Book stock rolled back
```

首先，线程 1 读取图书库存，然后睡眠。这时，线程 1 的事务尚未提交。当线程 1 睡眠时，线程 2 启动并增加图书库存。在 READ_COMMITTED 隔离级别下，线程 2 可以更新尚未提交的事务所读取的库存值。

不过，如果线程 1 再次读取图书库存，其值与第一次读取的值是不同的。该问题叫做不可重复读，因为事务会读到同一个字段的不同值。

为了避免不可重复读问题，应该将 checkStock()的隔离级别提高至 REPEATABLE_READ。

```
package com.apress.springrecipes.bookshop.spring;
...
import org.springframework.transaction.annotation.Isolation;
import org.springframework.transaction.annotation.Transactional;

public class JdbcBookShop extends JdbcDaoSupport implements BookShop {
    ...
    @Transactional(isolation = Isolation.REPEATABLE_READ)
    public int checkStock(String isbn) {
        ...
    }
}
```

再次运行该应用，直到线程 1 提交了事务，线程 2 才能更新图书库存。以这种方式，不可重复读问题通过防止一个事务更新已经被另一个未提交事务所读取的值得以避免。

```
Thread 1-Prepare to check book stock
Thread 1-Book stock is 10
Thread 1-Sleeping
Thread 2-Prepare to increase book stock
Thread 1-Wake up
Thread 2-Book stock increased by 5
Thread 2-Sleeping
Thread 2-Wake up
Thread 2-Book stock rolled back
```

对于支持 REPEATABLE_READ 隔离级别的底层数据库来说，它需要获取已读但尚未提交的行的读锁。接下来，其他事务在更新该行时需要等待，直到读锁释放为止。读锁的释放出现在加锁的事务提交或回滚的时候。

使用 SERIALIZABLE 隔离级别

在一个事务从一张表中读取了几行后，另一个事务向同一张表插入了几个新行。如果第一个事务再次读取同一张表，它会发现与第一次读时相比会多出来几行。这个问题叫做幻读。实际上，幻读问题非常类似于不可重复读，不过它涉及多行。

为了避免幻读问题，应该将隔离级别提高至最高级别：SERIALIZABLE。注意，该隔离级别是最慢的，因为它需要获取整张表的读锁。实际上，应该总是选择能够满足需求的最低的隔离级别。

10-8　设置回滚事务属性

问题提出

默认情况下，只有未检查异常（即 RuntimeException 或是 Error 类型）会导致事务回滚，而检查异常则不会。有时，你想要打破这个规则，可设置自己的回滚异常。

解决方案

可以通过回滚事务属性来指定哪些异常会导致事务回滚，哪些异常则不会。不在该属性中显式指定的任何异常都会按照默认回滚规则进行处理（即对未检查异常回滚，不对检查异常回滚）。

解释说明

可以通过@Transactional 注解的 rollbackFor 与 noRollbackFor 属性来定义事务的回滚规则。这两个属性声明为 Class[]类型，因此可以为每个属性指定多个异常。

```
package com.apress.springrecipes.bookshop.spring;
...
import org.springframework.transaction.annotation.Propagation;
import org.springframework.transaction.annotation.Transactional;
import java.io.IOException;

public class JdbcBookShop extends JdbcDaoSupport implements BookShop {
    ...
    @Transactional(
        propagation = Propagation.REQUIRES_NEW,
        rollbackFor = IOException.class,
        noRollbackFor = ArithmeticException.class)
    public void purchase(String isbn, String username) throws Exception {
        throw new ArithmeticException();
    }
}
```

10-9　设置超时与只读事务属性

问题提出

由于事务需要获取行和表的锁，因此长事务会占用资源并对整体性能产生影响。此外，如果事务只是读取而不更新数据，那么数据库引擎就可以优化事务。你可以指定这些属性来增强应用的性能。

解决方案

timeout 事务属性（一个整数，单位是秒）表示事务在强制回滚前会持续多长时间。这可以防止长事务始终占据着资源。read-only 属性表示事务只会读取而不会更新数据。read-only 标志只是一个让资源优化事务的提示，如果出现了写操作，资源不一定会产生失败。

解释说明

timeout 与 read-only 事务属性可以定义在@Transactional 注解中。注意，超时的单位是秒。

```
package com.apress.springrecipes.bookshop.spring;
...
import org.springframework.transaction.annotation.Isolation;
import org.springframework.transaction.annotation.Transactional;

public class JdbcBookShop extends JdbcDaoSupport implements BookShop {
    ...
    @Transactional(
        isolation = Isolation.REPEATABLE_READ,
        timeout = 30,
        readOnly = true)
    public int checkStock(String isbn) {
        ...
    }
}
```

10-10 使用加载期编织来管理事务

问题提出

默认情况下，Spring 的声明式事务管理是通过 AOP 框架开启的。不过，由于 Spring AOP 只能为声明在 IoC 容器中的 bean 的 public 方法添加通知，因此使用 Spring AOP 只能在这个范围内管理事务。有时，你想管理非 public 方法，或是在 Spring IoC 容器外创建的对象方法的事务。

解决方案

Spring 提供了一个名为 AnnotationTransactionAspect 的 AspectJ 切面，它可以管理任何对象的任何方法的事务，即便方法是非 public 的或者对象是在 Spring IoC 容器外创建的。该切面会管理使用了@Transactional 注解的任何方法的事务。可以选择 AspectJ 的编译期编织或是加载期编织来启用这个切面。

解释说明

要想将该切面在加载期织入到领域类中，你需要在配置类上使用@EnableLoadTimeWeaving 注解。要想开启 Spring 的 AnnotationTransactionAspect 进行事务管理，只需定义@EnableTransactionManagement 注解并将其 mode 属性设为 ASPECTJ 即可。@EnableTransactionManagement 注解的 mode 属性接收两个值：ASPECTJ 与 PROXY。ASPECTJ 规定容器应该使用加载期或编译期编织来启用事务通知。这要求 spring-instrument JAR 要位于类路径上，同时在加载期和编译期进行恰当的配置。

另外，PROXY 规定容器应该使用 Spring AOP 机制。值得注意的是，ASPECTJ 模式不支持在接口上使用@Transactional 注解。接下来，事务切面就会自动开启。你还需要为该切面提供一个事务管理器。默认情况下，它会寻找名字为 transactionManager 的事务管理器。

```
package com.apress.springrecipes.bookshop;

Configuration
@EnableTransactionManagement(mode = AdviceMode.ASPECTJ)
@EnableLoadTimeWeaving
public class BookstoreConfiguration { ... }
```

注意： 要想为 AspectJ 使用 Spring aspect 库，你需要将 spring-aspects 模块添加到类路径上。要想启用加载期编织，还需要引入一个 Java 代理（agent），该代理位于 spring-instrument 模块中。

对于简单的 Java 应用来说，你可以将 Spring 代理（agent）作为 VM 参数在加载期将该切面织入到类中。

```
java -javaagent:lib/spring-instrument-5.0.0.RELEASE.jar -jar recipe_10_10_i.jar
```

小结

本章介绍了事务以及为何应该使用事务。我们介绍了历史上 Java EE 所采取的事务管理方式，接下来介绍了 Spring 框架所提供的方式存在哪些不同。我们学习了如何在代码中显式使用事务，以及如何通过注解驱动的切面隐式使用事务。我们创建了数据库并使用事务保证数据库中的有效状态。

第 11 章将会介绍 Spring Batch。Spring Batch 提供了基础设施和组件，它们可以作为批处理任务的基础。

第 11 章

Spring Batch

批处理已有几十年的历史了。最早广泛使用的用于管理信息（信息技术）的技术应用就是批处理应用。这些环境中没有交互式会话，通常也无法在内存中加载多个应用。计算机相当昂贵，与今天的服务器没有可比性。而且机器是多租户的，并在白天使用（共享时间）。不过到了晚上，机器就闲置下来了，这是一种极大的浪费。于是，利用离线时间进行处理的业务越来越多。基于这种实践涌现出了批处理。

批处理解决方案通常是离线运行的，它们感知不到系统中的事件。过去，批处理离线运行是出于必要。不过，今天大多数批处理之所以离线运行，是因为对于大多数架构来说，让任务能在一个可预测的时间内完成并且将任务分块完成是一种需要。批处理解决方案通常不会对请求进行响应，尽管没有理由不因为消息或请求而无法启动它。批处理解决方案常常用在大规模数据集中，处理的持续时间对于架构和实现来说是关键因素。

一个处理可能会运行几分钟、几小时，甚至几天！任务的持续时间可能没有边界（比如说一直运行，直到所有任务完成，即便这意味着可能需要运行好几天），也可能有严格的边界（任务必须在固定时间内处理，无论边界是什么，每一个任务都要花费相同的时间，这可以预测给定的任务会在某个时间窗口内完成）。

批处理拥有很长的历史，甚至对现代批处理解决方案产生了影响。

大型机应用使用了批处理，当今最大的批处理环境之一的 z/OS 上的客户信息控制系统（CICS）依旧是个大型机操作系统。CICS 非常适合于这种类型的任务：接收输入、对其处理、将其写到输出。CICS 是个交易服务器，主要用在金融机构和政府中，可运行各种语言编写的程序（COBOL、C、PLI 等）。它可以轻松支持每秒钟成千上万笔的交易。虽然 CICS 本身是在 1969 年才粉墨登场的，但它是第一代容器（容器这个概念对于 Spring 和 Java EE 用户来说是耳熟能详的）。CICS 的安装费用非常昂贵，虽然 IBM 依旧在销售和安装 CICS，但很多其他的解决方案从那时起已经出现了。这些解决方案通常都特定于具体环境：大型机上的 COBOL/CICS、UNIX 上的 C，以及今天任何环境上的 Java。问题在于几乎没有标准化基础设施来处理这些批处理解决方案。几乎没有人意识到了他们错过了什么，因为 Java 平台上几乎没有对批处理提供原生支持。需要批处理解决方案的业务通常会由开发人员自己编写，这会产生脆弱、特定于领域的代码。

不过，问题在这里：事务支持、快速 I/O、调度器支持（如 Quartz）、健壮的线程支持，以及 Java EE 和 Spring 中强大的应用容器的概念。自然而然，Dave Syer 和团队一起构建了 Spring Batch，这是一个针对 Spring 平台的批处理解决方案。

在深入细节之前，我们有必要思考一下这个框架所解决的各种问题，这是非常重要的。技术是由它的解决方案空间所定义的。典型的 Spring Batch 应用通常会读取大量数据，然后将修改后的数据写回。关于事务屏障、输入规模、并发和处理步骤的顺序等决策都是一个典型的集成所涉及的维度。

一种常见的需求是从逗号分隔值（CSV）文件中加载数据，这也许是 B2B 的交易，也许是作为一种集成技术来处理老式遗留应用。另一种常见应用是对数据库中的记录进行重要处理。输出也许是对数据库记录本身的更新。一个例子就是调整文件系统中图片的大小，其元数据存储在数据库中，或是需要根据某些条件触发其他处理。

■ **注意**：固定宽度的数据是一种行和单元格格式，非常类似于 CSV 文件。CSV 文件的单元格是通过逗号或制表符分隔的，而固定宽度的数据则假定每个值有某些长度。第一个值可能是前 9 个字符，第二个值是接下来的 4 个字符，以此类推。

常常用在遗留或嵌入式系统中的固定宽度的数据很适合于批处理。对于非事务性资源（如 Web Service 或文件）的处理很适合使用批处理，因为批处理提供了大多数 Web Service 都缺乏的重试/跳过/失败功能。

理解 Spring Batch 不能做什么也是非常重要的。Spring Batch 是个灵活但并非包办一切的解决方案。就像 Spring 在可能的情况下不会重复发明轮子一样，Spring Batch 也将一些重要的方面留给了实现者自己来决断。比如，Spring Batch 提供了一种通用机制来加载任务，可以使用命令行、UNIX cron、操作系统服务、Quartz（第 13 章会介绍），或是对企业服务总线上的事件进行响应（比如说，Mule ESB 或 Spring 自己的类似于 ESB 的解决方案 Spring Integration，第 15 章会介绍）。再有就是 Spring Batch 管理批处理状态的方式。Spring Batch 需要一个持久化存储。JobRepository（Spring Batch 提供的一个接口，用于存储批元数据实体）唯一一个可用的实现需要用到数据库，因为数据库是事务性的，没必要再重复发明轮子。不过，应该部署到哪个数据库是没有指定的，当然，它已经为你提供了一些默认值。

■ **注意**：JEE7 规范包含了 JSR-352（针对 Java 平台的批应用）。Spring Batch 是该规范的参考实现。

运行时元数据模型

Spring Batch 使用到了 JobRepository，它是针对每个任务（包括组件部分，如 JobInstances、JobExecution 和 StepExecution）的所有知识和元数据的守护者。每个任务由一个或多个步骤组成，一个接着一个执行。借助于 Spring Batch，一个步骤可以有条件地跟着另一个步骤，这考虑到了原始的工作流。

这些步骤还可以是并发的；两个步骤可以同时运行。

当任务运行时，通常会有 JobParameter 来参数化任务本身的运行期行为。比如说，任务可能会接收一个 date 参数来确定要处理哪些记录。为了标识一个运行的任务，系统会创建一个 JobInstance。JobInstance 是唯一的，这是由与其关联的 JobParameters 决定的。每次当相同的 JobInstance（即同样的 Job 和 JobParameters）运行时，这就叫做一个 JobExecution。这是任务版本的一个运行时上下文。理想情况下，对于每个 JobInstance 来说，只会有一个 JobExecution：JobInstance 第一次运行时所创建的 JobExecution。不过，如果出现了任何错误，那么 JobInstance 就应该重启；随后的运行会创建另一个 JobExecution。对于原始任务中的每一步来说，在 JobExecution 中都会有一个 StepExecution。

这样，你会看到 Spring Batch 有一个镜像对象图：一个反映了任务的设计/构建期视图；另一个反映了任务的运行期视图。这种对原型和实例的划分类似于很多工作流引擎所采取的方式，包括 jBPM。

比如说，假设一份日报是在 2:00 生成的。任务的参数是日期（很可能是前一天的日期）。在这种情况下，任务会有加载步骤、总结步骤和输出步骤。每一天该任务都会运行，同时会创建新的 JobInstance 和 JobExecution。如果同一个 JobInstance 出现了任何重试，那么可以想象到的是会创建多个 JobExecution。

11-1 搭建 Spring Batch 基础设施

问题提出

Spring Batch 为应用提供了大量的灵活性与保证，不过它不能自己完成这些事情。为了完成这些工作，JobRepository 需要数据存储（可以是 SQL 数据库或是其他存储数据的方式）。此外，Spring Batch 还需要其他几个协作者一同完成它的工作。配置大部分都是样板式的。

解决方案

本节将会搭建 Spring Batch 数据库并创建一个 Spring 应用配置，该配置会在随后的解决方案中导入。这个配置涉及很多重复性工作且很无趣。它还会告诉 Spring Batch 将元数据存储到哪个数据库中。

解释说明

在搭建 Spring Batch 的过程中，首先需要处理 JobRepository 接口。通常不会在代码中处理它，而是通过 Spring 配置。该配置是让其他一切都能正确运行的关键所在。JobRepository 接口只有一个真正有用的实现，即 SimpleJobRepository，它会将关于批处理的状态信息存储到数据存储中。创建是通过 JobRepository-FactoryBean 完成的。另一个标准工厂 MapJobRepositoryFactoryBean 主要用于测试目的，因为它的状态并不是持久化的——它是一个内存实现。这两个工厂都会创建一个 SimpleJobRepository 实例。

由于这个 JobRepository 实例会使用你的数据库，因此需要创建模式让 Spring Batch 能够使用它。不同数据库的模式位于 Spring Batch 分发包中。初始化数据库最简单的方式是在 Java 配置中使用 DataSourceInitializer。该文件位于 org/springframework/batch/core 目录中；这里有几个.sql 文件，每个文件都包含了你所选择的数据库所需模式的数据定义语言（DDL，它是 SQL 的子集，用于定义和检查数据库的结构）。这些示例将会使用 H2（一个内存数据库），因此你需要使用针对 H2 的 DDL：schema-h2.sql。请确保将其配置好并告诉 Spring Batch，配置如下所示：

```
@Configuration
@ComponentScan("com.apress.springrecipes.springbatch")
@PropertySource("classpath:batch.properties")
public class BatchConfiguration {

    @Autowired
    private Environment env;

    @Bean
    public DataSource dataSource() {
        DriverManagerDataSource dataSource = new DriverManagerDataSource();
        dataSource.setUrl(env.getRequiredProperty("dataSource.url"));
        dataSource.setUsername(env.getRequiredProperty("dataSource.username"));
        dataSource.setPassword(env.getRequiredProperty("dataSource.password"));
        return dataSource;
    }

    @Bean
    public DataSourceInitializer dataSourceInitializer() {
        DataSourceInitializer initializer = new DataSourceInitializer();
        initializer.setDataSource(dataSource());
        initializer.setDatabasePopulator(databasePopulator());
        return initializer;
    }

    private DatabasePopulator databasePopulator() {
        ResourceDatabasePopulator databasePopulator = new ResourceDatabasePopulator();
        databasePopulator.setContinueOnError(true);
        databasePopulator.addScript(
            new ClassPathResource("org/springframework/batch/core/schema-h2.sql"));
        databasePopulator.addScript(
            new ClassPathResource("sql/reset_user_registration.sql"));
        return databasePopulator;
    }

    @Bean
    public DataSourceTransactionManager transactionManager() {
        return new DataSourceTransactionManager(dataSource());
    }

    @Bean
    public JobRepositoryFactoryBean jobRepository() {
        JobRepositoryFactoryBean jobRepositoryFactoryBean = new JobRepositoryFactoryBean();
        jobRepositoryFactoryBean.setDataSource(dataSource());
        jobRepositoryFactoryBean.setTransactionManager(transactionManager());
        return jobRepositoryFactoryBean;
```

```
    }

    @Bean
    public JobLauncher jobLauncher() throws Exception {
        SimpleJobLauncher jobLauncher = new SimpleJobLauncher();
        jobLauncher.setJobRepository(jobRepository().getObject());
        return jobLauncher;
    }

    @Bean
    public JobRegistryBeanPostProcessor jobRegistryBeanPostProcessor() {
        JobRegistryBeanPostProcessor jobRegistryBeanPostProcessor =
            new JobRegistryBeanPostProcessor();
        jobRegistryBeanPostProcessor.setJobRegistry(jobRegistry());
        return jobRegistryBeanPostProcessor;
    }

    @Bean
    public JobRegistry jobRegistry() {
        return new MapJobRegistry();
    }
}
```

前几个 bean 与配置紧密相关。对于 Spring Batch 来说没什么特殊的东西：一个数据源、一个事务管理器和一个数据源初始化器。

然后是 MapJobRegistry 实例的声明处。这是非常重要的——它是关于给定任务信息的中央化存储，控制着系统中所有任务的“全貌”，其他一切都会用到该实例。

接下来是 SimpleJobLauncher，其唯一目的就是提供一种启动批任务的机制，这里的“任务”指的就是批解决方案。jobLauncher 用于指定待运行的批解决方案的名字和所需的参数。11-2 节将会对其做进一步的介绍。

然后定义了一个 JobRegistryBeanPostProcessor。这个 bean 会扫描 Spring 上下文文件并将任何配置好的任务关联到 MapJobRegistry。

最后是 SimpleJobRepository（它是由 JobRepositoryFactoryBean 工厂创建的）。JobRepository 是“仓库”（使用企业应用架构模式的表述）的一个实现：它会处理各种步骤和任务中涉及的领域模型的持久化与获取。

@PropertySource 注解会让 Spring 加载 batch.properties 文件（位于 src/main/resource 下）。需要通过 Environment 类获取所需的属性。

> **提示：**还可以通过@Value 注解注入所有属性，不过在配置类中需要多个属性时，更简单的方式则是使用 Environment 对象。

batch.properties 文件的内容如下所示：

```
dataSource.password=sa
dataSource.username=
dataSource.url= jdbc:h2:~/batch
```

虽然这样做可行，但 Spring Batch 还可以通过@EnableBatchProcessing 注解以开箱即用的方式配置这些默认属性。这使得配置能够简化很多：

```
package com.apress.springrecipes.springbatch.config;

import org.apache.commons.dbcp2.BasicDataSource;
import org.springframework.batch.core.configuration.annotation.EnableBatchProcessing;
import org.springframework.beans.factory.annotation.Autowired;
import org.springframework.context.annotation.Bean;
import org.springframework.context.annotation.ComponentScan;
import org.springframework.context.annotation.Configuration;
import org.springframework.context.annotation.PropertySource;
import org.springframework.core.env.Environment;
import org.springframework.core.io.ClassPathResource;
import org.springframework.jdbc.datasource.init.DataSourceInitializer;
```

```
import org.springframework.jdbc.datasource.init.ResourceDatabasePopulator;
import javax.sql.DataSource;

@Configuration
@EnableBatchProcessing
@ComponentScan("com.apress.springrecipes.springbatch")
@PropertySource("classpath:/batch.properties")
public class BatchConfiguration {

    @Autowired
    private Environment env;

    @Bean
    public DataSource dataSource() {
        BasicDataSource dataSource = new BasicDataSource();
        dataSource.setUrl(env.getRequiredProperty("dataSource.url"));
        dataSource.setDriverClassName(
            env.getRequiredProperty("dataSource.driverClassName"));
        dataSource.setUsername(env.getProperty("dataSource.username"));
        dataSource.setPassword(env.getProperty("dataSource.password"));
        return dataSource;
    }

    @Bean
    public DataSourceInitializer databasePopulator() {
        ResourceDatabasePopulator populator = new ResourceDatabasePopulator();
        populator.addScript(
            new ClassPathResource("org/springframework/batch/core/schema-derby.sql"));
        populator.addScript(new ClassPathResource("sql/reset_user_registration.sql"));
        populator.setContinueOnError(true);
        populator.setIgnoreFailedDrops(true);

        DataSourceInitializer initializer = new DataSourceInitializer();
        initializer.setDatabasePopulator(populator);
        initializer.setDataSource(dataSource());
        return initializer;
    }
}
```

这个类只有两个 bean 定义：一个用于数据源；另一个用于初始化数据库。其他一切都由@EnableBatchProcessing 注解所处理。上面的配置类会使用一些默认值启动 Spring Batch。

默认配置会配置 JobRepository、JobRegistry 和 JobLauncher。

如果应用中有多个数据源，那就需要显式添加一个 BatchConfigurer 来选择数据源供应用中的批处理部分所用。

如下的 Main 类会使用基于 Java 的配置来运行批处理应用：

```
package com.apress.springrecipes.springbatch;

import com.apress.springrecipes.springbatch.config.BatchConfiguration;
import org.springframework.batch.core.configuration.JobRegistry;
import org.springframework.batch.core.launch.JobLauncher;
import org.springframework.batch.core.repository.JobRepository;
import org.springframework.context.ApplicationContext;
import org.springframework.context.annotation.AnnotationConfigApplicationContext;

public class Main {
    public static void main(String[] args) throws Throwable {
        ApplicationContext context =
            new AnnotationConfigApplicationContext(BatchConfiguration.class);

        JobRegistry jobRegistry = context.getBean("jobRegistry", JobRegistry.class);
        JobLauncher jobLauncher = context.getBean("jobLauncher", JobLauncher.class);
        JobRepository jobRepository = context.getBean("jobRepository", JobRepository.class);
```

```
        System.out.println("JobRegistry: " + jobRegistry);
        System.out.println("JobLauncher: " + jobLauncher);
        System.out.println("JobRepository: " + jobRepository);

    }
}
```

11-2 读写数据

问题提出

你想将 CSV 文件中的数据插入到数据库中。这个解决方案是最简单的一种，你可以借此机会全方位探索一个典型方案。

解决方案

我们将会构建一个解决方案，这个方案在工作量最小的前提下实现一个可行的应用。该解决方案会读取任意长度的文件，并将数据写入到数据库中。最终的结果是几乎不必编写代码。你将使用一个既有的模型类并编写一个类（包含 public static void main(String [] args)方法的类）来完成这个示例。这个模型类可以是 Hibernate 类或是 DAO 层的某个类；不过，在该方案中，它就是个普通的 POJO。该解决方案会使用到 11-1 节中配置的组件。

解释说明

该示例演示了 Spring Batch 最简单的用法：提供可伸缩性。这个程序只是从 CSV 文件中读取数据，其中的字段是由逗号分隔的，行是由换行符分隔的；接下来将记录插入到表中。利用 Spring Batch 提供的智能基础设施，无须担心可伸缩性问题。该应用可以手工轻松完成。现在还没有用上任何智能的事务功能，也没有考虑过重试问题。

这个解决方案就像 Spring Batch 解决方案一样简单。Spring Batch 使用 XML 模式建模解决方案，其抽象和术语与经典的批处理解决方案一脉相承，这样这些抽象和术语就可以从之前的技术移植到后面的技术上了。Spring Batch 提供了一些很有用的默认类，你既可以重写它们，也可以有选择地进行调整。在接下来的示例中，你会使用 Spring Batch 提供的大量辅助实现。从根本上来说，大多数解决方案看起来都是一样的，并组合使用了同样的接口集合。通常只要选择正确的接口就好了。

在运行这个程序时，它可以处理 20,000 行的文件，也可以处理 100 万行的文件。而且内存使用并没有增加，这意味着没有出现内存泄漏问题。显然，后者花费的时间会更长（对于 100 万行的插入来说，应用运行了几个小时）！

> **提示**：当然，在处理 100 万行时，如果在倒数第二条记录上失败了，那将是一场灾难。当事务回滚时，会丢失所有的工作！请阅读关于分块处理的示例。此外，还可以阅读第 10 章来复习事务相关的知识。

```
create table USER_REGISTRATION (
    ID BIGINT NOT NULL PRIMARY KEY GENERATED ALWAYS AS IDENTITY (START WITH 1, INCREMENT BY 1),
    FIRST_NAME VARCHAR(255) not null,
    LAST_NAME VARCHAR(255) not null,
    COMPANY VARCHAR(255) not null,
    ADDRESS VARCHAR(255) not null,
    CITY VARCHAR(255) not null,
    STATE VARCHAR(255) not null,
    ZIP VARCHAR(255) not null,
    COUNTY VARCHAR(255) not null,
    URL VARCHAR(255) not null,
    PHONE_NUMBER VARCHAR(255) not null,
    FAX VARCHAR(255) not null
) ;
```

这个表没有进行调优。比如说，除了主键外，任何列上都没有索引。这是为了避免让示例变得复杂。在重要的生产应用中请仔细审视这样的表。

Spring Batch 应用是高负载的应用，有可能揭示出应用中尚未发现的瓶颈。假设每 10 分钟向数据库插入 100 万条记录。你的数据库会慢慢宕机么？插入速度是应用速度的一个关键因素。软件开发人员会很理想地思考数据库模式，比如它是不是很好地强制了业务逻辑约束，是不是能够服务于整个业务模型。不过，在编写示例这样的应用时，从 DBA 的角度思考是很重要的。一种常见的解决方案是创建一个非规范化的表，当数据进入到数据库中时就一定是有效的数据，这可以通过插入操作上的触发器来实现。这在数据仓库中是非常典型的。接下来，你将会使用 Spring Batch 在插入之前对记录进行处理。这可以让开发人员验证或是重写进入到数据库中的输入。这种处理加上最适合于数据库的约束会使得应用变得健壮和快速。

任务配置

任务配置代码如下所示：

```
package com.apress.springrecipes.springbatch.config;

import com.apress.springrecipes.springbatch.UserRegistration;
import org.springframework.batch.core.Job;
import org.springframework.batch.core.Step;

import org.springframework.batch.core.configuration.annotation.JobBuilderFactory;
import org.springframework.batch.core.configuration.annotation.StepBuilderFactory;
import org.springframework.batch.item.ItemReader;
import org.springframework.batch.item.ItemWriter;
import org.springframework.batch.item.database.BeanPropertyItemSqlParameterSourceProvider;
import org.springframework.batch.item.database.JdbcBatchItemWriter;
import org.springframework.batch.item.file.FlatFileItemReader;
import org.springframework.batch.item.file.LineMapper;
import org.springframework.batch.item.file.mapping.BeanWrapperFieldSetMapper;
import org.springframework.batch.item.file.mapping.DefaultLineMapper;
import org.springframework.batch.item.file.transform.DelimitedLineTokenizer;
import org.springframework.beans.factory.annotation.Autowired;
import org.springframework.beans.factory.annotation.Value;
import org.springframework.context.annotation.Bean;
import org.springframework.context.annotation.Configuration;
import org.springframework.core.io.Resource;

import javax.sql.DataSource;

@Configuration
public class UserJob {

    private static final String INSERT_REGISTRATION_QUERY =
        "insert into USER_REGISTRATION (FIRST_NAME, LAST_NAME, COMPANY, ADDRESS,CITY,
        STATE,ZIP,COUNTY,URL,PHONE_NUMBER,FAX)" +
        " values " +
        "(:firstName,:lastName,:company,:address,:city,:state,:zip,:county,:url,:
        phoneNumber,:fax)";

    @Autowired
    private JobBuilderFactory jobs;

    @Autowired
    private StepBuilderFactory steps;

    @Autowired
    private DataSource dataSource;

    @Value("file:${user.home}/batches/registrations.csv")
    private Resource input;

    @Bean
    public Job insertIntoDbFromCsvJob() {
```

```
        return jobs.get("User Registration Import Job")
            .start(step1())
            .build();
    }

    @Bean
    public Step step1() {
        return steps.get("User Registration CSV To DB Step")
            .<UserRegistration,UserRegistration>chunk(5)
            .reader(csvFileReader())
            .writer(jdbcItemWriter())
            .build();
    }

    @Bean
    public FlatFileItemReader<UserRegistration> csvFileReader() {
        FlatFileItemReader<UserRegistration> itemReader = new FlatFileItemReader<>();
        itemReader.setLineMapper(lineMapper());
        itemReader.setResource(input);
        return itemReader;
    }

    @Bean
    public JdbcBatchItemWriter<UserRegistration> jdbcItemWriter() {
        JdbcBatchItemWriter<UserRegistration> itemWriter = new JdbcBatchItemWriter<>();
        itemWriter.setDataSource(dataSource);
        itemWriter.setSql(INSERT_REGISTRATION_QUERY);
        itemWriter.setItemSqlParameterSourceProvider(new BeanPropertyItemSql
        ParameterSourceProvider<>());
        return itemWriter;
    }

    @Bean
    public DefaultLineMapper<UserRegistration> lineMapper() {
        DefaultLineMapper<UserRegistration> lineMapper = new DefaultLineMapper<>();
        lineMapper.setLineTokenizer(tokenizer());
        lineMapper.setFieldSetMapper(fieldSetMapper());
        return lineMapper;
    }

    @Bean
    public BeanWrapperFieldSetMapper<UserRegistration> fieldSetMapper() {
        BeanWrapperFieldSetMapper<UserRegistration> fieldSetMapper = new
        BeanWrapperFieldSetMapper<>();
        fieldSetMapper.setTargetType(UserRegistration.class);
        return fieldSetMapper;
    }

    @Bean
    public DelimitedLineTokenizer tokenizer() {
        DelimitedLineTokenizer tokenizer = new DelimitedLineTokenizer();
        tokenizer.setDelimiter(",");
        tokenizer.setNames(new String[]{"firstName","lastName","company","address",
        "city","state","zip","county","url","phoneNumber","fax"});
        return tokenizer;
    }
}
```

如前所述，一个任务包含了若干步骤，这些步骤是给定任务真正需要完成的工作。根据需要，步骤既可以很复杂，也可以很简单。实际上，一个步骤可以看作是任务的最小工作单元。输入（读取的内容）会被传递给一个步骤并处理；然后输出（写入的内容）会在一个步骤中创建。这种处理使用了 Tasklet（这是 Spring Batch 提供的另一个接口）。你可以为不同的处理场景提供自己的 Tasklet 实现，或是使用预先配置好的配置。这些实现在配置方法方面是可用的。批处理最为重要的一个方面是面向块的处理，这里通过 chunk()配置方法使用了块。

在面向块的处理中，输入是从读取器读取的，经过可选的处理后进行聚合。最后，经过可配置的间隔（由 commit-interval 属性指定，用于配置在事务提交前需要处理多少条目）后，所有输入会发送给写出器。如果存在事务管理器，那么事务也会提交。在提交前，数据库中的元数据会更新以标记任务的进展。

当可以感知到事务的写出器（或处理器）回滚时，围绕着输入（读取）值的聚合会存在一些微小的差别。Spring Batch 会缓存它所读取的值，并将其写到写出器中。如果写出器组件是事务性的（就像数据库一样），而读取器不是，那么缓存读取的值就没什么问题，并且也许需要重试或是采取其他方式。如果读取器自身也是事务性的，那么从资源中所读取的值就会回滚，并且会发生变化，这会导致内存中缓存的值变成过时的。如果出现了这种情况，就可以通过在块元素上使用 reader-transactional-queue="true"让块不要缓存值。

输入

第一个职责是从文件系统中读取文件。该示例使用了一个既有实现。读取 CSV 文件是个常见的场景，Spring Batch 的支持也没有令人失望。org.springframework.batch.item.file.FlatFileItemReader<T>类将分隔文件中的字段与记录的任务委托给了 LineMapper<T>，它反过来又将识别记录中字段的任务委托给了 LineTokenizer。这里使用了 org.springframework.batch.item.file.transform.DelimitedLineTokenizer，它会提取出由逗号（,）字符分隔的字段。

DefaultLineMapper 还声明了一个 fieldSetMapper 属性，它需要一个 FieldSetMapper 实现。这个 bean 负责接收输入的名值对并生成一个类型，这个类型会赋给写入器组件。

该示例使用了一个 BeanWrapperFieldSetMapper，它会创建一个类型为 UserRegistration 的 JavaBean POJO。我们为字段起个名字，这样后续就可以在配置中引用它们了。这些名字不必非得是输入文件中某个标题行的值，它们只需要对应于输入文件中字段出现的顺序即可。这些名字还被 FieldSetMapper 用来匹配 POJO 中的属性。当每个记录被读取时，值就会应用到一个 POJO 的实例，然后返回这个 POJO。

```
@Bean
public FlatFileItemReader<UserRegistration> csvFileReader() {
    FlatFileItemReader<UserRegistration> itemReader = new FlatFileItemReader<>();
    itemReader.setLineMapper(lineMapper());
    itemReader.setResource(input);
    return itemReader;
}

@Bean
public DefaultLineMapper<UserRegistration> lineMapper() {
    DefaultLineMapper<UserRegistration> lineMapper = new DefaultLineMapper<>();
    lineMapper.setLineTokenizer(tokenizer());
    lineMapper.setFieldSetMapper(fieldSetMapper());
    return lineMapper;
}

@Bean
public BeanWrapperFieldSetMapper<UserRegistration> fieldSetMapper() {
    BeanWrapperFieldSetMapper<UserRegistration> fieldSetMapper =
        new BeanWrapperFieldSetMapper<>();
    fieldSetMapper.setTargetType(UserRegistration.class);
    return fieldSetMapper;
}

@Bean
public DelimitedLineTokenizer tokenizer() {
    DelimitedLineTokenizer tokenizer = new DelimitedLineTokenizer();
    tokenizer.setDelimiter(",");
    tokenizer.setNames(new String[]{"firstName","lastName","company","address","city",
    "state","zip","county","url","phoneNumber","fax"});
    return tokenizer;
}
```

从读取器返回的 UserRegistration 类就是个普通的 JavaBean。

```
package com.apress.springrecipes.springbatch;

public class UserRegistration implements Serializable {

    private String firstName;
    private String lastName;
    private String company;
    private String address;
    private String city;
    private String state;
    private String zip;
    private String county;
    private String url;
    private String phoneNumber;
    private String fax;

    //... accessor / mutators omitted for brevity ...

}
```

输出

下一个工作组件就是写入器，它负责接收从读取器所读取的条目的聚合集合。在该示例中会创建一个新的集合（java.util.List<UserRegistration>），然后写入，接下来当集合每次超出了 chunk 元素上的 commit-interval 属性时会重置。由于要写入到数据库中，因此需要使用 Spring Batch 的 org.springframework.batch.item.database.JdbcBatchItemWriter。这个类支持接收输入并将其写到数据库中。开发人员需要提供输入并指定对输入执行什么 SQL。它会运行 sql 属性所指定的 SQL，本质上会从数据库中进行读取，读取次数则是由块元素的 commit-interval 属性指定的，然后提交整个事务。这里执行的是简单的插入。具名参数的名字和值是由为 itemSqlParameterSourceProvider 属性所配置的 bean 创建的，它是个 BeanPropertyItemSqlParameterSourceProvider 的实例，其唯一的职责就是接收 JavaBean 属性并将其作为具名参数，使之对应于 JavaBean 中的属性名。

```
@Bean
public JdbcBatchItemWriter<UserRegistration> jdbcItemWriter() {
    JdbcBatchItemWriter<UserRegistration> itemWriter = new JdbcBatchItemWriter<>();
    itemWriter.setDataSource(dataSource);
    itemWriter.setSql(INSERT_REGISTRATION_QUERY);
    itemWriter.setItemSqlParameterSourceProvider(
        new BeanPropertyItemSqlParameterSourceProvider<>());
    return itemWriter;
}
```

一个可行的解决方案就是这些。仅通过极少的配置，同时没有自定义代码，就为接收大型 CSV 文件并将其发送到数据库构建了一个解决方案。这个解决方案是最为基本的，并未考虑大量的边界情况。比如说，你可以在读取完条目后对其进行处理（在插入到数据库之前）。

这演示了一个简单的任务。值得注意的是，有一些相似的类可以进行相反的转换：从数据库读取并写入到 CSV 文件中。

```
@Bean
public Job insertIntoDbFromCsvJob() {
    return jobs.get("User Registration Import Job")
        .start(step1())
        .build();
}

@Bean
public Step step1() {
    return steps.get("User Registration CSV To DB Step")
        .<UserRegistration,UserRegistration>chunk(5)
        .reader(csvFileReader())
```

```
            .writer(jdbcItemWriter())
            .build();
    }
```

为了对步骤进行配置，你为其起了名字 User Registration CSV To DB Step。你使用了基于块的处理，并且告诉它块的大小是 5。接下来提供一个读取器和一个写入器，最后告诉工厂构建这个步骤。配置好的这个步骤最终会作为任务的起始点，任务的名字叫做 User Registration Import Job，它只包含这一个步骤。

简化 ItemReader 和 ItemWriter 的配置

配置 ItemReader 和 ItemWriter 是个非常艰巨的任务。你需要对 Spring Batch 的内核有很好的理解与掌握（诸如该使用哪个类等）。从 Spring Batch 4 开始，读取器和写入器的配置变得轻松很多，因为现在有针对不同读取器和写入器的具体的构建器了。

要想配置 FlatFileItemReader，可以使用 FlatFileItemReaderBuilder，同时也不必再配置 4 个 bean 了，只需要 6 行代码就行（与示例中的代码格式相关）。

```
@Bean
public FlatFileItemReader<UserRegistration> csvFileReader() throws Exception {

    return new FlatFileItemReaderBuilder<UserRegistration>()
        .name(ClassUtils.getShortName(FlatFileItemReader.class))
        .resource(input)
        .targetType(UserRegistration.class)
        .delimited()
        .names(new String[]{"firstName","lastName","company","address","city","state",
        "zip","county","url","phoneNumber","fax"})
        .build();
}
```

这个构建器会自动创建 DefaultLineMapper、BeanWrapperFieldSetMapper 和 DelimitedLineTokenizer，你也不必知道它们在内部到底做了什么事情。现在可以基本上描述你的配置了，而不必再显式配置所有不同项。

使用 JdbcBatchItemWriterBuilder 配置 JdbcBatchItemWriter 也是一样的。

```
@Bean
public JdbcBatchItemWriter<UserRegistration> jdbcItemWriter() {
    return new JdbcBatchItemWriterBuilder<UserRegistration>()
        .dataSource(dataSource)
        .sql(INSERT_REGISTRATION_QUERY)
        .beanMapped()
        .build();
}
```

11-3 编写自定义 ItemWriter 与 ItemReader

问题提出

你想要使用某个资源（比如说 RSS 源或是其他自定义的数据格式），但 Spring Batch 不知道如何连接到这个资源。

解决方案

你可以轻松编写自己的 ItemWriter 和 ItemReader。接口非常简单，实现也不复杂。

解释说明

这个过程也很简单，只需重用既有的选择即可。对于编写 JMS（JmsItemWriter<T>）、JPA（JpaItemWriter<T>）、JDBC（JdbcBatchItemWriter<T>）、文件（FlatFileItemWriter<T>）、Hibernate（HibernateItemWriter<T>）等已经提供了支持。甚至还支持调用 bean（PropertyExtractingDelegatingItemWriter<T>）的方法，并将待写入的 Item 中的属性作为参数传递给该方法。其中一个更有用的写入器可以写入到一组带有编号的文件中。MultiResourceItemWriter<T>实现会委托给另一个恰当的 ItemWriter<T>实现来完成任务，不过会写入到多个文件而不是一个非常大的文件中。ItemReader 的实现要少一些，但表述性却很强。如果找不到，那就考虑自行编写。本节会编写自己的实现。

编写自定义的 ItemReader

ItemReader 示例很简单。创建一个 ItemReader，它知道如何从远程过程调用（RPC）端点获取到 UserRegistration 对象：

```
package com.apress.springrecipes.springbatch;

import org.springframework.batch.item.ItemReader;

import java.util.Collection;
import java.util.Date;

public class UserRegistrationItemReader implements ItemReader<UserRegistration> {

    private final UserRegistrationService userRegistrationService;

    public UserRegistrationItemReader(UserRegistrationService userRegistrationService) {
        this.userRegistrationService = userRegistrationService;
    }

    public UserRegistration read() throws Exception {
        final Date today = new Date();
        Collection<UserRegistration> registrations =
            userRegistrationService.getOutstandingUserRegistrationBatchForDate(1, today);
        return registrations.stream().findFirst().orElse(null);
    }
}
```

可以看到，接口很简单。在该示例中，大部分工作都交由远程服务完成了，这个远程服务会向你提供输入。接口要求返回一个记录。接口会被参数化为返回的对象（即“条目”）类型。所有读取的条目会聚合到一起，然后传递给 ItemWriter。

编写自定义的 ItemWriter

这个 ItemWriter 示例也很简单。你想要使用 Spring 提供的针对远程的一些选项来调用远程服务向其写入。ItemWriter<T>接口是通过你想要写入的条目类型进行参数化的。这里，我们期望从 ItemReader<T>获取到一个 UserRegistration 对象。该接口包含了一个方法，它接收一个类的参数化类型的 List。这些就是从 ItemReader<T>读取并聚合的对象。如果将 commit-interval 设为 10，那么 List 中的条目就是 10 个或不到 10 个。

```
package com.apress.springrecipes.springbatch;

import org.slf4j.Logger;
import org.slf4j.LoggerFactory;
import org.springframework.batch.item.ItemWriter;

import java.util.List;
public class UserRegistrationServiceItemWriter implements ItemWriter<UserRegistration> {

    private static final Logger logger = LoggerFactory.getLogger(UserRegistrationService
    ItemWriter.class);

    private final UserRegistrationService userRegistrationService;

    public UserRegistrationServiceItemWriter(UserRegistrationService userRegistrationService) {
        this.userRegistrationService = userRegistrationService;
    }

    public void write(List<?extends UserRegistration> items) throws Exception {
        items.forEach(this::write);
    }
```

```
        private void write(UserRegistration userRegistration) {
            UserRegistration registeredUserRegistration =
                userRegistrationService.registerUser(userRegistration);
            logger.debug("Registered: {}", registeredUserRegistration);

        }
    }
```

这里在服务的客户端接口中进行了装配。你只是遍历了 UserRegistration 对象并调用服务，它反过来返回一个同样的 UserRegistration 实例。如果删掉空格、花括号和日志输出，只需要编写两行代码就可以满足需求了。

UserRegistrationService 接口代码如下所示：

```
package com.apress.springrecipes.springbatch;

import java.util.Collection;
import java.util.Date;

public interface UserRegistrationService {

    Collection<UserRegistration> getOutstandingUserRegistrationBatchForDate(
        int quantity, Date date);

    UserRegistration registerUser(UserRegistration userRegistrationRegistration);

}
```

在该示例中，没有对该接口进行任何特定的实现，因为没必要：它可以是 Spring Batch 不知道的任何接口。

11-4 在写入前处理输入

问题提出

虽然直接从电子表格或是 CSV 存储中对数据进行转换会很有价值，不过还可以在数据写入前对其进行某种处理。CSV 文件中以及更为一般的任何来源的数据通常都不会刚好符合你的要求或是适合直接进行写入。虽然 Spring Batch 可以根据你的需要将其转换为 POJO，但这并非意味着数据的状态是正确的。在数据适合写入之前，你可能还需要推导出其他数据或是从其他服务中获取一些数据进行填充。

解决方案

Spring Batch 可以对读取器的输出进行处理。这种处理可以在输出传递给写入器之前对其进行任何操作，包括改变数据的类型。

解释说明

Spring Batch 可以让实现者对从读取器所读取的数据执行任何自定义的逻辑。块配置上的 processor 属性需要一个对接口 org.springframework.batch.item.ItemProcessor<I,O>的 bean 的引用。这样，重新定义的 11-3 节中的任务的代码如下所示：

```
@Bean
public Step step1() {
    return steps.get("User Registration CSV To DB Step")
        .<UserRegistration,UserRegistration>chunk(5)
        .reader(csvFileReader())
        .processor(userRegistrationValidationItemProcessor())
        .writer(jdbcItemWriter())
        .build();
}
```

目标是在将数据写到数据库之前对其进行某些验证。如果认为记录不合法，可以通过从 ItemProcessor<I,O>返回 null 来阻止进一步的处理。这是很重要的，并且提供了一种必要的防御措施。你希望确保数据格式是正确的（比如说，模式可能要求有效的两字母状态名，而不是更长的全状态名）。电话号码需要遵循某种格式，你可以使用该处理器去除掉电话号码中的任何无关字符，只保留有效的 10 位数字电话号码（在美国）。美国的邮编也一样，它包含了 5 个字符——1 个可选的连字符，后跟一个 4 位的数字代码。

最后，虽然重复性约束最好在数据库中实现，但对于记录来说，还会存在其他一些标准，且这些标准只能在插入前通过查询系统才能满足。

如下是 ItemProcessor 的配置：

```
@Bean
public ItemProcessor<UserRegistration, UserRegistration>
userRegistrationValidationItemProcessor() {
    return new UserRegistrationValidationItemProcessor();
}
```

为了让类的代码短一些，这里没有将类的全部代码都列出来，只是将重要部分呈现出来。

```
package com.apress.springrecipes.springbatch;
import java.util.Arrays;
import java.util.Collection;

import org.apache.commons.lang3.StringUtils;
import org.springframework.batch.core.StepExecution;
import org.springframework.batch.item.ItemProcessor;
import com.apress.springrecipes.springbatch.UserRegistration;

public class UserRegistrationValidationItemProcessor
    implements ItemProcessor<UserRegistration, UserRegistration> {

        private String stripNonNumbers(String input) { /* ... */ }

        private boolean isTelephoneValid(String telephone) { /* ... */ }

        private boolean isZipCodeValid(String zip) { /* ... */ }

        private boolean isValidState(String state) { /* ... */ }

        public UserRegistration process(UserRegistration input) throws Exception {
            String zipCode = stripNonNumbers(input.getZip());
            String telephone = stripNonNumbers(input.getPhoneNumber());
            String state = StringUtils.defaultString(input.getState());
            if (isTelephoneValid(telephone) && isZipCodeValid(zipCode) &&
            isValidState(state)) {
                input.setZip(zipCode);
                input.setPhoneNumber(telephone );
                return input;
            }
            return null;
        }
}
```

该类是个参数化类型。类型信息就是输入的类型以及输出的类型。输入是传递给方法要处理的内容，输出是从方法返回的数据。由于该示例没有转换任何内容，因此两个参数化类型是一样的。一旦处理完成，Spring Batch 元数据表中就会有大量有价值的信息。向数据库发出如下查询：

```
select * from BATCH_STEP_EXECUTION;
```

你会看到任务的退出状态、有多少个提交、读取了多少条目，以及过滤了多少条目等信息。如果上述任务运行在一个 100 行的批次上，每个条目都会被读取并通过处理器传递，那么它会找到 10 个不合法条目（返回 null 10 次），filter_count 列的值是 10。可以从 read_count 列看到读取了 100 个条目。write_count 列会反映出有 10 个条目不符合要求，其值显示为 90。

将处理器链接到一起

有时，你想要添加额外的处理，这个处理与已创建的处理器的目标并不一致。Spring Batch 提供了一个便捷类，名为 CompositeItemProcessor<I,O>，它会将过滤器的输出转发给后续过滤器的输入。通过这种方式，你可以编写很多 ItemProcessor<I,O>，每个都只关注一个方面，然后重用它们，必要时还可以将它们链接到一起。

```
@Bean
```

```
public CompositeItemProcessor<Customer, Customer> compositeBankCustomerProcessor() {
    List<ItemProcessor<Customer, Customer>> delegates = Arrays.asList(creditScoreValidation
    Processor(), salaryValidationProcessor(), customerEligibilityProcessor());
    CompositeItemProcessor<Customer, Customer> processor = new CompositeItemProcessor<>();
    processor.setDelegates(delegates);
    return processor;
}
```

该示例创建了一个简单的工作流。第一个 ItemProcessor<T>接收为任务所配置的来自于 ItemReader<T>的输入，假设为一个 Customer 对象。它会检查这个 Customer 的信用评分，如果通过就会将它转发给工资与收入验证处理器。如果验证通过，那么该 Customer 就会被转发给资格处理器，系统会检查重复性或是其他不合法的数据。它最终会被转发给写入器来添加到输出中。如果在这 3 个处理器中，任何一个 Customer 验证失败，那么 ItemProcessor 的执行就会返回 null 并停止处理。

11-5 通过事务增强健壮性

问题提出

你希望读写能够变得健壮一些。理想情况下，它们应该使用事务来恰当且正确地对异常做出反应。

解决方案

事务能力构建在核心 Spring 框架已经提供的一流支持之上。Spring Batch 提供了相关配置，这样你就可以控制它了。在面向块处理的上下文中，它还对提交频率、回滚语义等公开了大量控制。

解释说明

首先，我们将会介绍如何将步骤（或块）变成事务性的，然后介绍对步骤使用重试逻辑的配置。

事务

Spring 的核心框架为事务提供了一流的支持。你只需装配一个 PlatformTransactionManager 并给 Spring Batch 一个引用即可，就像在任何常规的 JdbcTemplate 或 HibernateTemplate 解决方案中所做的那样。在构建 Spring Batch 解决方案时，可以控制每个步骤在事务中的行为。之前我们已经见识过对事务的一些支持。

所有示例中所用的配置都会创建一个 DriverManagerDataSource 和 DataSourceTransactionManager bean。接下来，PlatformTransactionManager 和 DataSource 会装配到 JobRepository，它又会装配到 JobLauncher，可以使用 JobLauncher 启动所有任务。这会使得任务所创建的所有元数据以事务性的方式写到数据库中。

你可能好奇，在使用数据源引用来配置 JdbcItemWriter 时为何没有明确提到过事务管理器。尽管可以指定事务管理器引用，不过在你的解决方案中是不需要的，因为 Spring Batch 默认情况下会从上下文中寻找名为 transactionManager 的 PlatformTransactionManager，然后使用它。如果想要显式对其进行配置，可以在 tasklet 配置方法中指定 transactionManager 属性。针对 JDBC 的一个简单事务管理器的代码如下所示：

```
@Bean
protected Step step1() {
    return steps.get("step1")
        .<UserRegistration,UserRegistration>chunk(5)
        .reader(csvFileReader())
        .processor(userRegistrationValidationItemProcessor())
        .writer(jdbcItemWriter())
        .transactionManager(new DataSourceTransactionManager(dataSource))
        .build();
}
```

从 ItemReader<T>读取的条目会正常聚合。如果 ItemWriter<T>上的提交失败，那么聚合的条目会保留并再次提交。这个过程很高效，大多数时候都没什么问题。在有一个场景中它会破坏语义，那就是从事务性资源（比如说 JMS 队列或是数据库）中读取的时候。如果所在的事务（该示例中就是写入器的事务）失败了，那么从消息队列中的读取就可以，而且也应该回滚。

```
@Bean
protected Step step1() {
```

```
        return steps.get("step1")
            .<UserRegistration,UserRegistration>chunk(5)
            .reader(csvFileReader()).readerIsTransactionalQueue()
            .processor(userRegistrationValidationItemProcessor())
            .writer(jdbcItemWriter())
            .transactionManager(new DataSourceTransactionManager(dataSource))
            .build();
    }
```

回滚

处理简单的情况（“读取 X 个条目，每隔 Y 个条目提交一次数据库事务”）是很容易的。Spring Batch 在健壮性上做得非常棒，只需通过简单配置项就可以应对边界与失败的情况。

如果 ItemWriter 上的写入失败了，或是在处理过程中出现了其他异常，那么 Spring Batch 就会回滚事务。这是对于大多数情况的有效处理。在某些场景下，我们需要控制哪些异常情况会导致事务回滚。

在使用基于 Java 的配置进行回滚时，第一步是一个失败容错的步骤，它反过来可用于指定不回滚的异常。首先通过 faultTolerant()获取到一个失败容错的步骤，接下来使用 skipLimit()方法在实际停止执行任务前指定忽略的回滚数量，最后使用 noRollback()方法指定不会触发回滚的异常。要想指定多个异常，只需多次调用 noRollback()方法即可。

```
    @Bean
    protected Step step1() {
        return steps.get("step1")
            .<UserRegistration,UserRegistration>chunk(10)
                .faultTolerant()
                    .noRollback(com.yourdomain.exceptions.YourBusinessException.class)
            .reader(csvFileReader())
            .processor(userRegistrationValidationItemProcessor())
            .writer(jdbcItemWriter())
            .build();
    }
```

11-6 重试

问题提出

你要处理的一项功能需求可能会失败但又不是事务性的。也许它是事务性的，但并不可靠。在从资源中读取或者向其写入时可能会失败。失败的原因可能是网络连接问题、端点宕掉，或是其他原因。不过，你知道很快就会恢复，这时就需要重试。

解决方案

使用 Spring Batch 的重试功能系统性地对读和写进行重试。

解释说明

11-5 节讲到，使用 Spring Batch 可以轻松处理事务性资源。当遇到瞬态或是不可靠的资源时，我们就需要采取不同的方法。这种资源可能是分布式的，或是自己能够解决问题。有些资源（如 Web Service）是无法加入到事务中的，因为它们是分布式的。有些产品会在一台服务器上开启事务，然后将事务性上下文传播到分布式服务器上，并在那里结束，不过这种情况很少且低效。或者，如果可以使用分布式（“全局”或 XA）事务，那么 Spring Batch 也可以很好地支持分布式事务。不过，有时你要处理的资源并不是上面这两类。一个常见的例子就是调用远程服务，比如说 RMI 服务或是 REST 端点。一些调用会失败，但可以在事务性场景中进行重试，这样就有一定的成功可能性。比如说，对数据库进行更新所导致的 org.springframework.dao.DeadlockLoserDataAccess Exception 异常就可以进行重试。

配置一个步骤

在使用基于 Java 的配置开启重试时，第一个步骤需要是一个失败容错步骤，它反过来用于指定重试上限和重试异常。首先通过 faultTolerant()获取到一个失败容错步骤，接下来使用 retryLimit()方法指定重试的次数，最后使用 retry()方法指定触发重试的异常。要想指定多个异常，只需多次调用 retry()方法即可。

```
@Bean
public Step step1() {
    return steps.get("User Registration CSV To DB Step")
        .<UserRegistration,UserRegistration>chunk(10)
            .faultTolerant()
                .retryLimit(3).retry(DeadlockLoserDataAccessException.class)
        .reader(csvFileReader())
        .writer(jdbcItemWriter())
        .transactionManager(transactionManager)
        .build();
}
```

重试模板

此外，还可以在自己的代码中借助于 Spring Retry 支持来实现重试和恢复。比如说，你可能有一个自定义的 ItemWriter<T>，它需要重试功能，甚至整个服务接口都需要重试功能。

Spring Batch 通过 RetryTemplate 来支持这些场景，它（就像其他各种 Template 一样）将逻辑与重试隔离开来，使得在编写代码时就好像只会执行一次。让 Spring Batch 通过声明式配置来处理其他事情。

RetryTemplate 支持很多场景，它提供了便捷的 API 将冗余的重试/失败/恢复循环包装到简洁的单个方法调用之中。

我们来看看 11-4 节中那个简单的 ItemWriter<T>修改后的版本，了解一下如何编写自定义的 ItemWriter<T>。解决方案足够简单，理想情况下可以一直使用。不过，它无法处理服务中的错误情况。在使用 RPC 时，它总是会继续执行，就好像不可能让事情正常运转一样；服务本身可能会遇到违背语义和系统的情况。一种情况就是数据库键重复、信用卡卡号不合法等。当然，无论服务是分布式的还是在一个 VM 中，这些情况都会出现。

接下来，系统下面的 RPC 层也可能出现失败。下面是重写后的代码，这次考虑到了重试：

```
ppackage com.apress.springrecipes.springbatch;

import org.slf4j.Logger;
import org.slf4j.LoggerFactory;
import org.springframework.batch.item.ItemWriter;
import org.springframework.retry.RetryCallback;
import org.springframework.retry.support.RetryTemplate;

import java.util.List;

public class RetryableUserRegistrationServiceItemWriter implements
ItemWriter<UserRegistration> {

    private static final Logger logger = LoggerFactory.getLogger(RetryableUserRegistration
    ServiceItemWriter.class);

    private final UserRegistrationService userRegistrationService;
    private final RetryTemplate retryTemplate;

    public RetryableUserRegistrationServiceItemWriter(UserRegistrationService
    userRegistrationService, RetryTemplate retryTemplate) {
        this.userRegistrationService = userRegistrationService;
        this.retryTemplate = retryTemplate;
    }

    public void write(List<?extends UserRegistration> items)
        throws Exception {
        for (final UserRegistration userRegistration : items) {
            UserRegistration registeredUserRegistration = retryTemplate.execute(
                (RetryCallback<UserRegistration, Exception>) context ->
                userRegistrationService.registerUser(userRegistration));
```

```
            logger.debug("Registered: {}", registeredUserRegistration);
        }
    }
}
```

可以看到，代码修改并不多，结果也更加健壮。RetryTemplate 本身是在 Spring 上下文中配置的，不过在代码中创建也很容易。之所以在 Spring 上下文中声明，只是因为在创建对象时还需要做一些配置，我们让 Spring 处理这些配置。

RetryTemplate 更加有用的一个设置是这里所用的 BackOffPolicy。BackOffPolicy 表示 RetryTemplate 应该在两次重试之间回退多久。实际上，针对每次失败的尝试，它支持增加两次重试之间的延迟时间，这可以避免锁定尝试进行同样调用的其他客户端。当对相同的资源存在很多并发调用时可能会导致静态条件的出现，这种策略对于该情况来说很有意义。还有其他一些 BackOffPolicy 设置，包括将重试延迟设为固定时间的 FixedBackOffPolicy 策略。

```
@Bean
public RetryTemplate retryTemplate() {
    RetryTemplate retryTemplate = new RetryTemplate();
    retryTemplate.setBackOffPolicy(backOffPolicy());
    return retryTemplate;
}

@Bean
public ExponentialBackOffPolicy backOffPolicy() {
    ExponentialBackOffPolicy backOffPolicy = new ExponentialBackOffPolicy();
    backOffPolicy.setInitialInterval(1000);
    backOffPolicy.setMaxInterval(10000);
    backOffPolicy.setMultiplier(2);
    return backOffPolicy;
}
```

你已经配置了 RetryTemplate 的 backOffPolicy，这样 backOffPolicy 会在首次重试之前等待一秒钟（1000 毫秒）。随后的重试会将这个时间加倍（增长受到了 multiplier 的影响）。它会继续按照这种方式重试，直到达到 maxInterval 为止，之后所有的重试间隔时间就是一样的了。

基于 AOP 的重试

另一种方式是 Spring Batch 提供的 AOP 通知器，它会将在重试后成功无法得到保证的方法调用包装起来，就像在 RetryTemplate 中所做的那样。上一个示例重写了 ItemWriter<T>以使用模板。另一种方式只需要使用该重试逻辑通知整个 userRegistrationService 代理即可。在这种情况下，代码可以恢复到之前的样子，不使用 RetryTemplate！

要想做到这一点，需要为一个或多个需要重试的方式加上@Retryable 注解。为了达到与显式使用 RetryTemplate 的代码相同的效果，你需要添加如下内容：

```
@Retryable(backoff = @Backoff(delay = 1000, maxDelay = 10000, multiplier = 2))
public UserRegistration registerUser(UserRegistration userRegistrationRegistration) { … }
```

只添加这个注解还不够；还需要在配置上添加@EnableRetry 注解来启用注解处理。

```
@Configuration
@EnableBatchProcessing
@EnableRetry
@ComponentScan("com.apress.springrecipes.springbatch")
@PropertySource("classpath:/batch.properties")
public class BatchConfiguration { … }
```

11-7 控制步骤的执行

问题提出

你想要控制步骤的执行方式，目的也许是引入并发来减少不必要的时间浪费，也许是只有当某个条件为真时才执行步骤。

解决方案

有不同的方式可以修改任务的运行时信息，它们主要是通过控制步骤的执行方式来做到的：并发步骤、决策和串行步骤。

解释说明

到目前为止，你已经知道了如何在一个任务中运行一个步骤。不过，无论任务的复杂度如何，典型的任务都会有多个步骤。一个步骤对 bean 和它所封装的逻辑提供了一个边界（事务性或是非事务性的）。一个步骤可以有自己的读取器、写入器与处理器。每个步骤都会参与决策下一个步骤是什么。步骤之间是隔离的，每个步骤都会提供专门的功能，在复杂工作流中可以通过更新的模式与 Spring Batch 中的配置项对这些功能进行装配。实际上，如果对业务流程管理（Business Process Management，BPM）系统和工作流感兴趣的话，你会对接下来要介绍的一些概念和模式感到很熟悉。BPM 对流程和任务控制提供了很多支持，这与你现在看到的东西很相似。一个步骤通常对应于你在纸上勾勒出的任务定义的要点。比如说，加载每日销售额并生成报告的批任务可以描述为如下。

1. 从 CSV 文件中将客户信息加载到数据库中。
2. 计算每日统计数据并写到报告文件中。
3. 向消息队列发送消息，通知外部系统每个新加载的客户都登记成功了。

串行步骤

在上一个示例中，前两个步骤之间存在一个隐含的顺序；直到所有登记全部完成后才能写入到审计文件中。这种关系是两个步骤之间的默认关系。一个出现在另一个之后。每个步骤使用自己的执行上下文执行，多个步骤之间只共享父任务执行上下文与顺序号。

```
@Bean
public Job nightlyRegistrationsJob () {
    return jobs.get("nightlyRegistrationsJob ")
        .start(loadRegistrations())
        .next(reportStatistics())
        .next(...)
        .build();
    }
}
```

并发

Spring Batch 的第一个版本只会在同一个线程中执行批处理任务，通过某种调整，也可以在同一个虚拟机中执行。当然，尽管存在一些变通方法，不过情况还是不太理想。

在对这个示例任务的概述中，第一个步骤必须要在后两个步骤之前执行，这是因为后两个步骤依赖于第一个。不过，后两个步骤并不知道这种依赖关系。没有理由不能在传递 JMS 消息的同时写入审计日志。Spring Batch 提供了派生处理的能力，可以开启这种编排。

```
@Bean
public Job insertIntoDbFromCsvJob() {
    JobBuilder builder = jobs.get("insertIntoDbFromCsvJob");
    return builder
        .start(loadRegistrations())
        .split(taskExecutor())
            .add(
                builder.flow(reportStatistics()),
                builder.flow(sendJmsNotifications()))
        .build();
}
```

可以在任务构建器上使用 split()方法。要想将一个步骤纳入到一个流程中，可以使用任务构建器的 flow()方法；接下来，要想向流程添加更多的步骤，可以使用 next()方法。split()方法需要设置一个 TaskExecutor；

参见 2-23 节了解关于调度和并发的更多信息。

在该示例中，可以在 flow 元素中添加多个步骤，也可以在 split 元素后使用更多的步骤。就像 step 元素一样，split 元素也接收一个 next 属性。

Spring Batch 提供了一种机制来将处理转移到另一个进程当中。这种分发需要某种持久、可靠的连接。这正是 JMS 的用武之地，因为它可靠而且是事务性的，具有快速、可信赖等特点。Spring Batch 支持是在更高的层次上建模的，对于 Spring Integration 通道是在 Spring Integration 抽象之上的。该支持并非核心 Spring Batch 提供的；可以在 spring-batch-integration 项目中找到。远程分块使得每个步骤可以像通常一样在主线程中读取并聚合条目。这个步骤叫做 master。读取的条目会发送给运行在另一个进程（叫做 slave）中的 ItemProcessor<I,O>/ItemWriter<T>。如果 slave 是一个聚合消费者，那么你有一种简单、通用的机制实现可伸缩：工作可以在多个 JMS 客户端上立刻进行移交。这种聚合-消费者模式指的是多个 JMS 客户端都消费同一个队列中的消息这种情况。如果一个客户端消费消息并且忙于处理，那么其他空闲队列就会获取到消息。只要有空闲客户端，那么消息就会被立刻处理。

此外，Spring Batch 支持通过一种叫做分区的特性进行隐式伸缩。该特性很有趣，因为它是内建的且非常灵活。将步骤实例替换为它的一个子类 PartitionStep，该子类知道如何协调分布式执行器并维护步骤执行的元数据，这样在“远程分块”技术中就无须持久化的通信媒介了。

这里所提及的功能也是通用的。它可以与任何网格技术搭配使用，如 GridGain 或 Hadoop。Spring Batch 只自带了一个 TaskExecutorPartitionHandler，它可以通过 TaskExecutor 策略在多个线程中执行步骤。这个简单的实现对于该特性来说足够了！不过，如果解决不了你的问题，那可以对其进行扩展。

使用状态实现条件化步骤

使用给定任务或步骤的 ExitStatus 来决定下一个步骤是最简单的条件流形式。Spring Batch 通过“停止”“下一个”“失败”和“结束”元素对其提供了支持。默认情况下，Spring Batch 不会介入其中，一个步骤的 ExitStatus 会匹配其 BatchStatus，它是一个属性，其值定义在一个枚举中，可能为 COMPLETED、STARTING、STARTED、STOPPING、STOPPED、FAILED、ABANDONED 或是 UNKNOWN。

我们来看一个示例，它会根据上一个步骤成功与否来执行两个步骤其中的一个：

```
@Bean
public Job insertIntoDbFromCsvJob() {
    return jobs.get("User Registration Import Job")
        .start(step1())
            .on("COMPLETED").to(step2())
            .on("FAILED").to(failureStep())
        .build();
}
```

还可以使用通配符。如果想确保任意数量的 BatchStatus 值都有一个特定的行为，那么可以与只匹配一个 BatchStatus 值的更为具体的下一个元素相结合，这就会很有用。

```
@Bean
public Job insertIntoDbFromCsvJob() {
    return jobs.get("User Registration Import Job")
        .start(step1())
            .on("COMPLETED").to(step2())
            .on("*").to(failureStep())
        .build();
}
```

在该示例中，你让 Spring Batch 根据未明确指定的 ExitStatus 来执行某个步骤。另一种情况是只在 BatchStatus 值为 FAILED 时才停止处理。可以使用“失败”元素做到这一点。对上述示例进行些许改造，代码如下所示：

```
@Bean
public Job insertIntoDbFromCsvJob() {
    return jobs.get("User Registration Import Job")
        .start(step1())
            .on("COMPLETED").to(step2())
```

```
            .on("FAILED").fail()
        .build();
}
```

在所有这些示例中，你只对 Spring Batch 提供的标准 BatchStatus 值做出了响应。不过还可以使用自己的 ExitStatus。比如说，如果想让整个任务在出现 MAN DOWN 这个自定义的 ExitStatus 时失败，那就可以这么做：

```
@Bean
public Job insertIntoDbFromCsvJob() {
    return jobs.get("User Registration Import Job")
        .start(step1())
            .on("COMPLETED").to(step2())
            .on("FAILED").end("MAN DOWN")
        .build();
}
```

最后，如果想在 BatchStatus 为 COMPLETED 时结束处理，那就可以使用 end()方法。这是一种显式结束流程的方式，就好像它执行完了所有步骤且没有遇到错误一样。

```
@Bean
public Job insertIntoDbFromCsvJob() {
    return jobs.get("User Registration Import Job")
        .start(step1())
            .on("COMPLETED").end()
            .on("FAILED").to(errorStep())
        .build();
}
```

使用决策实现条件化步骤

如果想根据比任务的 ExitStatus 值更为复杂的逻辑来改变执行流，那么可以通过决策元素并使用 JobExecutionDecider 的实现来做到。

```
package com.apress.springrecipes.springbatch;

import org.springframework.batch.core.JobExecution;
import org.springframework.batch.core.StepExecution;
import org.springframework.batch.core.job.flow.FlowExecutionStatus;
import org.springframework.batch.core.job.flow.JobExecutionDecider;

public class HoroscopeDecider implements JobExecutionDecider {

    private boolean isMercuryIsInRetrograde () { return Math.random() > .9 ; }

    public FlowExecutionStatus decide(JobExecution jobExecution,
                                      StepExecution stepExecution) {
        if (isMercuryIsInRetrograde()) {
            return new FlowExecutionStatus("MERCURY_IN_RETROGRADE");
        }
        return FlowExecutionStatus.COMPLETED;
    }
}
```

剩下的就是配置，代码如下所示：

```
@Bean
public Job insertIntoDbFromCsvJob() {
    JobBuilder builder = jobs.get("insertIntoDbFromCsvJob");
    return builder
        .start(step1())
        .next((horoscopeDecider())
            .on("MERCURY_IN_RETROGRADE").to(step2())
            .on(("COMPLETED ").to(step3())
        .build();
}
```

11-8 启动任务

问题提出

Spring Batch 支持哪些部署场景？Spring Batch 如何启动？Spring Batch 如何使用系统调度器，比如说 cron 和 autosys，或是来自于一个 Web 应用？你想要理解这些内容。

解决方案

Spring Batch 在 Spring 所能运行的环境中都可以正常运行：public static void main、OSGi、Web 应用——任何地方！不过，有些情况颇具挑战的。比如说，让 Spring Batch 运行在与 HTTP 响应相同的线程中是不切实际的，因为这会延缓执行。Spring Batch 针对这种场景支持异步执行。它还提供了一个便捷类，可以与 cron 和 autosys 配合使用来支持任务的启动。此外，Spring 优秀的 scheduler 命名空间提供了很好的机制来调度任务。

解释说明

在创建解决方案之前，重要的是得知道有哪些方式可以部署并运行这些解决方案。所有解决方案都至少需要一个任务和一个 JobLauncher。你已经在前文中配置好了这些组件。任务是配置在 Spring 应用上下文中的，稍后将会看到。从 Java 代码启动一个 Spring Batch 解决方案的最简单的示例大约需要 5 行代码（如果已经获取到了 ApplicationContext，那就只需要 3 行）！

```
package com.apress.springrecipes.springbatch;

import org.springframework.batch.core.Job;
import org.springframework.batch.core.JobParameters;
import org.springframework.batch.core.JobParametersBuilder;
import org.springframework.batch.core.launch.JobLauncher;
import org.springframework.context.support.ClassPathXmlApplicationContext;

import java.util.Date;

public class Main {
    public static void main(String[] args) throws Throwable {
        ClassPathXmlApplicationContext ctx = new ClassPathXmlApplicationContext("solution2.xml");

        JobLauncher jobLauncher = ctx.getBean("jobLauncher", JobLauncher.class);
        Job job = ctx.getBean("myJobName", Job.class);
        JobExecution jobExecution = jobLauncher.run(job, new JobParameters());
    }
}
```

可以看到，你获取到之前配置的 JobLauncher 引用并通过它启动一个任务实例，其结果是个 JobExecution。可以通过 JobExecution 获取到任务的状态信息，包括退出状态和运行状态。

```
JobExecution jobExecution = jobLauncher.run(job, jobParameters);
BatchStatus batchStatus = jobExecution.getStatus();
while(batchStatus.isRunning()) {
    System.out.println( "Still running...");
    Thread.sleep( 10 * 1000 ); // 10 seconds
}
```

还可以得到 ExitStatus。

```
System.out.println( "Exit code: "+ jobExecution.getExitStatus().getExitCode());
```

JobExecution 还提供了其他很多有用的信息，如 Job 的创建时间、开始时间、上次更新日期与结束时间——所有这些时间都是 java.util.Date 实例。如果想要将任务关联到数据库中，那还需要任务实例与 ID。

```
JobInstance jobInstance = jobExecution.getJobInstance();
System.out.println( "job instance Id: "+ jobInstance.getId());
```

在这个简单的示例中，你使用了一个空的 JobParameters 实例。实际上，它只能使用一次。Spring Batch 会基于参数构建唯一的键，并使用它来区分不同 Job 的运行实例。下一节将会详细介绍如何参数化一个 Job。

从 Web 应用中启动

从 Web 应用中启动一个任务的方式稍有不同，因为客户端线程（比如说一个 HTTP 请求）通常不能等待一个批处理任务完成。理想的解决方案是从 Web 层的控制器或是行为（action）中启动时让任务异步执行，不由客户端线程处理。Spring Batch 借助于 Spring TaskExecutor 来支持这种场景。这需要对 JobLauncher 的配置进行一个简单的修改，不过 Java 代码不变。这里使用了一个 SimpleAsyncTaskExecutor，它会派生出一个执行线程并在不阻塞的情况下管理这个线程：

```
package com.apress.springrecipes.springbatch.config;

@Configuration
@EnableBatchProcessing
@ComponentScan("com.apress.springrecipes.springbatch")
@PropertySource("classpath:/batch.properties")
public class BatchConfiguration {

    @Bean
    public SimpleAsyncTaskExecutor taskExecutor() {
        return new SimpleAsyncTaskExecutor();
    }
}
```

由于不能再使用默认设置，因此需要添加自己的 BatchConfigurer 实现来配置 TaskExecutor 并将其添加到 SimpleJobLauncher 中。对于该实现来说，你使用 DefaultBatchConfigurer 作为参考；只需要重写 createJobLauncher 方法来添加 TaskExecutor 即可。

```
package com.apress.springrecipes.springbatch.config;

import org.springframework.batch.core.configuration.annotation.DefaultBatchConfigurer;
import org.springframework.batch.core.launch.JobLauncher;
import org.springframework.batch.core.launch.support.SimpleJobLauncher;
import org.springframework.core.task.TaskExecutor;
import org.springframework.stereotype.Component;

@Component
public class CustomBatchConfigurer extends DefaultBatchConfigurer {

    private final TaskExecutor taskExecutor;

    public CustomBatchConfigurer(TaskExecutor taskExecutor) {
        this.taskExecutor = taskExecutor;
    }

    @Override
    protected JobLauncher createJobLauncher() throws Exception {
        SimpleJobLauncher jobLauncher = new SimpleJobLauncher();
        jobLauncher.setJobRepository(getJobRepository());
        jobLauncher.setTaskExecutor(this.taskExecutor);
        jobLauncher.afterPropertiesSet();
        return jobLauncher;
    }
}
```

从命令行运行

另一种常见的场景是通过系统调度器（如 cron 或 autosys），甚至是 Windows 的事件调度器来部署批处理任务。Spring Batch 提供了一个便捷类，它接收 XML 应用上下文（包含了运行任务所需的一切）的名字和任务 bean 自身的名字作为参数。还可以通过其他一些额外的参数来定制任务。这些参数的形式必须为 name =value。如下示例展示了通过命令行（在 Linux/UNIX 系统上）来调用这个类，假设已经配置好了类路径：

```
java CommandLineJobRunner jobs.xml hourlyReport date='date +%m/%d/%Y time=date +%H'
```

CommandLineJobRunner 甚至可以返回系统错误码（0 表示成功，1 表示失败，2 表示加载批任务时出现了问题），这样 shell（被大多数系统调度器所用）就可以对失败进行响应和处理了。可以通过实现接口 ExitCodeMapper 创建并声明一个顶层的 bean 来返回更加复杂的返回码，这样就可以将退出状态消息转换为整数错误码了，shell 会在进程退出时看到它。

在调度中运行

Spring 支持调度框架（参见 3-22 节）。该框架非常适合于运行 Spring Batch。首先，修改既有的应用上下文配置，通过使用@EnableScheduling 注解并添加一个 ThreadPoolTaskScheduler 来启用调度。

```
package com.apress.springrecipes.springbatch.config;

@Configuration
@EnableBatchProcessing
@ComponentScan("com.apress.springrecipes.springbatch")
@PropertySource("classpath:/batch.properties")

@EnableScheduling
@EnableAsync
public class BatchConfiguration {

    @Bean
    public ThreadPoolTaskScheduler taskScheduler() {
        ThreadPoolTaskScheduler taskScheduler = new ThreadPoolTaskScheduler();
        taskScheduler.setThreadGroupName("batch-scheduler");
        taskScheduler.setPoolSize(10);
        return taskScheduler;
    }

}
```

这些导入对调度提供了最为简单的支持。上面的注解确保了包 com.apress.springrecipes. springbatch 下的所有 bean 都会根据需要得到配置和调度。调度器 bean 的代码如下所示：

```
package com.apress.springrecipes.springbatch.scheduler;

import org.springframework.batch.core.Job;
import org.springframework.batch.core.JobExecution;
import org.springframework.batch.core.JobParameters;
import org.springframework.batch.core.JobParametersBuilder;
import org.springframework.batch.core.launch.JobLauncher;
import org.springframework.scheduling.annotation.Scheduled;
import org.springframework.stereotype.Component;

import java.util.Date;

@Component
public class JobScheduler {

    private final JobLauncher jobLauncher;
    private final Job job;

    public JobScheduler(JobLauncher jobLauncher, Job job) {
        this.jobLauncher = jobLauncher;
        this.job = job;
    }

    public void runRegistrationsJob(Date date) throws Throwable {
        System.out.println("Starting job at " + date.toString());

        JobParametersBuilder jobParametersBuilder = new JobParametersBuilder();
        jobParametersBuilder.addDate("date", date);
        jobParametersBuilder.addString("input.file", "registrations");
```

```
        JobParameters jobParameters = jobParametersBuilder.toJobParameters();

        JobExecution jobExecution = jobLauncher.run(job, jobParameters);

        System.out.println("jobExecution finished, exit code: " + jobExecution.
        getExitStatus().getExitCode());
    }

    @Scheduled(fixedDelay = 1000 * 10)
    public void runRegistrationsJobOnASchedule() throws Throwable {
        runRegistrationsJob(new Date());
    }
}
```

这里没什么特别之处；它很好地说明了 Spring 框架的不同组件之间是如何协同工作的。由于使用了@Component 注解，因此 bean 会被识别出来并成为应用上下文的一部分，而@Component 注解的启用则是通过配置类上的@ComponentScan 注解来实现的。UserJob 类中只有一个 Job，并且只有一个 JobLauncher，这样就可以轻松将它们自动装配到你的 bean 中。最后，启动批处理运行的逻辑位于 runRegistrationsJob (java.util.Date date) 方法中。可以在任何地方调用该方法。该功能的唯一客户端就是调度方法 runRegistrationsJobOnASchedule。框架会根据@Scheduled 注解所指定的时间线来调用这个方法。

这类任务还有其他一些选择；传统上，在 Java 与 Spring 的世界中，这类问题非常适合于使用 Quartz。现在也是如此，因为 Spring 的调度支持不如 Quartz 的扩展性好。如果需要更为传统、易操作的调度工具，那么有一些经典的选择，如 cron、autosys 和 BMC 等。

11-9 参数化任务

问题提出

之前的示例运行得很好，不过在灵活性上还有待改进。要想将批处理代码应用到其他文件上，你需要编辑配置并在那里硬编码名字。参数化批处理解决方案的能力是非常有价值的。

解决方案

使用 JobParameters 来参数化任务，接下来可以通过 Spring Batch 的表达式语言或 API 调用供步骤使用。

解释说明

首先，我们将会介绍如何通过 JobParameters 来启动任务，接下来会介绍如何在 Job 和配置中使用并访问 JobParameters。

使用参数来启动任务

任务是 JobInstance 的原型。JobParameters 提供了一种方式来识别任务的唯一运行（一个 JobInstance）。这些 JobParameters 可以将输入传递给批处理过程，就像 Java 中的方法定义一样。之前的示例中已经出现过了 JobParameters，不过并未对其进行详细介绍。在使用 JobLauncher 启动任务时会创建 JobParameters 对象。要想启动名为 dailySalesFigures 的任务，同时向其传递一个需要处理的日期，请参见如下代码：

```
package com.apress.springrecipes.springbatch;

import com.apress.springrecipes.springbatch.config.BatchConfiguration;
import org.springframework.batch.core.*;
import org.springframework.batch.core.launch.JobLauncher;
import org.springframework.context.ApplicationContext;
import org.springframework.context.annotation.AnnotationConfigApplicationContext;

import java.util.Date;

public class Main {
    public static void main(String[] args) throws Throwable {

        ApplicationContext context =
```

```
            new AnnotationConfigApplicationContext(BatchConfiguration.class);

        JobLauncher jobLauncher = context.getBean(JobLauncher.class);
        Job job = context.getBean("dailySalesFigures", Job.class);

        jobLauncher.run(job, new JobParametersBuilder()
            .addDate( "date", new Date() ).toJobParameters());
    }
}
```

访问 JobParameters

从技术上来说，可以通过任意的 ExecutionContext（步骤与任务）获得 JobParameters。得到后，就可以通过调用 getLong()、getString()等方法以类型安全的方式访问参数了。一种简单的方式是绑定到@BeforeStep 事件，保存 StepExecution，然后遍历参数。接下来，你就可以查看参数并对其进行任何处理了。我们通过之前编写的 ItemProcessor<I,O>来看看。

```
// ...
private StepExecution stepExecution;

@BeforeStep
public void saveStepExecution(StepExecution stepExecution) {
    this.stepExecution = stepExecution;
}

public UserRegistration process(UserRegistration input) throws Exception {

    Map<String, JobParameter> params = stepExecution.getJobParameters().getParameters();
    for (String jobParameterKey : params.keySet()) {
        System.out.println(String.format("%s=%s", jobParameterKey,
    params.get(jobParameterKey).getValue().toString()));
    }

    Date date = stepExecution.getJobParameters().getDate("date");
    // etc ...
}
```

这么做价值有限。在 80%的情况下，你需要将任务启动的参数绑定到应用上下文中的 Spring bean 上。这些参数只在运行期才能使用，而 XML 应用上下文中的步骤则是在设计期配置的。这出现在很多地方。之前的示例演示了通过硬编码的路径来配置 ItemWriters<T>和 ItemReaders<T>。除非你想要参数化文件名，否则这种方式用起来没什么问题。如果不是想让任务只使用一次的话，这种方式令人难以接受！

核心的 Spring 框架提供了一种强大的表达式语言，Spring Batch 可以使用它将参数绑定推迟到正确的时间（在该示例中就是等到 bean 处于正确的作用域中时）。Spring Batch 的“步骤”作用域就是为了这个目的。我们来看看如何改写之前的示例，为 ItemReader 的资源使用参数化的文件名：

```
@Bean
@StepScope
public ItemReader<UserRegistration> csvFileReader(@Value("file:${user.home}/
batches/#{jobParameters['input.fileName']}.csv") Resource input) { … }
```

你所要做的就是将 bean（FlatFileItemReader<T>）的作用域设为一个步骤的生命周期（这时，JobParameters 将会正确解析完毕），然后使用 EL 语法来参数化路径。

小结

本章介绍了批处理的概念、一些历史，以及它适合于现代架构的原因所在。我们介绍了 Spring Batch（来自于 SpringSource 的批处理解决方案），以及如何在批任务中使用 ItemReader<T>和 ItemWriter<T>的实现进行读写。你根据需要编写了自己的 ItemReader<T>和 ItemWriter <T>实现，并掌握了如何控制步骤在一个任务中的执行。

第 12 章

Spring 与 NoSQL

大多数应用都会用到关系型数据库，如 Oracle、MySQL 或是 PostgreSQL；不过，除了 SQL 数据库之外，数据存储还有其他选择。现在有：

- 关系型数据库（Oracle、MySQL、PostgreSQL 等）；
- 文档存储（MongoDB、Couchbase）；
- 键值存储（Redis、Volgemort）；
- 列存储（Cassandra）；
- 图存储（Neo4j、Giraph）。

每一种技术（以及所有实现）的工作方式都不同，因此你需要花时间学习每种想要使用的技术。此外，还得编写大量重复的样板代码来处理事务和异常转换。

Spring Data 项目可以让你轻松一些；它可以通过样板代码来配置不同的技术。每个集成模块都支持将异常转换为 Spring 一致的 DataAccessException 层次体系并使用 Spring 的模板方式。Spring Data 还为一些技术提供了跨存储的解决方案，这意味着一部分模型可以通过 JPA 存储在关系型数据库中，另一部分则可以存储在图或是文档存储中。

■ **提示**：本章的每一节都会介绍如何下载和安装所需的持久化存储。不过，bin 目录包含的脚本可以为每一种持久化存储创建 Docker 容器。

12-1 使用 MongoDB

问题提出

你想要使用 MongoDB 存储和获取文档。

解决方案

下载并配置 MongoDB。

解释说明

在开始使用 MongoDB 前，你需要安装一个实例并运行它。当运行后，你需要连接它才能使用数据存储进行实际的存储。我们首先使用原始的 MongoDB 来了解如何存储和获取文档，接下来将会使用 Spring Data MongoDB，最后会使用反应式版本的仓库。

下载并启动 MongoDB

从 MongoDB 官网下载 MongoDB。选择适合于所用系统的版本并按照手册中的安装指令操作（http://docs.mongodb.org/manual/installation/）。安装完毕后就可以启动 MongoDB 了。要想启动 MongoDB，请在命令行上执行 mongodb 命令，如图 12-1 所示。这会在端口 27017 上启动一个 MongoDB 服务器。如果需要使用不同的端口，那么在启动服务器时可以通过在命令行上指定--port 选项来做到。

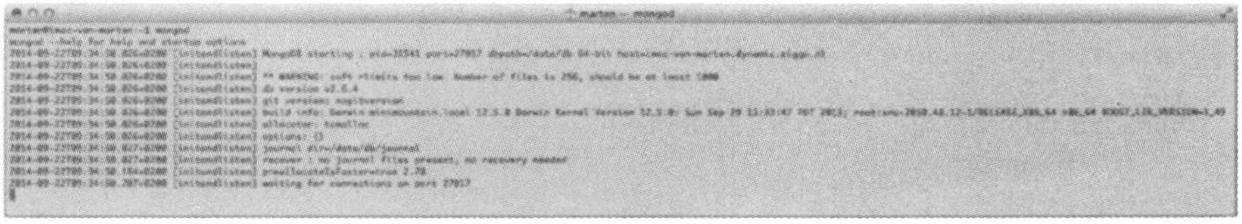

图 12-1 MongoDB 首次启动后的输出

默认的数据存储位置在\data\db（对于 Windows 用户来说，它在 MongoDB 安装的磁盘根目录处）。要想改变路径，请在命令行上使用--dbpath 选项。请确保该目录存在且 MongoDB 可以写入。

连接到 MongoDB

要想连接 MongoDB，你需要使用一个 Mongo 实例。可以通过该实例获取到要使用的数据库以及实际的底层集合。我们来创建一个小系统，它会创建一个对象并使用 MongoDB 将其存储起来。

```
package com.apress.springrecipes.nosql;

public class Vehicle {

    private String vehicleNo;
    private String color;
    private int wheel;
    private int seat;

    public Vehicle() {
    }

    public Vehicle(String vehicleNo, String color, int wheel, int seat) {
        this.vehicleNo = vehicleNo;
        this.color = color;
        this.wheel = wheel;
        this.seat = seat;
    }
    /// Getters and Setters have been omitted for brevity.
}
```

为了使用该对象，创建一个仓库接口。

```
package com.apress.springrecipes.nosql;

public interface VehicleRepository {

    long count();
    void save(Vehicle vehicle);
    void delete(Vehicle vehicle);
    List<Vehicle> findAll()
    Vehicle findByVehicleNo(String vehicleNo);
}
```

对于 MongoDB 来说，创建 VehicleRepository 的一个实现 MongoDBVehicleRepository。

```
package com.apress.springrecipes.nosql;

import com.mongodb.*;

import java.util.ArrayList;
import java.util.List;

public class MongoDBVehicleRepository implements VehicleRepository {

    private final Mongo mongo;
    private final String collectionName;
    private final String databaseName;
```

```
    public MongoDBVehicleRepository(Mongo mongo, String databaseName, String collectionName){
        this.mongo = mongo;
        this.databaseName=databaseName;
        this.collectionName = collectionName;
    }

    @Override
    public long count() {
        return getCollection().count();
    }

    @Override
    public void save(Vehicle vehicle) {
        BasicDBObject query = new BasicDBObject("vehicleNo", vehicle.getVehicleNo());
        DBObject dbVehicle = transform(vehicle);
        DBObject fromDB = getCollection().findAndModify(query, dbVehicle);
        if (fromDB == null) {
            getCollection().insert(dbVehicle);
        }
    }

    @Override
    public void delete(Vehicle vehicle) {
        BasicDBObject query = new BasicDBObject("vehicleNo", vehicle.getVehicleNo());
        getCollection().remove(query);
    }

    @Override
    public List<Vehicle> findAll() {
        DBCursor cursor = getCollection().find(null);
        List<Vehicle> vehicles = new ArrayList<>(cursor.size());
        for (DBObject dbObject : cursor) {
            vehicles.add(transform(dbObject));
        }
        return vehicles;
    }

    @Override
    public Vehicle findByVehicleNo(String vehicleNo) {
        BasicDBObject query = new BasicDBObject("vehicleNo", vehicleNo);
        DBObject dbVehicle = getCollection().findOne(query);
        return transform(dbVehicle);
    }

    private DBCollection getCollection() {
        return mongo.getDB(databaseName).getCollection(collectionName);
    }

    private Vehicle transform(DBObject dbVehicle) {
        return new Vehicle(
            (String) dbVehicle.get("vehicleNo"),
            (String) dbVehicle.get("color"),
            (int) dbVehicle.get("wheel"),
            (int) dbVehicle.get("seat"));
    }

    private DBObject transform(Vehicle vehicle) {
        BasicDBObject dbVehicle = new BasicDBObject("vehicleNo", vehicle.getVehicleNo())
            .append("color", vehicle.getColor())
            .append("wheel", vehicle.getWheel())
            .append("seat", vehicle.getSeat());
        return dbVehicle;
    }
}
```

首先注意到构造方法接收 3 个参数。第一个是实际的 MongoDB 客户端，第二个是将要使用的数据库名，最后一个是对象存储的集合名。MongoDB 中的文档存储在集合中，集合则属于数据库。

要想轻松访问所用的 DBCollection，可以使用 getCollection 方法，它会得到数据库的一个连接并返回配置好的 DBCollection。接下来，这个 DBCollection 可用于执行诸如存储、删除和更新文档等操作。

save 方法首先会尝试更新既有的文档。如果失败，那就会创建一个针对给定 Vehicle 的新文档。要想存储对象，首先需要将领域对象 Vehicle 转换为 DBObject，在该示例中就是 BasicDBObject。BasicDBObject 接收 Vehicle 对象不同属性的键值对。在查询文档时会使用同样的 DBObject，给定对象上的键值对被用于查找文档；仓库中的 findByVehicleNo 方法就说明了这一点。DBObject 对象与 Vehicle 对象之间的转换是通过两个 transform 方法实现的。

创建如下 Main 类来使用这个类：

```
package com.apress.springrecipes.nosql;

import com.mongodb.MongoClient;

import java.util.List;

public class Main {

    public static final String DB_NAME = "vehicledb";

    public static void main(String[] args) throws Exception {
        // Default monogclient for localhost and port 27017
        MongoClient mongo = new MongoClient();

        VehicleRepository repository = new MongoDBVehicleRepository(mongo, DB_NAME,
        "vehicles");

        System.out.println("Number of Vehicles: " + repository.count());

        repository.save(new Vehicle("TEM0001", "RED", 4, 4));
        repository.save(new Vehicle("TEM0002", "RED", 4, 4));

        System.out.println("Number of Vehicles: " + repository.count());

        Vehicle v = repository.findByVehicleNo("TEM0001");

        System.out.println(v);

        List<Vehicle> vehicleList = repository.findAll();

        System.out.println("Number of Vehicles: " + vehicleList.size());
        vehicleList.forEach(System.out::println);
        System.out.println("Number of Vehicles: " + repository.count());
        // Cleanup and close
        mongo.dropDatabase(DB_NAME);
        mongo.close();
    }
}
```

Main 类构建了一个 MongoClient 实例，它会连接到 localhost 上的 27017 端口来获取一个 MongoDB 实例。如果需要使用其他端口或主机，那么它还有一个将主机与端口作为参数的构造方法：new MongoClient("mongodb-server.local", 28018)。接下来构建了一个 MongoDBVehicleRepository 类的实例，之前构建的 MongoClient 会传递给它，同时还会传递数据库名（vehicledb）和集合名（vehicles）。

接下来的几行代码会向数据库插入两个 vehicle，然后查询它们，最后将其删除。Main 类的最后几行代码会先删除数据库，然后关闭 MongoClient。在使用生产数据库时请不要对数据库执行删除操作。

使用 Spring 进行配置

MongoClient 与 MongoDBVehicleRepository 的创建和配置可以轻松迁移到 Spring 配置中。

```
package com.apress.springrecipes.nosql.config;

import com.apress.springrecipes.nosql.MongoDBVehicleRepository;
import com.apress.springrecipes.nosql.VehicleRepository;
import com.mongodb.Mongo;
import com.mongodb.MongoClient;
import org.springframework.context.annotation.Bean;
import org.springframework.context.annotation.Configuration;

import java.net.UnknownHostException;

@Configuration
public class MongoConfiguration {

    public static final String DB_NAME = "vehicledb";

    @Bean
    public Mongo mongo() throws UnknownHostException {
        return new MongoClient();
    }

    @Bean
    public VehicleRepository vehicleRepository(Mongo mongo) {
        return new MongoDBVehicleRepository(mongo, DB_NAME, " vehicles");
    }
}
```

将被@PreDestroy 所注解的如下方法添加到 MongoDBVehicleRepository 中来负责数据库的清理。

```
@PreDestroy
public void cleanUp() {
    mongo.dropDatabase(databaseName);
}
```

最后，更新 Main 类来使用修改后的配置。

```
package com.apress.springrecipes.nosql;

...
import org.springframework.context.ApplicationContext;
import org.springframework.context.annotation.AnnotationConfigApplicationContext;
import org.springframework.context.support.AbstractApplicationContext;

import java.util.List;

public class Main {

    public static final String DB_NAME = "vehicledb";

    public static void main(String[] args) throws Exception {
        ApplicationContext ctx =
            new AnnotationConfigApplicationContext(MongoConfiguration.class);
        VehicleRepository repository = ctx.getBean(VehicleRepository.class);
        ...

         ((AbstractApplicationContext) ctx).close();

    }
}
```

配置是由 AnnotationConfigApplicationContext 加载的。从该上下文中加载 VehicleRepository bean 并执行操作。当运行上下文的代码关闭时，它会触发 MongoDBVehicleRepository 中 cleanUp 方法的调用。

使用 MongoTemplate 简化 MongoDB 代码

现在，MongoDBVehicleRepository 类使用的是原生的 MongoDB API。虽然并不复杂，但还是需要了解 API 才行。此外，这里有一些重复性任务，比如说 DBObject 与 Vehicle 对象之间的转换。使用 MongoTemplate 能够极大地简化仓库的操作。

注意：在使用 Spring Data Mongo 前，需要将相关的 JAR 添加到类路径下。如果使用 Maven，请添加如下依赖：

```
<dependency>
    <groupId>org.springframework.data</groupId>
    <artifactId>spring-data-mongodb</artifactId>
    <version>1.10.1.RELEASE</version>
</dependency>
```

如果使用 Gradle，请添加如下依赖：

```
compile 'org.springframework.data:spring-data-mongodb:1.10.1.RELEASE'
```

```java
package com.apress.springrecipes.nosql;

import org.springframework.data.mongodb.core.MongoTemplate;
import org.springframework.data.mongodb.core.query.Query;

import javax.annotation.PreDestroy;
import java.util.List;

import static org.springframework.data.mongodb.core.query.Criteria.where;

public class MongoDBVehicleRepository implements VehicleRepository {

    private final MongoTemplate mongo;
    private final String collectionName;

    public MongoDBVehicleRepository(MongoTemplate mongo, String collectionName) {
        this.mongo = mongo;
        this.collectionName = collectionName;
    }

    @Override
    public long count() {
        return mongo.count(null, collectionName);
    }

    @Override
    public void save(Vehicle vehicle) {
        mongo.save(vehicle, collectionName);
    }

    @Override
    public void delete(Vehicle vehicle) {
        mongo.remove(vehicle, collectionName);
    }

    @Override
    public List<Vehicle> findAll() {
        return mongo.findAll(Vehicle.class, collectionName);
    }

    @Override
    public Vehicle findByVehicleNo(String vehicleNo) {
        return mongo.findOne(new Query(where("vehicleNo").is(vehicleNo)), Vehicle.class,
```

```
            collectionName);
    }

    @PreDestroy
    public void cleanUp() {
        mongo.execute(db -> {
            db.drop();
            return null;
        });
    }
}
```

使用 MongoTemplate 后的代码看起来简洁了不少。它为几乎所有操作都提供了便捷方法：save、update 和 delete。此外，它还提供了非常棒的查询构建器方式（参见 findByVehicleNo 方法）。由于不再有与 MongoDB 类之间的映射，因此也就不需要再创建 DBObject 了。这些工作现在都由 MongoTemplate 来处理。要想将 Vehicle 对象转换为 MongoDB 类，可以使用 MongoConverter。默认情况下会使用 MappingMongoConverter。该映射器会将属性（properties）映射为属性（attribute）名，反之亦然，在映射时它还会尝试进行正确的数据类型转换。如果需要特定的映射，那么可以编写自己的 MongoConverter 实现并将其注册到 MongoTemplate 上。

由于使用了 MongoTemplate，配置需要做一些修改。

```
package com.apress.springrecipes.nosql.config;

import com.apress.springrecipes.nosql.MongoDBVehicleRepository;
import com.apress.springrecipes.nosql.VehicleRepository;
import com.mongodb.Mongo;
import org.springframework.context.annotation.Bean;
import org.springframework.context.annotation.Configuration;
import org.springframework.data.mongodb.core.MongoClientFactoryBean;
import org.springframework.data.mongodb.core.MongoTemplate;

@Configuration
public class MongoConfiguration {

    public static final String DB_NAME = "vehicledb";

    @Bean
    public MongoTemplate mongo(Mongo mongo) throws Exception {
        return new MongoTemplate(mongo, DB_NAME);
    }

    @Bean
    public MongoClientFactoryBean mongoFactoryBean() {
        return new MongoClientFactoryBean();
    }

    @Bean
    public VehicleRepository vehicleRepository(MongoTemplate mongo) {
        return new MongoDBVehicleRepository(mongo, "vehicles");
    }
}
```

注意，这里使用了 MongoClientFactoryBean。它可以轻松创建 MongoClient。它并非使用 MongoTemplate 的必须之选，但却有助于简化客户端的配置。另外一个好处就是不会再抛出 java.net.UnknownHostException，它会由 MongoClientFactoryBean 在内部所处理。

MongoTemplate 有多个构造方法。这里使用的方法接收了一个 Mongo 实例和要使用的数据库名。为了解析数据库，需要用到 MongoDbFactory 的一个实例；默认情况下，使用的是 SimpleMongoDbFactory。在大多数情况下这已足够，不过如果遇到一些特殊情况，比如说加密的集合，那么扩展默认实现也是相当简单的。最后，将 MongoTemplate 和集合名注入到 MongoDBVehicleRepository 中。

还需要为 Vehicle 对象再增加一个信息。它需要一个字段来存储生成的 ID。可以为其添加一个名为 id 的字段或是将一个字段标记为@Id 注解。

```
public class Vehicle {

    private String id;

...
}
```

使用注解来指定映射信息

现在，MongoDBVehicleRepository 需要知道待访问的集合名。如果可以在 Vehicle 对象上指定该信息，那就会变得更加轻松且灵活，就像 JPA @Table 注解一样。借助于 Spring Data Mongo，我们可以使用@Document 注解做到这一点。

```
package com.apress.springrecipes.nosql;

import org.springframework.data.mongodb.core.mapping.Document;

@Document(collection = "vehicles")
public class Vehicle { ... }
```

@Document 注解接收两个属性：collection 和 language。collection 属性指定要使用的集合名，language 属性指定该对象的语言。由于映射信息位于 Vehicle 类上，集合名就可以从 MongoDBVehicleRepository 中去掉了。

```
public class MongoDBVehicleRepository implements VehicleRepository {

    private final MongoTemplate mongo;

    public MongoDBVehicleRepository(MongoTemplate mongo) {
        this.mongo = mongo;
    }

    @Override
    public long count() {
        return mongo.count(null, Vehicle.class);
    }

    @Override
    public void save(Vehicle vehicle) {
        mongo.save(vehicle);
    }

    @Override
    public void delete(Vehicle vehicle) {
        mongo.remove(vehicle);
    }

    @Override
    public List<Vehicle> findAll() {
        return mongo.findAll(Vehicle.class);
    }

    @Override
    public Vehicle findByVehicleNo(String vehicleNo) {
        return mongo.findOne(new Query(where("vehicleNo").is(vehicleNo)), Vehicle.class);
    }
}
```

当然，集合名也可以从 MongoDBVehicleRepository 中去掉了。

```
@Configuration
public class MongoConfiguration {
...
    @Bean
```

```
    public VehicleRepository vehicleRepository(MongoTemplate mongo) {
        return new MongoDBVehicleRepository(mongo);
    }
}
```

当运行 Main 类时，结果应该与之前一样。

创建 Spring Data MongoDB 仓库

由于不需要对 MongoDB 类进行映射，以及不需要再传递集合名，因此代码已经精简了不少，不过还可以更进一步。借助于 Spring Data Mongo 的另一个特性，MongoDBVehicleRepository 的整个实现就可以移除了。

首先，需要修改一下配置。

```
package com.apress.springrecipes.nosql.config;

import com.mongodb.Mongo;
import org.springframework.context.annotation.Bean;
import org.springframework.context.annotation.Configuration;
import org.springframework.data.mongodb.core.MongoClientFactoryBean;
import org.springframework.data.mongodb.core.MongoTemplate;
import org.springframework.data.mongodb.repository.config.EnableMongoRepositories;

@Configuration
@EnableMongoRepositories(basePackages = "com.apress.springrecipes.nosql")
public class MongoConfiguration {

    public static final String DB_NAME = "vehicledb";

    @Bean
    public MongoTemplate mongoTemplate(Mongo mongo) throws Exception {
        return new MongoTemplate(mongo, DB_NAME);
    }

    @Bean
    public MongoClientFactoryBean mongoFactoryBean() {
        return new MongoClientFactoryBean();
    }
}
```

首先，注意到这里去掉了用于构建 MongoDBVehicleRepository 的@Bean 方法。其次，添加了@EnableMongoRepositories 注解。这开启了对继承 Spring Data CrudRepository 的接口的检测，并且可用于使用了注解@Document 的领域对象。

要想让 VehicleRepository 被 Spring Data 检测到，需要让其继承 CrudRepository 或是它的一个子接口，如 MongoRepository。

```
package com.apress.springrecipes.nosql;

import org.springframework.data.mongodb.repository.MongoRepository;

public interface VehicleRepository extends MongoRepository<Vehicle, String> {

    public Vehicle findByVehicleNo(String vehicleNo);

}
```

你可能想知道那些方法都去哪儿了。它们已经定义在了父接口中，因此该接口就不需要再定义它们了。findByVehicleNo 方法还在。该方法依然会根据 vehicleNo 属性查询 Vehicle。所有的 findBy 方法都会转换为一个 MongoDB 查询。findBy 后面的部分会被解释为一个属性名。还可以通过 and、or 和 between 等运算符编写更加复杂的查询。

再次运行 Main 类，结果应该是一样的；不过，使用 MongoDB 时实际编写的代码已经得到了最大的简化。

创建反应式 Spring Data MongoDB 仓库

相较于创建传统的 MongoDB 仓库，还可以创建反应式仓库，这是通过继承 ReactiveMongoRepository 类（或是其他反应式仓库接口）实现的。这会改变方法的返回类型，返回单个值的方法会变成返回一个 Mono<T>（如果没有返回值，则是 Mono<Void>），返回零或多个元素的方法会变成返回一个 Flux<T>。

注意：如果想要使用 RxJava 而非 Project Reactor，请继承 RxJava 的两个 Repository 接口之一，并用 Single 和 Observable 分别代替 Mono 和 Flux。

要想使用反应式仓库实现，首先需要使用 MongoDB 驱动的一个反应式实现并配置 Spring Data 来使用该驱动。为了简化，可以继承 AbstractReactiveMongoConfiguration 并实现两个必要的方法——getDatabaseName 和 mongoClient。

```
@Configuration
@EnableReactiveMongoRepositories(basePackages = "com.apress.springrecipes.nosql")
public class MongoConfiguration extends AbstractReactiveMongoConfiguration {

    public static final String DB_NAME = "vehicledb";

    @Bean
    @Override
    public MongoClient reactiveMongoClient() {
        return MongoClients.create();
    }

    @Override
    protected String getDatabaseName() {
        return DB_NAME;
    }
}
```

另一处变更是使用@EnableReactiveMongoRepositories 代替了@EnableMongoRepositories。数据库名依然还是需要的，你需要通过反应式驱动器连接到 MongoDB 实例。可以使用 MongoClients.create 方法；这里只需使用默认值即可。

接下来，修改 VehicleRepository，让其继承 ReactiveMongoRepository，这样它就成为反应式的了；还需要将 findByVehicleNo 方法的返回类型由普通的 Vehicle 改为 Mono<Vehicle>。

```
package com.apress.springrecipes.nosql;

import org.springframework.data.mongodb.repository.ReactiveMongoRepository;
import reactor.core.publisher.Mono;

public interface VehicleRepository extends ReactiveMongoRepository<Vehicle, String> {

    Mono<Vehicle> findByVehicleNo(String vehicleNo);

}
```

最后需要修改的是 Main 类。相比于阻塞调用，你希望以流式的方式来调用方法。

```
package com.apress.springrecipes.nosql;

import com.apress.springrecipes.nosql.config.MongoConfiguration;
import org.springframework.context.ApplicationContext;
import org.springframework.context.annotation.AnnotationConfigApplicationContext;
import org.springframework.context.support.AbstractApplicationContext;
import reactor.core.publisher.Flux;

import java.util.concurrent.CountDownLatch;

public class Main {
```

```
    public static void main(String[] args) throws Exception {
        ApplicationContext ctx =
            new AnnotationConfigApplicationContext(MongoConfiguration.class);
        VehicleRepository repository = ctx.getBean(VehicleRepository.class);

        CountDownLatch countDownLatch = new CountDownLatch(1);

        repository.count().doOnSuccess(cnt -> System.out.println("Number of Vehicles: " + cnt))
            .thenMany(repository.saveAll(
                Flux.just(
                    new Vehicle("TEM0001", "RED", 4, 4),
                    new Vehicle("TEM0002", "RED", 4, 4)))).last()
            .then(repository.count()).doOnSuccess(cnt -> System.out.println("Number of
            Vehicles: " + cnt))
            .then(repository.findByVehicleNo("TEM0001")).doOnSuccess(System.out::println)
            .then(repository.deleteAll())
                .doOnSuccess(x -> countDownLatch.countDown())
                .doOnError(t -> countDownLatch.countDown())
            .then(repository.count()).subscribe(cnt -> System.out.println
            ("Number of Vehicles: " + cnt.longValue()));

        countDownLatch.await();
        ((AbstractApplicationContext) ctx).close();

    }
}
```

流程从计算总数开始，如果成功，那就将 Vehicle 实例放到 MongoDB 中。当添加了 last() vehicle 后，再次计算总数。接下来是个查询，然后是 deleteAll。所有这些方法调用都以反应式的方式一个接着一个进行，由事件进行触发。由于不想使用 block()方法阻塞，因此使用一个 CountDownLatch 等待代码执行。当所有记录都删除后，计数器值会减少，接下来程序会继续执行。即便如此，这依然是阻塞的。当在完全的反应式栈中这么做时，你可能会从最后一个 then 中返回 Mono，然后进行进一步的组合或是通过 Spring WebFlux 控制器进行输出（见第 5 章）。

12-2　使用 Redis

问题提出

你想使用 Redis 来存储数据。

解决方案

下载并安装 Redis，使用 Spring 和 Spring Data 访问 Redis 实例。

解释说明

Redis 是个键值缓存或存储，它只会持有简单数据类型，如字符串和哈希表等。如果存储更为复杂的数据结构，那就需要对其进行转换。

下载并启动 Redis

可以从 Redis 官网下载 Redis 源码。在本书编写之际，3.2.8 是最新发布的稳定版。可以在 https://github.com/MSOpenTech/redis/releases 下载 Windows 的编译版本。官方下载站点只提供 UNIX 版本。Mac 用户可以通过 Homebrew（http://brew.sh）安装 Redis。

下载并安装 Redis 后，在命令行中使用 redis-server 命令启动。启动后，输出应该类似于图 12-2 所示。它会输出进程 ID（PID）和监听的端口号（默认是 6379）。

连接到 Redis

要想连接 Redis，需要一个客户端，就像连接到数据库所用的 JDBC 驱动一样。有几个客户端可以使用，可以在 Redis 网站（http://redis.io/clients）上看到可用客户端的完整列表。本节将会使用 Jedis 客户端，因为它相当活跃并且是 Redis 团队推荐的。

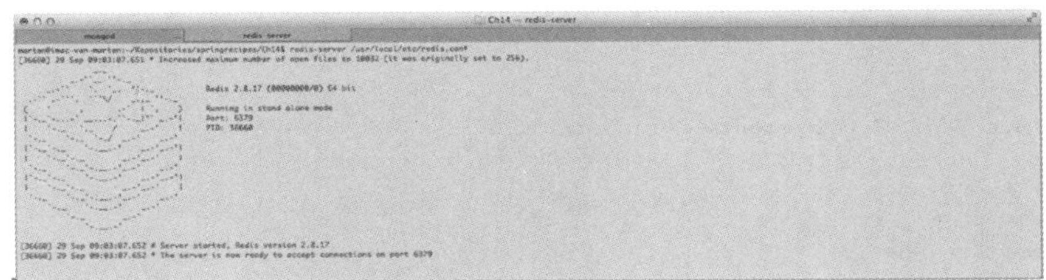

图 12-2　启动 Redis 后的输出

我们从一个简单的 Hello World 示例开始，看看如何创建到 Redis 的连接。

```
package com.apress.springrecipes.nosql;

import redis.clients.jedis.Jedis;

public class Main {

    public static void main(String[] args) {
        Jedis jedis = new Jedis("localhost");
        jedis.set("msg", "Hello World, from Redis!");
        System.out.println(jedis.get("msg"));
    }
}
```

上述代码创建了一个 Jedis 客户端并向其传递待连接的主机名，该示例中就是 localhost。Jedis 客户端的 set 方法会将一个消息置于存储中，通过 get 方法再将这条消息取出来。相较于简单对象来说，还可以让 Redis 模拟 List 或 Map。

```
package com.apress.springrecipes.nosql;

import redis.clients.jedis.Jedis;

public class Main {

    public static void main(String[] args) {
        Jedis jedis = new Jedis("localhost");
        jedis.rpush("authors", "Marten Deinum", "Josh Long", "Daniel Rubio", "Gary Mak");
        System.out.println("Authors: " + jedis.lrange("authors",0,-1));

        jedis.hset("sr_3", "authors", "Gary Mak, Danial Rubio, Josh Long, Marten Deinum");
        jedis.hset("sr_3", "published", "2014");

        jedis.hset("sr_4", "authors", "Josh Long, Marten Deinum");
        jedis.hset("sr_4", "published", "2017");

        System.out.println("Spring Recipes 3rd: " + jedis.hgetAll("sr_3"));
        System.out.println("Spring Recipes 4th: " + jedis.hgetAll("sr_4"));
    }
}
```

借助于 rpush 和 lpush，你可以向 List 中添加元素。rpush 会将元素添加到列表的末尾，lpush 则将它们添加到列表的开头。可以通过 lrange 和 rrange 方法取出元素。lrange 从左侧开始，接收一个起始与结束位置的索引。示例使用了-1，表示所有。

要想向 Map 添加元素，请使用 hset。它接收一个 key、一个 field 和一个 value。另一种方式是使用 hmset（多重集合），它接收一个 Map<String, String>或 Map<byte[], byte[]>作为参数。

使用 Redis 存储对象

Redis 是个键值存储，只能处理 String 和 byte[]，键也如此。因此，在 Redis 中存储对象并不像其他技术那么直接。对象存储前需要序列化为 String 或 byte[]。

我们重用 12-1 节中的 Vehicle 类，使用 Jedis 客户端存储并获取它。

```
package com.apress.springrecipes.nosql;

import java.io.Serializable;

public class Vehicle implements Serializable{

    private String vehicleNo;
    private String color;
    private int wheel;
    private int seat;

    public Vehicle() {
    }

    public Vehicle(String vehicleNo, String color, int wheel, int seat) {
        this.vehicleNo = vehicleNo;
        this.color = color;
        this.wheel = wheel;
        this.seat = seat;
    }
    // getters/setters omitted
}
```

注意，Vehicle 类实现了 Serializable。这样才能让这个 Java 对象实现序列化。在存储对象前，需要将对象转换为 Java 中的 byte[]。ObjectOutputStream 可以写入对象，ByteArrayOutputStream 可以将对象写到 byte[]中。要想将 byte[]转换回对象，可以使用 ObjectInputStream 和 ByteArrayInputStream。Spring 为此提供了一个名为 org.springframework.util.SerializationUtils 的辅助类，它拥有 serialize 和 deserialize 方法。

现在，在 Main 类中创建一个 Vehicle 并使用 Jedis 存储它。

```
package com.apress.springrecipes.nosql;

import org.springframework.util.SerializationUtils;
import redis.clients.jedis.Jedis;

public class Main {

    public static void main(String[] args) throws Exception {
        Jedis jedis = new Jedis("localhost");

        final String vehicleNo = "TEM0001";
        Vehicle vehicle = new Vehicle(vehicleNo, "RED", 4,4);

        jedis.set(vehicleNo.getBytes(), SerializationUtils.serialize(vehicle));

        byte[] vehicleArray = jedis.get(vehicleNo.getBytes());

        System.out.println("Vehicle: " + SerializationUtils.deserialize(vehicleArray));
    }
}
```

首先，创建了一个 Vehicle 实例。接下来，使用之前介绍的 SerializationUtils 将对象转换为 byte[]。在存储 byte[]时，键也需要是 byte[]；因此，键（这里是 vehicleNo）也需要转换。最后，使用相同的键再从存储中读取出序列化的对象，并将其转换回对象。这种方式的缺点是存储的每个对象都需要实现 Serializable 接口。如果做不到这一点，对象就可能丢失，或是在序列化过程中就可能出错。此外，byte[]是类的一种表示。如果类发生了变化，那么将其转换回对象时就很有可能会失败。

另一种方式是使用对象的 String 表示。将 Vehicle 对象转换为 XML 或 JSON，它们要比 byte[]更加灵活。我们看看如何使用优秀的 Jackson JSON 库将对象转换为 JSON：

```
package com.apress.springrecipes.nosql;
```

```
import com.fasterxml.jackson.databind.ObjectMapper;
import redis.clients.jedis.Jedis;

public class Main {

    public static void main(String[] args) throws Exception {
        Jedis jedis = new Jedis("localhost");
        ObjectMapper mapper = new ObjectMapper();
        final String vehicleNo = "TEM0001";
        Vehicle vehicle = new Vehicle(vehicleNo, "RED", 4,4);

        jedis.set(vehicleNo, mapper.writeValueAsString(vehicle));

        String vehicleString = jedis.get(vehicleNo);

        System.out.println("Vehicle: " + mapper.readValue(vehicleString, Vehicle.class));
    }
}
```

首先，需要一个 ObjectMapper 实例。它用于实现对象与 JSON 之间的转换。这里使用 writeValueAsString 方法将对象转换为了 JSON String。然后将这个 String 存储到 Redis 中。接下来，再次读取该 String 并传递给 ObjectMapper 的 readValue 方法。这里根据类型参数 Vehicle.class 构建出一个对象，JSON 会被映射为给定类的一个实例。

使用 Redis 存储对象并非那么直观，有人认为这不是 Redis 的本来意图（存储复杂的对象结构）。

配置并使用 RedisTemplate

由于用于连接到 Redis 的客户端库的不同，使用 Redis API 会更加困难一些。为了统一这一点，出现了 RedisTemplate。它支持大多数 Redis Java 客户端。除了提供统一的方式外，它还负责将异常转换为 Spring 的 DataAccessException 层次体系。这使得它可以很好地集成既有的数据访问并使用 Spring 的事务支持。

RedisTemplate 需要一个 RedisConnectionFactory 来获取连接。RedisConnectionFactory 是个接口，它有几个实现。该示例使用的是 JedisConnectionFactory。

```
package com.apress.springrecipes.nosql.config;

import com.apress.springrecipes.nosql.Vehicle;
import org.springframework.context.annotation.Bean;
import org.springframework.context.annotation.Configuration;
import org.springframework.data.redis.connection.RedisConnectionFactory;
import org.springframework.data.redis.connection.jedis.JedisConnectionFactory;
import org.springframework.data.redis.core.RedisTemplate;

@Configuration
public class RedisConfig {

    @Bean
    public RedisTemplate<String, Vehicle> redisTemplate(RedisConnectionFactory
    connectionFactory) {
        RedisTemplate template = new RedisTemplate();
        template.setConnectionFactory(connectionFactory);
        return template;
    }

    @Bean
    public RedisConnectionFactory redisConnectionFactory() {
        return new JedisConnectionFactory();
    }
}
```

注意 redisTemplate bean 方法的返回类型。RedisTemplate 是个泛型类，需要指定一个键值类型。在该示例中，键的类型是 String，值的类型是 Vehicle。在存储和获取对象时，RedisTemplate 负责进行转换。转换是通过 RedisSerializer 接口完成的，该接口有几个实现（见表 12-1）。默认的 RedisSerializer 实现 JdkSerializationRedisSerializer 使用标准的 Java 序列化实现对象与 byte[]之间的转换。

表 12-1 默认的 RedisSerializer 实现

名字	说明
GenericToStringSerializer	String 到 byte[]序列化器；使用 Spring ConversionService 先将对象转换为 String，然后再转换为 byte[]
Jackson2JsonRedisRedisSerializer	使用 Jackson 2 ObjectMapper 读写 JSON
JacksonJsonRedisRedisSerializer	使用 Jackson ObjectMapper 读写 JSON
JdkSerializationRedisSerializer	使用默认的 Java 序列化与反序列化，它是默认实现
OxmSerializer	使用 Spring 的 Marshaller 与 Unmarshaller 读写 XML
StringRedisSerializer	简单的 String 到 byte[]转换器

为了使用 RedisTemplate，需要修改一下 Main 类。在该类中会加载配置并获取到 RedisTemplate。

```
package com.apress.springrecipes.nosql;

import com.apress.springrecipes.nosql.config.RedisConfig;
import org.springframework.context.ApplicationContext;
import org.springframework.context.annotation.AnnotationConfigApplicationContext;
import org.springframework.data.redis.core.RedisTemplate;

public class Main {

    public static void main(String[] args) throws Exception {
        ApplicationContext context = new AnnotationConfigApplicationContext(RedisConfig.class);
        RedisTemplate<String, Vehicle> template = context.getBean(RedisTemplate.class);

        final String vehicleNo = "TEM0001";
        Vehicle vehicle = new Vehicle(vehicleNo, "RED", 4,4);
        template.opsForValue().set(vehicleNo, vehicle);
        System.out.println("Vehicle: " + template.opsForValue().get(vehicleNo));
    }
}
```

从 ApplicationContext 获取到 RedisTemplate 模板后，它就可以使用了。它最大的好处在于你可以使用对象了，模板会处理对象转换的繁重工作。注意，set 方法接收 String 和 Vehicle 作为参数，而非只接收 String 或是 byte[]。这使得代码的可读性更好，可维护性更强。默认情况下使用的是 JDK 序列化。要想使用 Jackson，需要配置一个不同的 RedisSerializer。

```
package com.apress.springrecipes.nosql.config;
...
import org.springframework.data.redis.serializer.Jackson2JsonRedisSerializer;

@Configuration
public class RedisConfig {

    @Bean
    public RedisTemplate<String, Vehicle> redisTemplate() {
        RedisTemplate template = new RedisTemplate();
        template.setConnectionFactory(redisConnectionFactory());
        template.setDefaultSerializer(new Jackson2JsonRedisSerializer(Vehicle.class));
        return template;
    }
...
}
```

RedisTemplate 模板现在会使用 Jackson ObjectMapper 对象进行序列化和反序列化。其余代码保持不变。再次运行程序，依然没问题，对象会使用 JSON 进行存储。当在事务中使用 Redis 时，它也可以参与到这个事务中。要想做到这一点，请将 RedisTemplate 模板的 enableTransactionSupport 属性设为 true。当事务提交时，它会负责在事务中执行 Redis 操作。

12-3 使用 Neo4j

问题提出

你想在应用中使用 Neo4j。

解决方案

使用 Spring Data Neo4j 库访问 Neo4j。

解释说明

在开始使用 Neo4J 前，需要安装它的一个实例并运行起来。运行后，你需要连接 Neo4J 才能使用数据库进行实际的存储。我们首先介绍普通的基于 Neo4J 的仓库来说明如何存储与获取对象，接下来介绍基于 Spring Data Neo4J 的仓库。

下载并运行 Neo4J

可以从 Neo4j 网站下载 Neo4j。对于本节来说，下载社区版就足够了；不过，使用 Neo4j 商业版也是可以的。Windows 用户可以运行安装器来安装 Neo4j。Mac 与 Linux 用户可以解压缩归档文件，进入创建的目录中，使用 bin/neo4j 来启动。Mac 用户还可以使用 Homebrew（http://brew.sh），通过 brew install neo4j 命令来安装 Neo4j。然后在命令行上执行命令 neo4j start 就可以启动 Neo4j 了。

在命令行启动 Neo4j 后，输出应该类似于图 12-3 所示。

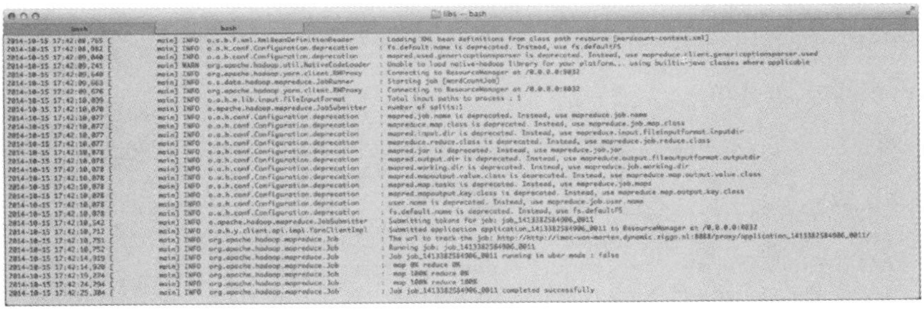

图 12-3　首次启动 Neo4j 的输出

开始使用 Neo4j

我们首先使用嵌入式的 Neo4j 服务器来创建一个简单的 Hello World 程序。创建一个 Main 类来启动嵌入式服务器，向 Neo4j 添加一些数据，然后再次获取这些数据。

```java
package com.apress.springrecipes.nosql;

import org.neo4j.graphdb.GraphDatabaseService;
import org.neo4j.graphdb.Node;
import org.neo4j.graphdb.Transaction;
import org.neo4j.graphdb.factory.GraphDatabaseFactory;

import java.nio.file.Paths;

public class Main {

    public static void main(String[] args) {
        final String DB_PATH = System.getProperty("user.home") + "/friends";

        GraphDatabaseService db = new GraphDatabaseFactory()
            .newEmbeddedDatabase(Paths.get(DB_PATH).toFile());
```

```
        Transaction tx1 = db.beginTx();

        Node hello = db.createNode();
        hello.setProperty("msg", "Hello");

        Node world = db.createNode();
        world.setProperty("msg", "World");
        tx1.success();

        db.getAllNodes().stream()
            .map(n -> n.getProperty("msg"))
            .forEach(m -> System.out.println("Msg: " + m));

        db.shutdown();
    }
}
```

这个 Main 类会启动一个嵌入式 Neo4j 服务器。接下来，它会开启一个事务并创建两个节点。然后，获取到所有节点并将 msg 属性值打印到控制台上。Neo4j 非常善于遍历节点间的关系，这一点得到了专门的优化（就像其他图数据存储一样）。

我们创建一些节点并在节点之间建立关系。

```
package com.apress.springrecipes.nosql;

import org.neo4j.graphdb.*;
import org.neo4j.graphdb.factory.GraphDatabaseFactory;

import java.nio.file.Paths;

import static com.apress.springrecipes.nosql.Main.RelationshipTypes.*;

public class Main {

    enum RelationshipTypes implements RelationshipType {FRIENDS_WITH, MASTER_OF, SIBLING, LOCATION}

    public static void main(String[] args) {
        final String DB_PATH = System.getProperty("user.home") + "/friends";
        final GraphDatabaseService db = new GraphDatabaseFactory()
            .newEmbeddedDatabase(Paths.get(DB_PATH).toFile());
        final Label character = Label.label("character");
        final Label planet = Label.label("planet");

        try (Transaction tx1 = db.beginTx()) {

            // Planets
            Node dagobah = db.createNode(planet);
            dagobah.setProperty("name", "Dagobah");

            Node tatooine = db.createNode(planet);
            tatooine.setProperty("name", "Tatooine");

            Node alderaan = db.createNode(planet);
            alderaan.setProperty("name", "Alderaan");

            // Characters
            Node yoda = db.createNode(character);
            yoda.setProperty("name", "Yoda");

            Node luke = db.createNode(character);
            luke.setProperty("name", "Luke Skywalker");

            Node leia = db.createNode(character);
            leia.setProperty("name", "Leia Organa");
```

```
            Node han = db.createNode(character);
            han.setProperty("name", "Han Solo");
            // Relations
            yoda.createRelationshipTo(luke, MASTER_OF);
            yoda.createRelationshipTo(dagobah, LOCATION);
            luke.createRelationshipTo(leia, SIBLING);
            luke.createRelationshipTo(tatooine, LOCATION);
            luke.createRelationshipTo(han, FRIENDS_WITH);
            leia.createRelationshipTo(han, FRIENDS_WITH);
            leia.createRelationshipTo(alderaan, LOCATION);

            tx1.success();
        }

        Result result = db.execute("MATCH (n) RETURN n.name as name");
        result.stream()
            .flatMap(m -> m.entrySet().stream())
            .map(row -> row.getKey() + " : " + row.getValue() + ";")
            .forEach(System.out::println);

        db.shutdown();
    }
}
```

上述代码反映了星战宇宙的一小部分。它有人物角色及其位置，实际上都是星球。人与人之间还存在着关系（图 12-4 展示了关系图）。

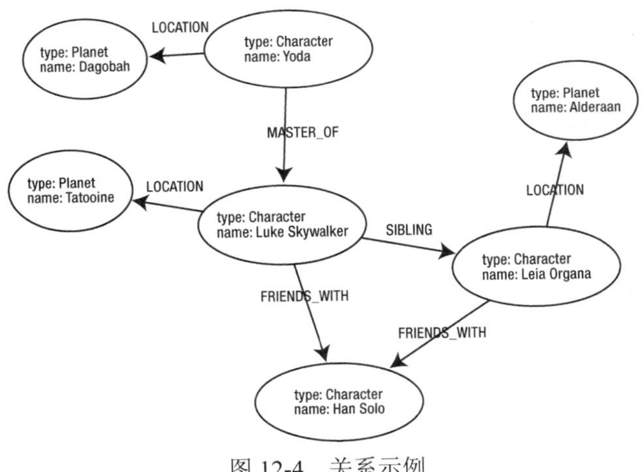

图 12-4　关系示例

代码中的关系是通过一个枚举来实现的，这个枚举实现了 Neo4j 的一个接口 RelationshipType。顾名思义，它用于区分不同类型的关系。节点类型是通过节点上的标签来区分的。所设置的名字作为节点的一个基本属性。当运行代码时，它会执行 cypher 查询：MATCH (n) RETURN n.name as name。这会选择所有节点并返回它们的 name 属性。

使用 Neo4j 映射对象

目前为止的代码都是低层次且绑定到 Neo4j 上的。创建和操纵节点是非常麻烦的。理想情况下，你应该使用 Planet 类与 Character 类，并可以将其存储到 Neo4j 以及从 Neo4j 中取回。首先，创建 Planet 与 Character 类。

```
package com.apress.springrecipes.nosql;

public class Planet {

    private long id = -1;
```

```
    private String name;
    // Getters and Setters omitted
}
package com.apress.springrecipes.nosql;

import java.util.ArrayList;
import java.util.Collections;
import java.util.List;

public class Character {

    private long id = -1;
    private String name;

    private Planet location;
    private final List<Character> friends = new ArrayList<>();
    private Character apprentice;

    public void addFriend(Character friend) {
        friends.add(friend);
    }

    // Getters and Setters omitted
}
```

Planet 类相当直观，它有 id 和 name 属性。Character 类稍微有点复杂。它也有 id 和 name 属性，此外还有针对于关系的其他一些属性：用于 LOCATION 关系的 location 属性、用于 FRIENDS_WITH 关系的 Character 集合属性，以及用于 MASTER_OF 关系的 apprentice 属性。

为了能够存储这些类，我们创建一个 StarwarsRepository 接口来封装保存操作。

```
package com.apress.springrecipes.nosql;

public interface StarwarsRepository {

    Planet save(Planet planet);
    Character save(Character character);

}
```

如下是针对 Neo4j 的实现：

```
package com.apress.springrecipes.nosql;

import org.neo4j.graphdb.GraphDatabaseService;
import org.neo4j.graphdb.Label;
import org.neo4j.graphdb.Node;
import org.neo4j.graphdb.Transaction;

import static com.apress.springrecipes.nosql.RelationshipTypes.*;

public class Neo4jStarwarsRepository implements StarwarsRepository {

    private final GraphDatabaseService db;

    public Neo4jStarwarsRepository(GraphDatabaseService db) {
        this.db = db;
    }

    @Override
    public Planet save(Planet planet) {
        if (planet.getId() != null) {
            return planet;
        }
        try (Transaction tx = db.beginTx()) {
            Label label = Label.label("planet");
```

```
            Node node = db.createNode(label);
            node.setProperty("name", planet.getName());
            tx.success();
            planet.setId(node.getId());
            return planet;
        }
    }

    @Override
    public Character save(Character character) {
        if (character.getId() != null) {
            return character;
        }

        try (Transaction tx = db.beginTx()) {
            Label label = Label.label("character");
            Node node = db.createNode(label);
            node.setProperty("name", character.getName());

            if (character.getLocation() != null) {
                Planet planet = character.getLocation();
                planet = save(planet);
                node.createRelationshipTo(db.getNodeById(planet.getId()), LOCATION);
            }

            for (Character friend : character.getFriends()) {
                friend = save(friend);
                node.createRelationshipTo(db.getNodeById(friend.getId()), FRIENDS_WITH);
            }

            if (character.getApprentice() != null) {
                save(character.getApprentice());
                node.createRelationshipTo(db.getNodeById(character.getApprentice().getId()),
                MASTER_OF);
            }

            tx.success();
            character.setId(node.getId());
            return character;
        }
    }

}
```

将对象转换为 Node 对象要做不少工作。对于 Planet 对象来说还是比较简单的。首先检查是否已经被持久化（如果是的话，ID 就会大于-1）；如果没有，那就开启事务，创建节点，设置 name 属性，将 id 值传递给 Planet 对象。不过，对于 Character 类来说就有些复杂了，因为需要考虑所有的关系。

需要修改一下 Main 类来反映这些变化。

```
package com.apress.springrecipes.nosql;

import org.neo4j.graphdb.GraphDatabaseService;
import org.neo4j.graphdb.Result;
import org.neo4j.graphdb.Transaction;
import org.neo4j.graphdb.factory.GraphDatabaseFactory;

import java.nio.file.Paths;

public class Main {

    public static void main(String[] args) {
        final String DB_PATH = System.getProperty("user.home") + "/starwars";
        final GraphDatabaseService db = new GraphDatabaseFactory().
        newEmbeddedDatabase(Paths.get(DB_PATH).toFile());
```

```
StarwarsRepository repository = new Neo4jStarwarsRepository(db);

try (Transaction tx = db.beginTx()) {

    // Planets
    Planet dagobah = new Planet();
    dagobah.setName("Dagobah");

    Planet alderaan = new Planet();
    alderaan.setName("Alderaan");

    Planet tatooine = new Planet();
    tatooine.setName("Tatooine");

    dagobah = repository.save(dagobah);
    repository.save(alderaan);
    repository.save(tatooine);

    // Characters
    Character han = new Character();
    han.setName("Han Solo");

    Character leia = new Character();
    leia.setName("Leia Organa");
    leia.setLocation(alderaan);
    leia.addFriend(han);

    Character luke = new Character();
    luke.setName("Luke Skywalker");
    luke.setLocation(tatooine);
    luke.addFriend(han);
    luke.addFriend(leia);

    Character yoda = new Character();
    yoda.setName("Yoda");
    yoda.setLocation(dagobah);
    yoda.setApprentice(luke);

    repository.save(han);
    repository.save(luke);
    repository.save(leia);
    repository.save(yoda);

    tx.success();
}

Result result = db.execute("MATCH (n) RETURN n.name as name");
result.stream()
        .flatMap(m -> m.entrySet().stream())
        .map(row -> row.getKey() + " : " + row.getValue() + ";")
        .forEach(System.out::println);

    db.shutdown();

    }
}
```

执行结果应该与之前一样。不过，主要的差别在于，现在的代码使用的是领域对象而非直接使用节点。将对象作为节点存储到 Neo4j 中是相当麻烦的。幸好，Spring Data Neo4j 简化了这一切。

使用 Neo4j OGM 映射对象

在转换为节点和关系时，属性的处理是相当麻烦的。如果只需要通过注解就能指定将什么存储到哪里岂不是很好？就像 JPA 一样！Neo4j OGM 提供了这些注解。要想让一个对象成为 Neo4j 映射的实体，请在类型上使用@NodeEntity 注解。可以通过@Relationship 注解对关系进行建模。要想标识作为 ID 使用的字段，请为其添加@GraphId 注解。将这些注解添加到 Planet 与 Character 类上，代码如下所示：

```
package com.apress.springrecipes.nosql;

import org.neo4j.ogm.annotation.GraphId;
import org.neo4j.ogm.annotation.NodeEntity;

@NodeEntity
public class Planet {

    @GraphId
    private Long id;
    private String name;

    // Getters/setters omitted
}
```

如下是 Character 类：

```
package com.apress.springrecipes.nosql;

import org.neo4j.ogm.annotation.GraphId;
import org.neo4j.ogm.annotation.NodeEntity;
import org.neo4j.ogm.annotation.Relationship;

import java.util.Collections;
import java.util.HashSet;
import java.util.Objects;
import java.util.Set;

@NodeEntity
public class Character {

    @GraphId
    private Long id;
    private String name;

    @Relationship(type = "LOCATION")
    private Planet location;
    @Relationship(type="FRIENDS_WITH")
    private final Set<Character> friends = new HashSet<>();
    @Relationship(type="MASTER_OF")
    private Character apprentice;

    // Getters / Setters omitted

}
```

在实体使用了注解后，现在就可以重写仓库，通过 SessionFactory 和 Session 来加快访问了。

```
package com.apress.springrecipes.nosql;

import org.neo4j.ogm.model.Result;
import org.neo4j.ogm.session.Session;
import org.neo4j.ogm.session.SessionFactory;
import org.neo4j.ogm.transaction.Transaction;
import org.springframework.beans.factory.annotation.Autowired;
import org.springframework.stereotype.Repository;

import javax.annotation.PreDestroy;
import java.util.Collections;

@Repository
public class Neo4jStarwarsRepository implements StarwarsRepository {
```

```
    private final SessionFactory sessionFactory;

    @Autowired
    public Neo4jStarwarsRepository(SessionFactory sessionFactory) {
        this.sessionFactory = sessionFactory;
    }

    @Override
    public Planet save(Planet planet) {

        Session session = sessionFactory.openSession();
        try (Transaction tx = session.beginTransaction()) {
            session.save(planet);
            return planet;
        }
    }

    @Override
    public Character save(Character character) {

        Session session = sessionFactory.openSession();
        try (Transaction tx = session.beginTransaction()) {
            session.save(character);
            return character;
        }
    }

    @Override
    public void printAll() {

        Session session = sessionFactory.openSession();
        Result result = session.query("MATCH (n) RETURN n.name as name",
            Collections.emptyMap(), true);
        result.forEach(m -> m.entrySet().stream()
            .map(row -> row.getKey() + " : " + row.getValue() + ";")
            .forEach(System.out::println));
    }

}
```

这里有许多地方值得关注：使用 SessionFactory 和 Session 时，代码变得更加简洁，因为大量样板代码已经帮你写好了，特别是对象与节点之间的映射。最后说一下增加的 printAll 方法。它将 Main 类中的一些代码转移到了仓库中。

接下来要修改的是 Main 类，我们需要在 Main 类中构建 SessionFactory。要想构建 SessionFactory，你需要指定它要扫描哪些包来寻找被@NodeEntity 所注解的类。

```
package com.apress.springrecipes.nosql;

import org.neo4j.ogm.session.SessionFactory;

public class Main {

    public static void main(String[] args) {
        SessionFactory sessionFactory = new SessionFactory("com.apress.springrecipes.nosql");

        StarwarsRepository repository = new Neo4jStarwarsRepository(sessionFactory);

        // Planets
        Planet dagobah = new Planet();
        dagobah.setName("Dagobah");

        Planet alderaan = new Planet();
```

```
        alderaan.setName("Alderaan");

        Planet tatooine = new Planet();
        tatooine.setName("Tatooine");

        dagobah = repository.save(dagobah);
        repository.save(alderaan);
        repository.save(tatooine);

        // Characters
        Character han = new Character();
        han.setName("Han Solo");

        Character leia = new Character();
        leia.setName("Leia Organa");
        leia.setLocation(alderaan);
        leia.addFriend(han);

        Character luke = new Character();
        luke.setName("Luke Skywalker");
        luke.setLocation(tatooine);
        luke.addFriend(han);
        luke.addFriend(leia);

        Character yoda = new Character();
        yoda.setName("Yoda");
        yoda.setLocation(dagobah);
        yoda.setApprentice(luke);

        repository.save(han);
        repository.save(luke);
        repository.save(leia);
        repository.save(yoda);

        repository.printAll();

        sessionFactory.close();
    }
}
```

上述代码创建了一个 SessionFactory，并使用它构建了一个 Neo4jStarwarsRepository。接下来准备数据，然后调用 printAll 方法。之前这个位置处的代码现在已经被放到了 printAll 方法中。最终结果应该与之前的类似。

使用 Spring 进行配置

到现在为止，一切都是手工配置并装配的。我们来创建一个 Spring 配置类。

```
package com.apress.springrecipes.nosql;

import org.neo4j.ogm.session.SessionFactory;
import org.springframework.context.annotation.Bean;
import org.springframework.context.annotation.Configuration;

@Configuration
public class StarwarsConfig {

    @Bean
    public SessionFactory sessionFactory() {
        return new SessionFactory("com.apress.springrecipes.nosql");
    }

    @Bean
    public Neo4jStarwarsRepository starwarsRepository(SessionFactory sessionFactory) {
```

```
        return new Neo4jStarwarsRepository(sessionFactory);
    }
}
```

现在，SessionFactory 和 Neo4jStarwarsRepository 都是 Spring 管理的 bean 了。现在可以让 Main 类使用该配置启动一个 ApplicationContext 并从中获取 StarwarsRepository。

```
package com.apress.springrecipes.nosql;

import org.springframework.context.annotation.AnnotationConfigApplicationContext;

public class Main {

    public static void main(String[] args) {
        AnnotationConfigApplicationContext context =
            new AnnotationConfigApplicationContext(StarwarsConfig.class);

        StarwarsRepository repository = context.getBean(StarwarsRepository.class);

        // Planets
        Planet dagobah = new Planet();
        dagobah.setName("Dagobah");

        Planet alderaan = new Planet();
        alderaan.setName("Alderaan");

        Planet tatooine = new Planet();
        tatooine.setName("Tatooine");

        dagobah = repository.save(dagobah);
        repository.save(alderaan);
        repository.save(tatooine);

        // Characters
        Character han = new Character();
        han.setName("Han Solo");

        Character leia = new Character();
        leia.setName("Leia Organa");
        leia.setLocation(alderaan);
        leia.addFriend(han);

        Character luke = new Character();
        luke.setName("Luke Skywalker");
        luke.setLocation(tatooine);
        luke.addFriend(han);
        luke.addFriend(leia);

        Character yoda = new Character();
        yoda.setName("Yoda");
        yoda.setLocation(dagobah);
        yoda.setApprentice(luke);

        repository.save(han);
        repository.save(luke);
        repository.save(leia);
        repository.save(yoda);

        repository.printAll();

        context.close();
    }
}
```

大体上还是相同的。主要的差别在于现在是由 Spring 控制着 bean 的生命周期。

Spring Data Neo4j 还提供了一个 Neo4jTransactionManager 实现，它负责开启和停止事务，就像其他的 PlatformTransactionManager 实现所做的那样。首先，修改配置引入它，并添加@EnableTransactionManagement 注解。

```
package com.apress.springrecipes.nosql;

import org.neo4j.ogm.session.SessionFactory;
import org.springframework.context.annotation.Bean;
import org.springframework.context.annotation.Configuration;
import org.springframework.data.neo4j.transaction.Neo4jTransactionManager;
import org.springframework.transaction.annotation.EnableTransactionManagement;

@Configuration
@EnableTransactionManagement
public class StarwarsConfig {

    @Bean
    public SessionFactory sessionFactory() {
        return new SessionFactory("com.apress.springrecipes.nosql");
    }

    @Bean
    public Neo4jStarwarsRepository starwarsRepository(SessionFactory sessionFactory) {
        return new Neo4jStarwarsRepository(sessionFactory);
    }

    @Bean
    public Neo4jTransactionManager transactionManager(SessionFactory sessionFactory) {
        return new Neo4jTransactionManager(sessionFactory);
    }
}
```

接下来，可以进一步清理 Neo4jStarwarsRepository 类了。

```
package com.apress.springrecipes.nosql;

import org.neo4j.ogm.model.Result;
import org.neo4j.ogm.session.Session;
import org.neo4j.ogm.session.SessionFactory;
import org.springframework.beans.factory.annotation.Autowired;
import org.springframework.stereotype.Repository;
import org.springframework.transaction.annotation.Transactional;

import javax.annotation.PreDestroy;
import java.util.Collections;

@Repository
@Transactional
public class Neo4jStarwarsRepository implements StarwarsRepository {

    private final SessionFactory sessionFactory;

    @Autowired
    public Neo4jStarwarsRepository(SessionFactory sessionFactory) {
        this.sessionFactory = sessionFactory;
    }

    @Override
    public Planet save(Planet planet) {
        Session session = sessionFactory.openSession();
        session.save(planet);
        return planet;
    }

    @Override
```

```
    public Character save(Character character) {
        Session session = sessionFactory.openSession();
        session.save(character);
        return character;
    }

    @Override
    public void printAll() {

        Session session = sessionFactory.openSession();
        Result result = session.query("MATCH (n) RETURN n.name as name",
            Collections.emptyMap(), true);
        result.forEach(m -> m.entrySet().stream()
            .map(row -> row.getKey() + " : " + row.getValue() + ";")
            .forEach(System.out::println));
    }

    @PreDestroy
    public void cleanUp() {
        Session session = sessionFactory.openSession();
        session.query("MATCH (n) OPTIONAL MATCH (n)-[r]-() DELETE n,r", null);
    }
}
```

Main 类保持不变，存储和查询的结果应该与之前一样。

使用 Spring Data Neo4j 仓库

代码已经得到了极大的简化。SessionFactory 和 Session 的使用使得 Neo4j 中对实体的处理变得更加轻松。还可以更进一步。就像 Spring Data JPA 或 Mongo 一样，它可以为你生成仓库。你所要做的唯一一件事就是编写接口。我们创建 PlanetRepository 与 CharacterRepository 类来操作实体。

```
package com.apress.springrecipes.nosql;

import org.springframework.data.repository.CrudRepository;

public interface CharacterRepository extends CrudRepository<Character, Long> {}
```

下面是 PlanetRepository：

```
package com.apress.springrecipes.nosql;

import org.springframework.data.repository.CrudRepository;

public interface PlanetRepository extends CrudRepository<Planet, Long> {}
```

这两个仓库都继承自 CrudRepository，不过也可以继承 PagingAndSortingRepository 或是专门的 Neo4jRepository 接口。对于本节来说，CrudRepository 就足够了。

接下来，将 StarwarsRepository 及其实现重命名为 StarwarsService，因为它不再是一个仓库了；实现还需要做一些修改，以操作仓库而非 SessionFactory。

```
package com.apress.springrecipes.nosql;

import org.springframework.stereotype.Service;
import org.springframework.transaction.annotation.Transactional;

import javax.annotation.PreDestroy;

@Service
@Transactional
public class Neo4jStarwarsService implements StarwarsService {

    private final PlanetRepository planetRepository;
    private final CharacterRepository characterRepository;
```

```
    Neo4jStarwarsService(PlanetRepository planetRepository,
                         CharacterRepository characterRepository) {
        this.planetRepository=planetRepository;
        this.characterRepository=characterRepository;
    }

    @Override
    public Planet save(Planet planet) {
        return planetRepository.save(planet);
    }

    @Override
    public Character save(Character character) {
        return characterRepository.save(character);
    }

    @Override
    public void printAll() {
        planetRepository.findAll().forEach(System.out::println);
        characterRepository.findAll().forEach(System.out::println);
    }

    @PreDestroy
    public void cleanUp() {
        characterRepository.deleteAll();
        planetRepository.deleteAll();
    }
}
```

现在，所有操作都是在具体的仓库接口中完成的。这些接口本身并不会创建实例。要想开启创建，请将@EnableNeo4jRepositories 注解添加到配置类上。此外，添加@ComponentScan 注解以便能够检测到 StarwarsService 并进行自动装配。

```
@Configuration
@EnableTransactionManagement
@EnableNeo4jRepositories
@ComponentScan
public class StarwarsConfig { ... }
```

注意@EnableNeo4jRepositories 注解。该注解会扫描配置的基础包（base package）以寻找仓库。找到后就会创建一个动态实现，该实现最终会委托给 SessionFactory。

最后，修改 Main 类来使用重构后的 StarwarsService。

```
package com.apress.springrecipes.nosql;

import com.apress.springrecipes.nosql.config.StarwarsConfig;
import org.springframework.context.annotation.AnnotationConfigApplicationContext;

public class Main {

    public static void main(String[] args) {
        AnnotationConfigApplicationContext context =
            new AnnotationConfigApplicationContext(StarwarsConfig.class);

        StarwarsService service = context.getBean(StarwarsService.class);
        ...
    }
}
```

现在，所有组件都更新为使用动态创建的 Spring Data Neo4j 仓库了。

连接到远程 Neo4j 数据库

到目前为止，针对 Neo4j 的所有代码使用的都是嵌入式 Neo4j 实例；之前已经下载并安装好了 Neo4j。我们对配置做一些修改，连接远程 Neo4j 实例。

```java
package com.apress.springrecipes.nosql;

import org.neo4j.ogm.session.SessionFactory;
import org.springframework.context.annotation.Bean;
import org.springframework.context.annotation.ComponentScan;
import org.springframework.context.annotation.Configuration;
import org.springframework.data.neo4j.repository.config.EnableNeo4jRepositories;
import org.springframework.data.neo4j.transaction.Neo4jTransactionManager;
import org.springframework.transaction.annotation.EnableTransactionManagement;

@Configuration
@EnableTransactionManagement
@EnableNeo4jRepositories
@ComponentScan
public class StarwarsConfig {

    @Bean
    public org.neo4j.ogm.config.Configuration configuration() {
        return new org.neo4j.ogm.config.Configuration.Builder().uri("bolt://localhost").
        build();
    }

    @Bean
    public SessionFactory sessionFactory() {
        return new SessionFactory(configuration(),"com.apress.springrecipes.nosql");
    }

    @Bean
    public Neo4jTransactionManager transactionManager(SessionFactory sessionFactory) {
        return new Neo4jTransactionManager(sessionFactory);
    }
}
```

现在有一个 Configuration 对象，它会被 SessionFactory 所使用；你既可以使用 new Configuration，也可以使用 Builder 来构建 Configuration 对象。需要指定 Neo4j 服务器的 URI。该示例使用的是 localhost。默认会使用 Bolt 驱动器，它通过一个二进制协议来传输数据。还可以使用 HTTP(S)，不过这就需要引入一个依赖。SessionFactory 会使用创建好的 Configuration 对象来配置自身。

12-4　使用 Couchbase

问题提出

你想在应用中使用 Couchbase 来存储文档。

解决方案

首先，请下载、安装并设置 Couchbase；接下来，使用 Spring Data Couchbase 项目存储文档并从数据存储中再获取文档。

解释说明

在开始使用 Couchbase 前，需要安装它的一个实例并运行起来。运行后，你需要连接 Couchbase 才能使用数据存储进行实际的存储。我们首先介绍普通的基于 Couchbase 的仓库来说明如何存储与获取文档，接下来会介绍 Spring Data Couchbase，最后会使用反应式版本的仓库。

下载、安装并设置 Couchbase

下载并启动 Couchbase 后，打开浏览器访问 http://localhost:8091。你会看到如图 12-5 所示的页面。在该页面上单击 Setup 按钮。

在下一个界面上（见图 12-6）可以配置集群。既可以开启新的集群，也可以加入到既有的集群中。本节将会开启新的集群。指定内存上限以及磁盘存储的路径（此为可选项）。对于本节来说，使用默认值即可。然后单击 Next 按钮。

注意： 如果在 Docker 中运行 Couchbase，那就需要减少数据 RAM 限额，因为这是有限的。

下一个界面（见图 12-7）可以选择示例数据来使用 Couchbase 的默认示例。由于本节并不需要，因此不要选择任何一项，单击 Next 按钮。

图 12-5 安装 Couchbase

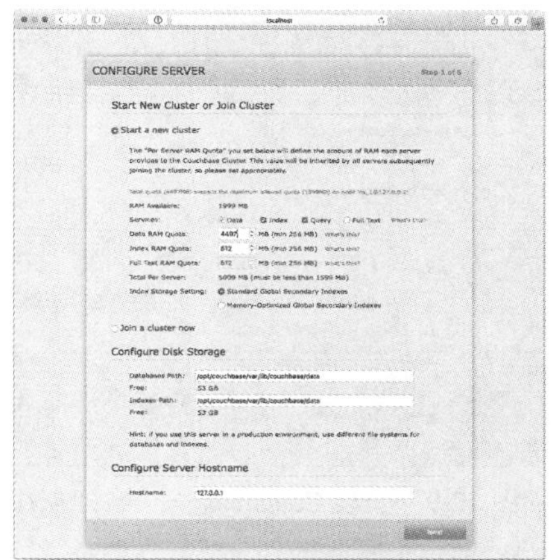

图 12-6 安装 Couchbase——集群设置

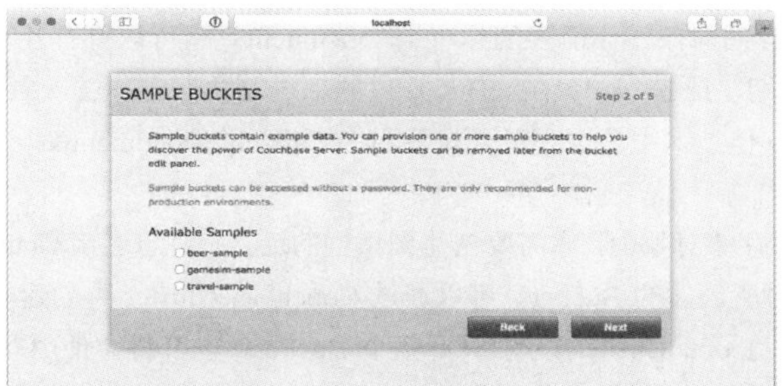

图 12-7 安装 Couchbase——示例 buckets

图 12-8 所示的界面可以创建默认 bucket。对于本节来说，保持默认设置即可，单击 Next 按钮。

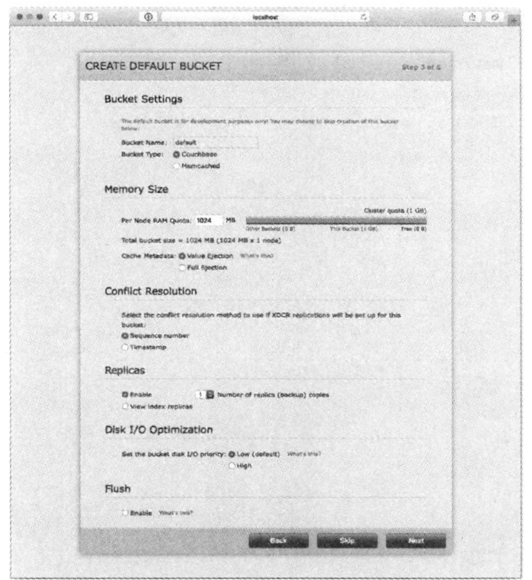

图 12-8 安装 Couchbase——创建默认 bucket

如果想要注册产品，请填写表单并确定是否想要接收关于软件更新的通知。至少勾选复选框来同意条款和条件（见图 12-9）。最后再单击一次 Next 按钮。

最后，你需要为服务器管理员账号指定一个用户名和密码。本节使用的用户名是 admin，密码是 sr4-admin，不过你可以根据需要设定，如图 12-10 所示。

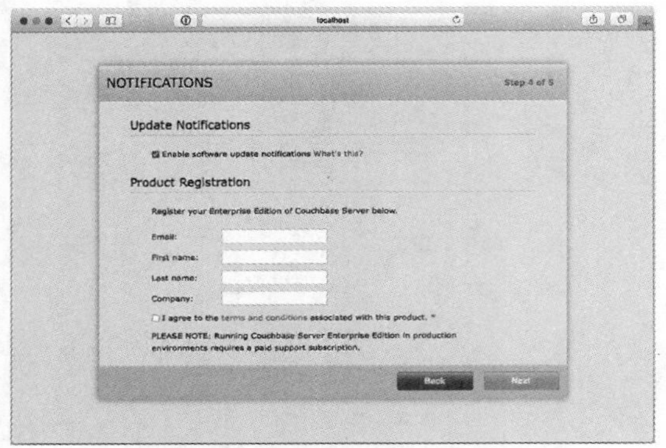

图 12-9　安装 Couchbase——通知与注册

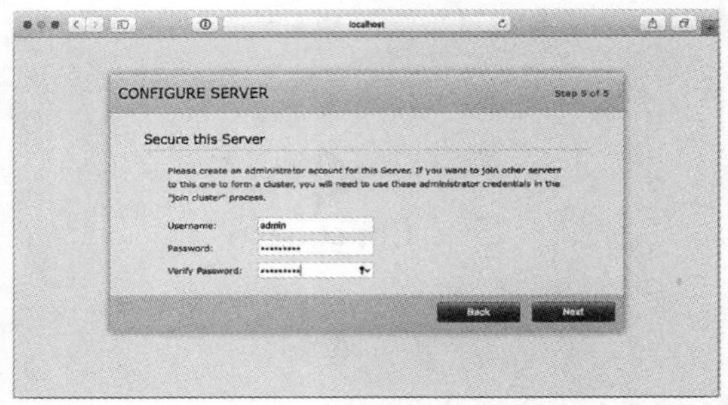

图 12-10　安装 Couchbase——创建一个 admin 用户

使用 Couchbase 存储并获取文档

要想在 Couchbase 中存储对象，你需要创建一个 Document，它可以持有各种类型的内容，如序列化的对象、JSON、字符串、日期，或是 Netty ByteBuf 形式的二进制数据。不过，主要的内容类型是 JSON。通过这种方式，你就可以在其他技术中使用它了。在使用 SerializableDocument 时，你只能使用基于 Java 的解决方案。

不过，在 Couchbase 中存储对象前，你需要创建到集群的连接。要想连接 Couchbase，你需要一个 Cluster 才能访问设置 Couchbase 时所创建的 Bucket。可以通过 CouchbaseCluster 类创建一个到之前所设置的集群的连接。得到的 Cluster 可通过 openBucket()方法打开一个 Bucket。本节将会使用默认 bucket 和最简单的集群设置。

首先，创建一个 Vehicle 类（或是使用 12-1 节中的），你要将它存储到 Couchbase 中。

```
package com.apress.springrecipes.nosql;

import java.io.Serializable;

public class Vehicle implements Serializable {

    private String vehicleNo;
    private String color;
    private int wheel;
    private int seat;

    public Vehicle() {
    }

    public Vehicle(String vehicleNo, String color, int wheel, int seat) {
        this.vehicleNo = vehicleNo;
        this.color = color;
        this.wheel = wheel;
        this.seat = seat;
    }

    public String getColor() {
        return color;
    }
```

```
    public int getSeat() {
        return seat;
    }

    public String getVehicleNo() {
        return vehicleNo;
    }

    public int getWheel() {
        return wheel;
    }

    public void setColor(String color) {
        this.color = color;
    }

    public void setSeat(int seat) {
        this.seat = seat;
    }

    public void setVehicleNo(String vehicleNo) {
        this.vehicleNo = vehicleNo;
    }

    public void setWheel(int wheel) {
        this.wheel = wheel;
    }

    @Override
    public String toString() {
        return "Vehicle [" +
                "vehicleNo='" + vehicleNo + '\'' +
                ", color='" + color + '\'' +
                ", wheel=" + wheel +
                ", seat=" + seat +
                ']';
    }
}
```

注意 implements Serializable 这部分。之所以需要实现 Serializable，是因为你要用 Couchbase 的 SerializableDocument 类来存储对象。

为了与 Couchbase 通信，你需要创建一个仓库。首先定义接口。

```
package com.apress.springrecipes.nosql;

public interface VehicleRepository {

    void save(Vehicle vehicle);

    void delete(Vehicle vehicle);

    Vehicle findByVehicleNo(String vehicleNo);

}
```

如下是实现，它会使用 SerializableDocument 存储 Vehicle 值：

```
package com.apress.springrecipes.nosql;

import com.couchbase.client.java.Bucket;
import com.couchbase.client.java.document.SerializableDocument;

class CouchBaseVehicleRepository implements VehicleRepository {

    private final Bucket bucket;
```

```
    public CouchBaseVehicleRepository(Bucket bucket) {
        this.bucket=bucket;
    }

    @Override
    public void save(Vehicle vehicle) {
        SerializableDocument vehicleDoc = SerializableDocument
            .create(vehicle.getVehicleNo(), vehicle);
        bucket.upsert(vehicleDoc);
    }

    @Override
    public void delete(Vehicle vehicle) {
        bucket.remove(vehicle.getVehicleNo());
    }

    @Override
    public Vehicle findByVehicleNo(String vehicleNo) {
        SerializableDocument doc = bucket.get(vehicleNo, SerializableDocument.class);
        if (doc != null) {
            return (Vehicle) doc.content();
        }
        return null;
    }
}
```

仓库需要一个 Bucket 来存储文档；它就像数据库表一样（Bucket 类似于表，Cluster 更像是整个数据库）。在存储 Vehicle 时，它会被包装到一个 SerializableDocument 中，vehicleNo 值会作为 ID；接下来会调用 upsert 方法。它会根据文档存在与否执行更新或是插入。

创建 Main 类，将 Vehicle 数据存储到 bucket 中，再从 bucket 中查询出来。

```
package com.apress.springrecipes.nosql;

import com.couchbase.client.java.Bucket;
import com.couchbase.client.java.Cluster;
import com.couchbase.client.java.CouchbaseCluster;

public class Main {

    public static void main(String[] args) {

        Cluster cluster = CouchbaseCluster.create();
         Bucket bucket = cluster.openBucket();

        CouchBaseVehicleRepository vehicleRepository = new CouchBaseVehicleRepository(bucket);
        vehicleRepository.save(new Vehicle("TEM0001", "GREEN", 3, 1));
        vehicleRepository.save(new Vehicle("TEM0004", "RED", 4, 2));

        System.out.println("Vehicle: " + vehicleRepository.findByVehicleNo("TEM0001"));
        System.out.println("Vehicle: " + vehicleRepository.findByVehicleNo("TEM0004"));

        bucket.remove("TEM0001");
        bucket.remove("TEM0004");

        bucket.close();
        cluster.disconnect();
    }
}
```

首先，通过 CouchbaseCluster.create()方法创建了一个到 Cluster 的连接。默认情况下，这会连接到 localhost 上的集群。当在生产环境中使用 Couchbase 时，你可能需要使用另一个 create 方法并传递一个要连接的主机列表，甚至需要使用 CouchbaseEnvironment 来设置更多属性（比如说设置 queryTimeout、searchTimeout 等）。

对于本节来说，使用默认 Cluster 就足够了。接下来需要指定用于存储和查询文档的 Bucket。由于使用默认配置，因此使用 cluster.openBucket()就足够了。还有其他几个重载的方法可以指定所用的特定 Bucket，以及到 bucket 的连接属性（比如说超时设置、用户名/密码等）。

Bucket 用于创建 CouchbaseVehicleRepository 的实例。接下来会保存两个 Vehicle，然后再将它们查询出来并删除（本节不会留下任何数据）。最后，关闭连接。

虽然现在将文档存储到了 CouchBase 中，但存在一个缺点：你使用的是 SerializableDocument，但 CouchBase 无法通过它处理索引。此外，只有基于 Java 的客户端才能读取它，其他语言（如 JavaScript）是无法读取的。推荐使用 JsonDocument。接下来重写 CouchbaseVehicleRepository 来反映出这一点。

```
package com.apress.springrecipes.nosql;

import com.couchbase.client.java.Bucket;
import com.couchbase.client.java.document.JsonDocument;
import com.couchbase.client.java.document.json.JsonObject;

class CouchbaseVehicleRepository implements VehicleRepository {

    private final Bucket bucket;

    public CouchbaseVehicleRepository(Bucket bucket) {
        this.bucket=bucket;
    }

    @Override
    public void save(Vehicle vehicle) {

        JsonObject vehicleJson = JsonObject.empty()
            .put("vehicleNo", vehicle.getVehicleNo())
            .put("color", vehicle.getColor())
            .put("wheels", vehicle.getWheel())
            .put("seat", vehicle.getSeat());

        JsonDocument vehicleDoc = JsonDocument.create(vehicle.getVehicleNo(), vehicleJson);
        bucket.upsert(vehicleDoc);
    }

    @Override
    public void delete(Vehicle vehicle) {
        bucket.remove(vehicle.getVehicleNo());
    }

    @Override
    public Vehicle findByVehicleNo(String vehicleNo) {

        JsonDocument doc = bucket.get(vehicleNo, JsonDocument.class);
        if (doc != null) {
            JsonObject result = doc.content();
            return new Vehicle(result.getString("vehicleNo"), result.getString("color"),
            result.getInt("wheels"), result.getInt("seat"));
        }
        return null;
    }
}
```

注意，上述代码使用了 JsonObject 对象并将 Vehicle 转换为了一个 JsonObject 对象，反之亦然。再次运行 Main 类，依然可以实现两个文档在 Couchbase 中的存储、查询和删除。

对于大型对象图来说，与 JSON 之间的转换是非常麻烦的，相较于手工进行转换来说，你可以通过 JSON 库实现与 JSON 之间的转换，比如说 Jackson。

```
package com.apress.springrecipes.nosql;
```

```
import com.couchbase.client.java.Bucket;
import com.couchbase.client.java.document.JsonDocument;
import com.couchbase.client.java.document.json.JsonObject;
import com.fasterxml.jackson.core.JsonProcessingException;
import com.fasterxml.jackson.databind.ObjectMapper;

import java.io.IOException;

class CouchbaseVehicleRepository implements VehicleRepository {

    private final Bucket bucket;
    private final ObjectMapper mapper;

    public CouchbaseVehicleRepository(Bucket bucket, ObjectMapper mapper) {
        this.bucket=bucket;
        this.mapper=mapper;
    }

    @Override
    public void save(Vehicle vehicle) {

        String json = null;
        try {
            json = mapper.writeValueAsString(vehicle);
        } catch (JsonProcessingException e) {
            throw new RuntimeException("Error encoding JSON.", e);
        }
        JsonObject vehicleJson = JsonObject.fromJson(json);
        JsonDocument vehicleDoc = JsonDocument.create(vehicle.getVehicleNo(), vehicleJson);
        bucket.upsert(vehicleDoc);
    }

    @Override
    public void delete(Vehicle vehicle) {
        bucket.remove(vehicle.getVehicleNo());
    }

    @Override
    public Vehicle findByVehicleNo(String vehicleNo) {
        JsonDocument doc = bucket.get(vehicleNo, JsonDocument.class);
        if (doc != null) {
            JsonObject result = doc.content();
            try {
                return mapper.readValue(result.toString(), Vehicle.class);
            } catch (IOException e) {
                throw new RuntimeException("Error decoding JSON.", e);
            }
        }
        return null;
    }
}
```

上述代码使用强大的 Jackson 库实现了与 JSON 之间的转换。

使用 Spring

到现在为止，一切都是在 Main 类中配置的。下面将配置部分迁移到 CouchbaseConfiguration 类中并通过它启动应用。

```
package com.apress.springrecipes.nosql;

import com.couchbase.client.java.Bucket;
import com.couchbase.client.java.Cluster;
import com.couchbase.client.java.CouchbaseCluster;
```

```
import com.fasterxml.jackson.databind.ObjectMapper;
import org.springframework.context.annotation.Bean;
import org.springframework.context.annotation.Configuration;

@Configuration
public class CouchbaseConfiguration {

    @Bean(destroyMethod = "disconnect")
    public Cluster cluster() {
        return CouchbaseCluster.create();
    }

    @Bean
    public Bucket bucket(Cluster cluster) {
        return cluster.openBucket();
    }

    @Bean
    public ObjectMapper mapper() {
        return new ObjectMapper();
    }

    @Bean
    public CouchbaseVehicleRepository vehicleRepository(Bucket bucket, ObjectMapper mapper) {
        return new CouchbaseVehicleRepository(bucket, mapper);
    }
}
```

注意 Cluster bean 上的 destroyMethod 方法。当应用关闭时会调用该方法。Bucket 中的 close 方法会被自动调用，因为它是会被自动检测到的诸多预先定义的方法之一。CouchbaseVehicleRepository 的构造保持不变，不过现在将 Spring 管理的两个 bean 传递给了它。

修改 Main 类以使用 CouchbaseConfiguration。

```
package com.apress.springrecipes.nosql;

import org.springframework.context.ApplicationContext;
import org.springframework.context.annotation.AnnotationConfigApplicationContext;

public class Main {

    public static void main(String[] args) {

        ApplicationContext context =
            new AnnotationConfigApplicationContext(CouchbaseConfiguration.class);
        VehicleRepository vehicleRepository =context.getBean(VehicleRepository.class);

        vehicleRepository.save(new Vehicle("TEM0001", "GREEN", 3, 1));
        vehicleRepository.save(new Vehicle("TEM0004", "RED", 4, 2));

        System.out.println("Vehicle: " + vehicleRepository.findByVehicleNo("TEM0001"));
        System.out.println("Vehicle: " + vehicleRepository.findByVehicleNo("TEM0004"));

        vehicleRepository.delete(vehicleRepository.findByVehicleNo("TEM0001"));
        vehicleRepository.delete(vehicleRepository.findByVehicleNo("TEM0004"));
    }
}
```

VehicleRepository 是从构建好的 ApplicationContext 中获取的，代码依然实现了 Vehicle 实例在 Couchbase 集群中的保存、获取和删除。

使用 Spring Data 的 CouchbaseTemplate

虽然在 Java 中使用 Couchbase 时通过 Jackson 来映射 JSON 是非常直接的，不过对于大型仓库或是使用特定的索引和 N1QL 进行查询来说，这么做会变得非常繁琐，更不必说将其集成到使用了其他方式来存储数据的应用中了。Spring Data Couchbase 项目包含了一个 CouchbaseTemplate 模板，它不仅去掉了你现在在仓库中所编写的部分样板代码，比如说与 JSON 之间的转换，而且还将异常转换为 DataAccessException。这使得它能与其他使用了 Spring 的数据访问技术轻松集成。

首先，修改仓库来使用 CouchbaseTemplate。

```
package com.apress.springrecipes.nosql;

import org.springframework.data.couchbase.core.CouchbaseTemplate;

public class CouchbaseVehicleRepository implements VehicleRepository {

    private final CouchbaseTemplate couchbase;

    public CouchbaseVehicleRepository(CouchbaseTemplate couchbase) {
        this.couchbase = couchbase;
    }

    @Override
    public void save(Vehicle vehicle) {
        couchbase.save(vehicle);
    }

    @Override
    public void delete(Vehicle vehicle) {
        couchbase.remove(vehicle);
    }

    @Override
    public Vehicle findByVehicleNo(String vehicleNo) {
        return couchbase.findById(vehicleNo, Vehicle.class);
    }
}
```

现在的仓库已经简化为只有十几行代码了。要想存储 Vehicle 对象，你需要为 Vehicle 添加注解；它需要知道哪个字段用作 ID。

```
package com.apress.springrecipes.nosql;

import com.couchbase.client.java.repository.annotation.Field;
import com.couchbase.client.java.repository.annotation.Id;

import java.io.Serializable;

public class Vehicle implements Serializable{

    @Id
    private String vehicleNo;
    @Field
    private String color;
    @Field
    private int wheel;
    @Field
    private int seat;

    // getters/setters omitted.
}
```

字段 vehicleNo 被添加了@Id 注解，其他字段则添加了@Field 注解。虽然后者并非强制要求，不过推荐使用。还可以通过@Field 注解为 JSON 属性名指定不同的名字，这对于将既有文档映射为 Java 对象会很有用。

最后，需要在配置类中配置一个 CouchbaseTemplate 模板。

```
package com.apress.springrecipes.nosql;
```

```
import com.couchbase.client.java.Bucket;
import com.couchbase.client.java.Cluster;
import com.couchbase.client.java.CouchbaseCluster;
import org.springframework.context.annotation.Bean;
import org.springframework.context.annotation.Configuration;
import org.springframework.data.couchbase.core.CouchbaseTemplate;

@Configuration
public class CouchbaseConfiguration {

    @Bean(destroyMethod = "disconnect")
    public Cluster cluster() {
        return CouchbaseCluster.create();
    }

    @Bean
    public Bucket bucket(Cluster cluster) {
        return cluster.openBucket();
    }

    @Bean
    public CouchbaseVehicleRepository vehicleRepository(CouchbaseTemplate couchbaseTemplate){
        return new CouchbaseVehicleRepository(couchbaseTemplate);
    }

    @Bean
    public CouchbaseTemplate couchbaseTemplate(Cluster cluster, Bucket bucket) {
        return new CouchbaseTemplate(cluster.clusterManager("default","").info(), bucket);
    }
}
```

CouchbaseTemplate 对象需要一个 Bucket，它需要访问 ClusterInfo，这可以通过 ClusterManager 来获取；这里传递了 Bucket 的名字，即 default，没有密码。另外，你还可以传递 admin/sr4-admin 作为用户名/密码组合。最后，配置的 CouchbaseVehicleRepository 实例是通过这个配置好的模板创建的。

当运行 Main 类时，Vehicle 对象的存储、查询与删除依然正常如故。

为了简化配置，Spring Data Couchbase 提供了一个基础配置类 AbstractCouchbaseConfiguration，你可以继承它，这样就不需要再配置 Cluster、Bucket 和 CouchbaseTemplate 对象了。

```
package com.apress.springrecipes.nosql;

import org.springframework.context.annotation.Bean;
import org.springframework.context.annotation.Configuration;
import org.springframework.data.couchbase.config.AbstractCouchbaseConfiguration;
import org.springframework.data.couchbase.core.CouchbaseTemplate;

import java.util.Collections;
import java.util.List;

@Configuration
public class CouchbaseConfiguration extends AbstractCouchbaseConfiguration {

    @Override
    protected List<String> getBootstrapHosts() {
        return Collections.singletonList("localhost");
    }

    @Override
    protected String getBucketName() {
        return "default";
    }

    @Override
    protected String getBucketPassword() {
```

```
        return "";
    }

    @Bean
    public CouchbaseVehicleRepository vehicleRepository(CouchbaseTemplate couchbaseTemplate){
        return new CouchbaseVehicleRepository(couchbaseTemplate);
    }
}
```

现在的配置类继承了 AbstractCouchbaseConfiguration 父类，你只需要提供 bucket 的名字、可选的密码以及主机列表即可。父配置类提供了 CouchbaseTemplate 及它所需要的全部对象。

使用 Spring Data 的 Couchbase 仓库

与其他技术一样，Spring Data Couchbase 可以指定接口并在运行期生成实际的仓库实现。通过这种方式，你只需要创建接口即可，无须创建具体的实现。要想做到这一点，就像其他 Spring Data 项目一样，你需要继承 CrudRepository。如果需要相应的功能，还可以继承 CouchbaseRepository 或 CouchbasePagingAndSortingRepository。本节将会使用 CrudRepository。

```
package com.apress.springrecipes.nosql;

import org.springframework.data.repository.CrudRepository;

public interface VehicleRepository extends CrudRepository<Vehicle, String> {}
```

可以看到，接口中没有其他方法，因为所有的 CRUD 方法已经提供了。接下来，需要在配置类上添加 @EnableCouchbaseRepositories 注解。

```
package com.apress.springrecipes.nosql;

import org.springframework.context.annotation.Configuration;
import org.springframework.data.couchbase.config.AbstractCouchbaseConfiguration;
import org.springframework.data.couchbase.repository.config.EnableCouchbaseRepositories;

import java.util.Collections;
import java.util.List;
@Configuration
@EnableCouchbaseRepositories(
public class CouchbaseConfiguration extends AbstractCouchbaseConfiguration { … }
```

最后，Main 类需要做一些小修改，因为不能再使用 findByVehicleNo 方法了，需要使用 findById 方法。

```
package com.apress.springrecipes.nosql;

import org.springframework.context.ApplicationContext;
import org.springframework.context.annotation.AnnotationConfigApplicationContext;

public class Main {

    public static void main(String[] args) {

        ApplicationContext context =
            new AnnotationConfigApplicationContext(CouchbaseConfiguration.class);
        VehicleRepository vehicleRepository =context.getBean(VehicleRepository.class);

        vehicleRepository.save(new Vehicle("TEM0001", "GREEN", 3, 1));
        vehicleRepository.save(new Vehicle("TEM0004", "RED", 4, 2));

        vehicleRepository.findById("TEM0001").ifPresent(System.out::println);
        vehicleRepository.findById("TEM0004").ifPresent(System.out::println);

        vehicleRepository.deleteById("TEM0001");
        vehicleRepository.deleteById("TEM0004");
    }
}
```

findById 方法返回一个 java.util.Optional 对象，这样就可以通过 ifPresent 方法将其打印到控制台上了。

使用 Spring Data 的反应式 Couchbase 仓库

除了阻塞式仓库外，我们还可以通过 ReactiveCouchbaseRepository 获取到反应式仓库。它会针对 findById 这样的拥有零个或一个返回结果的方法返回一个 Mono，会针对 findAll 这样的返回零个或多个元素的方法返回一个 Flux。默认的 Couchbase 驱动器已经内建了对反应式的支持。要想使用这个特性，需要在类路径上添加 RxJava 和 RxJava 反应式流。要想使用 ReactiveCouchbaseRepository 的反应式类型，还需要在类路径上添加 Pivotal Reactor。

为了配置 Couchbase 的反应式仓库，需要修改 CouchbaseConfiguration。让它继承 AbstractReactiveCouchbase-Configuration，这次不再使用@EnableCouchbaseRepositories，而是使用@EnableReactiveCouchbaseRepositories。

```
package com.apress.springrecipes.nosql;

import org.springframework.context.annotation.Configuration;
import org.springframework.data.couchbase.config.AbstractReactiveCouchbaseConfiguration;
import org.springframework.data.couchbase.repository.config.EnableReactiveCouchbaseRepositories;

import java.util.Collections;
import java.util.List;

@Configuration
@EnableReactiveCouchbaseRepositories
public class CouchbaseConfiguration extends AbstractReactiveCouchbaseConfiguration {

    @Override
    protected List<String> getBootstrapHosts() {
        return Collections.singletonList("localhost");
    }

    @Override
    protected String getBucketName() {
        return "default";
    }

    @Override
    protected String getBucketPassword() {
        return "";
    }
}
```

配置的其他部分与常规的 Couchbase 配置一样；依然需要连接到同样的 Couchbase 服务器并使用同样的 Bucket。

接下来，VehicleRepository 应该继承 ReactiveCrudRepository 而不是 CrudRepository。

```
package com.apress.springrecipes.nosql;

import org.springframework.data.repository.reactive.ReactiveCrudRepository;

public interface VehicleRepository extends ReactiveCrudRepository<Vehicle, String> {}
```

这基本上就是获取反应式仓库所需的全部工作了。为了测试，还需要修改 Main 类。

```
package com.apress.springrecipes.nosql;

import org.springframework.context.ApplicationContext;
import org.springframework.context.annotation.AnnotationConfigApplicationContext;
import reactor.core.publisher.Flux;
import reactor.core.publisher.Mono;

import java.util.Arrays;
import java.util.concurrent.CountDownLatch;
```

```
public class Main {

    public static void main(String[] args) throws InterruptedException {

        ApplicationContext context =
            new AnnotationConfigApplicationContext(CouchbaseConfiguration.class);
        VehicleRepository repository =context.getBean(VehicleRepository.class);

        CountDownLatch countDownLatch = new CountDownLatch(1);

        repository.saveAll(Flux.just(new Vehicle("TEM0001", "GREEN", 3, 1), //
            new Vehicle("TEM0004", "RED", 4, 2)))
             .last().log()
             .then(repository.findById("TEM0001")).doOnSuccess(System.out::println)
             .then(repository.findById("TEM0004")).doOnSuccess(System.out::println)
             .then(repository.deleteById(Flux.just("TEM0001", "TEM00004")))
                 .doOnSuccess(x -> countDownLatch.countDown())
                 .doOnError(t -> countDownLatch.countDown())
             .subscribe();

        countDownLatch.await();
    }
}
```

创建 ApplicationContext 并获取 VehicleRepository 与之前完全一样。不过，接下来有一个链式方法调用（一个跟着一个）。首先，向数据存储添加两个 Vehicle 实例。当最后一个保存好之后，向仓库查询每个实例。接下来，又将全部实例删除。为了让所有操作都能完成，既可以使用 block()阻塞，也可以自行等待。一般来说，需要避免在反应式系统中使用 block()。这正是使用 CountDownLatch 的原因所在；当 deleteById 方法执行完毕后，CountDownLatch 值会减少。countDownLatch.await()方法会等到计数器值变为 0 时完成程序的执行。

小结

本章介绍了不同类型的数据存储，包括如何使用它们以及如何通过 Spring Data 家族的不同模块来简化它们的使用。首先通过 MongoDB 介绍了文档驱动存储以及 Spring Data MongoDB 模块的使用。接下来介绍了键值存储；我们主要介绍了 Redis 并使用了 Spring Data Redis 模块。最后介绍了基于图的数据存储 Neo4j，我们学习了嵌入式 Neo4j 及其使用方法，以及如何构建仓库来存储实体。

对于其中的两种数据存储，我们还通过继承其反应式版本的接口以及为数据存储配置反应式驱动探索了它们的反应式特性。

第 13 章

Spring Java 企业服务与远程技术

本章将会介绍 Spring 对最常见的 Java 企业服务的支持：使用 Java Management Extensions（JMX）、使用 JavaMail 发送邮件，以及使用 Quartz 进行任务调度。此外，我们还会介绍 Spring 对各种远程技术的支持，如 RMI、Hessian、HTTP Invoker 与 SOAP Web Service。

JMX 是 Java SE 的一部分，它是一种用于管理和监控系统资源（如设备、应用、对象和服务驱动的网络等）的技术，这些资源表示为受管理的 bean（MBean）。Spring 通过将 Spring bean 导出为模型 MBeans 同时无须针对 JMX API 进行编程来支持 JMX。此外，Spring 还可以轻松访问远程 MBeans。

JavaMail 是 Java 中发送邮件的标准 API 和实现。Spring 进一步提供了一个抽象层，以一种独立于实现的方式来发送邮件。

Java 平台上主要有两种方式来调度任务：JDK Timer 与 Quartz Scheduler。JDK Timer 提供了简单的任务调度特性，它们都内建于 JDK 中。相较于 JDK Timer，Quartz 提供了更为强大的任务调度特性。对于这两种方式来说，Spring 提供了辅助类可以在 bean 配置文件中对调度任务进行配置，无须直接使用它们的 API。

远程是开发分布式应用，特别是多层企业应用的关键技术。它可以让运行在不同 JVM 或不同机器上的不同应用或组件能够通过特定的协议进行彼此间的通信。Spring 的远程支持能够跨越不同的远程技术保持一致性。在服务器端，Spring 可以通过服务导出器将任意一个 bean 公开为远程服务。在客户端，Spring 提供了各种代理工厂 bean 为远程服务创建本地代理，这样就可以像使用本地 bean 一样使用远程服务了。

我们将会介绍如何使用一系列的远程技术，包括 RMI、Hessian、HTTP Invoker 以及通过 Spring Web Services（Spring-WS）来使用 SOAP Web Service。

13-1　将 Spring POJO 注册为 JMX MBean

问题提出

你想要在 Java 应用中将对象注册为 JMX MBean，从而可以访问运行着的服务并在运行期操纵它们的状态。这样就可以执行很多任务了，比如说重新运行批任务、调用方法，以及修改配置元数据。

解决方案

Spring 可以将 IoC 容器中的任何 bean 导出为模型 MBean，通过这种方式来支持 JMX。这可以通过声明一个 MBeanExporter 实例轻松实现。借助于 Spring 的 JMX 支持，你无须直接使用 JMX API。此外，Spring 还可以声明 JSR-160（Java Management Extensions Remote API）连接器，并通过工厂 bean 为特定协议上的远程访问公开 MBean。Spring 为服务器端与客户端都提供了工厂 bean。

Spring 的 JMX 支持还提供了其他一些机制，你可以通过它们装配 MBean 的管理接口。这些选项包括根据方法名、接口和注解来使用导出的 bean。Spring 还可以对声明在 IoC 容器中的 bean 以及使用 Spring 定义的特定于 JMX 的注解的 bean 进行自动检测并导出为 MBean。

解释说明

假设你要开发这样一个功能：将文件从一个目录复制到另一个目录。我们为该功能设计如下接口：

```
package com.apress.springrecipes.replicator;
...
public interface FileReplicator {

    public String getSrcDir();
    public void setSrcDir(String srcDir);

    public String getDestDir();
    public void setDestDir(String destDir);

    public FileCopier getFileCopier();
    public void setFileCopier(FileCopier fileCopier);

    public void replicate() throws IOException;
}
```

源目录与目标目录被设计为 replicator 对象的属性而非方法参数。这意味着每个 file replicator 实例只会针对特定的源目录与目标目录进行复制。可以在应用中创建多个 replicator 实例。

不过在实现这个 replicator 前，我们再创建一个接口，将文件从一个目录复制到另一个目录，它会接收一个文件名。

```
package com.apress.springrecipes.replicator;
...
public interface FileCopier {

    public void copyFile(String srcDir, String destDir, String filename)
        throws IOException;
}
```

实现这个文件复制器的策略有很多。比如说，可以使用 Spring 提供的 FileCopyUtils 类。

```
package com.apress.springrecipes.replicator;
...
import org.springframework.util.FileCopyUtils;

public class FileCopierJMXImpl implements FileCopier {

    public void copyFile(String srcDir, String destDir, String filename)
        throws IOException {
        File srcFile = new File(srcDir, filename);
        File destFile = new File(destDir, filename);
        FileCopyUtils.copy(srcFile, destFile);
    }
}
```

借助于文件复制器的帮助，现在就可以实现 file replicator 了，代码如下所示：

```
package com.apress.springrecipes.replicator;

import java.io.File;
import java.io.IOException;

public class FileReplicatorJMXImpl implements FileReplicator {

    private String srcDir;
    private String destDir;
    private FileCopier fileCopier;

    // getters ommited for brevity

    public void setSrcDir(String srcDir) {
        this.srcDir = srcDir;
    }
```

```
    public void setDestDir(String destDir) {
        this.destDir = destDir;
    }

    public void setFileCopier(FileCopier fileCopier) {
        this.fileCopier = fileCopier;
    }

    public synchronized void replicate() throws IOException {
        File[] files = new File(srcDir).listFiles();
        for (File file : files) {
            if (file.isFile()) {
                fileCopier.copyFile(srcDir, destDir, file.getName());
            }
        }
    }
}
```

每次调用 replicate()方法时，源目录中的所有文件都会被复制到目标目录中。为了避免并发复制所导致的意外问题，将该方法声明为了 synchronized。

现在，可以在 Java 配置类中配置一个或多个 file replicator 实例。documentReplicator 实例需要引用两个目录：一个用于读取文件的源目录；另一个用于备份文件的目标目录。该示例中的代码会从操作系统中用户的主目录中读取名为 docs 的目录，然后将里面的文件复制到同一目录的 docs_backup 目录下。

当这个 bean 启动时，如果这两个目录不存在，则会创建。

提示：每个操作系统的“主目录”都是不同的，不过在 UNIX 上，通常它指的是~所解析的目录。在 Linux 上，该目录可能是/home/user。在 macOS 上，该目录可能是/Users/user。在 Windows 上，它类似于 C:\Documents and Settings\user。

```
package com.apress.springrecipes.replicator.config;
...

@Configuration
public class FileReplicatorConfig {

    @Value("#{systemProperties['user.home']}/docs")
    private String srcDir;
    @Value("#{systemProperties['user.home']}/docs_backup")
    private String destDir;

    @Bean
    public FileCopier fileCopier() {
        FileCopier fCop = new FileCopierJMXImpl();
        return fCop;
    }

    @Bean
    public FileReplicator documentReplicator() {
        FileReplicator fRep = new FileReplicatorJMXImpl();
        fRep.setSrcDir(srcDir);
        fRep.setDestDir(destDir);
        fRep.setFileCopier(fileCopier());
        return fRep;
    }

    @PostConstruct
    public void verifyDirectoriesExist() {
        File src = new File(srcDir);
        File dest = new File(destDir);
        if (!src.exists())
            src.mkdirs();
```

```
        if (!dest.exists())
            dest.mkdirs();
    }
}
```

一开始，有两个字段声明使用了@Value 注解来获取用户的主目录并定义源目录和目标目录。接下来，通过@Bean 注解创建了两个 bean 实例。注意 verifyDirectoriesExist()方法上的@PostConstuct 注解，它会确保源目录与目标目录都存在。

在有了应用的核心 bean 之后，我们来看看如何以 MBean 的形式注册并访问它们。

不借助于 Spring 的支持注册 MBean

首先，我们看看如何直接通过 JMX API 注册模型 MBean。在如下的 Main 类中，从 IoC 容器中获取到 FileReplicator bean 并将其注册为 MBean 用于管理和监控。所有属性和方法都包含在 MBean 的管理接口中。

```
package com.apress.springrecipes.replicator;
...
import java.lang.management.ManagementFactory;

import javax.management.Descriptor;
import javax.management.JMException;
import javax.management.MBeanServer;
import javax.management.ObjectName;
import javax.management.modelmbean.DescriptorSupport;
import javax.management.modelmbean.InvalidTargetObjectTypeException;
import javax.management.modelmbean.ModelMBeanAttributeInfo;
import javax.management.modelmbean.ModelMBeanInfo;
import javax.management.modelmbean.ModelMBeanInfoSupport;
import javax.management.modelmbean.ModelMBeanOperationInfo;
import javax.management.modelmbean.RequiredModelMBean;

import org.springframework.context.ApplicationContext;
import org.springframework.context.support.GenericXmlApplicationContext;

public class Main {

    public static void main(String[] args) throws IOException {
        ApplicationContext context =
            new AnnotationConfigApplicationContext("com.apress.
            springrecipes.replicator.config");

        FileReplicator documentReplicator = context.getBean(FileReplicator.class);
        try {
            MBeanServer mbeanServer = ManagementFactory.getPlatformMBeanServer();
            ObjectName objectName = new ObjectName("bean:name=documentReplicator");

            RequiredModelMBean mbean = new RequiredModelMBean();
            mbean.setManagedResource(documentReplicator, "objectReference");

            Descriptor srcDirDescriptor = new DescriptorSupport(new String[] {
                "name=SrcDir", "descriptorType=attribute",
                "getMethod=getSrcDir", "setMethod=setSrcDir" });
            ModelMBeanAttributeInfo srcDirInfo = new ModelMBeanAttributeInfo(
                "SrcDir", "java.lang.String", "Source directory",
                true, true, false, srcDirDescriptor);
            Descriptor destDirDescriptor = new DescriptorSupport(new String[] {
                "name=DestDir", "descriptorType=attribute",
                "getMethod=getDestDir", "setMethod=setDestDir" });
            ModelMBeanAttributeInfo destDirInfo = new ModelMBeanAttributeInfo(
                "DestDir", "java.lang.String", "Destination directory",
                true, true, false, destDirDescriptor);

            ModelMBeanOperationInfo getSrcDirInfo = new ModelMBeanOperationInfo(
```

```
                "Get source directory",
                FileReplicator.class.getMethod("getSrcDir"));
            ModelMBeanOperationInfo setSrcDirInfo = new ModelMBeanOperationInfo(
                "Set source directory",
                FileReplicator.class.getMethod("setSrcDir", String.class));
            ModelMBeanOperationInfo getDestDirInfo = new ModelMBeanOperationInfo(
                "Get destination directory",
                FileReplicator.class.getMethod("getDestDir"));
            ModelMBeanOperationInfo setDestDirInfo = new ModelMBeanOperationInfo(
                "Set destination directory",
                FileReplicator.class.getMethod("setDestDir", String.class));
            ModelMBeanOperationInfo replicateInfo = new ModelMBeanOperationInfo(
                "Replicate files",
                FileReplicator.class.getMethod("replicate"));

            ModelMBeanInfo mbeanInfo = new ModelMBeanInfoSupport(
                "FileReplicator", "File replicator",
                new ModelMBeanAttributeInfo[] { srcDirInfo, destDirInfo },
                null,
                new ModelMBeanOperationInfo[] { getSrcDirInfo, setSrcDirInfo,
                    getDestDirInfo, setDestDirInfo, replicateInfo },
                null);
            mbean.setModelMBeanInfo(mbeanInfo);

            mbeanServer.registerMBean(mbean, objectName);
        } catch (JMException e) {
            ...
        } catch (InvalidTargetObjectTypeException e) {
            ...
        } catch (NoSuchMethodException e) {
            ...
        }

        System.in.read();
    }
}
```

要想注册 MBean，你需要接口 javax.managment.MBeanServer 的一个实例。可以通过调用静态方法 ManagementFactory.getPlatformMBeanServer()来定位到一个平台 MBean 服务器。如果不存在，它会创建一个 MBean 服务器，然后将其注册供未来所用。每个 MBean 都需要一个 MBean 对象名，它包含了一个域名。之前的 MBean 注册在了名为 documentReplicator 的 domain bean 下。

从上述代码中可以看到，针对每个 MBean 属性和操作来说，你都需要创建一个 ModelMBeanAttributeInfo 对象和一个 ModelMBeanOperationInfo 对象来描述它。接下来，你需要创建一个 ModelMBeanInfo 对象，通过装配上述信息来描述 MBean 的管理接口。要想了解关于使用这些类的更多信息，请查看它们的 Javadocs。此外，在调用 JMX API 时，你需要处理特定于 JMX 的异常。这些异常是检查异常，因此必须要处理才行。注意，在使用 JMX 客户端工具查看内部信息之前，你必须要防止应用终止。使用 System.in.read()等待控制台输入是个不错的选择。

最后，需要添加 VM 参数-Dcom.sun.management.jmxremote 为该应用启用本地监控。如果使用本书的配套源代码，可以使用如下命令：

```
java -Dcom.sun.management.jmxremote -jar Recipe_13_1_i-4.0.0.jar
```

现在，你可以使用任意 JMX 客户端工具在本地监控 MBean 了。最简单的就是 JDK 自带的 JConsole。要想启动 JConsole，只需在 JDK 安装目录的 bin 目录下执行 jconsole 命令即可。

当 JConsole 启动后，你会在连接窗口的 Local 选项卡下看到一个启用了 JMX 的应用列表。选择与这个运行着的 Spring 应用相对应的进程（即 Recipe_13_1_i-1.0-SNAPSHOT.jar），如图 13-1 所示。

警告：如果使用的是 Windows，你可能在 JConsole 上看不到任何进程。这是一个已知 Bug，JConsole

无法检测到运行着的 Java 进程。为了解决这个问题，你需要确保用户在临时目录下存在 hsperfdata 这个目录。Java 与 JConsole 会使用该目录追踪运行着的进程，它可能并不存在。比如说，如果以用户 John.Doe 的身份运行应用，那么请确保如下路径是存在的：C:\Users\John.Doe\AppData\Local\Temp\hsperfdata_John.Doe\。

连接到 replicator 应用后，进入到 Mbeans 选项卡。接下来，单击左侧树的 Bean 目录，然后展开 operations 部分。你会在主界面上看到一系列按钮，它们可以调用 bean 的操作。要想调用 replicate()，只需单击 replicate 按钮即可。如图 13-2 所示。

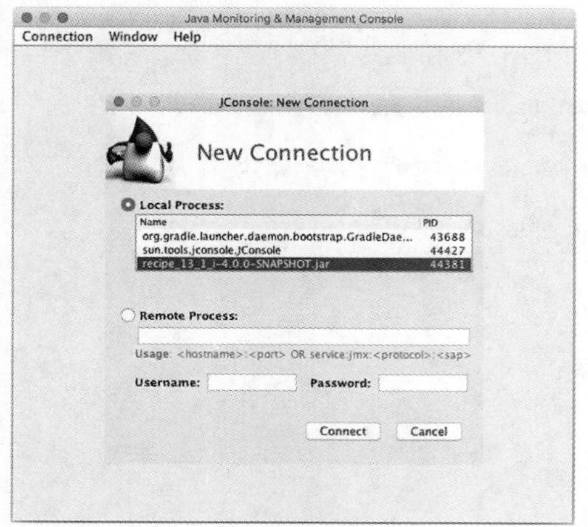

图 13-1 JConsole 启动窗口

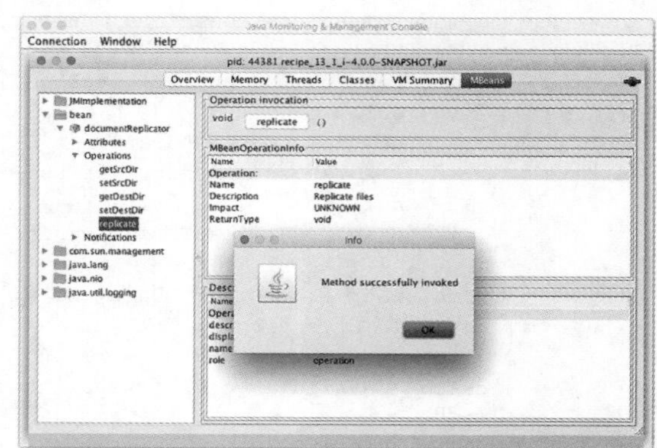

图 13-2 JConsole 模拟 Spring bean 操作

你会看到一个"Method successfully invoked"的弹出窗口。通过这个动作，源目录中的所有文件都会复制/同步到目标目录中。

借助 Spring 的支持注册 MBean

之前的应用直接使用了 JMX API。如在 Main 应用类中所看到的那样，代码非常多，难以编写、管理，有时也不好理解。要想将 Spring IoC 容器中配置的 bean 导出为 MBean，只需创建一个 MBeanExporter 实例并指定要导出的 bean 即可，将它们的 MBean 对象名作为键。这可以通过添加如下配置类来完成。注意，beansToExport map 中的键用作 bean 的 ObjectName，bean 则是由对应的值来引用的。

```
package com.apress.springrecipes.replicator.config;

import com.apress.springrecipes.replicator.FileReplicator;
import org.springframework.beans.factory.annotation.Autowired;
import org.springframework.context.annotation.Bean;
import org.springframework.context.annotation.Configuration;
import org.springframework.jmx.export.MBeanExporter;

import java.util.HashMap;
import java.util.Map;

@Configuration
public class JmxConfig {

    @Autowired
    private FileReplicator fileReplicator;

    @Bean
    public MBeanExporter mbeanExporter() {
        MBeanExporter mbeanExporter = new MBeanExporter();
```

```
        mbeanExporter.setBeans(beansToExport());
        return mbeanExporter;
    }

    private Map<String, Object> beansToExport() {
        Map<String, Object> beansToExport = new HashMap<>();
        beansToExport.put("bean:name=documentReplicator", fileReplicator);
        return beansToExport;
    }
}
```

上述配置将 FileReplicator bean 导出为一个 MBean，它位于 domain bean 之下，名字为 documentReplicator。默认情况下，所有 public 属性都会作为属性，所有 public 方法（除了 java.lang.Object 中的方法）都会作为 MBean 管理接口中的操作。借助于 Spring JMX MBeanExporter 的帮助，应用的 Main 类可以简化为如下几行代码：

```
package com.apress.springrecipes.replicator;
...
import org.springframework.context.annotation.AnnotationConfigApplicationContext;

public class Main {

    public static void main(String[] args) throws IOException {
        new AnnotationConfigApplicationContext("com.apress.springrecipes.replicator.config");
        System.in.read();
    }
}
```

使用多个 MBean Server 实例

Spring MBeanExporter 方式可以定位到一个 MBean 服务器实例并隐式将 MBeans 注册到它上面。当首次定位后，JDK 会创建一个 MBean 服务器，因此无须再显式创建了。如果应用运行在已经提供了 MBean 服务器的环境中（比如说 Java 应用服务器），那么情况也是一样的。

不过，如果有多个 MBean 服务器在运行，那就需要告诉 mbeanServer bean 要绑定到哪个服务器上。这是通过指定服务器的 agentId 值来实现的。要想在 JConsole 中查看给定服务器的 agentId 值，请进入到 Mbeans 选项卡，在左侧树中找到 JMImplementation/MBeanServerDelegate/Attributes/ MBeanServerId。你会在那里看到其字符串值。在这里的本地机器上，该值是 workstation_ 1253860476443。要想启用它，请配置 MBeanServer 的 agentId 属性。

```
@Bean
public MBeanServerFactoryBean mbeanServer() {
    MBeanServerFactoryBean mbeanServer = new MBeanServerFactoryBean();
    mbeanServer.setLocateExistingServerIfPossible(true);
    mbeanServer.setAgentId("workstation_1253860476443");
    return mbeanServer;
}
```

如果在上下文中有多个 MBean 服务器实例，那就可以显式指定一个特定的 MBean 服务器以让 MBeanExporter 将 MBeans 导入其中。在该示例中，MBeanExporter 并不会定位到一个 MBean 服务器；它会使用特定的 MBean 服务器实例。当存在多个 MBean 服务器时，你可以通过该属性指定特定的 MBean 服务器。

```
@Bean
public MBeanExporter mbeanExporter() {
    MBeanExporter mbeanExporter = new MBeanExporter();
    mbeanExporter.setBeans(beansToExport());
    mbeanExporter.setServer(mbeanServer().getObject());
    return mbeanExporter;
}

@Bean
public MBeanServerFactoryBean mbeanServer() {
    MBeanServerFactoryBean mbeanServer = new MBeanServerFactoryBean();
```

```
        mbeanServer.setLocateExistingServerIfPossible(true);
        return mbeanServer;
    }
```

注册 MBean 供 RMI 远程访问使用

如果想要远程访问 MBean，那就需要为 JMX 开启远程协议。JSR-160 通过一个 JMX 连接器为 JMX 远程访问定义了一个标准。Spring 可以通过 ConnectorServerFactoryBean 创建 JMX 连接器服务器。默认情况下，ConnectorServerFactoryBean 会创建并开启一个绑定到服务 URL service:jmx:jmxmp://localhost:9875 的 JMX 连接器服务器，它通过 JMX 消息协议（JMX Messaging Protocol，JMXMP）公开了 JMX 连接器。不过，大多数 JMX 实现并不支持 JMXMP。因此，你应该为 JMX 连接器选择一种广泛支持的远程协议，比如说 RMI。要想通过特定协议公开 JMX 连接器，只需为其提供服务 URL 即可。

```
@Bean
public FactoryBean<Registry> rmiRegistry() {
    return new RmiRegistryFactoryBean();
}

@Bean
@DependsOn("rmiRegistry")
public FactoryBean<JMXConnectorServer> connectorServer() {
    ConnectorServerFactoryBean connectorServerFactoryBean =
        new ConnectorServerFactoryBean();
    connectorServerFactoryBean
.setServiceUrl("service:jmx:rmi://localhost/jndi/rmi://localhost:1099/replicator");
    return connectorServerFactoryBean;
}
```

可以指定上述 URL 将 JMX 连接器绑定到监听 localhost 1099 端口的 RMI 注册器上。如果外部没有创建 RMI 注册器，那就应该通过 RmiRegistryFactoryBean 创建一个。该注册器的默认端口是 1099，不过可以在其 port 属性中指定其他端口。注意到 ConnectorServerFactoryBean 必须要在 RMI 注册器创建完毕并就绪后才能创建连接器服务器。你可以为其设置 depends-on 属性。

现在，MBean 可以通过 RMI 远程访问了。注意，现在没必要像之前那样通过 JMX -Dcom.sun.management. jmxremote 标记来启动开启了 RMI 访问的应用。当 JConsole 启动时，你可以在连接窗口中远程进程的服务 URL 中输入 service:jmx:rmi://localhost/jndi/rmi://localhost:1099/replicator，如图 13-3 所示。

连接创建后，就可以像之前那样调用 bean 方法了。

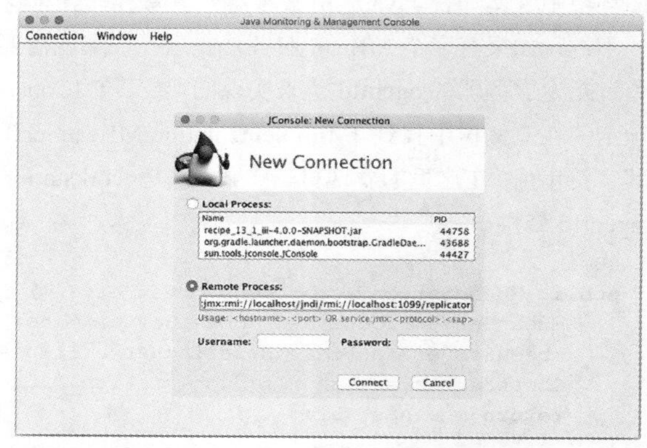

图 13-3　在 JConsole 中通过 RMI 连接到 MBean

装配 MBean 的管理接口

默认情况下，Spring MBeanExporter 会将 bean 的所有 public 属性导出为 MBean 属性，将所有 public 方法导出为 MBean 操作。不过，可以通过 MBean 装配器来装配 MBeans 的管理接口。Spring 中最简单的 MBean 装配器是 MethodNameBasedMBeanInfoAssembler，它可以指定要导出的方法名。

```
@Configuration
public class JmxConfig {
    ...
    @Bean
    public MBeanExporter mbeanExporter() {
        MBeanExporter mbeanExporter = new MBeanExporter();
        mbeanExporter.setBeans(beansToExport());
```

```
        mbeanExporter.setAssembler(assembler());
        return mbeanExporter;
    }

    @Bean
    public MBeanInfoAssembler assembler() {
        MethodNameBasedMBeanInfoAssembler assembler;
        assembler = new MethodNameBasedMBeanInfoAssembler();
        assembler.setManagedMethods(new String[] {"getSrcDir","setSrcDir","getDestDir",
        "setDestDir","replicate"});
        return assembler;
    }
}
```

另一个 MBean 装配器是 InterfaceBasedMBeanInfoAssembler，它会将你所指定的接口中定义的所有方法导出。

```
@Bean
public MBeanInfoAssembler assembler() {
    InterfaceBasedMBeanInfoAssembler assembler = new InterfaceBasedMBeanInfoAssembler();
    assembler.setManagedInterfaces(new Class[] {FileReplicator.class});
    return assembler;
}
```

Spring 还提供了 MetadataMBeanInfoAssembler，它会根据 bean 类中的元数据来装配 MBean 的管理接口。它支持两种类型的元数据：JDK 注解与 Apache Commons 属性（在背后，这是通过使用一个名为 JmxAttributeSource 的策略接口实现的）。对于使用了 JDK 注解的 bean 类来说，需要将 AnnotationJmxAttributeSource 实例指定为 MetadataMBeanInfoAssembler 的属性源。

```
@Bean
public MBeanInfoAssembler assembler() {
    MetadataMBeanInfoAssembler assembler = new MetadataMBeanInfoAssembler();
    assembler.setAttributeSource(new AnnotationJmxAttributeSource());
    return assembler;
}
```

接下来，为 bean 类与方法添加注解@ManagedResource、@ManagedAttribute 与@ManagedOperation 供 MetadataMBeanInfoAssembler 为 bean 装配管理接口。这些注解很容易理解。它们公开了被自己所注解的元素。如果有兼容于 JavaBeans 的属性，那么 JMX 就会将其作为自己的属性来用。类本身被看作是资源。在 JMX 中，方法叫做操作。了解了这些就比较好理解如下代码的功能了：

```
package com.apress.springrecipes.replicator;
...
import org.springframework.jmx.export.annotation.ManagedAttribute;
import org.springframework.jmx.export.annotation.ManagedOperation;
import org.springframework.jmx.export.annotation.ManagedResource;

@ManagedResource(description = "File replicator")
public class FileReplicatorJMXImpl implements FileReplicator {
    ...
    @ManagedAttribute(description = "Get source directory")
    public String getSrcDir() {
        ...
    }

    @ManagedAttribute(description = "Set source directory")
    public void setSrcDir(String srcDir) {
        ...
    }

    @ManagedAttribute(description = "Get destination directory")
    public String getDestDir() {
        ...
    }
```

```
    @ManagedAttribute(description = "Set destination directory")
    public void setDestDir(String destDir) {
        ...
    }

        ...

    @ManagedOperation(description = "Replicate files")
    public synchronized void replicate() throws IOException {
        ...
    }
}
```

使用注解注册 MBean

除了显式通过 MBeanExporter 导出 bean 外，还可以配置它的子类 AnnotationMBeanExporter 从声明在 IoC 容器中的 bean 中自动检测 MBean。无须为该导出器配置 MBean 装配器，因为默认情况下它会使用 MetadataMBeanInfoAssembler 和 AnnotationJmxAttributeSource。可以将之前注册时所用的 bean 和装配器属性删掉，只保留如下内容即可：

```
@Configuration
public class JmxConfig {

    @Bean
    public MBeanExporter mbeanExporter() {
        AnnotationMBeanExporter mbeanExporter = new AnnotationMBeanExporter();
        return mbeanExporter;
    }
}
```

AnnotationMBeanExporter 会检测到 IoC 容器中配置的使用了@ManagedResource 注解的 bean，并将其导出为 MBean。默认情况下，该导出器会将 bean 导出到名字与其包名相同的域下。此外，它会将 IoC 容器中的 bean 名字作为 MBean 的名字，并使用 bean 的简短类名作为其类型。因此，documentReplicator bean 会在 MBean 对象名 com.apress.springrecipes. replicator:name= documentReplicator,type=FileReplicatorJMXImpl 下被导出。

如果不想将包名作为域名，那可以通过添加 defaultDomain 属性为导出器设置默认域：

```
@Bean
public MBeanExporter mbeanExporter() {
    AnnotationMBeanExporter mbeanExporter = new AnnotationMBeanExporter();
    mbeanExporter.setDefaultDomain("bean");
    return mbeanExporter;
}
```

设置完 bean 的默认域后，documentReplicator bean 就会在如下 MBean 对象名下被导出：

```
bean:name=documentReplicator,type=FileReplicatorJMXImpl
```

此外，可以在@ManagedResource 注解的 objectName 属性中指定 bean 的 MBean 对象名。比如说，可以通过如下注解将文件复制器导出为 MBean：

```
package com.apress.springrecipes.replicator;
...
import org.springframework.jmx.export.annotation.ManagedOperation;
import org.springframework.jmx.export.annotation.ManagedOperationParameter;
import org.springframework.jmx.export.annotation.ManagedOperationParameters;
import org.springframework.jmx.export.annotation.ManagedResource;

@ManagedResource(
    objectName = "bean:name=fileCopier,type=FileCopierJMXImpl",
    description = "File Copier")
public class FileCopierImpl implements FileCopier {

    @ManagedOperation(
        description = "Copy file from source directory to destination directory")
```

```
        @ManagedOperationParameters( {
            @ManagedOperationParameter(
                name = "srcDir", description = "Source directory"),
            @ManagedOperationParameter(
                name = "destDir", description = "Destination directory"),
            @ManagedOperationParameter(
                name = "filename", description = "File to copy") })
        public void copyFile(String srcDir, String destDir, String filename)
            throws IOException {
            ...
        }
    }
```

不过，通过这种方式指定对象名只适用于 IoC 容器中的单例类（比如说 file copier），对于会创建多个实例的类来说并不适用（比如说 file replicator）。这是因为只能为一个类指定单个对象名。因此，不应该在不修改名字的情况下运行同一个服务器多次。

最后，另一种做法是依赖于 Spring 的类扫描机制来导出使用了@ManagedResource 注解的 MBean。如果 bean 是在 Java 配置类中初始化的，那就可以在配置类上使用@EnableMBeanExport 注解。这会告诉 Spring 导出被@EnableMBeanExport 所修饰且由@Bean 注解创建的所有 bean。

```
package com.apress.springrecipes.replicator.config;

...
import org.springframework.context.annotation.EnableMBeanExport;

@Configuration
@EnableMBeanExport
public class FileReplicatorConfig {
    ....
    @Bean
    public FileReplicatorJMXImpl documentReplicator() {
        FileReplicatorJMXImpl fRep = new FileReplicatorJMXImpl();
        fRep.setSrcDir(srcDir);
        fRep.setDestDir(destDir);
        fRep.setFileCopier(fileCopier());
        return fRep;
    }
    ...
}
```

由于使用了@EnableMBeanExport 注解，类型为 FileReplicatorJMXImpl 的 bean documentReplicatior 就会作为 MBean 导出，因为 FileReplicatorJMXImpl 类使用了注解@ManagedResource。

> **警告**：@EnableMBeanExport 注解用在了具体类的@Bean 实例上，而不像之前的示例那样用在接口上。基于接口的 bean 隐藏了目标类以及 JMX 管理的资源注解，MBean 是不会导出的。

13-2 发布并监听 JMX 通知

问题提出

你想从 MBean 发布 JMX 通知并通过 JMX 通知监听器监听它们。

解决方案

Spring 可以通过 NotificationPublisher 接口让 bean 发布 JMX 通知。还可以在 IoC 容器中注册标准的 JMX 通知监听器来监听 JMX 通知。

解释说明

要想发布事件，你需要访问 NotificationPublisher，这可以通过实现 NotificationPublisherAware 接口在 Spring 中获取到。要想监听事件，可以使用默认的 JMX 构造方式，实现 NotificationListener 接口并将实现注册到 JMX 上。

发布 JMX 通知

Spring IoC 容器支持导出为 MBean 的 bean 发布 JMX 通知。这些 bean 必须要实现 NotificationPublisherAware 接口来获取到 NotificationPublisher，这样才能发布通知。

```
package com.apress.springrecipes.replicator;
...
import javax.management.Notification;

import org.springframework.jmx.export.notification.NotificationPublisher;
import org.springframework.jmx.export.notification.NotificationPublisherAware;

@ManagedResource(description = "File replicator")
public class FileReplicatorImpl implements FileReplicator,
    NotificationPublisherAware {
    ...
    private int sequenceNumber;
    private NotificationPublisher notificationPublisher;

    public void setNotificationPublisher(
        NotificationPublisher notificationPublisher) {
        this.notificationPublisher = notificationPublisher;
    }

    @ManagedOperation(description = "Replicate files")
    public void replicate() throws IOException {
        notificationPublisher.sendNotification(
            new Notification("replication.start", this, sequenceNumber));
        ...
        notificationPublisher.sendNotification(
            new Notification("replication.complete", this, sequenceNumber));
        sequenceNumber++;
    }
}
```

在这个 file replicator 中，你会在复制开始和结束时发送一条 JMX 消息。可以在控制台的标准输出以及 JConsole Notifications 菜单中的 Mbeans 选项卡中看到通知，如图 13-4 所示。

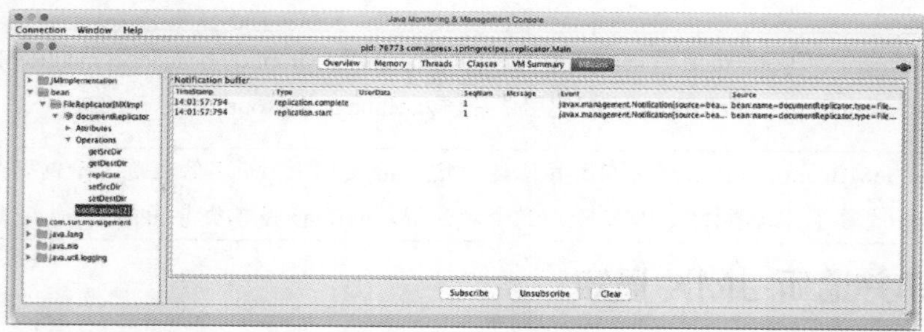

图 13-4　JConsole 中报告的 MBean 事件

要想在 JConsole 中查看通知，首先需要单击底部的 Subscribe 按钮，如图 13-4 所示。接下来，当在 MBean 操作一栏中使用 JConsole 按钮调用 replicate()方法时，你会看到两条通知。Notification 构造方法的第一个参数是通知类型，第二个参数是通知来源。

监听 JMX 通知

现在，我们创建一个通知监听器来监听 JMX 通知。由于监听器会收到多种不同类型的通知，比如说当 MBean 的属性发生变化时的 javax.management.AttributeChangeNotification，因此需要过滤那些想要处理的通知。

```
package com.apress.springrecipes.replicator;

import javax.management.Notification;
import javax.management.NotificationListener;

public class ReplicationNotificationListener implements NotificationListener {

    public void handleNotification(Notification notification, Object handback) {
        if (notification.getType().startsWith("replication")) {
            System.out.println(
                notification.getSource() + " " +
                notification.getType() + " #" +
                notification.getSequenceNumber());
        }
    }
}
```

接下来，可以将这个通知监听器注册到 MBean 导出器上来监听某些 MBean 发出的通知。

```
@Bean
public AnnotationMBeanExporter mbeanExporter() {
    AnnotationMBeanExporter mbeanExporter = new AnnotationMBeanExporter();
    mbeanExporter.setDefaultDomain("bean");
    mbeanExporter.setNotificationListenerMappings(notificationMappings());
    return mbeanExporter;
}

public Map<String, NotificationListener> notificationMappings() {
    Map<String, NotificationListener> mappings = new HashMap<>();
    mappings.put("bean:name=documentReplicator,type=FileReplicatorJMXImpl",
        new ReplicationNotificationListener());
    return mappings;
}
```

13-3 在 Spring 中访问远程 JMX MBean

问题提出

你想要访问 JMX 连接器所公开的远程 MBean 服务器上运行的 JMX MBean。当使用 JMX API 直接访问远程 MBean 时，你需要编写复杂的特定于 JMX 的代码。

解决方案

Spring 提供了两种方式来简化远程 MBean 的访问。首先，它提供了一个工厂 bean 以声明式的方式创建 MBean 服务器连接。借助于该服务器连接，你可以查询并更新 MBean 的属性，还可以调用它的操作。其次，Spring 提供了另一个工厂 bean，可以为远程 MBean 创建一个代理。借助于该代理，你可以像操纵本地 bean 一样操纵远程 MBean。

解释说明

为了简化 JMX 的操作，Spring 提供了两种方式。一种是通过配置到 MBean 服务器的连接来简化对普通 JMX 的处理。另一种更像是 Spring 所提供的其他远程技术，它提供了对远程 MBean 的代理。

通过 MBean 服务器连接来访问远程 MBean

JMX 客户端需要一个 MBean 服务器连接才能访问运行在远程 MBean 服务器上的 MBean。Spring 通过 org.springframework.jmx.support.MBeanServerConnectionFactoryBean 并以声明式的方式创建到启用了 JSR-160 的远程 MBean 服务器的连接。你只需提供服务 URL 就可以定位到 MBean 服务器。现在，我们在客户端 bean 配置类中声明该工厂 bean。

```
package com.apress.springrecipes.replicator.config;

import org.springframework.beans.factory.FactoryBean;
import org.springframework.context.annotation.Bean;
```

```
import org.springframework.context.annotation.Configuration;
import org.springframework.jmx.support.MBeanServerConnectionFactoryBean;

import javax.management.MBeanServerConnection;
import java.net.MalformedURLException;

@Configuration
public class JmxClientConfiguration {

    @Bean
    public FactoryBean<MBeanServerConnection> mbeanServerConnection()
    throws MalformedURLException {
        MBeanServerConnectionFactoryBean mBeanServerConnectionFactoryBean =
            new MBeanServerConnectionFactoryBean();
        mBeanServerConnectionFactoryBean
            .setServiceUrl("service:jmx:rmi://localhost/jndi/rmi://localhost:1099/
            replicator");
        return mBeanServerConnectionFactoryBean;
    }
}
```

借助于该工厂 bean 所创建的 MBean 服务器连接，你可以访问并操纵运行在 RMI 服务器（端口为 1099）上的 MBean。

> ■ 提示：可以使用 13-1 节中的 RMI 服务器，它公开了 MBean。如果使用本书的源代码，那么在使用 Gradle 构建完应用后，就可以通过命令 java -jar Recipe_14_1_iii-1.0- SNAPSHOT.jar 来启动服务器了。

建立好服务器与客户端之间的连接后，通过 getAttribute()和 setAttribute()方法并传递进 MBean 的对象名与属性名就可以查询并更新 MBean 的属性了。还可以通过 invoke()方法调用 MBean 的操作。

```
package com.apress.springrecipes.replicator;

import javax.management.Attribute;
import javax.management.MBeanServerConnection;
import javax.management.ObjectName;

import org.springframework.context.ApplicationContext;
import org.springframework.context.support.Generic XmlApplicationContext;

public class Client {

    public static void main(String[] args) throws Exception {
        ApplicationContext context =
            new AnnotationConfigApplicationContext("com.apress.springrecipes.
            replicator.config");

        MBeanServerConnection mbeanServerConnection =
            context.getBean(MBeanServerConnection.class);

        ObjectName mbeanName = new ObjectName("bean:name=documentReplicator");

        String srcDir = (String) mbeanServerConnection.getAttribute(mbeanName, "SrcDir");

        mbeanServerConnection.setAttribute(mbeanName,
            new Attribute("DestDir", srcDir + "_backup"));

        mbeanServerConnection.invoke(mbeanName, "replicate", new Object[] {}, new String[]{});
    }
}
```

此外，我们创建一个 JMX 通知监听器，这样就可以监听文件复制通知了。

```
package com.apress.springrecipes.replicator;

import javax.management.Notification;
```

```
import javax.management.NotificationListener;

public class ReplicationNotificationListener implements NotificationListener {

    public void handleNotification(Notification notification, Object handback) {
        if (notification.getType().startsWith("replication")) {
            System.out.println(
                notification.getSource() + " " +
                notification.getType() + " #" +
                notification.getSequenceNumber());
        }
    }
}
```

可以将该通知监听器注册到 MBean 服务器连接上，以监听该 MBean 服务器所发出的通知。

```
package com.apress.springrecipes.replicator;
...
import javax.management.MBeanServerConnection;
import javax.management.ObjectName;

public class Client {

    public static void main(String[] args) throws Exception {
        ...
        MBeanServerConnection mbeanServerConnection =
            (MBeanServerConnection) context.getBean("mbeanServerConnection");

        ObjectName mbeanName = new ObjectName(
            "bean:name=documentReplicator");

        mbeanServerConnection.addNotificationListener(
            mbeanName, new ReplicationNotificationListener(), null, null);
        ...
    }
}
```

运行该应用客户端后，打开 JConsole 查看该 RMI 服务器应用——使用远程进程设置 service: jmx:rmi://localhost/jndi/rmi://localhost:1099/replicator。在 Notifications 菜单的 Mbeans 选项卡下，你会看到新的通知类型 jmx.attribute.change，如图 13-5 所示。

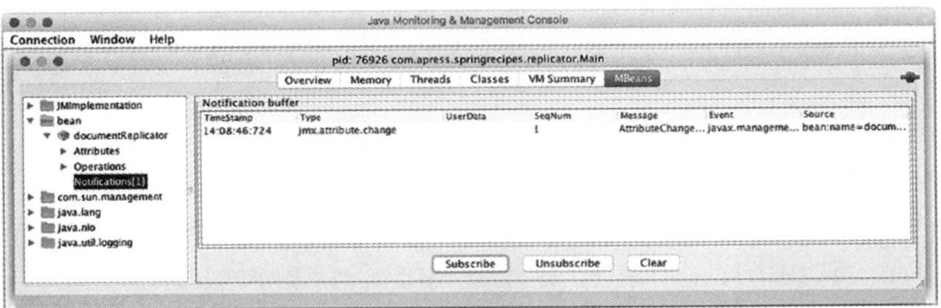

图 13-5 通过 RMI 调用的 JConsole 通知事件

通过 MBean 代理访问远程 MBean

Spring 提供的另一种远程 MBean 访问方式是通过 MBeanProxy 实现的，它可以由 MbeanProxyFactoryBean 所创建。

```
package com.apress.springrecipes.replicator.config;

...
import org.springframework.beans.factory.FactoryBean;
import org.springframework.jmx.access.MBeanProxyFactoryBean;
import org.springframework.jmx.support.MBeanServerConnectionFactoryBean;
```

```
import javax.management.MBeanServerConnection;
import java.net.MalformedURLException;

@Configuration
public class JmxClientConfiguration {

...

    @Bean
    public MBeanProxyFactoryBean fileReplicatorProxy() throws Exception {
        MBeanProxyFactoryBean fileReplicatorProxy = new MBeanProxyFactoryBean();
        fileReplicatorProxy.setServer(mbeanServerConnection().getObject());
        fileReplicatorProxy.setObjectName("bean:name=documentReplicator");
        fileReplicatorProxy.setProxyInterface(FileReplicator.class);
        return fileReplicatorProxy;
    }
}
```

你需要为想要代理的 MBean 指定对象名和服务器连接。最重要的是代理接口，其本地方法调用会在背后转换为远程 MBean 调用。现在，你可以通过这个代理操纵远程 MBean 了，它就像是个本地 bean 一样。之前直接在 MBean 服务器连接上调用的 MBean 操作可以简化为如下代码：

```
package com.apress.springrecipes.replicator;
...
public class Client {

    public static void main(String[] args) throws Exception {
        ...
        FileReplicator fileReplicatorProxy = context.getBean(FileReplicator.class);
        String srcDir = fileReplicatorProxy.getSrcDir();
        fileReplicatorProxy.setDestDir(srcDir + "_backup");
        fileReplicatorProxy.replicate();
    }
}
```

13-4　使用 Spring 的邮件支持来发送邮件

问题提出

很多应用都需要发送邮件。在 Java 应用中，可以通过 JavaMail API 发送邮件。不过，在使用 JavaMail 时，需要处理特定于 JavaMail 的会话与异常。这样，应用就会依赖于 JavaMail，难以切换至其他邮件 API。

解决方案

Spring 的邮件支持通过提供一个抽象且独立于实现的 API 来发送邮件，从而简化了邮件的发送。Spring 邮件支持的核心接口是 MailSender。JavaMailSender 是 MailSender 的子接口，包含了特定于 JavaMail 的特性，比如说多用途 Internet 邮件扩展（Multipurpose Internet Mail Extensions，MIME）消息支持。要想发送带有 HTML 内容、内联图片或是附件的邮件消息，你需要以 MIME 消息的形式发送。

解释说明

假设你希望前文中的 file replicator 应用能够将错误通知到管理员。首先，创建如下 ErrorNotifier 接口，它包含了一个方法用来通知文件复制错误：

```
package com.apress.springrecipes.replicator;

public interface ErrorNotifier {

    public void notifyCopyError(String srcDir, String destDir, String filename);
}
```

> **注意**：当有错误时，由你自己来完成这个通知器的调用。由于可以将错误处理看作是一个横切关注点，因此 AOP 将是该问题的一个理想解决方案。可以编写一个异常通知来调用这个通知器。

接下来，可以实现该接口，使其按照自己的方式来发送通知。最常见的方式是发送邮件。在按照这种方式实现接口前，你需要有一个支持简单邮件传输协议（Simple Mail Transfer Protocol，SMTP）的邮件服务器进行测试。我们推荐安装 Apache James Server，它很容易安装和配置。

注意：可以从 Apache James 网站下载 Apache James Server（比如说 2.3.2 版），将其解压缩到一个目录中完成安装。要想启动它，只需执行运行脚本即可（位于 bin 目录中）。

我们创建两个用户账户来使用该服务器进行邮件的发送和接收。默认情况下，James 的远程管理器服务会监听端口 4555。你可以通过控制台使用 telnet 登录到这个端口，并运行如下命令来添加用户 system 和 admin，其密码都是 12345：

```
> telnet 127.0.0.1 4555
JAMES Remote Administration Tool 2.3.2
Please enter your login and password
Login id:
Root
Password:
itroot
Welcome root. HELP for a list of commands
adduser system 12345
User system added
adduser admin 12345
User admin added
listusers
Existing accounts 2
user: admin
user: system
quit
Bye
```

使用 JavaMail API 发送邮件

现在，我们看看如何通过 JavaMail API 发送邮件。可以实现 ErrorNotifier 接口以在出错时发送邮件通知。

```
package com.apress.springrecipes.replicator;

import java.util.Properties;

import javax.mail.Message;
import javax.mail.MessagingException;
import javax.mail.Session;
import javax.mail.Transport;
import javax.mail.internet.InternetAddress;
import javax.mail.internet.MimeMessage;

public class EmailErrorNotifier implements ErrorNotifier {

    public void notifyCopyError(String srcDir, String destDir, String filename) {
        Properties props = new Properties();
        props.put("mail.smtp.host", "localhost");
        props.put("mail.smtp.port", "25");
        props.put("mail.smtp.username", "system");
        props.put("mail.smtp.password", "12345");
        Session session = Session.getDefaultInstance(props, null);
        try {

            Message message = new MimeMessage(session);
            message.setFrom(new InternetAddress("system@localhost"));
            message.setRecipients(Message.RecipientType.TO,
                    InternetAddress.parse("admin@localhost"));
            message.setSubject("File Copy Error");
            message.setText(
```

```
                "Dear Administrator,\n\n" +
                "An error occurred when copying the following file :\n" +
                "Source directory : " + srcDir + "\n" +
                "Destination directory : " + destDir + "\n" +
                "Filename : " + filename);
            Transport.send(message);
        } catch (MessagingException e) {
            throw new RuntimeException(e);
        }
    }
}
```

首先通过定义属性打开一个邮件会话并连接到 SMTP 服务器上。接下来，从该会话创建一个消息来构建邮件。然后调用 Transport.send() 来发送邮件。在使用 JavaMail API 时，你需要处理检查异常 MessagingException。注意，所有这些类、接口与异常都是由 JavaMail 定义的。

接下来，在 Spring IoC 容器中声明一个 EmailErrorNotifier 实例，以在文件复制出错时发送邮件通知。

```
package com.apress.springrecipes.replicator.config;

import com.apress.springrecipes.replicator.EmailErrorNotifier;
import com.apress.springrecipes.replicator.ErrorNotifier;
import org.springframework.context.annotation.Bean;
import org.springframework.context.annotation.Configuration;

@Configuration
public class MailConfiguration {

    @Bean
    public ErrorNotifier errorNotifier() {
        return new EmailErrorNotifier();
    }
}
```

编写如下 Main 类来测试 EmailErrorNotifier。运行后，你可以配置这个邮件应用，通过 POP3 从 James 服务器接收邮件。

```
package com.apress.springrecipes.replicator;

import org.springframework.context.ApplicationContext;
import org.springframework.context.support.GenericXmlApplicationContext;

public class Main {
    public static void main(String[] args) {
        ApplicationContext context =
            new AnnotationConfigApplicationContext("com.apress.springrecipes.replicator.
            config");

        ErrorNotifier errorNotifier = context.getBean(ErrorNotifier.class);
        errorNotifier.notifyCopyError("c:/documents", "d:/documents", "spring.doc");
    }
}
```

为了验证邮件已经发送，可以登录到 Apache James 自带的 POP 服务器上。你可以通过控制台使用 telnet 连接到 110 端口上并运行如下命令查看用户 admin 的邮件，其密码与创建时的一样：

```
> telnet 127.0.0.1 110
OK workstation POP3 server <JAMES POP3 Server 2.3.2> ready
USER admin
+OK
PASS 12345
+OK Welcome admin
LIST
+ OK 1 698
RETR 1
+OK Message follows
...
```

使用 Spring 的 MailSender 发送邮件

现在，我们来看看如何通过 Spring 的 MailSender 接口发送邮件。可以调用它的 send()方法发送 SimpleMailMessage。借助于该接口，程序代码将不会绑定到 JavaMail，并且也更易于测试。

```
package com.apress.springrecipes.replicator;

import org.springframework.mail.MailSender;
import org.springframework.mail.SimpleMailMessage;

public class EmailErrorNotifier implements ErrorNotifier {

    private MailSender mailSender;

    public void setMailSender(MailSender mailSender) {
        this.mailSender = mailSender;
    }

    public void notifyCopyError(String srcDir, String destDir, String filename) {
        SimpleMailMessage message = new SimpleMailMessage();
        message.setFrom("system@localhost");
        message.setTo("admin@localhost");
        message.setSubject("File Copy Error");
        message.setText(
            "Dear Administrator,\n\n" +
            "An error occurred when copying the following file :\n" +
            "Source directory : " + srcDir + "\n" +
            "Destination directory : " + destDir + "\n" +
            "Filename : " + filename);
        mailSender.send(message);
    }
}
```

接下来，需要在 bean 配置文件中配置一个 MailSender 实现并将其注入到 EmailErrorNotifier 中。在 Spring 中，该接口的唯一实现是 JavaMailSenderImpl，它使用 JavaMail 发送邮件。

```
@Configuration
public class MailConfiguration {

    @Bean
    public ErrorNotifier errorNotifier() {
        EmailErrorNotifier errorNotifier = new EmailErrorNotifier();
        errorNotifier.setMailSender(mailSender());
        return errorNotifier;
    }

    @Bean
    public JavaMailSenderImpl mailSender() {
        JavaMailSenderImpl mailSender = new JavaMailSenderImpl();
        mailSender.setHost("localhost");
        mailSender.setPort(25);
        mailSender.setUsername("system");
        mailSender.setPassword("12345");
        return mailSender;
    }
}
```

JavaMailSenderImpl 所用的默认端口是标准的 SMTP 端口 25，如果邮件服务器针对 SMTP 监听的就是这个端口，那就可以省略掉该属性。此外，如果 SMTP 服务器不需要用户认证，那就不需要设置用户名和密码。

如果在 Java 应用服务器中配置了 JavaMail 会话，可以借助于 JndiLocatorDelegate 来查找到它。

```
@Bean
public Session mailSession() throws NamingException {
    return JndiLocatorDelegate
```

```
        .createDefaultResourceRefLocator()
        .lookup("mail/Session", Session.class);
}
```

可以将 JavaMail session 注入到 JavaMailSenderImpl 中供其使用。在该示例中，无须再设置主机、端口、用户名与密码。

```
@Bean
public JavaMailSenderImpl mailSender() {
    JavaMailSenderImpl mailSender = new JavaMailSenderImpl();
    mailSender.setSession(mailSession());
    return mailSender;
}
```

定义邮件模板

在方法体中从头构建邮件模板并不高效，因为你需要硬编码邮件属性。此外，以 Java 字符串的形式编写邮件文本也是很困难的。可以考虑在 bean 配置文件中定义邮件消息模板并从中构建新的邮件消息。

```
@Configuration
public class MailConfiguration {
...
    @Bean
    public ErrorNotifier errorNotifier() {
        EmailErrorNotifier errorNotifier = new EmailErrorNotifier();
        errorNotifier.setMailSender(mailSender());
        errorNotifier.setCopyErrorMailMessage(copyErrorMailMessage());
        return errorNotifier;
    }

    @Bean
    public SimpleMailMessage copyErrorMailMessage() {
        SimpleMailMessage message = new SimpleMailMessage();
        message.setFrom("system@localhost");
        message.setTo("admin@localhost");
        message.setSubject("File Copy Error");
        message.setText("Dear Administrator,\n" +
            "\n" +
            "                   An error occurred when copying the following file :\n" +
            "\t\t       Source directory : %s\n" +
            "\t\t       Destination directory : %s\n" +
            "\t\t       Filename : %s");
        return message;
    }
}
```

注意，在上述消息文本中使用了占位符%s，这会通过 String.format()被消息参数替换掉。当然，还可以使用强大的模板语言，如 Velocity 或 FreeMarker，根据模板来生成消息文本。分离邮件消息模板与 bean 配置文件也是一种最佳实践。

每次发送邮件时，都可以通过这个注入的模板来构建新的 SimpleMailMessage 实例。接下来，可以使用 String.format()将%s 占位符替换为自己的消息参数来生成消息文本。

```
package com.apress.springrecipes.replicator;
...
import org.springframework.mail.SimpleMailMessage;

public class EmailErrorNotifier implements ErrorNotifier {
    ...
    private SimpleMailMessage copyErrorMailMessage;

    public void setCopyErrorMailMessage(SimpleMailMessage copyErrorMailMessage) {
        this.copyErrorMailMessage = copyErrorMailMessage;
    }
```

```
    public void notifyCopyError(String srcDir, String destDir, String filename) {
        SimpleMailMessage message = new SimpleMailMessage(copyErrorMailMessage);
        message.setText(String.format(
            copyErrorMailMessage.getText(), srcDir, destDir, filename));
        mailSender.send(message);
    }
}
```

发送带附件的邮件（MIME 消息）

到目前为止，你所使用的 SimpleMailMessage 类只能发送简单的普通文本邮件消息。要想发送带有 HTML 内容、内联图片或是附件的邮件，需要构建并发送 MIME 消息。JavaMail 通过 Javax.mail.internet.MimeMessage 类来支持 MIME。

首先，需要使用 JavaMailSender 接口而非它的父接口 MailSender。你所注入的 JavaMailSenderImpl 实例实现了该接口，因此无须修改 bean 配置。如下通知器会将 Spring 的 bean 配置文件作为邮件附件发送给管理员：

```
package com.apress.springrecipes.replicator;

import javax.mail.MessagingException;
import javax.mail.internet.MimeMessage;

import org.springframework.core.io.ClassPathResource;
import org.springframework.mail.MailParseException;
import org.springframework.mail.SimpleMailMessage;
import org.springframework.mail.javamail.JavaMailSender;
import org.springframework.mail.javamail.MimeMessageHelper;

public class EmailErrorNotifier implements ErrorNotifier {

    private JavaMailSender mailSender;
    private SimpleMailMessage copyErrorMailMessage;

    public void setMailSender(JavaMailSender mailSender) {
        this.mailSender = mailSender;
    }

    public void setCopyErrorMailMessage(SimpleMailMessage copyErrorMailMessage) {
        this.copyErrorMailMessage = copyErrorMailMessage;
    }

    public void notifyCopyError(String srcDir, String destDir, String filename) {
        MimeMessage message = mailSender.createMimeMessage();
        try {
            MimeMessageHelper helper = new MimeMessageHelper(message, true);
            helper.setFrom(copyErrorMailMessage.getFrom());
            helper.setTo(copyErrorMailMessage.getTo());

            helper.setSubject(copyErrorMailMessage.getSubject());
            helper.setText(String.format(
                copyErrorMailMessage.getText(), srcDir, destDir, filename));

            ClassPathResource config = new ClassPathResource("beans.xml");
            helper.addAttachment("beans.xml", config);
        } catch (MessagingException e) {
            throw new MailParseException(e);
        }
        mailSender.send(message);
    }
}
```

与 SimpleMailMessage 不同，MimeMessage 类是由 JavaMail 定义的，因此只能通过调用 mailSender.createMimeMessage()来实例化它。Spring 提供了辅助类 MimeMessageHelper 来简化 MimeMessage 的操作。

借助于 MimeMessageHelper，可以从 Spring Resource 对象中添加附件。不过，该辅助类的操作依然会抛出 JavaMail 的 MessagingException。你需要将该异常转换为 Spring 的邮件运行期异常以保持一致性。Spring 提供了另外一个方法来构建 MIME 消息，这是通过实现 MimeMessagePreparator 接口达成的。

```
package com.apress.springrecipes.replicator;
...
import javax.mail.internet.MimeMessage;

import org.springframework.mail.javamail.MimeMessagePreparator;

public class EmailErrorNotifier implements ErrorNotifier {
    ...
    public void notifyCopyError(
        final String srcDir, final String destDir, final String filename) {
        MimeMessagePreparator preparator = new MimeMessagePreparator() {

            public void prepare(MimeMessage mimeMessage) throws Exception {
                MimeMessageHelper helper =
                    new MimeMessageHelper(mimeMessage, true);
                helper.setFrom(copyErrorMailMessage.getFrom());
                helper.setTo(copyErrorMailMessage.getTo());
                helper.setSubject(copyErrorMailMessage.getSubject());
                helper.setText(String.format(
                    copyErrorMailMessage.getText(), srcDir, destDir, filename));

                ClassPathResource config = new ClassPathResource("beans.xml");
                helper.addAttachment("beans.xml", config);
            }
        };
        mailSender.send(preparator);
    }
}
```

在 prepare()方法中，可以准备 MimeMessage 对象，它是为 JavaMailSender 预先创建的。如果抛出了异常，那么异常就会自动转换为 Spring 的邮件运行期异常。

13-5　借助 Spring 的 Quartz 支持来调度任务

问题提出

应用有高级的调度需求，你想通过 Quartz 调度器来实现。这种需求可能会很复杂，比如说在任意指定的时间运行，或是在奇怪的间隔时间（“每两周的周二上午 10 点后和下午 2 点前”）运行。此外，你想以声明式的方式来配置调度任务。

解决方案

Spring 为 Quartz 提供了辅助类，无须使用 Quartz API 编写代码就能开启任务的调度。

解释说明

首先，我们将会介绍如何在不使用 Spring 提供的辅助类的情况下在 Spring 中使用 Quartz，然后介绍使用 Spring 提供的 Quartz 辅助类的方式。

不借助 Spring 的支持使用 Quartz

要想使用 Quartz 进行调度，首先需要实现 Job 接口来创建一个任务。比如说，如下任务执行了前文中所设计的文件复制器的 replicate()方法。它通过 JobExecutionContext 对象获取到一个任务数据 map（这是个 Quartz 概念，用于定义任务）。

```
package com.apress.springrecipes.replicator;
...
import org.quartz.Job;
import org.quartz.JobExecutionContext;
import org.quartz.JobExecutionException;
```

```
public class FileReplicationJob implements Job {

    public void execute(JobExecutionContext context)
        throws JobExecutionException {
        Map dataMap = context.getJobDetail().getJobDataMap();
        FileReplicator fileReplicator =
            (FileReplicator) dataMap.get("fileReplicator");
        try {
            fileReplicator.replicate();
        } catch (IOException e) {
            throw new JobExecutionException(e);
        }
    }
}
```

创建任务后，使用 Quartz API 配置并调度它。比如说，如下调度器每隔 60 秒且在首次执行时延迟 5 秒后运行文件复制任务：

```
package com.apress.springrecipes.replicator;
...
import org.quartz.JobDetail;
import org.quartz.JobDataMap;
import org.quartz.JobBuilder;
import org.quartz.Trigger;
import org.quartz.TriggerBuilder;
import org.quartz.SimpleScheduleBuilder;
import org.quartz.DateBuilder.IntervalUnit.*;
import org.quartz.Scheduler;
import org.quartz.impl.StdSchedulerFactory;

import org.springframework.context.ApplicationContext;
import org.springframework.context.support.GenericXmlApplicationContext;

public class Main {

    public static void main(String[] args) throws Exception {
        ApplicationContext context =
            new AnnotationConfigApplicationContext("com.apress.springrecipes.
            replicator.config");

        FileReplicator documentReplicator = context.getBean(FileReplicator.class);

        JobDataMap jobDataMap = new JobDataMap();
        jobDataMap.put("fileReplicator", documentReplicator);

        JobDetail job = JobBuilder.newJob(FileReplicationJob.class)
                                  .withIdentity("documentReplicationJob")
                                  .storeDurably()
                                  .usingJobData(jobDataMap)
                                  .build();
            Trigger trigger = TriggerBuilder.newTrigger()
        .withIdentity("documentReplicationTrigger")
        .startAt(new Date(System.currentTimeMillis() + 5000))
        .forJob(job)
        .withSchedule(SimpleScheduleBuilder.simpleSchedule()
    .withIntervalInSeconds(60)
    .repeatForever())
    .build();

        Scheduler scheduler = new StdSchedulerFactory().getScheduler();
        scheduler.start();
        scheduler.scheduleJob(job, trigger);
    }
}
```

在 Main 类中，首先创建一个任务 map。在该示例中只有单个任务，其键是个描述性的名字，值则是指向任务的对象引用。接下来，在 JobDetail 对象中定义文件复制任务的任务细节信息，并在其 jobDataMap 属性中准备任务数据。然后，创建了一个 SimpleTrigger 对象来配置调度属性。最后，创建一个调度器并使用该触发器来运行任务。

Quartz 支持不同类型的调度以不同的间隔时间来运行任务。调度被定义为触发器的一部分。在最新的版本中，Quartz 调度有 SimpleScheduleBuilder、CronScheduleBuilder、CalendarIntervalScheduleBuilder 与 DailyTimeIntervalScheduleBuilder。SimpleScheduleBuilder 通过设置诸如开始时间、结束时间、重复间隔与重复次数等属性来调度任务。CronScheduleBuilder 接受一个 UNIX cron 表达式来指定时间以运行任务。比如说，可以将之前的 SimpleScheduleBuilder 替换为如下的 CronScheduleBuilder 以在每天 17:30 运行任务：.withSchedule(CronScheduleBuilder. cronSchedule(" 0 30 17 * * ?"))。cron 表达式包含了 7 个字段（最后一个可选），中间用空格分开。表 13-1 所示为 cron 表达式的字段。

表 13-1　　cron 表达式的字段描述

位置	字段名	范围
1	秒	0～59
2	分钟	0～59
3	小时	0～23
4	每月的天数	1～31
5	月份	1～12 或 JAN～DEC
6	每周的天数	1～7 或 SUN～SAT
7	年（可选）	1970～2099

cron 表达式的每一部分都可以指定为一个具体的值（比如说 3）、一个范围（比如说 1～5）、一个列表（比如说 1,3,5）、一个通配符（*匹配所有值），或是一个问号（?用在“每月的天数”或“每周的天数”字段中，但不能同时指定，用于匹配这两个字段之一）。CalendarIntervalScheduleBuilder 可以根据日历时间（天、周、月、年）来调度任务，而 DailyTimeIntervalScheduleBuilder 则提供了便捷的方式来设置任务的结束时间（比如说 endingDailyAt()和 endingDailyAfterCount()等方法）。

借助 Spring 的支持使用 Quartz

在使用 Quartz 时，可以通过实现 Job 接口创建任务，通过 JobExecutionContext 从任务数据 map 中获取到任务数据。为了将任务类与 Quartz API 解耦，Spring 提供了 QuartzJobBean，你可以继承它并通过 setter 方法获取任务数据。QuartzJobBean 将任务数据 map 转换为属性，并通过 setter 方法将它们注入进来。

```
package com.apress.springrecipes.replicator;
...
import org.quartz.JobExecutionContext;
import org.quartz.JobExecutionException;
import org.springframework.scheduling.quartz.QuartzJobBean;

public class FileReplicationJob extends QuartzJobBean {

    private FileReplicator fileReplicator;

    public void setFileReplicator(FileReplicator fileReplicator) {
        this.fileReplicator = fileReplicator;
    }

    protected void executeInternal(JobExecutionContext context)
        throws JobExecutionException {
        try {
```

```
                fileReplicator.replicate();
            } catch (IOException e) {
                throw new JobExecutionException(e);
            }
        }
    }
```

接下来，可以通过 JobDetailBean 在 Spring 的 bean 配置文件中配置 Quartz JobDetail 对象。默认情况下，Spring 将该 bean 的名字作为任务名。可以通过设置其 name 属性进行修改。

```
@Bean
@Autowired
public JobDetailFactoryBean documentReplicationJob(FileReplicator fileReplicator) {
    JobDetailFactoryBean documentReplicationJob = new JobDetailFactoryBean();
    documentReplicationJob.setJobClass(FileReplicationJob.class);
    documentReplicationJob.setDurability(true);
    documentReplicationJob.setJobDataAsMap(
        Collections.singletonMap("fileReplicator", fileReplicator));
    return documentReplicationJob;
}
```

Spring 还提供了 MethodInvokingJobDetailFactoryBean 来定义一个任务，该任务执行特定对象中的单个方法。这样就无须再创建任务类了。可以通过如下任务信息来替换掉之前的信息：

```
@Bean
@Autowired
public MethodInvokingJobDetailFactoryBean documentReplicationJob(FileReplicator
fileReplicator) {
    MethodInvokingJobDetailFactoryBean documentReplicationJob =
        new MethodInvokingJobDetailFactoryBean();
    documentReplicationJob.setTargetObject(fileReplicator);
    documentReplicationJob.setTargetMethod("replicatie");
    return documentReplicationJob;
}
```

定义任务后，就可以配置 Quartz 触发器了。Spring 支持 SimpleTriggerFactoryBean 和 CronTriggerFactoryBean。SimpleTriggerFactoryBean 需要一个对 JobDetail 对象的引用，并为调度属性提供了常见值，比如说开始时间与重复次数。

```
@Bean
@Autowired
public SimpleTriggerFactoryBean documentReplicationTrigger(JobDetail documentReplicationJob)
{
    SimpleTriggerFactoryBean documentReplicationTrigger = new SimpleTriggerFactoryBean();
    documentReplicationTrigger.setJobDetail(documentReplicationJob);
    documentReplicationTrigger.setStartDelay(5000);
    documentReplicationTrigger.setRepeatInterval(60000);
    return documentReplicationTrigger;
}
```

还可以通过 CronTriggerFactoryBean 来配置 cron 风格的调度。

```
@Bean
@Autowired
public CronTriggerFactoryBean documentReplicationTrigger(JobDetail documentReplicationJob){
    CronTriggerFactoryBean documentReplicationTrigger = new CronTriggerFactoryBean();
    documentReplicationTrigger.setJobDetail(documentReplicationJob);
    documentReplicationTrigger.setStartDelay(5000);
    documentReplicationTrigger.setCronExpression("0/60 * * * * ?");
    return documentReplicationTrigger;
}
```

最后，有了 Quartz 任务和触发器后，可以配置 SchedulerFactoryBean 实例来创建一个 Scheduler 对象以运行触发器。可以在该工厂 bean 中指定多个触发器。

```
@Bean
@Autowired
public SchedulerFactoryBean scheduler(Trigger[] triggers) {
    SchedulerFactoryBean scheduler = new SchedulerFactoryBean();
```

```
        scheduler.setTriggers(triggers);
        return scheduler;
    }
```

现在，只需通过如下 Main 类开启调度器。通过这种方式，不需要编写一行代码就可以实现任务的调度。

```
package com.apress.springrecipes.replicator;

import org.springframework.context.annotation.AnnotationConfigApplicationContext;

public class Main {

    public static void main(String[] args) throws Exception {
        new AnnotationConfigApplicationContext("com.apress.springrecipes.replicator.config");
    }
}
```

13-6　使用 Spring 的调度支持来调度任务

问题提出

你想以一致的方式调度方法调用（使用 cron 表达式、间隔时间或是一个比率），但又不想使用 Quartz 来做这件事。

解决方案

Spring 支持配置 TaskExecutors 和 TaskSchedulers。这个能力再加上可通过@Scheduled 注解来调度方法可使得 Spring 调度支持变得非常简单；你只需一个方法和一个注解，并在注解扫描器上进行切换即可。

解释说明

我们再来看一下上一节中的示例：你想要通过 cron 表达式对 bean 的复制方法的调用进行调度。配置类的代码如下所示：

```
package com.apress.springrecipes.replicator.config;

import org.springframework.context.annotation.Bean;
import org.springframework.context.annotation.Configuration;
import org.springframework.scheduling.annotation.EnableScheduling;
import org.springframework.scheduling.annotation.SchedulingConfigurer;
import org.springframework.scheduling.config.ScheduledTaskRegistrar;

import java.util.concurrent.Executor;
import java.util.concurrent.Executors;

@Configuration
@EnableScheduling
public class SchedulingConfiguration implements SchedulingConfigurer {

    @Override
    public void configureTasks(ScheduledTaskRegistrar taskRegistrar) {
        taskRegistrar.setScheduler(scheduler());
    }

    @Bean
    public Executor scheduler() {
        return Executors.newScheduledThreadPool(10);
    }

}
```

通过指定@EnableScheduling 开启了注解驱动的调度支持。这会注册一个 bean，这个 bean 会扫描应用上下文中带有@Scheduled 注解的 beans。你还实现了接口 SchedulingConfigurer，因为想要对调度器进行一些额外的配置。你创建了一个拥有 10 个线程的线程池来执行调度任务。

```
package com.apress.springrecipes.replicator;

import org.springframework.scheduling.annotation.Scheduled;
```

```
import java.io.File;
import java.io.IOException;

public class FileReplicatorImpl implements FileReplicator {

    @Scheduled(fixedDelay = 60 * 1000)
    public synchronized void replicate() throws IOException {
        File[] files = new File(srcDir).listFiles();

        for (File file : files) {
            if (file.isFile()) {
                fileCopier.copyFile(srcDir, destDir, file.getName());
            }
        }
    }
}
```

注意到你对 replicate()方法使用了@Scheduled 注解。这是告诉调度器每 60 秒执行一次方法。此外，还可以对@Scheduled 注解指定 fixedRate 值，这会指定连续开始的时间间隔，然后触发另一次运行。

```
@Scheduled(fixedRate = 60 * 1000)
public synchronized void replicate() throws IOException {
    File[] files = new File(srcDir).listFiles();

    for (File file : files) {
        if (file.isFile()) {
            fileCopier.copyFile(srcDir, destDir, file.getName());
        }
    }
}
```

最后，你可能想对方法的执行进行更为复杂一些的控制。在该示例中，你可以使用 cron 表达式，就像在 Quartz 示例中所做的那样。

```
@Scheduled( cron = "0/60 * * * * ? " )
public synchronized void replicate() throws IOException {
    File[] files = new File(srcDir).listFiles();

    for (File file : files) {
        if (file.isFile()) {
            fileCopier.copyFile(srcDir, destDir, file.getName());
        }
    }
}
```

我们也可以在 Java 中配置这一切。如果不想或是不能向既有的 bean 方法上添加注解，那么这就很有价值了。如下展示了如何通过 Spring ScheduledTaskRegistrar 来重新创建之前以注解为中心的示例：

```
package com.apress.springrecipes.replicator.config;

import com.apress.springrecipes.replicator.FileReplicator;
import org.springframework.beans.factory.annotation.Autowired;
import org.springframework.context.annotation.Bean;
import org.springframework.context.annotation.Configuration;
import org.springframework.scheduling.annotation.EnableScheduling;
import org.springframework.scheduling.annotation.SchedulingConfigurer;
import org.springframework.scheduling.config.ScheduledTaskRegistrar;

import java.io.IOException;
import java.util.concurrent.Executor;
import java.util.concurrent.Executors;

@Configuration
@EnableScheduling
public class SchedulingConfiguration implements SchedulingConfigurer {
```

```
    @Autowired
    private FileReplicator fileReplicator;

    @Override
    public void configureTasks(ScheduledTaskRegistrar taskRegistrar){
        taskRegistrar.setScheduler(scheduler());
        taskRegistrar.addFixedDelayTask(() -> {
            try {
                fileReplicator.replicate();
            } catch (IOException e) {
                e.printStackTrace();
            }
        }, 60000);
    }

    @Bean
    public Executor scheduler() {
        return Executors.newScheduledThreadPool(10);
    }

}
```

13-7 通过 RMI 公开和调用服务

问题提出

你想公开 Java 应用中的服务供其他 Java 客户端远程调用。由于二者都运行在 Java 平台上，因此可以选择纯 Java 解决方案而无须考虑跨平台的移植性问题。

解决方案

远程方法调用（Remote Method Invocation，RMI）是个基于 Java 的远程技术，可以实现运行在不同 JVM 上的两个 Java 应用之间的通信。借助于 RMI，对象可以调用远程对象的方法。RMI 依赖于对象序列化来对方法参数和返回值进行编组与反编组。

要想通过 RMI 公开服务，你需要创建服务接口，它需要继承 java.rmi.Remote，接口中的方法会抛出 java.rmi.RemoteException。接下来创建接口的服务实现。然后，启动 RMI 注册器并将服务注册上来。因此，即便公开一个简单的服务也需要很多步骤。

要想通过 RMI 调用服务，首先需要在 RMI 注册器中查找远程服务引用，然后调用其方法。不过，要想调用远程服务的方法，需要处理 java.rmi.RemoteException 以防止远程服务抛出任何异常。

Spring 的远程设施可以极大简化服务器端与客户端的 RMI 使用。在服务器端，你可以通过 RmiServiceExporter 将 Spring POJO 导出为 RMI 服务，其方法就可以被远程调用了。这只需要几行 bean 配置即可，无须编写任何代码。通过这种方式导出的 beans 无须实现 java.rmi.Remote，也无须抛出 java.rmi.RemoteException。在客户端，只需通过 RmiProxyFactoryBean 为远程服务创建代理即可。借助于代理，可以像使用本地 bean 一样使用远程服务。此外，也无须编写额外的代码。

解释说明

假设你要为运行在不同平台上的客户端构建一个天气 Web 服务。该服务包含了一个查询某城市多天的温度的操作。首先，创建 TemperatureInfo 类，表示特定城市与日期的最低、最高与平均温度。

```
package com.apress.springrecipes.weather;
...
public class TemperatureInfo implements Serializable {

    private String city;
    private Date date;
    private double min;
    private double max;
    private double average;
```

```
    // Constructors, Getters and Setters
    ...
}
```

接下来，定义服务接口，它有一个 getTemperatures()操作，该操作会返回一个城市多天的温度。

```
package com.apress.springrecipes.weather;
...
public interface WeatherService {

    List<TemperatureInfo> getTemperatures(String city, List<Date> dates);
}
```

你需要为该接口提供实现。在生产应用中，你需要查询数据库来实现该服务接口。不过，这里将会硬编码温度进行测试。

```
package com.apress.springrecipes.weather;
...
public class WeatherServiceImpl implements WeatherService {

    public List<TemperatureInfo> getTemperatures(String city, List<Date> dates) {
        List<TemperatureInfo> temperatures = new ArrayList<TemperatureInfo>();
        for (Date date : dates) {
            temperatures.add(new TemperatureInfo(city, date, 5.0, 10.0, 8.0));
        }
        return temperatures;
    }
}
```

公开 RMI 服务

接下来，将天气服务公开为 RMI 服务。要想使用 Spring 的远程设施，需要构建一个 Java 配置类来创建必要的 bean，并通过 RmiServiceExporter 将天气服务导出为 RMI 服务。

```
package com.apress.springrecipes.weather.config;

...
import com.apress.springrecipes.weather.WeatherService;
import com.apress.springrecipes.weather.WeatherServiceImpl;

import org.springframework.remoting.rmi.RmiServiceExporter;

@Configuration
public class WeatherConfig {

    @Bean
    public WeatherService weatherService() {
        return new WeatherServiceImpl();
    }

    @Bean
    public RmiServiceExporter rmiService() {
        RmiServiceExporter rmiService = new RmiServiceExporter();
        rmiService.setServiceName("WeatherService");
        rmiService.setServiceInterface(com.apress.springrecipes.weather.
        WeatherService.class);
        rmiService.setService(weatherService());
        return rmiService;
    }
}
```

RmiServiceExporter 实例有几个属性必须要配置，包括服务名、服务接口和待导出的服务对象。你可以将配置在 IoC 容器中的任何 bean 导出为 RMI 服务。RmiServiceExporter 会创建一个 RMI 代理来包装这个 bean 并将其绑定到 RMI 注册器上。当代理接收到来自 RMI 注册器的调用请求时，它会调用 bean 中对应的方法。默认情况下，RmiServiceExporter 会寻找 localhost 上端口为 1099 的 RMI 注册器。如果找不到，那么它会启动一个新的。不过，如果想将服务绑定到另一个运行着的 RMI 注册器上，那就可以在 registryHost 与 registryPort 属性中指定注册器的主机和端口号。注意，一旦指定了注册器主机，RmiServiceExporter 就不会再启动新的注册器，即便指定的注册器不存在亦如此。运行如下 RmiServer 类来创建一个应用上下文：

```
package com.apress.springrecipes.weather;

import com.apress.springrecipes.weather.config.WeatherConfigServer;
import org.springframework.context.annotation.AnnotationConfigApplicationContext;

public class RmiServer {

    public static void main(String[] args) {
        new AnnotationConfigApplicationContext(WeatherConfigServer.class);
    }
}
```

在该配置中，服务器会启动；在输出中，会看到一条消息，指明无法找到既有的 RMI 注册器。

调用 RMI 服务

借助于 Spring 的远程设施，可以像调用本地 bean 一样调用远程服务。比如说，可以创建一个客户端，通过接口来引用天气服务。

```
package com.apress.springrecipes.weather;

import java.util.Arrays;
import java.util.Date;
import java.util.List;

public class WeatherServiceClient {

    private final WeatherService weatherService;

    public WeatherServiceClient(WeatherService weatherService) {
        this.weatherService = weatherService;
    }

    public TemperatureInfo getTodayTemperature(String city) {
        List<Date> dates = Arrays.asList(new Date());
        List<TemperatureInfo> temperatures =
            weatherService.getTemperatures(city, dates);
        return temperatures.get(0);
    }
}
```

注意，weatherService 字段是通过构造方法进行装配的，因此需要创建该 bean 的一个实例。weatherService 会通过 RmiProxyFactoryBean 为远程服务创建一个代理。接下来就可以使用该服务了，就好像它是个本地 bean 一样。如下 Java 配置类展示了该 RMI 客户端所需的必要 bean：

```
package com.apress.springrecipes.weather.config;

import com.apress.springrecipes.weather.WeatherService;
import com.apress.springrecipes.weather.WeatherServiceClient;

import org.springframework.context.annotation.Bean;
import org.springframework.context.annotation.Configuration;
import org.springframework.remoting.rmi.RmiProxyFactoryBean;

@Configuration
public class WeatherConfigClient {

    @Bean
    public RmiProxyFactoryBean weatherService() {
        RmiProxyFactoryBean rmiProxy = new RmiProxyFactoryBean();
```

```
        rmiProxy.setServiceUrl("rmi://localhost:1099/WeatherService");
        rmiProxy.setServiceInterface(WeatherService.class);
        return rmiProxy;
    }

    @Bean
    public WeatherServiceClient weatherClient(WeatherService weatherService) {
        return new WeatherServiceClient(weatherService);
    }
}
```

RmiProxyFactoryBean 实例有两个属性是必须要配置的。服务 URL 属性指定了 RMI 注册器的主机和端口号以及服务名。服务接口可以让这个工厂 bean 针对已知、共享的 Java 接口为远程服务创建代理。代理会透明地将调用请求转移给远程服务。除了 RmiProxyFactoryBean 实例外，你还需要创建一个名为 weatherClient 的 WeatherServiceClient 实例。

接下来，运行如下 RmiClient 主类：

```
package com.apress.springrecipes.weather;

import com.apress.springrecipes.weather.config.WeatherConfigClient;
import org.springframework.context.ApplicationContext;
import org.springframework.context.annotation.AnnotationConfigApplicationContext;

public class RmiClient {

    public static void main(String[] args) {
        ApplicationContext context =
            new AnnotationConfigApplicationContext(WeatherConfigClient.class);

        WeatherServiceClient client = context.getBean(WeatherServiceClient.class);

        TemperatureInfo temperature = client.getTodayTemperature("Houston");
        System.out.println("Min temperature : " + temperature.getMin());
        System.out.println("Max temperature : " + temperature.getMax());
        System.out.println("Average temperature : " + temperature.getAverage());
    }
}
```

13-8　通过 HTTP 公开和调用服务

问题提出

RMI 通过自己的协议进行通信，这可能无法穿越防火墙。理想情况下，你想要通过 HTTP 进行通信。

解决方案

Hessian 是由 Caucho Technology 公司开发的一款简单、轻量级的远程技术。它通过 HTTP 使用私有的消息进行通信，并且拥有自己的序列化机制，不过要比 RMI 简单不少。除了 Java 外，其他平台也支持 Hessian 的消息格式，如 PHP、Python、C#和 Ruby。这样，Java 应用就可以与运行在其他平台上的应用进行通信了。

除了之前介绍的技术外，Spring 框架还提供了一个名为 HTTP Invoker 的远程技术。它也通过 HTTP 进行通信，不过使用了 Java 的对象序列化机制来序列化对象。与 Hessian 不同，HTTP Invoker 要求服务的两端都运行在 Java 平台上且都需要使用 Spring 框架。不过，它可以序列化所有类型的 Java 对象，其中有一些类型是无法通过 Hessian 的私有机制进行序列化的。

Spring 的远程设施在使用这些技术公开和调用远程服务时是一致的。在服务器端，你可以创建如 HessianServiceExporter 或 HttpInvokerServiceExporter 这样的服务导出器将 Spring bean 导出为远程服务，服务的方法可以远程进行调用。这只需要几行 bean 配置即可，无须编写其他代码。在客户端，你还可以配置如 HessianProxyFactoryBean 或 HttpInvokerProxyFactoryBean 这样的代理工厂 bean 来为远程服务创建代理。这样就可以像使用本地 bean 一样使用远程服务了。此外，它也不需要编写其他代码。

解释说明

要想导出服务，需要通过 HessianServiceExporter 和 HttpInvokerServiceExporter 来实现。要想消费公开出来的服务，可以使用一些辅助类，如 HessianProxyFactoryBean 和 HttpInvokerProxyFactorybean。这里会介绍针对 Hessian 与 HTTP 的这两个解决方案。

公开 Hessian 服务

我们使用 13.7 节中所用的同样的天气服务，并通过 Spring 将其导出为 Hessian 服务。我们将会使用 Spring MVC 来创建一个简单的 Web 应用以部署服务。首先，创建一个 WeatherServiceInitializer 类来启动 Web 应用和 Spring 应用上下文。

```
package com.apress.springrecipes.weather.config;

import org.springframework.web.servlet.support.
AbstractAnnotationConfigDispatcherServletInitializer;

public class WeatherServiceInitializer extends
AbstractAnnotationConfigDispatcherServletInitializer {

    @Override
    protected String[] getServletMappings() {
        return new String[] {"/*"};
    }

    @Override
    protected Class<?>[] getRootConfigClasses() {
        return null;
    }

    @Override
    protected Class<?>[] getServletConfigClasses() {
        return new Class[] {WeatherConfigHessianServer.class};
    }
}
```

WeatherServiceInitializer 类创建了一个 DispatcherServlet servlet，以映射根路径（/*）下的所有 URL，并使用 WeatherConfigHessianServer 类进行配置。

```
package com.apress.springrecipes.weather.config;

import com.apress.springrecipes.weather.WeatherService;
import com.apress.springrecipes.weather.WeatherServiceImpl;
import org.springframework.context.annotation.Bean;
import org.springframework.context.annotation.Configuration;
import org.springframework.remoting.caucho.HessianServiceExporter;

@Configuration
public class WeatherConfigHessianServer {

    @Bean
    public WeatherService weatherService() {
        WeatherService wService = new WeatherServiceImpl();
        return wService;
    }

    @Bean(name = "/weather")
    public HessianServiceExporter exporter() {
        HessianServiceExporter exporter = new HessianServiceExporter();
        exporter.setService(weatherService());
        exporter.setServiceInterface(WeatherService.class);
        return exporter;
    }
}
```

扫描组件可以让 Spring 检测到实例化了 weatherService bean 的 Java 配置类，这个 bean 包含了将由 HessianServiceExporter 实例所公开的操作。该示例中的 weatherService bean 等价于 13.7 节中所用的 bean。可以查看本书源代码找到预先构建好的应用。

对于 HessianServiceExporter 实例来说，你需要配置一个待导出的服务对象及其服务接口。可以将任何 Spring bean 导出为 Hessian 服务。HessianServiceExporter 会创建一个代理来包装这个 bean。

当代理接收到调用请求时，它会调用该 bean 上对应的方法。默认情况下，BeanNameUrlHandlerMapping 会针对 Spring MVC 应用预先配置好，这就意味着 bean 会被映射到模式为 bean 名字的 URL 上。上述配置会将 URL 模式/weather 映射到该导出器上。接下来，可以将该 Web 应用部署到 Web 容器中（比如说 Apache Tomcat)。默认情况下，Tomcat 会监听 8080 端口，因此如果将应用部署到 Hessian 上下文路径下，那就可以通过 URL http://localhost:8080/ hessian/weather 访问服务了。

调用 Hessian 服务

借助于 Spring 的远程设施，你可以像使用本地 bean 一样调用远程服务。在客户端应用中，你可以在 Java 配置类中创建一个 HessianProxyFactoryBean 实例来为远程 Hessian 服务创建代理。接下来就可以使用该服务了，就好像它是个本地 bean 一样。

```
@Bean
public HessianProxyFactoryBean weatherService() {
    HessianProxyFactoryBean factory = new HessianProxyFactoryBean();
    factory.setServiceUrl("http://localhost:8080/hessian/weather");
    factory.setServiceInterface(WeatherService.class);
    return factory;
}
```

对于 HessianProxyFactoryBean 实例来说，你需要配置两个属性。服务 URL 属性指定了目标服务的 URL。服务接口属性供工厂 bean 为远程服务创建本地代理所用。代理会透明地将调用请求发送给远程服务。

公开 HTTP Invoker 服务

使用 HTTP Invoker 公开服务的配置类似于 Hessian，只不过需要使用 HttpInvokerServiceExporter。

```
@Bean(name = "/weather")
public HttpInvokerServiceExporter exporter() {
    HttpInvokerServiceExporter exporter = new HttpInvokerServiceExporter();
    exporter.setService(weatherService());
    exporter.setServiceInterface(WeatherService.class);
    return exporter;
}
```

调用 HTTP Invoker 服务

调用通过 HTTP Invoker 公开的服务也类似于 Hessian 和 Burlap。这次需要使用 HttpInvokerProxyFactoryBean。

```
@Bean
public HttpInvokerProxyFactoryBean weatherService() {
    HttpInvokerProxyFactoryBean factory = new HttpInvokerProxyFactoryBean();
    factory.setServiceUrl("http://localhost:8080/httpinvoker/weather");
    factory.setServiceInterface(WeatherService.class);
    return factory;
}
```

13-9 使用 JAX-WS 公开和调用 SOAP Web Service

问题提出

SOAP 是个企业标准和跨平台的应用通信技术。大多数现代与任务关键性的软件远程任务（如银行服务和库存应用）通常都会使用该标准。你想在 Java 应用中调用第三方 SOAP Web Service，还想在 Java 应用中公开 Web Service，这样不同平台上的第三方应用就可以通过 SOAP 调用它们了。

解决方案

使用 JAX-WS 的@WebService 和@WebMethod 注解以及 Spring 的 SimpleJaxWsServiceExporter 可以通过 SOAP 访问 bean 业务逻辑。还可以使用 Apache CXF 和 Spring 在 Tomcat 等 Java 服务器中公开 SOAP 服务。要想访问 SOAP 服务，可以使用 Apache CXF 与 Spring 或是利用 Spring 的 JaxWsPortProxyFactoryBean。

解释说明

JAX-WS 2.0(Java API for XML-based Web Service)是 JAX-RPC 1.1 的后继。如果想在 Java 中使用 SOAP，那么 JAX-WS 就是最新的标准，Java EE 与标准 JDK 都对其提供了支持。

通过 JDK 中的 JAX-WS 端点支持公开 Web Service

可以依赖于 Java JDK JAX-WS 运行时支持来公开 JAX-WS 服务。这意味着无须将 JAX-WS 服务作为 Java Web 应用的一部分进行部署了。默认情况下，如果没有其他运行时，那就会使用 JDK 中的 JAX-WS 支持。我们使用 JDK 通过 JAX-WS 来实现前文中的天气服务应用。你需要对天气服务添加注解来标识应该将其公开给客户端。修改后的 WeatherServiceImpl 实现需要使用@WebService 与@WebMethod 注解。Main 类使用了@WebService 注解，服务所公开的方法需要使用@WebMethod 注解。

```
package com.apress.springrecipes.weather;

import javax.jws.WebMethod;
import javax.jws.WebService;
import java.util.ArrayList;
import java.util.Date;
import java.util.List;

@WebService(serviceName = "weather", endpointInterface = " com.apress.springrecipes. weather.
WeatherService ")
public class WeatherServiceImpl implements WeatherService {

    @WebMethod(operationName = "getTemperatures")
    public List<TemperatureInfo> getTemperatures(String city, List<Date> dates) {
        List<TemperatureInfo> temperatures = new ArrayList<TemperatureInfo>();

        for (Date date : dates) {
            temperatures.add(new TemperatureInfo(city, date, 5.0, 10.0, 8.0));
        }

        return temperatures;
    }
}
```

注意，无须为注解提供任何参数，比如说 endpointInterface 和 serviceName，不过这里加上的原因在于让生成的 SOAP 契约可读性更好一些。与之类似，也无须在@WebMethod 注解上提供 operationName。不过这通常是个最佳实践，因为它将对 Java 实现所做的任何重构与 SOAP 端点的客户端隔离开了。

接下来，为了让 Spring 能够检测到使用了@WebService 注解的 bean，你需要使用 Spring 的 SimpleHttpServerJaxWsServiceExporter。如下是 Java 配置类中该类的@Bean 定义：

```
@Bean
public SimpleHttpServerJaxWsServiceExporter jaxWsService() {
    SimpleHttpServerJaxWsServiceExporter simpleJaxWs =
        new SimpleHttpServerJaxWsServiceExporter();
    simpleJaxWs.setPort(8888);
    simpleJaxWs.setBasePath("/jaxws/");
    return simpleJaxWs;
}
```

注意，bean 定义调用了 setPort 与 setBasePath，并将其设置为 8888 与/jaxws/。这是应用 JAX-WS 服务的端点。使用@WebService 注解定义的所有 bean 都会位于该地址下（JDK 所创建的独立服务器）。因此，如果有个名为 weather 的@Webservice，那就可以通过 http://localhost:8888/jaxws/weather 来访问。

注意：如果将其作为 Web 部署的一部分，那么请使用 SimpleJaxWsServiceExporter，因为 SimpleHttpServerJaxWsServiceExporter 会启动一个内建的 HTTP 服务器，这在 Web 部署中通常是不可行的。

启动浏览器并通过 http://localhost:8888/jaxws/weather?wsdl 查看结果，你会看到如下生成的 SOAP WSDL 契约：

```
<?xml version="1.0" encoding="UTF-8"?>
<!--
 Published by JAX-WS RI (http://jax-ws.java.net). RI's version is JAX-WS RI 2.2.9-
b130926.1035 svn-revision#5f6196f2b90e9460065a4c2f4e30e065b245e51e.
-->
<!--
 Generated by JAX-WS RI (http://jax-ws.java.net). RI's version is JAX-WS RI 2.2.9-
b130926.1035 svn-revision#5f6196f2b90e9460065a4c2f4e30e065b245e51e.
-->
<definitions xmlns="http://schemas.xmlsoap.org/wsdl/" xmlns:soap="http://schemas.
xmlsoap. org/wsdl/soap/" xmlns:tns="http://weather.springrecipes.apress.com/"
xmlns:wsam="http:// www.w3.org/2007/05/addressing/metadata" xmlns:wsp=
"http://www.w3.org/ns/ws-policy" xmlns:wsp1_2="http://schemas.xmlsoap.org/ws/2004/
09/policy" xmlns:wsu="http://docs.oasis-open.org/wss/2004/01/oasis-200401-wss-
wssecurity-utility-1.0.xsd" xmlns:xsd="http:// www.w3.org/2001/XMLSchema"
targetNamespace="http://weather.springrecipes.apress.com/" name="weather">
    <types>
        <xsd:schema>
            <xsd:import namespace="http://weather.springrecipes.apress.com/"
            schemaLocation="http://localhost:8888/jaxws/weather?xsd=1" />
        </xsd:schema>
    </types>
    <message name="getTemperatures">
        <part name="parameters" element="tns:getTemperatures" />
    </message>
    <message name="getTemperaturesResponse">
        <part name="parameters" element="tns:getTemperaturesResponse" />
    </message>
    <portType name="WeatherService">
        <operation name="getTemperatures">
            <input wsam:Action="http://weather.springrecipes.apress.com/WeatherService/
            getTemperaturesRequest" message="tns:getTemperatures" />
            <output wsam:Action="http://weather.springrecipes.apress.com/WeatherService/
            getTemperaturesResponse" message="tns:getTemperaturesResponse" />
        </operation>
    </portType>
    <binding name="WeatherServiceImplPortBinding" type="tns:WeatherService">
        <soap:binding transport="http://schemas.xmlsoap.org/soap/http" style="document" />
        <operation name="getTemperatures">
            <soap:operation soapAction="" />
        <input>
            <soap:body use="literal" />
        </input>
        <output>
        <soap:body use="literal" />
        </output>
        </operation>
    </binding>
    <service name="weather">
        <port name="WeatherServiceImplPort" binding="tns:WeatherServiceImplPortBinding">
            <soap:address location="http://localhost:8888/jaxws/weather" />
        </port>
    </service>
</definitions>
```

客户端会通过 SOAP WSDL 契约来访问服务。如果查看生成的 WSDL，你会发现其内容很简单（描述了天气服务的方法），不过更为重要的是，它是独立于编程语言的。这种中立性就是 SOAP 的全部目标：能够跨越可以解释 SOAP 的各种平台访问服务。

使用 CXF 公开 Web Service

使用 JAX-WS 服务导出器和 JAX-WS JDK 支持公开独立的 SOAP 端点是很简单的。不过，这个解决方案忽略了这样一个事实：真实环境中的大多数 Java 应用都会在 Java 应用运行时上运行，比如 Tomcat。Tomcat 本身并不支持 JAX-WS，因此需要使用 JAX-WS 运行时来装配应用。

选择有很多，你可以自由选取。一种流行的选择是 CXF，它是个 Apache 项目。对于该示例来说，我们会使用 CXF，因为它健壮，而且经过了良好测试，并且对其他重要标准提供了支持，如 JAX-RS，这是个针对 RESTful 端点的 API。

首先，我们来看看 Initializer 类，它会在兼容于 Servlet 3.1 的服务器（如 Apache Tomcat 8.5）下启动应用。

```
package com.apress.springrecipes.weather.config;

import org.springframework.web.WebApplicationInitializer;
import org.springframework.web.servlet.DispatcherServlet;
import org.springframework.web.context.ContextLoaderListener;

import org.springframework.web.context.support.XmlWebApplicationContext;

import org.apache.cxf.transport.servlet.CXFServlet;

import javax.servlet.ServletRegistration;

import javax.servlet.ServletContext;
import javax.servlet.ServletException;

public class Initializer implements WebApplicationInitializer {
    public void onStartup(ServletContext container) throws ServletException {
        XmlWebApplicationContext context = new XmlWebApplicationContext();
        context.setConfigLocation("/WEB-INF/appContext.xml");

        container.addListener(new ContextLoaderListener(context));

        ServletRegistration.Dynamic cxf = container.addServlet("cxf", new CXFServlet());
        cxf.setLoadOnStartup(1);
        cxf.addMapping("/*");
    }
}
```

这个 Initializer 类看起来与所有 Spring MVC 应用所做的事情一样。唯一的差别就是配置了一个 CXFServlet，它会处理公开服务所要做的很多工作。在 Spring MVC 配置文件中，你会使用 CXF 提供的用于配置服务的 Spring 命名空间支持。Spring 上下文文件很简单；大部分都是样板 XML 命名空间和 Spring 上下文文件导入。这里只显示了两个突出的部分，你首先像往常一样配置服务本身，然后，使用 CXF jaxws:endpoint 命名空间来配置端点。

```
package com.apress.springrecipes.weather.config;

import com.apress.springrecipes.weather.WeatherService;
import com.apress.springrecipes.weather.WeatherServiceImpl;
import org.apache.cxf.Bus;
import org.apache.cxf.jaxws.EndpointImpl;
import org.springframework.context.annotation.Bean;
import org.springframework.context.annotation.Configuration;
import org.springframework.context.annotation.ImportResource;

@Configuration
@ImportResource("classpath:META-INF/cxf/cxf.xml")
public class WeatherConfig {

    @Bean
```

```
    public WeatherService weatherService() {
        return new WeatherServiceImpl();
    }

    @Bean(initMethod = "publish")
    public EndpointImpl endpoint(Bus bus) {
        EndpointImpl endpoint = new EndpointImpl(bus, weatherService());
        endpoint.setAddress("/weather");
        return endpoint;
    }
}
```

这里使用 EndpointImpl 类来注册一个端点。它需要 CXF Bus，这是在导入的 cxf.xml 文件中配置的（由 Apache CXF 提供），同时它会使用 weatherService Spring bean 作为实现。你通过 address 属性告诉它在什么地址发布服务。在该示例中，由于 Initializer 将 CXF servlet 挂接到了根目录（/）下，因此 CXF weatherService 端点可以通过/cxf/weather 来访问（因为应用部署在了/cxf 下）。

注意，publish 方法用作为 initMethod。除了这种方式和 setAddress 外，还可以使用 endpoint.publish ("/weather")。不过，使用 initMethod 可以在端点发布前使用回调来增强/配置实际的 EndpointImpl。比如说，如果有多个端点需要配置为 SSL，那么这种方式就很方便了。

注意，weatherServiceImpl 中的 Java 代码与之前一样，依然使用了@WebService 和@WebMethod 注解。启动应用和 Web 容器，在浏览器中访问它。在本书的源代码中，该应用被构建为一个名为 cxf.war 的 WAR，由于 CXF 部署在/cxf，因此 SOAP WSDL 契约的地址为 http://localhost:8080/cxf/ weather。如果访问 http://localhost:8080/cxf，你会看到可用服务及其操作的一个列表。单击服务的 WSDL 链接（或是在服务端点后加上 wsdl），你会看到该服务的 WSDL。该 WSDL 契约与上节中使用 JAX-WS JDK 支持时的契约是非常类似的。唯一的差别在于这个 WSDL 契约是在 CXF 的帮助下生成的。

使用 Spring 的 JaxWsPortProxyFactoryBean 调用 Web Service

Spring 提供了访问 SOAP WSDL 契约以及与底层服务进行通信的功能，就好像它们是常规的 Spring bean 一样。这个功能是由 JaxWsPortProxyFactoryBean 提供的。如下是个简单的 bean 定义，它使用 JaxWsPortProxyFactoryBean 访问 SOAP 天气服务：

```
@Bean
public JaxWsPortProxyFactoryBean weatherService() throws MalformedURLException {
    JaxWsPortProxyFactoryBean weatherService = new JaxWsPortProxyFactoryBean();
    weatherService.setServiceInterface(WeatherService.class);
    weatherService.setWsdlDocumentUrl(new URL("http://localhost:8080/cxf/weather?WSDL"));
    weatherService.setNamespaceUri("http://weather.springrecipes.apress.com/");
    weatherService.setServiceName("weather");
    weatherService.setPortName("WeatherServiceImplPort");
    return weatherService;
}
```

bean 实例的名字是 weatherService。你可以通过该引用调用底层 SOAP 服务的方法，就好像它们运行在本地一样（比如说 weatherService.getTemperatures(city, dates)）。JaxWsPortProxyFactoryBean 需要几个属性，下面会介绍。

serviceInterface 属性为 SOAP 定义了服务接口。对于天气服务来说，你可以使用服务器端实现代码，这是一样的。如果所访问的 SOAP 服务没有服务器端代码，那么可以通过 java2wsdl 等工具从 WSDL 契约中创建这个 Java 接口。注意，客户端所用的 serviceInterface 需要与服务器端实现使用相同的 JAX-WS 注解（比如说@WebService）。

wsdlDocumentUrl 属性表示 WSDL 契约的位置。在该示例中，它会指向本节中的 CXF SOAP 端点，不过你也可以定义该属性，访问本节中的 JAX-WS JDK 端点或是任意 WSDL 契约。

namesapceUrl、serviceName 与 portName 属性都与 WSDL 契约本身有关。由于 WSDL 契约中可能会有多种命名空间、服务与端口，因此你需要告诉 Spring 访问服务需要使用哪些值。这里所列出的值可以通过手工查看天气 WSDL 契约轻松进行验证。

使用 CXF 调用 Web Service

现在使用 CXF 定义 Web Service 客户端。该客户端与前文中的一样，无须特殊的 Java 配置或是代码。你只需将服务接口放置于类路径下即可。完成后就可以使用 CXF 命名空间支持来创建客户端了。

```
package com.apress.springrecipes.weather.config;

import com.apress.springrecipes.weather.WeatherService;
import com.apress.springrecipes.weather.WeatherServiceClient;
import org.apache.cxf.jaxws.JaxWsProxyFactoryBean;
import org.springframework.context.annotation.Bean;
import org.springframework.context.annotation.Configuration;
import org.springframework.context.annotation.ImportResource;

@Configuration
@ImportResource("classpath:META-INF/cxf/cxf.xml")
public class WeatherConfigCxfClient {

    @Bean
    public WeatherServiceClient weatherClient(WeatherService weatherService) {
        return new WeatherServiceClient(weatherService);
    }

    @Bean
    public WeatherService weatherServiceProxy() {
        JaxWsProxyFactoryBean factory = new JaxWsProxyFactoryBean();
        factory.setServiceClass(WeatherService.class);
        factory.setAddress("http://localhost:8080/cxf/weather");
        return (WeatherService) factory.create();
    }
}
```

注意@ImportResource("classpath:META-INF/cxf/cxf.xml")的使用，它加载了由 Apache CXF 提供的基础设施 bean。

要想创建客户端，可以使用 JaxWsProxyFactoryBean 并将所需的服务类和需要连接的地址传递进来。接下来可以通过 create 方法创建代理，这个代理就像是个常规的 WeatherService 一样。这就是要做的全部工作了。前文中的示例没有发生变化：将客户端注入到 WeatherServiceClient 中并通过 weatherService 引用调用它（如 weatherService.getTemperatures (city, dates)）。

13-10　使用契约优先的 SOAP Web Service

问题提出

你想开发契约优先的 SOAP Web Service 而非上一节中所做的代码优先的 SOAP Web Service。

解决方案

有两种方式可以开发 SOAP Web Service。一种叫做代码优先，这意味着先编写 Java 类，然后构建 WSDL 契约。另一种叫做契约优先，这意味着先编写 XML 数据契约（比 WSDL 简单一些），然后构建 Java 类来实现服务。要想为契约优先的 SOAP Web Service 创建数据契约，你需要一个 XSD 文件或是 XML Schema 文件，它描述了服务所支持的操作与数据。之所以需要 XSD 文件是因为 SOAP 服务的客户端与服务器之间的“底层”通信是通过定义在 XSD 文件中的 XML 来实现的。不过，由于正确编写 XSD 文件有点困难，因此推荐的做法是先创建一个示例 XML 消息，然后通过它来生成 XSD 文件。接下来，借助于这个 XSD 文件，你就可以使用诸如 Spring-WS 等技术通过它来构建 SOAP Web Service 了。

解释说明

开始契约优先的 Web Service 的最简单的方式是编写示例 XML 消息，编写好之后就可以使用工具从中提取出契约、XSD 文件了。有了 XSD 文件后，你就可以使用它们和另外一个工具来构建 Web Service 客户端了。

创建样例 XML 消息

我们实现与前文中相同的天气服务，不过这次使用 SOAP 契约优先的方式。你要编写一个 SOAP 服务，它可以根据城市与日期获取天气信息，返回最低、最高与平均温度。相较于上一节中所做的编写代码来支持这些功能，我们这次使用契约优先的方式和 XML 消息来描述特定城市与日期的温度，如下所示：

```
<TemperatureInfo city="Houston" date="2013-12-01">
    <min>5.0</min>
    <max>10.0</max>
    <average>8.0</average>
</TemperatureInfo>
```

这是以 SOAP 契约优先的方式为天气服务编写数据契约的第一步。现在来定义一些操作。你想让客户端能够查询特定城市几天中的温度。每个请求都包含了一个城市元素和多个日期元素。还需要为请求指定命名空间以免与其他 XML 文档产生命名冲突。我们来创建这个 XML 消息并将其保存到名为 request.xml 的文件中。

```
<GetTemperaturesRequest
    xmlns="http://springrecipes.apress.com/weather/schemas">
    <city>Houston</city>
    <date>2013-12-01</date>
    <date>2013-12-08</date>
    <date>2013-12-15</date>
</GetTemperaturesRequest>
```

上面这个请求类型的响应会包含多个 TemperatureInfo 元素，每个都代表了特定城市与日期的温度，这与所请求的日期是相对应的。我们创建这个 XML 消息并将其保存到名为 response.xml 的文件中。

```
<GetTemperaturesResponse
    xmlns="http://springrecipes.apress.com/weather/schemas">
    <TemperatureInfo city="Houston" date="2013-12-01">
        <min>5.0</min>
        <max>10.0</max>
        <average>8.0</average>
    </TemperatureInfo>
    <TemperatureInfo city="Houston" date="2007-12-08">
        <min>4.0</min>
        <max>13.0</max>
        <average>7.0</average>
    </TemperatureInfo>
    <TemperatureInfo city="Houston" date="2007-12-15">
        <min>10.0</min>
        <max>18.0</max>
        <average>15.0</average>
    </TemperatureInfo>
</GetTemperaturesResponse>
```

从示例 XML 消息生成 XSD 文件

现在，可以从上述示例 XML 消息生成 XSD 文件了。大多数 XML 工具和企业级 Java IDE 都可以从一组 XML 文件生成 XSD 文件。这里使用 Apache XMLBeans 来生成 XSD 文件。

> **注意：** 可以从 Apache XMLBeans 网站下载 Apache XMLBeans（比如说 v2.6.0），然后将其解压缩到指定目录中来完成安装。

Apache XMLBeans 提供了一个名为 inst2xsd 的工具，可用于从 XML 文件生成 XSD 文件。它支持几种设计类型来生成 XSD 文件。最简单的一种叫做俄罗斯套娃设计（Russian doll design），它会为目标 XSD 文件生成本地元素与本地类型。由于 XML 消息中没有使用枚举类型，因此可以禁用掉枚举生成特性。可以执行如下命令从之前的 XML 文件生成 XSD 文件：

```
inst2xsd -design rd -enumerations never request.xml response.xml
```

生成的 XSD 文件有默认的名字 schema0.xsd，位于相同目录中。我们将其重命名为 temperature.xsd。

```
<?xml version="1.0" encoding="UTF-8"?>
<xs:schema attributeFormDefault="unqualified"
```

```
    elementFormDefault="qualified"
    targetNamespace=http://springrecipes.apress.com/weather/schemas
    xmlns:xs="http://www.w3.org/2001/XMLSchema">

    <xs:element name="GetTemperaturesRequest">
        <xs:complexType>
            <xs:sequence>
                <xs:element type="xs:string" name="city" />
                <xs:element type="xs:date" name="date"
                    maxOccurs="unbounded" minOccurs="0" />
            </xs:sequence>
        </xs:complexType>
    </xs:element>

    <xs:element name="GetTemperaturesResponse">
        <xs:complexType>
            <xs:sequence>
                <xs:element name="TemperatureInfo"
                    maxOccurs="unbounded" minOccurs="0">
                    <xs:complexType>
                        <xs:sequence>
                            <xs:element type="xs:float" name="min" />
                            <xs:element type="xs:float" name="max" />
                            <xs:element type="xs:float" name="average" />
                        </xs:sequence>
                        <xs:attribute type="xs:string" name="city"
                            use="optional" />
                        <xs:attribute type="xs:date" name="date"
                            use="optional" />
                    </xs:complexType>
                </xs:element>
            </xs:sequence>
        </xs:complexType>
    </xs:element>
</xs:schema>
```

优化生成的 XSD 文件

可以看到，生成的 XSD 文件可以让客户端查询无限天数的温度。如果想对最大与最小的查询天数进行限制，可以修改 maxOccurs 与 minOccurs 属性。

```
<?xml version="1.0" encoding="UTF-8"?>
<xs:schema attributeFormDefault="unqualified"
    elementFormDefault="qualified"
    targetNamespace=http://springrecipes.apress.com/weather/schemas
    xmlns:xs="http://www.w3.org/2001/XMLSchema">

    <xs:element name="GetTemperaturesRequest">
        <xs:complexType>
            <xs:sequence>
                <xs:element type="xs:string" name="city" />
                <xs:element type="xs:date" name="date"
                    maxOccurs="5" minOccurs="1" />
            </xs:sequence>
        </xs:complexType>
    </xs:element>

    <xs:element name="GetTemperaturesResponse">
        <xs:complexType>
            <xs:sequence>
                <xs:element name="TemperatureInfo"
                    maxOccurs="5" minOccurs="1">
                    ...
                </xs:element>
```

```
            </xs:sequence>
        </xs:complexType>
    </xs:element>
</xs:schema>
```

预览生成的 WSDL 文件

Spring-WS 可以自动从 XSD 文件生成 WSDL 契约，接下来将会对此进行详细介绍。如下代码片段展示了针对于此目的的 Spring bean 配置——下一节将会介绍如何使用这个代码片段，以及如何通过 Spring-WS 构建 SOAP Web Service。

```
<sws:dynamic-wsdl id="temperature" portTypeName="Weather" locationUri="/">
    <sws:xsd location="/WEB-INF/temperature.xsd"/>
</sws:dynamic-wsdl>
```

现在预览生成的 WSDL 文件以更好地理解服务契约。出于简单性的考虑，这里省略了不重要的部分。

```
<?xml version="1.0" encoding="UTF-8" ?>
<wsdl:definitions ...
    targetNamespace="http://springrecipes.apress.com/weather/schemas">
    <wsdl:types>
        <!-- Copied from the XSD file -->
        ...
    </wsdl:types>
    <wsdl:message name="GetTemperaturesResponse">
        <wsdl:part element="schema:GetTemperaturesResponse"
            name="GetTemperaturesResponse">
        </wsdl:part>
    </wsdl:message>
    <wsdl:message name="GetTemperaturesRequest">
        <wsdl:part element="schema:GetTemperaturesRequest"
            name="GetTemperaturesRequest">
        </wsdl:part>
    </wsdl:message>
    <wsdl:portType name="Weather">
        <wsdl:operation name="GetTemperatures">
            <wsdl:input message="schema:GetTemperaturesRequest"
                name="GetTemperaturesRequest">
            </wsdl:input>
            <wsdl:output message="schema:GetTemperaturesResponse"
                name="GetTemperaturesResponse">
            </wsdl:output>
        </wsdl:operation>
    </wsdl:portType>
    ...
    <wsdl:service name="WeatherService">
        <wsdl:port binding="schema:WeatherBinding" name="WeatherPort">
            <soap:address
                location="http://localhost:8080/weather/services" />
        </wsdl:port>
    </wsdl:service>
</wsdl:definitions>
```

在 Weather 端口类型中定义了一个 GetTemperatures 操作，其名字来自于输入与输出消息的前缀（即<GetTemperaturesRequest>和<GetTemperaturesResponse>）。这两个元素的定义位于<wsdl:types>部分，定义在数据契约中。

现在已经有了 WSDL 契约，你可以生成必要的 Java 接口，然后为与 XML 消息对应的每个操作编写实现代码。下一节将会介绍完整的实现技术，其中会使用 Spring-WS。

13-11　使用 Spring-WS 公开和调用 SOAP Web Service

问题提出

你有一个 XSD 文件，用于开发契约优先的 SOAP Web Service，但不知道如何以及该使用什么工具实现契约优先的 SOAP 服务。

Spring-WS 从设计伊始就支持契约优先的 SOAP Web Service。不过，这并不意味着 Spring-WS 是 Java 中创建 SOAP Web Service 的唯一方式。CXF 这样的 JAX-WS 实现也支持该技术。然而，Spring-WS 是在 Spring 应用上下文中实现契约优先的 SOAP Web Service 的更为成熟和自然的方式。使用其他契约优先的 SOAP Java 技术会导致脱离 Spring 框架的范围。

解决方案

Spring-WS 提供了一套设施来开发契约优先的 SOAP Web Service。构建 Spring-WS Web Service 的核心任务如下所示。

1. 为 Spring-WS 创建并配置 Spring MVC 应用。
2. 将 Web Service 请求映射到端点。
3. 创建服务端点来处理请求消息并返回响应消息。
4. 为 Web Service 发布 WSDL 文件。

创建 Spring-WS 应用

要想通过 Spring-WS 实现 Web Service，首先需要创建一个 Web 应用初始化器类来启动提供 SOAP Web Service 的 Web 应用。你需要配置 MessageDispatcherServlet servlet，它是 Spring-WS 的一部分。该 servlet 专门用于将 Web Service 消息分发到恰当的端点并检测 Spring-WS 的框架设施。

```
package com.apress.springrecipes.weather.config;

import org.springframework.ws.transport.http.support.
AbstractAnnotationConfigMessageDispatcherServletInitializer;

public class Initializer extends AbstractAnnotationConfigMessageDispatcherServletInitializer
{

    @Override
    protected Class<?>[] getRootConfigClasses() {
        return null;
    }

    @Override
    protected Class<?>[] getServletConfigClasses() {
        return new Class<?>[] {SpringWsConfiguration.class};
    }

}
```

为了简化配置，可以继承 AbstractAnnotationConfigMessageDispatcherServletInitializer 父类。你需要向其提供包含了 rootConfig 与 servletConfig 的配置类。前者可以为 null，后者则是必需的。

上述配置会使用 SpringWsConfiguration 类启动 MessageDispatcherServlet 并将其注册到/services/*与*.wsdl URL 上。

```
package com.apress.springrecipes.weather.config;

import org.springframework.context.annotation.Bean;
import org.springframework.context.annotation.ComponentScan;
import org.springframework.context.annotation.Configuration;
import org.springframework.core.io.ClassPathResource;
import org.springframework.ws.config.annotation.EnableWs;
import org.springframework.ws.wsdl.wsdl11.DefaultWsdl11Definition;
import org.springframework.xml.xsd.SimpleXsdSchema;
import org.springframework.xml.xsd.XsdSchema;

@Configuration
@EnableWs
@ComponentScan("com.apress.springrecipes.weather")
```

```
public class SpringWsConfiguration {

    ...
}
```

SpringWsConfiguration 类使用了@EnableWs 注解，它会注册必要的 bean 让 MessageDispatcherServlet 能够正常使用。还使用了@ComponentScan 注解来扫描@Service 与@Endpoint bean。

创建服务端点

Spring-WS 支持通过@Endpoint 注解将任意一个类作为服务端点，这样就可以作为服务访问它了。除了@Endpoint 注解外，你还需要对处理器方法使用注解@PayloadRoot，以将其映射到服务请求上。每个处理器方法还依赖于@ResponsePayload 与@RequestPayload 注解来处理进出的服务数据。

```
package com.apress.springrecipes.weather;

import org.dom4j.Document;
import org.dom4j.DocumentHelper;
import org.dom4j.Element;
import org.dom4j.XPath;
import org.dom4j.xpath.DefaultXPath;
import org.springframework.ws.server.endpoint.annotation.Endpoint;
import org.springframework.ws.server.endpoint.annotation.PayloadRoot;
import org.springframework.ws.server.endpoint.annotation.RequestPayload;
import org.springframework.ws.server.endpoint.annotation.ResponsePayload;

import java.text.DateFormat;
import java.text.SimpleDateFormat;
import java.util.*;

@Endpoint
public class TemperatureEndpoint {

    private static final String namespaceUri = "http://springrecipes.apress.com/weather/
    schemas";
    private XPath cityPath;
    private XPath datePath;

    private final WeatherService weatherService;

    public TemperatureEndpoint(WeatherService weatherService) {
        this.weatherService = weatherService;
        // Create the XPath objects, including the namespace
        Map<String, String> namespaceUris = new HashMap<String, String>();
        namespaceUris.put("weather", namespaceUri);
        cityPath = new DefaultXPath("/weather:GetTemperaturesRequest/weather:city");
        cityPath.setNamespaceURIs(namespaceUris);
        datePath = new DefaultXPath("/weather:GetTemperaturesRequest/weather:date");
        datePath.setNamespaceURIs(namespaceUris);
    }
    @PayloadRoot(localPart = "GetTemperaturesRequest", namespace = namespaceUri)
    @ResponsePayload
    public Element getTemperature(@RequestPayload Element requestElement) throws Exception
    {
        DateFormat dateFormat = new SimpleDateFormat("yyyy-MM-dd");
        // Extract the service parameters from the request message
        String city = cityPath.valueOf(requestElement);
        List<Date> dates = new ArrayList<Date>();
        for (Object node : datePath.selectNodes(requestElement)) {
            Element element = (Element) node;
            dates.add(dateFormat.parse(element.getText()));
        }
```

```
        // Invoke the back-end service to handle the request
        List<TemperatureInfo> temperatures =
            weatherService.getTemperatures(city, dates);

        // Build the response message from the result of back-end service
        Document responseDocument = DocumentHelper.createDocument();
        Element responseElement = responseDocument.addElement(
            "GetTemperaturesResponse", namespaceUri);
        for (TemperatureInfo temperature : temperatures) {
            Element temperatureElement = responseElement.addElement(
                "TemperatureInfo");
            temperatureElement.addAttribute("city", temperature.getCity());
            temperatureElement.addAttribute(
                "date", dateFormat.format(temperature.getDate()));
            temperatureElement.addElement("min").setText(
                Double.toString(temperature.getMin()));
            temperatureElement.addElement("max").setText(
                Double.toString(temperature.getMax()));
            temperatureElement.addElement("average").setText(
                Double.toString(temperature.getAverage()));
        }
        return responseElement;
    }
}
```

在@PayloadRoot 注解中，指定了要处理的负载根元素的本地名（getTemperaturesRequest）和命名空间（http://springrecipes.apress.com/weather/schemas）。接下来，为方法添加了@ResponsePayload 注解，标识方法的返回值是服务响应数据。此外，对方法的输入参数使用了@RequestPayload 注解，标识它是服务输入值。

接下来，在处理器方法中，首先从请求消息中提取出服务参数。这里使用 XPath 来帮助定位元素。XPath 对象是在构造方法中创建的，这样就可以在后续的请求处理中重用了。注意，还需要在 XPath 表达式中包含命名空间，否则它们将无法正确定位元素。在提取出服务参数后，调用后端服务来处理请求。由于该端点是在 Spring IoC 容器中配置的，因此可以通过依赖注入轻松引用其他 bean。最后，从后端服务的结果构建出响应消息。该示例使用了 dom4j 库，它提供了丰富的 API 来构建 XML 消息。不过，也可以使用其他 XML 处理 API 或是 Java 解析器（如 DOM）。

由于已经在 SpringWsConfiguration 类中定义了@ComponentScan 注解，因此 Spring 会自动找到所有 Spring-WS 注解并将端点部署到 servlet。

发布 WSDL 文件

完成 SOAP Web Service 的最后一步是发布 WSDL 文件。在 Spring-WS 中，没必要手工编写 WSDL 文件；只需向 SpringWsConfiguration 类添加一个 bean 即可。

```
@Bean
public DefaultWsdl11Definition temperature() {
    DefaultWsdl11Definition temperature = new DefaultWsdl11Definition();
    temperature.setPortTypeName("Weather");
    temperature.setLocationUri("/");
    temperature.setSchema(temperatureSchema());
    return temperature;
}

@Bean
public XsdSchema temperatureSchema() {
    return new SimpleXsdSchema(new ClassPathResource("/META-INF/xsd/temperature.xsd"));
}
```

DefaultWsdl11Definition 类需要指定两个属性：类的 portTypeName 以及部署最终 WSDL 的 locationUri。此外，还需要指定 XSD 文件的位置，通过它来创建 WSDL——请参见上一节了解如何创建 XSD 文件。在该示例中，XSD 文件位于应用的 META-INF 目录下。由于在 XSD 文件中定义了<GetTemperaturesRequest>与

<GetTemperaturesResponse>，并将端口类型名指定为 Weather，因此 WSDL 构建器会生成如下 WSDL 端口类型与操作。如下代码片段来自于生成的 WSDL 文件：

```
<wsdl:portType name="Weather">
    <wsdl:operation name="GetTemperatures">
        <wsdl:input message="schema:GetTemperaturesRequest"
            name="GetTemperaturesRequest" />
        <wsdl:output message="schema:GetTemperaturesResponse"
            name="GetTemperaturesResponse" />
    </wsdl:operation>
</wsdl:portType>
```

最后，可以通过连接定义的 bean 名字与.wsdl 后缀来访问这个 WSDL 文件。假设 Web 应用打包到名为 springws 的 WAR 文件中，那么服务就会部署在 http://localhost:8080/springws/（这是因为初始化器中的 Spring-WS servlet 部署在了/服务路径下），WSDL 文件的 URL 就是 http://localhost:8080/springws/services/weather/temperature.wsdl，因为 WSDL 定义的 bean 名字是 temperature。

使用 Spring-WS 调用 SOAP Web Service

现在，我们创建 Spring-WS 客户端，根据发布的契约来调用天气服务。可以通过解析请求和响应 XML 消息来创建 Spring-WS 客户端。作为示例，可以使用 dom4j 来实现，不过也可以使用任何其他的 XML 解析 API。

为了将客户端与底层调用细节隔离开来，你创建了一个本地代理来调用 SOAP Web Service。该代理也实现了 WeatherService 接口，它会将本地方法调用转换为远程 SOAP Web Service 调用。

```
package com.apress.springrecipes.weather;

import org.dom4j.Document;
import org.dom4j.DocumentHelper;
import org.dom4j.Element;
import org.dom4j.io.DocumentResult;
import org.dom4j.io.DocumentSource;
import org.springframework.ws.client.core.WebServiceTemplate;

import java.text.DateFormat;
import java.text.ParseException;
import java.text.SimpleDateFormat;
import java.util.ArrayList;
import java.util.Date;
import java.util.List;

public class WeatherServiceProxy implements WeatherService {

    private static final String namespaceUri = "http://springrecipes.apress.com/weather/
    schemas";
    private final WebServiceTemplate webServiceTemplate;

    public WeatherServiceProxy(WebServiceTemplate webServiceTemplate) throws Exception {
        this.webServiceTemplate = webServiceTemplate;
    }

    public List<TemperatureInfo> getTemperatures(String city, List<Date> dates) {
        private DateFormat dateFormat = new SimpleDateFormat("yyyy-MM-dd");

        Document requestDocument = DocumentHelper.createDocument();
        Element requestElement = requestDocument.addElement(
                "GetTemperaturesRequest", namespaceUri);
        requestElement.addElement("city").setText(city);
        for (Date date : dates) {
            requestElement.addElement("date").setText(dateFormat.format(date));
        }

        DocumentSource source = new DocumentSource(requestDocument);
```

```
        DocumentResult result = new DocumentResult();
        webServiceTemplate.sendSourceAndReceiveToResult(source, result);

        Document responsetDocument = result.getDocument();
        Element responseElement = responsetDocument.getRootElement();
        List<TemperatureInfo> temperatures = new ArrayList<TemperatureInfo>();
        for (Object node : responseElement.elements("TemperatureInfo")) {
            Element element = (Element) node;
            try {
                Date date = dateFormat.parse(element.attributeValue("date"));
                double min = Double.parseDouble(element.elementText("min"));
                double max = Double.parseDouble(element.elementText("max"));
                double average = Double.parseDouble(
                    element.elementText("average"));
                temperatures.add(
                    new TemperatureInfo(city, date, min, max, average));
            } catch (ParseException e) {
                throw new RuntimeException(e);
            }
        }
        return temperatures;
    }
}
```

在 getTemperatures()方法中，首先通过 dom4j API 构建请求消息。WebServiceTemplate 提供了一个 sendSourceAndReceiveToResult()方法，它接收一个 java.xml.transform.Source 对象和一个 java.xml.transform.Result 对象作为参数。需要构建一个 dom4j DocumentSource 对象来包装请求文档，并为方法创建一个新的 dom4j DocumentResult 对象来将响应文档写入其中。最后，获取响应消息并从中提取出结果。

借助于所编写的服务代理，你可以在配置类中声明它，后面使用独立类来调用它。

```
package com.apress.springrecipes.weather.config;

import com.apress.springrecipes.weather.WeatherService;
import com.apress.springrecipes.weather.WeatherServiceClient;
import com.apress.springrecipes.weather.WeatherServiceProxy;
import org.springframework.context.annotation.Bean;
import org.springframework.context.annotation.Configuration;
import org.springframework.ws.client.core.WebServiceTemplate;

@Configuration
public class SpringWsClientConfiguration {

    @Bean
    public WeatherServiceClient weatherServiceClient(WeatherService weatherService)
    throws Exception {
        return new WeatherServiceClient(weatherService);
    }

    @Bean
    public WeatherServiceProxy weatherServiceProxy(WebServiceTemplate webServiceTemplate)
    throws Exception {
        return new WeatherServiceProxy(webServiceTemplate);
    }

    @Bean
    public WebServiceTemplate webServiceTemplate() {
        WebServiceTemplate webServiceTemplate = new WebServiceTemplate();
        webServiceTemplate.setDefaultUri("http://localhost:8080/springws/services");
        return webServiceTemplate;
    }
}
```

注意到 webServiceTemplate 将其 defaultUri 值设置为前文中为 Spring-WS 所定义的端点。当配置被应用加载后，就可以通过如下类调用 SOAP 服务了：

```
package com.apress.springrecipes.weather;
```

```
import org.springframework.context.ApplicationContext;
import org.springframework.context.support.GenericXmlApplicationContext;

public class SpringWSInvokerClient {

    public static void main(String[] args) {
        ApplicationContext context =
            new AnnotationConfigApplicationContext("com.apress.springrecipes.weather.config");

        WeatherServiceClient client = context.getBean(WeatherServiceClient.class);
        TemperatureInfo temperature = client.getTodayTemperature("Houston");
        System.out.println("Min temperature : " + temperature.getMin());
        System.out.println("Max temperature : " + temperature.getMax());
        System.out.println("Average temperature : " + temperature.getAverage());
    }
}
```

13-12 使用 Spring-WS 与 XML 编组来开发 SOAP Web Service

问题提出

要想通过契约优先的方式开发 Web Service，你需要处理请求和响应 XML 消息。如果直接使用 XML 解析 API 来解析 XML 消息，那就需要通过底层 API 一个接一个地处理 XML 元素，这是非常繁琐且低效的做法。

解决方案

Spring-WS 支持使用 XML 编组技术将对象编组到 XML 文档以及将 XML 文档反编组到对象。通过这种方式，你可以处理对象属性而非 XML 元素。这项技术也叫做对象/XML 映射（OXM），因为你实际上做的是对象与 XML 文档之间的映射。要想通过 XML 编组技术来实现端点，你可以为其配置一个 XML 编组器。表 13-2 所示为 Spring 为不同 XML 编组 API 所提供的编组器。

表 13-2　　针对不同 XML 编排 API 的编组器

API	编组器
JAXB 2.0	org.springframework.oxm.jaxb.Jaxb2Marshaller
Castor	org.springframework.oxm.castor.CastorMarshaller
XMLBeans	org.springframework.oxm.xmlbeans.XmlBeansMarshaller
JiBX	org.springframework.oxm.jibx.JibxMarshaller
Xstream	org.springframework.oxm.xstream.XStreamMarshaller

类似地，Spring WS 客户端也可以使用同样的编组与反编组技术来简化 XML 数据的处理。

解释说明

可以对端点和客户端使用编组与反编组。首先，我们将会介绍如何创建端点并使用 Spring OXM 编组器，然后介绍如何在客户端使用它。

使用 XML 编组创建服务端点

Spring-WS 支持各种 XML 编组 API，包括 JAXB 2.0、Castor、XMLBeans、JiBX 和 XStream。作为示例，你将使用 Castor 作为编组器来创建服务端点。使用其他 XML 编组 API 也是类似的。使用 XML 编组的第一步是根据 XML 消息格式创建对象模型。该模型通常可以通过编组 API 生成。对于某些编组 API 来说，对象模型必须要由它们来生成，这样它们才能插入特定于编组的信息。由于 Castor 支持 XML 消息与任意 Java 对象之间的编组，因此可以先创建如下类：

```
package com.apress.springrecipes.weather;
...
public class GetTemperaturesRequest {
```

```
    private String city;
    private List<Date> dates;

    // Constructors, Getters and Setters
    ...
}

package com.apress.springrecipes.weather;
...
public class GetTemperaturesResponse {

    private List<TemperatureInfo> temperatures;

    // Constructors, Getters and Setters
    ...
}
```

创建好对象模型后，你可以轻松在任意端点上集成编组。我们将这项技术应用到上一节中的端点上。

```
package com.apress.springrecipes.weather;

import org.springframework.ws.server.endpoint.annotation.Endpoint;
import org.springframework.ws.server.endpoint.annotation.PayloadRoot;
import org.springframework.ws.server.endpoint.annotation.RequestPayload;
import org.springframework.ws.server.endpoint.annotation.ResponsePayload;

import java.util.List;

@Endpoint
public class TemperatureMarshallingEndpoint {

    private static final String namespaceUri = "http://springrecipes.apress.com/weather/
    schemas";

    private final WeatherService weatherService;

    public TemperatureMarshallingEndpoint(WeatherService weatherService) {
        this.weatherService = weatherService;
    }

    @PayloadRoot(localPart = "GetTemperaturesRequest", namespace = namespaceUri)
    public @ResponsePayload GetTemperaturesResponse getTemperature(@RequestPayload
    GetTemperaturesRequest request) {
        List<TemperatureInfo> temperatures =
            weatherService.getTemperatures(request.getCity(), request.getDates());
        return new GetTemperaturesResponse(temperatures);
    }
}
```

注意，在这个新方法端点中你所要做的全部事情就是处理请求对象和返回响应对象。接下来，它会被编组为响应 XML 消息。除了对这个端点的修改外，编组端点还需要设置 marshaller 与 unmarshaller 属性。通常来说，可以为这两个属性指定单个编组器。对于 Castor 来说，声明一个 CastorMarshaller bean 作为编组器。除了编组器外，还需要注册 MethodArgumentResolver 与 MethodReturnValueHandler 来实际处理方法参数与返回类型的编组。为了做到这一点，需要继承 WsConfigurerAdapter 并重写 addArgumentResolvers 与 addReturnValueHandlers 方法，并将 MarshallingPayloadMethodProcessor 添加到这两个列表中。

```
@Configuration
@EnableWs
@ComponentScan("com.apress.springrecipes.weather")
public class SpringWsConfiguration extends WsConfigurerAdapter {

    @Bean
    public MarshallingPayloadMethodProcessor marshallingPayloadMethodProcessor() {
        return new MarshallingPayloadMethodProcessor(marshaller());
```

```
    }

    @Bean
    public Marshaller marshaller() {
        CastorMarshaller marshaller = new CastorMarshaller();
        marshaller.setMappingLocation(new ClassPathResource("/mapping.xml"));
        return marshaller;
    }

    @Override
    public void addArgumentResolvers(List<MethodArgumentResolver> argumentResolvers) {
        argumentResolvers.add(marshallingPayloadMethodProcessor());
    }

    @Override
    public void addReturnValueHandlers(List<MethodReturnValueHandler> returnValueHandlers)
    {
        returnValueHandlers.add(marshallingPayloadMethodProcessor());
    }
}
```

注意，Castor 需要一个映射配置文件来知晓如何对对象和 XML 文档进行映射。你可以在类路径根目录下创建该文件并在 mappingLocation 属性中指定它（比如说 mapping.xml）。如下 Castor 映射文件为 GetTemperaturesRequest、GetTemperaturesResponse 与 TemperatureInfo 类定义了映射：

```
<!DOCTYPE mapping PUBLIC "-//EXOLAB/Castor Mapping DTD Version 1.0//EN"
    "http://castor.org/mapping.dtd">

<mapping>
    <class name="com.apress.springrecipes.weather.GetTemperaturesRequest">
        <map-to xml="GetTemperaturesRequest"
            ns-uri="http://springrecipes.apress.com/weather/schemas" />
        <field name="city" type="string">
            <bind-xml name="city" node="element" />
        </field>
        <field name="dates" collection="arraylist" type="string"
            handler="com.apress.springrecipes.weather.DateFieldHandler">
            <bind-xml name="date" node="element" />
        </field>
    </class>

    <class name="com.apress.springrecipes.weather.
    GetTemperaturesResponse">
        <map-to xml="GetTemperaturesResponse"
            ns-uri="http://springrecipes.apress.com/weather/schemas" />
        <field name="temperatures" collection="arraylist"
            type="com.apress.springrecipes.weather.TemperatureInfo">
            <bind-xml name="TemperatureInfo" node="element" />
        </field>
    </class>

    <class name="com.apress.springrecipes.weather.TemperatureInfo">
        <map-to xml="TemperatureInfo"
            ns-uri="http://springrecipes.apress.com/weather/schemas" />
        <field name="city" type="string">
            <bind-xml name="city" node="attribute" />
        </field>
        <field name="date" type="string"
            handler="com.apress.springrecipes.weather.DateFieldHandler">
            <bind-xml name="date" node="attribute" />
        </field>
        <field name="min" type="double">
            <bind-xml name="min" node="element" />
        </field>
```

```
        <field name="max" type="double">
            <bind-xml name="max" node="element" />
        </field>
        <field name="average" type="double">
            <bind-xml name="average" node="element" />
        </field>
    </class>
</mapping>
```

此外，需要在所有的日期字段中指定处理器来使用特定的日期格式对日期进行转换。处理器代码如下所示：

```
package com.apress.springrecipes.weather;
...
import org.exolab.castor.mapping.GeneralizedFieldHandler;

public class DateFieldHandler extends GeneralizedFieldHandler {

    private DateFormat format = new SimpleDateFormat("yyyy-MM-dd");

    public Object convertUponGet(Object value) {
        return format.format((Date) value);
    }

    public Object convertUponSet(Object value) {
        try {
            return format.parse((String) value);
        } catch (ParseException e) {
            throw new RuntimeException(e);
        }
    }

    public Class getFieldType() {
        return Date.class;
    }
}
```

使用 XML 编组调用 Web Service

Spring-WS 客户端还可以实现请求和响应对象与 XML 消息之间的编组与反编组。作为示例，你会使用 Castor 创建一个客户端作为编组器，这样就可以重用对象模型 GetTemperaturesRequest、GetTemperaturesResponse 与 TemperatureInfo，以及服务端点中的映射配置文件 mapping.xml 了。我们通过 XML 编组来实现服务代理。WebServiceTemplate 提供了一个 marshalSendAndReceive()方法，它接收一个请求对象作为方法参数，该参数会被编组为请求消息。该方法需要返回一个能从响应消息中进行反编组的响应对象。

```
package com.apress.springrecipes.weather;

import org.springframework.ws.client.core.WebServiceTemplate;

import java.util.Date;
import java.util.List;

public class WeatherServiceProxy implements WeatherService {

    private WebServiceTemplate webServiceTemplate;

    public WeatherServiceProxy(WebServiceTemplate webServiceTemplate) throws Exception {
        this.webServiceTemplate = webServiceTemplate;
    }

    public List<TemperatureInfo> getTemperatures(String city, List<Date> dates) {

        GetTemperaturesRequest request = new GetTemperaturesRequest(city, dates);
        GetTemperaturesResponse response = (GetTemperaturesResponse)
            this.webServiceTemplate.marshalSendAndReceive(request);
```

```
        return response.getTemperatures();
    }
}
```

在使用 XML 编组时，WebServiceTemplate 需要设置 marshaller 与 unmarshaller 属性。如果继承了 WebServiceGatewaySupport 以实现 WebServiceTemplate 的自动创建，那么就可以将属性设置为 WebServiceGatewaySupport。通常来说，可以为这两个属性指定单个编组器。对于 Castor 来说，声明一个 CastorMarshaller bean 作为编组器。

```
package com.apress.springrecipes.weather.config;

import com.apress.springrecipes.weather.WeatherService;
import com.apress.springrecipes.weather.WeatherServiceClient;
import com.apress.springrecipes.weather.WeatherServiceProxy;
import org.springframework.context.annotation.Bean;
import org.springframework.context.annotation.Configuration;
import org.springframework.core.io.ClassPathResource;
import org.springframework.oxm.castor.CastorMarshaller;
import org.springframework.ws.client.core.WebServiceTemplate;

@Configuration
public class SpringWsClientConfiguration {

    @Bean
    public WeatherServiceClient weatherServiceClient(WeatherService weatherService)
    throws Exception {
        return new WeatherServiceClient(weatherService);
    }

    @Bean
    public WeatherServiceProxy weatherServiceProxy(WebServiceTemplate webServiceTemplate)
    throws Exception {
        return new WeatherServiceProxy(webServiceTemplate);
    }

    @Bean
    public WebServiceTemplate webServiceTemplate() {
        WebServiceTemplate webServiceTemplate = new WebServiceTemplate(marshaller());
        webServiceTemplate.setDefaultUri("http://localhost:8080/springws/services");
        return webServiceTemplate;
    }

    @Bean
    public CastorMarshaller marshaller() {
        CastorMarshaller marshaller = new CastorMarshaller();
        marshaller.setMappingLocation(new ClassPathResource("/mapping.xml"));
        return marshaller;
    }
}
```

小结

本章介绍了 JMX 以及相关的一些规范。介绍了如何将 Spring bean 导出为 JMX MBean 以及如何通过 Spring 代理在远程和本地从客户端中使用这些 MBean，以及从 Spring 中发布并监听 JMX 服务器上的通知事件。还介绍了如何通过 Spring 来发送邮件、如何通过 Quartz Scheduler 以及 Spring 的任务命名空间来调度任务。本章介绍了 Spring 所支持的各种远程技术，还介绍了如何发布和消费 RMI 服务，以及如何通过 3 种不同的技术/协议构建服务并通过 HTTP 进行操作，这些技术协议分别是 Burlap、Hessian 和 HTTP Invoker。接下来，介绍了 SOAP Web Service 以及如何通过 JAX-WS 和 Apache CXF 框架构建并消费这些类型的服务。最后，介绍了契约优先的 SOAP Web Service 以及如何通过 Spring-WS 创建和消费这些类型的服务。

第 14 章

Spring 消息机制

本章将会介绍 Spring 对消息机制的支持。消息机制是个非常强大的技术，可以实现应用的可扩展性。它可以对服务所要处理的大量任务进行排队，还鼓励一种解耦的架构。比如说，一个组件可能只会消费基于 java.util.Map 的键值对的消息。这种松耦合的契约使得它可以成为多个不同系统的通信枢纽。

本章将会多次提及主题和队列。消息解决方案被设计为解决两种架构需求：消息从一个应用中的一个点进入到另一个已知点；消息从一个应用中的一个点进入到多个未知点。这些模式相当于一种中间件：面对面跟一个人说话，以及通过扬声器跟一屋子人说话。

如果想通过消息队列将消息广播给"监听"该消息的多个未知客户端（类似于使用扬声器），那就可以将消息发送给一个主题。如果想将消息发送给单个已知的客户端，那就可以通过队列发送。

学习完本章内容后，你可以通过 Spring 来创建并访问基于消息的中间件。本章还会对消息机制进行总体介绍，这会对下一章所介绍的 Spring Integration 起到帮助作用。

我们将会介绍消息抽象以及如何在 JMS、AMQP 和 Apache Kafka 中使用消息。对于每一种技术来说，Spring 都通过基于模板的方式简化了使用，以轻松实现消息的发送和接收。此外，Spring 还可以让声明在 IoC 容器中的 bean 监听消息并对其作出响应。对于每一种技术来说，方式都是一样的。

注意：在 ch14\bin 目录中有几个脚本，它们用于启动不同消息提供者的 Docker 版本：针对 JMS 的 ActiveMQ、针对 AMQP 的 RabbitMQ，最后是 Apache Kafka。

14-1 使用 Spring 发送和接收 JMS 消息

问题提出

要想发送和接收 JMS 消息，你需要执行如下任务。

1．创建消息代理的 JMS 连接工厂。

2．创建 JMS 目的地，可以是队列或是主题。

3．从连接工厂中打开 JMS 连接。

4．从连接中获取 JMS 会话。

5．使用消息生产者和消费者来发送和接收 JMS 消息。

6．处理 JMSException，它是一个检查异常，所以必须要处理。

7．关闭 JMS 会话与连接。

可以看到，发送或是接收一条简单的 JMS 消息需要编写大量代码。实际上，大多数任务都是样板式的，需要在每次使用 JMS 时都要重复做一遍。

解决方案

Spring 提供了基于模板的解决方案来简化 JMS 编码。借助于 JMS 模板（Spring 框架类 JmsTemplate），通过很少的代码就可以完成 JMS 消息的发送与接收。模板会处理样板任务，还会将 JMS API 的 JMSException

层次体系转换为 Spring 的运行期异常 org.springframework.jms.JmsException 层次体系。

解释说明

假设你要开发一个邮局系统，它包含两个子系统：前端桌面子系统与后端邮局子系统。当前端桌面收到邮件时，它会将邮件传递给邮局进行分类和传送。同时，前端桌面子系统会向后端邮局子系统发送一条 JMS 消息，通知它有新邮件到来。邮件信息是通过如下类来表示的：

```
package com.apress.springrecipes.post;
public class Mail {

    private String mailId;
    private String country;
    private double weight;

    // Constructors, Getters and Setters
    ...
}
```

用于发送和接收邮件信息的方法定义在 FrontDesk 与 BackOffice 接口中，如下所示：

```
package com.apress.springrecipes.post;

public interface FrontDesk {

    public void sendMail(Mail mail);
}
package com.apress.springrecipes.post;

public interface BackOffice {

    public Mail receiveMail();
}
```

不借助于 Spring 的 JMS 模板支持来发送和接收消息

我们看看在不借助于 Spring 的 JMS 模板支持的情况下如何发送和接收 JMS 消息。如下 FrontDeskImpl 类会直接使用 JMS API 来发送 JMS 消息。

```
package com.apress.springrecipes.post;

import javax.jms.Connection;
import javax.jms.ConnectionFactory;
import javax.jms.Destination;
import javax.jms.JMSException;
import javax.jms.MapMessage;
import javax.jms.MessageProducer;
import javax.jms.Session;

import org.apache.activemq.ActiveMQConnectionFactory;
import org.apache.activemq.command.ActiveMQQueue;

public class FrontDeskImpl implements FrontDesk {

    public void sendMail(Mail mail) {
        ConnectionFactory cf =
            new ActiveMQConnectionFactory("tcp://localhost:61616");
        Destination destination = new ActiveMQQueue("mail.queue");

        Connection conn = null;
        try {
            conn = cf.createConnection();
            Session session =
                conn.createSession(false, Session.AUTO_ACKNOWLEDGE);
            MessageProducer producer = session.createProducer(destination);
```

```
            MapMessage message = session.createMapMessage();
            message.setString("mailId", mail.getMailId());
            message.setString("country", mail.getCountry());
            message.setDouble("weight", mail.getWeight());
            producer.send(message);

            session.close();
        } catch (JMSException e) {
            throw new RuntimeException(e);
        } finally {
            if (conn != null) {
                try {
                    conn.close();
                } catch (JMSException e) {
                }
            }
        }
    }
}
```

在上述的 sendMail()方法中，首先通过 ActiveMQ 所提供的类创建了特定于 JMS 的 ConnectionFactory 与 Destination 对象。如果在本机运行，那么代码中所使用的消息代理 URL 就是它的默认值。在 JMS 中有两类目的地：队列与主题。

如本章一开始所说的那样，队列针对的是点对点通信模型，而主题则用于发布-订阅通信模型。由于要点对点地从前端桌面向后端邮局发送 JMS 消息，因此应该使用消息队列。可以通过 ActiveMQTopic 类轻松创建一个主题来作为目的地。

接下来，在发送消息前需要创建连接、会话和消息生产者。JMS API 中定义了几类消息，包括 TextMessage、MapMessage、BytesMessage、ObjectMessage 和 StreamMessage。MapMessage 中的消息内容是键值对形式的，就像一个 map 一样。它们都是接口，其父接口是 Message。同时，你还需要处理 JMSException，JMS API 会抛出该异常。最后，要记得关闭会话与连接来释放系统资源。每次关闭 JMS 连接时，在该连接上打开的所有会话都会自动关闭。因此，只需要在 finally 块中确保恰当地关闭掉 JMS 连接即可。

另一方面，如下 BackOfficeImpl 类会直接通过 JMS API 接收 JMS 消息：

```
package com.apress.springrecipes.post;

import javax.jms.Connection;
import javax.jms.ConnectionFactory;
import javax.jms.Destination;
import javax.jms.JMSException;
import javax.jms.MapMessage;
import javax.jms.MessageConsumer;
import javax.jms.Session;

import org.apache.activemq.ActiveMQConnectionFactory;
import org.apache.activemq.command.ActiveMQQueue;

public class BackOfficeImpl implements BackOffice {

    public Mail receiveMail() {
        ConnectionFactory cf =
            new ActiveMQConnectionFactory("tcp://localhost:61616");
        Destination destination = new ActiveMQQueue("mail.queue");

        Connection conn = null;
        try {
            conn = cf.createConnection();
            Session session =
                conn.createSession(false, Session.AUTO_ACKNOWLEDGE);
            MessageConsumer consumer = session.createConsumer(destination);
```

```
                conn.start();
                MapMessage message = (MapMessage) consumer.receive();
                Mail mail = new Mail();
                mail.setMailId(message.getString("mailId"));
                mail.setCountry(message.getString("country"));
                mail.setWeight(message.getDouble("weight"));
                session.close();
                return mail;
            } catch (JMSException e) {
                throw new RuntimeException(e);
            } finally {
                if (conn != null) {
                    try {
                        conn.close();
                    } catch (JMSException e) {
                    }
                }
            }
        }
    }
```

该方法中的大部分代码都类似于发送 JMS 消息时的代码，只不过这里创建的是消息消费者并从它接收 JMS 消息。注意，这里使用了连接的 start()方法，但在之前的 FrontDeskImpl 示例中并未使用它。

在使用 Connection 接收消息时，可以向连接添加监听器以便在接收到消息时得到调用，或是同步阻塞，以等待消息的到来。容器并不知道你采取的是哪种方式，因此直到显式调用 start()后才会开始轮询消息。如果添加监听器或是进行任何配置，那么请在调用 start()之前进行。

最后，创建两个配置类，一个用于前端桌面子系统（即 FrontOfficeConfiguration），另一个用于后端邮局子系统（即 BackOfficeConfiguration）。

```
package com.apress.springrecipes.post.config;

import com.apress.springrecipes.post.FrontDeskImpl;
import org.springframework.context.annotation.Bean;
import org.springframework.context.annotation.Configuration;

@Configuration
public class FrontOfficeConfiguration {

    @Bean
    public FrontDeskImpl frontDesk() {
        return new FrontDeskImpl();
    }
}
package com.apress.springrecipes.post.config;

import com.apress.springrecipes.post.BackOfficeImpl;
import org.springframework.context.annotation.Bean;
import org.springframework.context.annotation.Configuration;

@Configuration
public class BackOfficeConfiguration {

    @Bean
    public BackOfficeImpl backOffice() {
        return new BackOfficeImpl();
    }
}
```

现在，前端桌面与后端邮局子系统几乎可以发送和接收 JMS 消息了。不过在进入到最后一步之前，请先启动 ActiveMQ 消息代理（如果还没有启动的话）。

可以轻松监控 ActiveMQ 消息代理的活动。在默认安装下，可以打开 http://localhost:8161/admin/queueGraph.jsp 查看示例中所用的队列 mail.queue 的情况。此外，ActiveMQ 还通过 JMX 公开了非常有价值的 bean 与统计数据。只需运行 jconsole 并在 Mbeans 选项卡下进入到 org.apache.activemq 即可。

接下来，创建两个主类运行这个消息系统，一个用于前端桌面子系统（FrontDeskMain 类），另一个用于后端邮局子系统（BackOfficeMain 类）。

```
package com.apress.springrecipes.post;

import com.apress.springrecipes.post.config.FrontOfficeConfiguration;
import org.springframework.context.ApplicationContext;
import org.springframework.context.annotation.AnnotationConfigApplicationContext;

public class FrontDeskMain {

    public static void main(String[] args) {

        ApplicationContext context =
            new AnnotationConfigApplicationContext(FrontOfficeConfiguration.class);

        FrontDesk frontDesk = context.getBean(FrontDesk.class);
        frontDesk.sendMail(new Mail("1234", "US", 1.5));
    }
}
package com.apress.springrecipes.post;

import com.apress.springrecipes.post.config.BackOfficeConfiguration;
import org.springframework.context.ApplicationContext;
import org.springframework.context.annotation.AnnotationConfigApplicationContext;

public class BackOfficeMain {

    public static void main(String[] args) {

        ApplicationContext context =
            new AnnotationConfigApplicationContext(BackOfficeConfiguration.class);

        BackOffice backOffice = context.getBean(BackOffice.class);
        Mail mail = backOffice.receiveMail();
        System.out.println("Mail #" + mail.getMailId() + " received");
    }
}
```

每次使用上述 FrontDeskMain 类运行前端桌面应用时，就会将一条消息发送至代理；每次使用上述 BackOfficeMain 类运行后端邮局应用时，都会从代理中获取一条消息。

使用 Spring 的 JMS 模板来发送和接收消息

Spring 提供了一个 JMS 模板，可以极大简化 JMS 代码。要想通过该模板发送 JMS 消息，只需调用 send() 方法并提供消息目的地即可，此外还有一个 MessageCreator 对象，可用于创建待发送的消息。MessageCreator 对象通常会被实现为一个匿名内部类。

```
package com.apress.springrecipes.post;

import javax.jms.Destination;
import javax.jms.JMSException;
import javax.jms.MapMessage;
import javax.jms.Message;
import javax.jms.Session;

import org.springframework.jms.core.JmsTemplate;
import org.springframework.jms.core.MessageCreator;

public class FrontDeskImpl implements FrontDesk {

    private JmsTemplate jmsTemplate;
```

```
    private Destination destination;

    public void setJmsTemplate(JmsTemplate jmsTemplate) {
        this.jmsTemplate = jmsTemplate;
    }

    public void setDestination(Destination destination) {
        this.destination = destination;
    }

    public void sendMail(final Mail mail) {
        jmsTemplate.send(destination, new MessageCreator() {
            public Message createMessage(Session session) throws JMSException {
                MapMessage message = session.createMapMessage();
                message.setString("mailId", mail.getMailId());
                message.setString("country", mail.getCountry());
                message.setDouble("weight", mail.getWeight());
                return message;
            }
        });
    }
}
```

注意，内部类只能访问外部方法中声明为 final 的参数和变量。MessageCreator 接口只声明了一个 createMessage()方法供你实现。在该方法中，需要通过提供的 JMS 会话来创建并返回 JMS 消息。

JMS 模板可以帮助你获取并释放 JMS 连接与会话，它会发送由 MessageCreator 对象所创建的 JMS 消息。此外，它还会将 JMS API 的 JMSException 层次体系转换为 Spring 的 JMS 运行期异常层次体系，其父类是 org.springframework.jms.JmsException。你可以捕获该 send 方法以及另外一个 send 方法所抛出的 JmsException，然后在 catch 块中进行处理。

前端桌面子系统的 bean 配置文件中声明了一个 JMS 模板，它引用了 JMS 连接工厂来开启连接。接下来，将该模板和消息目的地注入到前端桌面 bean 中。

```
package com.apress.springrecipes.post.config;

import com.apress.springrecipes.post.FrontDeskImpl;
import org.apache.activemq.ActiveMQConnectionFactory;
import org.apache.activemq.command.ActiveMQQueue;
import org.springframework.context.annotation.Bean;
import org.springframework.context.annotation.Configuration;
import org.springframework.jms.core.JmsTemplate;

import javax.jms.ConnectionFactory;
import javax.jms.Queue;

@Configuration
public class FrontOfficeConfiguration {

    @Bean
    public ConnectionFactory connectionFactory() {
        return new ActiveMQConnectionFactory("tcp://localhost:61616");
    }

    @Bean
    public Queue destination() {
        return new ActiveMQQueue("mail.queue");
    }

    @Bean
    public JmsTemplate jmsTemplate() {
        JmsTemplate jmsTemplate = new JmsTemplate();
        jmsTemplate.setConnectionFactory(connectionFactory());
        return jmsTemplate;
```

```
    }

    @Bean
    public FrontDeskImpl frontDesk() {
        FrontDeskImpl frontDesk = new FrontDeskImpl();
        frontDesk.setJmsTemplate(jmsTemplate());
        frontDesk.setDestination(destination());
        return frontDesk;
    }
}
```

要想通过 JMS 模板接收 JMS 消息，可以调用 receive()方法并向其提供消息目的地。该方法会返回一个 JMS 消息 javax.jms.Message，其类型是父 JMS 消息类型（也就是接口），因此在进一步处理前需要将其转换为恰当的类型。

```
package com.apress.springrecipes.post;

import javax.jms.Destination;
import javax.jms.JMSException;
import javax.jms.MapMessage;

import org.springframework.jms.core.JmsTemplate;
import org.springframework.jms.support.JmsUtils;

public class BackOfficeImpl implements BackOffice {

    private JmsTemplate jmsTemplate;
    private Destination destination;

    public void setJmsTemplate(JmsTemplate jmsTemplate) {
        this.jmsTemplate = jmsTemplate;
    }

    public void setDestination(Destination destination) {
        this.destination = destination;
    }

    public Mail receiveMail() {
        MapMessage message = (MapMessage) jmsTemplate.receive(destination);
        try {
            if (message == null) {
                return null;
            }
            Mail mail = new Mail();
            mail.setMailId(message.getString("mailId"));
            mail.setCountry(message.getString("country"));
            mail.setWeight(message.getDouble("weight"));
            return mail;
        } catch (JMSException e) {
            throw JmsUtils.convertJmsAccessException(e);
        }
    }
}
```

不过，在从接收到的 MapMessage 对象中提取信息时，依然需要处理 JMS API 的 JMSException。这与框架的默认行为形成了鲜明的对比，因为在调用 JmsTemplate 的方法时，框架会自动映射异常。为了保持该方法抛出的异常类型的一致性，你需要调用 JmsUtils.convertJmsAccessException()将 JMS API 的 JMSException 转换为 Spring 的 JmsException。

后端邮局子系统的 bean 配置文件声明了一个 JMS 模板，并将其和消息目的地注入到了后端邮局 bean 中。

```
package com.apress.springrecipes.post.config;

import com.apress.springrecipes.post.BackOfficeImpl;
import org.apache.activemq.ActiveMQConnectionFactory;
```

```
import org.apache.activemq.command.ActiveMQQueue;
import org.springframework.context.annotation.Bean;
import org.springframework.context.annotation.Configuration;
import org.springframework.jms.core.JmsTemplate;

import javax.jms.ConnectionFactory;
import javax.jms.Queue;

@Configuration
public class BackOfficeConfiguration {

    @Bean
    public ConnectionFactory connectionFactory() {
        return new ActiveMQConnectionFactory("tcp://localhost:61616");
    }

    @Bean
    public Queue destination() {
        return new ActiveMQQueue("mail.queue");
    }

    @Bean
    public JmsTemplate jmsTemplate() {
        JmsTemplate jmsTemplate = new JmsTemplate();
        jmsTemplate.setConnectionFactory(connectionFactory());
        jmsTemplate.setReceiveTimeout(10000);
        return jmsTemplate;
    }

    @Bean
    public BackOfficeImpl backOffice() {
        BackOfficeImpl backOffice = new BackOfficeImpl();
        backOffice.setDestination(destination());
        backOffice.setJmsTemplate(jmsTemplate());
        return backOffice;
    }
}
```

请注意 JMS 模板的 receiveTimeout 属性，它指定了等待的毫秒数。默认情况下，该模板会在目的地永远等待 JMS 消息，与此同时调用线程也会阻塞。为了避免等待消息如此长的时间，你应该为该模板指定一个超时时间。如果目的地在这段时间内没有消息，那么 JMS 模板的 receive()方法就会返回一条 null 消息。

在你的应用中，接收消息的主要用途可能是因为要对某个东西进行响应或是在一个间隔时间内检查消息、处理消息然后直到下一个间隔时间的到来。如果想要接收消息并以服务的形式对其响应，可能得使用本章后面将会介绍的消息驱动的 POJO 功能。这里，我们探讨这样一种机制，它会一直等待消息，当消息到来时通过回调应用对其进行处理。

从默认目的地发送和接收消息

相较于为每个 JMS 模板的 send()与 receive()方法调用都指定消息目的地，你可以为 JMS 模板指定默认目的地。接下来，就无须再将其注入到消息发送者与接收者 beans 中了。

```
@Configuration
public class FrontOfficeConfiguration {
...
    @Bean
    public JmsTemplate jmsTemplate() {
        JmsTemplate jmsTemplate = new JmsTemplate();
        jmsTemplate.setConnectionFactory(connectionFactory());
        jmsTemplate.setDefaultDestination(mailDestination());
        return jmsTemplate;
    }
```

```
    @Bean
    public FrontDeskImpl frontDesk() {
        FrontDeskImpl frontDesk = new FrontDeskImpl();
        frontDesk.setJmsTemplate(jmsTemplate());
        return frontDesk;
    }
}
```

对于后端邮局来说，配置如下所示：

```
@Configuration
public class BackOfficeConfiguration {
...
    @Bean
    public JmsTemplate jmsTemplate() {
        JmsTemplate jmsTemplate = new JmsTemplate();
        jmsTemplate.setConnectionFactory(connectionFactory());
        jmsTemplate.setDefaultDestination(mailDestination());
        jmsTemplate.setReceiveTimeout(10000);
        return jmsTemplate;
    }

    @Bean
    public BackOfficeImpl backOffice() {
        BackOfficeImpl backOffice = new BackOfficeImpl();
        backOffice.setJmsTemplate(jmsTemplate());
        return backOffice;
    }
}
```

为 JMS 模板指定了默认目的地后，可以从消息发送者与接收者类中将消息目的地的 setter 方法删除。现在，当调用 send()和 receive()方法时就无须再指定消息目的地了。

```
package com.apress.springrecipes.post;
...
import org.springframework.jms.core.MessageCreator;

public class FrontDeskImpl implements FrontDesk {
    ...
    public void sendMail(final Mail mail) {
        jmsTemplate.send(new MessageCreator() {
            ...
        });
    }
}
package com.apress.springrecipes.post;
...
import javax.jms.MapMessage;
...

public class BackOfficeImpl implements BackOffice {
    ...
    public Mail receiveMail() {
        MapMessage message = (MapMessage) jmsTemplate.receive();
        ...
    }
}
```

此外，相较于为 JMS 模板指定 Destination 接口的实例，还可以指定目的地名来让 JMS 模板解析它，这样就可以从两个 bean 配置类中将目的地属性声明删除掉了。这是通过添加 defaultDestinationName 属性做到的。

```
@Bean
    public JmsTemplate jmsTemplate() {
        JmsTemplate jmsTemplate = new JmsTemplate();
        ...
```

```
        jmsTemplate.setDefaultDestinationName("mail.queue");
        return jmsTemplate;
    }
```

继承 JmsGatewaySupport 类

JMS 发送者与接收者类还可以继承 JmsGatewaySupport 以获取 JMS 模板。对于继承了 JmsGatewaySupport 的类来说，有如下两种选择来创建 JMS 模板。

- 为 JmsGatewaySupport 注入 JMS 连接工厂来自动创建 JMS 模板。不过，如果按照这种方式来做，将无法配置 JMS 模板的详细信息。
- 为 JmsGatewaySupport 注入你所创建和配置的 JMS 模板。

对于这两种方式来说，如果需要自己配置 JMS 模板，那么第二种方式会更加适合。你可以从发送者和接收者类中将私有字段 jmsTemplate 及其 setter 方法删除。在需要访问 JMS 模板时，只需调用 getJmsTemplate() 即可。

```
package com.apress.springrecipes.post;

import org.springframework.jms.core.support.JmsGatewaySupport;
...

public class FrontDeskImpl extends JmsGatewaySupport implements FrontDesk {
    ...
    public void sendMail(final Mail mail) {
        getJmsTemplate().send(new MessageCreator() {
            ...
        });
    }
}
package com.apress.springrecipes.post;
...

import org.springframework.jms.core.support.JmsGatewaySupport;

public class BackOfficeImpl extends JmsGatewaySupport implements BackOffice {
    public Mail receiveMail() {
        MapMessage message = (MapMessage) getJmsTemplate().receive();
        ...
    }
}
```

14-2 转换 JMS 消息

问题提出

应用从消息队列接收到了消息，不过需要将这些消息从特定于 JMS 的类型转换为特定于业务的类型。

解决方案

Spring 提供了 SimpleMessageConverter 的一个实现，可以将接收到的 JMS 消息转换为业务对象以及将业务对象转换为 JMS 消息。你可以使用默认实现，也可以提供自己的实现。

解释说明

前文中处理的是原始的 JMS 消息。Spring 的 JMS 模板可以通过消息转换器实现 JMS 消息与 Java 对象之间的转换。默认情况下，JMS 模板会使用 SimpleMessageConverter 实现 TextMessage 与字符串之间、BytesMessage 与字节数组之间、MapMessage 与 map 之间，以及 ObjectMessage 与序列化对象之间的转换。

对于上一节中的前端桌面与后端邮局类来说，你可以通过 convertAndSend()和 receiveAndConvert()方法来发送和接收 map，这个 map 会与 MapMessage 进行转换。

```
package com.apress.springrecipes.post;
...
public class FrontDeskImpl extends JmsGatewaySupport implements FrontDesk {
```

```
    public void sendMail(Mail mail) {
        Map<String, Object> map = new HashMap<String, Object>();
        map.put("mailId", mail.getMailId());
        map.put("country", mail.getCountry());
        map.put("weight", mail.getWeight());
        getJmsTemplate().convertAndSend(map);
    }
}
package com.apress.springrecipes.post;
...
public class BackOfficeImpl extends JmsGatewaySupport implements BackOffice {
    public Mail receiveMail() {
        Map map = (Map) getJmsTemplate().receiveAndConvert();
        Mail mail = new Mail();
        mail.setMailId((String) map.get("mailId"));
        mail.setCountry((String) map.get("country"));
        mail.setWeight((Double) map.get("weight"));
        return mail;
    }
}
```

还可以通过实现 MessageConverter 接口，创建自定义的消息转换器来转换 mail 对象。

```
package com.apress.springrecipes.post;

import javax.jms.JMSException;
import javax.jms.MapMessage;
import javax.jms.Message;
import javax.jms.Session;

import org.springframework.jms.support.converter.MessageConversionException;
import org.springframework.jms.support.converter.MessageConverter;

public class MailMessageConverter implements MessageConverter {

    public Object fromMessage(Message message) throws JMSException,
        MessageConversionException {
        MapMessage mapMessage = (MapMessage) message;
        Mail mail = new Mail();
        mail.setMailId(mapMessage.getString("mailId"));
        mail.setCountry(mapMessage.getString("country"));
        mail.setWeight(mapMessage.getDouble("weight"));
        return mail;
    }

    public Message toMessage(Object object, Session session) throws JMSException,
        MessageConversionException {
        Mail mail = (Mail) object;
        MapMessage message = session.createMapMessage();
        message.setString("mailId", mail.getMailId());
        message.setString("country", mail.getCountry());
        message.setDouble("weight", mail.getWeight());
        return message;
    }
}
```

要想使用该消息转换器，你需要在两个 bean 配置类中声明它，并将其注入到 JMS 模板中。

```
@Configuration
public class BackOfficeConfiguration {
    ...
    @Bean
    public JmsTemplate jmsTemplate() {
        JmsTemplate jmsTemplate = new JmsTemplate();
        jmsTemplate.setMessageConverter(mailMessageConverter());
        ...
        return jmsTemplate;
```

```
    }

    @Bean
    public MailMessageConverter mailMessageConverter() {
        return new MailMessageConverter();
    }
}
```

当显式为 JMS 模板设置消息转换器时，它会覆盖默认的 SimpleMessageConverter。现在，可以调用 JMS 模板的 convertAndSend()与 receiveAndConvert()方法来发送和接收 mail 对象了。

```
package com.apress.springrecipes.post;
...
public class FrontDeskImpl extends JmsGatewaySupport implements FrontDesk {
    public void sendMail(Mail mail) {
        getJmsTemplate().convertAndSend(mail);
    }
}
package com.apress.springrecipes.post;
...
public class BackOfficeImpl extends JmsGatewaySupport implements BackOffice {
    public Mail receiveMail() {
        return (Mail) getJmsTemplate().receiveAndConvert();
    }
}
```

14-3　管理 JMS 事务

问题提出

你想要参与到 JMS 的事务中，这样消息的接收与发送就都是事务性的了。

解决方案

可以使用与其他 Spring 组件同样的事务策略。根据需要使用 Spring 的 TransactionManager 实现并将行为装配到 bean 中。

解释说明

当在一个方法中生产或是消费多条 JMS 消息时，如果中间出现了错误，那么目的地处所生产和消费的 JMS 消息就可能会处于不一致的状态。你需要使用事务将方法包围起来以避免这个问题。

在 Spring 中，JMS 的事务管理与其他数据访问策略是一致的。比如说，你可以对需要事务管理的方法使用注解@Transactional。

```
package com.apress.springrecipes.post;

import org.springframework.jms.core.support.JmsGatewaySupport;
import org.springframework.transaction.annotation.Transactional;
...
public class FrontDeskImpl extends JmsGatewaySupport implements FrontDesk {

    @Transactional
    public void sendMail(Mail mail) {
        ...
    }
}
package com.apress.springrecipes.post;

import org.springframework.jms.core.support.JmsGatewaySupport;
import org.springframework.transaction.annotation.Transactional;
...
public class BackOfficeImpl extends JmsGatewaySupport implements BackOffice {

    @Transactional
    public Mail receiveMail() {
        ...
```

```
    }
}
```

接下来，在两个 Java 配置类中，添加@EnableTransactionManagement 注解并声明事务管理器。与本地 JMS 事务对应的事务管理器是 JmsTransactionManager，它需要引用 JMS 连接工厂。

```
package com.apress.springrecipes.post.config;
...
import org.springframework.jms.connection.JmsTransactionManager;
import org.springframework.transaction.annotation.EnableTransactionManagement;

import javax.jms.ConnectionFactory;

@Configuration
@EnableTransactionManagement
public class BackOfficeConfiguration {

    @Bean
    public ConnectionFactory connectionFactory() { ... }

    @Bean
    public PlatformTransactionManager transactionManager() {
        return new JmsTransactionManager(connectionFactory());
    }
}
```

如果需要跨多个资源的事务管理，比如说数据源和 ORM 资源工厂，或是需要分布式事务管理，那就需要在应用服务器中配置 JTA 事务并使用 JtaTransactionManager。注意，为了支持多个资源事务，JMS 连接工厂必须要兼容于 XA（即必须要支持分布式事务）。

14-4　在 Spring 中创建消息驱动的 POJO

问题提出

当调用 JMS 消息消费者的 receive()方法来接收消息时，调用线程会阻塞，直到消息可用为止。线程除了等待之外什么事情也做不了。这种类型的消息接收叫做同步接收，因为应用必须要等待消息的到来后才能完成其工作。你可以创建消息驱动的 POJO（MDP）来支持异步的 JMS 消息接收。MDP 会使用注解@MessageDriven。

> **注意：** 本节上下文中的消息驱动 POJO 或 MDP 指的是没有任何特殊的运行时要求，只是监听 JMS 消息的 POJO。它指的并非消息驱动 bean（MDB），这种 beans 对应于 EJB 规范，需要 EJB 容器才行。

解决方案

Spring 可以让声明在其 IoC 容器中的 bean 按照与 MDB 相同的方式监听 JMS 消息，而 MDB 的方式则是基于 EJB 规范的。由于 Spring 向 POJO 添加了消息监听能力，因此它们被称作消息驱动的 POJO（MDP）。

解释说明

假设你想为邮局的后勤办公室增加一个电子板，当邮件从前端桌面传过来时能够实时显示它的信息。当前端桌面随邮件发送 JMS 消息时，后勤办公室子系统能够监听到这些消息并在电子板上展示出来。为了实现更好的系统性能，你应该使用异步 JMS 接收方式以避免阻塞接收这些 JMS 消息的线程。

使用消息监听器监听 JMS 消息

首先，创建一个消息监听器来监听 JMS 消息。消息监听器提供了与前文中使用 JmsTemplate 的 BackOfficeImpl 不同的方式。监听器也可以消费来自于代理的消息。比如说，如下 MailListener 会监听包含了邮件信息的 JMS 消息：

```
package com.apress.springrecipes.post;

import javax.jms.JMSException;
```

```
import javax.jms.MapMessage;
import javax.jms.Message;
import javax.jms.MessageListener;

import org.springframework.jms.support.JmsUtils;

public class MailListener implements MessageListener {

    public void onMessage(Message message) {
        MapMessage mapMessage = (MapMessage) message;
        try {
            Mail mail = new Mail();
            mail.setMailId(mapMessage.getString("mailId"));
            mail.setCountry(mapMessage.getString("country"));
            mail.setWeight(mapMessage.getDouble("weight"));
            displayMail(mail);
        } catch (JMSException e) {
            throw JmsUtils.convertJmsAccessException(e);
        }
    }

    private void displayMail(Mail mail) {
        System.out.println("Mail #" + mail.getMailId() + " received");
    }
}
```

消息监听器必须要实现 javax.jms.MessageListener 接口。当 JMS 消息到来时，onMessage()方法会被调用，同时消息会作为方法参数。在该示例中，你只是将邮件信息打印到了控制台上。注意，在从 MapMessage 对象提取消息信息时，你需要处理 JMS API 的 JMSException。可以调用 JmsUtils.convertJmsAccessException()将其转换为 Spring 的运行期异常 JmsException。

接下来，需要在后勤办公室的配置中配置这个监听器。仅仅声明这个监听器还不足以监听 JMS 消息。你需要一个消息监听器容器在消息目的地监控 JMS 消息，当消息到来时触发消息监听器。

```
package com.apress.springrecipes.post.config;

import com.apress.springrecipes.post.MailListener;
import org.apache.activemq.ActiveMQConnectionFactory;
import org.springframework.context.annotation.Bean;
import org.springframework.context.annotation.Configuration;
import org.springframework.jms.listener.SimpleMessageListenerContainer;

import javax.jms.ConnectionFactory;

@Configuration
public class BackOfficeConfiguration {

    @Bean
    public ConnectionFactory connectionFactory() { ... }

    @Bean
    public MailListener mailListener() {
        return new MailListener();
    }

    @Bean
    public Object container() {
        SimpleMessageListenerContainer smlc = new SimpleMessageListenerContainer();
        smlc.setConnectionFactory(connectionFactory());
        smlc.setDestinationName("mail.queue");
        smlc.setMessageListener(mailListener());
        return smlc;
    }
}
```

Spring 在 org.springframework.jms.listener 包中提供了几种消息监听器容器供你选择，其中的 SimpleMessageListenerContainer 与 DefaultMessageListenerContainer 是最常使用的。SimpleMessageListenerContainer 是最简单的一个，它不支持事务。如果在接收消息时有事务上的需求，那就要使用 SimpleMessageListenerContainer。

现在，可以启动消息监听器了。由于不需要调用 bean 来触发消息消费（监听器会帮你做），因此如下 main 类只会启动 Spring IoC 容器，这足够了：

```
package com.apress.springrecipes.post;

import org.springframework.context.support.GenericXmlApplicationContext;

public class BackOfficeMain {

    public static void main(String[] args) {
        new AnnotationConfigApplicationContext(BackOfficeConfiguration.class);
    }
}
```

当启动这个后勤办公室应用时，它就会监听代理（即 ActiveMQ）上的消息。当前端桌面应用向代理发送消息时，后勤办公室应用就会对其进行响应并将消息打印到控制台上。

使用 POJO 监听 JMS 消息

实现了 MessageListener 接口的监听器可以监听消息，声明在 Spring IoC 容器中的任意 bean 也可以。这么做意味着 beans 会与 Spring 框架接口和 JMS MessageListener 接口解耦。要想在消息到来时触发 bean 的方法，方法必须要接收如下类型之一作为其唯一的参数。

- 原始的 JMS 消息类型：针对 TextMessage、MapMessage、BytesMessage 与 ObjectMessage。
- String：只针对 TextMessage。
- Map：只针对 MapMessage。
- byte[]：只针对 BytesMessage。
- Serializable：只针对 ObjectMessage。

比如说，要想监听 MapMessage，你声明一个方法，它接受 Map 作为参数并使用注解@JmsListener。这个类无须再实现 MessageListener 接口。

```
package com.apress.springrecipes.post;

import org.springframework.jms.annotation.JmsListener;

import java.util.Map;

public class MailListener {

    @JmsListener(destination = "mail.queue")
    public void displayMail(Map map) {
        Mail mail = new Mail();
        mail.setMailId((String) map.get("mailId"));
        mail.setCountry((String) map.get("country"));
        mail.setWeight((Double) map.get("weight"));
        System.out.println("Mail #" + mail.getMailId() + " received");
    }
}
```

要想检测@JmsListener 注解，需要将@EnableJms 注解加在配置类上，还需要注册 JmsListenerContainerFactory，默认情况下会使用名字 jmsListenerContainerFactory 来检测它。

POJO 是通过 JmsListenerContainerFactory 注册到监听器容器上的。该工厂会创建和配置一个 MessageListenerContainer，并将被注解的方法作为消息监听器注册到上面。你可以实现自己的 JmsListener-ContainerFactory，不过通常情况下使用已有的就足够了。SimpleJmsListenerContainerFactory 会创建 SimpleMessageListenerContainer 的实例，而 DefaultJmsListenerContainerFactory 则会创建 DefaultMessage-

ListenerContainer 实例。

你现在使用的是 SimpleJmsListenerContainerFactory。如果有更多的需求，那么可以轻松切换至 DefaultMessageListenerContainer，比如说，需要事务或是 TaskExecutor 的异步处理。

```
package com.apress.springrecipes.post.config;

import com.apress.springrecipes.post.MailListener;
import org.apache.activemq.ActiveMQConnectionFactory;
import org.springframework.context.annotation.Bean;
import org.springframework.context.annotation.Configuration;
import org.springframework.jms.annotation.EnableJms;
import org.springframework.jms.config.SimpleJmsListenerContainerFactory;
import org.springframework.jms.listener.SimpleMessageListenerContainer;
import org.springframework.jms.listener.adapter.MessageListenerAdapter;

import javax.jms.ConnectionFactory;

@Configuration
@EnableJms
public class BackOfficeConfiguration {

    @Bean
    public ConnectionFactory connectionFactory() {
        return new ActiveMQConnectionFactory("tcp://localhost:61616");
    }

    @Bean
    public MailListener mailListener() {
        return new MailListener();
    }

    @Bean
    public SimpleJmsListenerContainerFactory jmsListenerContainerFactory() {
        SimpleJmsListenerContainerFactory listenerContainerFactory =
            new SimpleJmsListenerContainerFactory();
        listenerContainerFactory.setConnectionFactory(connectionFactory());
        return listenerContainerFactory;
    }
}
```

转换 JMS 消息

还可以创建消息转换器来转换包含了邮件信息的 JMS 消息中的 mail 对象。由于消息监听者只接收消息，因此方法 toMessage()并不会被调用，这样它只需返回 null 即可。不过，如果还使用该消息转换器发送消息，那就需要实现该方法。如下示例再次展示了之前编写的 MailMessageConverter 类。

```
package com.apress.springrecipes.post;

import javax.jms.JMSException;
import javax.jms.MapMessage;
import javax.jms.Message;
import javax.jms.Session;

import org.springframework.jms.support.converter.MessageConversionException;
import org.springframework.jms.support.converter.MessageConverter;

public class MailMessageConverter implements MessageConverter {

    public Object fromMessage(Message message) throws JMSException,
        MessageConversionException {
        MapMessage mapMessage = (MapMessage) message;
        Mail mail = new Mail();
        mail.setMailId(mapMessage.getString("mailId"));
```

```
        mail.setCountry(mapMessage.getString("country"));
        mail.setWeight(mapMessage.getDouble("weight"));
        return mail;
    }

    public Message toMessage(Object object, Session session) throws JMSException,
        MessageConversionException {
        ...
    }
}
```

消息转换器应当应用到监听器容器工厂上，使其能在调用 POJO 方法前将消息转换为对象。

```
package com.apress.springrecipes.post.config;

import com.apress.springrecipes.post.MailListener;
import com.apress.springrecipes.post.MailMessageConverter;
import org.apache.activemq.ActiveMQConnectionFactory;
import org.springframework.context.annotation.Bean;
import org.springframework.context.annotation.Configuration;
import org.springframework.jms.annotation.EnableJms;
import org.springframework.jms.config.SimpleJmsListenerContainerFactory;

import javax.jms.ConnectionFactory;

@Configuration
@EnableJms
public class BackOfficeConfiguration {

    @Bean
    public ConnectionFactory connectionFactory() {
        return new ActiveMQConnectionFactory("tcp://localhost:61616");
    }

    @Bean
    public MailListener mailListener() {
        return new MailListener();
    }

    @Bean
    public MailMessageConverter mailMessageConverter() {
        return new MailMessageConverter();
    }

    @Bean
    public SimpleJmsListenerContainerFactory jmsListenerContainerFactory() {
        SimpleJmsListenerContainerFactory listenerContainerFactory =
            new SimpleJmsListenerContainerFactory();
        listenerContainerFactory.setConnectionFactory(connectionFactory());
        listenerContainerFactory.setMessageConverter(mailMessageConverter());
        return listenerContainerFactory;
    }
}
```

借助于该消息转换器，POJO 的监听器方法就可以接受一个 mail 对象作为方法参数了。

```
package com.apress.springrecipes.post;

import org.springframework.jms.annotation.JmsListener;

public class MailListener {

    @JmsListener(destination = "mail.queue")
    public void displayMail(Mail mail) {
        System.out.println("Mail #" + mail.getMailId() + " received");
    }
}
```

管理 JMS 事务

如前所述，SimpleMessageListenerContainer 不支持事务。因此，如果消息监听器方法需要事务，那就得使用 DefaultMessageListenerContainer。对于本地 JMS 事务来说，只需开启其 sessionTransacted 属性即可，这样监听器方法就会运行在本地 JMS 事务中（与 XA 事务相反）。要想使用 DefaultMessageListenerContainer，请将 SimpleJmsListenerContainerFactory 改为 DefaultJmsListenerContainerFactory，并配置方才所说的 sessionTransacted 属性。

```
@Bean
public DefaultJmsListenerContainerFactory jmsListenerContainerFactory() {
    DefaultJmsListenerContainerFactory listenerContainerFactory =
        new DefaultJmsListenerContainerFactory();
    listenerContainerFactory.setConnectionFactory(cachingConnectionFactory());
    listenerContainerFactory.setMessageConverter(mailMessageConverter());
    listenerContainerFactory.setSessionTransacted(true);
    return listenerContainerFactory;
}
```

不过，如果想让监听器参与到 JTA 事务中，那就需要声明一个 JtaTransactionManager 实例并将其注入到监听器容器工厂中。

14-5 缓存与池化 JMS 连接

问题提出

在本章中，出于简化的目的，我们通过 org.apache.activemq.ActiveMQConnectionFactory 的一个简单实例作为连接工厂来探索 Spring 的 JMS 支持。这在实际开发中并非最佳选择，因为存在着性能上的考量。

问题的关键在于，JmsTemplate 会在每次调用时关闭会话和消费者。这意味着它会清理所有这些对象并释放内存。这么做是“安全的”，但并不高效，因为创建的一些对象（比如说消费者）需要长期存活。该行为源自于应用服务器环境下 JmsTemplate 的使用，在这种环境下通常会使用应用服务器的连接工厂，其内部会提供连接池。在这种环境下，归还所有对象只是将其放到池中，这是我们期望的行为。

解决方案

这个问题并没有“一刀切”的解决方案。你需要进行权衡并作出恰当的反应。

解释说明

一般来说，你需要一个连接工厂，在使用 JmsTemplate 发布消息时会提供某种池化和缓存的功能。首先你应该在所用的应用服务器中寻找池化的连接工厂。默认情况下，它会提供一个。

在本章的示例中，你在独立配置中使用了 ActiveMQ。就像很多厂商一样，ActiveMQ 提供了一个池化的连接工厂类替代方案。实际上，它提供了两个：一个在使用 JCA 连接器消费消息时所用；另一个则在 JCA 容器外使用。在发送消息时，你可以使用它们来缓存生产者和会话。如下配置会在独立配置中池化一个连接工厂。在发布消息时，可以使用它来代替前文中所用的连接工厂。

```
@Bean(destroyMethod = "stop")
public ConnectionFactory connectionFactory() {
    ActiveMQConnectionFactory connectionFactoryToUse =
        new ActiveMQConnectionFactory("tcp://localhost:61616");
    PooledConnectionFactory connectionFactory = new PooledConnectionFactory();
    connectionFactory.setConnectionFactory(connectionFactoryToUse);
    return connectionFactory;
}
```

如果是接收消息，则依然可以提升效率，因为 JmsTemplate 每次都会构建一个新的 MessageConsumer。在这种情况下，你有几种选择：使用 Spring 的各种*MessageListenerContainer 实现（MDP），因为它会正确缓存消费者；使用 Spring 的 ConnectionFactory 实现。第一个实现 org.springframework.jms.connection.SingleConnectionFactory 每次都会返回同样的底层 JMS 连接（根据 JMS API，它是线程安全的），并且会忽略掉对 close()方法的调用。

一般来说，该实现与 JMS API 搭配得非常好。org.springframework.jms.connection. CachingConnectionFactory 则是更新的一种选择。首先，显而易见的好处就是它提供了缓存多个实例的能力。其次，它会缓存会话、消息生产者和消息消费者。最后，无论 JMS 连接工厂实现是什么都不会对其产生影响。

```
@Bean
public ConnectionFactory cachingConnectionFactory() {
    return new CachingConnectionFactory(connectionFactory());
}
```

14-6　使用 Spring 发送和接收 AMQP 消息

问题提出

你想使用 RabbitMQ 发送和接收消息。

解决方案

Spring AMQP 项目提供了轻松访问 AMQP 协议的能力。它所提供的支持类似于 Spring JMS。它带有一个 RabbitTemplate，可以提供基本的发送与接收选项；它还有一个模拟 Spring JMS 的 MessageListenerContainer 选项。

解释说明

我们看看如何通过 RabbitTemplate 发送消息。要想访问 RabbitTemplate，继承 RabbitGatewaySupport 是最简单的方式。这里会使用 FrontDeskImpl，它用到了 RabbitTemplate。

不借助 Spring 的模板支持来发送和接收消息

我们看看如何在不借助于 Spring 模板支持的情况下发送和接收消息。如下的 FrontDeskImpl 类会使用普通的 API 向 RabbitMQ 发送一条消息：

```
package com.apress.springrecipes.post;

import com.fasterxml.jackson.databind.ObjectMapper;
import com.rabbitmq.client.Channel;
import com.rabbitmq.client.Connection;
import com.rabbitmq.client.ConnectionFactory;
import org.springframework.scheduling.annotation.Scheduled;

import java.io.IOException;
import java.util.Locale;
import java.util.Random;
import java.util.concurrent.TimeoutException;

public class FrontDeskImpl implements FrontDesk {

    private static final String QUEUE_NAME = "mail.queue";

    public void sendMail(final Mail mail) {
        ConnectionFactory connectionFactory = new ConnectionFactory();
        connectionFactory.setHost("localhost");
        connectionFactory.setUsername("guest");
        connectionFactory.setPassword("guest");
        connectionFactory.setPort(5672);

        Connection connection = null;
        Channel channel = null;
        try {

            connection = connectionFactory.newConnection();
            channel = connection.createChannel();
            channel.queueDeclare(QUEUE_NAME, true, false, false, null);
            String message = new ObjectMapper().writeValueAsString(mail);
```

```
            channel.basicPublish("", QUEUE_NAME, null, message.getBytes("UTF-8"));

        } catch (IOException | TimeoutException e) {
            throw new RuntimeException(e);
        } finally {
            if (channel != null) {
                try {
                    channel.close();
                } catch (IOException | TimeoutException e) {
                }
            }

            if (connection != null) {
                try {
                    connection.close();
                } catch (IOException e) {
                }
            }
        }
    }
}
```

首先，创建一个 ConnectionFactory 来获取到 RabbitMQ 的连接；这里配置为 localhost 并提供了用户名/密码组合。接下来需要获取到一个 Channel 来最终创建一个队列。然后，使用 Jackson ObjectMapper 将传递进来的 Mail 消息转换为 JSON 并将其发送到队列。在创建连接和发送消息时，你需要处理出现的不同异常，发送后还需要再关闭并释放 Connection，这也会抛出异常。

在发送和接收 AMQP 消息前，你需要安装一个 AMQP 消息代理。

注意：在 bin 目录中有一个 rabbitmq.sh 文件，它会在一个 Docker 容器中下载并启动 RabbitMQ 代理。

如下 BackOfficeImpl 类会通过普通的 RabbitMQ API 接收消息：

```
package com.apress.springrecipes.post;

import com.fasterxml.jackson.databind.ObjectMapper;
import com.rabbitmq.client.*;
import org.springframework.stereotype.Service;

import javax.annotation.PreDestroy;
import java.io.IOException;
import java.util.concurrent.TimeoutException;

@Service
public class BackOfficeImpl implements BackOffice {

    private static final String QUEUE_NAME = "mail.queue";

    private MailListener mailListener = new MailListener();
    private Connection connection;

    @Override
    public Mail receiveMail() {

        ConnectionFactory connectionFactory = new ConnectionFactory();
        connectionFactory.setHost("localhost");
        connectionFactory.setUsername("guest");
        connectionFactory.setPassword("guest");
        connectionFactory.setPort(5672);

        Channel channel = null;
        try {

            connection = connectionFactory.newConnection();
```

```
            channel = connection.createChannel();
            channel.queueDeclare(QUEUE_NAME, true, false, false, null);

            Consumer consumer = new DefaultConsumer(channel) {
                @Override
                public void handleDelivery(String consumerTag, Envelope envelope,
                AMQP.BasicProperties properties, byte[] body)
                    throws IOException {
                    Mail mail = new ObjectMapper().readValue(body, Mail.class);
                    mailListener.displayMail(mail);
                }
            };
            channel.basicConsume(QUEUE_NAME, true, consumer);

        } catch (IOException | TimeoutException e) {
            throw new RuntimeException(e);
        }

        return null;
    }

    @PreDestroy
    public void destroy() {
        if (this.connection != null) {
            try {
                this.connection.close();
            } catch (IOException e) {
            }
        }
    }
}
```

上述代码与 FrontDeskImpl 大体相同，只不过现在是注册了一个 Consumer 对象来接收消息。在该消费者中，使用 Jackson 将消息映射为 Mail 对象并将其传递给 MailListener，它又会将转换后的消息打印到控制台。在使用通道时，可以添加一个消费者，使之在接收到消息时被调用。使用 basicConsume 方法将消费者注册到通道后，该消费者就可以使用了。

如果运行了 FrontDeskImpl，那就会立刻看到消息到来。

借助 Spring 的模板支持来发送消息

FrontDeskImpl 类继承了 RabbitGatewaySupport，它会根据所传递的配置来配置一个 RabbitTemplate。为了发送消息，你使用 getRabbitOperations 方法获取模板，然后转换并发送消息。这里使用到了 convertAndSend 方法。该方法首先会使用 MessageConverter 将消息转换为 JSON，然后将其发送到配置好的队列。

```
package com.apress.springrecipes.post;

import org.springframework.amqp.rabbit.core.RabbitGatewaySupport;

public class FrontDeskImpl extends RabbitGatewaySupport implements FrontDesk {

    public void sendMail(final Mail mail) {
        getRabbitOperations().convertAndSend(mail);
    }
}
```

我们来看一下配置：

```
package com.apress.springrecipes.post.config;

import com.apress.springrecipes.post.FrontDeskImpl;
import org.springframework.amqp.rabbit.connection.CachingConnectionFactory;
import org.springframework.amqp.rabbit.connection.ConnectionFactory;
import org.springframework.amqp.rabbit.core.RabbitTemplate;
```

```
import org.springframework.amqp.support.converter.Jackson2JsonMessageConverter;
import org.springframework.context.annotation.Bean;
import org.springframework.context.annotation.Configuration;

@Configuration
public class FrontOfficeConfiguration {

    @Bean
    public ConnectionFactory connectionFactory() {
        CachingConnectionFactory connectionFactory =
            new CachingConnectionFactory("127.0.0.1");
        connectionFactory.setUsername("guest");
        connectionFactory.setPassword("guest");
        connectionFactory.setPort(5672);
        return connectionFactory;
    }

    @Bean
    public RabbitTemplate rabbitTemplate() {
        RabbitTemplate rabbitTemplate = new RabbitTemplate();
        rabbitTemplate.setConnectionFactory(connectionFactory());
        rabbitTemplate.setMessageConverter(new Jackson2JsonMessageConverter());
        rabbitTemplate.setRoutingKey("mail.queue");
        return rabbitTemplate;
    }

    @Bean
    public FrontDeskImpl frontDesk() {
        FrontDeskImpl frontDesk = new FrontDeskImpl();
        frontDesk.setRabbitOperations(rabbitTemplate());
        return frontDesk;
    }
}
```

该配置非常类似于 JMS 配置。你需要一个 ConnectionFactory 来连接 RabbitMQ 代理，使用 CachingConnectionFactory 以便可以重用连接。接下来有个 RabbitTemplate，它会使用连接并通过 MessageConverter 转换消息。消息会通过 Jackson2 库转换为 JSON，这正是要配置 Jackson2JsonMessageConverter 的原因所在。最后，将 RabbitTemplate 传递给 FrontDeskImpl 类，这样 FrontDeskImpl 就可以使用了。

```
package com.apress.springrecipes.post;

import com.apress.springrecipes.post.config.FrontOfficeConfiguration;
import org.springframework.context.ConfigurableApplicationContext;
import org.springframework.context.annotation.AnnotationConfigApplicationContext;

public class FrontDeskMain {

    public static void main(String[] args) throws Exception {
        ConfigurableApplicationContext context =
            new AnnotationConfigApplicationContext(FrontOfficeConfiguration.class);

        FrontDesk frontDesk = context.getBean(FrontDesk.class);
        frontDesk.sendMail(new Mail("1234", "US", 1.5));

        System.in.read();

        context.close();
    }
}
```

使用消息监听器监听 AMQP 消息

Spring AMQP 支持通过 MessageListenerContainers 来获取消息，方式与 Spring JMS 针对 JMS 所做的一样。

Spring AMQP 提供了@RabbitListener 注解来标识基于 AMQP 的消息监听器。我们来看看所用的 MessageListener。

```
package com.apress.springrecipes.post;

import org.springframework.amqp.rabbit.annotation.RabbitListener;

public class MailListener {

    @RabbitListener(queues = "mail.queue")
    public void displayMail(Mail mail) {
        System.out.println("Received: " + mail);
    }
}
```

MailListener 与 14-4 节中创建的用于接收 JMS 消息的监听器一样。区别在于配置。

```
package com.apress.springrecipes.post.config;

import com.apress.springrecipes.post.MailListener;
import org.springframework.amqp.rabbit.annotation.EnableRabbit;
import org.springframework.amqp.rabbit.config.SimpleRabbitListenerContainerFactory;
import org.springframework.amqp.rabbit.connection.CachingConnectionFactory;
import org.springframework.amqp.rabbit.connection.ConnectionFactory;
import org.springframework.amqp.rabbit.listener.RabbitListenerContainerFactory;
import org.springframework.amqp.support.converter.Jackson2JsonMessageConverter;
import org.springframework.context.annotation.Bean;
import org.springframework.context.annotation.Configuration;

@Configuration
@EnableRabbit
public class BackOfficeConfiguration {

    @Bean
    public RabbitListenerContainerFactory rabbitListenerContainerFactory() {
        SimpleRabbitListenerContainerFactory containerFactory =
            new SimpleRabbitListenerContainerFactory();
        containerFactory.setConnectionFactory(connectionFactory());
        containerFactory.setMessageConverter(new Jackson2JsonMessageConverter());
        return containerFactory;
    }

    @Bean
    public ConnectionFactory connectionFactory() {
        CachingConnectionFactory connectionFactory = new
        CachingConnectionFactory("127.0.0.1");
        connectionFactory.setUsername("guest");
        connectionFactory.setPassword("guest");
        connectionFactory.setPort(5672);
        return connectionFactory;
    }

    @Bean
    public MailListener mailListener() {
        return new MailListener();
    }

}
```

为了启用 AMQP 基于注解的监听器，需要向配置类添加@EnableRabbit 注解。由于每个监听器都需要一个 MessageListenerContainer，因此需要配置一个 RabbitListenerContainerFactory，它负责创建这些容器。默认情况下，@EnableRabbit 的逻辑会查找名为 rabbitListenerContainerFactory 的 bean。

RabbitListenerContainerFactory 需要一个 ConnectionFactory，因此使用了 CachingConnectionFactory。在 MailListener.displayMail 方法被 MessageListenerContainer 调用前，它需要通过 Jackon2JsonMessageConverter

将 JSON 类型的消息负载转换为 Mail 对象。

要想监听消息，请创建一个带有 main 方法的类，它只需要构建应用上下文即可。

```
package com.apress.springrecipes.post;

import com.apress.springrecipes.post.config.BackOfficeConfiguration;
import org.springframework.context.annotation.AnnotationConfigApplicationContext;

public class BackOfficeMain {

    public static void main(String[] args) {
        new AnnotationConfigApplicationContext(BackOfficeConfiguration.class);
    }
}
```

14-7 使用 Spring Kafka 发送和接收消息

问题提出

你想使用 Apache Kafka 发送和接收消息。

解决方案

Spring Kafka 项目提供了轻松访问 Apache Kafka 的方式。它所提供的支持类似于使用了 Spring 消息抽象的 Spring JMS。它带有 KafkaTemplate，提供了基本的发送选项；还带有一个模仿 Spring JMS 的 MessageListenerContainer 选项，可以通过@EnableKafka 开启。

解释说明

首先，我们会介绍如何创建 KafkaTemplate 来发送消息，以及如何通过 KafkaListener 监听消息，然后介绍如何通过 MessageConverters 将对象转换为消息负载。

借助 Spring 的模板支持来发送消息

首先，重写 FrontOfficeImpl 类来使用 KafkaTemplate 发送消息。为了做到这一点，需要一个实现了 KafkaOperations 的对象，KafkaOperations 是 KafkaTemplate 实现的一个接口。

```
package com.apress.springrecipes.post;

import com.fasterxml.jackson.core.JsonProcessingException;
import com.fasterxml.jackson.databind.ObjectMapper;
import org.springframework.kafka.core.KafkaOperations;
import org.springframework.kafka.support.SendResult;
import org.springframework.util.concurrent.ListenableFuture;
import org.springframework.util.concurrent.ListenableFutureCallback;
public class FrontDeskImpl implements FrontDesk {

    private final KafkaOperations<Integer, String> kafkaOperations;

    public FrontDeskImpl(KafkaOperations<Integer, String> kafkaOperations) {
        this.kafkaOperations = kafkaOperations;
    }

    public void sendMail(final Mail mail) {

        ListenableFuture<SendResult<Integer, String>> future =
            kafkaOperations.send("mails", convertToJson(mail));
        future.addCallback(new ListenableFutureCallback<SendResult<Integer, String>>() {

            @Override
            public void onFailure(Throwable ex) {
                ex.printStackTrace();
            }

            @Override
```

```
            public void onSuccess(SendResult<Integer, String> result) {
                System.out.println("Result (success): " + result.getRecordMetadata());
            }
        });
    }

    private String convertToJson(Mail mail) {
        try {
            return new ObjectMapper().writeValueAsString(mail);
        } catch (JsonProcessingException e) {
            throw new IllegalArgumentException(e);
        }
    }
}
```

注意 kafkaOperations 字段，其类型是 KafkaOperations<Integer, String>。这意味着所发送的消息的键（发送消息时生成的）是 Integer 类型，消息本身则是 String 类型的。基于此，你需要将传递进来的 Mail 实例转换为 String。这是由 convertToJson 方法通过 Jackson2 ObjectMapper 处理的。消息将会发送到 mails 主题，它是 send 方法的第一个参数；第二个参数是待发送的负载（转换后的 Mail 消息）。

使用 Kafka 发送消息通常是个异步操作，KafkaOperations.send 方法所返回的 ListenableFuture 就能够反映出这一点。它是个普通的 Future，因此可以调用 get()使其成为一个阻塞操作，或是注册 ListenableFutureCallback，使其在操作成功或失败时获得通知。

接下来，需要创建一个配置类来配置 KafkaTemplate，以在 FrontDeskImpl 中使用。

```
package com.apress.springrecipes.post.config;

import com.apress.springrecipes.post.FrontDeskImpl;
import org.apache.kafka.clients.producer.ProducerConfig;
import org.apache.kafka.common.serialization.IntegerSerializer;
import org.apache.kafka.common.serialization.StringSerializer;
import org.springframework.context.annotation.Bean;
import org.springframework.context.annotation.Configuration;
import org.springframework.kafka.core.DefaultKafkaProducerFactory;
import org.springframework.kafka.core.KafkaTemplate;
import org.springframework.kafka.core.ProducerFactory;

import java.util.HashMap;
import java.util.Map;

@Configuration
public class FrontOfficeConfiguration {

    @Bean
    public KafkaTemplate<Integer, String> kafkaTemplate() {
        KafkaTemplate<Integer, String> kafkaTemplate = new KafkaTemplate<>(producerFactory());
        return kafkaTemplate;
    }

    @Bean
    public ProducerFactory<Integer, String> producerFactory() {
        DefaultKafkaProducerFactory producerFactory = new DefaultKafkaProducerFactory<>
        (producerFactoryProperties());
        return producerFactory;
    }

    @Bean
    public Map<String, Object> producerFactoryProperties() {
        Map<String, Object> properties = new HashMap<>();
        properties.put(ProducerConfig.BOOTSTRAP_SERVERS_CONFIG, "localhost:9092");
        properties.put(ProducerConfig.KEY_SERIALIZER_CLASS_CONFIG, IntegerSerializer.class);
        properties.put(ProducerConfig.VALUE_SERIALIZER_CLASS_CONFIG, StringSerializer.class);
        return properties;
```

```
    }

    @Bean
    public FrontDeskImpl frontDesk() {
        return new FrontDeskImpl(kafkaTemplate());
    }

}
```

上面的配置创建了一个最小化配置的 KafkaTemplate。你需要配置 KafkaTemplate 所用的 ProducerFactory；它至少需要一个连接的 URL，还需要知道将消息序列化成的键与值的类型。URL 是通过 ProducerConfig.BOOTSTRAP_SERVERS_CONFIG 指定的，它可以接受要连接的一个或多个服务器。ProducerConfig.KEY_SERIALIZER_CLASS_CONFIG 与 ProducerConfig.VALUE_SERIALIZER_CLASS_CONFIG 分别配置所使用的键、值序列化器。由于键使用了 Integer，值使用了 String，因此分别使用 IntegerSerializer 与 StringSerializer 对其进行了配置。

最后，将构造好的 KafkaTemplate 传递给 FrontDeskImpl。为了运行这个前端桌面应用，我们只需要如下的 Main 类：

```
package com.apress.springrecipes.post;

import org.springframework.context.ConfigurableApplicationContext;
import org.springframework.context.annotation.AnnotationConfigApplicationContext;

import com.apress.springrecipes.post.config.FrontOfficeConfiguration;

public class FrontDeskMain {

    public static void main(String[] args) throws Exception {
        ConfigurableApplicationContext context =
            new AnnotationConfigApplicationContext(FrontOfficeConfiguration.class);
        context.registerShutdownHook();

        FrontDesk frontDesk = context.getBean(FrontDesk.class);
        frontDesk.sendMail(new Mail("1234", "US", 1.5));

        System.in.read();

    }
}
```

这会启动前端桌面应用并通过 Kafka 发送一条消息。

使用 Spring Kafka 监听消息

Spring Kafka 也提供了消息监听器容器来监听主题上的消息，就像 Spring JMS 与 Spring AMQP 一样。要想开启这些容器，需要将@EnableKafka 注解添加到配置类上并通过@KafkaListener 创建和配置 Kafka 消费者。

首先，我们来创建监听器。这非常简单，只需对方法添加一个单参数的@KafkaListener 注解即可。

```
package com.apress.springrecipes.post;

import org.springframework.kafka.annotation.KafkaListener;

public class MailListener {

    @KafkaListener(topics = "mails")
    public void displayMail(String mail) {
        System.out.println(" Received: " + mail);
    }
}
```

现在，你会关注原始的 String 负载，因为它是真正发送的内容。

接下来需要配置监听器容器。

```java
package com.apress.springrecipes.post.config;

import com.apress.springrecipes.post.MailListener;
import org.apache.kafka.clients.consumer.ConsumerConfig;
import org.apache.kafka.common.serialization.IntegerDeserializer;
import org.apache.kafka.common.serialization.StringDeserializer;
import org.springframework.context.annotation.Bean;
import org.springframework.context.annotation.Configuration;
import org.springframework.kafka.annotation.EnableKafka;
import org.springframework.kafka.config.ConcurrentKafkaListenerContainerFactory;
import org.springframework.kafka.config.KafkaListenerContainerFactory;
import org.springframework.kafka.core.ConsumerFactory;
import org.springframework.kafka.core.DefaultKafkaConsumerFactory;
import org.springframework.kafka.listener.ConcurrentMessageListenerContainer;

import java.util.HashMap;
import java.util.Map;

@Configuration
@EnableKafka
public class BackOfficeConfiguration {

    @Bean
    KafkaListenerContainerFactory<ConcurrentMessageListenerContainer<Integer, String>>
    kafkaListenerContainerFactory() {
        ConcurrentKafkaListenerContainerFactory factory =
            new ConcurrentKafkaListenerContainerFactory();
        factory.setConsumerFactory(consumerFactory());
        return factory;
    }

    @Bean
    public ConsumerFactory<Integer, String> consumerFactory() {
        return new DefaultKafkaConsumerFactory<>(consumerConfiguration());
    }

    @Bean
    public Map<String, Object> consumerConfiguration() {
        Map<String, Object> properties = new HashMap<>();
        properties.put(ConsumerConfig.BOOTSTRAP_SERVERS_CONFIG, "localhost:9092");
        properties.put(ConsumerConfig.KEY_DESERIALIZER_CLASS_CONFIG, IntegerDeserializer.class);
        properties.put(ConsumerConfig.VALUE_DESERIALIZER_CLASS_CONFIG,
                       StringDeserializer.class);
        properties.put(ConsumerConfig.GROUP_ID_CONFIG, "group1");
        return properties;
    }

    @Bean
    public MailListener mailListener() {
        return new MailListener();
    }

}
```

该配置与客户端类似；需要传递一个或多个 URL 来连接到 Apache Kafka 上，由于需要反序列化消息，因此需要指定键值的反序列化器。最后，需要添加一个 group ID，否则将无法连接到 Kafka 上。URL 是通过 ConsumerConfig.BOOTSTRAP_SERVERS_CONFIG 传递的；对于键值序列化器来说，键使用的是 IntegerDeserializer（因为它是个整数）；由于负载是 String，因此需要使用 StringDeserializer。最后，设置了 group 属性。

借助于这些属性，就可以配置 KafkaListenerContainerFactory 了，它是个工厂，用于创建基于 Kafka 的 MessageListenerContainer。该容器在内部会被添加@EnableKafka 注解所启用的功能所使用。对于每个使用了@KafkaListener 注解的方法来说，都会创建一个 MessageListenerContainer。

要想运行该后端邮局应用，需要加载这个配置并让其监听：

```
package com.apress.springrecipes.post;

import com.apress.springrecipes.post.config.BackOfficeConfiguration;
import org.springframework.context.annotation.AnnotationConfigApplicationContext;

public class BackOfficeMain {

    public static void main(String[] args) {
        new AnnotationConfigApplicationContext(BackOfficeConfiguration.class);
    }
}
```

启动前端邮局应用后，Main 消息会被转换为 String 并通过 Kafka 发送给后端邮局，然后生成如下结果：

```
Received: {"mailId":"1234","country":"US","weight":1.5}
```

使用MessageConverter将负载转换为对象

监听器现在会接收 String，不过如果能自动将其转换为 Mail 对象就更好了。只需要修改一些配置就可以轻松做到这一点。这里使用的 KafkaListenerContainerFactory 接受一个 MessageConverter，要想自动将 String 转换为期望的对象，可以将一个 StringJsonMessageConverter 传递给它。它会接收 String 并根据被@KafkaListener 注解的方法所指定的参数将其转换为对象。

首先，更新一下配置。

```
package com.apress.springrecipes.post.config;

import org.springframework.kafka.support.converter.StringJsonMessageConverter;

@Configuration
@EnableKafka
public class BackOfficeConfiguration {

    @Bean
    KafkaListenerContainerFactory<ConcurrentMessageListenerContainer<Integer, String>>
    kafkaListenerContainerFactory() {
        ConcurrentKafkaListenerContainerFactory factory =
            new ConcurrentKafkaListenerContainerFactory();
        factory.setMessageConverter(new StringJsonMessageConverter());
        factory.setConsumerFactory(consumerFactory());
        return factory;
    }
    ...
}
```

接下来需要修改 MailListener，使其使用 Mail 对象来代替普通的 String。

```
package com.apress.springrecipes.post;

import org.springframework.kafka.annotation.KafkaListener;

public class MailListener {

    @KafkaListener(topics = "mails")
    public void displayMail(Mail mail) {
        System.out.println("Mail #" + mail.getMailId() + " received");
    }
}
```

运行后端邮局与前端邮局应用，消息会被发送并接收到。

将对象转换为负载

在前端邮局中，Mail 实例是手工转换为 JSON 字符串的。虽然这并不难，不过如果框架能够透明地进行转换就更好了。这可以通过配置 JsonSerializer（而不是 StringSerializer）来做到。

```
package com.apress.springrecipes.post.config;

import com.apress.springrecipes.post.FrontDeskImpl;
import org.apache.kafka.clients.producer.ProducerConfig;
import org.apache.kafka.common.serialization.IntegerSerializer;
import org.springframework.context.annotation.Bean;
import org.springframework.context.annotation.Configuration;
import org.springframework.kafka.core.DefaultKafkaProducerFactory;
import org.springframework.kafka.core.KafkaTemplate;
import org.springframework.kafka.core.ProducerFactory;
import org.springframework.kafka.support.serializer.JsonSerializer;

import java.util.HashMap;
import java.util.Map;

@Configuration
public class FrontOfficeConfiguration {

    @Bean
    public KafkaTemplate<Integer, Object> kafkaTemplate() {
        return new KafkaTemplate<>(producerFactory());
    }

    @Bean
    public ProducerFactory<Integer, Object> producerFactory() {
        return new DefaultKafkaProducerFactory<>(producerFactoryProperties());
    }

    @Bean
    public Map<String, Object> producerFactoryProperties() {
        Map<String, Object> properties = new HashMap<>();
        properties.put(ProducerConfig.BOOTSTRAP_SERVERS_CONFIG, "localhost:9092");
        properties.put(ProducerConfig.KEY_SERIALIZER_CLASS_CONFIG, IntegerSerializer.class);
        properties.put(ProducerConfig.VALUE_SERIALIZER_CLASS_CONFIG, JsonSerializer.class);
        return properties;
    }

    @Bean
    public FrontDeskImpl frontDesk() {
        return new FrontDeskImpl(kafkaTemplate());
    }
}
```

相较于 KafkaTemplate<Integer, String>，现在使用的是 KafkaTemplate<Integer, Object>，因为现在可以将对象序列化为 String 发送给 Kafka 了。

FrontOfficeImpl 类现在也不需要了，因为转换为 JSON 的工作已经由 KafkaTemplate 来处理了。

```
package com.apress.springrecipes.post;

import org.springframework.kafka.core.KafkaOperations;
import org.springframework.kafka.support.SendResult;
import org.springframework.util.concurrent.ListenableFuture;
import org.springframework.util.concurrent.ListenableFutureCallback;

public class FrontDeskImpl implements FrontDesk {

    private final KafkaOperations<Integer, Object> kafkaOperations;

    public FrontDeskImpl(KafkaOperations<Integer, Object> kafkaOperations) {
        this.kafkaOperations = kafkaOperations;
    }

    public void sendMail(final Mail mail) {
```

```
        ListenableFuture<SendResult<Integer, Object>> future = kafkaOperations.send("mails",
        mail);
        future.addCallback(new ListenableFutureCallback<SendResult<Integer, Object>>() {

            @Override
            public void onFailure(Throwable ex) {
                ex.printStackTrace();
            }

            @Override
            public void onSuccess(SendResult<Integer, Object> result) {
                System.out.println("Result (success): " + result.getRecordMetadata());
            }
        });
    }
}
```

小结

本章介绍了 Spring 的消息支持，以及如何通过它来构建面向消息的架构。我们介绍了如何通过不同的消息解决方案来生产和消费消息。对于不同的消息解决方案来说，我们介绍了如何通过 MessageListenerContainer 构建消息驱动的 POJO。

我们介绍了 JMS 与 AMQP 及 ActiveMQ（这是一个可靠的开源消息队列），同时还简要介绍了 Apache Kafka。

下一章将会介绍 Spring Integration，这是个类似于 ESB 的框架，用于构建应用集成解决方案，类似于 Mule ESB 和 ServiceMix。你将利用本章学到的知识并借助于 Spring Integration 将面向消息的应用提升到一个新的高度。

第 15 章

Spring Integration

本章将会介绍企业应用集成（Enterprise Application Integration，EAI）背后的原理，很多现代化应用都使用它来解耦组件之间的依赖。Spring 框架提供了一个名为 Spring Integration 的强大且可扩展的框架。就像核心 Spring 框架为应用中的组件所提供的解耦能力一样，Spring Integration 为不同的系统与数据也提供了同样的解耦能力。本章将会介绍为理解 EAI 涉及的各种模式而必须要掌握的知识，从而理解何为企业服务总线（Enterprise Service Bus，ESB），最后将会介绍如何通过 Spring Integration 构建解决方案。如果之前曾使用过 EAI 服务器或是 ESB，那么你会发现 Spring Integration 要比以前使用过的那些简单很多。

学习完本章后，你将可以编写相当复杂的 Spring Integration 解决方案来集成应用，从而共享服务与数据。你还会学习到 Spring Integration 的诸多配置选项。如果喜欢的话，Spring Integration 可以完全在标准的 XML 命名空间中进行配置，不过你会发现混合式的方式（使用注解与 XML）会更加自然。你还会了解到为何 Spring Integration 对于那些拥有传统企业应用集成背景的人来说会有强大的吸引力。如果你曾经使用过 ESB（如 Mule 或 ServiceMix）或是传统的 EAI 服务器（如 Axway 的 Integrator 或 TIBCO 的 ActiveMatrix），则会对这里介绍的术语很熟悉，同时配置也是非常直观的。

15-1 使用 EAI 进行系统集成

问题提出

你有两个应用，它们需要通过外部接口进行彼此间的通信。你需要在应用的服务和数据间建立一个连接。

解决方案

你需要使用 EAI，它是使用一套已知模式进行应用与数据集成的一门学科。这些模式都被总结和呈现在 Gregor Hohpe 与 Bobby Woolf 所著的 *Enterprise Integration Patterns* 这本里程碑之作中。时至今日，这些模式已经成为标准，并且是现代 ESB 的通用语。

解释说明

有几种不同的集成方式可供使用，你可以使用文件系统、数据库、消息，甚至执行远程过程调用。接下来，你会探索如何实现不同的集成方式，以及 Spring Integration 提供了哪些选择。

选择一种集成方式

有多种集成方式，每种方式都有适合的应用与需求类型。基本前提很简单：应用不能像在一个系统中使用本地机制那样与其他系统进行通信。因此，可以设计一种桥接连接，在上面构建一些抽象的东西或是对其他系统的特性进行一些变通，使之便于调用系统。对于每个应用来说，抽象的东西是不同的。有时是位置，有时是调用的同步或异步特性，有时又会是消息协议。选择集成方式有多种标准，这与你希望应用之间的耦合程度有关、与服务器的亲和度有关、与需要的消息格式有关，等等。从某种程度上来说，TCP/IP 是最为知名的集成技术，因为它将一个应用与另一个服务器进行了解耦。

你可能已经使用了一些或是所有如下的集成方式（依旧使用了 Spring）构建了应用。比如说，共享数据库可以通过 Spring 的 JDBC 支持轻松实现；远程过程调用可以通过 Spring 的导出器功能轻松实现。

4 种集成方式如下所示。

- 文件传输：每个应用都可以生产共享数据的文件供其他应用消费，它也可以消费其他应用生产的文件。
- 共享数据库：应用将想要共享的数据存储到一个共同数据库中。这通常会采取数据库的形式来让不同的应用访问。一般来说，这并非很好的一种方式，因为需要将数据公开给不同客户端，但可能没有遵从你所规定的限制（没有外显出来）。使用视图与存储过程可以实现这个功能，但并非理想之选。并没有特定的支持来访问数据库自身，不过你可以构建一个端点将 SQL 数据库中的新结果处理为消息负载。与数据库集成无法做到细粒度或是面向消息的，相反，它是面向批处理的。毕竟，数据库中 100 万的新行并不是一个事件，而是一个批处理！因此，Spring Batch（见第 11 章）提供了面向 JDBC 输入与输出的令人惊叹的支持就不足为奇了。
- 远程过程调用：每个应用公开它的部分过程，这样它们就可以被远程调用了，其他应用可以调用它们来发起动作并交换数据。Spring Integration 为优化 RPC 交换提供了专门的支持（包括远程过程调用，如 SOAP、RMI 和 HTTP Invoker）。
- 消息：每个应用都连接到一个公共的消息系统并通过消息来交换数据和调用行为。这种方式（在 JEE 世界中大多数是通过 JMS 开启的）还描述了其他异步或是多路广播的发布-订阅架构。从某种程度上来说，诸如 Spring Integration 等 ESB 或是 EAI 容器可以让你处理大多数其他方式，就好像处理的是消息队列一样：请求进入队列中并被管理起来，然后对其进行响应或是转发给其他队列。

基于 ESB 解决方案进行构建

在知道该如何处理集成后，现在就来实现它。时至今日，有多种选择可供选取。如果需求足够常见，那么大多数中间件或是框架都会以某种方式提供。JEE、.NET 和其他方案都可以通过 SOAP、XML-RPC、二进制层（如 EJB）或二进制远程、JMS 或是 MQ 抽象来处理常见场景。不过，如果需求比较少见或是有大量配置要做，那么也许就需要使用 ESB 了。ESB 是个中间件，提供了高层次的方式来对集成进行建模，它使用了 EAI 所描述的各种模式。ESB 提供了可管理的配置格式，以一种简单的高层次格式来对集成的不同部分进行编排。

作为 SpringSource 系列产品中的一个 API，Spring Integration 提供了健壮的机制来对大量集成场景进行建模，它能与 Spring 完美结合。相较于大量其他的 ESB 来说，Spring Integration 拥有很多优势，特别是框架本身的轻量级特性。新兴的 ESB 市场充满了选择。一些是以前的 EAI 服务器，重新处理后强调了以 ESB 为中心的架构。一些则是真正的 ESB，以 ESB 的理念进行的构建。还有一些只不过是带有适配器的消息队列。

实际上，如果寻找超级强大的 EAI 服务器（几乎集成了 JEE 平台，但价格也非常昂贵），那就可以考虑 Axway Integrator。它几乎无所不能。TIBCO 和 WebMethods（后来被收购）等厂商也是榜上有名的，因为它们提供了优秀的工具来处理企业中的集成问题。这些选择虽然强大，但常常也价格高昂且以中间件为中心；你的集成需要部署到中间件上。

诸如 Java 业务集成（Java Business Integration，JBI）等标准化尝试在一定程度上证明了自身的成功，基于这些标准有很多不错的兼容于它的 ESB（如 OpenESB 和 ServiceMix）。ESB 市场上一个有着深谋远虑的领导者——Mule ESB，它有很好的名声，它是免费的，对开源很友好，对社区也很友好，并且是轻量级的。这些特质也使得 Spring Integration 非常有吸引力。很多时候，你只是想与另一个开放系统进行通信，并不想购买那么贵的，甚至比几套房子还贵的中间件。

每个 Spring Integration 应用都是完全嵌入式的，无须服务器基础设施。实际上，你可以在另一个应用中部署集成，也许是在 Web 应用端点中。Spring Integration 改变了大多数 ESB 的部署模式。Spring Integration 应该部署到应用中，而不是将应用部署到 Spring Integration 中。Spring Integration 没有启动与停止脚本，也没有监听端口号。最简单的可运行的 Spring Integration 应用就是个启动 Spring 上下文的 Java public static void main()方法而已。

```
package com.apress.springrecipes.springintegration;

import org.springframework.context.annotation.AnnotationConfigApplicationContext;

public class Main {
    public static void main(String [] args){
        ApplicationContext applicationContext =
            new AnnotationConfigApplicationContext(IntegrationConfiguration.class);
    }
}
```

你创建了一个标准的 Spring 应用上下文并启动。Spring 应用上下文的内容会在随后小节中介绍，不过可以看到它是多么简单。你可能想将上下文提升到 Web 应用、EJB 容器，或是任何想使用的地方。实际上，可以通过 Spring Integration 在 Swing/JavaFX 应用中增强邮件轮询功能。它可以做到非常轻量级。在后续示例中，展示的配置应该被放到 XML 文件中，在运行这个类时将 XML 文件作为第一个参数进行引用。当 main 方法执行完毕时，上下文会开启 Spring Integration 总线并开始对配置在应用上下文 XML 中的组件上的请求进行响应。

15-2 使用 JMS 集成两个系统

问题提出

你想要构建一个集成，使用 JMS 将一个应用连接到另一个应用，它为 Java 应用提供了现代化中间件之上的定位与临时性解耦的功能。你想要应用更为复杂的路由，并且想将代码与具体的消息来源隔离开来（在该示例中就是 JMS 队列或主题）。

解决方案

虽然可以通过常规的 JMS 代码、EJB 对消息驱动 bean（MDB）的支持或是核心 Spring 的消息驱动 POJO（MDP）的支持来实现，但所有这些都需要编写代码来处理来自于 JMS 的消息。代码会与 JMS 绑定到一起。使用 ESB 可以从处理代码中将消息来源隐藏掉。你将该解决方案作为一种简单的方式来了解 Spring Integration 解决方案是如何构建的。Spring Integration 提供了一种简单的方式来使用 JMS，就像在核心 Spring 容器中使用 MDP 一样。不过，可以将 JMS 中间件替换为邮件，而响应消息的代码则无须改变。

解释说明

回忆一下第 14 章的内容，Spring 可以通过 MDP 来取代 EJB 的 MDB 功能。这对于想要处理消息队列中的消息的应用来说是一个强大的解决方案。你要构建一个 MDP，不过可以通过 Spring Integration 更为简洁的配置来进行，同时提供一个非常基本的集成示例。该集成要做的就是接收到来的 JMS 消息（其负载类型是 Map<String,Object>）并将其写到日志中。

正如一个标准的 MDP 一样，我们需要 ConnectionFactory 类的配置。如下是个配置类。在创建时，你可以将其作为参数传递给 Spring ApplicationContext 实例（就像在上一节中的 Main 类一样）。

```
package com.apress.springrecipes.springintegration;

import org.apache.activemq.ActiveMQConnectionFactory;
import org.springframework.context.annotation.Bean;
import org.springframework.context.annotation.ComponentScan;
import org.springframework.context.annotation.Configuration;
import org.springframework.integration.config.EnableIntegration;
import org.springframework.integration.dsl.IntegrationFlow;
import org.springframework.integration.dsl.IntegrationFlows;
import org.springframework.integration.jms.dsl.Jms;
import org.springframework.jms.connection.CachingConnectionFactory;
import org.springframework.jms.core.JmsTemplate;

import javax.jms.ConnectionFactory;
```

```
@Configuration
@EnableIntegration
@ComponentScan
public class IntegrationConfiguration {

    @Bean
    public CachingConnectionFactory connectionFactory() {
        ActiveMQConnectionFactory connectionFactory =
            new ActiveMQConnectionFactory("tcp://localhost:61616");
        return new CachingConnectionFactory(connectionFactory);
    }

    @Bean
    public JmsTemplate jmsTemplate(ConnectionFactory connectionFactory) {
        return new JmsTemplate(connectionFactory);
    }

    @Bean
    public InboundHelloWorldJMSMessageProcessor messageProcessor() {
        return new InboundHelloWorldJMSMessageProcessor();
    }

    @Bean
    public IntegrationFlow jmsInbound(ConnectionFactory connectionFactory) {
        return return IntegrationFlows
            .from(Jms.messageDrivenChannelAdapter(connectionFactory)
            .extractPayload(true)
            .destination("recipe-15-2"))
            .handle(messageProcessor())
            .get();
    }
}
```

可以看到，内容最多的部分就是模式导入！其余代码都是标准的样板代码。你定义了一个 connectionFactory，就像配置一个标准的 MDP 一样。

接下来，定义特定于该解决方案的 bean，在该示例中定义的 bean 会响应来自于消息队列 messageProcessor 并进入到总线的消息。服务激活器是 Spring Integration 中的一种通用端点，用于调用功能（无论是服务中的操作、普通 POJO 中的某个程序，还是你想要的任何东西）来响应输入通道中所发送过来的消息。虽然后面会详细介绍它，不过这里有趣的是你要使用它来响应消息。这些 bean 一起构成了解决方案中的协作者，该示例能够很好地展现出大多数集成是什么样子的。你定义了协作组件，然后通过 Spring Integration Java DSL（用于配置解决方案自身）定义了流程。

提示：还有 Spring Integration Groovy DSL。

配置始于 IntegrationFlows，它用于定义消息如何流经系统。流程开始于 messageDrivenChannelAdapter 的定义，它会接收来自于 recipe-15-2 目的地的消息并将其传递给一个 Spring Integration 通道。顾名思义，messageDrivenChannelAdapter 是个适配器。适配器是这样一种组件，它知道如何与特定类型的子系统进行通信，并能将该子系统中的消息转换为 Spring Integration 总线可以使用的形式。适配器还可以做相反的工作，即接收 Spring Integration 总线上的消息，并将其转换为具体的子系统可以理解的形式。这与服务激活器（后面会介绍）是不同的，因为适配器是总线与外部端点之间的一种通用连接。不过，服务激活器只会在接收到消息时调用应用的业务逻辑。业务逻辑中做什么（是否连接到其他系统）完全取决于你自己。

下一个组件是服务激活器，它会监听进入到通道的消息并通过 handle 方法调用被引用的 bean，在该示例中就是之前定义的 messageProcessor bean。由于组件的方法使用了@ServiceActivator 注解，因此 Spring Integration 知道该调用哪个方法。

```
package com.apress.springrecipes.springintegration;
```

```
import org.slf4j.Logger;
import org.slf4j.LoggerFactory;
import org.springframework.integration.annotation.ServiceActivator;
import org.springframework.messaging.Message;

import java.util.Map;

public class InboundHelloWorldJMSMessageProcessor {

    private final Logger logger =
        LoggerFactory.getLogger(InboundHelloWorldJMSMessageProcessor.class);

    @ServiceActivator
    public void handleIncomingJmsMessage(Message<Map<String, Object>> inboundJmsMessage)
        throws Throwable {
        Map<String, Object> msg = inboundJmsMessage.getPayload();
        logger.info("firstName: {}, lastName: {}, id: {}", msg.get("firstName"),
                                                            msg.get("lastName"),
                                                            msg.get("id"));
    }
}
```

注意，这里有个注解@ServiceActivator，它告诉 Spring 要配置这个组件，同时将这个方法作为来自于通道的消息负载的接收者，消息负载会以 Message<Map<String, Object>> inboundJmsMessage 的形式传递给方法。在之前的配置中，extract-payload="true"告诉 Spring Integration 从 JMS 队列中获取消息负载（该示例中就是 Map<String,Object>），将其提取出来并以 org.springframework.messaging.Message<T>的类型作为流经 Spring Integration 通道的消息负载。不要将 Spring Message 接口与 JMS Message 接口搞混了，虽然它们存在一些相似性。如果没有指定 extractPayload 选项，那么 Spring Message 接口中的负载类型就会是 javax.jms.Message。是否提取负载取决于开发者，不过有时访问该信息是很有用的。重写上述代码，处理未展开的 javax.jms.Message 接口，示例稍有不同，代码如下所示：

```
package com.apress.springrecipes.springintegration;

import org.slf4j.Logger;
import org.slf4j.LoggerFactory;
import org.springframework.integration.annotation.ServiceActivator;
import org.springframework.messaging.Message;

import javax.jms.MapMessage;

public class InboundHelloWorldJMSMessageProcessor {

    private final Logger logger =
        LoggerFactory.getLogger(InboundHelloWorldJMSMessageProcessor.class);

    @ServiceActivator
    public void handleIncomingJmsMessageWithPayloadNotExtracted(
        Message<javax.jms.Message> msgWithJmsMessageAsPayload) throws Throwable {
        javax.jms.MapMessage jmsMessage =
            (MapMessage) msgWithJmsMessageAsPayload.getPayload();
        logger.debug("firstName: {}, lastName: {}, id: {}", jmsMessage.
                                                             getString("firstName"),
                                                             jmsMessage.
                                                             getString("lastName"),
                                                             jmsMessage.getLong("id"));
    }
}
```

可以将负载类型作为传递给方法的参数类型。比如说，如果来自于 JMS 的消息负载的类型是 Cat，那么方法原型就是 public void handleIncomingJmsMessageWithPayload NotExtracted(Cat inboundJmsMessage) throws Throwable。Spring Integration 知道该如何做。在该示例中，我们使用了 Spring Message<T>，它包含了

消息头的值，这对于获取信息来说很有帮助。

此外，不需要指定 throws Throwable。在 Spring Integration 中，错误处理可以是通用的，也可以是具体的（取决于自己的需要）。

在该示例中，你使用@ServiceActivator 注解所调用的功能中结束了集成过程。不过，可以通过从方法中返回值来将一个激活中的响应传递给下一个通道。返回值的类型用来决定系统中所发送的下一条消息是什么。如果返回 Message<T>，那就会直接发送。如果返回的不是 Message<T>，那么值会在 Message<T>实例中被包装为一个负载，它会成为下一条消息并发送给处理管道中的下一个组件。该 Message<T>接口会在服务激活器上配置的输出通道中被发送出去。输出通道中所发送的消息与输入通道中进来的消息类型可以是不同的；这是一种转换消息类型的有效方式。服务激活器是一种非常灵活的组件，可以在其中放置系统钩子来辅助集成。

该解决方案非常直接，就针对一个 JMS 队列的配置来说，它并不比直接的 MDP 有优势，因为还要处理一个额外的间接层。Spring Integration 设施使得构建复杂集成要比 Spring Core 或 EJB3 更加轻松，因为配置是中心化的。你可以总览整个集成，借助于路由与处理的中心化，能更好地重新定位集成中的组件。不过，可以看到，Spring Integration 并不是要与 EJB 和 Spring Core 进行竞争；它所实现的解决方案是 EJB3 和 Spring Core 无法自然而然构建的，而这一点正是它最擅长的。

15-3　查询 Spring Integration 消息以获取上下文信息

问题提出

除了消息类型隐式带给你的信息之外，你还想获取与进入到 Spring Integration 处理管道中的消息相关的更多信息。

解决方案

查询 Spring Integration Message<T>接口以获取消息中的消息头信息。这些值以消息头值的形式存在于 map（Map<String,Object>类型）中。

解释说明

Spring Message<T>接口是个通用包装器，包含了一个指针，该指针指向实际的消息负载，还指向提供了上下文消息元数据的消息头。你可以操纵或是增强该元数据以启用/增强下游组件的功能。比如说，在通过邮件发送消息时，指定 TO/FROM 头信息就很有用。

无论何时，只要你向框架公开一个类来处理某个需求（比如说向服务激活器组件或是转换器组件提供的功能），那就有机会与 Message<T>和消息头进行交互。记住，Spring Integration 会通过处理管道推送一个 Message<T>实例。与 Message<T>实例关联的每个组件都需要对其进行处理、做一些事情，或是对其进行转发。向这些组件提供信息，获取到该点处关于组件上所发生的事情的信息的一种方式就是查询 MessageHeaders。

你应该知道在使用 Spring Integration 时的几个值（见表 15-1 和表 15-2）。这些常量是在 org.springframework.messaging.MessageHeaders 接口和 org.springframework.integration.IntegrationMessageHeaderAccessor 上公开的。

表 15-1　核心 Spring Messaging 上的常见头信息

常量	说明
ID	这是 Spring Integration 引擎赋给消息的唯一值
TIMESTAMP	这是赋给消息的时间戳
REPLY_CHANNEL	这是当前组件应该将输出发送给的通道的 String 名字；可以覆盖
ERROR_CHANNEL	这是运行期出现异常时，当前组件应该将输出发送给的通道的 String 名字；可以覆盖
CONTENT_TYPE	这是消息的内容类型（MIME 类型），主要用于 Web Socket 消息

除了 Spring 消息所定义的头信息外，还有一些 Spring Integration 中常用的头信息；它们定义在 org.springframework.integration.IntegrationMessageHeaderAccessor 中，如表 15-2 所示。

表 15-2　Spring Integration 中常见的头信息

常量	说明
CORRELATION_ID	这是可选的，一些组件（如聚合器）使用它在某个处理管道中对消息进行分组
EXPIRATION_DATE	一些组件将其用作一个阈值，在一个组件等待了这个时间后才开始继续处理
PRIORITY	这是消息的优先级；更大的数字表示更高的优先级
SEQUENCE_NUMBER	这是消息序列的顺序；通常作为序列器来使用
SEQUENCE_SIZE	这是序列的大小，这样聚合器就知道何时停止等待更多消息的到来并继续进行。这在实现联合功能时很有用
ROUTING_SLIP	该消息头包含了关于何时使用 Routing Slip 模式的信息
CLOSEABLE_RESOURCE	这是可选的，一些组件使用它来决定是否可以/应该关闭消息负载（比如说 File 或是 InputStream）

一些消息头值是特定于源消息负载类型的；比如说，来自于文件系统中文件的负载就与来自于 JMS 队列中的负载不同，它们与来自于邮件系统中的消息也不同。这些不同的组件通常都会打包到自己的 JAR 中，一般来说会有一个类提供访问这些头信息的常量。org.springframework.integration.file.FileHeaders 中的 FILENAME 与 PREFIX 常量是针对文件的，它们就是特定于组件的头信息。如果不确定，那就可以手工枚举这些值，因为头信息只不过是一个 java.util.Map 实例而已。

```
package com.apress.springrecipes.springintegration;

import org.slf4j.Logger;
import org.slf4j.LoggerFactory;
import org.springframework.integration.annotation.ServiceActivator;
import org.springframework.messaging.Message;
import org.springframework.messaging.MessageHeaders;

import java.io.File;
import java.util.Map;

public class InboundFileMessageServiceActivator {
    private final Logger logger = LoggerFactory.getLogger(InboundFileMessageService
    Activator.class);

    @ServiceActivator
    public void interrogateMessage(Message<File> message) {
        MessageHeaders headers = message.getHeaders();
        for (Map.Entry<String, Object> header : headers.entrySet()) {
            logger.debug("{} : {}", header.getKey(), header.getValue() );
        }
    }
}
```

借助于这些头信息，可以在不依赖于具体接口的情况下获悉这些消息的具体特性。还可以通过它们进行处理并为下游组件指定自定义元数据。为下游组件提供额外数据的行为叫做消息增强。消息增强指的是获取到给定消息的消息头，然后将其添加到下游处理管道的组件上。大家可以想象这样一种场景：处理一条消息以向客户关系管理（CRM）系统添加一个客户，该系统会调用第三方网站来建立信用评级。该信用会被添加到消息头中，这样下游组件（其任务是添加或是拒绝客户）就可以基于这些消息头值进行判断了。

访问消息头元数据的另一种方式是将其作为参数传递给组件的方法。你只需为参数添加@Header 注解即可，Spring Integration 会负责其他的一切。

```
package com.apress.springrecipes.springintegration;

import java.io.File;

import org.slf4j.Logger;
import org.slf4j.LoggerFactory;
import org.springframework.integration.annotation.ServiceActivator;
import org.springframework.integration.file.FileHeaders;
import org.springframework.messaging.MessageHeaders;
import org.springframework.messaging.handler.annotation.Header;

public class InboundFileMessageServiceActivator {
    private final Logger logger = LoggerFactory.getLogger(InboundFileMessageService
    Activator.class);

    @ServiceActivator
    public void interrogateMessage(
        @Header(MessageHeaders.ID) String uuid,
        @Header(FileHeaders.FILENAME) String fileName, File file) {
        logger.debug("the id of the message is {}, and name of the file payload is {}",
        uuid, fileName);
    }
}
```

还可以让 Spring Integration 只是传递 Map<String,Object>。

```
package com.apress.springrecipes.springintegration;

import java.io.File;
import java.util.Map;

import org.slf4j.Logger;
import org.slf4j.LoggerFactory;
import org.springframework.integration.annotation.ServiceActivator;
import org.springframework.integration.file.FileHeaders;
import org.springframework.messaging.MessageHeaders;
import org.springframework.messaging.handler.annotation.Header;

public class InboundFileMessageServiceActivator {
    private final Logger logger = LoggerFactory.getLogger(InboundFileMessageService
    Activator.class);

    @ServiceActivator
    public void interrogateMessage(
        @Header(MessageHeaders.ID) Map<String, Object> headers, File file) {
        logger.debug("the id of the message is {}, and name of the file payload is {}",
            headers.get(MessageHeaders.ID), headers.get(FileHeaders.FILENAME));
    }
}
```

15-4 使用文件系统来集成两个系统

问题提出

你想要构建这样一个解决方案：获取一个已知、共享的文件系统中的文件，并将其作为通道与另一个系统进行集成。一个例子就是应用每一小时就为添加到系统中的所有客户生成一个逗号分隔值（CSV）转储文件。公司的第三方财务系统会根据这些销售数据进行更新，方式是检查一个共享目录（通过网络文件系统进行挂载）并处理 CSV 记录。我们所需要的是将新文件的出现作为总线上的一个事件。

解决方案

你知道如何通过标准技术来构建解决方案，不过想要使用更加优雅的方式。让 Spring Integration 将你与文件系统的事件驱动的本性以及文件输入/输出的要求隔离开来。相反，我们会通过它专注于编写处理 java.io.File 负载本身的代码。通过这种方式，你可以编写能够进行单元测试的代码，接受一个输入并通过将客户添加到财务系统中进行响应。当功能编写完成后，你在 Spring Integration 管道中对其进行配置，当文件系统识别到有新的文件时让 Spring Integration 调用你所编写的功能。这是事件驱动架构（Event-Driven Architecture，EDA）的一个示例。EDA 可以让你忽略掉事件是如何产生的，并专注于对其的反应上。这与事件驱动的 GUI 非常类似，它让你将代码的关注点由控制用户如何触发动作变为对调用本身的实际反应。Spring Integration 使得松耦合的解决方案成为一种自然的方式。实际上，代码应该与为 JMS 队列所构建的解决方案是类似的，因为它只不过是接收一个参数（一个 Spring Integration Message<T>接口，与消息负载相同类型的参数）的另一个类而已。

解释说明

构建与 JMS 进行通信的解决方案在之前已经介绍过。相反，我们来考虑构建一个使用共享文件系统的解决方案。想象一下，如何在不使用 ESB 方案的情况下进行构建。你需要一种机制来周期性轮询文件系统并检测新文件。使用 Quartz 或其他缓存？你需要快速读取这些文件，然后将负载高效传递给处理逻辑。最后，系统需要处理该负载。

Spring Integration 将你从所有这些基础设施代码中解放出来；你只需要配置即可。不过，对基于文件系统的处理存在一些问题，这需要你自己来解决。在背后，Spring Integration 依然会轮询文件系统并检测新文件。当文件“写完”时，它不知道应用要做什么，因此需要你自己来设置。

存在几种解决方式。你可以写入一个文件，然后再写入另一个零字节的文件。该文件的存在意味着我们可以安全地假设真正的负载是存在的。配置 Spring Integration 来寻找这个文件。如果找到了，它就知道还有另外一个文件（也许同名但文件扩展名不同？），并且可以开始读取/处理它。类似思路下的另一个解决方案是让客户端（“生产者”）将文件写入到目录中，这次使用的文件名是 Spring Integration 用于轮询目录时所使用的 glob 模式不会检测到的名字。接下来，当写入完成时，如果你相信文件系统会做正确的事情，那就执行一个 mv 命令。

我们重新看一下第一个解决方案，不过这次使用的是一个基于文件的适配器。配置看起来与之前一样，只不过适配器的配置发生了变化，去除了针对 JMS 适配器的大量配置，比如说连接工厂等。相反，你告诉 Spring Integration 另一个源（文件系统），消息将会来自这里。

```java
package com.apress.springrecipes.springintegration;

import org.springframework.beans.factory.annotation.Value;
import org.springframework.context.annotation.Bean;
import org.springframework.context.annotation.ComponentScan;
import org.springframework.context.annotation.Configuration;
import org.springframework.integration.config.EnableIntegration;
import org.springframework.integration.dsl.IntegrationFlow;
import org.springframework.integration.dsl.IntegrationFlows;
import org.springframework.integration.dsl.Pollers;
import org.springframework.integration.file.dsl.Files;

import java.io.File;
import java.util.concurrent.TimeUnit;

@Configuration
@EnableIntegration
@ComponentScan
public class IntegrationConfiguration {

    @Bean
    public InboundHelloWorldFileMessageProcessor messageProcessor() {
        return new InboundHelloWorldFileMessageProcessor();
    }

    @Bean
    public IntegrationFlow inboundFileFlow(@Value("${user.home}/inboundFiles/new/") File
    directory) {
```

```
        return IntegrationFlows
            .from(
                Files.inboundAdapter(directory).patternFilter("*.csv"),
                c -> c.poller(Pollers.fixedRate(10, TimeUnit.SECONDS)))
            .handle(messageProcessor())
            .get();
    }
}
```

代码之前都见过。Files.inboundAdapter 是唯一新增的元素。@ServiceActivator 注解的代码发生了变化，从而反映出这样一个事实——你期望接收的消息中包含了类型为 Message<java.io.File>的消息。

```
package com.apress.springrecipes.springintegration;

import org.slf4j.Logger;
import org.slf4j.LoggerFactory;
import org.springframework.integration.annotation.ServiceActivator;
import org.springframework.messaging.Message;

import java.io.File;

public class InboundHelloWorldFileMessageProcessor {
    private final Logger logger = LoggerFactory.getLogger(InboundHelloWorldFileMessage
    Processor.class);

    @ServiceActivator
    public void handleIncomingFileMessage(Message<File> inboundJmsMessage)
        throws Throwable {
        File filePayload = inboundJmsMessage.getPayload();
        logger.debug("absolute path: {}, size: {}", filePayload.getAbsolutePath(),
        filePayload.length());
    }
}
```

15-5　将消息由一种类型转换为另一种类型

问题提出

你想要向总线发送一条消息，在进一步处理前对其进行转换。通常情况下，这是根据下游组件的要求对消息进行适配来做到的。你还想通过消息增强来对其进行转换——添加额外的头信息或是增强负载，这样处理管道中的下游组件就可以从中获益了。

解决方案

使用转换器组件来接收负载的 Message<T>接口并使用不同类型的负载将 Message<T>发送出去。还可以使用转换器添加额外的头信息或是根据处理管道中下游组件的要求更新头信息的值。

解释说明

Spring Integration 提供了一个转换器消息端点来增强消息头信息或是对消息本身进行转换。在 Spring Integration 中，组件是链接到一起的，一个组件的输出是通过调用该组件的方法来返回的。方法的返回值会在组件的“依赖通道”中传递给下一个组件，下一个组件将它作为输入参数。转换器组件可以让你改变返回的对象类型或是添加额外的头信息，更新后的对象会传递给链中的下一个组件。

修改消息负载

转换器组件的配置与之前看到的非常相似。

```
package com.apress.springrecipes.springintegration;

import org.springframework.integration.annotation.Transformer;
import org.springframework.messaging.Message;

import java.util.Map;
```

```
public class InboundJMSMessageToCustomerTransformer {

    @Transformer
    public Customer transformJMSMapToCustomer(
        Message<Map<String, Object>> inboundSprignIntegrationMessage) {
        Map<String, Object> jmsMessagePayload =
            inboundSprignIntegrationMessage.getPayload();
        Customer customer = new Customer();
        customer.setFirstName((String) jmsMessagePayload.get("firstName"));
        customer.setLastName((String) jmsMessagePayload.get("lastName"));
        customer.setId((Long) jmsMessagePayload.get("id"));
        return customer;
    }
}
```

没什么复杂的东西：传递进一个类型为 Map<String,Object>的 Message<T>接口。值会被手工提取出来并用于构建类型为 Customer 的对象。Customer 对象会返回，这相当于在该组件的依赖通道上进行传递。配置中的下一个组件会接收到该对象并作为其输入 Message<T>。

该解决方案与之前看到过的几乎一样，不过有了一个新的转换器元素。

```
package com.apress.springrecipes.springintegration;

import javax.jms.ConnectionFactory;

import org.apache.activemq.ActiveMQConnectionFactory;
import org.springframework.context.annotation.Bean;
import org.springframework.context.annotation.ComponentScan;
import org.springframework.context.annotation.Configuration;
import org.springframework.integration.config.EnableIntegration;
import org.springframework.integration.dsl.IntegrationFlow;
import org.springframework.integration.dsl.IntegrationFlows;
import org.springframework.integration.jms.dsl.Jms;
import org.springframework.jms.connection.CachingConnectionFactory;
import org.springframework.jms.core.JmsTemplate;

@Configuration
@EnableIntegration
@ComponentScan
public class IntegrationConfiguration {

    @Bean
    public CachingConnectionFactory connectionFactory() {
        ActiveMQConnectionFactory connectionFactory =
            new ActiveMQConnectionFactory("tcp://localhost:61616");
        return new CachingConnectionFactory(connectionFactory);
    }

    @Bean
    public JmsTemplate jmsTemplate(ConnectionFactory connectionFactory) {
        return new JmsTemplate(connectionFactory);
    }

    @Bean
    public InboundJMSMessageToCustomerTransformer customerTransformer() {
        return new InboundJMSMessageToCustomerTransformer();
    }

    @Bean
    public InboundCustomerServiceActivator customerServiceActivator() {
        return new InboundCustomerServiceActivator();
    }

    @Bean
    public IntegrationFlow jmsInbound(ConnectionFactory connectionFactory) {
```

```
        return IntegrationFlows
            .from(Jms.messageDrivenChannelAdapter(connectionFactory).
             extractPayload(true).destination("recipe-15-5"))
            .transform(customerTransformer())
            .handle(customerServiceActivator())
            .get();
    }
}
```

这里指定了一个 messageDrivenChannelAdapter 组件，它将传入的内容传递给 InboundJMSMessageTo-CustomerTransformer，它会将其转换为一个 Customer，该 Customer 会被发送给 InboundCustomerServiceActivator。

下一个组件的代码现在可以声明对 Customer 接口的依赖了。借助于转换器，你可以从任意的源中接收消息并将它们转换为一个 Customer，这样就可以重用 InboundCustomerServiceActivator 实例了。

```
package com.apress.springrecipes.springintegration;

import org.slf4j.Logger;
import org.slf4j.LoggerFactory;
import org.springframework.integration.annotation.ServiceActivator;
import org.springframework.messaging.Message;

public class InboundCustomerServiceActivator {
    private static final Logger logger = LoggerFactory.getLogger(InboundCustomerService
    Activator.class);

    @ServiceActivator
    public void doSomethingWithCustomer(Message<Customer> customerMessage) {
        Customer customer = customerMessage.getPayload();
        logger.debug("id={}, firstName: {}, lastName: {}",
            customer.getId(), customer.getFirstName(), customer.getLastName());
    }
}
```

修改消息的头信息

有时，仅仅修改消息负载还不够。你可能想要更新负载和头信息。这么做会很有趣，因为需要用到 MessageBuilder<T>类，它可以让你使用任意负载和任意头信息数据来创建新的 Message<T>对象。该示例中的 XML 配置是一样的。

```
package com.apress.springrecipes.springintegration;

import org.springframework.integration.annotation.Transformer;
import org.springframework.integration.core.Message;
import org.springframework.integration.message.MessageBuilder;

import java.util.Map;

public class InboundJMSMessageToCustomerTransformer {
    @Transformer
    public Message<Customer> transformJMSMapToCustomer(
        Message<Map<String, Object>> inboundSpringIntegrationMessage) {
        Map<String, Object> jmsMessagePayload =
           inboundSpringIntegrationMessage.getPayload();
        Customer customer = new Customer();
        customer.setFirstName((String) jmsMessagePayload.get("firstName"));
        customer.setLastName((String) jmsMessagePayload.get("lastName"));
        customer.setId((Long) jmsMessagePayload.get("id"));
        return MessageBuilder.withPayload(customer)
            .copyHeadersIfAbsent( inboundSpringIntegrationMessage.getHeaders())
            .setHeaderIfAbsent("randomlySelectedForSurvey", Math.random() > .5)
            .build();
    }
}
```

与之前一样，代码只不过是一个带有输入和输出的方法而已。输出是通过 MessageBuilder<T>动态构建的，创建一条消息，该消息与输入消息拥有相同的负载，同时复制既有的头信息并添加了一个新的头信息：randomlySelectedForSurvey。

15-6　使用 Spring Integration 进行错误处理

问题提出

Spring Integration 将分布在不同节点、计算机、服务、协议和语言栈上的系统集成到了一起。实际上，当启动后，Spring Integration 解决方案可能无法在同一时间段内完成工作。那么，异常处理就不像在单个线程中拥有异步行为的组件那样使用语言级的 try/catch 块那么简单了。这意味着你所构建的很多解决方案（使用了通道和队列等）都需要一种方式来发出分布式的且对创建它的组件来说是很自然的错误。这样，错误可能会经由不同的 JMS 队列，或是在不同线程的队列中进行处理。

解决方案

使用 Spring Integration 对错误通道的支持，既可以隐式使用，也可以通过代码显式处理。

解释说明

Spring Integration 提供了捕获异常并将其发送到所选错误通道的能力。默认情况下，这是个名为 errorChannel 的全局通道。Spring Integration 在默认情况下会向该通道注册一个名为 LoggingHandler 的处理器，该处理器只会记录异常日志和堆栈信息。要想让其发挥作用，你需要告诉消息驱动的通道适配器你要将错误发送给 errorChannel；这可以通过配置错误通道属性来实现。

```
@Bean
public IntegrationFlow jmsInbound(ConnectionFactory connectionFactory) {
    return IntegrationFlows
        .from(Jms.messageDrivenChannelAdapter(connectionFactory).extractPayload(true).
         destination("recipe-15-6").errorChannel("errorChannel"))
        .transform(customerTransformer())
        .handle(customerServiceActivator())
        .get();
}
```

使用自定义处理器来处理异常

当然，还可以让组件订阅来自于该通道的消息以覆盖异常处理行为。你可以创建一个类，当消息进入到 errorChannel 通道时调用该类。

```
@Bean
public IntegrationFlow errorFlow() {
    return IntegrationFlows
        .from("errorChannel")
        .handle(errorHandlingServiceActivator())
        .get();
}
```

上述 Java 代码正是我们所期望的。当然，接收来自于 errorChannel 通道错误消息的组件不必非得是一个服务激活器。这里使用它只是为了方便而已。如下服务激活器代码描述了为 errorChannel 构建处理器的具体方式：

```
package com.apress.springrecipes.springintegration;

import org.slf4j.Logger;
import org.slf4j.LoggerFactory;
import org.springframework.integration.annotation.ServiceActivator;
import org.springframework.messaging.Message;
import org.springframework.messaging.MessagingException;

public class DefaultErrorHandlingServiceActivator {
    private static final Logger logger = LoggerFactory.getLogger(DefaultErrorHandlingService
    Activator.class);
```

```
    @ServiceActivator
    public void handleThrowable(Message<Throwable> errorMessage)
        throws Throwable {
        Throwable throwable = errorMessage.getPayload();
        logger.debug("Message: {}", throwable.getMessage(), throwable);

        if (throwable instanceof MessagingException) {
            Message<?> failedMessage = ((MessagingException) throwable).getFailedMessage();

            if (failedMessage != null) {
                // do something with the original message
            }
        } else {
            // it's something that was thrown in the execution of code in some component you created
        }
    }
}
```

Spring Integration 组件抛出的所有错误都是 MessagingException 的子类。MessagingException 持有一个指向导致错误的原始 Message 的指针，你可以通过它了解更多上下文信息。在该示例中，你使用了比较丑陋的 instanceof。显然，根据异常类型委托给自定义异常处理器是很好的做法。

根据异常类型路由到自定义处理器

有时，我们需要更为具体的错误处理。在如下代码中，我们将一个路由器配置为异常类型路由器，它反过来会监听 errorChannel。接下来，它会根据异常类型来确定哪个通道会接收到结果。

```
@Bean
public ErrorMessageExceptionTypeRouter exceptionTypeRouter() {
    ErrorMessageExceptionTypeRouter router = new ErrorMessageExceptionTypeRouter();
    router.setChannelMapping(MyCustomException.class.getName(), "customExceptionChannel");
    router.setChannelMapping(RuntimeException.class.getName(), "runtimeExceptionChannel");
    router.setChannelMapping(MessageHandlingException.class.getName(),
    "messageHandlingExceptionChannel");
    return router;
}

@Bean
public IntegrationFlow errorFlow() {
    return IntegrationFlows
        .from("errorChannel")
        .route(exceptionTypeRouter())
        .get();
}
```

使用多个错误通道构建解决方案

上述示例对于简单场景来说没有问题，不过不同的集成常常需要不同的错误处理方式，这意味着将所有错误都发给同一个通道最终会导致一个类中充满了 switch 语句，复杂到难以维护。与之相反，更好的做法是有选择地将错误消息路由到最适合的错误通道中。这可以避免集中处理所有的错误。一种方式就是显式指定给定集成的错误应该进入到哪个通道中。如下示例展示了一个组件（服务激活器），它会在接收到消息时添加一个消息头，标识错误通道的名字。Spring Integration 会使用这个头信息并将消息处理过程中遇到的错误转发给该通道。

```
package com.apress.springrecipes.springintegration;

import org.apache.log4j.Logger;
import org.springframework.integration.annotation.ServiceActivator;
import org.springframework.integration.core.Message;
import org.springframework.integration.core.MessageHeaders;
import org.springframework.integration.message.MessageBuilder;
```

```
public class ServiceActivatorThatSpecifiesErrorChannel {
    private static final Logger logger = Logger.getLogger(
        ServiceActivatorThatSpecifiesErrorChannel.class);

    @ServiceActivator
    public Message<?> startIntegrationFlow(Message<?> firstMessage)
        throws Throwable {
        return MessageBuilder.fromMessage(firstMessage).
            setHeaderIfAbsent( MessageHeaders.ERROR_CHANNEL,
                "errorChannelForMySolution").build();
    }
}
```

这样，使用了该组件的集成所产生的所有错误都会被重定向给 customErrorChannel，你可以向其订阅任何组件。

15-7 派生集成控制：分割器与聚合器

问题提出

你想将处理流程从一个组件派生为多个，要么同时执行，要么根据条件汇总为单个结果。

解决方案

可以使用分割器组件（也许还有它的搭档——聚合器组件）来派生与合并流程控制。

解释说明

ESB 的一个根基就是路由。你已经看到了如何将组件链接到一起来创建序列，其中的处理几乎是线性的。一些解决方案需要将一条消息分割为多个组成部分。其中一个原因就是有些问题在本质上是并行的，无须依赖彼此就可以完成。你应该在可能的情况下实现并行的效果。

使用分割器

将大的负载划分为具有单独处理流程的独立消息通常是很有用的。在 Spring Integration 中，这是通过使用分割器组件来实现的。分割器会接收一条输入消息，并询问你组件的用户是谁，以及根据什么来分割 Message<T>：你负责提供分割功能。一旦告诉 Spring Integration 如何分割 Message<T>后，它就会将分割器组件输出通道的每个结果转发出去。在一些场景下，Spring Integration 自带了很有用的分割器，无须再定制。比如说 XPathMessageSplitter 分割器，它会根据 XPath 查询来分割 XML 负载。

下面展示一个示例来说明分割器的价值：一个文本文件有若干数据行，每一行都需要处理。你的目标是将每一行都提交给一个服务进行处理。我们所要做的就是提取出每一行并将其作为新的 Message<T>进行转发。该解决方案的配置如下所示：

```
@Configuration
@EnableIntegration
public class IntegrationConfiguration {

    @Bean
    public CustomerBatchFileSplitter splitter() {
        return new CustomerBatchFileSplitter();
    }

    @Bean
    public CustomerDeletionServiceActivator customerDeletionServiceActivator() {
        return new CustomerDeletionServiceActivator();
    }

    @Bean
    public IntegrationFlow fileSplitAndDelete(@Value("file:${user.home}/customerstoremove/
    new/") File inputDirectory) throws Exception {
```

```
        return IntegrationFlows.from(
            Files.inboundAdapter(inputDirectory).patternFilter("customerstoremove-*.txt"),
            c -> c.poller(Pollers.fixedRate(1, TimeUnit.SECONDS)))
            .split(splitter())
            .handle(customerDeletionServiceActivator())
            .get();
    }
}
```

配置与之前的解决方案大致相同。Java 代码几乎一样，只不过被@Splitter 所注解的方法的返回类型是 java.util.Collection。

```
package com.apress.springrecipes.springintegration;

import org.springframework.integration.annotation.Splitter;

import java.io.File;
import java.io.IOException;
import java.nio.file.Files;
import java.util.Collection;

public class CustomerBatchFileSplitter {

    @Splitter
    public Collection<String> splitAFile(File file) throws IOException {
        System.out.printf("Reading %s....%n", file.getAbsolutePath());
        return Files.readAllLines(file.toPath());
    }
}
```

消息负载以 java.io.File 组件的形式传递进来，其内容会被读取并返回结果（一个集合或是数组值；在该示例中是个 Collection<String>集合）。Spring Integration 会对结果执行一种 foreach，将集合中的每个值从为分割器所配置的输出通道中发送出去。通常会对消息进行分割，这样就可以将每一部分转发到专门的处理中。由于消息更容易管理，因此处理上的要求就简化了很多（在很多不同的架构中都是这样的）。在 map/reduce 解决方案中，任务会被分割，然后并行处理，BPM 系统中的 fork/join 结构可以控制流程并行进行，这样就可以更快得到最终的结果。

使用聚合器

有时，你要做相反的事情：将多条消息合并为一条，创建单个结果并从输出通道中返回。@Aggregator 会收集一系列消息（基于为 Spring Integration 设置的消息间的相关性）并向下游组件发布单条消息。假设你要从系统中的 22 个角色获取到 22 条不同的消息，但不知道消息何时会到来。比如说有公司要进行拍卖，在选择最终的供应商之前，它会收集不同供应商的报价，这与方才说的是很类似的。公司只有在收到所有供应商的报价后才会接受其中一个报价。否则就会出现签署的合同并不是最佳选择的结果。聚合器非常适合于构建这种逻辑。

Spring Integration 有多种方式对到来的消息进行关联。为了确定停止前需要读取多少条消息，它使用了 SequenceSizeCompletionStrategy 类，它会读取一个已知的头信息值（聚合器常常用在分割器之后。这样，默认的头信息值就由分割器提供了，不过也完全可以自行创建头信息参数）。SequenceSizeCompletionStrategy 类会计算需要读取的消息数量，并且会观察相较于期望总数的消息索引（如 3/22）。

对于相关性来说，如果你不知道总数，但却知道要处理的消息会在一个已知的时间内共享一个公共的消息头值，Spring Integration 提供了 HeaderAttributeCorrelationStrategy。在这种方式下，它知道拥有该值的所有消息都来自于同一个分组，其方式与按照你的姓氏将你标识在一个更大的分组内一样。

我们再来看看上述示例。假设文件被分割了（按照行分割，每一行属于一个新的消费者），随后进行处理。你现在想重新聚合消费者，使其同时执行一些清理工作。在该示例中，你使用了默认的完成策略与相关性策略，这样就可以在集成流配置中使用默认的 aggregate()了。结果会被传递给另一个服务激活器，它会打印出一个汇总。

```
package com.apress.springrecipes.springintegration;

import org.springframework.beans.factory.annotation.Value;
import org.springframework.context.annotation.Bean;
import org.springframework.context.annotation.Configuration;
import org.springframework.integration.config.EnableIntegration;
import org.springframework.integration.dsl.IntegrationFlow;
import org.springframework.integration.dsl.IntegrationFlows;
import org.springframework.integration.dsl.Pollers;
import org.springframework.integration.file.dsl.Files;

import java.io.File;
import java.util.concurrent.TimeUnit;

@Configuration
@EnableIntegration
public class IntegrationConfiguration {

    @Bean
    public CustomerBatchFileSplitter splitter() {
        return new CustomerBatchFileSplitter();
    }

    @Bean
    public CustomerDeletionServiceActivator customerDeletionServiceActivator() {
        return new CustomerDeletionServiceActivator();
    }

    @Bean
    public SummaryServiceActivator summaryServiceActivator() {
        return new SummaryServiceActivator();
    }

    @Bean
    public IntegrationFlow fileSplitAndDelete(@Value("file:${user.home}/customerstoremove/
    new/") File inputDirectory) throws Exception {

        return IntegrationFlows.from(
            Files.inboundAdapter(inputDirectory).patternFilter("customerstoremove-*.txt"),
            c -> c.poller(Pollers.fixedRate(1, TimeUnit.SECONDS)))
            .split(splitter())
            .handle(customerDeletionServiceActivator())
            .aggregate()
            .handle(summaryServiceActivator())
            .get();
    }
}
```

SummaryServiceActivator 的 Java 代码相当简单。

```
package com.apress.springrecipes.springintegration;

import org.springframework.integration.annotation.ServiceActivator;

import java.util.Collection;

public class SummaryServiceActivator {

    @ServiceActivator
    public void summary(Collection<Customer> customers) {
        System.out.printf("Removed %s customers.%n", customers.size());
    }
}
```

15-8 使用路由器实现条件路由

问题提出

你想基于一些标准根据条件让消息通过不同的流程。这是 EAI 之于 if/else 分支。

解决方案

可以使用路由器组件，基于断言来调整处理流程。还可以使用路由器向多个订阅者广播消息（就像使用分割器时所做的那样）。

解释说明

借助于路由器，可以指定一个已知的通道列表，让传入的 Message 对象都会通过该列表。这有一些强大的暗示，意味着你可以根据条件来改变处理流程，可以根据需要将 Message 对象转发给更多或更少的通道。有一些方便、默认的路由器可以满足常见需求，比如说基于负载类型的路由（PayloadTypeRouter）以及路由至一个分组或是通道列表（RecipientListRouter）。

比如说，有一个处理管道，它会将信用分高的客户路由到一个服务，将信用分低的客户路由到另一个流程，在这个流程中会对信息进行排队处理，进行人工审计和验证。与往常一样，配置是很直接的。如下示例展示了配置。路由器元素（反过来会将路由逻辑委托给一个类）就是 CustomerCreditScoreRouter。

```
@Bean
public IntegrationFlow fileSplitAndDelete(@Value("file:${user.home}/customerstoimport/new/")
File inputDirectory) throws Exception {

    return IntegrationFlows.from(
        Files.inboundAdapter(inputDirectory).patternFilter("customers-*.txt"), c ->
        c.poller(Pollers.fixedRate(1, TimeUnit.SECONDS)))
        .split(splitter())
        .transform(transformer())
        .<Customer, Boolean>route(c -> c.getCreditScore() > 770,
            m -> m
                .channelMapping(Boolean.TRUE, "safeCustomerChannel")
                .channelMapping(Boolean.FALSE, "riskyCustomerChannel").applySequence(false)
        ).get();
}
```

如果一个类中的方法使用了@Router 注解，那么你也可以使用它。它看起来像是工作流引擎的条件元素，甚至像是 JSF backing-bean 的方法，因为它会将路由逻辑放置到 XML 配置中，使之从代码中剥离出来，这会将决策延迟到运行时。在该示例中，返回的 String 就是 Message 组件应该传递进去的通道名字。

```
package com.apress.springrecipes.springintegration;

import org.springframework.integration.annotation.Router;

public class CustomerCreditScoreRouter {

    @Router
    public String routeByCustomerCreditScore(Customer customer) {
        if (customer.getCreditScore() > 770) {
            return "safeCustomerChannel";
        } else {
            return "riskyCustomerChannel";
        }
    }
}
```

如果不想传递 Message<T>并且要停止处理，那就可以返回 null 而非 String。

15-9 使用 Spring Batch 发起事件

问题提出

一个文件有 100 万行记录。这个文件太大了，无法按照一个事件来处理；更自然的方式是将每一行作为一个事件进行响应。

解决方案

Spring Batch 能够非常好地支持这些类型的解决方案。它可以让你接收一个输入文件或是负载，并且可靠、系统地将其分解为 ESB 能够处理的事件。

解释说明

Spring Integration 支持将文件读取到总线，Spring Batch 支持为数据提供定制化、唯一的端点。然而，就像妈妈总说的那样："你有能力去做这件事并不代表你应该去做。"虽然两者看起来好像存在很多重叠，但实际上是有区别的（还是个很重要的区别）。虽然这两个系统都能够处理文件和消息队列，或是你能够想象到的可以通过代码与之通信的东西，但 Spring Integration 却并不擅长大型负载的处理，这是因为很难将拥有 100 万行的大型文件的处理（需要好几个小时）作为一个事件。这对于 ESB 来说是个巨大的负担。在这种情况下，术语"事件"就失去了它的意义。CSV 文件中的 100 万行记录并不是总线上的一个事件；它是个拥有 100 万行记录的文件，每一行最终都是一个事件。这是个细微的差别。

拥有 100 万行的文件需要分解为更小的事件。Spring Batch 可以做到这一点：它可以让你系统化地进行读取并应用验证，然后跳过（可选的）或是重试不合法的记录。处理可以在 ESB 上开始，比如说 Spring Integration。我们可以联合使用 Spring Batch 与 Spring Integration 来构建真正可伸缩的解耦系统。

多阶段事件驱动架构（Staged Event Driven Architecture，SEDA）是一种能够处理这种问题的架构风格。在 SEDA 中，你将架构组件的负载按照阶段放置到队列中，并且只放置那些下游组件可以处理的负载。换言之，想象一下视频处理。如果你运行一个站点，有 100 万个用户上传视频，并且视频需要转码，但却只有 10 台服务器；如果系统在接收到上传的视频后就对其进行转码，那么系统就会宕掉。转码可能需要几个小时，当系统在工作时会一直占用 CPU 资源。最明智的做法则是先存储文件，然后在容量允许的情况下处理每一个视频。通过这种方式，处理转码的节点的负载就得到了管理。总会有足够的工作让机器处理，但并不会超载。

与之类似，没有处理系统（如 ESB）能够高效地同时处理 100 万条记录。请将大的事件与消息分解为更小粒度的事件。想象这样一个解决方案，它会容纳一些批文件，这些文件代表每小时的销售额。这些批文件会放到 Spring Integration 监控的分区中。Spring Integration 在看到新文件到来时就会开始处理。Spring Integration 会告诉 Spring Batch 文件信息并异步启动 Spring Batch 任务。

Spring Batch 会读取文件，将记录转换为对象，并将输出写到 JMS 主题中，通过键将原始的批信息关联到 JMS 消息上。这需要半天才能完成，不过确实能完成。Spring Integration 完全意识不到半天前它所启动的任务现在完成了，它会逐一处理主题上的消息。对记录的处理现在就开始了。简单的处理会涉及多个组件，从 ESB 开始。

如果执行是个长期的过程，涉及很多角色的长期的会话状态，那么对每个记录的执行就可以交由 BPM 引擎来完成。BPM 引擎会贯通不同的角色与工作列表，并且可以让工作持续很久而非 Spring Integration 适合的毫秒级别时间。在该示例中，我们提到使用 Spring Batch 作为跳板来控制下游组件的负载。在这种情况下，下游组件又是一个 Spring Integration 流程，它接收任务并创建好后经由 BPM 引擎进行处理，这是最终流程开始的地方。Spring Integration 可以使用目录轮询作为触发器来启动批任务并将文件名提供给流程。为了从 Spring Integration 启动任务，Spring Batch 提供了 JobLaunchingMessageHandler 类。该类接收一个 JobLaunchRequest 实例来确定使用哪些参数启动哪个任务。你需要创建一个转换器将传入的 Message<File> 转换为 JobLaunchRequest 实例。

转换器代码如下所示：

```
package com.apress.springrecipes.springintegration;

import org.springframework.batch.core.Job;
import org.springframework.batch.core.JobParametersBuilder;
import org.springframework.batch.integration.launch.JobLaunchRequest;
import org.springframework.integration.annotation.Transformer;
```

```
import java.io.File;

public class FileToJobLaunchRequestTransformer {

    private final Job job;
    private final String fileParameterName;

    public FileToJobLaunchRequestTransformer(Job job, String fileParameterName) {
        this.job=job;
        this.fileParameterName=fileParameterName;
    }

    @Transformer
    public JobLaunchRequest transform(File file) throws Exception {

        JobParametersBuilder builder = new JobParametersBuilder();
        builder.addString(fileParameterName, file.getAbsolutePath());
        return new JobLaunchRequest(job, builder.toJobParameters());
    }
}
```

转换器需要一个 Job 对象和一个 filename 参数进行构建；Spring Batch 任务会使用该参数确定加载哪个文件。传入的消息会在 JobLaunchRequest 中得到转换（使用文件的全名作为参数值）。该请求可用于启动一个批任务。

要想将这一切装配起来，你可以使用如下配置（注意，这里没有提供 Spring Batch 的配置；请参阅第 11 章了解关于配置 Spring Batch 的信息）：

```
package com.apress.springrecipes.springintegration;

import org.springframework.batch.core.Job;
import org.springframework.batch.core.launch.JobLauncher;
import org.springframework.batch.integration.launch.JobLaunchingMessageHandler;
import org.springframework.beans.factory.annotation.Value;
import org.springframework.context.annotation.Bean;
import org.springframework.integration.dsl.IntegrationFlow;
import org.springframework.integration.dsl.IntegrationFlows;
import org.springframework.integration.dsl.Pollers;
import org.springframework.integration.file.dsl.Files;

import java.io.File;
import java.util.concurrent.TimeUnit;

public class IntegrationConfiguration {

    @Bean
    public FileToJobLaunchRequestTransformer transformer(Job job) {
        return new FileToJobLaunchRequestTransformer(job, "filename");
    }

    @Bean
    public JobLaunchingMessageHandler jobLaunchingMessageHandler(JobLauncher jobLauncher)
    {
        return new JobLaunchingMessageHandler(jobLauncher);
    }

    @Bean
    public IntegrationFlow fileToBatchFlow(@Value("file:${user.home}/customerstoimport/
    new/") File directory, FileToJobLaunchRequestTransformer transformer,
    JobLaunchingMessageHandler handler) {
        return IntegrationFlows
            .from(Files.inboundAdapter(directory).patternFilter("customers-*.txt"),
            c -> c.poller(Pollers.fixedRate(10, TimeUnit.SECONDS)))
            .transform(transformer)
```

```
                .handle(handler)
            .get();

    }

}
```

FileToJobLaunchRequestTransformer 的配置方式与 JobLaunchingMessageHandler 一样。这里使用文件入站通道适配器来轮询文件。当检测到文件时，会在通道中放置一条消息。这里配置了一个链来监听该通道。当接收到消息时，先转换消息，然后再传递给 JobLaunchingMessageHandler。

现在，会启动一个批任务来处理文件。典型的任务会使用一个 FlatFileItemReader 来根据 filename 参数实际读取传递过来的文件。使用 JmsItemWriter 基于读取的每一行向主题写入消息。在 Spring Integration 中，JMS 入站通道适配器可用于接收消息并对其进行处理。

15-10　使用网关

问题提出

你想向客户端公开服务接口，同时又不想让客户端知道服务是通过消息中间件实现的。

解决方案

使用网关（该模式来源于 Gregor Hohpe 与 Bobby Woolf 合著的 *Enterprise Integration Patterns* 一书），Spring Integration 对其提供了强有力的支持。

解释说明

网关本身很特别，类似于其他很多模式，但与其他模式还是存在着本质的区别。之前的示例使用适配器实现外部、松耦合的中间件组件之间的通信。这个外部组件可以是任何东西：文件系统、JMS 队列/主题、Twitter 等。

你还知道什么是门面，它将其他组件的功能抽象到了一个简化的接口中，提供细粒度的功能。你可能使用门面构建过面向接口的度假计划，它抽象了使用汽车租赁、酒店预订以及航空公司预订系统等的细节信息。

另一方面，你通过构建网关为系统提供了一个接口，它将客户端与系统中的中间件或是消息隔离开来，这样它们就不再依赖于 JMS 或是 Spring Integration API 了。借助于网关，我们可以对系统输入和输出的编译期进行限制。

这么做存在几个原因。首先是更加整洁。其次，如果要求客户端遵循一个接口，那么它就是提供接口的一个很好的方式。你对中间件的使用就是实现细节了。也许架构的消息中间件可以借助于异步消息实现性能上的提升，不过你不能以精确、显式、外部化的接口为代价来获取这种提升。

该特性（将消息隐藏到 POJO 接口之后的能力）很有趣，也成为其他几个项目的关注点。Lingo（来自于 Codehaus 的项目，后者不再处于活跃开发状态）就提供了特定于 JMS 与 Java EE Connector Architecture（JCA）的这种特性，它原来用于 Java 加密架构（Java Cryptography Architecture），但现在在 Java EE 连接器架构（Connector Architecture）中得到了更为广泛的应用。从那时起，开发者就开始转向了 Apache Camel。

本节将会探索 Spring Integration 对消息网关的核心支持，以及它对于消息交换模式的支持。接下来，我们将会介绍如何完全从客户端接口中移除实现细节。

SimpleMessagingGateway

对网关最基本的支持来自于 Spring Integration 类 SimpleMessagingGateway。该类可以指定请求发向的通道以及接收响应的通道，还可以指定回应发向的通道。这样，你就可以基于既有的消息系统表达进-出模式以及纯进的模式。该类可以根据负载进行处理，这样就可以将你与发送和接收的消息细节信息隔离开来。这已经是一层抽象了。你可以使用 SimpleMessagingGateway 和 Spring Integration 的通道概念实现与文件系统、JMS、邮件以及其他任何系统进行交互，只处理负载和通道即可。还有一些实现能够支持常见的端点，如 Web Service 和 JMS。

我们看看如何使用通用的消息网关。在该示例中，你会向服务激活器发送消息，然后接收响应。我们手工与 SimpleMessageGateway 进行交互，这样就可以见识到它的方便之处了。

```
package com.apress.springrecipes.springintegration;

import org.springframework.context.ConfigurableApplicationContext;
import org.springframework.context.annotation.AnnotationConfigApplicationContext;
import org.springframework.messaging.MessageChannel;

public class Main {
    public static void main(String[] args) {
        ConfigurableApplicationContext ctx =
            new AnnotationConfigApplicationContext(AdditionConfiguration.class);
        MessageChannel request = ctx.getBean("request", MessageChannel.class);
        MessageChannel response = ctx.getBean("response", MessageChannel.class);

        SimpleMessagingGateway msgGateway = new SimpleMessagingGateway();
        msgGateway.setRequestChannel(request);
        msgGateway.setReplyChannel(response);
        msgGateway.setBeanFactory(ctx);
        msgGateway.afterPropertiesSet();
        msgGateway.start();

        Number result = msgGateway.convertSendAndReceive(new Operands(22, 4));

        System.out.printf("Result: %f%n", result.floatValue());

        ctx.close();

    }
}
```

接口很直观。SimpleMessagingGateway 需要一个请求和一个响应通道，它会协调其他事宜。在该示例中，你只是将请求转发给服务激活器，它反过来添加了操作数并将其发送给响应通道。配置信息并不多，因为大多数工作都通过下面这 5 行 Java 代码完成了。

```
package com.apress.springrecipes.springintegration;

import org.springframework.context.annotation.Bean;
import org.springframework.context.annotation.Configuration;
import org.springframework.integration.config.EnableIntegration;
import org.springframework.integration.dsl.IntegrationFlow;
import org.springframework.integration.dsl.IntegrationFlows;

@Configuration
@EnableIntegration
public class AdditionConfiguration {

    @Bean
    public AdditionService additionService() {
        return new AdditionService();
    }

    @Bean
    public IntegrationFlow additionFlow() {

        return IntegrationFlows
            .from("request")
            .handle(additionService(), "add")
            .channel("response")
            .get();
    }
}
```

打破接口依赖

上述示例揭示了背后发生的事情。你只使用了 Spring Integration 接口，并且与不同端点之间的细微差别隔离开来。不过，依然会有很多客户端无法轻松遵循的约束存在。最简单的解决方案就是将消息隐藏在接口后面。我们来构建一个虚拟的酒店预订搜索引擎。搜索酒店可能需要很长时间，理想情况下，处理应该在单独的服务器上进行。理想的解决方案是使用 JMS，因为可以实现主动的消费者模式，只需添加更多的消费者就可以轻松实现扩容。在该示例中，客户端依然会阻塞来等待结果，不过服务器却并不会超载或是处于阻塞状态中。

下面构建两个 Spring Integration 解决方案：一个针对客户端（它会包含网关）；另一个针对服务本身，它位于独立的主机上，只通过已知的消息队列连接到客户端。

首先来看看客户端配置。客户端配置要做的第一件事就是声明一个 ConnectionFactory。接下来声明流程，它从 VacationService 接口的网关开始。网关元素存在的目的只是用来标识组件和接口，指定代理转向的目标并将这些提供给客户端，jms-outbound-gateway 是完成大部分工作的组件。它接收创建的消息，并将其发送给请求 JMS 目的地，设置响应头信息等。最后，声明一个通用的网关元素，它会完成大多数工作。

```
package com.apress.springrecipes.springintegration;

import com.apress.springrecipes.springintegration.myholiday.VacationService;
import org.apache.activemq.ActiveMQConnectionFactory;
import org.springframework.context.annotation.Bean;
import org.springframework.context.annotation.Configuration;
import org.springframework.integration.config.EnableIntegration;
import org.springframework.integration.dsl.IntegrationFlow;
import org.springframework.integration.dsl.IntegrationFlows;
import org.springframework.integration.jms.dsl.Jms;
import org.springframework.jms.connection.CachingConnectionFactory;

import java.util.Arrays;

@Configuration
@EnableIntegration
public class ClientIntegrationContext {

    @Bean
    public CachingConnectionFactory connectionFactory() {
        ActiveMQConnectionFactory connectionFactory =
            new ActiveMQConnectionFactory("tcp://localhost:61616");
        connectionFactory.setTrustAllPackages(true);
        return new CachingConnectionFactory(connectionFactory);
    }

    @Bean
    public IntegrationFlow vacationGatewayFlow() {
        return IntegrationFlows
            .from(VacationService.class)
            .handle(
                Jms.outboundGateway(connectionFactory())
                    .requestDestination("inboundHotelReservationSearchDestination")
                    .replyDestination("outboundHotelReservationSearchResultsDestination"))
            .get();
    }

}
```

要想将 VacationService 作为网关使用，需要为其添加@MessagingGateway 注解，作为入口点的方法需要使用注解@Gateway。

```
package com.apress.springrecipes.springintegration.myholiday;
```

```
import org.springframework.integration.annotation.Gateway;
import org.springframework.integration.annotation.MessagingGateway;

import java.util.List;

@MessagingGateway
public interface VacationService {

    @Gateway
    List<HotelReservation> findHotels(HotelReservationSearch hotelReservationSearch);
}
```

这是客户端面对的接口。由网关组件公开的客户端面对的接口与最终处理消息的服务接口之间没有任何耦合。服务与客户端都使用了该接口，从而简化了名字的使用，而名字则是理解一切的基础。传统的同步远程技术需要服务接口和客户端接口匹配，但该示例并不是这样的。

在该示例中，我们使用两个非常简单的对象进行说明：HotelReservationSearch 与 HotelReservation。这两个对象没什么值得说的；它们只是实现了 Serializable 并包含一些访问器/设值器来填充信息的简单 POJO 而已。

如下的客户端 Java 代码展示了如何将它们连接到一起：

```
package com.apress.springrecipes.springintegration;

import com.apress.springrecipes.springintegration.myholiday.HotelReservation;
import com.apress.springrecipes.springintegration.myholiday.HotelReservationSearch;
import com.apress.springrecipes.springintegration.myholiday.VacationService;
import org.springframework.context.ConfigurableApplicationContext;
import org.springframework.context.annotation.AnnotationConfigApplicationContext;

import java.time.LocalDate;
import java.time.ZoneId;
import java.util.Date;
import java.util.List;

public class Main {
    public static void main(String[] args) throws Throwable {
        // Start server
        ConfigurableApplicationContext serverCtx =
            new AnnotationConfigApplicationContext(ServerIntegrationContext.class);

        // Start client and issue search
        ConfigurableApplicationContext clientCtx =
            new AnnotationConfigApplicationContext(ClientIntegrationContext.class);

        VacationService vacationService = clientCtx.getBean(VacationService.class);
        LocalDate now = LocalDate.now();
        Date start = Date.from(now.plusDays(1).atStartOfDay(ZoneId.systemDefault()).
        toInstant());
        Date stop = Date.from(now.plusDays(8).atStartOfDay(ZoneId.systemDefault()).
        toInstant());
        HotelReservationSearch hotelReservationSearch =
            new HotelReservationSearch(200f, 2, start, stop);
        List<HotelReservation> results = vacationService.findHotels(hotelReservationSearch);

        System.out.printf("Found %s results.%n", results.size());
        results.forEach(r -> System.out.printf("\t%s%n", r));

        serverCtx.close();
        clientCtx.close();
    }
}
```

代码非常简洁！完全看不到 Spring Integration 接口。你发出一个请求，搜索就会开始，当处理完毕后你会拿到结果。服务实现很有趣，原因并不是你添加的东西，而是缺少的东西。

```
package com.apress.springrecipes.springintegration;
```

```
import com.apress.springrecipes.springintegration.myholiday.VacationServiceImpl;
import org.apache.activemq.ActiveMQConnectionFactory;
import org.springframework.context.annotation.Bean;
import org.springframework.context.annotation.Configuration;
import org.springframework.integration.config.EnableIntegration;
import org.springframework.integration.dsl.IntegrationFlow;
import org.springframework.integration.dsl.IntegrationFlows;
import org.springframework.integration.jms.dsl.Jms;
import org.springframework.jms.connection.CachingConnectionFactory;

import java.util.Arrays;

@Configuration
@EnableIntegration
public class ServerIntegrationContext {

    @Bean
    public CachingConnectionFactory connectionFactory() {
        ActiveMQConnectionFactory connectionFactory =
            new ActiveMQConnectionFactory("tcp://localhost:61616");
        connectionFactory.setTrustAllPackages(true);
        return new CachingConnectionFactory(connectionFactory);
    }

    @Bean
    public VacationServiceImpl vacationService() {
        return new VacationServiceImpl();
    }

    @Bean
    public IntegrationFlow serverIntegrationFlow() {
        return IntegrationFlows.from(
            Jms.inboundGateway(connectionFactory())
                .destination("inboundHotelReservationSearchDestination"))
            .handle(vacationService())
            .get();
    }
}
```

这里定义了一个入站 JMS 网关。来自入站 JMS 网关的消息会被放到通道中，通道中的消息会被转发给服务激活器，这是意料之中的。服务激活器会进行实际的处理。有趣的是这里并未提及响应通道，无论是服务激活器还是入站 JMS 网关都如此。服务激活器会查找响应通道，但找不到，因此它会使用入站 JMS 网关组件所创建的响应通道，它反过来会根据入站 JMS 消息中的头信息元数据来创建响应通道。这样，一切就都可以正常工作，无须指定。

这个实现是个接口实现，简单但没什么实际用途。

```
package com.apress.springrecipes.springintegration.myholiday;

import org.springframework.integration.annotation.ServiceActivator;

import javax.annotation.PostConstruct;
import java.util.Arrays;
import java.util.List;

public class VacationServiceImpl implements VacationService {
    private List<HotelReservation> hotelReservations;

    @PostConstruct
    public void afterPropertiesSet() throws Exception {
        hotelReservations = Arrays.asList(
                new HotelReservation("Bilton", 243.200F),
                new HotelReservation("East Western", 75.0F),
```

```
                new HotelReservation("Thairfield Inn", 70F),
                new HotelReservation("Park In The Inn", 200.00F));
    }

    @ServiceActivator
    public List<HotelReservation> findHotels(HotelReservationSearch searchMsg) {
        try {
            Thread.sleep(1000);
        } catch (Throwable th) {
        }

        return this.hotelReservations;
    }
}
```

小结

本章介绍了如何通过 Spring Integration（构建在 Spring 框架之上的类似于 ESB 的框架）构建集成解决方案。我们学习了企业应用集成的核心概念，以及如何处理一些集成场景，包括 JMS 与文件轮询。

下一章将会介绍 Spring 在测试领域所提供的能力。

第 16 章

Spring 测试

本章将会介绍用于测试 Java 应用的基础技术以及 Spring 框架所提供的测试支持特性。这些特性可以简化你的测试任务，促成更好的应用设计。一般来说，使用 Spring 框架与依赖注入模式所开发的应用都很容易测试。

在软件开发中，测试是确保质量的一项关键活动。测试有很多类型，包括单元测试、集成测试、功能测试、系统测试、性能测试与验收测试。Spring 对测试的支持聚焦在单元与集成测试上，不过它也对其他类型的测试提供了支持。测试可以手工进行，也能自动化完成。不过，由于自动化测试可以在开发过程的不同阶段重复且持续运行，因此是高度推荐的，特别是在敏捷开发过程中。Spring 框架是个敏捷框架，非常适合这些类型的过程。

Java 平台上有很多测试框架。目前，JUnit 与 TestNG 是最流行的。JUnit 拥有悠久的历史，在 Java 社区中有大量的用户群组。TestNG 则是另一个流行的 Java 测试框架。相较于 JUnit 来说，TestNG 提供了额外的强大特性，比如说测试组、依赖测试方法和数据驱动的测试。

Spring 的测试支持特性是由 Spring TestContext 框架提供的，它通过如下概念抽象了底层的测试框架。

- 测试上下文：封装了测试执行的上下文，包括应用上下文、测试类、当前的测试实例、当前的测试方法以及当前的测试异常。
- 测试上下文管理器：管理着测试的测试上下文并在预先定义的测试执行点上触发测试执行，包括在准备测试实例时、在执行测试方法前（在任何特定于框架的初始化方法前），以及在执行测试方法后（在任何特定于框架的清理方法后）。
- 测试执行监听器：定义了监听器接口，通过实现该接口，你可以监听测试执行事件。TestContext 框架为常见的测试特性提供了几个测试执行监听器，不过也可以自行创建。

Spring 为 JUnit 和 TestNG 提供了便捷的 TestContext 支持类，特定的测试执行监听器已经提前注册好了。你只需继承这些支持类就可以使用 TestContext 框架了，无须了解框架的细节信息。

学习完本章后，你将会掌握测试的基本概念与技术，同时学会流行的 Java 测试框架 JUnit 和 TestNG。你还可以使用 Spring TestContext 框架创建单元测试与集成测试。

16-1 使用 JUnit 与 TestNG 创建测试

问题提出

你想为 Java 应用创建自动化测试，这样它们就可以重复运行来确保应用的正确性了。

解决方案

Java 平台上最流行的单元测试框架是 JUnit 与 TestNG。无论使用哪个，你都可以为测试方法添加@Test 注解，这样任何 public 方法都可以作为测试用例来运行。

解释说明

假设你要为银行开发一个系统。为了确保系统的质量，你需要测试每一部分。首先，我们考虑一个利息计算器，其接口定义代码如下所示：

```
package com.apress.springrecipes.bank;

public interface InterestCalculator {

    void setRate(double rate);
    double calculate(double amount, double year);
}
```

每个利息计算器都需要设置一个固定利率。现在，可以使用一个简单的利息公式来实现该计算器，代码如下所示：

```
package com.apress.springrecipes.bank;

public class SimpleInterestCalculator implements InterestCalculator {

    private double rate;

    public void setRate(double rate) {
        this.rate = rate;
    }

    public double calculate(double amount, double year) {
        if (amount < 0 || year < 0) {
            throw new IllegalArgumentException("Amount or year must be positive");
        }
        return amount * year * rate;
    }
}
```

接下来，我们使用流行的测试框架 JUnit 与 TestNG（版本 5）来测试这个简单的利息计算器。

提示：通常情况下，测试及其目标类应该位于相同的包下，不过测试的源文件是存储在单独的目录下的（如 test），与其他类的源文件不在同一个目录中（如 src）。

使用 JUnit 进行测试

测试用例只不过是使用了@Test 注解的 public 方法而已。为了创建数据，你可以使用@Before 来注解一个方法。为了清理资源，你可以使用@After 来注解一个方法。还可以使用@BeforeClass 或@AfterClass 来注解 public static 方法，这样它们就会在该类中所有测试用例运行前后各运行一次。

你需要直接调用 org.junit.Assert 类中声明的静态断言方法。不过，可以通过静态 import 声明来导入所有断言方法。可以创建如下 JUnit 测试用例来测试这个简单的利息计算器。

注意：要想编译和运行针对 JUnit 创建的测试用例，需要将 JUnit 放到 CLASSPATH 中。如果使用 Maven，请向项目中添加如下依赖：

```
<dependency>
    <groupId>junit</groupId>
    <artifactId>junit</artifactId>
    <version>4.12</version>
</dependency>
```

如果使用的是 Gradle，请添加如下依赖：

```
testCompile 'junit:junit:4.12'
```

```
package com.apress.springrecipes.bank;

import static org.junit.Assert.*;
```

```
import org.junit.Before;
import org.junit.Test;

public class SimpleInterestCalculatorJUnit4Tests {

    private InterestCalculator interestCalculator;

    @Before
    public void init() {
        interestCalculator = new SimpleInterestCalculator();
        interestCalculator.setRate(0.05);
    }

    @Test
    public void calculate() {
        double interest = interestCalculator.calculate(10000, 2);
        assertEquals(interest, 1000.0, 0);
    }

    @Test(expected = IllegalArgumentException.class)
    public void illegalCalculate() {
        interestCalculator.calculate(-10000, 2);
    }
}
```

JUnit 提供了一个强大的特性，使得你可以期望测试用例中抛出的异常。只需在@Test 注解的 expected 属性中指定异常类型即可。

使用 TestNG 进行测试

TestNG 测试看起来与 JUnit 测试很相似，只不过需要使用 TestNG 框架所定义的类和注解类型。

■ **注意**：要想编译和运行针对 TestNG 创建的测试用例，需要将 TestNG 放到 CLASSPATH 中。如果使用 Maven，请向项目中添加如下依赖：

```
<dependency>
    <groupId>org.testng</groupId>
    <artifactId>testng</artifactId>
    <version>6.11</version>
</dependency>
```

如果使用的是 Gradle，请添加如下依赖：

```
testCompile 'org.testng:testng:6.11'
```

```
package com.apress.springrecipes.bank;

import static org.testng.Assert.*;

import org.testng.annotations.BeforeMethod;
import org.testng.annotations.Test;

public class SimpleInterestCalculatorTestNGTests {

    private InterestCalculator interestCalculator;

    @BeforeMethod
    public void init() {
        interestCalculator = new SimpleInterestCalculator();
        interestCalculator.setRate(0.05);
    }

    @Test
```

```
    public void calculate() {
        double interest = interestCalculator.calculate(10000, 2);
        assertEquals(interest, 1000.0);
    }

    @Test(expectedExceptions = IllegalArgumentException.class)
    public void illegalCalculate() {
        interestCalculator.calculate(-10000, 2);
    }
}
```

提示：如果你使用的是 Eclipse，那么可以从 TestNG 官网下载并安装 TestNG Eclipse 插件以在 Eclipse 中运行 TestNG 测试。如果所有测试均通过，你会看到一个绿条，否则就是红条。

TestNG 的一个强大特性就是针对数据驱动测试提供了内建的支持。TestNG 能够清晰地将测试数据与测试逻辑分开，这样就可以针对不同数据集运行一个测试方法多次。在 TestNG 中，测试数据集是由数据提供者来提供的，它们是使用了@DataProvider 注解的方法。

```
package com.apress.springrecipes.bank;

import org.testng.annotations.BeforeMethod;
import org.testng.annotations.DataProvider;
import org.testng.annotations.Test;

import static org.testng.Assert.assertEquals;

public class SimpleInterestCalculatorTestNGTests {

    private InterestCalculator interestCalculator;

    @BeforeMethod
    public void init() {
        interestCalculator = new SimpleInterestCalculator();
        interestCalculator.setRate(0.05);
    }

    @DataProvider(name = "legal")
    public Object[][] createLegalInterestParameters() {
        return new Object[][]{new Object[]{10000, 2, 1000.0}};
    }

    @DataProvider(name = "illegal")
    public Object[][] createIllegalInterestParameters() {
        return new Object[][]{
            new Object[]{-10000, 2},
            new Object[]{10000, -2},
            new Object[]{-10000, -2}
        };
    }

    @Test(dataProvider = "legal")
    public void calculate(double amount, double year, double result) {
        double interest = interestCalculator.calculate(amount, year);
        assertEquals(interest, result);
    }

    @Test(
        dataProvider = "illegal",
        expectedExceptions = IllegalArgumentException.class)
    public void illegalCalculate(double amount, double year) {
        interestCalculator.calculate(amount, year);
    }
}
```

如果使用 TestNG 运行上述测试，那么 calculate()方法只会执行一次，而 illegalCalculate() 方法则会执行 3 次，因为 illegal 数据提供者返回了 3 个数据集。

16-2　创建单元测试与集成测试

问题提出

常见的测试技术是单独测试应用的每个模块，然后再组合起来测试。你想按照这种方式来测试你的 Java 应用。

解决方案

单元测试用于测试单个程序单元。在面向对象的语言中，单元通常指的是类或是方法。单元测试的范围是单个单元，不过在真实世界中，大多数单元并非孤立的。它们通常需要与其他单元协作来完成任务。当测试的单元依赖于其他单元时，可以使用的常见技术就是通过桩（stub）和模拟（mock）对象来模拟单元依赖，它们都可以降低由依赖所导致的单元测试的复杂度。

桩指的是模拟依赖对象的对象，它拥有测试所需的最少数量的方法。方法是以一种预先确定好的方式来实现的，通常使用硬编码的数据。桩还会为测试公开一些方法以验证桩的内部状态。与桩相反，模拟对象通常知道其方法在测试中是如何被调用的。接下来，模拟对象会针对期望被调用的方法对实际调用的方法进行验证。在 Java 中有一些库可以帮助我们创建模拟对象，比如说 Mockito、EasyMock 和 jMock。桩对象与模拟对象之间的主要差别在于桩对象通常用于状态验证，而模拟对象则用于行为验证。

与之相反，集成测试用于对几个单元的组合进行整体上的测试。它们会测试单元之间的集成与交互是否正确。其中每个单元都应该已经通过了单元测试，因此集成测试通常会在单元测试之后进行。

最后，注意到使用接口与实现分离原则以及依赖注入模式开发的应用是很容易测试的，对于单元测试与集成测试来说均如此。这是因为该原则与模式可以降低应用中不同单元之间的耦合。

解释说明

首先，我们来探索如何为单个类编写单元测试，接下来扩展到使用模拟与桩协作者对类进行测试。最后，我们会介绍如何编写集成测试。

为孤立的类创建单元测试

这个银行系统的核心功能应该围绕着客户账户进行设计。首先，创建如下领域类 Account 并自定义 equals()与 hashCode()方法：

```
package com.apress.springrecipes.bank;

public class Account {

    private String accountNo;
    private double balance;

    // Constructors, Getters and Setters
    ...

    @Override
    public boolean equals(Object o) {
        if (this == o) return true;
        if (o == null || getClass() != o.getClass()) return false;
        Account account = (Account) o;
        return Objects.equals(this.accountNo, account.accountNo);
    }

    @Override
    public int hashCode() {
        return Objects.hash(this.accountNo);
    }
}
```

接下来，定义如下 DAO 接口以在银行系统的持久层中持久化账户对象：

```
package com.apress.springrecipes.bank;

public interface AccountDao {

    public void createAccount(Account account);
    public void updateAccount(Account account);
    public void removeAccount(Account account);
    public Account findAccount(String accountNo);
}
```

为了揭示单元测试的概念，我们使用一个 map 来存储账户对象以实现该接口。AccountNotFoundException 与 DuplicateAccountException 类是 RuntimeException 的子类，你需要自行创建。

```
package com.apress.springrecipes.bank;

import java.util.Collections;
import java.util.HashMap;
import java.util.Map;

public class InMemoryAccountDao implements AccountDao {

    private Map<String, Account> accounts;

    public InMemoryAccountDao() {
        accounts = Collections.synchronizedMap(new HashMap<String, Account>());
    }

    public boolean accountExists(String accountNo) {
        return accounts.containsKey(accountNo);
    }

    public void createAccount(Account account) {
        if (accountExists(account.getAccountNo())) {
            throw new DuplicateAccountException();
        }
        accounts.put(account.getAccountNo(), account);
    }

    public void updateAccount(Account account) {
        if (!accountExists(account.getAccountNo())) {
            throw new AccountNotFoundException();
        }
        accounts.put(account.getAccountNo(), account);
    }

    public void removeAccount(Account account) {
        if (!accountExists(account.getAccountNo())) {
            throw new AccountNotFoundException();
        }
        accounts.remove(account.getAccountNo());
    }

    public Account findAccount(String accountNo) {
        Account account = accounts.get(accountNo);
        if (account == null) {
            throw new AccountNotFoundException();
        }
        return account;
    }
}
```

显然，这个简单的 DAO 实现并不支持事务。不过，为了确保线程安全，你可以将存储账户的 map 包装为一个同步 map，这样就只能串行访问了。

现在，我们使用 JUnit 为这个 DAO 实现创建单元测试。由于这个类并不直接依赖于其他类，因此是很容易测试的。为了确保这个类在异常情况与正常情况下都能正确运行，还应该为其创建异常测试用例。通常，异常测试用例会期望抛出一个异常。

```
package com.apress.springrecipes.bank;

import static org.junit.Assert.*;

import org.junit.Before;
import org.junit.Test;

public class InMemoryAccountDaoTests {

    private static final String EXISTING_ACCOUNT_NO = "1234";
    private static final String NEW_ACCOUNT_NO = "5678";

    private Account existingAccount;
    private Account newAccount;
    private InMemoryAccountDao accountDao;

    @Before
    public void init() {
        existingAccount = new Account(EXISTING_ACCOUNT_NO, 100);
        newAccount = new Account(NEW_ACCOUNT_NO, 200);
        accountDao = new InMemoryAccountDao();
        accountDao.createAccount(existingAccount);
    }

    @Test
    public void accountExists() {
        assertTrue(accountDao.accountExists(EXISTING_ACCOUNT_NO));
        assertFalse(accountDao.accountExists(NEW_ACCOUNT_NO));
    }

    @Test
    public void createNewAccount() {
        accountDao.createAccount(newAccount);
        assertEquals(accountDao.findAccount(NEW_ACCOUNT_NO), newAccount);
    }

    @Test(expected = DuplicateAccountException.class)
    public void createDuplicateAccount() {
        accountDao.createAccount(existingAccount);
    }

    @Test
    public void updateExistedAccount() {
        existingAccount.setBalance(150);
        accountDao.updateAccount(existingAccount);
        assertEquals(accountDao.findAccount(EXISTING_ACCOUNT_NO), existingAccount);
    }

    @Test(expected = AccountNotFoundException.class)
    public void updateNotExistedAccount() {
        accountDao.updateAccount(newAccount);
    }

    @Test
    public void removeExistedAccount() {
        accountDao.removeAccount(existingAccount);
        assertFalse(accountDao.accountExists(EXISTING_ACCOUNT_NO));
    }
```

```
    @Test(expected = AccountNotFoundException.class)
    public void removeNotExistedAccount() {
        accountDao.removeAccount(newAccount);
    }

    @Test
    public void findExistedAccount() {
        Account account = accountDao.findAccount(EXISTING_ACCOUNT_NO);
        assertEquals(account, existingAccount);
    }

    @Test(expected = AccountNotFoundException.class)
    public void findNotExistedAccount() {
        accountDao.findAccount(NEW_ACCOUNT_NO);
    }
}
```

使用桩与模拟对象为依赖类创建单元测试

测试没有依赖的类是很容易的，因为不需要考虑依赖的使用方式，也不需要创建依赖。不过，如果一个类依赖于其他类或服务（如数据库服务和网络服务）的结果，那么测试就会变得有点困难了。比如说，我们考虑服务层中的如下 AccountService 接口：

```
package com.apress.springrecipes.bank;

public interface AccountService {

    void createAccount(String accountNo);
    void removeAccount(String accountNo);
    void deposit(String accountNo, double amount);
    void withdraw(String accountNo, double amount);
    double getBalance(String accountNo);
}
```

该服务接口的实现依赖于持久层中的 AccountDao 对象来持久化账户对象。InsufficientBalanceException 也是你需要创建的 RuntimeException 子类。

```
package com.apress.springrecipes.bank;

public class AccountServiceImpl implements AccountService {

    private AccountDao accountDao;

    public AccountServiceImpl(AccountDao accountDao) {
        this.accountDao = accountDao;
    }

    public void createAccount(String accountNo) {
        accountDao.createAccount(new Account(accountNo, 0));
    }

    public void removeAccount(String accountNo) {
        Account account = accountDao.findAccount(accountNo);
        accountDao.removeAccount(account);
    }

    public void deposit(String accountNo, double amount) {
        Account account = accountDao.findAccount(accountNo);
        account.setBalance(account.getBalance() + amount);
        accountDao.updateAccount(account);
    }

    public void withdraw(String accountNo, double amount) {
        Account account = accountDao.findAccount(accountNo);
```

```
        if (account.getBalance() < amount) {
            throw new InsufficientBalanceException();
        }
        account.setBalance(account.getBalance() - amount);
        accountDao.updateAccount(account);
    }

    public double getBalance(String accountNo) {
        return accountDao.findAccount(accountNo).getBalance();
    }
}
```

在单元测试中，降低依赖所导致的复杂性的常见技术是使用桩。桩必须要实现与目标对象相同的接口，这样它就可以代替目标对象了。比如说，你可以为 AccountDao 创建一个桩，它会存储单个客户账户并且只实现 findAccount()与 updateAccount()方法，因为 deposit()与 withdraw()需要用到它们。

```
package com.apress.springrecipes.bank;

import static org.junit.Assert.*;

import org.junit.Before;
import org.junit.Test;

public class AccountServiceImplStubTests {

    private static final String TEST_ACCOUNT_NO = "1234";
    private AccountDaoStub accountDaoStub;
    private AccountService accountService;

    private class AccountDaoStub implements AccountDao {

        private String accountNo;
        private double balance;

        public void createAccount(Account account) {}
        public void removeAccount(Account account) {}

        public Account findAccount(String accountNo) {
            return new Account(this.accountNo, this.balance);
        }

        public void updateAccount(Account account) {
            this.accountNo = account.getAccountNo();
            this.balance = account.getBalance();
        }
    }

    @Before
    public void init() {
        accountDaoStub = new AccountDaoStub();
        accountDaoStub.accountNo = TEST_ACCOUNT_NO;
        accountDaoStub.balance = 100;
        accountService = new AccountServiceImpl(accountDaoStub);
    }

    @Test
    public void deposit() {
        accountService.deposit(TEST_ACCOUNT_NO, 50);
        assertEquals(accountDaoStub.accountNo, TEST_ACCOUNT_NO);
        assertEquals(accountDaoStub.balance, 150, 0);
    }

    @Test
    public void withdrawWithSufficientBalance() {
```

```
        accountService.withdraw(TEST_ACCOUNT_NO, 50);
        assertEquals(accountDaoStub.accountNo, TEST_ACCOUNT_NO);
        assertEquals(accountDaoStub.balance, 50, 0);
    }

    @Test(expected = InsufficientBalanceException.class)
    public void withdrawWithInsufficientBalance() {
        accountService.withdraw(TEST_ACCOUNT_NO, 150);
    }
}
```

不过，自己编写桩需要很多代码。更具效率的技术是使用模拟对象。Mockito 可以动态创建模拟对象，以记录/回放的机制工作。

注意：要想使用 Mockito 进行测试，需要将其添加到 CLASSPATH 中。如果使用 Maven，请向项目中添加如下依赖：

```
<dependency>
    <groupId>org.mockito</groupId>
    <artifactId>mockito-core</artifactId>
    <version>2.7.20</version>
    <scope>test</scope>
</dependency>
```

如果使用的是 Gradle，请添加如下依赖：

```
testCompile 'org.mockito:mockito-core:2.7.20'
```

```
package com.apress.springrecipes.bank;

import org.junit.Before;
import org.junit.Test;

import static org.mockito.Mockito.*;

public class AccountServiceImplMockTests {

    private static final String TEST_ACCOUNT_NO = "1234";

    private AccountDao accountDao;
    private AccountService accountService;

    @Before
    public void init() {
        accountDao = mock(AccountDao.class);
        accountService = new AccountServiceImpl(accountDao);
    }

    @Test
    public void deposit() {
        // Setup
        Account account = new Account(TEST_ACCOUNT_NO, 100);
        when(accountDao.findAccount(TEST_ACCOUNT_NO)).thenReturn(account);

        // Execute
        accountService.deposit(TEST_ACCOUNT_NO, 50);

        // Verify
        verify(accountDao, times(1)).findAccount(any(String.class));
        verify(accountDao, times(1)).updateAccount(account);

    }

    @Test
```

```
    public void withdrawWithSufficientBalance() {
        // Setup
        Account account = new Account(TEST_ACCOUNT_NO, 100);
        when(accountDao.findAccount(TEST_ACCOUNT_NO)).thenReturn(account);

        // Execute
        accountService.withdraw(TEST_ACCOUNT_NO, 50);

        // Verify
        verify(accountDao, times(1)).findAccount(any(String.class));
        verify(accountDao, times(1)).updateAccount(account);

    }

    @Test(expected = InsufficientBalanceException.class)
    public void testWithdrawWithInsufficientBalance() {
        // Setup
        Account account = new Account(TEST_ACCOUNT_NO, 100);
        when(accountDao.findAccount(TEST_ACCOUNT_NO)).thenReturn(account);

        // Execute
        accountService.withdraw(TEST_ACCOUNT_NO, 150);
    }
}
```

借助于 Mockito，你可以为任意接口或类动态创建模拟对象。这种模拟可以为方法调用设定某些行为，你可以通过它有选择地验证某些事情是否发生。在测试中，你期望 findAccount 方法会返回某个 Account 对象。为此，你使用了 Mockito.when 方法，接下来既可以返回一个值，也可以抛出异常，还可以通过 Answer 进行更为详尽的处理。模拟的默认行为是返回 null。你可以使用 Mockito.verify 方法对应该发生的动作有选择地进行验证。你要确保 findAccount 方法被调用了，并且账户得到了更新。

创建集成测试

集成测试用于对几个单元进行整体测试，确保单元集成没问题且能正常交互。比如说，你可以创建集成测试，使用 InMemoryAccountDao 作为 DAO 实现来测试 AccountServiceImpl。

```
package com.apress.springrecipes.bank;

import static org.junit.Assert.*;

import org.junit.After;
import org.junit.Before;
import org.junit.Test;

public class AccountServiceTests {

    private static final String TEST_ACCOUNT_NO = "1234";
    private AccountService accountService;

    @Before
    public void init() {
        accountService = new AccountServiceImpl(new InMemoryAccountDao());
        accountService.createAccount(TEST_ACCOUNT_NO);
        accountService.deposit(TEST_ACCOUNT_NO, 100);
    }

    @Test
    public void deposit() {
        accountService.deposit(TEST_ACCOUNT_NO, 50);
        assertEquals(accountService.getBalance(TEST_ACCOUNT_NO), 150, 0);
    }
```

```
    @Test
    public void withDraw() {
        accountService.withdraw(TEST_ACCOUNT_NO, 50);
        assertEquals(accountService.getBalance(TEST_ACCOUNT_NO), 50, 0);
    }

    @After
    public void cleanup() {
        accountService.removeAccount(TEST_ACCOUNT_NO);
    }
}
```

16-3　为 Spring MVC 控制器实现单元测试

问题提出

在 Web 应用中，你想测试使用 Spring MVC 框架开发的 Web 控制器。

解决方案

Spring MVC 控制器是由 DispatcherServlet 调用的，并且传递进一个 HTTP 请求对象和一个 HTTP 响应对象。在处理完请求后，控制器会将其返回给 DispatcherServlet 来渲染视图。对 Spring MVC 控制器进行单元测试的主要挑战在于，就像其他 Web 应用框架中的 Web 控制器一样，需要在单元测试环境中模拟 HTTP 请求对象和响应对象。幸好，Spring 通过为 Servlet API 提供一套模拟对象（包括 MockHttpServletRequest、MockHttpServletResponse 与 MockHttpSession）来支持 Web 控制器的测试。

要想测试 Spring MVC 控制器的输出，你需要检查返回给 DispatcherServlet 的对象是否是正确的。Spring 还提供了一套断言功能来检查对象的内容。

解释说明

在这个银行系统中，假设你要为银行职员开发一个 Web 界面，用于输入账号数字与存款金额。你使用 Spring MVC 提供的技术创建了一个名为 DepositController 的控制器。

```
package com.apress.springrecipes.bank.web;

import org.springframework.beans.factory.annotation.Autowired;
import org.springframework.stereotype.Controller;
import org.springframework.ui.ModelMap;
import org.springframework.web.bind.annotation.RequestMapping;
import org.springframework.web.bind.annotation.RequestParam;

@Controller
public class DepositController {

    private AccountService accountService;

    @Autowired
    public DepositController(AccountService accountService) {
        this.accountService = accountService;
    }

    @RequestMapping("/deposit.do")
    public String deposit(
        @RequestParam("accountNo") String accountNo,
        @RequestParam("amount") double amount,
        ModelMap model) {
        accountService.deposit(accountNo, amount);
        model.addAttribute("accountNo", accountNo);
        model.addAttribute("balance", accountService.getBalance(accountNo));
        return "success";
    }
}
```

由于这个控制器没有使用到Servlet API，因此很容易测试。可以像简单的Java类一样测试它。

```
package com.apress.springrecipes.bank.web;

import static org.junit.Assert.*;

import com.apress.springrecipes.bank.AccountService;
import org.junit.Before;
import org.junit.Test;
import org.mockito.Mockito;
import org.springframework.ui.ModelMap;

public class DepositControllerTests {

    private static final String TEST_ACCOUNT_NO = "1234";
    private static final double TEST_AMOUNT = 50;
    private AccountService accountService;
    private DepositController depositController;

    @Before
    public void init() {
        accountService = Mockito.mock(AccountService.class);
        depositController = new DepositController(accountService);
    }

    @Test
    public void deposit() {
        //Setup
        Mockito.when(accountService.getBalance(TEST_ACCOUNT_NO)).thenReturn(150.0);
        ModelMap model = new ModelMap();

        //Execute
        String viewName =
            depositController.deposit(TEST_ACCOUNT_NO, TEST_AMOUNT, model);

        assertEquals(viewName, "success");
        assertEquals(model.get("accountNo"), TEST_ACCOUNT_NO);
        assertEquals(model.get("balance"), 150.0);
    }
}
```

16-4　在集成测试中管理应用上下文

问题提出

在为 Spring 应用创建集成测试时，你需要访问声明在应用上下文中的 bean。如果没有 Spring 的测试支持，你需要在测试的初始化方法中手工加载应用上下文，在 JUnit 中就是使用了@Before 或@BeforeClass 注解的方法。不过，当初始化方法在每个测试方法或测试类前被调用时，同样的应用上下文可能会被加载多次。在拥有很多 bean 的大型应用中，加载应用上下文可能需要很长时间，这会导致测试运行很慢。

解决方案

Spring 的测试支持设施可以帮你管理测试的应用上下文，包括从一个或多个 bean 配置文件中加载上下文，跨越多个测试执行获取上下文等。在单个 JVM 中，应用上下文会跨越所有测试被缓存起来，使用配置文件位置作为键。这样，测试的运行速度就会很快，无须重新加载同样的应用上下文多次。

TestContext 框架提供了一些测试执行监听器，它们默认就会注册，如表 16-1 所示。

表 16-1　　默认的测试执行监听器

TestExecutionListener	说明
DependencyInjectionTestExecutionListener	它会将依赖（包括托管的应用上下文）注入到测试中
DirtiesContextTestExecutionListener，DirtiesContextBeforeModesTestExecutionListener	处理@DirtiesContext 注解，必要时重新加载应用上下文
TransactionalTestExecutionListener	处理测试用例中的@Transactional 注解，在测试结束时进行回滚
SqlScriptsTestExecutionListener	检测测试中的@Sql 注解，在测试开始前执行 SQL
ServletTestExecutionListener	在检测到@WebAppConfiguration 注解时处理 Web 应用上下文的加载

要想让 TestContext 框架管理应用上下文，测试类需要在内部与测试上下文管理器进行集成。为了方便，TestContext 框架提供了支持类来做到这一点，如表 16-2 所示。这些类集成了测试上下文管理器并实现了 ApplicationContextAware 接口，这样它们就可以通过受保护的字段 applicationContext 访问托管的应用上下文了。

表 16-2　　针对上下文管理的 TestContext 支持类

测试框架	TestContext 支持类
Junit	AbstractJUnit4SpringContextTests
TestNG	AbstractTestNGSpringContextTests

测试类只需继承所用测试框架中相应的 TestContext 支持类即可。

这些 TestContext 支持类只启用了 DependencyInjectionTestExecutionListener、DirtiesContextTestExecution-Listener 与 ServletTestExecutionListener。

如果使用 JUnit 或是 TestNG，那就可以自行集成测试类与测试上下文管理器，并直接实现 ApplicationContextAware 接口，无须继承 TestContext 支持类。通过这种方式，测试类就不会绑定到 TestContext 框架类层次体系上，因此可以继承自己的父类。在 JUnit 中，只需使用测试运行器 SpringRunner 来运行测试就可以实现与测试上下文管理器的集成。不过在 TestNG 中，需要手工集成测试上下文管理器。

解释说明

首先，在配置类中声明一个 AccountService 实例和一个 AccountDao 实例。接下来为它们创建集成测试。

```
package com.apress.springrecipes.bank.config;

import com.apress.springrecipes.bank.AccountServiceImpl;
import com.apress.springrecipes.bank.InMemoryAccountDao;
import org.springframework.context.annotation.Bean;
import org.springframework.context.annotation.Configuration;

@Configuration
public class BankConfiguration {

    @Bean
    public InMemoryAccountDao accountDao() {
        return new InMemoryAccountDao();
    }

    @Bean
    public AccountServiceImpl accountService() {
        return new AccountServiceImpl(accountDao());
    }
}
```

在 JUnit 中使用 TestContext 框架访问上下文

如果使用 JUnit 通过 TestContext 框架创建测试，那么有两种方式来访问托管应用上下文。第一种是实现

ApplicationContextAware 接口或是在 ApplicationContext 类型的字段上使用@Autowired 注解。对于这种方式来说，你需要显式指定一个特定于 Spring 的测试运行器来运行你的测试 SpringRunner。可以在类级别的@RunWith 注解中指定。

```
package com.apress.springrecipes.bank;

import com.apress.springrecipes.bank.config.BankConfiguration;
import org.junit.After;
import org.junit.Before;
import org.junit.Test;
import org.junit.runner.RunWith;
import org.springframework.beans.factory.annotation.Autowired;
import org.springframework.test.context.ContextConfiguration;
import org.springframework.test.context.junit4.SpringRunner;

import static org.junit.Assert.assertEquals;

@RunWith(SpringRunner.class)
@ContextConfiguration(classes = BankConfiguration.class)
public class AccountServiceJUnit4ContextTests implements ApplicationContextAware {

    private static final String TEST_ACCOUNT_NO = "1234";
    private ApplicationContext applicationContext;
    private AccountService accountService;

    @Override
    public void setApplicationContext(ApplicationContext applicationContext) throws
    BeansException {
        this.applicationContext=applicationContext;
    }

    @Before
    public void init() {
        accountService = applicationContext.getBean(AccountService.class);
        accountService.createAccount(TEST_ACCOUNT_NO);
        accountService.deposit(TEST_ACCOUNT_NO, 100);
    }

    @Test
    public void deposit() {
        accountService.deposit(TEST_ACCOUNT_NO, 50);
        assertEquals(accountService.getBalance(TEST_ACCOUNT_NO), 150, 0);
    }

    @Test
    public void withDraw() {
        accountService.withdraw(TEST_ACCOUNT_NO, 50);
        assertEquals(accountService.getBalance(TEST_ACCOUNT_NO), 50, 0);
    }

    @After
    public void cleanup() {
        accountService.removeAccount(TEST_ACCOUNT_NO);
    }

}
```

可以在类级别注解@ContextConfiguration 的 classes 属性中指定配置类。当使用基于 XML 的配置时，可以使用 locations 属性。如果没有指定任何测试配置，那么 TestContext 就会尝试进行检测。它首先会从与测试类相同的包中加载名字为测试类名加上-context.xml 后缀的文件（如 AccountServiceJUnit4Testscontext.xml）。接下来，它会扫描测试类，寻找使用了@Configuration 注解的 public static 内部类。如果检测到文件或是类，它们就会用于加载测试配置。

默认情况下，应用上下文会被缓存起来，每个测试方法可以重用它；不过，如果想在特定的测试方法执行完毕后重新加载它，可以对测试方法使用@DirtiesContext 注解，这样应用上下文就会重新加载供接下来的测试方法使用。

访问托管应用上下文的第二种方式是继承特定于 JUnit 的 TestContext 支持类：AbstractJUnit4SpringContextTests。这个类实现了 ApplicationContextAware 接口，这样就可以通过受保护的字段 applicationContext 访问托管应用上下文了。不过，首先需要删除私有字段 applicationContext 及其 setter 方法。注意，如果继承了这个支持类，那就无须在@RunWith 注解中指定 SpringRunner 了，因为该注解会从父类中继承下来。

```
package com.après.springrecipes.bank;
...
import org.springframework.test.context.ContextConfiguration;
import org.springframework.test.context.junit4.AbstractJUnit4SpringContextTests;

@ContextConfiguration(classes = BankConfiguration.class)
public class AccountServiceJUnit4ContextTests extends AbstractJUnit4SpringContextTests {

    private static final String TEST_ACCOUNT_NO = "1234";
    private AccountService accountService;

    @Before
    public void init() {
        accountService = applicationContext.getBean(AccountService.class);
        accountService.createAccount(TEST_ACCOUNT_NO);
        accountService.deposit(TEST_ACCOUNT_NO, 100);
    }
    ...
}
```

在 TestNG 中使用 TestContext 框架访问上下文

要想在 TestNG 中使用 TestContext 框架访问托管应用上下文，可以继承 TestContext 支持类 AbstractTestNGSpringContextTests。该类也实现了 ApplicationContextAware 接口。

```
package com.apress.springrecipes.bank;

import com.apress.springrecipes.bank.config.BankConfiguration;
import org.springframework.test.context.ContextConfiguration;
import org.springframework.test.context.testng.AbstractTestNGSpringContextTests;
import org.testng.annotations.AfterMethod;
import org.testng.annotations.BeforeMethod;
import org.testng.annotations.Test;

import static org.testng.Assert.assertEquals;

@ContextConfiguration(classes = BankConfiguration.class)
public class AccountServiceTestNGContextTests extends AbstractTestNGSpringContextTests {

    private static final String TEST_ACCOUNT_NO = "1234";
    private AccountService accountService;

    @BeforeMethod
    public void init() {
        accountService = applicationContext.getBean(AccountService.class);
        accountService.createAccount(TEST_ACCOUNT_NO);
        accountService.deposit(TEST_ACCOUNT_NO, 100);
    }

    @Test
    public void deposit() {
        accountService.deposit(TEST_ACCOUNT_NO, 50);
```

```
        assertEquals(accountService.getBalance(TEST_ACCOUNT_NO), 150, 0);
    }

    @Test
    public void withDraw() {
        accountService.withdraw(TEST_ACCOUNT_NO, 50);
        assertEquals(accountService.getBalance(TEST_ACCOUNT_NO), 50, 0);
    }

    @AfterMethod
    public void cleanup() {
        accountService.removeAccount(TEST_ACCOUNT_NO);
    }

}
```

如果不想让 TestNG 测试类继承 TestContext 支持类，那么可以实现 ApplicationContextAware 接口，就像在 JUnit 中所做的那样。不过，你需要自行集成测试上下文管理器。请参阅 AbstractTestNGSpringContextTests 源代码了解详情。

16-5 向集成测试注入测试构件

问题提出

Spring 应用集成测试的测试构件几乎都是声明在应用上下文中的 bean。你希望通过依赖注入让 Spring 能够自动注入测试构件，这样就无须手工从应用上下文中获取它们了。

解决方案

Spring 的测试支持设施可以自动将托管应用上下文中的 beans 注入到测试中作为测试构件。你只需为测试的 setter 方法或是字段添加 Spring 的@Autowired 注解或是 JSR-250 的@Resource 注解就可以自动注入构件。对于@Autowired 来说，构件是根据类型注入的；对于@Resource 来说，构件是根据名字注入的。

解释说明

我们将会使用 JUnit 与 TestNG 来展示如何注入测试构件。

在 JUnit 中使用 TestContext 框架注入测试构件

在使用 TestContext 框架创建测试时，可以通过为字段或 setter 方法应用@Autowired 或@Resource 注解从托管应用上下文中注入测试构件。在 JUnit 中，可以将 SpringRunner 指定为测试运行器，无须继承支持类。

```
package com.apress.springrecipes.bank;
...
import org.springframework.beans.factory.annotation.Autowired;
import org.springframework.test.context.ContextConfiguration;
import org.springframework.test.context.junit4.SpringRunner;

@RunWith(SpringRunner.class)
@ContextConfiguration(classes = BankConfiguration.class)
public class AccountServiceJUnit4ContextTests {

    private static final String TEST_ACCOUNT_NO = "1234";

    @Autowired
    private AccountService accountService;

    @Before
    public void init() {
        accountService.createAccount(TEST_ACCOUNT_NO);
        accountService.deposit(TEST_ACCOUNT_NO, 100);
    }
    ...
}
```

如果对测试的字段或是 setter 方法应用了@Autowired 注解，那么它会使用根据类型自动装配进行注入。可以在@Qualifier 注解中提供 bean 的名字来进一步指定自动装配的候选 bean。不过，如果想让字段或是 setter 方法根据名字进行自动装配，那么可以使用@Resource 注解。

通过继承 TestContext 支持类 AbstractJUnit4SpringContextTests，也可以从托管应用上下文中注入测试构件。在这种情况下，无须为测试指定 SpringRunner，因为它会从父类中继承下来。

```
package com.apress.springrecipes.bank;
...
import org.springframework.beans.factory.annotation.Autowired;
import org.springframework.test.context.ContextConfiguration;
import org.springframework.test.context.junit4.AbstractJUnit4SpringContextTests;

@ContextConfiguration(classes = BankConfiguration.class)
public class AccountServiceJUnit4ContextTests extends AbstractJUnit4SpringContextTests {

    private static final String TEST_ACCOUNT_NO = "1234";

    @Autowired
    private AccountService accountService;
    ...
}
```

在 TestNG 中使用 TestContext 框架注入测试构件

在 TestNG 中，可以通过继承 TestContext 支持类 AbstractTestNGSpringContextTests 从托管应用上下文中注入测试构件。

```
package com.apress.springrecipes.bank;
...
import org.springframework.beans.factory.annotation.Autowired;
import org.springframework.test.context.ContextConfiguration;
import org.springframework.test.context.testng.AbstractTestNGSpringContextTests;

@ContextConfiguration(classes = BankConfiguration.class)
public class AccountServiceTestNGContextTests extends AbstractTestNGSpringContextTests {

    private static final String TEST_ACCOUNT_NO = "1234";

    @Autowired
    private AccountService accountService;

    @BeforeMethod
    public void init() {
        accountService.createAccount(TEST_ACCOUNT_NO);
        accountService.deposit(TEST_ACCOUNT_NO, 100);
    }
    ...
}
```

16-6　在集成测试中管理事务

问题提出

在为访问数据库的应用创建集成测试时，通常需要在初始化方法中准备测试数据。在每个测试方法运行后，它可能会修改了数据库中的数据。因此，还需要清理数据库以确保接下来的测试方法会在一个一致的状态下运行。这样，你需要处理很多数据库清理任务。

解决方案

Spring 的测试支持设施可以为每个测试方法创建和回滚事务，这样在一个测试方法中所做的变更就不会影响到下一个。此外，你也无须再处理清理任务来清理数据库了。

TestContext 框架提供了一个与事务管理器关联的测试执行监听器。如果没有显式自行指定的话，那么默认情况下它会被注册到测试上下文管理器上。

TransactionalTestExecutionListener 会处理类级别与方法级别上的@Transactional 注解，让方法自动运行在事务中。

测试类可以继承与所用测试框架对应的 TestContext 支持类，如表 16-3 所示，从而让测试方法运行在事务中。这些类与测试上下文管理器进行了集成，并在类级别上启用了@Transactional 注解。注意，bean 配置文件中也需要事务管理器。

表 16-3　　用于事务管理的 TestContext 支持类

测试框架	TestContext 支持类
Junit	AbstractTransactionalJUnit4SpringContextTests
TestNG	AbstractTransactionalTestNGSpringContextTests

除了 DependencyInjectionTestExecutionListener 与 DirtiesContextTestExecutionListener 之外，这些 TestContext 支持类还启用了 TransactionalTestExecutionListener 与 SqlScriptsTestExecutionListener。

在 JUnit 与 TestNG 中，只需在类级别或方法级别上使用注解@Transactional 就可以让测试方法运行在事务中，无须继承 TestContext 支持类。不过，要想集成测试上下文管理器，你需要使用测试运行器 SpringRunner 运行 JUnit 测试，对于 TestNG 测试来说则需要手工操作。

解释说明

我们将银行系统的账户存储到关系型数据库中。可以选择支持事务的任何兼容于 JDBC 的数据库引擎，然后执行如下 SQL 语句来创建 ACCOUNT 表。出于测试的目的，我们使用了内存数据库 H2。

```
CREATE TABLE ACCOUNT (
    ACCOUNT_NO     VARCHAR(10)    NOT NULL,
    BALANCE        DOUBLE         NOT NULL,
    PRIMARY KEY (ACCOUNT_NO)
);
```

接下来，创建一个新的 DAO 实现，它会使用 JDBC 访问数据库。可以使用 JdbcTemplate 来简化操作。

```
package com.apress.springrecipes.bank;

import org.springframework.jdbc.core.support.JdbcDaoSupport;

public class JdbcAccountDao extends JdbcDaoSupport implements AccountDao {

    public void createAccount(Account account) {
        String sql = "INSERT INTO ACCOUNT (ACCOUNT_NO, BALANCE) VALUES (?, ?)";
        getJdbcTemplate().update(
            sql, account.getAccountNo(), account.getBalance());
    }

    public void updateAccount(Account account) {
        String sql = "UPDATE ACCOUNT SET BALANCE = ? WHERE ACCOUNT_NO = ?";
        getJdbcTemplate().update(
            sql, account.getBalance(), account.getAccountNo());
    }

    public void removeAccount(Account account) {
        String sql = "DELETE FROM ACCOUNT WHERE ACCOUNT_NO = ?";
        getJdbcTemplate().update(sql, account.getAccountNo());
    }

    public Account findAccount(String accountNo) {
        String sql = "SELECT BALANCE FROM ACCOUNT WHERE ACCOUNT_NO = ?";
        double balance =
            getJdbcTemplate().queryForObject(sql, Double.class, accountNo);
        return new Account(accountNo, balance);
    }
}
```

在创建集成测试来测试 AccountService 实例（它使用该 DAO 持久化账户对象）前，你需要在配置类中将该 DAO 替换为 InMemoryAccountDao 并配置目标数据源。

注意： 要想使用 H2，需要将其作为依赖添加到类路径中：

```
<dependency>
    <groupId>com.h2database:</groupId>
    <artifactId>h2</artifactId>
    <version>1.4.194</version>
</dependency>
```

如果使用的是 Gradle，请添加如下依赖：

```
testCompile 'com.h2database:h2:1.4.194'
```

```
@Configuration
public class BankConfiguration {

    @Bean
    public DataSource dataSource() {
        DriverManagerDataSource dataSource = new DriverManagerDataSource();
        dataSource.setUrl("jdbc:h2:mem:bank-testing");
        dataSource.setUsername("sa");
        dataSource.setPassword("");
        return dataSource;
    }

    @Bean
    public AccountDao accountDao() {
        JdbcAccountDao accountDao = new JdbcAccountDao();
        accountDao.setDataSource(dataSource());
        return accountDao;
    }

    @Bean
    public AccountService accountService() {
        return new AccountServiceImpl(accountDao());
    }
}
```

在 JUnit 中使用 TestContext 框架管理事务

在使用 TestContext 框架创建测试时，可以通过在类级别或方法级别上使用@Transactional 注解让测试方法运行在事务中。在 JUnit 中，可以为测试类指定 SpringRunner，这样就无须继承支持类了。

```
package com.apress.springrecipes.bank;
...
import org.springframework.beans.factory.annotation.Autowired;
import org.springframework.test.context.ContextConfiguration;
import org.springframework.test.context.junit4.SpringRunner;
import org.springframework.transaction.annotation.Transactional;

@RunWith(SpringRunner.class)
@ContextConfiguration(classes = BankConfiguration.class)
@Transactional
public class AccountServiceJUnit4ContextTests {

    private static final String TEST_ACCOUNT_NO = "1234";

    @Autowired
    private AccountService accountService;

    @Before
```

```
        public void init() {
            accountService.createAccount(TEST_ACCOUNT_NO);
            accountService.deposit(TEST_ACCOUNT_NO, 100);
        }

        // Don't need cleanup() anymore
        ...
}
```

如果对测试类使用@Transactional 注解，那么其所有测试方法都将运行在事务中。另一种方式是对单个方法而非整个类应用@Transactional 注解。

默认情况下，测试方法的事务会在方法执行结束后回滚。可以通过禁用@TransactionConfiguration 的 defaultRollback 属性来改变该行为，@TransactionConfiguration 注解需要在类级别上使用。此外，可以在方法级别上使用@Rollback 注解来覆盖类级别的回滚行为，它需要一个 Boolean 值。

注意：被@Before 或@After 注解修饰的方法会与测试方法运行在相同的事务中。如果有方法需要在事务前后执行初始化或清理任务，那就需要使用@BeforeTransaction 或@AfterTransaction 注解。

最后，还需要在 bean 配置文件中配置一个事务管理器。默认情况下会使用类型为 PlatformTransactionManager 的 bean，不过也可以在@TransactionConfiguration 注解的 transactionManager 属性中通过名字来指定其他的事务管理器。

```
@Bean
public DataSourceTransactionManager transactionManager(DataSource dataSource) {
    return new DataSourceTransactionManager(dataSource);
}
```

在 JUnit 中，管理测试方法事务的另一种方式是继承传统的 TestContext 支持类 AbstractTransactionalJUnit4SpringContextTests，它在类级别上开启了@Transactional，这样就无须再次开启。通过继承这个支持类，我们就无须再为测试指定 SpringRunner 了，因为它会从父类中继承下来。

```
package com.apress.springrecipes.bank;
...
import org.springframework.test.context.ContextConfiguration;
import org.springframework.test.context.junit4.
AbstractTransactionalJUnit4SpringContextTests;

@ContextConfiguration(classes = BankConfiguration.class)
public class AccountServiceJUnit4ContextTests extends
AbstractTransactionalJUnit4SpringContextTests {
    ...
}
```

在 TestNG 中使用 TestContext 框架管理事务

要想创建在事务中运行的 TestNG 测试，测试类可以继承 TestContext 支持类 AbstractTransactionalTestNGSpringContextTests，使其方法运行在事务中。

```
package com.apress.springrecipes.bank;
...
import org.springframework.beans.factory.annotation.Autowired;
import org.springframework.test.context.ContextConfiguration;
import org.springframework.test.context.testng.
AbstractTransactionalTestNGSpringContextTests;

@ContextConfiguration(classes = BankConfiguration.class)
public class AccountServiceTestNGContextTests extends
    AbstractTransactionalTestNGSpringContextTests {

    private static final String TEST_ACCOUNT_NO = "1234";

    @Autowired
```

```
        private AccountService accountService;

        @BeforeMethod
        public void init() {
            accountService.createAccount(TEST_ACCOUNT_NO);
            accountService.deposit(TEST_ACCOUNT_NO, 100);
        }

        // Don't need cleanup() anymore
       ...
    }
```

16-7 在集成测试中访问数据库

问题提出

在为访问数据库，特别是使用了 ORM 框架的应用创建集成测试时，你可能想要直接访问数据库，以准备测试数据并在测试方法运行后验证数据。

解决方案

Spring 的测试支持设施可以为你创建并提供 JDBC 模板以在测试中执行数据库相关的任务。测试类可以继承一个事务性 TestContext 支持类来访问事先创建好的 JdbcTemplate 实例。这些类还需要在 bean 配置文件中配置好数据源和事务管理器。

解释说明

在使用 TestContext 框架创建测试时，你可以继承对应的 TestContext 支持类以通过一个 protected 字段来使用 JdbcTemplate 实例。对于 JUnit 来说，该类是 AbstractTransactionalJUnit4SpringContextTests，它提供了类似的便捷方法来计算表中的行数、从表中删除行，以及执行 SQL 脚本。

```
package com.apress.springrecipes.bank;
...
import org.springframework.test.context.ContextConfiguration;
import org.springframework.test.context.junit4.
AbstractTransactionalJUnit4SpringContextTests;

@ContextConfiguration(classes = BankConfiguration.class)
public class AccountServiceJUnit4ContextTests extends AbstractTransactionalJUnit4Spring
ContextTests {
    ...
    @Before
    public void init() {
        executeSqlScript("classpath:/bank.sql",true);
        jdbcTemplate.update(
            "INSERT INTO ACCOUNT (ACCOUNT_NO, BALANCE) VALUES (?, ?)",
            TEST_ACCOUNT_NO, 100);
    }

    @Test
    public void deposit() {
        accountService.deposit(TEST_ACCOUNT_NO, 50);
        double balance = jdbcTemplate.queryForObject(
            "SELECT BALANCE FROM ACCOUNT WHERE ACCOUNT_NO = ?",
            Double.class, TEST_ACCOUNT_NO);
        assertEquals(balance, 150.0, 0);
    }

    @Test
    public void withDraw() {
        accountService.withdraw(TEST_ACCOUNT_NO, 50);
        double balance = jdbcTemplate.queryForObject(
            "SELECT BALANCE FROM ACCOUNT WHERE ACCOUNT_NO = ?",
            Double.class, TEST_ACCOUNT_NO);
```

```
        assertEquals(balance, 50.0, 0);
    }
}
```

相较于使用 executeSqlScript 方法，还可以将@Sql 注解放到类或是测试方法上来执行 SQL 或是脚本。

```
@ContextConfiguration(classes = BankConfiguration.class)
@Sql(scripts="classpath:/bank.sql")
public class AccountServiceJUnit4ContextTests extends
AbstractTransactionalJUnit4SpringContextTests {

    private static final String TEST_ACCOUNT_NO = "1234";

    @Autowired
    private AccountService accountService;

    @Before
    public void init() {
        jdbcTemplate.update(
            "INSERT INTO ACCOUNT (ACCOUNT_NO, BALANCE) VALUES (?, ?)",
            TEST_ACCOUNT_NO, 100);
    }
}
```

借助于@Sql 方法，可以执行脚本，脚本可以在 scripts 属性中指定，也可以直接在注解的 statements 属性中写入 SQL 语句。最后，可以指定何时执行指定的指令，比如是在测试方法执行前还是执行后。可以在类或方法上添加多个@Sql 注解，这样就可以在测试运行前后执行语句了。

在 TestNG 中，可以继承 AbstractTransactionalTestNGSpringContextTests 来使用 JdbcTemplate 实例。

```
package com.apress.springrecipes.bank;
...
import org.springframework.beans.factory.annotation.Autowired;
import org.springframework.test.context.ContextConfiguration;
import org.springframework.test.context.testng.
AbstractTransactionalTestNGSpringContextTests;

@ContextConfiguration(classes = BankConfiguration.class)
public class AccountServiceTestNGContextTests extends
AbstractTransactionalTestNGSpringContextTests {
    ...
    @BeforeMethod
    public void init() {
        executeSqlScript("classpath:/bank.sql",true);
        jdbcTemplate.update(
            "INSERT INTO ACCOUNT (ACCOUNT_NO, BALANCE) VALUES (?, ?)",
            TEST_ACCOUNT_NO, 100);
    }

    @Test
    public void deposit() {
        accountService.deposit(TEST_ACCOUNT_NO, 50);
        double balance = jdbcTemplate.queryForObject(
            "SELECT BALANCE FROM ACCOUNT WHERE ACCOUNT_NO = ?",
            Double.class, TEST_ACCOUNT_NO);
        assertEquals(balance, 150, 0);
    }

    @Test
    public void withDraw() {
        accountService.withdraw(TEST_ACCOUNT_NO, 50);
        double balance = jdbcTemplate.queryForObject(
            "SELECT BALANCE FROM ACCOUNT WHERE ACCOUNT_NO = ?",
            Double.class, TEST_ACCOUNT_NO);
        assertEquals(balance, 50, 0);
    }
}
```

16-8 使用 Spring 常见的测试注解

问题提出

你常常需要手工实现一些常见的测试任务，比如说期望抛出异常、重复执行一个测试方法多次、确保测试方法会在特定的时间周期内完成等。

解决方案

Spring 的测试支持提供了一套常见的测试注解来简化测试的创建。这些注解是 Spring 专有的，不过独立于底层的测试框架。表 16-4 列出的注解对于常见的测试任务来说很有帮助。不过，它们只能用于 JUnit。

表 16-4 Spring 的测试注解

注解	说明
@Repeat	表示测试方法会运行多次，运行的次数是通过注解值来指定的
@Timed	表示测试方法必须要在指定的时间周期（单位是毫秒）内完成。否则，测试就会失败。注意，时间周期包含了测试方法的重复执行以及初始化和清理方法
@IfProfileValue	表示测试方法只能运行在指定的测试环境中。只有当实际的 profile 值与指定的 profile 值匹配时，测试方法才会运行。还可以指定多个值，这样当任何值匹配时，测试方法都会执行。默认情况下会使用 SystemProfileValueSource 来获取系统属性作为 profile 值，不过也可以创建自己的 ProfileValueSource 实现并在 @ProfileValueSourceConfiguration 注解中指定

可以通过继承一个 TestContext 支持类来使用 Spring 的测试注解。如果没有继承支持类但却使用测试运行器 SpringRunner 来运行 JUnit 测试，那也可以使用这些注解。

解释说明

在使用 TestContext 框架创建 JUnit 测试时，如果使用 SpringRunner 运行测试或是继承了 JUnitTestContext 支持类，那就可以使用 Spring 的测试注解。

```
package com.apress.springrecipes.bank;
...
import org.springframework.test.annotation.Repeat;
import org.springframework.test.annotation.Timed;
import org.springframework.test.context.ContextConfiguration;
import org.springframework.test.context.junit4.
AbstractTransactionalJUnit4SpringContextTests;

@ContextConfiguration(locations = "/beans.xml")
public class AccountServiceJUnit4ContextTests extends
AbstractTransactionalJUnit4SpringContextTests {
    ...
    @Test
    @Timed(millis = 1000)
    public void deposit() {
        ...
    }
    @Test
    @Repeat(5)
    public void withDraw() {
        ...
    }
}
```

16-9 为 Spring MVC 控制器实现集成测试

问题提出

在 Web 应用中，你想对使用 Spring MVC 框架开发的 Web 控制器进行集成测试。

解决方案

Spring MVC 控制器是由 DispatcherServlet 调用的，同时会传递一个 HTTP 请求对象和一个 HTTP 响应对象。在处理完请求后，控制器会将其返回给 DispatcherServlet 来渲染视图。对 Spring MVC 控制器（以及其他 Web 应用框架中的 Web 控制器）进行集成测试的主要挑战在于要在单元测试环境中模拟 HTTP 请求和响应对象，还要为单元测试创建模拟环境。幸好，Spring 在测试支持中提供了 mock MVC 的功能。这可以让我们轻松创建一个模拟的 servlet 环境。

Spring Test Mock MVC 会根据配置创建一个 WebApplicationContext。接下来，可以使用 MockMvc API 模拟 HTTP 请求并验证结果。

解释说明

在银行应用中，你想要对 DepositController 进行集成测试。在测试前，你需要创建一个配置类来配置 Web 相关的 bean。

```
package com.apress.springrecipes.bank.web.config;

import org.springframework.context.annotation.Bean;
import org.springframework.context.annotation.ComponentScan;
import org.springframework.context.annotation.Configuration;
import org.springframework.web.servlet.ViewResolver;
import org.springframework.web.servlet.config.annotation.EnableWebMvc;
import org.springframework.web.servlet.view.InternalResourceViewResolver;

@Configuration
@EnableWebMvc
@ComponentScan(basePackages = "com.apress.springrecipes.bank.web")
public class BankWebConfiguration {

    @Bean
    public ViewResolver viewResolver() {
        InternalResourceViewResolver viewResolver = new InternalResourceViewResolver();
        viewResolver.setPrefix("/WEB-INF/views/");
        viewResolver.setSuffix(".jsp");
        return viewResolver;

    }
}
```

上述配置通过使用@EnableWebMvc 注解启用了基于注解的控制器；接下来，通过使用@ComponentScan 注解可以自动获取被@Controller 注解修饰的 bean。最后，InternalResourceViewResolver 会将视图名转换为 URL，正常情况下它会由浏览器渲染，这是你在控制器中验证的内容。

在基于 Web 的配置就绪后，现在可以开始创建集成测试了。单元测试需要加载 BankWebConfiguration 类，还需要使用@WebAppConfiguration 注解来告诉 TestContext 框架你需要的是 WebApplicationContext 而非常规的 ApplicationContext。

使用 JUnit 对 Spring MVC 控制器进行集成测试

在 JUnit 中，继承一个父类是最简单的，在该示例中我们继承了 AbstractTransactionalJUnit4SpringContextTests，因为你想插入一些测试数据并在测试完成后回滚。

```
package com.apress.springrecipes.bank.web;

import com.apress.springrecipes.bank.config.BankConfiguration;
import com.apress.springrecipes.bank.web.config.BankWebConfiguration;
import org.junit.Before;
import org.junit.Test;
import org.springframework.beans.factory.annotation.Autowired;
import org.springframework.test.context.ContextConfiguration;
import org.springframework.test.context.junit4.AbstractTransactionalJUnit4Spring
ContextTests;
```

```
import org.springframework.test.context.web.WebAppConfiguration;
import org.springframework.test.web.servlet.MockMvc;
import org.springframework.test.web.servlet.setup.MockMvcBuilders;
import org.springframework.web.context.WebApplicationContext;

import static org.springframework.test.web.servlet.request.MockMvcRequestBuilders.get;
import static org.springframework.test.web.servlet.result.MockMvcResultHandlers.print;
import static org.springframework.test.web.servlet.result.MockMvcResultMatchers.forwardedUrl;
import static org.springframework.test.web.servlet.result.MockMvcResultMatchers.status;

@ContextConfiguration(classes= { BankWebConfiguration.class, BankConfiguration.class})
@WebAppConfiguration
public class DepositControllerJUnit4ContextTests extends
AbstractTransactionalJUnit4SpringContextTests {

    private static final String ACCOUNT_PARAM = "accountNo";
    private static final String AMOUNT_PARAM = "amount";

    private static final String TEST_ACCOUNT_NO = "1234";
    private static final String TEST_AMOUNT = "50.0";

    @Autowired
    private WebApplicationContext webApplicationContext;

    private MockMvc mockMvc;

    @Before
    public void init() {
        executeSqlScript("classpath:/bank.sql", true);
        jdbcTemplate.update(
            "INSERT INTO ACCOUNT (ACCOUNT_NO, BALANCE) VALUES (?, ?)",
            TEST_ACCOUNT_NO, 100);
        mockMvc = MockMvcBuilders.webAppContextSetup(webApplicationContext).build();

    }

    @Test
    public void deposit() throws Exception {
        mockMvc.perform(
            get("/deposit.do")
                .param(ACCOUNT_PARAM, TEST_ACCOUNT_NO)
                .param(AMOUNT_PARAM, TEST_AMOUNT))
            .andDo(print())
            .andExpect(forwardedUrl("/WEB-INF/views/success.jsp"))
            .andExpect(status().isOk());
    }
}
```

在 init 方法中，你通过使用便捷的 MockMvcBuilders 准备了 MockMvc 对象。借助于工厂方法 webAppContextSetup，你可以使用已经加载好的 WebApplicationContext 来初始化 MockMvc 对象。基本上，MockMvc 对象会模拟 DispatcherServlet 的行为，这在基于 Spring MVC 的应用中会用到。它会使用传进来的 WebApplicationContext 来配置处理器映射和视图解析策略，还会应用配置好的拦截器。

这里还创建了一个测试账户，这样就可以使用它了。

在 deposit 测试方法中，初始化的 MockMvc 对象用于模拟发向/deposit.do URL 的请求，它有两个参数，分别是 accountNo 与 amount。MockMvcRequestBuilders.get 工厂方法会生成一个 RequestBuilder 实例，它会传递给 MockMvc.perform 方法。

perform 方法返回一个 ResultActions 对象，它可用于执行断言并对返回结果执行某些动作。测试方法会通过 andDo(print())打印出创建的请求和返回的响应信息，这对于测试的调试来说很有帮助。最后，有两个断言来验证一切如期望的一样。DepositController 返回 success 作为视图名，根据 ViewResolver 的配置，它应该转发至/WEB-INF/views/success.jsp。请求的返回码应该是 200（OK），这可以通过 status().isOk()或 status().is(200)来测试。

使用 TestNG 对 Spring MVC 控制器进行集成测试

Spring Mock MVC 还可与 TestNG 搭配使用来继承恰当的父类 AbstractTransactionalTestNGSpringContextTests 并添加@WebAppConfiguration 注解。

```
@ContextConfiguration(classes= { BankWebConfiguration.class, BankConfiguration.class})
@WebAppConfiguration
public class DepositControllerTestNGContextTests
    extends AbstractTransactionalTestNGSpringContextTests {

    @BeforeMethod
    public void init() {
        executeSqlScript("classpath:/bank.sql", true);
        jdbcTemplate.update(
            "INSERT INTO ACCOUNT (ACCOUNT_NO, BALANCE) VALUES (?, ?)",
            TEST_ACCOUNT_NO, 100);
        mockMvc = MockMvcBuilders.webAppContextSetup(webApplicationContext).build();
    }

}
```

16-10 为 REST 客户端编写集成测试

问题提出

你想为基于 RestTemplate 的客户端编写集成测试。

解决方案

在为基于 REST 的客户端编写集成测试时，你不想依赖于外部服务的存在与否。可以使用模拟服务器来编写集成测试，使其返回期望的结果而非调用真实的端点。

解释说明

在银行工作时，你需要验证人们输入的账号；可以实现自己的验证或是重用既有的验证。你打算实现一个 IBAN 验证服务，它会用到 http://openiban.com 提供的 API。

首先，编写接口，定义契约。

```
package com.apress.springrecipes.bank.web;

public interface IBANValidationClient {

    IBANValidationResult validate(String iban);
}
```

IBANValidationResult 包含了调用验证端点的结果。

```
package com.apress.springrecipes.bank.web;

import java.util.ArrayList;
import java.util.HashMap;
import java.util.List;
import java.util.Map;

public class IBANValidationResult {

    private boolean valid;
    private List<String> messages = new ArrayList<>();
    private String iban;

    private Map<String, String> bankData = new HashMap<>();

    public boolean isValid() {
```

```
        return valid;
    }

    public void setValid(boolean valid) {
        this.valid = valid;
    }

    public List<String> getMessages() {
        return messages;
    }

    public void setMessages(List<String> messages) {
        this.messages = messages;
    }

    public String getIban() {
        return iban;
    }

    public void setIban(String iban) {
        this.iban = iban;
    }

    public Map<String, String> getBankData() {
        return bankData;
    }

    public void setBankData(Map<String, String> bankData) {
        this.bankData = bankData;
    }

    @Override
    public String toString() {
        return "IBANValidationResult [" +
               "valid=" + valid +
               ", messages=" + messages +
               ", iban='" + iban + '\'' +
               ", bankData=" + bankData +
               ']';
    }
}
```

接下来，编写 OpenIBANValidationClient 类，它会使用 RestTemplate 实例与 API 通信。为了能够轻松访问 RestTemplate 实例，这里继承了 RestGatewaySupport。

```
package com.apress.springrecipes.bank.web;

import org.springframework.stereotype.Service;
import org.springframework.web.client.support.RestGatewaySupport;

@Service
public class OpenIBANValidationClient extends RestGatewaySupport implements
IBANValidationClient {

    private static final String URL_TEMPLATE = "https://openiban.com/validate/{IBAN_NUMBER}?
    getBIC=true&validateBankCode=true";

    @Override
    public IBANValidationResult validate(String iban) {

        return getRestTemplate().getForObject(URL_TEMPLATE, IBANValidationResult.class, iban);
    }
}
```

接下来，创建一个测试，它会为 OpenIBANValidationClient 类构建一个 MockRestServiceServer 类，对其

进行配置，使之返回 JSON 格式的针对期望请求的具体结果。

```
package com.apress.springrecipes.bank.web;

import com.apress.springrecipes.bank.config.BankConfiguration;
import org.junit.Before;
import org.junit.Test;
import org.junit.runner.RunWith;
import org.springframework.beans.factory.annotation.Autowired;
import org.springframework.core.io.ClassPathResource;
import org.springframework.http.MediaType;
import org.springframework.test.context.ContextConfiguration;
import org.springframework.test.context.junit4.SpringRunner;
import org.springframework.test.web.client.MockRestServiceServer;

import static org.junit.Assert.assertFalse;
import static org.junit.Assert.assertTrue;
import static org.springframework.test.web.client.match.MockRestRequestMatchers.requestTo;
import static org.springframework.test.web.client.response.MockRestResponseCreators.
withSuccess;

@RunWith(SpringRunner.class)
@ContextConfiguration(classes= { BankConfiguration.class})
public class OpenIBANValidationClientTest {

    @Autowired
    private OpenIBANValidationClient client;

    private MockRestServiceServer mockRestServiceServer;

    @Before
    public void init() {
        mockRestServiceServer = MockRestServiceServer.createServer(client);
    }

    @Test
    public void validIban() {

        mockRestServiceServer
            .expect(requestTo("https://openiban.com/validate/NL87TRIO0396451440?getBIC=
            true&validateBankCode=true"))
                .andRespond(withSuccess(new ClassPathResource("NL87TRIO0396451440-result. json"),
                MediaType.APPLICATION_JSON));

        IBANValidationResult result = client.validate("NL87TRIO0396451440");
        assertTrue(result.isValid());
    }

    @Test
    public void invalidIban() {

        mockRestServiceServer
            .expect(requestTo("https://openiban.com/validate/NL28XXXX389242218?getBIC=
            true&validateBankCode=true"))
                .andRespond(withSuccess(new ClassPathResource("NL28XXXX389242218-result. json"),
                MediaType.APPLICATION_JSON));

        IBANValidationResult result = client.validate("NL28XXXX389242218");
        assertFalse(result.isValid());
    }
}
```

该测试类有两个测试方法，它们非常相似。在 init 方法中，你使用 OpenIBANValidationClient 类创建了一个 MockRestServiceService 类（这是可以的，因为它继承了 RestGatewaySupport；如果没有继承的话，就需要使用配置好的 RestTemplate 类来创建模拟服务器）。在测试方法中，你使用 URL 创建了期望，现在当 URL 调用后，来自类路径的 JSON 响应就会作为结果返回。

对于测试来说，你可能想要使用来自服务器的一些已知的响应。为此，你可以使用来自于实时系统的一些记录好的结果，或许它们已经提供了供测试使用的结果。

小结

本章介绍了测试 Java 应用所需的基本概念与技术。JUnit 与 TestNG 是 Java 平台上最为流行的测试框架。单元测试用于测试单个程序单元，在面向对象的语言中，它通常是一个类或是方法。在对依赖于其他单元的单元进行测试时，你可以使用桩和模拟对象来模拟其依赖，这会简化测试。与之相反，集成测试用于整体测试几个单元。

在 Web 层，控制器通常是难以测试的。Spring 为 Servlet API 提供了模拟对象，这样就可以轻松模拟 Web 请求和响应对象来测试 Web 控制器了。Servlet Mock MVC 可以简化对控制器的集成测试。适用于控制器的同样适用于基于 REST 的客户端。为了帮助你测试这些客户端，Spring 提供了 MockRestServiceServer，可以使用它模拟外部系统。

Spring 的测试支持设施可以为测试管理应用上下文，这是通过从 bean 配置文件中加载 bean 并跨越多个测试执行将其缓存起来做到的。你可以在测试中访问托管应用上下文，并且从应用上下文中自动注入测试构件。此外，如果测试涉及数据库更新，Spring 则可以为其管理事务，这样一个测试方法中的变更就会回滚，因此不会影响到接下来的测试方法。Spring 还可以为你创建 JDBC 模板来准备并验证数据库中的测试数据。

Spring 提供了一套常见的测试注解来简化测试的创建。这些注解是 Spring 专有的，但独立于底层的测试框架。不过，其中一些只能用于 JUnit。

第 17 章

Grails

在着手创建 Java Web 应用时，你需要将一系列 Java 类放到一起，创建配置文件并建立特定的布局，但所有这一切与应用所要解决的问题并没有什么关系。这些东西通常叫做样板代码或是样板步骤，因为它们只不过是达到目的的手段而已，而目的则是应用实际要完成的。

Grails 是一个框架，旨在减少 Java 应用中需要执行的样板步骤的数量。它基于 Groovy 语言，Groovy 是一种兼容于 Java 虚拟机的语言，Grails 基于约定将很多需要在 Java 应用中执行的步骤进行了自动处理。

比如说，在创建应用控制器时，除了需要某种配置文件以正常工作外，它们最终都会伴随着一系列视图（如 JavaServer Pages [JSP]页面）。如果使用 Grails 生成控制器，那么 Grails 会通过约定自动执行大量步骤（比如说创建视图和配置文件）。接下来可以针对更为具体的场景修改 Grails 生成的文件，不过毫无疑问，Grails 减少了开发时间，因为无须再从头编写一切了（比如说编写 XML 配置文件以及准备项目目录结构等）。

Grails 完全集成了 Spring，因此可以使用它来创建 Spring 应用，从而降低开发成本。

17-1 获取并安装 Grails

问题提出

你想要开始创建 Grails 应用，但不知道从哪里获取 Grails 以及如何安装。

解决方案

可以从 Grails 官网下载 Grails。请确保下载 Grails 3.2 或更高版本。Grails 是个独立的框架，带有很多脚本来自动化 Java 应用的创建。从这个意义上来说，你只需解压缩分发包并执行几步安装就可以在自己的工作站创建 Java 应用了。

解释说明

在工作站解压缩 Grails 后，请在操作系统中定义两个环境变量：GRAILS_HOME 与 PATH。这样就可以在工作站的任何地方调用 Grails 操作了。如果使用 Linux 工作站，那么可以编辑全局 bashrc 文件，它位于/etc/目录下；或是编辑用户的.bashrc 文件，它位于用户的主目录下。注意，根据 Linux 发行版的不同，这些文件名可能会发生变化（如 bash.bashrc）。这两个文件使用了相同的语法来定义环境变量，只不过一个文件定义的变量对所有用户有效，另一个文件只对单个用户有效。请将如下内容添加到其中一个文件中：

```
GRAILS_HOME=/<installation_directory>/grails
export GRAILS_HOME
export PATH=$PATH:$GRAILS_HOME/bin
```

如果使用的是 Windows 工作站，那么请打开“控制面板”，单击“系统”图标。在打开的窗口中，单击“高级选项”选项卡。接下来，单击“环境变量”打开环境变量编辑器。可以在这里为单个用户或是所有用户添加和修改环境变量，请按照如下步骤进行。

1. 单击“新建”按钮。

2. 创建一个环境变量，名字为 GRAILS_HOME，值为 Grails 的安装目录（如/<installation_

directory>/grails）。

3．选中 PATH 环境变量，单击“修改”按钮。

4．将;%GRAILS_HOME%\bin 值添加到 PATH 环境变量的末尾。

警告：请确保在最后添加这个值，并且不要以任何形式修改 PATH 环境变量，因为这可能会导致某些应用停止运行。

在 Windows 或是 Linux 工作站上执行这些步骤后，你就可以开始创建 Grails 应用了。在工作站的任意目录中执行命令 grails help，你会看到 Grails 丰富的命令。

17-2 创建 Grails 应用

问题提出

你想创建 Grails 应用。

解决方案

要想创建 Grails 应用，请在创建应用的地方调用如下命令：grails create-app <grailsappname>。这会创建一个 Grails 应用目录，其项目结构与框架设计保持一致。如果命令执行失败，请参阅 17-1 节。如果 Grails 安装正确，那么 grails 命令应该可以在任意控制台或是终端中使用。

解释说明

比如说，输入 grails create-app court 会在名为 court 的目录下创建一个 Grails 应用。在该目录中，你会看到由 Grails 根据约定生成的一系列文件和目录。图 17-1 所示为 Grails 应用的初始项目结构。

注意：除了这个布局外，Grails 还为应用创建了一系列工作目录与文件（不要直接修改）。这些工作目录与文件位于用户主目录下的.grails/<grails_version>/目录中。

```
court
  gradle
  grails-app
    assets
    conf
    controllers
    domain
    i18n
    init
    services
    taglib
    utils
    views
  src
    integration-test
    main
      groovy
      webapp
    test
      groovy
  .gitignore
  build.gradle
  gradle.properties
  gradlew
  gradlew.bat
  grails-wrapper.jar
  grailsw
  grailsw.bat
  settings.gradle
```

图 17-1　Grails 应用项目结构

从图 17-1 中可以看到，Grails 会生成大多数 Java 应用中常见的一系列文件与目录。目录 src\main\groovy 用于存放源代码文件，目录 src\main\web-app 包含了 Java Web 应用的常见布局（如/WEB-INF/、/META-INF/、css、images 和 js）。这都是开箱即用的，Grails 通过单个命令将这些常见的 Java 应用组件放到了一起，从而节省你的时间。

探索 Grails 应用的文件与目录结构

由于一些文件与目录是 Grails 特有的，因此我们来介绍一下其背后的目的。

- gradle.properties：用于定义应用的构建属性，包括 Grails 版本、servlet 版本和应用名。
- grails-app：一个目录，包含了应用的核心，它又包含了如下目录。
  - assets：包含应用的静态资源的目录（如.css 和.js 文件）。
  - conf：包含应用的配置源的目录。
  - controllers：包含应用的控制器文件的目录。
  - domain：包含应用的领域文件的目录。
  - i18n：包含应用的国际化（i18n）文件的目录。
  - services：包含应用的服务文件的目录。
  - taglib：包含应用的标签库的目录。
  - utils：包含应用的公共文件的目录。
  - views：包含应用的视图文件的目录。

- src\main：存放应用的源代码文件的目录；包含了一个名为 groovy 的子目录，用于放置使用该语言编写的源文件（可以添加一个 java 子目录来编写 Java 文件）。
- src\test：存放应用的单元测试文件的目录。
- src\integration-test：存放应用的集成测试文件的目录。
- web-app：存放应用的部署结构的目录；包含了标准的 Web 归档文件（WAR）和目录结构（如 /WEB-INF/、/META-INF/、css、images 和 js）。

运行应用

Grails 已经预先配置为可以在 Apache Tomcat Web 容器中运行应用。类似于 Grails 应用的创建，运行 Grails 应用的过程也是高度自动化的。

进入到 Grails 应用的根目录下，调用 grails run-app。如果需要，该命令会触发应用的构建过程，同时还会启动 Apache Tomcat Web 容器并部署应用。

由于 Grails 是根据约定来操作的，应用会被部署到项目名后跟着的上下文中。比如说，名为 court 的应用会被部署到 URL http://localhost:8080/。图 17-2 所示为 Grails 应用默认的主界面。

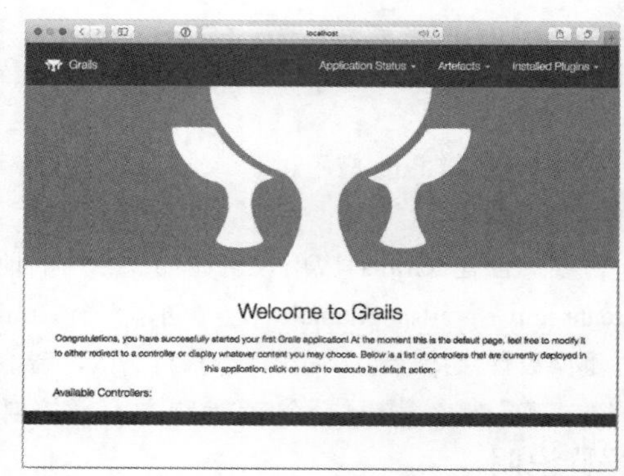

图 17-2　court Grails 应用默认的主界面

应用依旧处于开箱即用的状态。接下来，我们将会介绍如何创建第一个 Grails 组件，看看它是如何节省我们的时间的。

创建第一个 Grails 应用组件

在见识到创建一个 Grails 应用有多么轻松后，我们来创建一个应用组件——控制器。这会进一步展示出 Grails 是如何自动处理 Java 应用开发过程中的一系列步骤的。

进入到 Grails 应用的根目录下，调用 grails create-controller welcome 命令。该命令会执行如下步骤。

1．在应用目录 grails-app/controllers 下创建一个名为 WelcomeController.groovy 的控制器。
2．在应用目录 grails-app/views 下创建一个名为 welcome 的目录。
3．在应用目录 src/u 下创建一个名为 WelcomeControllerSpec.groovy 的测试类。

我们来分析一下 Grails 生成的控制器的内容。WelcomeController.groovy 控制器的内容如下所示：

```
class WelcomeController {
    def index {}
}
```

如果不熟悉 Groovy，那么会觉得其语法有些笨拙。不过它只不过是个名为 WelcomeController 的类，里面有个名为 index 的方法。其目的与第 3 章中创建的 Spring MVC 控制器一样。WelcomeController 表示一个控制器类，而方法 index 表示一个处理器方法。不过，现在这个控制器什么都没有做。修改一下：

```
class WelcomeController {
    Date now = new Date()
    def index = {[today:now]}
}
```

首先添加的是个 Date 对象，并将其赋值给 now 类字段，它表示系统日期。def index {}表示一个处理器方法，添加的[today:now]用作返回值。在该示例中，返回值表示一个使用 now 类字段定义的名为 today 的变量，该变量的值会被传递给与处理器方法关联的视图中。

拥有了控制器以及返回当前日期的处理器方法后，你可以创建相应的视图。如果在 grails-app/views/welcome 目录下，则找不到视图。不过，Grails 会尝试在该目录中寻找与 WelcomeController 控制器对应的视图，视图名与处理器方法的名字保持一致；这又是 Grails 所用的诸多约定之一。

因此，在该目录下创建一个名为 index.gsp 的 GSP 页面，其内容如下所示：

```
<!DOCTYPE html>
<html>
<head>
    <title>Welcome</title>
</head>

<body>
<h2>Welcome to Court Reservation System</h2>
Today is <g:formatDate format="yyyy-MM-dd" date="${today}"/>
</body>
</html>
```

这是个标准的 GSP 页面，它使用了表达式和标签库。在编写 GSP 页面时，默认的标签库可以通过 g 标签来访问。formatDate 标签会渲染一个名为${today}的变量，它就是名为 index 的控制器处理器方法所返回的变量名。

接下来，在 Grails 应用的根目录下调用命令 grails run-app。这会自动构建应用，编译控制器类并将文件复制到需要的地方，同时会启动 Apache Tomcat Web 容器并部署应用。

遵循着同样的 Grails 约定流程，可以通过上下文路径 http://localhost:8080/welcome/访问 WelcomeController 及其处理器方法和视图。由于 index 是上下文路径所用的默认页面，如果打开浏览器并访问 http://localhost:8080/welcome/或显式的 URL http://localhost:8080/welcome/index，你会看到之前的 JSP 页面，它渲染了控制器返回的当前日期。注意，URL 中没有视图扩展名（比如说.html）。默认情况下，Grails 隐藏了视图技术；这么做的原因在更加高级的 Grails 场景中是显而易见的。

在重复创建应用控制器与视图所需的简单步骤时，请记住一点，即没必要创建或是修改任何配置文件、手工将文件复制到别的地方，或是创建 Web 容器来运行应用。随着应用不断推进，在 Java Web 应用中避免这些常见的样板步骤会极大减少开发时间。

将 Grails 应用导出为 WAR

上述步骤都是在 Grails 环境范围内执行的。也就是说，你依赖于 Grails 来启动 Web 容器并运行应用。不过，当在生产环境中运行 Grails 应用时，你需要生成一种格式将应用部署到外部 Web 容器中，对于 Java 应用来说就是 WAR 文件。

进入到 Grails 应用的根目录中，调用 grails war 命令。执行该命令会在根目录下生成一个 WAR 文件，其形式为<application-name>-<application-version>.war。这个 WAR 是个独立的文件，拥有在任意 Java 标准的 Web 容器中运行 Grails 应用所需的全部必要元素。对于 court 应用来说，会在 Grails 应用的根目录下生成一个名为 court-0.1.war 的文件，应用的版本号来自于 application.properties 文件中定义的 app.version 参数。

与 Apache Tomcat 的部署约定一致，名为 court-0.1.war 的 WAR 要通过 URL http://localhost:8080/court-0.1/访问。根据 Java Web 容器的不同，WAR 部署到 URL 的约定可能会发生变化（如 Jetty 或 Oracle WebLogic）。

17-3 获取 Grails 插件

问题提出

你想在 Grails 应用中使用 Java 框架或是 Java API 的功能，同时又想利用同样的 Grails 技术来去除样板性工作。这个问题并不只是在应用中使用 Java 框架或是 Java API；只要向应用的 lib 目录中放入相应的 JAR 就可以解决该问题。不过，不需要将 Java 框架或是 Java API 紧紧集成到 Grails 中，这是通过 Grails 插件的形式来提供的。

通过紧紧集成到 Grails 中，我们可以拥有简洁命令的能力（如 grails <plug-in-task>）来执行特定的 Java 框架或是 Java API 任务，也可以使用应用中的类与配置文件，同时无须重复那些样板步骤。

解决方案

Grails 实际上带有一些预先安装好的插件，不过如果你坚持使用 Grails 开箱即用的能力，这一点就不是那么明显了。然而，有很多 Grails 插件可以使用特定的 Java 框架或是 Java API，使用过程就像使用 Grails 核心功能那样高效。下面列出一些流行的 Grails 插件：

- App Engine：将 Google 的 App Engine SDK 与工具集成到了 Grails 中。
- Quartz：集成了 Quartz Enterprise Job Scheduler（企业级任务调度器）来调度任务，并使用专门的间隔或是 cron 表达式来执行它们。
- Spring WS：基于 Spring Web Services 项目集成并支持 Web Service 的能力。
- Clojure：集成 Clojure 并且可以在 Grails 构件中执行 Clojure 代码。

要想获取完整的 Grails 插件列表，可以执行命令 grails list-plugins。这个命令会连接到 Grails 插件仓库并列出所有可用的 Grails 插件。此外，命令 grails plugin-info <plugin_name>可用于获取特定插件的详细信息。还可以访问 Grails 插件页面 http://grails.org/plugin/home 来查看插件信息。

安装 Grails 插件很简单，只需向 build.gradle 文件中添加依赖就可以；卸载则与之相反，从 build.gradle 文件中将依赖移除即可。

解释说明

Grails 插件遵循了一系列约定，使其能够将特定的 Java 框架或是 Java API 紧紧集成到 Grails 中。默认情况下，Grails 预先安装了 Apache Tomcat 与 Hibernate 插件。

除了这些默认插件外，还可以根据具体应用安装其他插件。比如说，要想安装 Clojure 插件，需要向 build.gradle 文件中添加如下依赖：

```
dependencies {
    compile "org.grails.plugins:clojure:2.0.0.RC4"
}
```

17-4　Grails 环境中的开发、生产与测试

问题提出

你想根据应用运行的环境（如开发、生产和测试）为相同的应用使用不同的参数。

解决方案

Grails 预料到 Java 应用会经历各种阶段，需要不同的参数。这些阶段（在 Grails 中叫做环境）可能有开发、生产和测试等。

最显而易见的场景涉及数据源，你要为开发、生产和测试环境使用不同的持久化存储系统。由于每一个存储系统都会使用不同的连接参数，因此很容易为多个环境配置参数，让 Grails 根据应用的操作来连接到每一个存储系统。

除了数据源之外，Grails 也为其他参数提供了同样的特性，可以在应用环境间进行切换，比如说用于创建应用的绝对链接的服务器 URL。

Grails 应用环境的配置参数是在应用的/grails-app/conf/目录下的文件中指定的。

解释说明

根据你所执行的操作，Grails 会自动选择最适合的环境：开发、生产或是测试。比如说，当调用命令 grails run-app 时，这表示你依然在本地开发应用，因此会使用开发环境。实际上，在执行这个命令时，可以从输出中看到下面这一行：

```
Environment set to development
```

这意味着针对开发环境设定的参数会用于构建、配置和运行应用。再比如说 grails war 命令。由于将 Grails 应用导出为一个独立的 WAR 意味着你要在外部 Web 容器中运行它，Grails 就会使用生产环境。在该命令生成的输出中，你会看到如下这一行：

```
Environment set to production
```

这意味着针对生产环境设定的参数会用于构建、配置和运行应用。最后，如果运行 grails test-app 命令，那么 Grails 就会使用测试环境。这意味着针对测试环境设定的参数会用于构建、配置和运行测试。在该命令生成的输出中，你会看到如下这一行：

```
Environment set to test
```

在应用目录/grails-app/conf/下的 application.yml 文件中，你会看到如下内容：

```
environments:
    development:
        dataSource:
            dbCreate: create-drop
            url: jdbc:h2:mem:devDb;MVCC=TRUE;LOCK_TIMEOUT=10000;DB_CLOSE_ON_EXIT=FALSE
    test:
        dataSource:
            dbCreate: update
            url: jdbc:h2:mem:testDb;MVCC=TRUE;LOCK_TIMEOUT=10000;DB_CLOSE_ON_EXIT=FALSE
    production:
        dataSource:
            dbCreate: none
            url: jdbc:h2:./prodDb;MVCC=TRUE;LOCK_TIMEOUT=10000;DB_CLOSE_ON_EXIT=FALSE
```

在上面的代码中，根据应用环境为持久化存储系统指定了不同的连接参数。这样，应用就可以操作不同的数据集了，你肯定不想在开发环境下对数据所做的修改发生在生产环境下。值得注意的是，这并不意味着根据应用环境只能配置这些参数。你可以在对应的环境部分添加任意参数。这些示例只是表示了应用环境间最有可能变化的参数。

还可以根据给定的应用环境执行程序逻辑（比如说在类或脚本中）。这是通过 grails.util.Environment 类实现的。如下代码展示了这一处理：

```
import grails.util.Environment
...
...
switch(Environment.current) {
    case Environment.DEVELOPMENT:
        // Execute development logic
    break
    case Environment.PRODUCTION:
        // Execute production logic
    break
}
```

在上述代码片段中，类首先导入了 grails.util.Environment 类。接下来，基于 Environment.current 值（包含了应用运行的环境），代码使用一个 switch 条件根据该值执行逻辑。

这种场景在一些领域中是具有共通性的，如发送邮件或是执行地理定位。在开发环境中发送邮件或是确定用户的位置是没有意义的，因为除了不需要应用的邮件通知外，开发团队的位置与此并没有什么关系。

最后，值得注意的是，你可以覆盖任何 Grails 命令所用的默认环境。

比如说，默认情况下，grails run-app 命令会使用针对于开发环境的参数。如果出于某种原因，你想使用针对生产环境的参数来运行该命令，那就可以使用命令 grails prod run-app。如果想要使用针对于测试环境的参数，那就可以使用命令 grails test run-app。

出于同样的原因，对于诸如 grails test-app 这样的命令来说，它使用了针对测试环境的参数，你可以通过命令 grails dev test-app 使用针对开发环境的参数。这种方式适用于所有其他命令，只需在 grails 命令后插入 prod、test 或是 dev 关键字即可。

17-5 创建应用的领域类

问题提出

你需要定义应用的领域类。

解决方案

领域类用于描述应用的主要元素和特征。如果设计一个应用来接受预订，那么就需要提供一个领域类来保存预订。同样地，如果预订与人相关联，那么应用就需要提供一个领域类来保存人。

在 Web 应用中，领域类常常是首先要定义的，因为这些类表示存储起来供后续使用的数据（在持久化的存储系统中），这样它就可以与控制器交互了；还可以表示在视图上展示的数据。

在 Grails 中，领域类放在/grails-app/domain/目录下。就像 Grails 中的其他内容一样，领域类的创建可以通过执行如下形式的简单命令来进行：

```
grails create-domain-class <domain_class_name>
```

上述命令会在/grails-app/domain/目录下生成一个名为<domain_class_name>.groovy 的骨架领域类文件。

解释说明

Grails 会创建骨架领域类，不过你依然需要修改每个领域类来反映出应用的目的。

我们创建一个预订系统，类似于第 4 章介绍 Spring MVC 时创建的那个。创建两个领域类，一个名为 Reservation，另一个名为 Player。请执行如下命令来创建：

```
grails create-domain-class Player
grails create-domain-class Reservation
```

通过执行这些命令，名为 Player.groovy 和 Reservation.groovy 的类文件就会在应用的/grails-app/domain/下创建出来。此外，针对每个领域类的单元测试文件会在应用的 src/test/groovy 目录下生成出来，不过测试会在 17-10 节中介绍。接下来，打开 Player.groovy 类编辑其内容，代码如下所示：

```
class Player {
    static hasMany = [ reservations : Reservation ]
    String name
    String phone
    static constraints = {
        name(blank:false)
        phone(blank:false)
    }
}
```

添加的第一行 static hasMany = [reservations : Reservation]表示领域类之间的关系。这条语句表示 Player 领域类有一个 reservations 字段，它会关联多个 Reservation 对象。接下来的语句表示 Player 领域类还有两个 String 字段，一个叫做 name，另一个叫做 phone。

剩下的元素 static constraints = { }在领域类上定义了约束。在该示例中，声明 name(blank:false)表示 Player 对象的 name 字段不能为空。声明 phone(blank:false)表示只有当 phone 字段有值后才可以创建 Player 对象。修改好 Player 领域类后，请打开 Reservation.groovy 类来编辑其内容。

```
package court

import java.time.DayOfWeek
import java.time.LocalDateTime

class Reservation {

    static belongsTo = Player
    String courtName;
    LocalDateTime date;
    Player player;
    String sportType;
    static constraints = {
        sportType(inList: ["Tennis", "Soccer"])
        date(validator: { val, obj ->
            if (val.getDayOfWeek() == DayOfWeek.SUNDAY && (val.getHour() < 8 || val.
            getHour() > 22)) {
                return ['invalid.holidayHour']
            } else if (val.getHour() < 9 || val.getHour() > 21) {
                return ['invalid.weekdayHour']
            }
```

```
        })
    }
}
```

添加到 Reservation 领域类的第一条语句 static belongsTo = Player 表示 Reservation 对象永远都属于 Player 对象。接下来的语句表示 Reservation 领域类有一个类型为 String，名字为 courtName 的字段；一个类型为 LocalDateTime，名字为 date 的字段；一个类型为 Player，名字为 player 的字段，以及一个类型为 String，名字为 sportType 的字段。

Reservation 领域类的约束要比 Player 复杂一些。第一个约束 sportType(inList:["Tennis", "Soccer"])将 Reservation 对象的 sportType 字段限制为一个字符串值，要么是 Tennis，要么是 Soccer。第二个约束是个自定义的验证器，确保 Reservation 对象的 date 字段会出现在某个时间范围之内（取决于一周中的哪一天）。

在有了应用的领域类后，就可以为应用创建对应的视图和控制器了。

不过在继续前，我们先说一下关于 Grails 的领域类。虽然本节创建的领域类让你对定义 Grails 领域类的语法有了一个基本的了解，不过这里只涉及 Grails 领域类中的部分特性。

随着领域类之间的关系变得愈发复杂，我们需要更为复杂的构件来定义 Grails 领域类。这是因为 Grails 依赖于领域类来实现各种应用功能。

比如说，如果领域对象从应用持久化存储系统中更新或是删除了，那么领域类之间的关系就需要建立好。如果没有建立好关系，那就有可能出现应用中数据不一致的情况（比如说，如果删除了一个 person 对象，那么它对应的关系也需要删除，以避免应用预订出现不一致的状态）。

同样地，各种约束可用于限制领域类的结构。在某些情况下，如果约束太复杂，那么通常会在创建领域类对象前将其纳入到应用的控制器中。不过，对于本节来说，模型约束用于展示 Grails 领域类的设计。

17-6　为应用的领域类生成 CRUD 控制器与视图

问题提出

你需要为应用的领域类生成创建、读取、更新与删除（CRUD）控制器与视图。

解决方案

应用的领域类本身是没什么用的。映射到领域类的数据依然需要创建，并展示给最终用户；可能还需要保存到持久化存储系统中供后续使用。

在持久化存储系统所支撑的 Web 应用中，这些针对领域类的操作通常叫做 CRUD 操作。在大多数 Web 框架中，生成 CRUD 控制器与视图还需要不少工作量。这需要控制器能够在持久化存储系统中创建、读取、更新和删除领域对象，同时还要为最终用户创建相应的视图（如 JSP 页面）使之能够对相同的对象执行创建、读取、更新和删除操作。

不过，既然 Grails 基于约定来操作，因此为应用的领域类生成 CRUD 控制器和视图的机制就是非常简单的了。可以执行如下命令为应用的领域类生成相应的 CRUD 控制器和视图：

```
grails generate-all <domain_class_name>
```

解释说明

Grails 可以审视应用的领域类并生成必要的控制器与视图来创建、读取、更新和删除属于应用领域类的实例。

比如说，考虑之前创建的 Player 领域类。要想生成其 CRUD 控制器与视图，只需在应用的根目录下执行如下命令：

```
grails generate-all court.Player
```

类似的命令也适用于 Reservation 领域类。只需执行如下命令就可以生成其 CRUD 控制器与视图：

```
grails generate-all court.Reservation
```

那么，执行这些步骤实际会生成哪些内容呢？如果看一下这些命令的输出就知道了，不过这里还是要重述一下过程。

1. 编译应用的类。

2．在目录 grails-app/i18n 下生成 12 个属性文件以支持应用的国际化（如 messages_ <language>.properties）。

3．创建一个名为<domain_class>Controller.groovy 的控制器，它位于应用的 grails-app/controllers 目录下，拥有针对于 RDBMS 的 CRUD 操作。

4．创建与控制器类的 CRUD 操作对应的 4 个视图，名字分别为 create.gsp、edit.gsp、index.gsp 与 show.gsp。注意，.gsp 名扩展表示"Groovy Server Pages"，它相当于 JavaServer Pages，只不过使用 Groovy 声明语句而非 Java。这些视图位于应用的 grails-app/views/<domain_class>目录下。

完成这些步骤后就可以使用 grails run-app 启动 Grails 应用并让最终用户来使用应用了。你没有看错，在执行完这些简单命令后，应用就可以供最终用户使用了。这正是 Grails 基于约定的强大之处，它通过一个命令简化了样板代码的创建。当应用启动后，可以通过如下 URL 对 Player 领域类执行 CRUD 操作。

- 创建：http://localhost:8080/player/create。
- 读取：http://localhost:8080/player/list（所有 player）或 http://localhost:8080/court/player/ show/<player_id>。
- 更新：http://localhost:8080/player/edit/<player_id>。
- 删除：http://localhost:8080/player/delete/<player_id>。

视图间的页面导航要比这些 URL 还要直观，不过这里将会通过一些截图来简要解释一下。关于这些 URL，值得注意的重要一点在于它们的约定。注意模式<domain>/<app_name>/<domain_ class>/<crud_action>/<object_id>，根据操作，其中的<object_id>是可选的。

除了用于定义 URL 模式外，这些约定在整个应用构件中都得到了使用。比如说，如果查看 PlayerController.groovy 控制器，你会发现有命名为各种<crud_action>值的处理器方法。与之类似，如果查看应用的后端 RDBMS，你会发现领域类对象是使用 URL 中所用的相同<player_id>进行保存的。

在认识到 CRUD 操作在 Grails 应用中是如何组织的后，现在就访问地址 http://localhost:8080/player/create 来创建一个 Player 对象。访问该页面后，你会看到一个 HTML 表单，它拥有与 Player 领域类定义时同样的字段值。

为 name 与 phone 字段输入任意两个值并提交表单，将一个 Player 对象持久化到 RDBMS 中。默认情况下，Grails 预先配置为使用 HSQLDB，这是个内存 RDBMS。后文会介绍如何将其切换至其他 RDBMS 现在 HSQLDB 就足够了。

接下来，再次提交相同的表单，不过这次不输入任何值。Grails 不会再持久化 Player 对象；相反，它会展示两个警告消息，表示 name 与 phone 字段不能为空，如图 17-3 所示。

这个验证过程是由 Player 领域类中的语句 name(blank:false)和 phone(blank:false)所强制的。你不需要修改应用的控制器和视图，甚至不需要为这些错误消息创建属性文件；一切都是由 Grails 基于约定的方法来处理的。

■ **注意：**在使用支持 HTML5 的浏览器时，你将无法提交表单。这两个输入元素都是必填的，这会使得表单在这些浏览器中无法提交。这些规则也是基于之前提及的语句添加的。

体验一下 Player 领域类的其他页面，直接通过 Web 浏览器创建、读取、更新和删除对象来感受 Grails 是如何处理这些任务的。

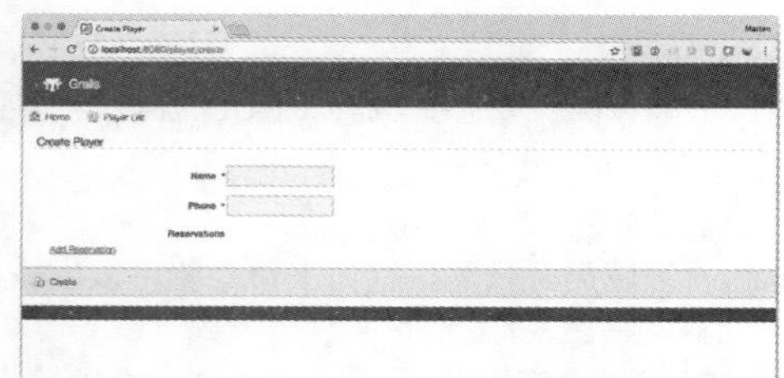

图 17-3　Grails 领域类验证出现在视图中（在该示例中是个 HTML 表单）

还可以通过如下 URL 对 Reservation 领域类执行 CRUD 操作：

- 创建：http://localhost:8080/reservation/create。
- 读取：http://localhost:8080/reservation/list（所有 reservation）或 http://localhost:8080/court/ reservation/show/<reservation_id>。
- 更新：http://localhost:8080/reservation/edit/<reservation_id>。
- 删除：http://localhost:8080/reservation/delete/<reservation_id>。

这些 URL 的作用与 Player 领域类的那些一样：从 Web 界面上创建、读取、更新和删除属于 Reservation 领域类的对象。

接下来，我们分析一下用于创建 Reservation 对象的 HTML 表单，它位于 URL http://localhost:8080/reservation/create。图 17-4 所示为该表单。

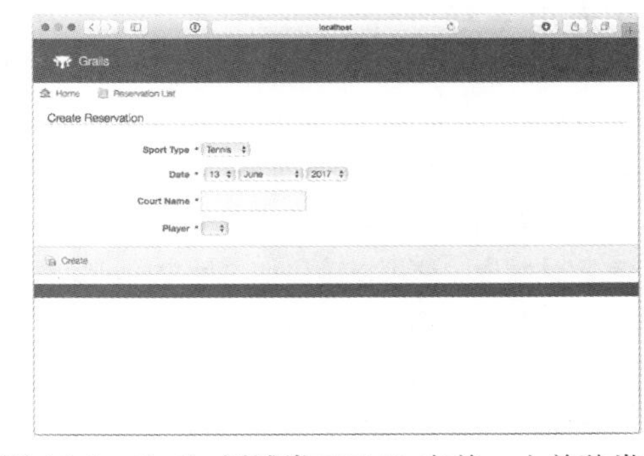

图 17-4 Grails 领域类 HTML 表单，由单独类的领域对象装配而成

图 17-4 很有意思。虽然 HTML 表单依然是根据 Reservation 领域类的字段创建的，就像 Player 领域类的 HTML 表单一样，但它拥有各种预先装配好的 HTML 选择菜单。

第一个选择菜单属于 sportType 字段。由于这个特殊字段有一个定义约束，其值只能是字符串 Soccer 或 Tennis，因此 Grails 会自动将这些选项提供给用户而不是让用户随意输入再做验证。

第二个选择菜单属于 date 字段。在该示例中，Grails 会生成多个 HTML 选择菜单来表示日期，使得日期选择过程更容易一些，而不是让用户随意输入再做验证。

第三个选择菜单属于 player 字段。这个选择菜单的选项是不同的，它们来自于你为应用所创建的 Player 对象，其值是查询应用的 RDBMS 所获得的；如果再添加另一个 Player 对象，那么它就会自动出现在该选择菜单中。

此外，还会对 date 字段执行验证。如果所选的日期不符合某个范围，那么 Reservation 对象就无法持久化，一条警告消息会出现在表单上。

现在，你无法提交合法的 Reservation，因为 date 字段只接收日期而不是时间。为了解决这个问题，打开 views/reservation 目录下的 create.gsp（以及 edit.gsp）文件。它里面有个<f:all bean="reservation" />标签。该标签负责创建 HTML 表单，它会根据字段的类型渲染出一个 HTML input 元素。为了能够输入时间，添加一个<g:datePicker />标签并从默认表单中排除掉 date 字段。

```
<fieldset class="form">
    <f:all bean="reservation" except="date" />
    <div class="fieldcontain required">
        <label for="date">Date</label>
        <g:datePicker name="date" value="${reservation?.date}" precision="minute"/>
    </div>
</fieldset>
```

<g:datePicker />标签可以指定一个精确值。当设为 minute 时，那就可以在日期旁设置小时与分钟了。现在，当选择恰当的范围后，Reservation 就可以存储起来了。

体验一下 Reservation 领域类的其他视图，直接通过 Web 浏览器创建、读取、更新和删除对象。

最后，我们再来看看在这几步中应用到底做了什么事情：验证输入，从 RDBMS 中创建、读取、更新和删除对象；根据 RDBMS 中的数据完成 HTML 表单；支持国际化。你甚至没有修改一个配置文件，没有编写 HTML，也没有使用 SQL 或是对象关系映射（ORM）。

17-7 为消息属性实现国际化（I18n）

问题提出

你需要对 Grails 应用所使用的值进行国际化。

解决方案

默认情况下，所有 Grails 应用都支持国际化。在应用的/grails-app/i18n/目录下，可以找到一系列*.properties 文件，它们以 12 种语言定义了消息。这些*.properties 文件中声明的值可以让 Grails 应用基于用户的语言首选项或是应用的默认语言来显示消息。在 Grails 应用中，声明在*.properties 文件中的值可以从包含视图（JSP 或 GSP 页面）或应用上下文的地方访问。

解释说明

Grails 基于两个标准来确定为用户使用哪个地域（即使用哪个国际化属性文件）：

- 应用的/grails-app/conf/spring/resource.groovy 文件中的显式配置；
- 用户的浏览器语言首选项。

由于显式的应用地域配置的优先级要高于用户的浏览器语言首选项，因此应用的 resource.groovy 文件中并没有默认配置。这可以确保如果用户的浏览器语言首选项设置为 Spanish（es）或 German（de），那么用户就会使用来自于 Spanish 或 German 属性文件的消息（即 messages_es.properties 或 messages_ de.properties）。另一方面，如果应用的 resource.groovy 文件配置为使用 Italian（it），那么无论用户的浏览器语言首选项是什么，用户总是会使用来自于 Italian 属性文件的消息（即 messages_it.properties）。

因此，只有当强制用户使用特定的语言地域时，才应该在应用的/grails-app/conf/spring/resource.groovy 文件中定义显式的配置。比如说，也许你不想更新某些国际化属性文件或是只想保持一样的值。

由于 Grails 国际化以 Spring 的 Locale Resolver 为基础，因此需要将如下内容放到应用的/grails-app/conf/spring/resource.groovy 文件中，以强制对用户使用特定的语言：

```
import org.springframework.web.servlet.i18n.SessionLocaleResolver

beans = {
    localeResolver(SessionLocaleResolver) {
        defaultLocale= Locale.ENGLISH
        Locale.setDefault (Locale.ENGLISH)
    }
}
```

通过使用上述声明，无论浏览器的语言首选项是什么，任何访问者都会使用来自于 English 属性文件的消息（即 messages_en.properties）。值得一提的是，如果指定的地域没有对应的属性文件，那么 Grails 就会使用默认的 messages.properties 文件，默认情况下，该文件是使用英文编写的，不过如果愿意的话，也可以轻松修改其值来使用其他语言。当用户的浏览器语言首选项确定了选择标准时，这种场景也是适用的（比如说，如果用户的浏览器语言首选项设置为中文，但并没有中文属性文件，那么 Grails 就会回退为使用默认的 messages.properties 文件）。

在知道了 Grails 是如何确定所用的属性文件来满足地域化内容后，我们来看看 Grails *.properties 文件的语法：

```
default.paginate.next=Next
typeMismatch.java.net.URL=Property {0} must be a valid URL
default.blank.message=Property [{0}] of class [{1}] cannot be blank
default.invalid.email.message=Property [{0}] of class [{1}] with value [{2}] is not a valid
e-mail address
default.invalid.range.message=Property [{0}] of class [{1}] with value [{2}] does not fall
within the valid range from [{3}] to [{4}]
```

第一行是*.properties 文件中最简单的声明形式。如果 Grails 在应用中遇到了名为 default.paginate.next 的属性，那么它就会将其替换为值 Next，或是以用户决定的地域为基础使用针对于该属性的其他值。

在某些情况下，有必要提供更为明确的消息，消息内容是在地域文件被调用时确定的。这正是键{0}、{1}、{2}、{3}和{4}的目的。这些参数要与地域属性搭配使用。在这种方式下，展示给用户的地域消息会表达出更为详尽的信息。图 17-5 所示为 court 应用所用的地域与参数化消息，它们是根据用户浏览器的语言首选项来确定的。

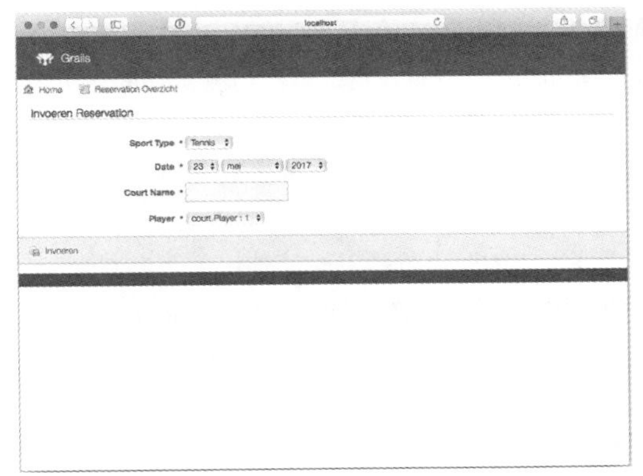

图 17-5　Grails 地域与参数化消息，根据用户浏览器的语言首选项来确定

了解了这些内容后，在 Grails message.properties 文件中定义如下 4 个属性：

```
invalid.holidayHour=Invalid holiday hour
invalid.weekdayHour=Invalid weekday hour
welcome.title=Welcome to Grails
welcome.message=Welcome to Court Reservation System
```

接下来，是时候看看如何在 Grails 应用中定义属性占位符了。

在 17-5 节中，你可能还没有认识到这一点，不过为 Reservation 领域类声明了一个地域属性。在验证部分（static constraints = { }），你创建了如下形式的语句：

```
return ['invalid.weekdayHour']
```

如果遇到该语句，Grails 就会尝试在属性文件中寻找名为 invalid.weekdayHour 的属性，并根据用户确定的地域替换它的值。还可以向应用的视图引入地域属性。比如说，你可以修改 17-2 节创建的 /court/grails-app/views/welcome/index.gsp GSP 文件，代码如下所示：

```
<html>
<!DOCTYPE html>
<html>
<head>
    <title><g:message code="welcome.title"/></title>
</head>

<body>
<h2><g:message code="welcome.message"/></h2>
Today is <g:formatDate format="yyyy-MM-dd" date="${today}"/>
</body>
</html>
```

该 GSP 页面使用了<g:message/>标签。接下来使用了 code 属性，定义了属性 welcome.title 与 welcome.messsage，当渲染 JSP 时，它们都会被替换为相应的地域化值。

17-8　变更持久化存储系统

问题提出

你想将 Grails 应用的持久化存储系统改为自己喜欢的 RDBMS。

解决方案

Grails 使用 RDBMS 作为持久化存储系统。默认情况下，Grails 预先配置为使用 HSQLDB。在部署应用时，Grails 会自动启动 HSQLDB 数据库（即执行 grails run-app 时）。

不过，HSQLDB 的简单性也是它的主要缺点。应用每次在开发和测试环境中重启时，HSQLDB 都会丢失所有数据，因为它是在内存操作的。虽然在生产环境中，Grails 应用的 HSQLDB 配置为将数据持久化存储到文件中，但对于某些应用需求来说，HSQLDB 特性集还是存在限制的。

可以配置 Grails 使用其他 RDBMS，方式是修改应用的 application.yml 文件，它位于 grails-app/conf 目录

下。在该文件中，一共可以配置 3 个 RDBMS，每个 RDBMS 针对应用的一种环境，分别是开发、生产和测试环境。参见 17-4 节了解关于 Grails 应用中的开发、生产与测试环境的更多信息。

解释说明

Grails 依赖于标准的 Java JDBC 来指定 RDBMS 连接参数，同时还依赖于每个 RDBMS 厂商提供的相应 JDBC 驱动来创建、读取、更新和删除信息。

如果变更 RDBMS，那么需要注意的一点就是 Grails 使用了名为 Groovy 对象关系映射（Groovy Object Relational Mapper，GORM）的 ORM 来与 RDBMS 进行交互。GORM 的目的与其他 ORM 解决方案一样，可以让你专注于应用的业务逻辑，无须关心具体的 RDBMS 实现细节（从数据类型的差别到直接使用 SQL）。借助于 GORM，我们可以设计应用的领域类并将设计映射到你所选择的 RDBMS 上。

创建 RDBMS 驱动

变更 Grails 默认 RDBMS 的第一步就是将所选择的 RDBMS 的 JDBC 驱动添加到 gradle.build 文件中。这样，应用就可以访问 JDBC 类将对象持久化到特定的 RDBMS 中了。

配置 RDBMS 实例

第二步需要修改应用 grails-app/conf 目录下的 application.yml 文件。在该文件中有 3 个部分用于定义 RDBMS 实例。

每个 RDBMS 实例对应于不同的应用环境：开发、生产与测试环境。根据所采取的动作，Grails 会选择其中一个实例执行应用需要做的持久化存储操作。参见 17-4 节了解 Grails 应用中关于开发、生产与测试环境的更多信息。

不过，每一部分中用于声明 RDBMS 的语法是一样的。表 17-1 包含了配置 RDBMS 所需的 dataSource 中可用的各种属性。

表 17-1　配置 RDBMS 的 dataSource 属性

<table>
<tr><th>属性</th><th colspan="2">定义</th></tr>
<tr><td>driverClassName</td><td colspan="2">JDBC 驱动的类名</td></tr>
<tr><td>username</td><td colspan="2">建立到 RDBMS 连接的用户名</td></tr>
<tr><td>password</td><td colspan="2">建立到 RDBMS 连接的密码</td></tr>
<tr><td>url</td><td colspan="2">RDBMS 的 URL 连接参数</td></tr>
<tr><td>pooled</td><td colspan="2">表示是否为 RDBMS 使用连接池；默认为 true</td></tr>
<tr><td>jndiName</td><td colspan="2">表示数据源的 JNDI 连接字符串（这是除了在 Grails 中直接配置 driverClassName、username、password 与 url 外的另一种方式，需要依赖在 Web 容器中配置的数据源）</td></tr>
<tr><td>logSql</td><td colspan="2">表示是否启用 SQL 日志</td></tr>
<tr><td>dialect</td><td colspan="2">表示执行操作的 RDBMS 方言</td></tr>
<tr><td>properties</td><td colspan="2">用于指定 RDBMS 操作的额外参数</td></tr>
<tr><td>dbCreate</td><td colspan="2">表示自动生成 RDBMS 数据定义语言（DDL）</td></tr>
<tr><td></td><th>dbCreate 值</th><th>定义</th></tr>
<tr><td></td><td>create-drop</td><td>当 Grails 运行时删除并重新创建 RDBMS DDL（警告：这会删除 RDBMS 中所有的数据）</td></tr>
<tr><td></td><td>create</td><td>如果不存在则创建 RDBMS DDL，如果存在则不会修改（警告：这会删除 RDBMS 中所有的数据）</td></tr>
<tr><td></td><td>update</td><td>如果不存在会创建 RDBMS DDL，如果存在则更新</td></tr>
</table>

如果使用过 Java ORM，如 Hibernate 或 EclipseLink，那么你会对表 17-1 中介绍的参数感到很熟悉。如下代码演示了针对于 MySQL RDBMS 的 dataSource 定义：

```
dataSource:
    dbCreate: update
    pooled: true
    jmxExport: true
    driverClassName: com.mysql.jdbc.Driver
    username: grails
    password: groovy
```

在上面定义的属性中，最应该注意的是 dbCreate，因为它会销毁 RDBMS 中的数据。在这种情况下，update 值是所有 3 个可用值中最保守的，如表 17-1 所解释的那样。

如果使用生产 RDBMS，那么 dbCreate="update"显然是首选策略，因为它并不会销毁 RDBMS 中的任何数据。另一方面，如果 Grails 应用正在测试，那么你可能希望每次测试运行后 RDBMS 中的数据都会被清理掉；这样，诸如 dbCreate="create"或 dbCreate="create-drop"这样的值就更适合了。对于开发 RDBMS 来说，这些选项中哪一个才是更好的策略则取决于 Grails 应用的开发是如何进行的。

Grails 还可以使用配置在 Web 容器中的 RDBMS。在这种情况下，诸如 Apache Tomcat 这样的 Web 容器就会使用相应的 RDBMS 连接参数进行配置，并且通过 JNDI 来访问 RDBMS。如下代码展示了通过 JNDI 访问 RDBMS 时的 dataSource 定义：

```
dataSource:
    jndiName: java:comp/env/grailsDataSource
```

最后，值得一提的是，可以配置一个 dataSource 定义使之在应用的各种环境下都生效，同时为每个具体环境进一步指定属性。如下代码展示了该配置：

```
dataSource:
    driverClassName: com.mysql.jdbc.Driver
    username: grails

environments:
    production:
        dataSource:
            url: jdbc:mysql://localhost/grailsDBPro
            password: production
    development:
        dataSource:
            url: jdbc:mysql://localhost/grailsDBDev
            password: development
```

在上述代码中，dataSource 的 driverClassName 与 username 属性是全局定义的，对所有环境生效，而其他的 dataSource 属性则是针对每个环境单独声明的。

17-9 定制日志输出

问题提出

你想定制 Grails 应用生成的日志输出。

解决方案

Grails 依赖于 Logback 执行日志操作。为了做到这一点，所有配置参数都是在应用的/grails-app/conf 目录下的 logback.groovy 文件中指定的。如果更熟悉 Logback 的 XML 配置的话，可以将其替换为 logback.xml 文件。

考虑到 Logback 日志功能的多样性，Grails 应用日志可以通过多种方式进行配置。这包括创建自定义追加器、日志级别、控制台输出、根据构件输出日志，以及自定义日志布局。

解释说明

Grails 预先配置了一套基本的参数。这些参数定义在应用/grails-app/conf 目录下的 logback.groovy 文件中，参数如下所示：

```
import grails.util.BuildSettings
import grails.util.Environment
```

```
import org.springframework.boot.logging.logback.ColorConverter
import org.springframework.boot.logging.logback.WhitespaceThrowableProxyConverter

import java.nio.charset.Charset

conversionRule 'clr', ColorConverter
conversionRule 'wex', WhitespaceThrowableProxyConverter

// See http://logback.qos.ch/manual/groovy.html for details on configuration
appender('STDOUT', ConsoleAppender) {
    encoder(PatternLayoutEncoder) {
        charset = Charset.forName('UTF-8')

        pattern =
            '%clr(%d{yyyy-MM-dd HH:mm:ss.SSS}){faint} ' + // Date
                '%clr(%5p) ' + // Log level
                '%clr(---){faint} %clr([%15.15t]){faint} ' + // Thread
                '%clr(%-40.40logger{39}){cyan} %clr(:){faint} ' + // Logger
                '%m%n%wex' // Message
    }
}

def targetDir = BuildSettings.TARGET_DIR
if (Environment.isDevelopmentMode() && targetDir != null) {
    appender("FULL_STACKTRACE", FileAppender) {
        file = "${targetDir}/stacktrace.log"
        append = true
        encoder(PatternLayoutEncoder) {
            pattern = "%level %logger - %msg%n"
        }
    }
    logger("StackTrace", ERROR, ['FULL_STACKTRACE'], false)
}
root(ERROR, ['STDOUT'])
```

在日志的术语中，每个包叫做一个日志器。Logback 支持如下日志级别：error、warn、info、debug 与 trace。error 是最严重的。Grails 通过对大多数包使用 error 级别，从而使用了一种保守、默认的日志策略。指定严重程度低一些的级别（如 debug）会生成更多的日志信息，这对于大多数情况来说并不实际。

默认情况下，所有日志消息都会发送给应用根目录下的 stacktrace.log 文件，如果可能还会发送给正在运行的应用的标准输出（如控制台）。在执行 Grails 命令时，你会看到日志消息会被发送给标准输出。

配置自定义追加器与日志器

Logback 依赖于追加器与日志器来提供各种日志功能。追加器指的是日志信息发往的地方（比如说文件或是标准输出），而日志器指的是日志信息生成的地方（比如说类或是包）。

Grails 配置了一个根日志器，所有其他的日志器都会继承它的行为。可以通过如下语句在应用的 Logback.groovy 文件中定制 Grails 应用的默认日志器：

```
root(ERROR, ['STDOUT'])
```

上述语句定义了一个日志器，这样 error 或是更严重级别的消息就会进入到标准输出。这就是你会看到其他日志器（如类或是包）的日志消息被发送到标准输出的原因所在；除了指定自己的日志级别外，它们都继承了根日志器的行为。另一方面，Logback 追加器提供了一种方式将日志消息发送到各种地方。默认情况下有 4 类追加器。

- jdbc：将日志输出到 JDBC 连接的追加器。
- console：将日志输出到标准输出的追加器。
- file：将日志输出到文件的追加器。
- rollingFile：将日志输出到一组卷动文件的追加器。

若想在 Grails 应用中定义追加器，你需要在应用的 Logback.groovy 文件中声明它们，如下所示：

```
def USER_HOME = System.getProperty("user.home")

appender('customlogfile', FileAppender) {
    encoder(PatternLayoutEncoder) {
        Pattern = "%d %level %thread %mdc %logger - %m%n"
    }
    file = '${USER_HOME}/logs/grails.log'
}

appender('rollinglogfile', RollingFileAppender) {
    encoder(PatternLayoutEncoder) {
        Pattern = "%d %level %thread %mdc %logger - %m%n"
    }

    rollingPolicy(TimeBasedRollingPolicy) {
        FileNamePattern = "${USER_HOME}/logs/rolling-grails-%d{yyyy-MM}.log"
    }
}
```

要想使用追加器，只需将其添加到能够接收输入的相应的日志器即可。

如下声明展示了如何将追加器、日志器与日志级别放在一起使用：

```
root(DEBUG, ['STDOUT','customlogfile'])
```

上述配置覆盖了默认的根日志器。它使用 debug 级别将日志消息输出到 stdout 追加器（如标准输出或是控制台）和 customlogfile 追加器，后者表示定义在追加器部分中的一个文件。请注意，debug 级别会生成大量日志消息。

17-10 运行单元与集成测试

问题提出

为了确保应用的类能够按照指定的方式运行，你需要对其执行单元与集成测试。

解决方案

Grails 对应用的单元与集成测试提供了内建的支持。之前在生成 Grails 构件时（比如说应用的领域类），你会看到一系列测试类会自动生成出来。

在 Grails 应用中，测试位于应用的 src/test 或 src/integration-test 目录下。类似于 Grails 提供的其他功能，创建与配置应用测试的大量工作都被 Grails 处理掉了。你只需聚焦在设计测试上即可。

设计好应用的测试后，在 Grails 中运行测试是非常简单的，只需在应用的根目录下执行 grails test-app 命令即可。

解释说明

Grails 会启动一个必要的环境来执行应用测试。这个环境包含了库（如 JAR）与持久化存储系统（如 RDBMS），还有其他对于执行单元与集成测试来说必要的构件。

首先分析一下执行 grails test-app 命令的输出，如图 17-6 所示。

第一部分表示测试的执行，它来自于应用根目录下的 src/test/groovy 目录。在该示例中，有 13 个单元测试失败，4 个成功，它们对应于应用领域类创建时所生成的骨架测试类。由于这些测试类包含了一个测试骨架，因此大多数都失败了。

第二部分表示是成功还是失败了，在该示例中是失败了。结果可以通过 HTML 报告看到，并且给出了链接；可以在项目的 build/reports/tests/test 目录中找到。

在知道 Grails 是如何执行测试的后，我们现在来修改既有的单元测试类，根据领域类的逻辑来引入单元测试。考虑到 Grails 测试基于 JUnit 单元测试框架，如果不熟悉，建议查阅其文档掌握它的语法和方法。后续章节假设你已经对 JUnit 有了基本的了解（还可以参阅第 16 章）。

```
court — marten@iMac-van-Marten — ..th/ch17/court — -zsh — 160×68
→ court git:(master) ✗ grails test-app
:compileJava UP-TO-DATE
:compileGroovy
:buildProperties UP-TO-DATE
:processResources
:classes
:compileTestJava UP-TO-DATE
:compileTestGroovy
:processTestResources UP-TO-DATE
:testClasses
:test

court.PlayerControllerSpec > Test the save action correctly persists an instance FAILED
    org.spockframework.runtime.ConditionNotSatisfiedError at PlayerControllerSpec.groovy:15

court.PlayerControllerSpec > Test that the show action returns the correct model FAILED
    org.spockframework.runtime.ConditionNotSatisfiedError at PlayerControllerSpec.groovy:15

court.PlayerControllerSpec > Test that the edit action returns the correct model FAILED
    org.spockframework.runtime.ConditionNotSatisfiedError at PlayerControllerSpec.groovy:15

court.PlayerControllerSpec > Test the update action performs an update on a valid domain instance FAILED
    org.spockframework.runtime.ConditionNotSatisfiedError at PlayerControllerSpec.groovy:15

court.PlayerControllerSpec > Test that the delete action deletes an instance if it exists FAILED
    org.spockframework.runtime.ConditionNotSatisfiedError at PlayerControllerSpec.groovy:15

court.PlayerSpec > test something FAILED
    org.spockframework.runtime.SpockComparisonFailure at PlayerSpec.groovy:20

court.ReservationControllerSpec > Test the save action correctly persists an instance FAILED
    org.spockframework.runtime.ConditionNotSatisfiedError at ReservationControllerSpec.groovy:15

court.ReservationControllerSpec > Test that the show action returns the correct model FAILED
    org.spockframework.runtime.ConditionNotSatisfiedError at ReservationControllerSpec.groovy:15

court.ReservationControllerSpec > Test that the edit action returns the correct model FAILED
    org.spockframework.runtime.ConditionNotSatisfiedError at ReservationControllerSpec.groovy:15

court.ReservationControllerSpec > Test the update action performs an update on a valid domain instance FAILED
    org.spockframework.runtime.ConditionNotSatisfiedError at ReservationControllerSpec.groovy:15

court.ReservationControllerSpec > Test that the delete action deletes an instance if it exists FAILED
    org.spockframework.runtime.ConditionNotSatisfiedError at ReservationControllerSpec.groovy:15

court.ReservationSpec > test something FAILED
    org.spockframework.runtime.SpockComparisonFailure at ReservationSpec.groovy:20

court.WelcomeControllerSpec > test something FAILED
    org.spockframework.runtime.SpockComparisonFailure at WelcomeControllerSpec.groovy:20

17 tests completed, 13 failed
:test FAILED

FAILURE: Build failed with an exception.

* What went wrong:
Execution failed for task ':test'.
> There were failing tests. See the report at: file:///Users/marten/Repositories/spring-recipes-4th/ch17/court/build/reports/tests/test/index.html

* Try:
Run with --stacktrace option to get the stack trace. Run with --info or --debug option to get more log output.

BUILD FAILED

Total time: 18.166 secs
| Tests FAILED Test execution failed
→ court git:(master) ✗
```

图 17-6　测试输出

向应用 src/test/groovy 目录下的 PlayerSpec.groovy 类中添加如下方法（即单元测试），并删除“test something”方法：

```
void "A valid player is constructed"() {
    given:
        def player = new Player(name: 'James', phone: '120-1111')
    when: "validate is called"
        def result = player.validate();
    then: "it should be valid"
        result
}

void "A player without a name is constructed"() {
    given:
        def player = new Player(name: '', phone: '120-1111')
    when: "validate is called"
        def result = player.validate();
    then: "The name should be rejected"
        !result
        player.errors['name'].codes.contains('nullable')
}

void "A player without a phone is constructed"() {
    given:
        def player = new Player(name: 'James', phone: '')
    when: "validate is called"
        def result = player.validate()
    then: "The phone number should be rejected."
        !result
        player.errors['phone'].codes.contains('nullable')
}
```

第一个单元测试创建了一个 Player 对象并通过 name 字段和 phone 字段将其实例化。与声明在 Player 领域类中的约束一致，该实例总是合法的。因此，语句 assertTrue player.validate()会确认对该对象的验证总是 true。

后两个单元测试也创建了一个 Player 对象。不过，注意在一个测试中，Player 对象的实例化使用了一个空的 name 字段；在另一个测试中，Player 对象的实例化使用了一个空的 phone 字段。与声明在 Player 领域类中的约束一致，这两个实例总是非法的。因此，then:块中的!result 语句会确认对这些对象的验证总是 false。player.errors['phone'].codes.contains ('nullable')部分会检查验证是否包含了验证异常对应的验证码。

接下来，向应用 src/test/groovy 目录下的 ReservationSpec.groovy 类添加如下方法（即单元测试）：

```
void testReservation() {
    given:
    def calendar = LocalDateTime.of(2017, 10, 13, 15, 00)
        .toInstant(ZoneOffset.UTC)
    def validDateReservation = Date.from(calendar)
    def reservation = new Reservation(
        sportType:'Tennis', courtName:'Main',
        date:validDateReservation,player:new Player(name:'James',phone:'120-1111'))

    expect:
        reservation.validate()
}

void testOutOfRangeDateReservation() {
    given:
    def calendar = LocalDateTime.of(2017, 10, 13, 23, 00)
        .toInstant(ZoneOffset.UTC)

    def invalidDateReservation = Date.from(calendar)
    def reservation = new Reservation(
        sportType:'Tennis',courtName:'Main',
        date:invalidDateReservation,player:new Player(name:'James',phone:'120-1111'))

    expect:
        !reservation.validate()
        reservation.errors['date'].code == 'invalid.weekdayHour'
}

void testOutOfRangeSportTypeReservation() {
    given:
    def calendar = LocalDateTime.of(2017, 10, 13, 15, 00)
        .toInstant(ZoneOffset.UTC)
    def validDateReservation = Date.from(calendar)
    def reservation = new Reservation(
        sportType:'Baseball',courtName:'Main',
        date:validDateReservation,player:new Player(name:'James',phone:'120-1111'))

    expect:
        !reservation.validate()
        reservation.errors['sportType'].codes.contains('not.inList')
}
```

上述代码包含了 3 个单元测试，用于验证 Reservation 对象的完整性。第一个测试创建了一个 Reservation 对象实例并确认相应的值能够通过 Reservation 领域类的约束。第二个测试创建了一个 Reservation 对象，它违背了领域类的 date 约束并确认该实例是不合法的。第三个测试创建了一个 Reservation 对象，它违背了领域类的 sportType 约束并确认该实例是不合法的。

如果执行 grails test-app 命令，Grails 会自动执行所有测试并将测试结果输出到应用的 build 目录下。

还有失败的测试，特别是 PlayerControllerSpec 与 ReservationControllerSpec。打开文件，你会看到有个名为 populateValidParams 的方法包含了一个@TODO。

```
def populateValidParams(params) {
    assert params != null

    // TODO: Populate valid properties like...
    //params["name"] = 'someValidName'
    assert false, "TODO: Provide a populateValidParams() implementation for this generated
    test suite"
}
```

为了修复测试，你需要修改它，向控制器提交恰当的值。对于 PlayerControllerSpec 来说，修改 populateValidParams，使之包含 params["name"]与 params["phone"]。

```
def populateValidParams(params) {
    assert params != null

    params["name"] = 'J. Doe'
    params["phone"] = '555-123-4567'
}
```

向 ReservationControllerSpec 添加如下内容：

```
def populateValidParams(params) {
    assert params != null

    def calendar = LocalDateTime.of(2017, 10, 13, 12, 00)
        .toInstant(ZoneOffset.UTC)

    params["courtName"] = 'Tennis Court #1'
    params["sportType"] = "Tennis"
    params["date"] = Date.from(calendar)
    params["player"] = new Player(name: "J. Doe", phone: "555-432-1234")
}
```

现在，剩下的唯一失败的测试是 WelcomeControllerSpec，你可以将其删除，以实现成功的构建。

在为 Grails 应用创建了单元测试后，我们现在来创建集成测试。

与单元测试不同，集成测试会验证应用执行过的更为复杂的逻辑。各种领域类之间的交互或对 RDBMS 执行的操作都属于集成测试的领域。在这种情况下，Grails 通过自动启动 RDBMS 和其他应用属性来执行集成测试，从而辅助集成测试的进行。

单元测试旨在验证单个领域类中的逻辑。由于这个事实，除了自动化执行这种测试，Grails 并未提供启动属性来执行这类测试。

集成测试旨在验证跨越一系列应用类的更为复杂的逻辑。因此，Grails 不仅会启动 RDBMS 来针对这种持久化存储系统运行测试，还会启动领域类的动态方法简化这种测试的创建。相较于单元测试来说，执行这种测试肯定会增加额外的成本。

住的注意的是，如果仔细查看 Grails 为单元和集成测试所生成的骨架测试类，就会发现几乎没有任何区别。唯一的差别在于位于 integration 目录下的测试可以访问方才提及的一系列功能，而位于 unit 目录下的测试则不行。可以将单元测试放到 integration 目录中，不过需要自己在便捷性和成本之间做出权衡。

接下来，执行命令 grails create-integration-test CourtIntegrationTest 为应用创建集成类。这会在应用的 src/integration-test/groovy 目录下生成一个集成测试类。

将如下方法（即集成测试）添加到上面的类中，验证应用所执行的 RDBMS 操作：

```
void testQueries() {
    given: "2 Existing Players"
        // Define and save players
        def players = [ new Player(name:'James',phone:'120-1111'),
                        new Player(name:'Martha',phone:'999-9999')]
        players*.save()

        // Confirm two players are saved in the database
        Player.list().size() == 2
    when: "Player James is retrieved"
        // Get player from the database by name
        def testPlayer = Player.findByName('James')
    then: "The phone number should match"
        // Confirm phone
        testPlayer.phone == '120-1111'
    when: "Player James is Updated"
        // Update player name
        testPlayer.name = 'Marcus'
        testPlayer.save()

    then: "The name should be updated in the DB"
        // Get updated player from the database, but now by phone
```

```
        def updatedPlayer = Player.findByPhone('120-1111')

        // Confirm name
        updatedPlayer.name == 'Marcus'

    when: "The updated player is deleted"
        // Delete player
        updatedPlayer.delete()

    then: "The player should be removed from the DB."
        // Confirm one player is left in the database
        Player.list().size() == 1

        // Confirm updatedPlayer is deleted
        def nonexistantPlayer = Player.findByPhone('120-1111')
        nonexistantPlayer == null
}
```

上述代码针对应用的 RDBMS 执行了一系列操作，它首先保存两个 Player 对象，然后从数据库查询、更新并删除它们。在每个操作之后会执行一个验证步骤来确保逻辑（在该示例中，位于 PlayerController 控制器类中）如期望的一样（即控制器的 list()方法返回正确的 RDBMS 中的对象号）。

默认情况下，Grails 会针对 HSQLDB 执行 RDBMS 测试操作。不过，你可以使用喜欢的任何 RDBMS。请参见 17-8 节了解变更 Grails RDBMS 的详细信息。

最后，值得提及的是，如果想要执行一种类型的测试（即单元测试或集成测试），你可以使用命令标记-unit 或-integration。执行 grails test-app -unit 命令只会执行应用的单元测试，而执行 grails test-app-integration 命令则只会执行应用的集成测试。如果这两种测试的数量有很多，那么就很有用了，因为它能减少执行测试的总时间。

17-11 使用自定义布局与模板

问题提出

你需要自定义布局与模板来展示应用的内容。

解决方案

默认情况下，Grails 会使用一个全局布局来展示应用的内容。这样，视图就可以使用最少量的展示元素（即 HTML、CSS 与 JavaScript）并从单独的位置继承它们的布局行为。

这种继承过程可以让应用设计者与图形设计者单独完成他们的工作，应用设计者专注于通过必要的数据来创建视图，而图形设计者则聚焦于这些数据的布局（即美学）。

你可以创建自定义的布局来包含复杂的 HTML 展示以及自定义的 CSS 或 JavaScript 库。Grails 还支持模板的概念，其目的与布局一样，只不过应用在更为细粒度的层面上。此外，还可以使用模板渲染控制器的输出，从而代替大多数控制器所使用的视图。

解释说明

在应用的/grails-app/view/目录下，你会看到一个名为 layouts 的子目录，它包含了应用可用的布局。默认情况下，有一个名为 main.gsp 的文件，其内容如下所示：

```
<!doctype html>
<html lang="en" class="no-js">
<head>
    <meta http-equiv="Content-Type" content="text/html; charset=UTF-8"/>
    <meta http-equiv="X-UA-Compatible" content="IE=edge"/>
    <title>
        <g:layoutTitle default="Grails"/>
    </title>
    <meta name="viewport" content="width=device-width, initial-scale=1"/>

    <asset:stylesheet src="application.css"/>
```

```
    <g:layoutHead/>
</head>
<body>

    <div class="navbar navbar-default navbar-static-top" role="navigation">
        <div class="container">
            <div class="navbar-header">
                <button type="button" class="navbar-toggle" data-toggle="collapse"
                data-target=".navbar-collapse">
                    <span class="sr-only">Toggle navigation</span>
                    <span class="icon-bar"></span>
                    <span class="icon-bar"></span>
                    <span class="icon-bar"></span>
                </button>
                <a class="navbar-brand" href="/#">
                    <i class="fa grails-icon">
                        <asset:image src="grails-cupsonly-logo-white.svg"/>
                    </i> Grails
                </a>
            </div>
            <div class="navbar-collapse collapse" aria-expanded="false" style="height: 0.8px;">
                <ul class="nav navbar-nav navbar-right">
                    <g:pageProperty name="page.nav" />
                </ul>
            </div>
        </div>
    </div>

    <g:layoutBody/>

    <div class="footer" role="contentinfo"></div>

    <div id="spinner" class="spinner" style="display:none;">
        <g:message code="spinner.alt" default="Loading&hellip;"/>
    </div>

    <asset:javascript src="application.js"/>

</body>
</html>
```

显然，这是个简单的 HTML 文件，它包含了一些作为占位符的元素，可以让应用视图（如 JSP 和 GSP 页面）继承同样的布局。

第一个元素是<g:*>命名空间中的 Groovy 标签。<g:layoutTitle>标签用于定义布局的标题部分的内容。如果视图从该布局继承了行为并且没有提供这个值，那么 Grails 就会自动赋予 Grails 值，这是由 default 属性提供的。另一方面，如果继承的视图提供了该值，那么它就会展示出来。

<g:layoutHead>标签用于定义布局的头部分的内容。如果视图继承了该布局，那么声明在视图头部分的值在渲染时就会展示出来。

<asset:javascript src="application">标签可以让继承了该布局的视图能够自动访问 JavaScript 库。在渲染时，该元素会转换为<script type="text/javascript" src="/court/assets/application.js"> </script>。请记住，需要将 JavaScript 库放在 Grails /<app-name>/web-app/assets/javascripts 子目录下；在该示例中，<app-name>就是 court。

继续往下看，你还会看到几处${assetPath*}形式的声明，它们有一个 src 属性。Grails 会转换这种元素以反映出包含在应用中的资源。比如说，语句${assetPath(src: 'favicon.ico')}会转换为/court/assets/images /favicon.ico。注意，应用名（即上下文路径）会添加到转换后的值中。这样，布局可以在多个应用中重用，同时还可以引用相同的图片，其中图片需要放在 Grails /court/web-app/assets/images 子目录下。

在知道了 Grails 布局的组织后，我们来看看视图是如何继承其行为的。打开之前创建的应用控制器所生成的任何视图（player、reservation 或 welcome，它们也位于 views 目录下），你会看到用于从 Grails 布局中继承行为的如下语句：

```
<meta name="layout" content="main"/>
```

<meta>标签是个标准的 HTML 标签，它对页面的展示没有影响，不过 Grails 会使用它检测布局，视图则会从该布局继承其行为。通过使用该元素，视图会自动使用名为 main 的布局进行渲染，它就是上面介绍的模板。

进一步查看视图的结构，你会发现所有生成的视图都是独立的 HTML 页面；它们包含了<html>、<body>和其他 HTML 标签，类似于布局模板。不过，这并不意味着页面在渲染时会包含重复的 HTML 标签。Grails 会自动对替换过程进行分类，方式是将视图的<title>内容放到<g:layoutTitle>标签中，将视图的<body>内容放到<g:layoutBody />标签中，以此类推。

如果从 Grails 视图中删除了<meta>标签会发生什么呢？表面看，这个问题的答案是显而易见的：渲染视图时不会使用任何布局，这也意味着不会有可视化元素被渲染出来（如图片、菜单和 CSS 边框等）。不过，由于 Grails 是基于约定进行操作的，因此它总是会根据控制器的名字来尝试应用布局。

比如说，即便对应于 reservation 控制器的视图没有<meta name="layout">标签声明，如果名为 reservation.gsp 的文件位于应用的 layout 目录中，那么它就会应用到对应于控制器的所有视图上。

虽然布局提供了一种很好的基础来模块化应用视图，但它们只能应用到视图的整个页面。为了提供更为精细化的方式，模板可以重用视图页面的某些部分。

就拿展示运动员预订信息的 HTML 部分来说。你想在对应于该控制器的所有视图上展示这个信息以作为提醒。将这个 HTML 片段显式放到所有视图中不仅会导致大量的初始化工作，而且当发生变化时还需要进行多处修改。为了简化这种包含过程，我们可以使用模板。如下代码展示了名为_reservationList.gsp 的模板内容：

```
<table>
    <g:each in="${reservationInstanceList}" status="i" var="reservationInstance">
        <tr class="${(i % 2) == 0 ? 'odd' : 'even'}">
            <td><g:link action="show" id="${reservationInstance.id}">
                ${fieldValue(bean:reservationInstance, field:'id')}</g:link></td>
            <td>${fieldValue(bean:reservationInstance, field:'sportType')}</td>
            <td>${fieldValue(bean:reservationInstance, field:'date')}</td>
            <td>${fieldValue(bean:reservationInstance, field:'courtName')}</td>
            <td>${fieldValue(bean:reservationInstance, field:'player')}</td>
        </tr>
    </g:each>
</table>
```

上述模板会根据 Groovy 标签<g:each>生成一个 HTML 表格，展示一个预订列表。用于命名该文件的下划线（_）前缀是 Grails 的一个符号，用于区分模板和独立视图；模板总是会使用一个下划线作为前缀。

要想在视图中使用该模板，需要使用<g:render>标签，如下所示：

```
<g:render template="reservationList" model="[reservationList:reservationInstanceList]" />
```

在该示例中，<g:render>标签接收两个属性：template 属性标识了模板名；model 属性会传进模板所需的引用数据。另一种形式的<g:render>标签包含了模板的相对与绝对位置。通过声明 template="reservationList"，Grails 会尝试在与视图相同的目录下寻找模板。为了方便重用，模板可以从一个公共目录下加载，这时需要使用绝对目录。比如说，视图中使用了 template="/common/reservationList"形式的语句，那么它就会在应用的 grails-app/views/common 目录下寻找名为_reservationList.gsp 的模板。

最后，值得注意的是，模板还可以由控制器使用来渲染其输出。比如说，大多数控制器会使用如下语法将控制返回给视图：

```
render view:'reservations', model:[reservationList:reservationList]
```

还可以使用如下语法将控制返回给模板：

```
render template:'reservationList', model:[reservationList:reservationList]
```

通过使用上述渲染语句，Grails 会尝试根据名字_reservationList.gsp 来寻找模板。

17-12 使用 GORM 查询

问题提出

你想对应用的 RDBMS 执行查询。

解决方案

Grails 使用 GORM 执行 RDBMS 操作。GORM 基于流行的 Java ORM Hibernate，可以让 Grails 应用使用 Hibernate 查询语言（Hibernate Query Language，HQL）执行查询。不过，除了支持使用 HQL 外，GORM 还有一系列内建的功能来简化 RDBMS 查询。

解释说明

在 Grails 中，针对 RDBMS 的查询通常都是在控制器中执行的。如果查看 court 应用的控制器，则最简单的查询如下所示：

```
Player.get(id)
```

该查询用于使用特定的 ID 获取一个 Player 对象。不过在某些情况下，应用需要根据其他标准来执行查询。比如说，court 应用中的 Player 对象有 name 和 phone 字段，它们定义在 Player 领域类中。GORM 支持根据字段名来查询领域对象。这是通过 findBy<field_name>形式的方法提供的，代码如下所示：

```
Player.findByName('Henry')
Player.findByPhone('120-1111')
```

这两条语句用于查询 RDBMS 并根据名字和手机号获取一个 Player 对象。这些方法叫做动态查找器，因为它们是由 GORM 根据领域类的字段提供的。

与之类似，拥有自己的字段名的 Reservation 领域类也有动态查找器，如 findByPlayer()、findByCourtName() 和 findByDate()。可以看到，这个过程简化了 Java 应用中针对 RDBMS 的查询创建。

此外，动态查找器方法还可以使用比较器来进一步精化查询结果。如下代码片段展示了如何使用比较器按照特定的日期范围来提取出 Reservation 对象：

```
def now = new Date()
def tomorrow = now + 1
def reservations = Reservation.findByDateBetween( now, tomorrow )
```

除了 Between 比较器外，court 应用中还可以使用比较器 Like。如下代码片段展示了如何通过 Like 比较器提取出名字以字母 A 开头的 Player 对象：

```
def letterAPlayers = Player.findByNameLike('A%')
```

表 17-2 介绍了可用于动态查找器方法的各种比较器。

表 17-2 GORM 动态查找比较器

GORM 比较器	查询
InList	如果值位于给定的值列表中
LessThan	小于给定值的对象
LessThanEquals	小于或等于给定值的对象
GreaterThan	大于给定值的对象
GreaterThanEquals	大于或等于给定值的对象
Like	模糊匹配给定值的对象
Ilike	模糊匹配给定值的对象（不区分大小写）
NotEqual	不等于给定值的对象
Between	介于两个给定值之间的对象
IsNotNull	不为 null 的对象；不使用参数
IsNull	为 null 的对象；不使用参数

GORM 还支持在动态查找器方法的构建过程中使用布尔逻辑（与 / 或）。如下代码片段展示了如何查询符合某个 court 名和未来日期的 Reservation 对象。

```
def reservations = Reservation.findAllByCourtNameLikeAndDateGreaterThan("%main%",
new Date()+7)
```

与之类似，Or 语句（而不是 And）也可以用在上述动态查找器方法中，以提取出至少满足一个条件的 Reservation 对象。

最后，动态查找器方法还支持分页和排序以进一步精化查询。这是通过对动态查找器方法使用映射来实现的。如下代码片段展示了如何限制查询结果的数量，并且定义了其分类与排序属性：

```
def reservations = Reservation.findAllByCourtName("%main%", [ max: 3, sort: "date",
order: "desc"] )
```

如本节一开始所述，GORM 还支持使用 HQL 对 RDBMS 执行查询。不过要比上述做法更麻烦且容易出错，如下代码展示了使用 HQL 实现的相同功能的查询：

```
def letterAPlayers = Player.findAll("from Player as p where p.name like 'A%'")
def reservations = Reservation.findAll("from Reservation as r
    where r.courtName like '%main%' order by r.date desc", [ max: 3] )
```

17-13 创建自定义标签

问题提出

你想在 Grails 视图中执行逻辑，但内建的 GSP 或 JSTL 标签并未提供该功能，同时也不想在视图中使用代码。

解决方案

Grails 视图可以包含展示元素（即 HTML 标签）、业务逻辑元素（即 GSP 或 JSTL 标签）或是直接的 Groovy 或 Java 代码来实现展示。在某些情况下，视图需要特定的展示元素与业务逻辑的整合。比如说，按照月份展示特定运动员的预订信息需要编写自定义代码。为了简化这种组合并能在多个视图中重用，可以使用自定义标签。

解释说明

要想创建自定义标签，可以使用 grails create-taglib <tag-lib-name>命令。该命令会为应用/grails-app/tag-lib/目录下的自定义标签库创建一个骨架类。

了解了这一点后，我们为 court 应用创建一个自定义标签库，用于展示特殊的预订信息。第一个自定义标签会检测当前日期并根据这个信息展示特殊的预订信息。最后的结果是可以在应用的视图中使用<g:promoDailyAd/>这样的标签，而不是将代码放到视图中或是在控制器中执行逻辑。

执行 grails create-taglib DailyNotice 命令创建自定义标签库类。接下来，打开应用/grails-app/taglib/目录下生成的 DailyNoticeTagLib.groovy 类，添加如下方法（即自定义标签）：

```
def promoDailyAd = { attrs, body ->
    def dayoftheweek = Calendar.getInstance().get(Calendar.DAY_OF_WEEK)
    out << body() << (dayoftheweek == 7 ?
        "We have special reservation offers for Sunday!": "No special offers")
}
```

该方法的名字定义了自定义标签的名字。方法的第一个声明（attrs, body）表示自定义标签的输入值——其属性和体。接下来，使用一个 Calendar 对象确定一周的哪一天。

接下来会看到基于一周的哪一天的条件语句。如果天数是 7（周六），那么条件语句就会解析为字符串"We have special reservation offers for Saturday!"否则会解析为"No special offers"。

该字符串是通过<<输出的，并且首先通过 body()方法赋值，它表示自定义标签的主体（body），然后是自定义标签的输出。通过这种方式，可以使用如下语法在应用的视图中声明自定义标签：

```
<h3><g:promoDailyAd /></h3>
```

当渲染包含了该自定义标签的视图时，Grails 会执行支撑类方法中的逻辑并使用结果来代替它。这样，只需通过简单的声明就可以让视图基于更为复杂的逻辑展示结果了。

注意： 这种标签自动可以在 GSP 页面中使用，但无法在 JSP 页面中使用。要想在 JSP 中使用这个自定义标签来完成功能，需要将其添加到名为 grails.tld 的对应的标签库定义中（TLD）。TLD 位于应用的 /web-app/WEB-INF/tld/目录中。

自定义标签还可以根据作为标签属性传递进来的输入参数来执行支撑类的逻辑。如下代码展示了另一个自定义标签，它根据名为 offerdate 的属性来确定结果：

```
def upcomingPromos  = { attrs, body ->
    def dayoftheweek = attrs['offerdate']
    out << body() << (dayoftheweek == 7 ?
        "We have special reservation offers for Saturday!": "No special offers")
}
```

虽然与之前的自定义标签类似，不过上述代码使用了语句 attrs['offerdate']来确定一周中的哪一天。在该示例中，attrs 表示向类方法（即声明在视图中的方法）传递进来的作为输入参数的属性。因此，要想使用该自定义标签，需要使用如下声明：

```
<h3><g:upcomingPromos offerdate='saturday'/></h3>
```

这种自定义标签更为灵活，因为其逻辑是根据视图中所提供的数据来执行的。还可以使用变量来表示控制器向视图传递的数据，代码如下所示：

```
<h3><g:upcomingPromos offerdate='${promoDay}'/></h3>
```

最后，说一下 Grails 自定义标签所使用的命名空间。默认情况下，Grails 会将自定义标签赋予<g:>命名空间。要想使用自定义命名空间，需要在自定义标签库类的顶部声明 namespace 字段。

```
class DailyNoticeTagLib {
    static namespace = 'court'
    def promoDailyAd = { attrs, body ->
    ...
    }
    def upcomingPromos = { attrs, body ->
    ...
    }
}
```

通过上述声明，类的自定义标签就会被赋予名为 court 的自定义命名空间。对于该自定义标签来说，视图中的声明就需要改成下面的样子：

```
<h3><court:promoDailyAd/></h3>
<h3><court:upcomingPromos offerdate='${promoDay}'/></h3>
```

17-14 添加安全

问题提出

你想使用 Spring Security 保护应用。

解决方案

使用 Grails Spring Security 插件（参见 17-3 节了解关于插件的信息）为应用添加安全。

解释说明

要想保护应用，你需要向应用添加 Grails 插件 spring-security-core。打开 build.gradle 文件，添加插件。

```
dependencies {
    compile "org.grails.plugins:spring-security-core:3.1.2"
}
```

添加完插件后，运行 grails compile 会下载并安装插件。

安装好插件后，可以通过 s2-quickstart 命令建立安全。该命令接收一个包名和类名来表示用户与授权。

```
grails s2-quickstart court SecUser SecRole
```

运行上述命令会创建 SecUser 与 SecRole 领域对象。它还会修改(或创建)grails-app/conf/application.groovy 文件。这会添加安全部分的内容。

```
// Added by the Spring Security Core plugin:
grails.plugin.springsecurity.userLookup.userDomainClassName = 'court.SecUser'
```

```
grails.plugin.springsecurity.userLookup.authorityJoinClassName = 'court.SecUserSecRole'
grails.plugin.springsecurity.authority.className = 'court.SecRole'
grails.plugin.springsecurity.controllerAnnotations.staticRules = [
    [pattern: '/',                access: ['permitAll']],
    [pattern: '/error',           access: ['permitAll']],
    [pattern: '/index',           access: ['permitAll']],
    [pattern: '/index.gsp',       access: ['permitAll']],
    [pattern: '/shutdown',        access: ['permitAll']],
    [pattern: '/assets/**',       access: ['permitAll']],
    [pattern: '/**/js/**',        access: ['permitAll']],
    [pattern: '/**/css/**',       access: ['permitAll']],
    [pattern: '/**/images/**',    access: ['permitAll']],
    [pattern: '/**/favicon.ico', access: ['permitAll']]
]
grails.plugin.springsecurity.filterChain.chainMap = [
    [pattern: '/assets/**',       filters: 'none'],
    [pattern: '/**/js/**',        filters: 'none'],
    [pattern: '/**/css/**',       filters: 'none'],
    [pattern: '/**/images/**',    filters: 'none'],
    [pattern: '/**/favicon.ico', filters: 'none'],
    [pattern: '/**',              filters: 'JOINED_FILTERS']
]
```

当使用 grails run-app 运行应用时，应用会启动并添加安全。如果访问应用，你会看到一个登录界面，要求输入用户名与密码，如图 17-7 所示。

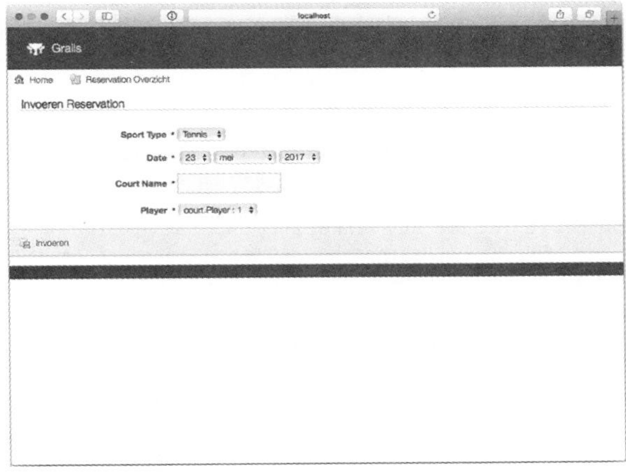

图 17-7　添加安全后的登录界面

现在的系统中还没有用户和角色，因此目前无法登录到系统。

启动安全

要想使用系统，需要应用中存在拥有密码和角色的用户。这可以来自数据库、LDAP 目录，或是文件系统中的文件。我们在应用的启动脚本中添加一些用户。打开 grails-app/init 目录下的 BootStrap.groovy 文件，向系统添加两个用户和两个角色。

```
class BootStrap {
def init = { servletContext ->
def adminRole = new court.SecRole(authority: 'ROLE_ADMIN').save(flush: true)
def userRole = new court.SecRole(authority: 'ROLE_USER').save(flush: true)
def testUser = new court.SecUser(username: 'user', password: 'password')
testUser.save(flush: true)
def testAdmin = new court.SecUser(username: 'admin', password: 'secret')
testAdmin.save(flush: true)
court.SecUserSecRole.create testUser, userRole, true
court.SecUserSecRole.create testAdmin, adminRole, true
```

```
        }
        ...
    }
```

前两个角色 ROLE_ADMIN 与 ROLE_USER 添加到了系统中。接下来添加的两个用户都有用户名和密码。最后，建立好用户与角色之间的链接。

现在一切就绪，重启应用并登录到系统，如图 17-8 所示。

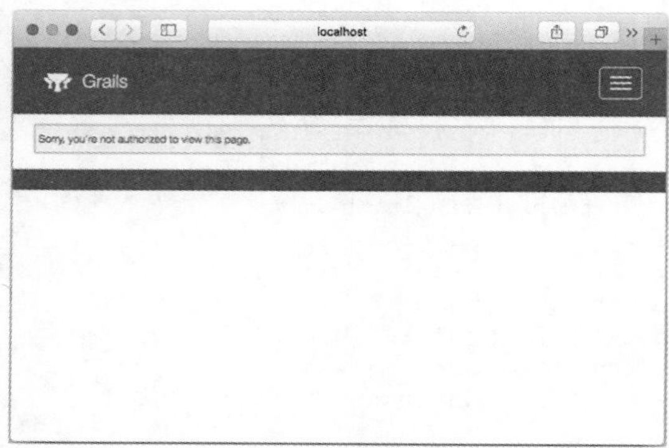

图 17-8 用户登录后的系统

虽然可以登录，但却出现一个页面，告诉你不允许访问所请求的页面。虽然系统中已经有了用户，但系统却不知道他们是可以访问页面的。基于此，你需要添加一些配置，用来表示哪些 URL 是可以访问的。

保护 URL

创建好安全配置后，只添加了一些默认 URL，这并不包含应用特定的 URL。出于这个原因，请打开 grails-app/conf 目录下的 application.groovy 文件。

```
grails.plugin.springsecurity.controllerAnnotations.staticRules = [
    [pattern: '/',               access: ['permitAll']],
    [pattern: '/error',          access: ['permitAll']],
    [pattern: '/index',          access: ['permitAll']],
    [pattern: '/index.gsp',      access: ['permitAll']],
    [pattern: '/shutdown',       access: ['permitAll']],
    [pattern: '/assets/**',      access: ['permitAll']],
    [pattern: '/**/js/**',       access: ['permitAll']],
    [pattern: '/**/css/**',      access: ['permitAll']],
    [pattern: '/**/images/**',   access: ['permitAll']],
    [pattern: '/**/favicon.ico', access: ['permitAll']],
    [pattern: '/player/**',      access: ['isAuthenticated()']],
    [pattern: '/reservation/**', access: ['isAuthenticated()']]
]
```

注意新添加的两处，一个是网站的 player 部分，另一个是 reservation 部分。表达式/player/**就是所谓的 Ant 风格的模式，它会匹配以/player 开头的地址。permitAll 表示所有人，甚至是不存在的用户都可以访问网站的这部分内容（对于静态与公开内容很有用）。借助于 isAuthenticated()，只有认证用户才允许访问网站。要想了解关于允许表达式的更多信息，请参考第 7 章。

重新构建并启动应用后，应该可以再次访问 player 与 reservations 界面了。

使用注解来实现安全

除了保护 URL 外，还可以保护方法；可以使用@Secured 注解实现。我们来保护 create 方法，这样只有管理员可以创建新的运动员。

```
import grails.plugin.springsecurity.annotation.Secured

class PlayerController {
```

```
...

    @Secured(['ROLE_ADMIN'])
    def create() {
        respond new Player(params)
    }
}
```

注意这里为 create 方法添加的@Secured 注解。该注解接收一个允许访问的角色数组。这里指定的是 ROLE_ADMIN，将访问限制为管理员。@Secured 注解还可用于类级别上，这会对类中的所有方法都添加安全。以普通用户的身份登录（使用 user/password）后，在创建一个新的运动员时会出现一个访问拒绝页面（与图 17-8 一样）。当以管理员（使用 admin/secret）的身份创建时，就可以输入新的运动员。

小结

本章介绍了如何使用 Grails 框架开发 Java Web 应用。首先学习了 Grails 应用的结构，然后快速了解了一个示例应用，它展示了如何自动化 Web 应用构建的步骤。

我们介绍了 Grails 如何在其自动化过程中借由约定来创建应用的视图、控制器、模型与配置文件。此外，我们介绍了如何使用 Grails 插件在 Grails 上下文中自动处理相关的 Java API 或框架的任务。接下来，我们介绍了 Grails 如何分离配置参数以及如何基于应用的运行环境执行任务，环境可以是开发、测试与生产环境。

然后，我们介绍了 Grails 如何通过应用的领域类来生成对应的控制器和视图，用于针对 RDBMS 执行 CRUD 操作。接下来，我们介绍了 Grails 国际化、日志与测试设施。

我们还介绍了 Grails 布局与模板，它们用于对应用的展示进行模块化；接下来介绍了 Grails ORM 设施以及自定义标签的创建。

最后，我们介绍了如何向 Grails 项目添加 Spring Security，以及如何配置和使用它来保护 URL 与方法。